Integral Vehicle Safety

Rodolfo Schöneburg
Editor

Integral Vehicle Safety

Biomechanics – Accident Prevention – Occupant Protection – Sensors – Vehicle Safety in the Development Process

Springer

Editor
Rodolfo Schöneburg
CEO, RSC Safety Engineering
Hechingen, Baden-Württemberg, Germany

ISBN 978-3-658-48099-8 ISBN 978-3-658-48100-1 (eBook)
https://doi.org/10.1007/978-3-658-48100-1

This book is a translation of the original German edition "Integrale Sicherheit von Kraftfahrzeugen," 5th edition, by Rodolfo Schöneburg, published by Springer Fachmedien Wiesbaden GmbH in 2023. The translation was done with the help of an artificial intelligence machine translation tool. A subsequent human revision was done primarily in terms of content, so that the book will read stylistically differently from a conventional translation. Springer Nature works continuously to further the development of tools for the production of books and on the related technologies to support the authors.

The translation was done with the help of artificial intelligence (machine translation by the service DeepL.com). A subsequent human revision was done primarily in terms of content.

This Springer imprint is published by the registered company Springer Fachmedien Wiesbaden GmbH, part of Springer Nature.
The registered company address is: Abraham-Lincoln-Str. 46, 65189 Wiesbaden, Germany

Preface to the Fifth Edition

Exactly 25 years ago, the first edition of this book on vehicle safety by Prof. Dr.-Ing. Florian Kramer, which has since become a standard work in this field, was published in Germany. I have been professionally and personally connected with him for many years, since my time at the Technical University of Berlin in the 1980s. From the second edition onwards, I was honored to contribute the chapter "Integral Safety in the Vehicle Development Process", a topic that accompanied me at Audi and later as a director at Mercedes-Benz, responsible for vehicle safety and other vehicle functions, for many years.

With great sorrow, we had to say goodbye to Florian Kramer in June 2019. He was unable to complete the fifth edition of his book, which we had been discussing intensively since 2017. The world of vehicle safety is rapidly evolving, and new rating programs, automated driving, and digitization in the development process are occupying the automotive industry to an unprecedented extent. After 10 years, a new edition of this book had become necessary. I am very grateful to Florian's children, who gave me and the publisher permission to continue the work of my long-time companion. The fifth edition of this book is now also available in English for the first time.

Since the turn of the millennium, efforts by vehicle manufacturers and the supplier industry to completely avoid accidents or mitigate their severity have been increasing. Indeed, the 'Vision Zero', i.e., no more fatalities or serious injuries in accidents, has now been anchored in many traffic safety programs across Europe. In addition, development conditions are changing and becoming more complex at ever shorter intervals due to new laws, global rating programs, and CAE methods.

A major focus and innovation priority in passenger car safety development is the increasing automation of driving tasks and a new aspect in the integral safety approach, the consideration of the pre-accident phase. This phase, also referred to as preventive safety, combines active with passive safety and offers many potentials. Systems based on this concept can be found in many new automobiles of safety-conscious vehicle manufacturers today.

The fourth edition of the book in 2013 marked an important milestone with its expansion from a purely occupant and partner protection related “Passive Safety” to the “Integral Safety of Motor Vehicles”. This milestone is enriched with many facets in this fifth edition. A completely new chapter on active safety, with in-depth discussions on driver assistance and automated driving, takes this into account, as do new sections on telematics and radar systems in sensor technology.

Except for a few updates and corrections, I have left the comprehensive, scientifically derived chapter on accident research as it was written by Florian Kramer. The biomechanics are also excellently explained and easy for engineers to understand. However, there have been developments in the protection criteria, e.g. with the introduction of the THOR dummy in test operations and in the assessment of neck loads, which have been added to this edition.

The chapters on “Evaluation of Vehicle Safety” and “Occupant and Partner Protection” have been extensively revised and updated according to the latest findings. In addition, the contents of the computational and experimental simulation have been integrated into the chapter “Integral Safety in the Vehicle Development Process” and supplemented with new methods. Thus, there is now a brief introduction to explicit time integration in the CAE section, as well as applications of the now established human body model (HBM), which allows the loads and kinematics of the human body to be increasingly accurately represented and evaluated in computational simulations.

I have adjusted the structure of the book for this new edition: After describing the basic **safety requirements** for motor vehicles based on accident research, biomechanics as well as laws and rating programs (safety assessment) in Chaps. 2 to 4, **measures of integral safety** regarding accident prevention, occupant and partner protection, and sensor technology are derived and explained in Chaps. 5 to 7. The last chapter deals in detail with the **vehicle development process**, including how and when to appropriately use the ‘development tools’ of computational and experimental simulation, and the importance of manufacturer-specific safety requirements in the context of product liability.

A newly added glossary in the book’s preface also helps to clarify the seemingly endless number of abbreviations in the field of integral vehicle safety. Finally, the now mostly colored illustrations not only enhance the book, but also help readers to grasp the complex content better and faster.

I wish you much success in your work, your teaching activities, or your studies related to vehicle safety. Let us do everything we can to make the roads and traffic even safer and to get closer to ‘Vision Zero’ every day.

Hechingen, Germany
October 2025

Prof. Dr. Rodolfo Schöneburg
RSC Safety Engineering

Acknowledgment

I would like to thank my colleagues and friends from industry and science, whom I was able to win for this present work and who contributed as co-authors in this extensively revised edition. In addition to their daily workload, they provided their competent support and further developed the book in many discussions.

These are the gentlemen (in the order of the chapters):

Dr. med. Wolfram Hell, President of the Society for Medical and Technical Trauma Biomechanics gmttb and owner of the Expert Office for Accident Analysis in Gruenwald, Germany. He updated and expanded the (admittedly quite specialized for vehicle engineers) chapter "Biomechanics" with new protection criteria and added historical notes and cross-references at appropriate places.

Dipl.-Wirtsch.-Ing. Ralf Reuter, Author of the well-known SafetyCompanions and employee at carhs.training GmbH in Alzenau, Germany. He revised the chapter "Evaluation of Vehicle Safety" (formerly "Investigation and Assessment of Safety") extensively in terms of the changed requirements in law and rating. Since the last edition, many requirements and evaluation programs for vehicle safety worldwide have been tightened or added.

Univ.-Prof. Dr.-Ing. Lutz Eckstein, Director of the Institute for Motor Vehicles (ika), RWTH Aachen University, and President of the Association of German Engineers e. V. in Düsseldorf, Germany, as well as Prof. Dr.-Ing. Adrian Zlocki, Head of the Automated Driving Department at fka GmbH in Aachen, Germany. With their new chapter "Accident Prevention and Severity Reduction", the book is enriched with many aspects of active safety and driver assistance. In addition, they expanded the methods of experimental simulation with the new section "Scenario-based Testing of Active Safety and Automated Driving", without which the vehicle development process would no longer be complete today.

Dr.-Ing. Frank Laakmann, Former employee at the ZF Group in the Passive Safety Technology Division in Alfdorf, Germany. He thoroughly revised and updated the chapter "Occupant and Partner Protection" (formerly "Safety Measures") completely and brought it up to date. He was supported by Dr. Christian Fischer, Bernd Issler, Dr. Detlef Last, Dr. Kai-Ulrich Machens, Marc Schledorn, Dr. Martin Seyffert, Marco Wahl, Lothar Zink and other colleagues. Dr. Swen Schaub, Head of Engineering Strategies & Communication at ZF LIFETEC, supported with the translation of the English edition. The many examples and new illustrations will be of great value to readers interested in passive safety, the development of restraint systems, and integral safety technologies.

Dipl.-Wirtsch.-Ing. Johannes Clemm, Managing Director of Continental Safety Engineering International GmbH in Alzenau, Germany, and Dr.-Ing. Stephan Zecha, Head of Integrated Safety System Development, along with his colleagues Dipl.-Ing. Andreas Forster and Dr. rer. nat. Marc Menzel, also from Continental. They updated and expanded the chapter "Sensor Technology for Accident Detection", which was already contributed by Thomas Görnig from Continental in Ingolstadt in the second edition and significantly expanded in the fourth edition by James Remfrey, former Head of Technology Intelligence at Continental in Frankfurt. I would particularly like to mention the new sections on telematics and radar technology for environment detection, without which a book on the integral safety of motor vehicles would no longer be complete today.

Dipl.-Math. Dipl.-Ing. (BA) Ulrich Franz, Managing Director of DYNAmore GmbH, a subsidiary of ANSYS Inc., and Prof. Dr.-Ing. André Haufe, Director LS-DYNA Method Development and Homologation, also at DYNAmore. They revised the extensive section "Computational Simulation", part of the chapter "Integral Safety in the Vehicle Development Process", and brought it up to a highly current state, which has been necessary since the fourth edition more than ten years ago. This section was supplemented with an introduction to explicit time integration. I would also like to thank Dr.-techn. Thomas Kinsky, Director of Business Development at Humanetics Europe, for the important revision of the subsection on dummies in the "Experimental Simulation" and the meaningful images to go with it.

During the creation of this book, I was able to experience the support of friendly and interested colleagues with great gratitude, listing whom would exceed the scope of this acknowledgment.

Finally, I would like to thank Ms. **Ulrike Butz**—Project Manager, Ms. **Mangayarkarassi Karthikeyan** and Mr. **Prasenjit Das**—Production, especially manuscript editing and formal coordination, Ms. **Angela Schulze-Thomin**—Project Coordinator Germany, Ms. **Julia Bauer**—Project Coordinator German Language and Autotranslations, Ms. **Juliane Wagner**—Project Coordinator Autotranslations, as well as Ms. **Stella Müller** and Mr. **Axel Garbers**—Editorial team of the automotive sector at Springer Vieweg publishing house in

Berlin and Heidelberg, Germany, for the many coordination meetings and the support in the implementation of the work. I also extend my thanks to them and many of their colleagues in Germany and India for proofreading, preparing the script, printing, and finally marketing the book on integral vehicle safety.

Hechingen, Germany
Oktober 2025

Prof. Dr.-Ing. Rodolfo Schöneburg
RSC Safety Engineering

Contents

Glossary of Abbreviations

3PGA	automatic three-point seat belt
ABS	anti-lock braking system (German: Antiblockiersystem)
ABS	acrylonitrile-butadiene-styrene copolymer (plastic material)
ABL	active buckle lifter (see also active control buckle)
ACB	active control buckle
ACC	adaptive cruise control
ACEA	European Automobile Manufacturer Association
ACR	active control retractor (active seat belt pretensioner)
ACRS	air cushion restraint system
AD	automated driving
ADAS	advanced driver assistance system
AEB	automated emergency braking
AEBS	advanced emergency braking system
AIS	abbreviated injury scale (simplified injury scale)
ALE	arbitrary Lagrangian Eulerian
ALR	automatic locking retractor (child seat safety function)
AkSIx	index for active safety
ALS	advanced level support (in rescue services)
AP	anchor pretensioner
APEAL	Auto Performance, Execution and Layout Study (USA, J.D. Power Study)
API	application programming interface
aPLI	advanced pedestrian legform impactor
ASIC	application specific integrated circuit
ATD	anthropometric test device (dummy)
AWS	anti-whiplash system
BAB	federal highway
BAS	brake assist system (emergency brake assistant)
BASt	Federal Highway Research Institute (Germany)
BEV	battery electric vehicle
BF	confirmation vehicle
BG	Employer's Liability Insurance Association (Germany)

BIR	bag-in-roof
BIW	body-in-white (vehicle body shell)
BLS	basic level support (in rescue services)
BMS	battery management system
BP	buckle pretensioner
BSM	blind spot monitoring
BSM	basic safety message (in telematics)
BWS	thoracic spine
CAB	curtain airbag module (windowbag)
CAD	computer aided design
CAE	computer aided engineering
CAM	cooperative awareness message
CAN	controller area network
Canfix	other/regional name for ISOFIX (Canada)
Car-to-X	vehicle real-time communication
CATARC	China Automotive Technology and Research Center
CDD	cyclic delay diversity
CDS	Crashworthiness Data System (USA)
C-IASI	China Insurance Automobile Safety Index
CE	Conformité Européenne (european conformity)
CeAB	center airbag module (middle-airbag module)
CFC	channel frequency class
CFD	computational fluid dynamics
CFK	carbon fiber-reinforced plastic
CFL	characteristic shoulder belt force level
CIREN	Crash Injury Research and Engineering Network (USA)
CLL	constant load limiter
CMVSS	Canada Motor Vehicle Safety Standards
CoP	conformity of production
COP	child occupant protection
CPM	cooperative perception message
CPU	central processing unit
CRABI	Child Restraint and Airbag Interaction (child dummy)
CRS	child restraint system
DAB	driver airbag module
DAMAGE	diffuse axonal multi-axis general evaluation
DARPA	Defense Advanced Research Projects Agency
DENM	decentralized environmental notification message
DGU	German Society for Trauma Surgery
DIVI	German Interdisciplinary Association for Intensive and Emergency Medicine
DLL	degressive load limiter (degressive force limitation)
DLT	dynamic locking tongue
DMS	driver monitoring system
DPT	digital prototype (also VP)

DOT	Department of Transportation (USA)
EAC	equivalent accident characteristics
EC	Enrollment Certificate
eCall	emergency call (automatic emergency call)
ECDSA	elliptic curve digital signature algorithm
ECE	Economic Commission for Europe
EDR	event data recorder
EEBL	emergency brake light
EEVC	European Enhanced Vehicle Safety Committee
EG	European Union (German: Europäische Gemeinschaft)
EKG	electrocardiogram
eKF	micro electric vehicle
ELK	emergency lane keeping
ELR	emergency locking retractor
EMV	electromagnetic compatibility
ES-2	EuroSID 2 – European side impact dummy, 2nd generation
ESA	evasive steering assist
ESD	electrostatic discharge
ESG	single-pane safety glass (German: Einscheiben-Sicherheitsglas)
ESP	electronic stability program
ETSC	European Transport Safety Council
ETSI	European Telecommunications Standards Institute
EU	European Union
EUC	equipment under control
FAS	driver assistance systems (also ADAS)
FAR	functional validation of the body-in-white (German: Funktionsabsicherung Rohbau)
FAT	Research Association for Automotive Technology e.V.
FCEV	vehicle with fuel cell drive
FCW	forward collision warning
FEM	finite element method
Flex-PLI	flexible pedestrian legform impactor
FMVSS	Federal Motor Vehicle Safety Standards (USA)
FPM	finite point-set method
Fps	frames per second
FUPD	front underride protection device (truck underride protection device)
FWDB	full width deformable barrier
GB-Standard	Guobiao (Chinese for "National Standard")
GFK	glass fiber-reinforced plastic
GG	gas generator (airbag inflator)
GIDAS	German In-Depth Accident Study (accident database)
GNSS	global navigation satellite system (GPS/Galileo/etc.)
GPS	global positioning system
GPU	graphical processing unit
GSR	General Safety Regulations (EU)

GTR	Global Technical Regulations
GuNi	guanidine nitrate
HANS	DYNAmore human body model
HBM	human body model
HIII	Hybrid III (dummy)
HIII-50M	HIII - 50th percentile (male)
HIII-5F	HIII - 5th percentile (female)
HIC	head injury criterion
HMI	human machine interface
HPC	head performance criterion
HPM	H-point machine
HPR	high penetration resistant
HUMOS	Human Model for Safety
HUS	rear underride protection
HWS	cervical spine (German: Halswirbelsäule)
I-Shaft	intermediate shaft (steering system)
i-Size	child seat standard (approval standard for child seats)
IBRL	internal bumper reference line (for pedestrian protection)
ICS	injury cost scale
ICW	intersection collision warning
IFM	Institute for Vehicle Safety Munich
IIHS	Insurance Institute for Highway Safety (USA)
IMA	intersection movement assist
IMU	inertial measurement unit
IoT	internet of things
IQS	Initial Quality Study (USA, J.D. Power Study)
IRS	inflatable restraint systems
ISOFIX:	ISO-International Standardisation Organization, FIX-fixation (attachment system for child seats)
ISS	injury severity score
IVIM	in-vehicle information message
LDW	lane departure warning
ITS	intelligent transportation system
KBA	German Federal Motor Transport Authority
Kbps	kilobit per second
KFZ	motor vehicle (German: Kraftfahrzeug)
KISI	see ALR
KMVSS	Korean Motor Vehicle Safety Standards
KnAB	knee airbag
KOM	bus, coach (German: Kraftomnibus)
LATCH	other/regional name for ISOFIX (USA)
LEM	light electric vehicles
LiDAR	light detection and ranging
LDW	lane departure warning
LKA	lane keeping assistance

LKW	truck, comercial vehicle, CV (German: Lastkraftwagen)
LL	load limiter
LMM	low-mount module (airbag)
LPD	lateral protection device (truck underride protection device)
LUAS	lower anchors and tethers for children (LATCH, USA)
LWS	lumbar spine
MaaS	mobility as a service
MAIS	maximum AIS value
MAP	intersection MAP message (road topology)
MCM	maneuver coordination message
MDB	mobile deformable barrier
MGG	micro gas generator (airbag inflator)
MIND	minimum emergency data set
MIPS	multi-directional impact protection
MMM	mid-mount module (airbag)
MPDB	mobile progressive deformable barrier
MKS	multi-body system (German: Mehrkoerpersystem)
MSD	minimum set of data
MZR	motorized two-wheeler
NASS	National Automotive Sampling System (USA)
NBW	normalized load value
NC	nitrocellulose
NCAP	New Car Assessment Programme
NFZ	commercial vehicle, CV (German: Nutzfahrzeug)
NHTSA	National Highway Traffic Safety Administration (USA)
NIDA	emergency, information and documentation assistant
NiGu	nitroguanidine
NVH	noise, vibration, harshness
OBD	onboard diagnostic connector
OCS	occupant classification system
ODB	offset deformable barrier
ODS	occupant detection system (occupant presence sensing system)
OFDM	orthogonal frequency division multiplexing
OoP	out-of-position
ODD	operational design domain
OLC	occupant load criterion
OPW	one piece woven
PA	polyamide (plastic material)
PAB	passenger airbag module (front passenger airbag module)
PAS	peripheral acceleration sensor
PaSIx	index for passive safety
PC	polycarbonate (plastic material)
PdB	Partnership for Dummy Technology and Biomechanics (Germany)
PES	polyester (textile)
PET	polyethylene terephthalate (plastic material)

PHEV	plug-in hybrid vehicle
PKI	public key infrastructure
PKW	passenger car (German: Personenkraftwagen)
PLL	progressive load limiter
PODS	probability of death score
PRE-SAFE®	preventive safety system (Mercedes-Benz)
PSI	peripheral sensor interface
PT	prototype
PTW	powered two-wheeler
PVB	polyvinyl butyral (plastic material)
PWM	pulse width modulation
QPSK	quadrature phase-shift keying or quaternary phase-shift keying
Radar	radio detection and ranging
RAM	random access memory
RCA	Root Certification Authority
RCAR	Research Council for Automobile Repairs
RDB	ride-down benefit
RDE	ride-down effect
RNS	latching cam buckle (seat belt buckle)
ROM	read only memory
R/P	retractor pretensioner
RSAB	rear seat airbag module
RSM	roadside safety message
RUPD	rear underride protection device (truck underride protection device)
RWW	road works warning
SAB	side airbag module
SAE	Society of Automotive Engineers
SbW	steer-by-wire
SDM	management of simulation data
SDM	sensing and diagnostic module (in airbag control unit)
SDSM	sensor data sharing message
SE	simultaneous engineering
SG	safety level (German: Sicherheitsgrad)
SID	side impact dummy
SiKriS	system of safety criteria
SIX	safety index
SLL	switchable load limiter
SPAT	signal phase and timing
SPR	snake pretensioner
SPUL	specific accident power
SRAB	seat ramp airbag module
SRP	seat reference point
SRS	supplemental restraint system (airbag, additional restraint system)
StVZO	Road Traffic Licensing Regulations (Germany)
TA	partial assembly (German: Teilaufbau)

TAI	international atomic time
TAU	tether activation unit (control of different airbag sizes)
THOR	**T**est device for **H**uman **O**ccupant **R**estraint (frontal impact dummy)
THUMS	**T**otal **Hu**man **M**odel for **S**afety
TI	tibia index
TOR	takeover request
TPE	thermoplastic elastomers (plastic material)
TTI	thoracic trauma index
TTA	time to arrival
TTC	time to collision
TTL	time to lane crossing
UCSSS	universal child safety seat system, other/regional name for ISOFIX (USA)
UDS	unified diagnostic services
UNECE	United Nations Economic Commission for Europe
UN-R	UN-Regulation
V2X	vehicle to everything
VC	viscous criterion
VDA	Association of the Automotive Industry e.V. (Germany)
VDS	Vehicle Dependability Study (USA, J.D. Power Study)
VP	virtual prototype (also: DPT)
VRU	vulnerable road user
VS	vehicle sensing
VSG	laminated safety glass (German: Verbund-Sicherheitsglas)
WAD	wrap around distance (unwinding length for determining the impact areas in pedestrian protection tests)
WCU	wheelchair user
WHO	World Health Organization
WS	WorldSID (side impact dummy)
WS	web sense (webbing sensing)
WSU	Wayne State University
ZFZR	National Vehicle Register (German: Zentrales Fahrzeugregister)

1 The Integral Safety

Rodolfo Schöneburg and Florian Kramer

Motor vehicles already demonstrate a high level of safety for occupants and for other road users (e.g., pedestrians and cyclists); further improvement presents the automotive industry with significant challenges that can only be addressed through an expansion of objectives and close collaboration between manufacturers, suppliers, and engineering service providers towards a common solution.

While passive safety deals with accident consequence-reducing measures to limit the effects of accidents, active safety refers to the prevention of accidents and the reduction of their frequency. Integral safety now links both areas with the aim of further increasing the protection potential of all road users. In this context, the entire accident sequence from the cause of the accident, through the collision, to the rescue services is considered holistically, and protective measures are developed, tested, and implemented as standard in vehicles. The efforts of safety engineers aim to recognize dangerous situations during normal driving conditions early on, interpret them, and consequently initiate measures for accident prevention and minimization of accident consequences.

Assistance systems, conditionable and reversible occupant protection systems, or automatic emergency call systems, which will be discussed in more detail below, are already state of the art at many car manufacturers and are available to users of newer vehicles. Viewed individually, these systems undoubtedly improve the potential for accident prevention. However, by using the information and data obtained through the assistance systems, if a collision appears unavoidable, occupant protection systems can be

R. Schöneburg (✉)
RSC Safety Engineering, Hechingen, Germany
e-mail: rsc@rschoeneburg.de

F. Kramer

R. Schöneburg (ed.), *Integral Vehicle Safety*,
https://doi.org/10.1007/978-3-658-48100-1_1

activated early, even before the impact, so that the consequences of the accident can be reduced, or at least the accident severity can be minimized, leading to an improvement in vehicle safety. This link between active and passive safety through the pre-accident or pre-crash phase characterizes "integral safety".

1.1 Basic Concepts in Safety Science

When using the term "safety", one is immediately faced with the question of whether there can be absolute safety, i.e., "freedom from danger" [8] for road traffic. Given the annual accident consequences, this question must probably be answered in the negative. So what does "safety" mean?

Firstly, the uncertainty of road traffic undoubtedly results from the risk, from a probabilistic (probability-theoretical) perspective, of suffering a damage of a certain extent to persons and/or property. The accident (as a synonym for the non-normal case, the disturbance case) can therefore be defined as follows:

Accident is an event in which the deviation between the given driving task and its fulfillment exceeds an acceptable level (unmanaged control task) and as a direct consequence, a damage of a certain type and severity occurs.

According to [2], a **damage** (or an injury) is a disadvantage due to injury to legal goods as a result of a specific technical process or condition.

The extent of the damage can be indicated on the involved motor vehicle by property damage (e.g., repair costs, replacement value, loss of value due to damage) and on the involved person by personal injury (e.g., degrees of injury severity, injury-related costs).

Following the definition also established in [2], **risk**, which is associated with a specific technical process or condition and is summarized by a probability statement, is

- the expected frequency of the occurrence of an event leading to damage and
- the expected extent of damage in the event of an occurrence.

For example, this risk definition is used by Schmid, where the product $R = W \cdot S$ of probability W and damage amount S is used for the comparative juxtaposition of different accident scenarios. The product is referred to as a "compatibility parameter" because it is suitable "for describing the benefits of safety measures" [10]. However, the adoption of this risk concept will not be used in the following, because the term "accident consequences" or, as a monetized parameter for the product of frequency and injury severity, the term "injury-related costs" is more appropriate [5]. Furthermore, the product formation contradicts the risk function to be shown later, with which the probability of a certain injury severity level is to be represented depending on the protection criterion. Risk is therefore understood in its broader definition as a probability statement that describes the expected value of a frequency for a certain damage extent in the event of its occurrence.

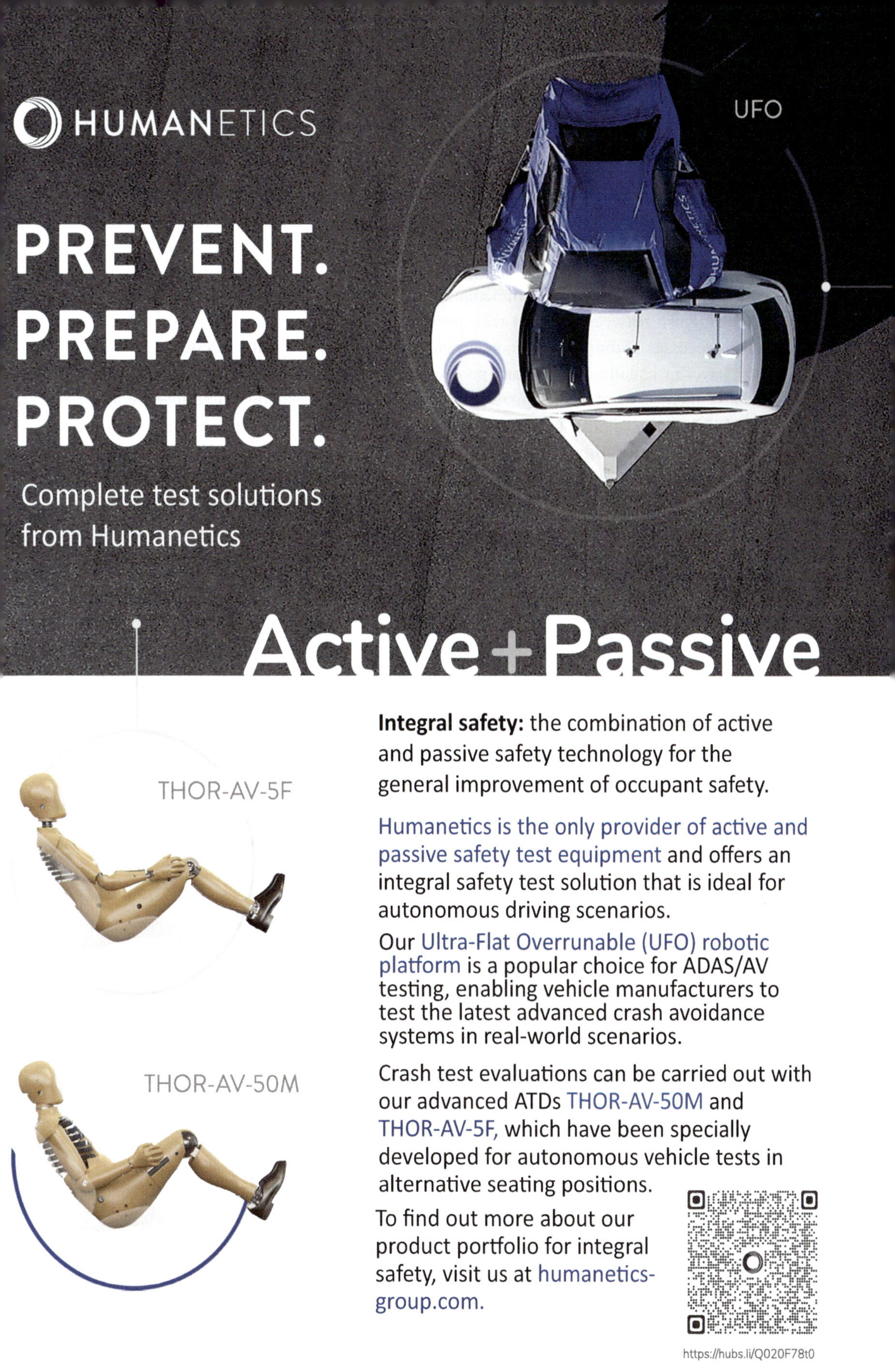
HUMANETICS
UFO
PREVENT.
PREPARE.
PROTECT.
Complete test solutions from Humanetics
Active + Passive
THOR-AV-5F
THOR-AV-50M
Integral safety: the combination of active and passive safety technology for the general improvement of occupant safety.
Humanetics is the only provider of active and passive safety test equipment and offers an integral safety test solution that is ideal for autonomous driving scenarios.
Our Ultra-Flat Overrunable (UFO) robotic platform is a popular choice for ADAS/AV testing, enabling vehicle manufacturers to test the latest advanced crash avoidance systems in real-world scenarios.
Crash test evaluations can be carried out with our advanced ATDs THOR-AV-50M and THOR-AV-5F, which have been specially developed for autonomous vehicle tests in alternative seating positions.
To find out more about our product portfolio for integral safety, visit us at humanetics-group.com.
https://hubs.li/Q020F78t0

With the introduction and explanation of the risk definition, the danger can be formulated according to [2] as follows:

Danger is a situation in which the risk is greater than the greatest still acceptable risk (border risk) of a certain technical process or state, and according to [3] is **hazard** a spatially and temporally as well as in terms of type, size and direction specific danger to a person, thing or function.

Thus, the accident, which immediately results in a damage, presupposes a hazard. To distinguish between accident-relevant driving maneuvers with and without damage, those processes without damage occurrence are referred to as near-accidents. Although they have an equally high hazard potential as accidents, they do not lead to damage because there is no damage-triggering circumstance (for example, a collision counterparty). Therefore, danger can be equated with uncertainty. As a result, safety is understood as the probability with which no danger emanates from a unit of observation during a certain time. This safety concept proposed by Meyna in [8], which has an interdisciplinary character from a safety science perspective, can be applied regardless of the respective field of expertise. However, it requires a certain precision for application in the field of motor vehicle safety: Uncertainty and safety are complementary terms, so that with the above equating of danger and uncertainty, the standardized sizes "safety" and "danger" can be supplemented to one. Safety can thus be formulated as

$$\text{safety} = 1 - \text{danger}$$

In reversal of the definition for danger, including the risk concept, safety (in reference to [2]) can be expressed as: **Safety** is a situation in which the risk is smaller than the greatest still acceptable risk (border risk) of a certain technical process or state.

Since the "acceptable risk" can only be quantified in the rarest cases, for example, for time series representation, safety can be considered sufficient using non-standardized parameters. The increase in safety is essentially achieved by reducing the risk, either by reducing the probability of an event leading to damage (accident prevention) or by reducing the extent of damage (mitigation of accident consequences).

1.2 Basics of Integral Safety

Traffic encompasses equally the people participating in traffic—regardless of their nature (driver, passenger, pedestrian)—as well as the vehicles and the traffic space. The traffic safety is accordingly oriented towards the road user, the means of transportation and the traffic routes. The safety of road traffic is aimed at the human, the vehicle, and the environment (Fig. 1.1).

Measures to improve traffic safety are distinguished into

- accident-avoiding measures to reduce the frequency of accidents and
- accident consequence-reducing measures to limit the expected damage.

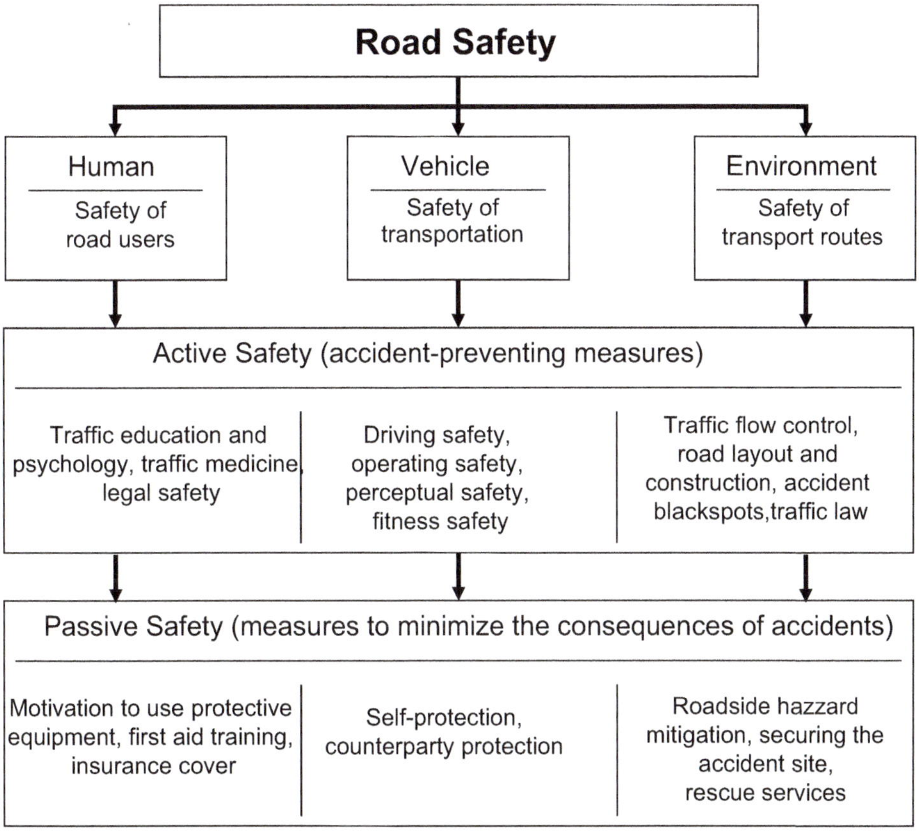

Fig. 1.1 Subareas of road safety and examples of safety measures

The accident-avoiding measures are assigned to the area of active safety and the accident consequence-reducing measures to the area of passive safety. They can be divided, as shown in Fig. 1.1 by some examples, into measures that influence the human, the vehicle, and the environment in terms of accident prevention or mitigation of accident consequences. Although the terms primary and secondary safety, which are frequently used in English, have a more accurate meaning—since the aim is first to avoid accidents and, if they do occur, to mitigate the consequences in the second line—these terms are not used further in the following, due to the established use of the terms "active" and "passive" safety.

To clarify the question of how the temporal development of the two safety areas (active and passive safety) has been shaped in the past, the usual safety indices, such as fatalities or injuries per registered or crashed vehicle or per kilometer number, are not very meaningful. For one thing, the relationship between accident numbers and vehicle inventory (Fig. 1.2) is highly non-linear and shows phases dependent on economic development in Germany: The restoration phase (up to about 1961) is characterized by a rapid

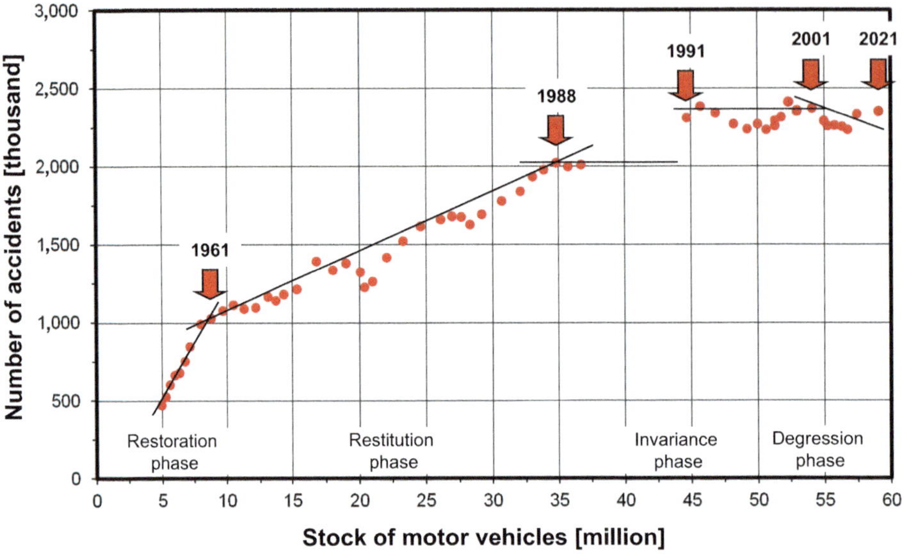

Fig. 1.2 Development of accidents in relation to the registration numbers in the period 1953 to 2021

increase in accident development depending on the number of registered motor vehicles, while the increase in the restitution phase (up to 1988) is significantly flatter. From the end of the 1980s, a stagnation sets in, i.e. the number of accidents remains largely constant despite an increasing vehicle inventory. This invariance phase continues after the reunification of the two parts of Germany at a significantly higher level and is replaced at the turn of the millennium by the regression phase. It is characterized by a slight decline in the number of accidents with a still increasing vehicle inventory; it should be taken into account that since 2008, temporarily decommissioned motor vehicles are no longer included in the vehicle inventory.

In such representations, however, the quite different safety advancements in the individual areas cannot be separated from each other. On the other hand, when forming quotients of injured and traffic or accident numbers, there are no indications of absolute numbers. And finally, the statements focused solely on killed road users are too vague, as they do not take into account the consequences of injuries in their entirety. The latter is particularly true considering the fact that over an observation period of about 60 years, the proportion of national economic injury-related costs for fatally injured has settled at about 18%, for severely injured at about 52%, and for slightly injured at 30% [6].

For separate representation, first, as an indicator of active safety, the approved vehicles involved in accidents are considered. Thus, the degree of active safety, expressed by the quotient of motor vehicle inventory and the number of accident vehicles, increases with a decreasing number of accident vehicles at the same inventory. Active safety

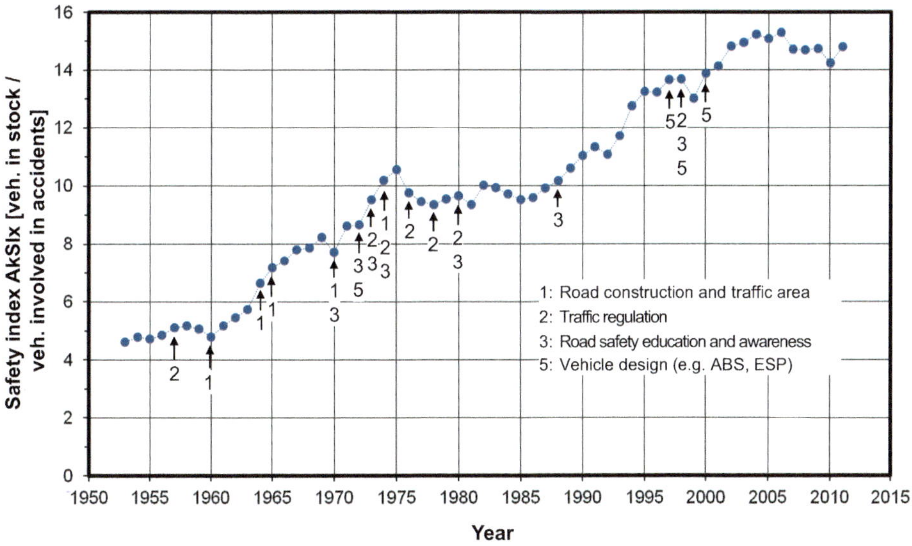

Fig. 1.3 Temporal development of active safety in the period 1953 to 2011 and introduction of measures for its improvement

also increases with the inventory of approved vehicles at the same number of accident vehicles. The criterion for passive safety is defined by the quotient of the number of all police-reported accidents and the injury-related costs (for scale reasons in millions of €). Accordingly, passive safety increases with decreasing consequential costs for fatalities and injuries or with the increase in accidents without simultaneous cost increase.

The derivation of the metrics for active and passive safety is detailed in [7], so here only the result in the form of the temporal development over a period of about 60 years should be shown (Figs. 1.3 and 1.4). For active safety, a tripling can be stated for the period from 1953 to 2011, and for passive safety even an eightfold increase. The two developments are each assigned to legally defined regulations in the field of active and passive safety. Even if the individual legal regulations and rules—with the exception of the reinforcement of fines for not wearing seat belts in August 1984 (compare increase from 1983 to 1985 in Fig. 1.4)—do not show immediate, clearly assignable effects, the increase of the two curves is undoubtedly due to the overall effect of interlocking individual measures [1].

These were measures:

- in road construction and through the design of the road area (1),
- in traffic regulation (2),
- in traffic education and awareness (3),
- in the form of protective devices (4),

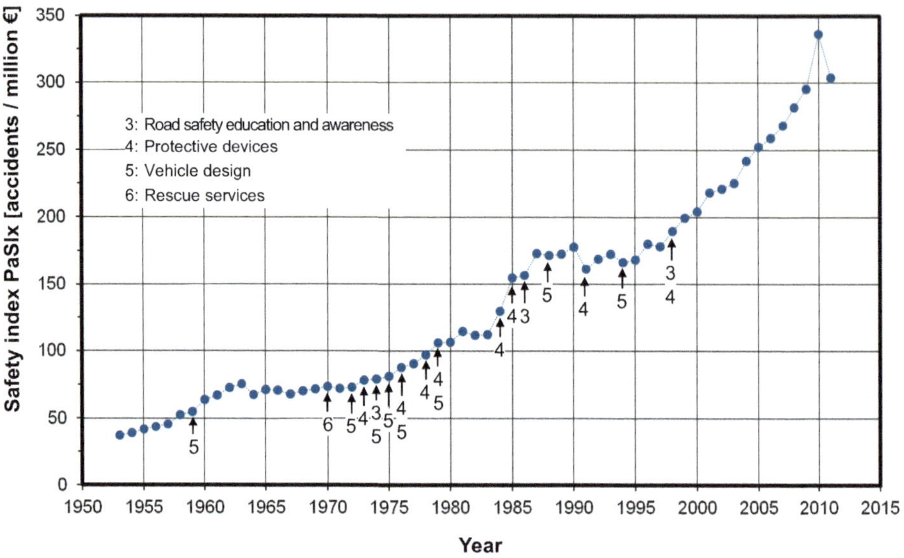

Fig. 1.4 Temporal development of passive safety in the period 1953 to 2011 and introduction of measures for its improvement

- in the design of the vehicles (5) and
- in rescue services (6).

In the field of vehicle safety, the priorities for action are primarily derived from accident research; it consists of accident data collection and statistics, accident reconstruction and accident analysis.

The passive vehicle safety and the associated impact areas are shown in Fig. 1.5. Using knowledge about injury causation from the field of biomechanics and safety assessment based on laws and ratings, safety measures for the mitigation of accident consequences are developed, designed, and tested in vehicles. Statements about the effectiveness of these measures (e.g., seat belt systems, airbags, etc.) can be made in the early development phases primarily using computational simulation.

According to this exemplary process, the present book is structured. After describing the basic **safety requirements** for the motor vehicle based on accident research, biomechanics as well as laws and ratings (safety assessment) in Chaps. 2 to 4, **measures of integral safety** are derived and conveyed in Chaps. 5 to 7. The last chapter deals extensively with the **vehicle development process**, including how and when the 'development tools' computational and experimental simulation should be used appropriately.

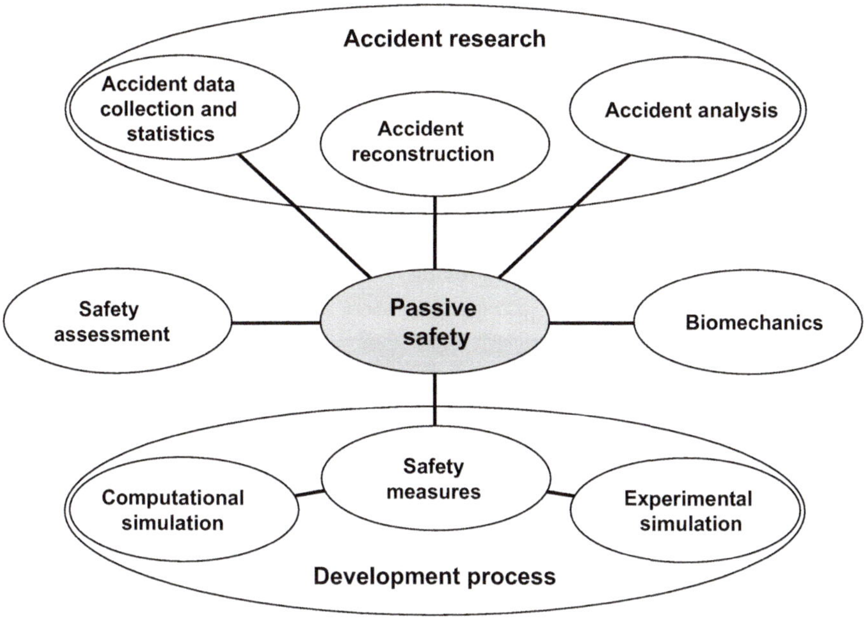

Fig. 1.5 The passive safety of motor vehicles and their impact areas

1.3 New Field of Action Pre-Accident Phase and the Integral Safety Approach

The traditional division of vehicle safety into active and passive safety is no longer sufficient [13]. An important safety phase is missing, which is now the focus of development for many new safety systems: preventive safety. What vehicle measures can be initiated as a precaution when a risk of accident is detected to improve occupant and partner protection?

Figure 1.6 shows the escalation stages of an accident and describes the phases of integral safety with 'safe driving', 'act preventively', 'protect as needed' and 'rescue, secure'. Every accident begins with a critical situation and ends with the securing of the vehicle and rescue of the accident participants.

This representation of an accident process leads to the integral safety approach [11], as it is now used by many vehicle manufacturers and suppliers in the field of vehicle safety and is shown in Fig. 1.7. It is not only divided into active safety—accident prevention and accident severity mitigation—and passive safety—mitigation of accident consequences—but particular importance is given to the transition phase shortly before the collision.

According to this safety strategy, protective measures are initiated by the vehicle in the event of an acute risk of accident. These can serve accident prevention or prepare

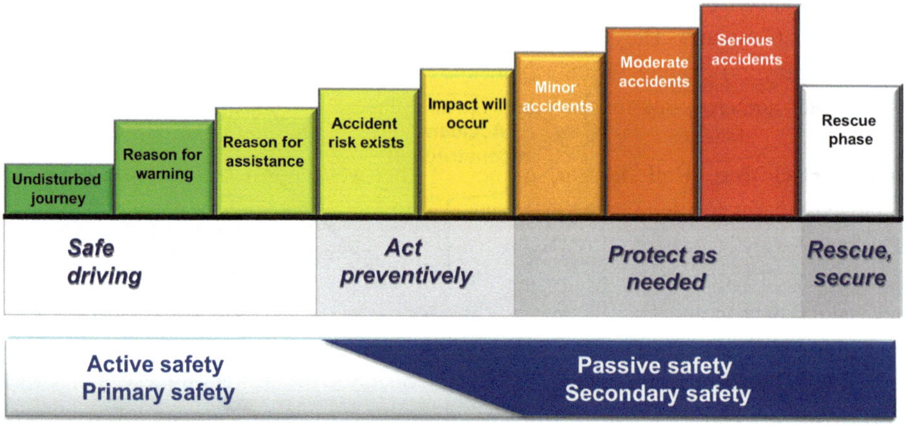

Fig. 1.6 Escalation of the accident process and the four safety phases

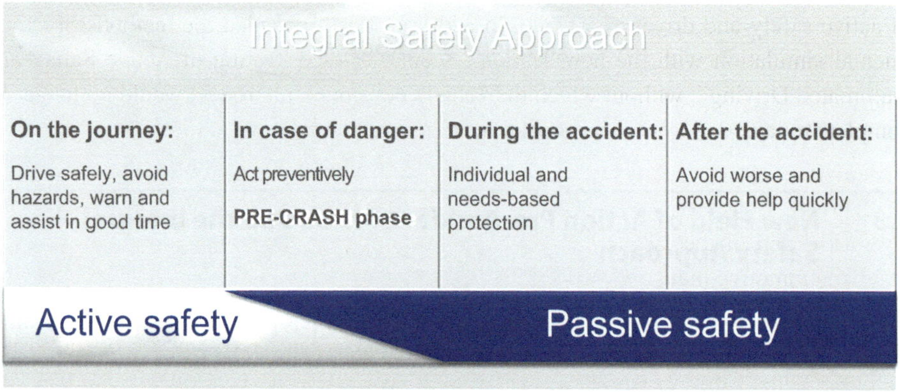

Fig. 1.7 The four phases of the integral safety approach

the vehicle occupants for the impending collision. The transition from active and passive safety is therefore fluid in this phase. Ideally, this phase, often referred to as pre-accident or pre-crash phase, is preceded by a warning to prompt the driver to take action.

The integral safety approach is closely associated with the automobile brand Mercedes-Benz. In 2000, the approach of integral safety was first presented at the Int. Automobile Exhibition IAA in Frankfurt and in 2001 finally PRE-SAFE®, the world's first preventive passenger car safety system, was shown at the ESV conference in Amsterdam [12]. The first series introduction took place as early as 2002, with reversible seat belts for the front occupants, seat conditioning as well as side window and sunroof closing in the model update of the then S-Class W220.

The holistic understanding of integral safety and the use of preventive measures unleashed innovative forces and led to a dynamic development of new safety systems. While in a first step primarily reversible protective measures for preparing the occupants for a possible accident were the focus of development, it quickly became clear that the new solutions could not stop at driving dynamic interventions. The goal was to reduce energy through autonomous braking interventions just before impact in order to reduce the own accident severity and that of a possible impact opponent—it is basically a virtual or electronical 'crumple zone' in front of the vehicle. This approach can be described as the central innovation of accident prevention and as the birth of future, accident-avoiding safety systems.

1.4 The 'Vision Zero'

The automotive industry faces significant challenges. In addition to the central competence and responsibility of making individual transport climate-neutral, and alongside new technological possibilities for communication and networking of the entire environment, it is also expected that the user behavior of vehicle drivers and all other people involved in traffic will change. Among all these aspects, a prominent goal remains to make road traffic safer. The fact that the number of traffic fatalities has almost stagnated since 2010 in many western countries is a clear signal that all efforts need to be intensified.

At the end of the 1990s, parallel to the development of the integral safety approach in the automotive industry, the idea of 'Vision Zero' [15] was first raised by the Swedish government as a strategy for traffic safety. Presumably, the steady decrease in the number of traffic fatalities in Europe during these years fueled the hope of an foreseeable achievement of this vision. In many parts of the world, the idea of this approach was adopted, but sometimes interpreted differently: 'Zero traffic fatalities', 'No serious injuries' or even 'No traffic accidents'.

'Vision Zero' is based on four principles [4]:

- Life is non-negotiable
- Humans are fallible
- Tolerable limits lie in the physical resilience of humans
- People have a right to a safe transport system.

Every type of mobility carries its specific risks, this has been the case since the invention and spread of the automobile. For example, in 1903 there were 215 killed road users [9] in London. The causes were naturally different than today, back then more than 90% of the accidents were caused by horses, horse-drawn carts and carriages. Whenever people move at finite speed on a plane, no matter how, it is associated with risks. And so it is not surprising that in 1886 the first automobile was also in focus in terms of safety. Thus,

Karl Benz advertised in 1888 for his 'Patent-Motorwagen' that the new automobile could be operated 'comfortably and absolutely safely!', and that 'steering, stopping and braking are easier and safer than with ordinary carriages' [14].

But even with an optimistic view of the vehicle-side possibilities, it was clear from the outset that road traffic is too complex to solve the safety problem with the vehicle alone. The risk of an accident is significantly influenced by the way road traffic takes place now and in the future. Compared to the very safe means of transportation, such as train or airplane, road traffic is much more complex and associated with much greater individual freedoms:

- There is no spatial and temporal separation of road users and traffic flows in the common traffic area.
- The speed differences of the road users in a confined space are very high and the movements take place on one plane.
- The most diverse road users are individually on the move in the traffic area (from trucks, to cars, to pedestrians and two-wheel riders).
- The technological standards of the vehicles that meet in traffic are highly different.
- The individual experiences and the condition of the road users are very different.
- Every road user acts individually.

In addition to the properties of the automobile itself, the design of the transport infrastructure, the traffic environment, as well as human influences and traffic rules, significantly determine the accident risk. Similar to the integral safety approach (Fig. 1.7), which is based on the premise that accident prevention, protective measures already before an impending accident with preventive safety systems, occupant and partner protection during the accident itself as well as accident aftercare must have equal importance in vehicle development. We can only get closer to 'Vision Zero' if all influencing factors are examined from an integrated perspective of 'road safety' (Fig. 1.1). All areas of traffic safety must be addressed, i.e., vehicle, infrastructural safety measures and the human factor in road traffic, all of which must contribute to getting closer to 'Vision Zero'.

References

1. Brühning, E. et al.: Zum Rückgang der Getötetenzahlen im Straßenverkehr der Bundesrepublik Deutschland von 1970 bis 1984. Zeitschrift für Verkehrssicherheit, Nr. 3 (1986)
2. DIN-VDE 31000: Allgemeine Leitsätze für das sicherheitsgerechte Gestalten technischer Erzeugnisse – Begriffe der Sicherheitstechnik, Grundbegriffe (1987)
3. DIN 31004 (Vornorm): Sicherheit und Schutz im Arbeitssystem – Begriffe, Wertzusammensetzungen (1982)

4. DVR Deutscher Verkehrssicherheitsrat: Vision Zero – Grundlagen und Strategien. Schriftenreihe Verkehrssicherheit Nr. 16, Bonn (2012)
5. Kramer, F.: Schutzkriterien für den Fahrzeug-Insassen im Falle sagittaler Belastung. Fortschritt-Berichte, VDI-Reihe 12. Bd 137, VDI Verlag, Düsseldorf (1989)
6. Kramer, F.: Passive Sicherheit/Biomechanik I und II. Script for the course of the same name at the Hochschule für Technik und Wirtschaft (HTW), Dresden (2011)
7. Kramer, F.: Zur Quantifizierung der Straßenverkehrssicherheit. Research report Nr. 323/88. Technische Universität Berlin, Berlin (1988)
8. Meyna, A.: Beitrag zur Entwicklung einer allgemeinen probabilistischen Sicherheitstheorie. Habilitationsschrift, Fachbereich 14 (Sicherheitstechnik). Gesamthochschule Wuppertal, Wuppertal (1980)
9. Niemann, H.: Béla Barényi – Sicherheitstechnik made by Mercedes-Benz. ISBN 978–3–613–04237–7. Motorbuch Verlag, Stuttgart (2002)
10. Schmid, W.: Vermindertes Verletzungsrisiko bei Verkehrsunfällen durch Fahrzeug-Kompatibilität. XVIII. FISITA Congress, Hamburg (1980)
11. Schöneburg, R.: Auf dem Weg zum unfallfreien Fahren – Sicherheitsstrategie Mercedes-Benz. 9. GMTTB Jahrestagung, Konstanz (2019)
12. Schöneburg, R., Baumann, K.-H., Justen, R..: A Vision for an Integrated Safety Concept. 17th ESV-Conference, Paper Number 01–493-O. Amsterdam, Netherlands (2001)
13. Schöneburg, R.: Integrale Fahrzeugsicherheit I und II. Script for the joint course of the same name at the TU and HTW Dresden (2019/2020)
14. Stolle, S.: Das Thema Sicherheit in der deutschen Anzeigenwerbung für Automobile. Betriebswirtschaftliche Schriften, Heft 159. ISBN 3–428–11283–0. Duncker&Humblot, Berlin (2004)
15. Tingvall, C. et al.: Vision Zero: An ethical approach to safety and mobility. 6th ITE Int. Conf. on Road Safety and Traffic Enforcement: Beyond 2000. Melbourne (1999)

Accident Research

2

Rodolfo Schöneburg and Florian Kramer

This chapter explores the extent and manner in which road traffic accidents occur and under what circumstances road users suffer damage. However, before that, the significance of traffic accident victims in the cause of death statistics will be highlighted.

According to federal statistics [20], the causes of death of the 858,768 people who died in Germany in 2010 can be divided into carcinomas (25.5%), diseases of the circulatory system (41.1%), heart muscle diseases, e.g., myocardial infarctions (6.5%), respiratory diseases (7.0%), digestive diseases (5.0%) and other natural causes of death, e.g., infections, tuberculosis, diabetes, etc., (11.1%) as well as deaths due to accidents and violence (3.9%). These main groups are shown in Fig. 2.1.

The last main group mentioned consists of a total of 33,312 cases with unnatural cause of death, i.e., injuries and poisonings. These can be divided into two groups, accidents and violence, namely traffic accidents (11.0%), home and leisure accidents (32.5%), work accidents (1.0%) and other accidents, such as deaths from falls, fire, weapons, chemical and biological substances as well as late effects of accidents (17.0%). The other group includes suicide and self-harm (30.1%) as well as other violence (8.4%), including murder and manslaughter. The respective proportions are shown in Fig. 2.2.

Of those who died in 2010 (858,768 deaths), a total of 3,648 people died in road traffic accidents, which is 0.42%. This was significantly fewer than suicide victims, but significantly more than violently killed persons.

Accident research includes the three areas of accident survey and statistics, accident reconstruction and accident analysis (Fig. 2.3). The competence and responsibility of

R. Schöneburg (✉)
RSC Safety Engineering, Hechingen, Germany
e-mail: rsc@rschoeneburg.de

F. Kramer

R. Schöneburg (ed.), *Integral Vehicle Safety*,
https://doi.org/10.1007/978-3-658-48100-1_2

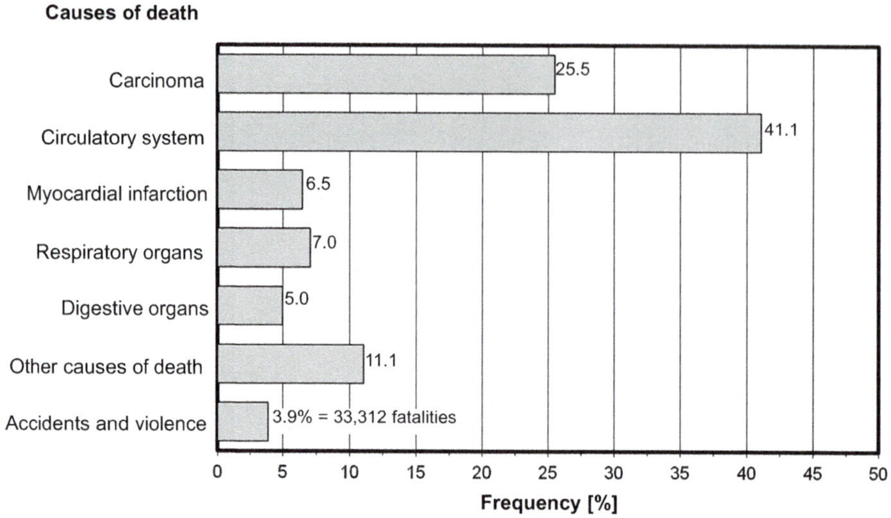

Fig. 2.1 Main groups of causes of death (n = 858,768) in 2010

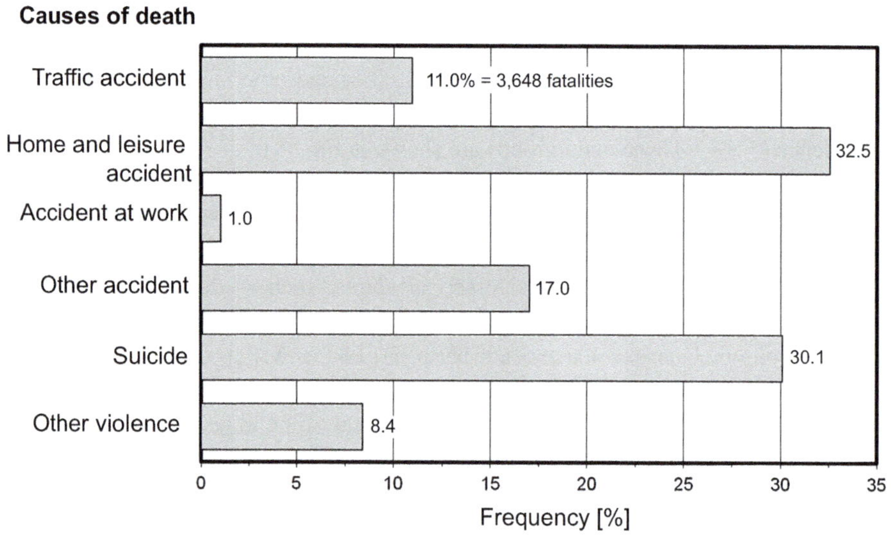

Fig. 2.2 Accidents and violence as causes of death (n = 33,312) in 2010

accident research is to clarify the causes that lead to an accident. From this explanation, insights can be derived, the implementation of which in technology, medicine, and psychology contributes to improving road traffic safety [19].

In the technical area of accident research, attempts are made to obtain characteristic quantities that describe the behavior of humans, vehicles and the environment during the

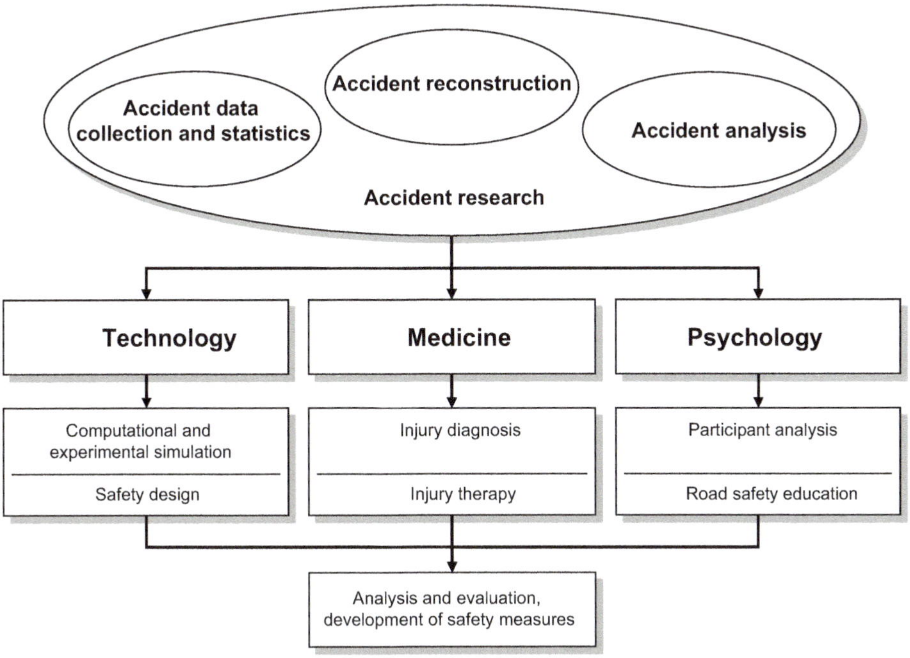

Fig. 2.3 Accident research and implementation of findings in various disciplines

accident using simulations, experiments with volunteers, corpses and dummies, or mathematical models of the respective road users (occupants, pedestrians, etc.). These quantities represent an equivalent to the injuries to humans and the damage to or in the vehicle as well as the influence of the environment. From them, a safety design to improve road traffic safety can be formulated.

In the medical area of accident research, hospital reports, medical diagnoses, and pathological findings of accident victims are evaluated with regard to injury patterns, frequency, and severity in order to gain insight into the causes and circumstances of the injuries [16].

In the psychological field of accident research, the causes leading to an accident are determined and analyzed in the three directions of driver, vehicle, and environment by interviewing the accident participants and witnesses. The risk of accidents in road traffic is usually operationalized by determining accident rates [18] or by means of relative accident frequencies, as shown in Fig. 1.3 using the measure to quantify active safety [13]. Safety measures aim at the traffic educational aspect, namely traffic education and enlightenment. With the development and introduction of assistance systems, the goal is also pursued to inform and warn the driver in critical driving situations so that he can adapt his behavior to such situations.

2.1 Accident Data Collection and Statistics

Traffic accidents contribute significantly to unnatural causes of death (see Fig. 2.2). Collecting and evaluating information about these types of accidents provides accident researchers with an opportunity to understand accident and injury mechanics and can suggest solutions for accident prevention and injury reduction. In contrast to simulation in the laboratory or on the computer, where all variables can be measured, calculated, kept constant or checked, most physical quantities in vehicle accidents cannot be measured and also change within a very short time.

The problem of accident researchers is to obtain sufficient information to understand the mechanisms of injury causation. Vehicle accidents account for the largest proportion of injuries and are the most difficult to document. Currently, there are two types of publicly accessible accident materials, which essentially differ in their level of detail on the one hand and their case number on the other: Police collected accident data (summarized in federal statistics) and so-called "in-depth" investigations conducted by professional accident teams.

2.1.1 Objective of Accident Statistics

The purpose of accident statistics is to process the information obtained during accident data collection. This is done using statistical methods, applying descriptive or analytical procedures. In descriptive, or explanatory statistics, frequency distributions and their summations, the distribution functions, are usually plotted to identify certain characteristics or to represent probabilities. With analytical statistics, previously formulated prediction models or correlation hypotheses are tested and possibly confirmed. Examples of both statistical procedures are the content of the following chapters.

2.1.2 Police-Collected Accident Data

The accident data from the **Federal Statistical Office** refers to police accident report forms, which are primarily used to clarify the question of fault (Fig. 2.4). All reported accidents are recorded, but they provide only limited specific information about injuries. The evaluation is subject to the assessment of the respective police officers, who often have sufficient experience regarding the causes and origins of injuries, but the accident data forms do not allow for the entry of such information (see, for example, [21]). Since all police-recorded accidents are recorded, this accident data provides a representative overview of accident occurrence in Germany. However, it does not allow statements about the accident constellations (impact points, damage areas, etc.) nor about the external accident severity parameters (speed, mass, stiffness, design, etc.). On the other hand, the data material contains no information about the seating position of the occupants;

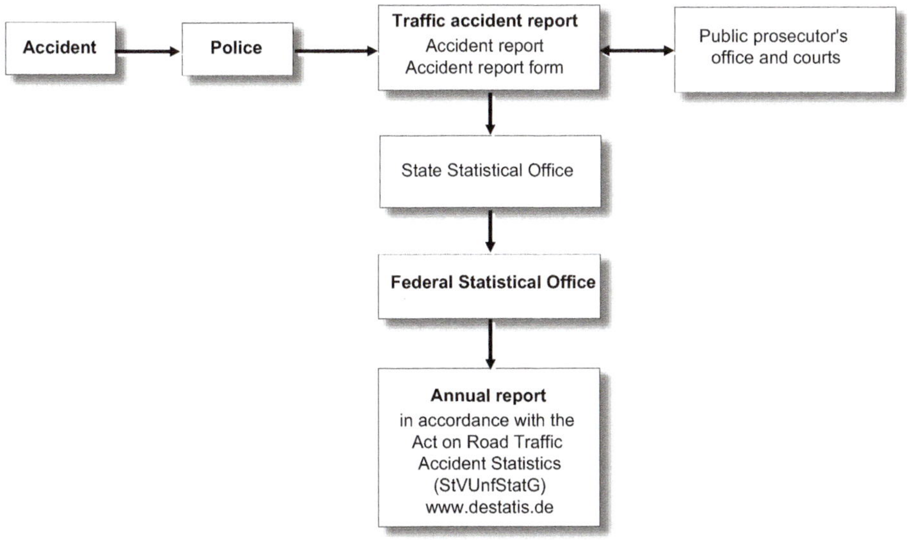

Fig. 2.4 Data collection for the provision of the federal statistics accident data

only rough estimates of the overall injuries in terms of light, severe and fatally injured are possible [14].

Due to the fact that only police-reported accidents are included in the federal statistics, there is a certain number of unreported cases compared to the real accident occurrence. These include, for example, the often-cited "disco accidents", which often occur under the influence of alcohol at night after a local stay of the occupants and where there are no seriously injured. For obvious reasons, no police are called to record such accidents.

The **Fatal Accident Reporting System (FARS)**, which has been in existence in the USA since 1975, was founded by the US National Highway Traffic Safety Administration (NHTSA). The most thoroughly investigated accidents are those with fatal outcomes for one or more participants, hence the information collected in such accidents is more complete than in lighter accidents. The US traffic safety authority finances the central provision of fatal accident data in the FARS database. Accident specialists trained by the NHTSA in all states compile a joint, complete report of all fatal accidents that occur in their states, which is stored in Washington in a data-technical manner and is publicly accessible via the NHTSA [5]. The FARS database contains data from police accident reports, from the registration authority and from medical reports (including pathological findings); it thus represents the most comprehensive and detailed accident data material on severe traffic accidents in the USA.

2.1.3 In-Depth Investigations

For in-depth investigations, accident research teams collect a large amount of information about environmental conditions, vehicle condition and injuries of the participants and their causes usually immediately after an accident. This is done by reviewing police and medical reports, checking the vehicles and interviewing accident participants. Due to the high costs of this time-consuming accident data collection, the number of such investigations is very limited (see Fig. 2.5).

The accident data material of the **German Insurance Association (GDV)** includes in its newer study 140,000 insurance cases with personal injuries. Such accident data collections have been carried out at periodic intervals since 1969. The accident data is suitable for analyzing the accident events. For the retrospective evaluation by engineers and doctors, the entire insurance damage files with all documents about cause of accident, accident sequence, vehicle damage and injury consequences were available [7].

The data material from the **German In-depth Accident Study (GIDAS)**, which comes from the Dresden University of Technology (TUD) and until 2019 from the Medical School Hanover (MHH), contains data from surveys directly at the accident site with a multitude of information about each accident, which allow a more in-depth analysis regarding active and passive safety. These are accidents with personal injury, because the focus of the research project funded by the Federal Highway Research Institute Germany (BASt) and the German Forschungsvereinigung Automobiltechnik e. V. (FAT) is on the medical sector. As part of the data collection for active safety, interviews with those involved and witnesses at the accident site, sometimes also later in the hospital, are conducted. These are intended to provide information about the personality of the

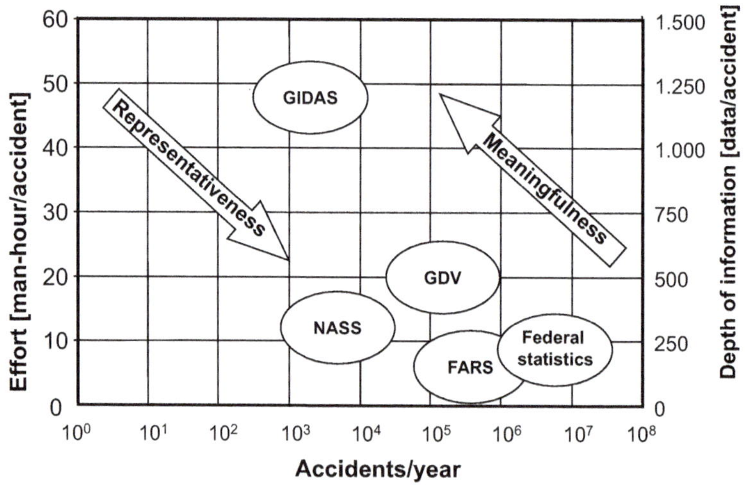

Fig. 2.5 Effort, detailing and number of cases in accident data collection

interviewee, the cause and origin of the accident, the course of the journey, the purpose of the journey, as well as reactions to the accident occurrence and its history [8]. In order to sustainably improve the statistical reliability of the local accident survey, the accident data collection has been carried out since the beginning of 1985 according to a sampling procedure with the aim of achieving better agreement with federal statistics and thus achieving general validity of the statements for the accident occurrence within Germany [14].

In the accident data surveys of various **car manufacturers** (e.g., Audi, BMW, Mercedes-Benz or VW), accidents in which their own vehicles were involved as collision objects are specifically investigated. The aim here is to improve the safety of their own products. However, the availability of data and information poses a problem for the non-internal accident researcher.

The in-depth investigation program conducted by the NHTSA in the USA since 1979, the **National Automotive Sampling System (NASS)** Crashworthiness Data System (CDS), which, like GIDAS, creates studies on traffic accidents, contains just 4,648 accidents in the 1989 data material (see Fig. 2.5). In earlier data collections, emphasis was placed on special accident constellations, such as rollovers, fires, and fatal accidents, in order to investigate causes and circumstances, identify characteristic sizes, and initiate measures for accident prevention, i.e., to improve active safety. The NASS system randomly selects accidents in randomly selected areas of individual states in the hope that the relatively small number of cases is statistically representative of the accident occurrence in the USA. The information contained in the NASS data material is therefore dependent on the collection location. The CDS program launched in 1988 focused on passive vehicle safety; the parallel data collection program, the General Estimates System (GES), incorporates information from police reports to complete the CDS data, so that insights into vehicle identification, damage, window destruction, intrusions, and accident severity in the form of speed change are available. In addition, the data material contains information about the size and weight of the occupants, the seating position, the use of occupant protection systems, etc. [5].

Finally, there is the **Highway Loss Data Institute (HLDI)**, an institution of insurers, whose accident research results are incorporated into US safety legislation in cooperation with the NHTSA via the relevant authorities.

Figure 2.5 shows the relationship between the effort of data collection and analysis and the number of data for each accident on the one hand, and the achieved number of cases per year on the other hand.

2.2 Accident Mechanics and Reconstruction

The reconstruction of accidents fundamentally aims to capture the course of an accident in its details, i.e., to clarify accident sequences and to determine and evaluate accident and damage causes. This allows the individual phases of the accident sequence to be determined and assessed in their spatial and temporal allocation. For the alignment of road traffic safety to the human, the vehicle, and the environment (see Fig. 1.1) from the perspective of the involved disciplines of technology, medicine, and psychology (see Fig. 2.3) in the individual accident phases initiation (pre-crash), collision (in-crash) and consequences (post-crash), the entire causal and effect chain of an accident must be understood; mutual dependencies and influences must be demonstrated [1]. Regardless of the specific question, the kinematic sequence of the accident must first be determined in whole or in part. This includes information about

- times (e.g., reaction time),
- velocities (e.g., approach, collision, and run-out speed) and
- local conditions (e.g., approach directions, reaction point, collision site, and run-out direction).

These details are of interest either on their own or in combination, e.g., using the distance/time diagram [3]. In the forensic field, questions are also important, such as

- the velocity of the vehicle before the accident,
- the spatial and temporal avoidability of the accident,
- the impact of a reduced (e.g., prescribed) speed regarding the avoidability of the accident and thus of injuries and damages,
- the possible injury reduction by using safety devices such as fastening the seat belt or by wearing helmets for motorcyclists, and
- the seating position of occupants to clarify who was driving the vehicle [9].

And in the safety-specific area, questions arise about

- the avoidability of the accident,
- the accident causes under psychological and physiological aspects,
- the movements of the vehicles and the movements of the occupants relative to the vehicle (accident mechanics),
- the causes of injury or the injury-inducing vehicle parts,
- the behavior of occupant protection systems and their components, and finally
- the injury mechanics and the injury pattern [9].

The questions in the forensic and safety engineering fields may well be different, but the methods and procedures initially run parallel for both areas [1]. They will be presented in the following in the necessary brevity.

2.2.1 Reconstruction of Accidents

The quality of the accident reconstruction and thus the legal certainty as well as the meaningfulness of derived sizes to the accident occurrence depend crucially on the quality and scope of the recorded accident data. Immediately after the accident, data can be determined that retrospectively can either not be obtained at all or only inadequately.

Within the scope of accident reconstruction, one can distinguish between [9]

- **measurable data:**
  Final position of the parties involved (vehicle and persons), throwing distance of pedestrians and cyclists, fixation of collision points by trace drawing (brake marks) and deposits on the road, tire abrasion marks according to type, length and direction, tires (condition, tread depth), wheelbase and track widths, static load conditions and extraordinary loads (e.g. roof racks), center of gravity height, visibility obstructions due to parked vehicles or roadside vegetation, traffic regulations (traffic light control, -times), brightness (time of day, street lighting), road condition and surface;
- **describable data:**
  Weather, road condition (dry or wet), visibility conditions (fogged or dirty windows), lighting condition (parking, low beam or high beam switched on), traffic volume, driving dynamics influencing conditions (engaged gear, locking wheels), impairment of the driver/operator due to poor vision, alcohol/drugs, illness or other passengers, number of passengers and their seating position, occupant protection systems;
- **experimentally or computationally determinable data:** Reaction time of the parties involved, driving speed of cyclists, walking speed of pedestrians, brake response time, brake force distribution, tire characteristics, friction values, acceleration and deceleration values, masses and moments of inertia, dynamic load conditions during collision and run-out, tipping behavior, visibility distance for the driver/operator.

The application of accident reconstruction up to quantitative results requires, in addition to the availability of specific accident data, the availability of basic data that can and must be referred to. This basic data has been and still is the subject of research.

In principle, two methods can be distinguished in accident reconstruction: experimental and computational simulation. In the case of computational reconstruction, three different procedures can be applied, namely kinematic analysis, impact calculation, and force calculation.

The **kinematic considerations** (starting and braking processes, distance/time diagram) may be sufficient in simple cases where the actual impact phase is not relevant [1].

The **impact calculation** is based on three conservation laws: the impulse, the angular momentum, and the law of conservation of energy. It is assumed that the impulse as well as the energy before and after the in-crash phase is the same and that the impact duration is infinitely short. Based on the collision point, the run-out movement (via trace drawing), the final positions as well as the deformations of the vehicles, the run-out impulse and deformation energy are determined. From this, under certain assumptions, the incoming impulses and thus the collision speeds of the parties involved can be calculated. Since the run-out movement is used to infer the incoming speed, this approach is called the **backward calculation**.

In the case of the **force calculation**, the incoming speed of the vehicles is assumed in the range of the probable solution. Using motion equations, all occurring forces and from this the movement of the vehicles are calculated in small time steps. This requires above all the most accurate possible knowledge of the deformation behavior of the vehicles involved. Since the motion sequence of the vehicles is calculated starting from the incoming speed, this approach is referred to as **forward computation**. This method gained significantly in importance with the application of computers [15]. The calculation can be repeated with changed initial conditions until the calculated results sufficiently match the final positions, trace courses and damage patterns on the vehicles determined during the accident data collection.

2.2.2 Accident Severity

When considering accident severity, not to be confused with accident consequences, it is assumed that the moving accident participant, the collision object, is exposed to an external load and thereby experiences a change in its movement behavior. At the same time, the external load causes a deformation of the collision object due to imposed forces. The deformation, for example in the area of the front structure during the collision phase, is quite desirable (except in minor cases), as the deformation characteristics can specifically influence one's own movement behavior, but also that of the collision opponent. However, the external deformation also results in the displacement of structural parts into the vehicle interior (intrusion), so that the forward displacement of the occupants can be restricted. The external load of the collision object during the in-crash phase thus has three decisive effects on occupant load, namely

- the change in movement behavior,
- the deformation of vehicle structures and as a result of this
- the intrusion.

However, since the intrusion can essentially only be determined qualitatively and only in the rarest cases quantitatively in the accident data collection, the accident parameter for characterizing the accident severity describes the movement behavior of the collision object. For this purpose, the following briefly described evaluation sizes are to be distinguished:

EBS (Equivalent Barrier Speed)
Impact speed on a flat rigid barrier, at which the same deformation energy is converted and the same damage pattern as in the real accident is caused in the considered vehicle.

ETS (Equivalent Test Speed)
Impact speed on a suitable, fixed or movable obstacle, at which the same deformation energy is converted and the same damage pattern as in the real accident is caused.

EES (Energy Equivalent Speed)
Impact speed on any fixed obstacle, at which the same deformation work as in the real accident is converted. The deformation energy is calculated as follows

$$W_{\text{def}} = \frac{m}{2} \cdot (EES)^2 \tag{2.1}$$

Change in Kinetic Energy ΔE
The change in kinetic energy is the energy loss related to the collision object during the in-crash phase; it corresponds in value to the deformation work.

Shock Drive $S = \Delta I$
The shock drive characterizes the impulse difference at two distinguished points in time, namely at the beginning and at the end of the in-crash phase.

Speed Change Δv
The speed change is the difference between the collision and the run-out speed; it can also be formulated as a specific shock drive by relating the shock drive acting on the collision object to the mass of the collision object [14]:

$$\Delta v_1 = v_1' - v_1 = \frac{m_2}{m_1 + m_2} \cdot (v_1 - v_2) \cdot (1 + \varepsilon) \tag{2.2}$$

It thus depends on the relative speed v_{rel}, the mass ratio $\mu = m_1/m_2$ and the energy absorption capacity. The restitution coefficient ε characterizes the energy absorption, i.e., $\varepsilon = 0$: plastic impact; $\varepsilon = 1$: elastic impact.

Specific Accident Power SPUL
The specific accident power is the product of the speed change Δv and the average vehicle deceleration $\bar{a}$ and is proportional to the change in kinetic energy acting on the collision object during the impact [14]. If this energy difference is divided by the collision duration, a power, namely the accident power, is present. If this is again related to the mass of the collision object, one obtains, apart from a proportionality factor, the specific accident power. It is calculated as follows

$$\text{SPUL} = 2 \cdot \frac{v_1 - v_1'}{v_1 + v_1'} \cdot \frac{\Delta E_1}{\tau \cdot m_1} = c \cdot \frac{\Delta E_1}{\tau \cdot m_1} = \Delta v_1 \cdot \bar{a}_1 \tag{2.3}$$

The energy-equivalent speed EES, the speed change Δv and the specific accident power were examined in [14] and [22] with regard to a correlation with the loads on car occupants. A clear preference for one of the accident parameters could not be shown. However, it should be noted that the speed change Δv and the energy-equivalent speed EES are currently the most frequently used parameters for characterizing frontal, but also side and rear collisions.

2.3 Accident Analysis

The aim of the accident analysis is to investigate the circumstances that led to the accident, taking into account various aspects, and to infer the causes of the accident in the technical, medical and/or psychological field from the effects, i.e., the consequences of the accident (injuries, property damage). In this case, the accident analysis is limited to passive safety with the aim of identifying accident consequence-reducing effects in order to initiate measures to reduce the expected damage.

The accident analysis can refer to an individual case that has been sufficiently investigated by a comprehensive reconstruction, but statements based on statistical studies of a sufficiently large number of cases are also possible.

2.3.1 Investigation of Accident Causes

The investigation of the cause of accident is the basis of accident analysis, from which a need for development and improvement of safety measures can be derived.

Human error can be considered as the main cause of road traffic accidents. In contrast, environmental causes such as road and weather conditions due to rain, snow, and ice only make up a subordinate part, and technical defects occur only to a vanishingly small extent (Fig. 2.6). It is likely that this pleasingly low proportion of technical defects can be attributed to adapted technology and continuous control through regular main inspections. The reduction of human inadequacies also requires improved, long-term stable training and further education of vehicle drivers as well as driver-supporting measures.

Upon further detailing, in accidents with personal injury, where human failure was observed, the most common cause of accident can be identified as errors when turning (Fig. 2.7). This group of causes is followed by accidents due to right-of-way errors, inappropriate speed, and distance errors. In accidents caused by incorrect road use, there is a significant difference in the frequency of occurrence between all vehicle drivers and car drivers. This can be attributed to the more prevalent misbehavior of other road users such as commercial vehicles and motorized two-wheelers.

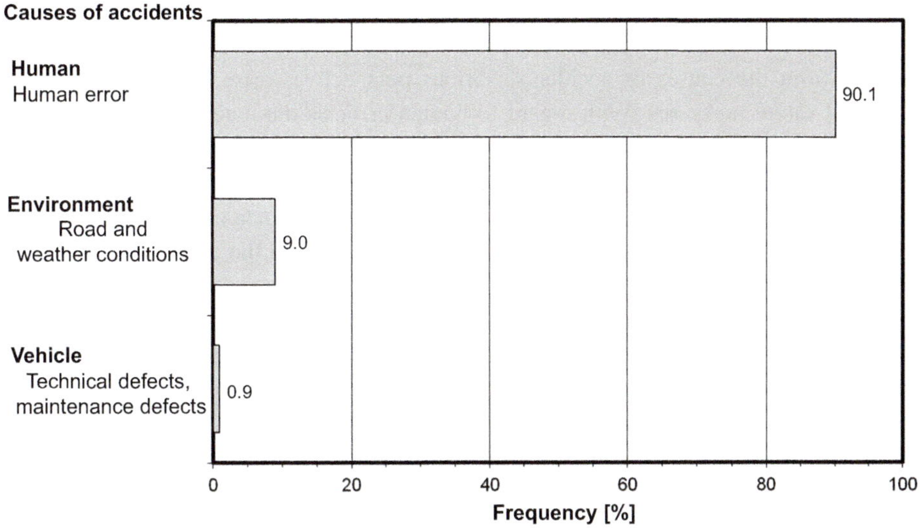

Fig. 2.6 Overview of causes of road traffic accidents with personal injury in 2011 (n = 424,461 accidents [21])

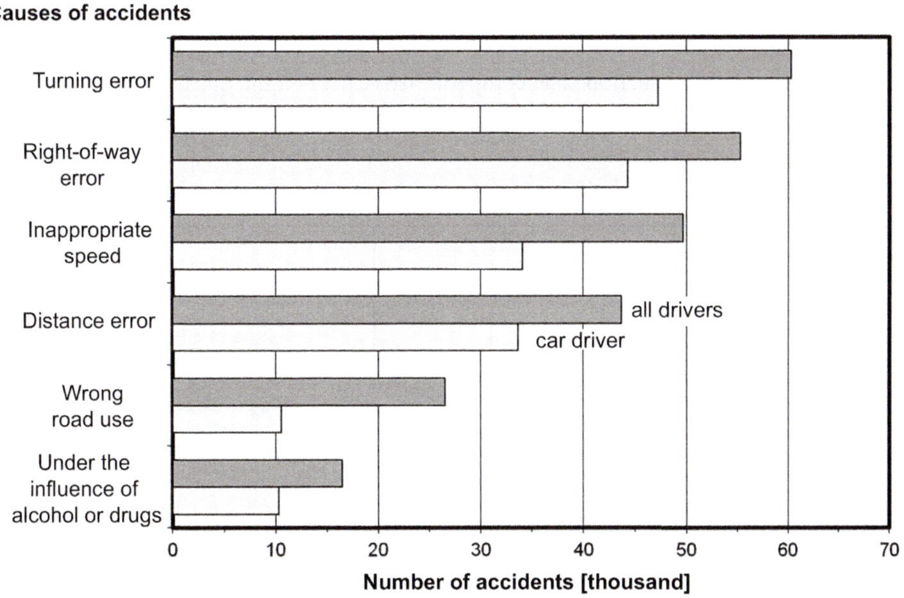

Fig. 2.7 Classification of selected examples of human failure in accidents with personal injury in 2011 (all vehicle drivers: 251,958, of which car drivers: 180,202 accidents [21])

2.3.2 Derivation of Improvement Measures

Starting from the causes of accidents, various perspectives arise for the provision of improved safety measures. With regard to human error as the main cause of accidents, this aspect is currently being pushed forward. The focus is on advanced assistance systems that support the driver with technical devices in such a way that he can recognize and react early to dangerous situations in terms of speed, distance to the vehicle in front, and crossing of lane markings; an independent intervention in the engine or brake management is being increasingly promoted.

In addition, the goal is to minimize the severity of the accident and its consequences, in addition to accident prevention. As examples of measures, the requirements of which result from the occurrence of accidents, some justified analysis results will be presented below:

The Occupant Protection Depending on the Vehicle Mass

The vehicle mass demonstrably plays a significant role in the injury risk of car occupants. Based on the safety figure, which was calculated by dividing the number of examined car occupants by the monetized injury consequences, it is evident that occupants in car-to-car collisions are safer from injuries in larger cars than in smaller ones. However, the injury risk in car/obstacle collisions is nearly independent of the mass of the collision object, and is as high, or the safety as low, as in lighter cars (Fig. 2.8). Therefore, small and micro cars must be equipped with the highest level of technical safety measures to improve self-protection. The compatibility issue, i.e., the question of counterparty

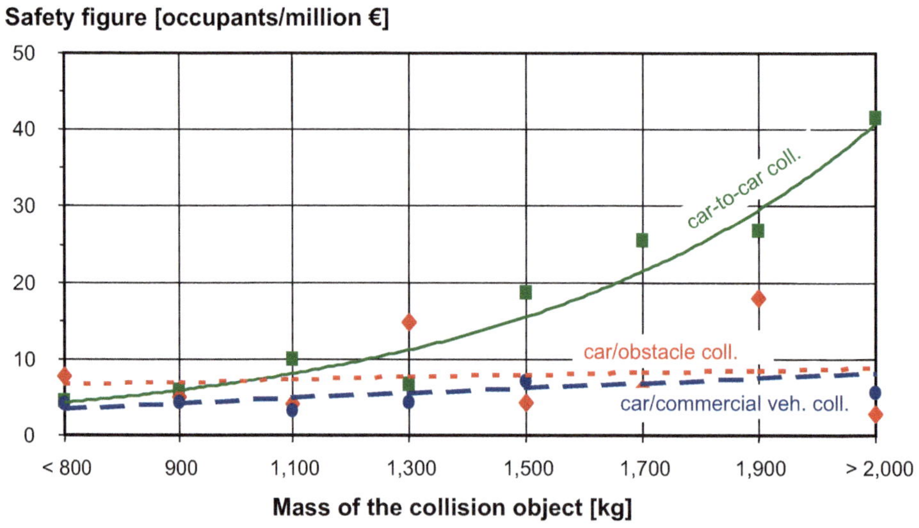

Fig. 2.8 Safety of car occupants depending on the vehicle mass in different types of collisions

protection, can be solved in such a way that larger cars are equipped with a yielding front structure to compensate for the Δv disadvantage of smaller cars, and accordingly smaller cars with a stiffer deformation structure, especially in the area of the passenger compartment [17].

Side Protection Measures

The examination of the types of impacts (see Sect. 2.4.5) will show that side collisions are of high importance, especially when not only the frequency but also the injury-related costs are used as evaluation criteria. For this, two test configurations were demanded in [7] (and are now legally prescribed in the regulations UN-R 95 and R 135):

- Large-scale impact on the driver's side against the passenger compartment with a yielding barrier and a speed of at least 50 km/h and
- Impact against a rigid obstacle (pole) at an angle of 75° and a speed of 32 km/h, to account for the frequently occurring obstacle collisions.

Pedestrian Protection Measures

Another important topic is pedestrian protection. In accidents involving pedestrians, they are the least protected and therefore exposed to a significantly higher injury risk. For this reason, it is important to take precautions here as well to minimize the consequences of injury in an accident or to avoid injuries altogether. This topic will be further elaborated and explained in detail.

European Safety Standards

Based on the examination of the types of impacts, it will be shown (see Sect. 2.4.6) that the safety requirements cannot be adequately described by a single crash test in frontal collisions. On the one hand, the EU Commission has issued regulations for crash tests (UN-R 94 and R 137 as well as R 95 and R 135). On the other hand, car manufacturers, system developers, and consumer protection associations carry out additional tests according to Euro NCAP with speeds (frontal up to 64 km/h) and obstacle shapes (offset barrier, pole) derived from accident occurrence (see Chap. 4).

Adapted Occupant Protection Systems

In most accidents, only the driver is in the vehicle. For reasons of repair costs, the ignition of the airbag and the belt tensioner on the unoccupied seats should be avoided using seat occupancy detection sensors. Moreover, it has been determined from the accident occurrence that there is a certain risk from the airbags, especially for smaller people and children [10]. To minimize the risk potential, the safety standard FMVSS 208 (since May 2000) was modified and intelligent occupant protection systems and components were developed to ensure compliance with the more restrictive criteria (so-called smart restraint systems).

Since the mass motorization and the associated injury risks, protective measures have been developed and improved. These developments were mostly driven by the challenge of compensating for human failure or technical inadequacies. In newer research and development projects, the goal is to create and design a **"fault-tolerant traffic system"**—keyword 'Vision Zero' –, as explained in more detail in Chap. 1.4. In addition to the road users, a greater degree of responsibility must also be demanded in the planning of infrastructure measures.

The list of results and findings could be continued indefinitely, but here the attempt should only be made to clarify the objective and competence and responsibility of accident analysis.

2.4 Structuring the Accident

The classification of the accident event is not a natural given, according to which the epidemiological manifestation of accidents would align. Rather, the structuring started by Appel in [1] and [2] and consistently continued in [11] follows a more or less obvious distinction for reasons of a uniform terminology, but also for reasons of comparability of similar events. For example, how would one compare the accident and injury mechanics of car occupants and those of cyclists? The following will show the classification of the accident event and present it in relation to accident figures from federal statistics or from the data material of the accident research group of the MHH; this is an updated excerpt from [11]. An overview of the structuring elements explained below is shown in Fig. 2.9.

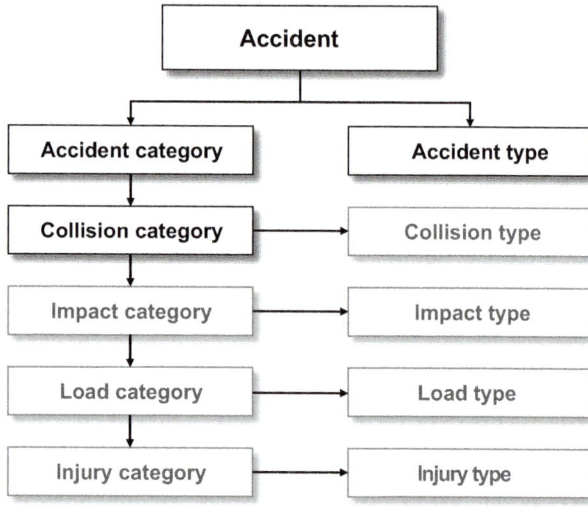

Fig. 2.9 Structure of the accident event; the darker structuring elements are of particular importance

2.4.1 Accident Category

In the federal statistics [21], accidents are divided according to their involvement in road traffic. This division is considered necessary there to account for the diversity of individual road users in terms of their collision characteristics. However, as similarities can be seen when looking at individual accidents, this historically grown, but here inappropriate division is to be summarized for the sake of simplification in treatment and with the aim of better clarity in favor of a sufficiently accurate accident category division and applied as follows: The categories of accidents, which serve to distinguish the accident participants according to their participation in road traffic, thus include:

- Commercial vehicle accidents (NFZ):
  Delivery cars and trucks, semi-trailers, agricultural and other tractors, trucks with special bodies, motor buses and trolleybuses as well as trams,
- Passenger car accidents (PKW):
  Passenger cars, station wagons and convertibles,
- Accidents with motorized two-wheelers (MZ):
  Motorcycles, mopeds and scooters,
- Accidents with bicycles (FR),
- Accidents with pedestrians (FG) as well as
- Single accidents (obstacle).

This allows all those accidents to be summarized under a category of accident in which at least one affected person can be assigned to a specific road traffic participation. Thus, the accident category PKW consists of accidents that occur between passenger cars, and between passenger cars and commercial vehicles, motorized two-wheelers, bicycles, pedestrians and obstacles. It should be noted that in the category of accident NFZ, the collisions between passenger cars and commercial vehicles occur with the same frequency and consequence of accident as in the accident category PKW; the same applies to the PKW/MZ accidents, which are represented both in the accident category PKW and in MZ, so that multiple counts occur when considering the categories of accidents.

By summing up the injured persons and assigning the respective injury-related costs (see Sect. 3.2.4), costs can be assigned to each accident category. Despite the multiple counts that occur, the percentage shares allow statements about the significance of the individual categories of accidents (Fig. 2.10): The accident category PKW takes by far the largest share, due to the high traffic participation and the associated high collision probability, with about 76 %, followed by the accidents that occur at obstacles; these are involved in almost 29 % of the resulting injury-related costs. Approximately equal shares of about 20 % are shown by the categories of accidents that are oriented towards motorized two-wheelers and bicycles. The accident category NFZ is still represented with about 16 % of the injury-related costs. The last place in the ranking is taken by accidents with pedestrians at about 12 %. The total sum of the percentages results in more than 100 % due to the multiple counts.

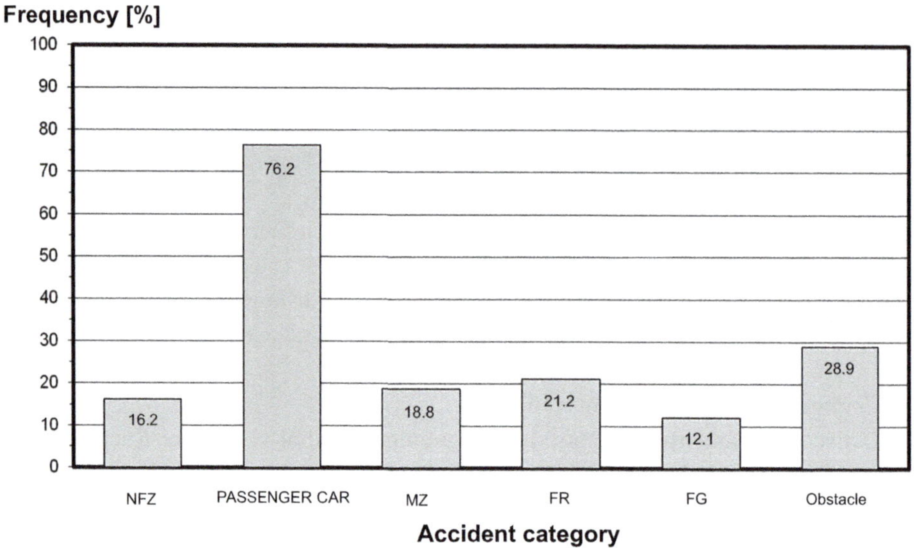

Fig. 2.10 Categories of accidents and percentage shares of injury-related costs, 2011 (according to [21])

2.4.2 Accident Type

In the first structuring level (Fig. 2.9), the accident type describes the conflict situation that precedes the accident, i.e., the accident genesis. In total, seven types of accidents are distinguished according to the definition in [21]:

Driving accident: A driving accident occurs in all those cases where the driver loses control of the vehicle, for example due to a misjudgment of the speed in relation to the course of the road or the condition of the road, but not "as a result of a conflict with another road user, an animal or an obstacle on the road or due to sudden incapacity or sudden damage to the vehicle" [21]. The conflict-triggering moment thus results from the mismatch between the demand and the supply of traffic space. During the course of the driving accident, there may be an impact on an obstacle, but also a collision with other road users.

Turning accident: A turning accident occurs when the "accident was triggered by a conflict between a turner and a road user coming from the same or opposite direction" [21].

Turn-in / Crossing accident: Such an accident occurs when the triggering conflict can be traced back to a critical situation "between a turning or crossing vehicle with a duty to wait and a vehicle with right of way" [21].

Crossing accident: In contrast to the other types of accidents mentioned, the collision opponents in a crossing accident are more specifically defined: Such an accident type occurs when a "conflict between a pedestrian crossing the road and a vehicle" [21] occurs; the accident is also assigned to this accident type if there is no collision between the accident-causing pedestrian and the vehicle.

Accident caused by stationary traffic: This accident type includes all those accidents where the conflict occurred between a "vehicle in flowing traffic and a vehicle stopping or parking on the road" [21], but without traffic-related waiting.

Accident in longitudinal traffic: An accident in longitudinal traffic occurs "when the accident was triggered by a conflict between road users moving in the same or opposite direction" [21]. This also includes accidents where a pedestrian moves in the longitudinal direction of the road; they are not attributed to crossing accidents.

Other accident: "This includes all accidents that cannot be assigned to any other accident type" [21].

The relevance of these seven accident types is shown due to their different characteristics in Fig. 2.11 for accidents that occurred within and in Fig. 2.12 for accidents outside built-up areas for the year 2011.

In urban accidents, as expected, turning/crossing accidents show a pronounced dominance in terms of injury-related costs (Fig. 2.11), while in interurban accidents, driving accidents and accidents in longitudinal traffic predominate (Fig. 2.12).

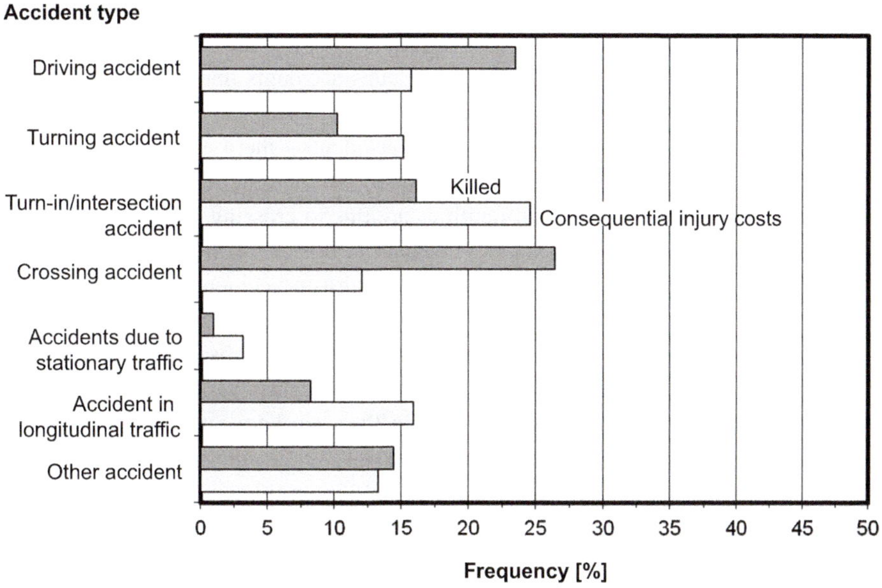

Fig. 2.11 Percentage distribution of accident types in 2011, urban

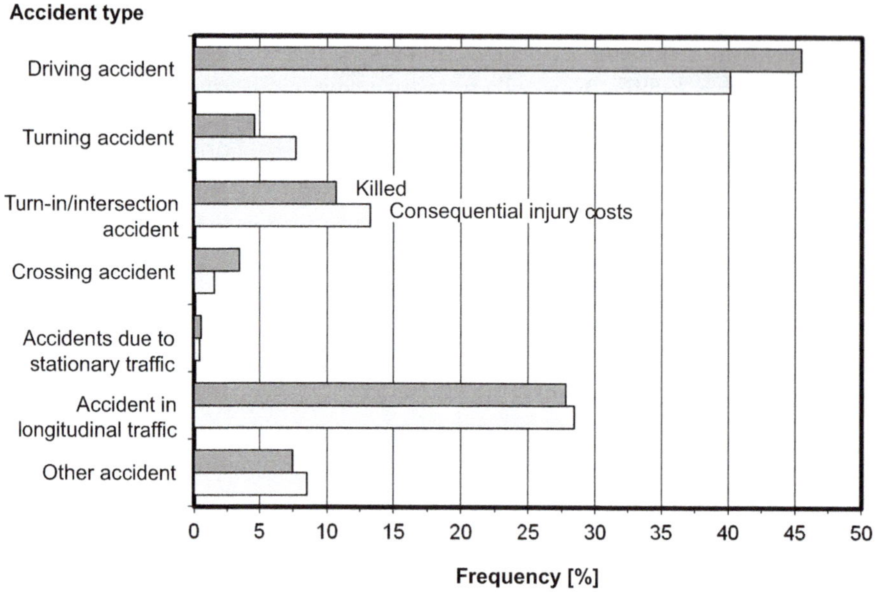

Fig. 2.12 Percentage distribution of accident types in 2011, interurban

The distribution of accident types can also illustrate that it is not sufficient for a statement of relevance to comparatively present the number of fatalities as a distinguishing feature. Rather, seriously and slightly injured people must also be included and charged with the respective injury-related costs in order to arrive at an objective risk statement. The comparison of the subtotals alone, with 1,115 fatalities among urban road users and 2,894 fatalities interurban, but approximately equal injury-related costs (13.05 billion € urban versus 11.53 billion € interurban), overvalues accidents that occur outside urban areas if only the number of fatalities is taken as a basis. But also in the distribution of accident types—Fig. 2.11 shows this for urban accidents—there are sometimes significantly different proportions depending on the distinguishing criterion:

Thus, in urban areas, 26.5% of all fatalities are due to crossing accidents, while only 12.1% of the injury-related costs can be attributed to this accident type. In contrast, the two accident types turning/crossing accident and turn-in accident, based on the injury-related costs, have a significantly higher share of 39.8% together than in the distribution of fatalities, which is 26.4%. These shifts could not be observed to this extent in interurban accidents (Fig. 2.12), because the higher accident severity led to more severe injuries and thus a smaller range of variation in the ratio between the number of seriously and fatally injured; this ratio is significantly lower in interurban accidents compared to accidents in urban areas (on average 11-fold interurban compared to the 33-fold value urban).

A conceivable statistical correlation between accident type and point of impact, for example on a passenger car, is possible, but does not seem appropriate, as the conflict situation does not necessarily lead to a collision between the road users. However, the relevance consideration of the accident types can be used as a decision-making aid for measures in the field of active safety, as it offers the possibility to initiate measures in the traffic environment with the description of the conflict triggering and thus contribute to a defusing of the conflict situation. In addition, possibilities for measures on the vehicle can be developed with the aim of positively influencing the vehicle characteristics in the interaction between driver and vehicle.

2.4.3 Collision Category

The term collision categories is used to describe accidents between individual collision counterparties, regardless of the assignment of causation and order of involvement, as reported in the federal statistics. By considering a single collision category, e.g., by summarizing all commercial vehicle/car or all car/car accidents, the significance of the collision category can be stated without contradiction, as this avoids multiple counting, unlike in the case of accident category studies. The type of presentation in the form of the so-called collision category matrix was already demonstrated in [2] with the frequencies of collisions and the number of fatalities. It serves to classify the accident events; strictly speaking, however, the collision category matrix only includes obstacle collisions and collisions with two involved collision counterparties (which account for approximately 90% of all accidents), not multiple collisions. The matrix, which is schematically represented in Fig. 2.13, is populated in such a way that each matrix element is assigned the respective interesting numerical values of a combination between two (same or different) collision counterparties. On the left side of the matrix, the road users are listed as row elements, who can collide with other road users and obstacles (column elements). The subdivision is made, according to the groups involved already used in

	NFZ	PASSENGER CAR	MZ	FR	FG	Obstacle
NFZ	n_{11}	n_{12}	n_{13}	n_{14}	n_{15}	n_{16}
PASSENGER CAR	---	n_{22}	n_{23}	n_{24}	n_{25}	n_{26}
MZ	---	---	n_{33}	n_{34}	n_{35}	n_{36}
FR	---	---	---	n_{44}	n_{45}	n_{46}
FG	---	---	---	---	n_{55}	n_{56}

Accident category PASSENGER CAR

Fig. 2.13 Schematic of the collision category matrix

the accident category, into commercial vehicles (NFZ), passenger cars (PKW), motorized two-wheelers (MZ), bicycles (FR), and pedestrians (FG). In this way, all single accidents, i.e., accidents with obstacles, and accidents with two participants are recorded.

Using the accident and injury numbers from the 2011 federal statistics [21], Fig. 2.14 shows the number of obstacle accidents and accidents with two collision counterparties with personal and severe property damage for each collision category entered into the corresponding matrix elements. However, with this collision category assignment, only 91% of all accidents can be recorded; the unconsidered portion includes multiple collisions.

As expected, the most common collision category is formed by car/car collisions, the second most common accident combination is car/obstacle collisions, followed in order of occurrence by the collision categories car/FR, NFZ/car, car/MZ, and car/FG.

For these collision categories, the number of fatally, severely, and lightly injured road users can be determined from the data material of the 2011 federal statistics [21]. If these numbers are linked with the determined injury costs (see Sect. 3.2.4) from an economic perspective, the result is—similar to what has already been done in connection with the accident types—the injury-related costs for each collision category, which are shown in Fig. 2.15 [12].

If one establishes a ranking based on these costs, which in turn is led by the collision category car/car and followed by car/obstacle, car/FR, car/MZ, car/FG, and truck/car accidents, a redistribution within the ranking becomes apparent compared to the frequency of accidents: Measured by the injury-related costs, for example, car/FG accidents have a significantly greater importance than the mere number of accidents suggests. But not only the ranking, but also the absolute amounts of the number of accidents and the injury-related costs (cf. e.g. car/car and car/obstacle accidents) show considerable differences to each other, which are to be clarified by the representation in Fig. 2.16.

It has already been shown that about 73% of the injury-related costs are caused by car accidents (see "category of accident car" in Fig. 2.10); this is about 16.473 billion €.

	NFZ	PASSENGER CAR	MZ	FR	FG	Obstacle
NFZ	3,599	32,276	3,839	5,828	3,685	4,974
PASSENGER CAR	---	114,314	27,763	45,744	22,259	50,792
MZ	---	---	789	1,216	850	12,270
FR	---	---	---	5,069	3,857	12,839
FG	---	---	---	---	0	0

The matrix shows 351,963 accidents with property damage and personal injury (= 90.8 %) compared to a total of 387,753 corresponding accidents

Fig. 2.14 Collision category matrix: Number of obstacle accidents and accidents between two collision counterparties with personal and severe property damage, 2011

	NFZ	PASSENGER CAR	MZ	FR	FG	Obstacle
NFZ	258,115	1.670,545	348,947	426,237	424,999	379,852
PASSENGER CAR	---	5.036,999	2.111,215	2.349,708	1.872,180	3.432,146
MZ	---	---	99,501	95,264	71,454	1.327,740
FR	---	---	---	353,304	236,881	1.112,550
FG	---	---	---	---	0,000	0,000

The matrix shows € 21,608 million in consequential injury costs (= 87.9 %) compared to a total of € 24,585 million due to all persons injured in road traffic accidents

Fig. 2.15 Collision category matrix: Injury-related costs in million € for obstacle collisions and accidents between two collision counterparties, 2011

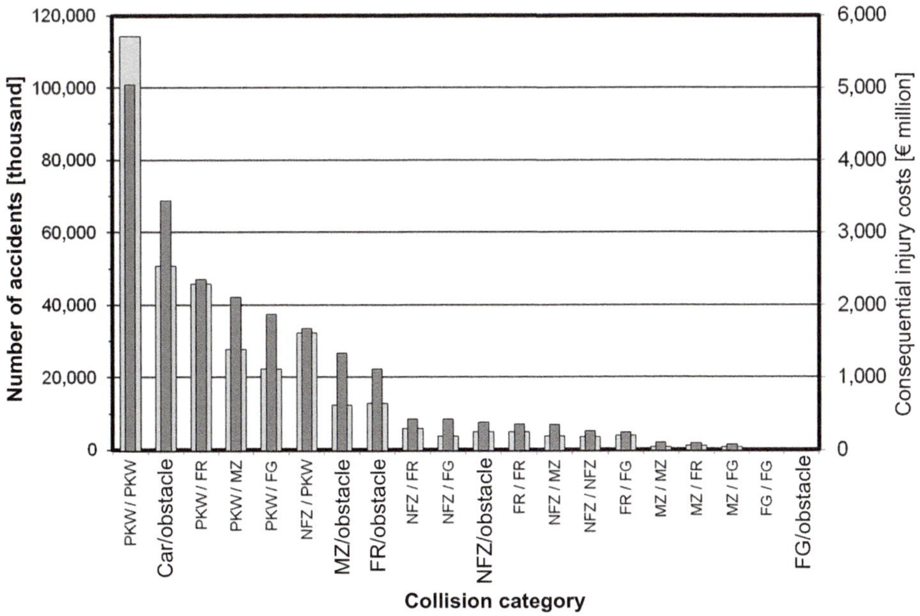

Fig. 2.16 Ranking of collision categories according to injury-related costs and comparison with the frequency of accidents, 2011

Another way to represent the safety for the individual collision categories is the **safety index**. With the help of this key figure, the number of accidents with personal and material damage is related to 1 million € injury-related costs for scale reasons. If one now compares the ranking shown in Fig. 2.17, it can be seen that the car/car collision has the highest safety. In contrast, car/obstacle collisions—with accident frequency and absolute injury-related costs each in second place—have only about 35% lower safety. This

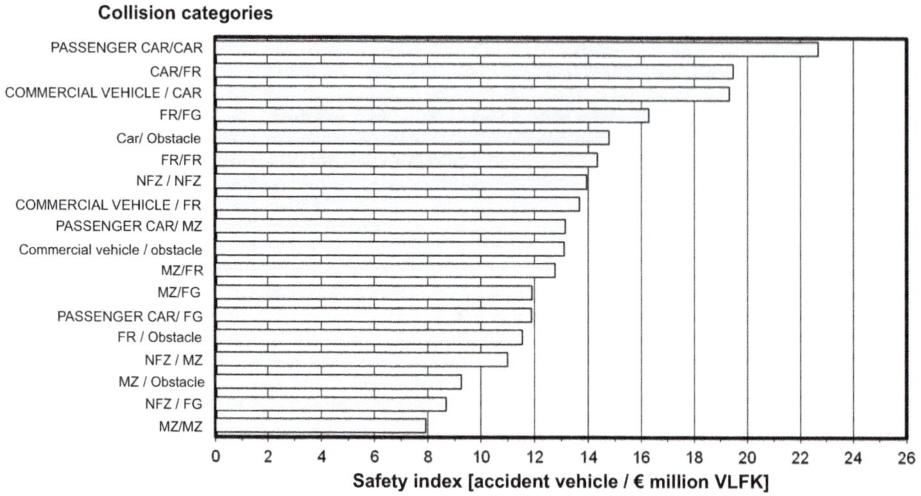

Fig. 2.17 Ranking of collision categories measured by the number of accidents per million € injury-related costs, 2011

surprising result is remarkable insofar as the safety of cars is designed and checked in crash tests using the barrier impact. In general, it can be stated that car collisions with other cars, trucks and obstacles have a relatively high safety (safety indices between about 15 and 23 accidents/million € injury-related costs). For the truck/obstacle and truck/truck collisions, from which conclusions can be drawn about the occupant protection for truck drivers, the safety index is only about 13 to 14 accidents/million € injury-related costs. Upon closer inspection, it is also noticeable that car/FR collisions result in higher safety than collisions between cyclists. The reason for this is likely to be found in the low injury severity of car occupants and the more favorable ejection kinematics of cyclists. The collision categories with the lowest safety index are, as expected, accidents involving pedestrians (FG) and those with motorized two-wheelers (MZ) due to the lack of deformation structures and, at least in the case of motorcycles, the high collision velocities.

2.4.4 Collision Type

With the help of collision types, the accident situation is qualitatively described in terms of the geometric conditions (such as the location of the collision opponents, the damage areas, etc.) and the deformation behavior of the involved structures. The classification of collision types thus allows for an initial necessary subdivision of accidents within a collision type with the aim of providing comparable accident and injury mechanisms of the occupants for the accident analysis. It will be shown later that further distinctions, e.g. regarding the used restraint system, are necessary, which focus more on the contact system, i.e. the relation between occupants and interior. In contrast, the classification of collision types can be seen as an "external" distinction possibility of the accident situation.

The definition of collision types for truck/car, car/car, and car/obstacle accidents were developed in [17] and [6] and applied to identify and verify safety measures. The collision types thus introduced are shown in Figs. 2.18, 2.19, 2.20. The collision types resulting for collided passenger cars are summarized in Table 2.1.

The material from the federal statistics is not sufficient in its depth to derive the collision types. This derivation, as well as the further numerical structuring, is based on the data material of the accident research at the Medical School Hannover (MHH) from the year 2010. The provided data material consists of 1,080 car accidents with a total of 2,330 injured car occupants. If accidents with external road users are neglected, 1,028 accidents remain from the original material with:

- 189 truck/car accidents,
- 639 car/car accidents (with 742 cars) and
- 200 car/obstacle accidents,

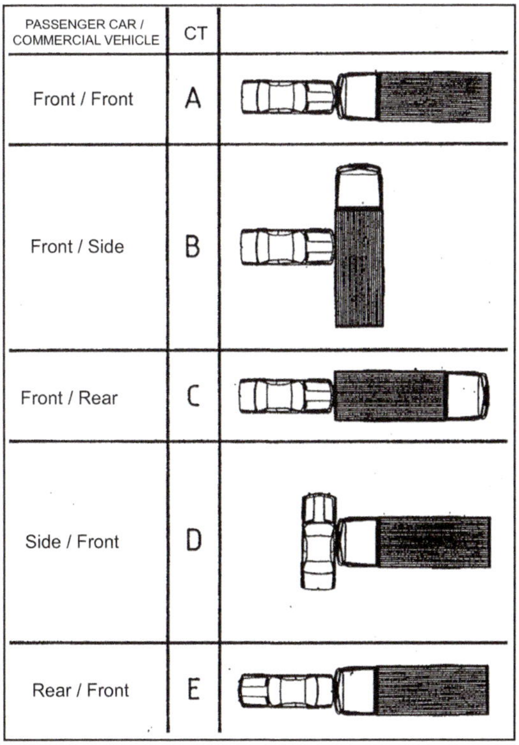

Fig. 2.18 Collision types (CT) in car/truck accidents

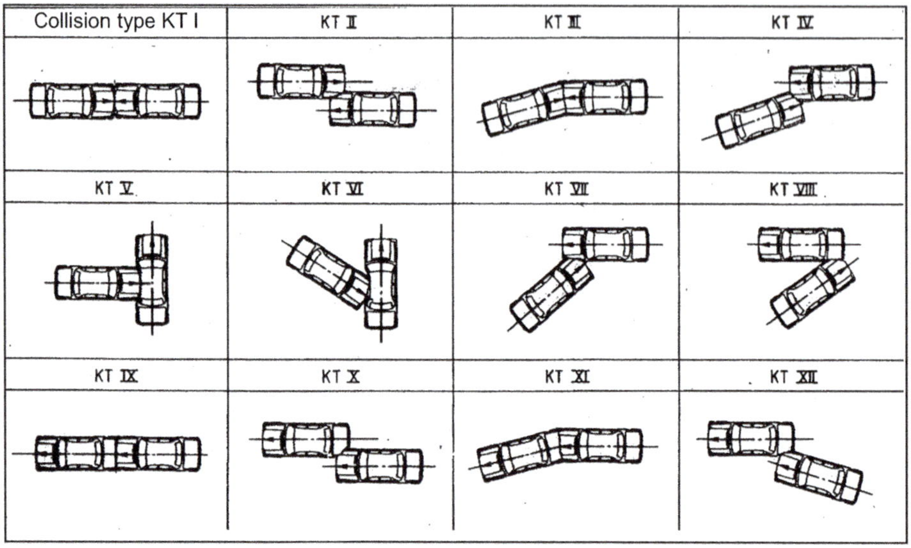

Fig. 2.19 Collision types in car/car accidents

in which a total of 1,131 passenger cars were involved. The distribution of the collision types resulting from truck/car collisions is shown in Fig. 2.21 based on 104 cases taking into account clear accident descriptions.

The distribution of collision types occurring in car/car accidents was determined based on 524 passenger cars, it is shown in Fig. 2.22.

The distinction between narrow (collision type 1 to 3) and wide obstacles (collision type 4 to 6) was made based on the obstacle description, with uncertain assignments (e.g., objects with a mass of less than 200 kg, fence, bushes, etc.) and obviously non-simulatable accidents (e.g., ditch, earth wall) being disregarded. The distribution of collision types for the remaining 167 frontal car/obstacle accidents is shown in Fig. 2.23.

Due to the inconsistent definitions of collision types, which in one case only describe the impact areas (in the case of commercial vehicle collisions), but in the other case at least partially include the deformation properties of both the collision object and the opponent (in the case of car/obstacle collisions), a consistent application of collision types is of little help and here only has a documenting character.

2.4.5 Impact Category

The impact category describes the accident situation of the individual collision object, i.e., the accident participant of interest. In the present case, this is the collided passenger car, which can be subjected to a frontal, lateral, and rear impact; in addition, the rollover is introduced as a distinguishing feature. The thus defined impact category implies

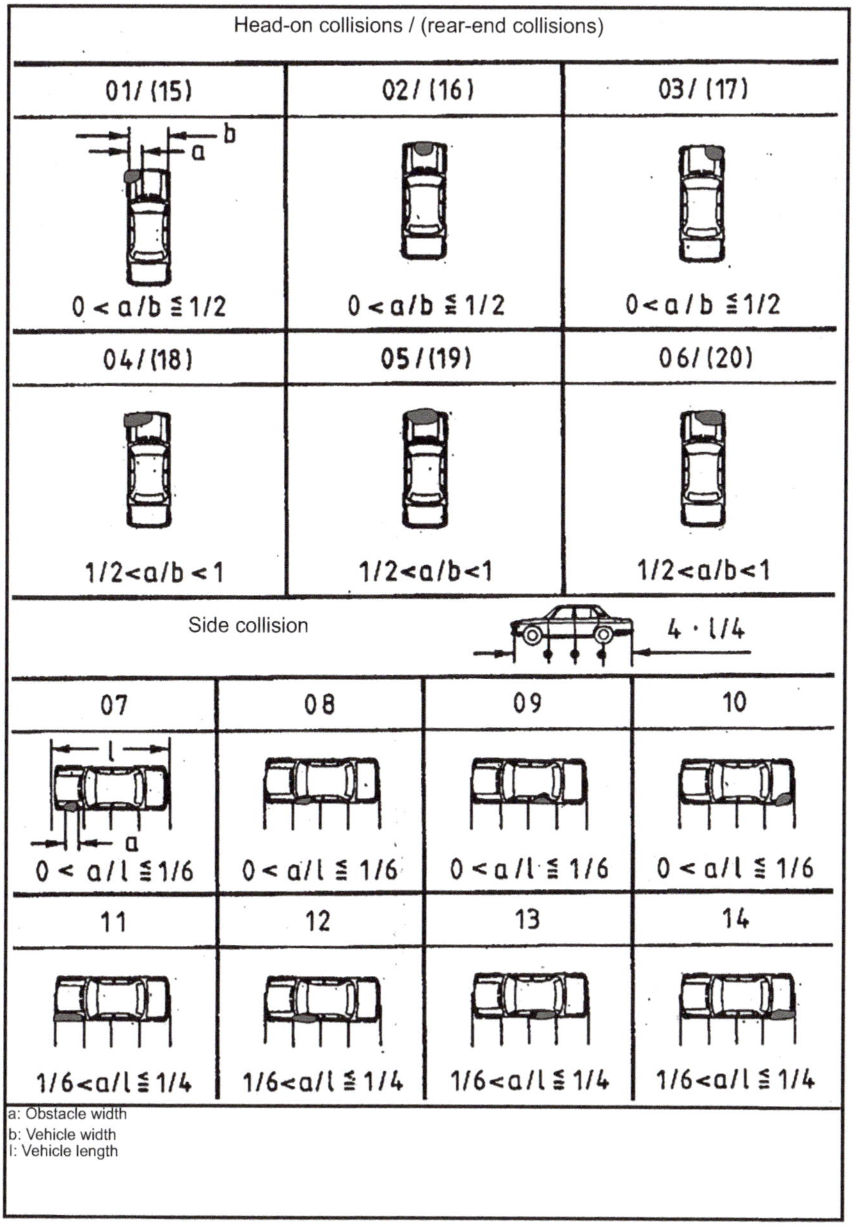

Fig. 2.20 Collision types in car/obstacle accidents

the completely different kinematics and the resulting load on the car occupants. To determine the impact category from the data provided by the accident research MHH, it can be assumed that 1,231 cars have collided with other passenger cars or commercial vehicles

Table 2.1 Collision types for car/truck, car/car, and car/obstacle collisions

Collision type	Number of collision types	Accident constellation for collision object		
		Front	Side	Rear
PASSENGER CAR / COMMERCIAL VEHICLE	5	A	D	E
CAR / PASSENGER CAR	12	I to IV	V to VII	IX to XII
Car / Obstacle	20	1 to 6	7 to 14	15 to 20

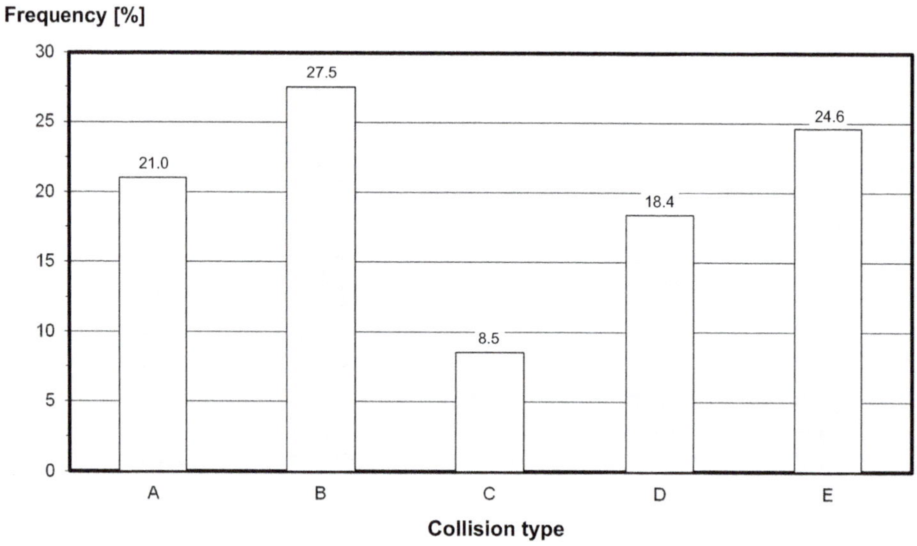

Fig. 2.21 Frequency distribution of collision types in truck/car accidents

or have hit an obstacle. Since the injury-related costs are to be included in the relevance study, accidents where the injuries of the occupants are unknown were neglected, so that only 1,211 passenger cars were available for the distribution of the impact category.

The extent of damage in the form of injury-related costs depends to a large extent on the category of impact; this is reflected in the numerical values of Table 2.2. Here, the number and percentage distribution of car impact types in collisions with the collision opponents commercial vehicles, cars, and obstacles are listed; the rollover is somewhat a special case of obstacle collision and does not occur due to the exclusion of multiple collisions in commercial vehicle and car collisions. Noteworthy is the relatively high proportion of frontal impacts in car collisions (70.8%) compared to other collision opponents (commercial vehicles: 56.5% or obstacle: 58.2%). On the other hand, car occupants are significantly more at risk from side impacts with commercial vehicles (37.5%) due to the pronounced architectural incompatibility between passenger cars and commercial vehicles.

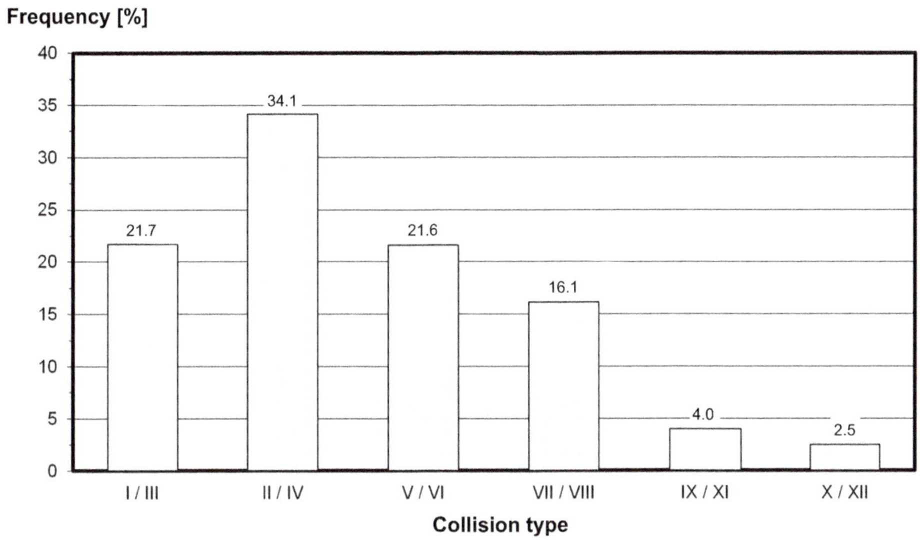

Fig. 2.22 Frequency distribution of collision types in car/car accidents

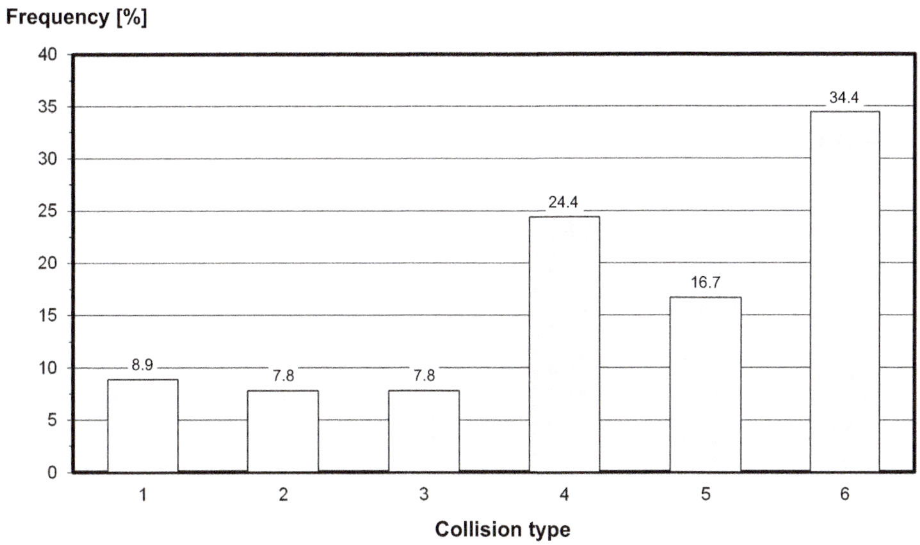

Fig. 2.23 Frequency distribution of collision types in frontal car/obstacle accidents

Table 2.2 Car impact category (data from the accident research MHH)

Category of impact PASSENGER CAR	Collision counterparties NFZ	PASSENGER CAR	Obstacle	Total number [-]	Consequential injury costs [€ million]	Safety figure [acc./million €]
Front	104 56.5%	524 70.8%	167 58.2%	795 65.7%	79.67 58.3%	5.10
Side	69 37.5%	179 24.2%	83 29.0%	331 27.3%	54.25 39.7%	3.12
Rear	11 6.0%	37 5.0%	3 1.0%	51 4.2%	0.76 0.6%	34.46
Rollover	0 0.0%	0 0.0%	34 11.8%	34 2.8%	1.97 1.4%	8.81
Total	184 100.0%	740 100.0%	287 100.0%	1.211 100.0%	136.65 100.0%	---

In the total number of impact categories, the frontal impact dominates as expected with 65.7%, followed by the side impact (27.3%) and the rear impact (4.2%); the rollover only accounts for 2.8%. This is also vividly illustrated by the comparison of impact types shown in Fig. 2.24.

In the present study of impacts categories, the aim is not so much to provide proof of representativeness based on accident numbers. Rather, the emphasis here is on the necessity of using injury-related costs in relevance studies: The distribution of injury-related costs listed in Table 2.2 shows a noteworthy shift towards side impact, which accounts

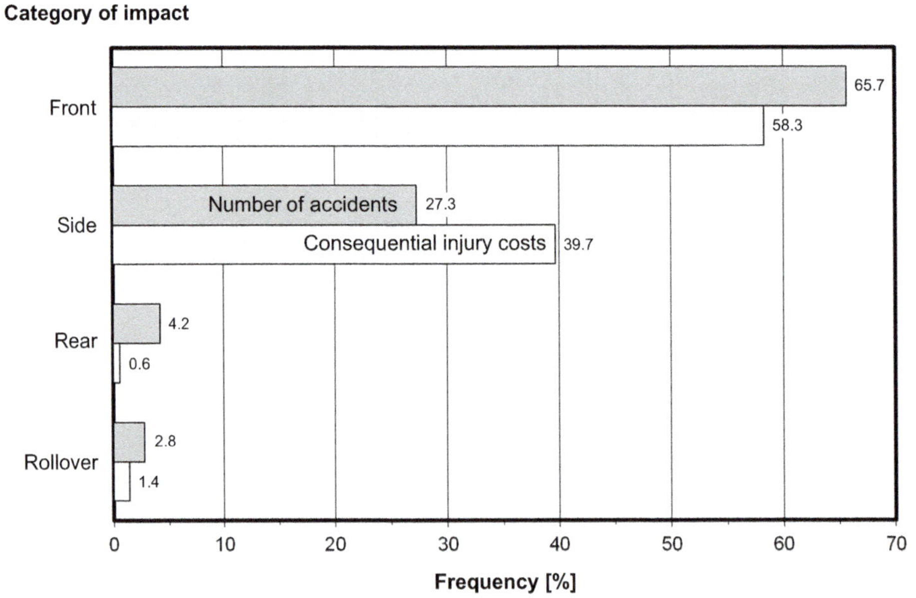

Fig. 2.24 Distribution of the frequency and the injury-related costs of impact categories in car collisions

for 27.3% in the frequency distribution, but 39.7% in the distribution of injury-related costs, thus a significantly higher injury risk (Fig. 2.24).

This is expressed, albeit in a different range of values, by the safety figure defined in [13]—applied here to the categories of impact: If the respective number of accidents is related to the determined consequential costs of occupant injuries, the safety figures listed in the right column of Table 2.2 are obtained, which characterize the safety of car occupants in different impact categories. Here too, the larger the safety figure and thus the level of passive safety, the lower the injury risk. As a result, rear impact has the highest occupant safety, as large deformation paths are available between the occupants and the collision opponent to achieve a gradual speed adjustment. The safety during rollover is significantly lower, where the belted occupants are largely protected, provided the passenger compartment retains its shape. There is less safety or a higher risk for the occupants of frontally collided passenger cars. Compared to this, the injury risk for side-impacted cars is considerably higher, as here the occupant is brought almost to the speed of the striking collision opponent within a very short time from a certain accident severity. However, when interpreting the safety figures shown, it should not be overlooked that the values of accident severity (e.g. impact velocity), that differ considerably in some cases for the different categories of impact, are implicitly included. A comparison of the distributions of the impact categories, based on the number of accidents on the one hand and the injury-related costs on the other, is shown in Fig. 2.24. The frontal impact is represented with significantly more than half both in terms of frequency and in terms of injury-related costs.

2.4.6 Impact Type

For the sequence of motion of the vehicles involved in the accident, but also for the kinematics and the load of the occupants, it is crucial in which area and to what extent the front-deformation structure of the passenger car is stressed during the impact. The distinction of deformation areas can be achieved using the impact type; it characterizes the degree of overlap and the location of the damage. This typification thus classifies the translational and rotational motion components of the vehicle during an impact and the corresponding occupant movement relative to the vehicle interior. It is easy to see that a car hitting frontally with full centric overlap experiences the highest translational deceleration with negligible rotational movement. The deceleration in a central pole impact is usually lower than in an impact with full overlap. With an off-center partial overlap, the translational deceleration also shows lower values, but the rotational component increases, the more eccentric it becomes.

In the data material of the accident research MHH, the vehicle deformation index (VDI: vehicle deformation index according to collision deformation classification [4]) is used to distinguish the impact types, which is given for both the collision object and the collision opponent; in the case of single accidents, the obstacle description is sufficient.

Starting from 795 frontally collided cars, 716 vehicles usable for impact type classification are obtained by excluding situations not described in sufficient detail. The damages of these cars can be divided into horizontal (*x*/*y* plane) and vertical (*x*/*z* plane) impact types based on the deformation indices; these are shown in Fig. 2.25.

The vast majority of cases in terms of their frequency can be assigned to impact type A (43%), where the deformation extends over the entire vertical front area. The impact types E, B and M, which include the vehicle front end (see Fig. 2.25), account for about 50% in terms of their frequency.

The distribution of impact types in the horizontal vehicle plane is shown in Table 2.3, with only minor shifts between the percentage shares of frequency and the injury-related costs.

For the asymmetrically occurring collisions, a strong predominance of the left-sided impact (Y: 2/3 or L: 1/3 coverage) can be observed compared to the right-sided (Z: 2/3 or R: 1/3 coverage), with the left-sided impact types having approximately twice as high shares as the right-sided ones (Fig. 2.26).

With the help of the structuring elements shown so far, the accidents were classified and differentiated according to external criteria. However, for the assessment of the injury situation of the occupants, it seems advantageous to also apply structuring criteria inside the vehicle. This allows for a summary of similar injury mechanisms (kinematics, contact, and injury causation), which allows for the greatest possible restriction of the scattering to occupant-specific distinguishing features (age, sex, constitution). However, the structuring has the consequence with increasing depth that the number of observations becomes smaller with the division of the accident material. The balance between

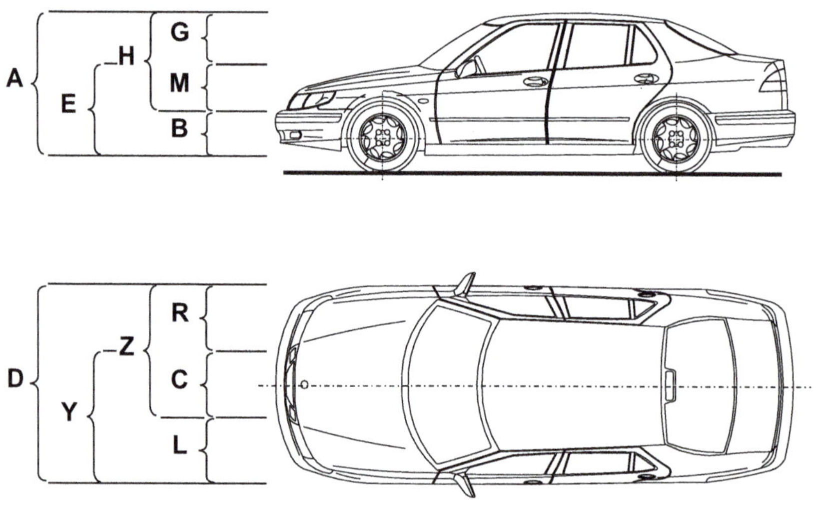

Fig. 2.25 Impact types in frontally collided cars (according to SAE J224)

Table 2.3 Distribution of frequency and injury-related costs of impact types (car front) in horizontal plane. (Source: Data material of the accident research MHH)

Impact type (horizontal level)	Frequency		Consequential injury costs	
	[-]	[%]	[€ million]	[%]
D	420	58.66	51.703	69.13
Y	115	16.06	10.512	14.06
Z	62	8.66	3.518	4.70
L	58	8.10	4.872	6.51
G	23	3.21	1.822	2.44
R	38	5.31	2.359	3.15
Total	**716**	**100.00**	**74.786**	**100.00**

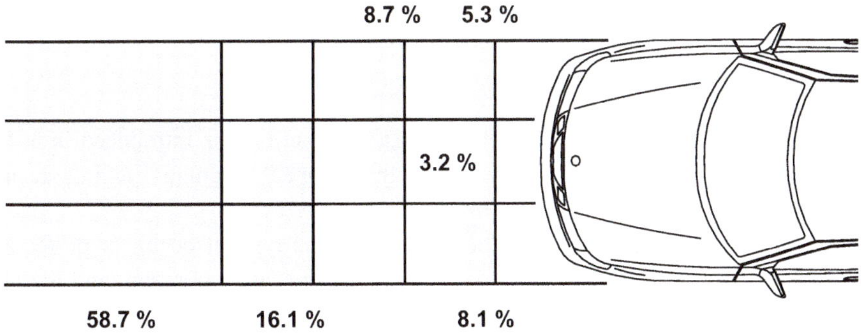

Fig. 2.26 Frequency distribution of impact types (car front)

both objectives, on the one hand the restriction to similar injury mechanisms to reduce the scattering and on the other hand the provision of a statistically sufficient number of injuries, requires a balanced approach when applying the selection criteria.

2.4.7 Load Category and Load Type

The **load category** refers to the body regions stressed during the accident. In the 1,288 car occupants examined here, a total of 6,091 different severe individual injuries were identified, which can be attributed to the head with a share of 33%, to the legs with 29%, and to the thorax and arms with approximately 14% each (Fig. 2.27). However, since very different injuries were found in drivers and passengers as well as in occupants with

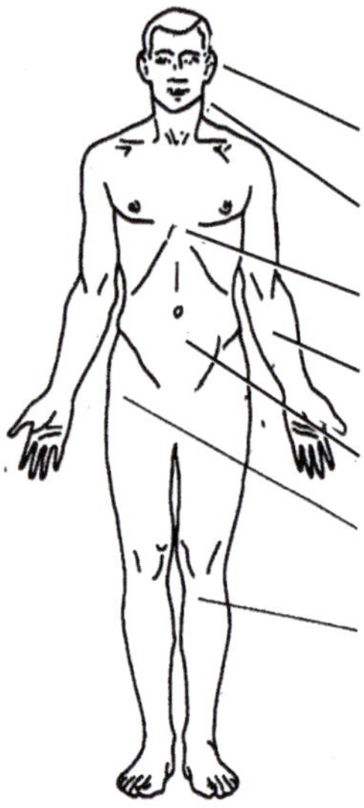

Injuries		
Body part	Frequency [-]	Frequency [%]
Head	2,033	33.4
Neck	132	2.2
Thorax	875	14.4
Arms	825	13.5
Abdomen	199	3.3
Pelvis	243	4.0
Legs	1,777	29.1
Unknown	7	0.1
	6,091	100

Fig. 2.27 Load category: Distribution of injuries of occupants in frontally collided cars

and without restraint systems (seat belt, airbag), the load type is introduced in addition to the load category for the characterization of the occupant kinematics and the resulting injury causation.

With the help of the **load type**, the question of the type of occupant protection is actually answered. This involves the load direction, the seating position, and the use of restraint systems; essentially, the provision of load types results in a classification of the occupant restraint. Another distinguishing feature within the load types is the seating position, as this determines the available forward displacement and, in the event of an impact, the contacted structure (steering wheel, dashboard, airbag, etc.) and its deformation properties.

Starting from 716 frontal car collisions, a total of 1,288 occupants are distributed among drivers (56%), passengers (26%), and rear occupants (18%). The frequency distribution for seating positions of the occupants is shown in Fig. 2.28.

Finally, the load type can be used to distinguish seat belt use among front occupants, neglecting cases marked as "unknown". For the remaining 1,022 front occupants, it was assumed that the seat belt system was used by drivers in 89% and by passengers in 88% of cases. The seat belt use rate of the presented older accident data was extrapolated to newer figures from 2003 [10]. Based on the determined injury-related costs, assuming a same accident severity distribution, a significant reduction in costs due to seat belt use can be noted (see Fig. 2.29): While unbelted front occupants are involved on average with about 6% and have an injury-related costs share of 11%, the share of injury costs

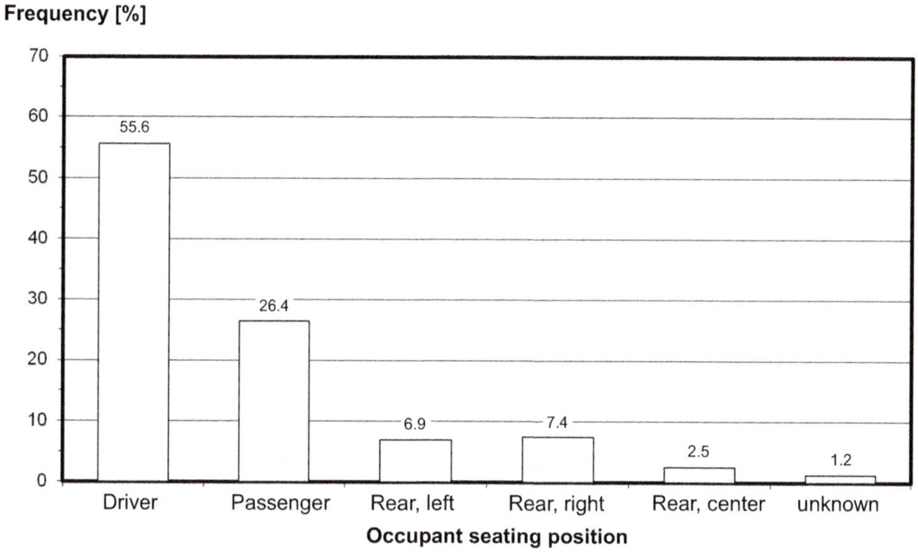

Fig. 2.28 Load type: Distribution of seating positions in frontally collided cars

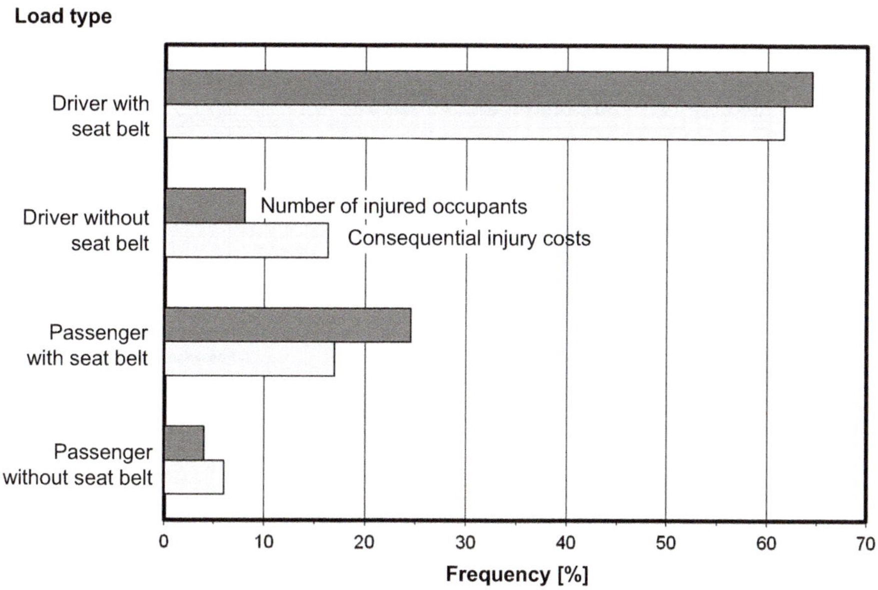

Fig. 2.29 Load type: Percentage distribution of seat belt use by injured occupants in frontally collided cars compared with injury-related costs

for belted occupants with a frequency of 45% reduces to 39%; the reduction in the cost share can be interpreted as the benefit of seat belt use.

Now that load category (body regions) and load type (seating position and occupant protection system) have been introduced, the load category related to front occupants can be meaningfully expanded by the load type. In addition to the frequency distribution of injuries to individual body regions for all occupants (see Fig. 2.27), Fig. 2.30 shows the distribution of individual injuries depending on the load type; for reasons of clarity, this was limited to drivers and passengers without and with seat belt use.

The comparison of injury frequency without/with seat belt use shows a reduction in head and leg injuries for both drivers and passengers, with head injuries of passengers decreasing significantly more due to seat belt use than those of drivers. This circumstance can be attributed to the greater forward displacement on the passenger side. The proportion of thorax and abdomen/pelvis injuries increases considerably due to the belt forces, although the injuries are less severe. The increase in neck injuries is likely due to the restraining effect of the chest, while the head is only restrained by the neck. In addition, the relatively high proportion of lower extremity injuries in belted occupants compared to, for example, reduced head injuries is striking. The importance of foot and leg injuries in terms of frequency, but not injury severity, increases even further with the use of airbags.

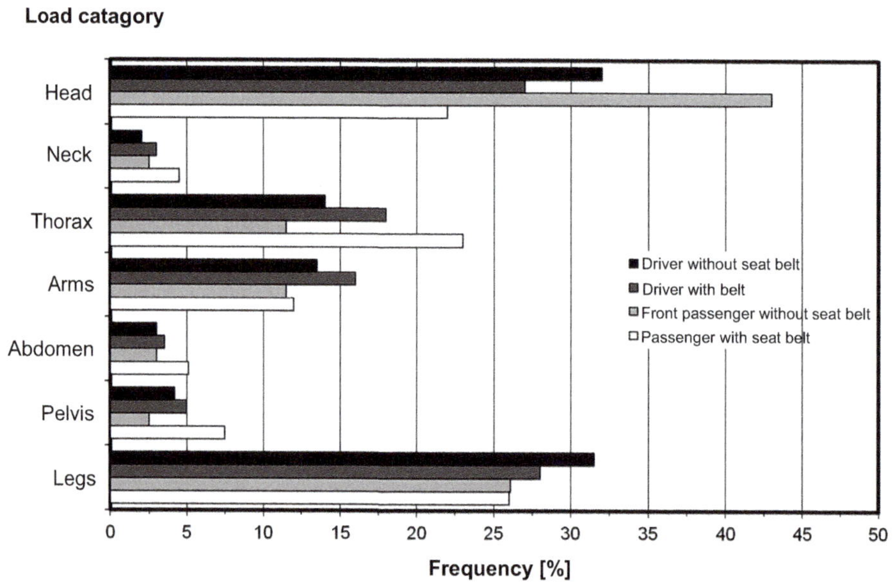

Fig. 2.30 Load category and load type: Percentage distribution of body part-specific injuries of occupants in frontally collided cars considering seat belt use

2.4.8 Category of Injury and Injury Type

The **category of injury** is classified as soft tissue, organ, vascular, ligament and tendon injuries as well as fractures. The **injury type**, on the other hand, characterizes the stress from a mechanical point of view. In the data material of the accident research MHH, the injury type is only specified in a few cases and thus eludes an exact statistical evaluation, so that it could only be estimated based on the injury description on the body part and the associated injury severity.

For numerical representation, the frequency distribution of the category of injury based on 1,022 front passengers is shown in Fig. 2.31. The most common category of injury is expectedly formed by soft tissue injuries with a share of 68%, where the injury severity and their consequential costs are lower compared to fractures, organ and vascular injuries, which occur with a frequency of 20 or 9%.

With the injury type, four fundamentally different injury-causing mechanisms are introduced, namely injuries due to

- direct force,
- indirect force,
- inertia, and
- hyperextension and hyperflexion (overstretching, overbending).

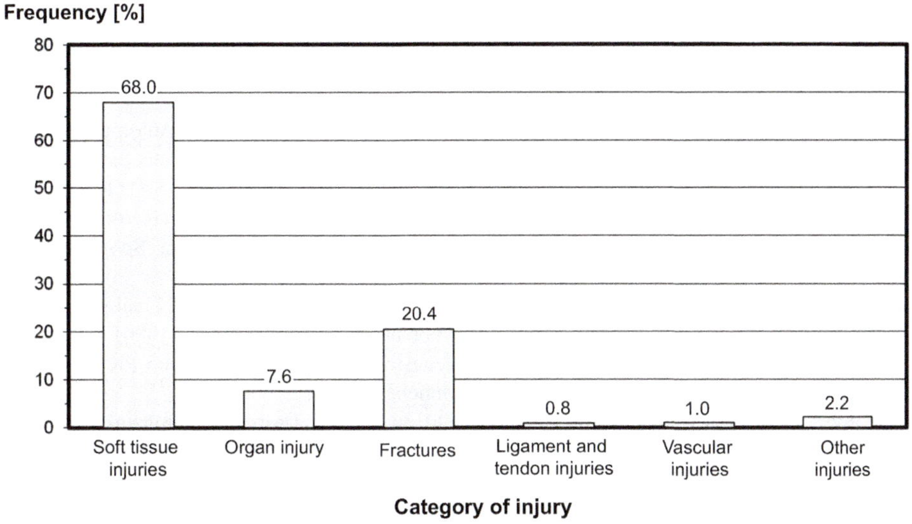

Fig. 2.31 Category of injury: Percentage distribution of injuries of occupants in frontally collided cars

A direct force is present in all those cases where injuries are caused by direct contact between body part and restraint system (vehicle interior, seat, seat belt system, airbag); these can be superficial soft tissue injuries as well as fractures to the head, thorax, pelvis and extremities or organ injuries in the abdominal area. In contrast, injuries caused by indirect force occur when they are not on the stressed (external) body part, but have arisen due to the forces introduced there. Such an injury can, for example, be a lung laceration (tear) as a result of thoracic instability. However, if the bony body part remains stable, displacements of organs can occur due to their inertia and, with sufficiently high intensity, lead to injuries. Such injuries can be, for example, aortic ruptures, pericardial contusions (without thoracic instability) or brain contusions due to contact on the skull bone or contre-coup injuries on the opposite (shock-averted) side of the brain. Finally, injuries can be traced back to hyperextensions and hyperflexions of the joints or the spine. This injury type is generally characterized by ligament or tendon injuries or, in severe cases, by dislocations and fractures of the joints and vertebrae. The four injury types described here are difficult to distinguish retrospectively based on the injury pattern.

References

1. Appel, H.: Unfallaufklärung aus technischer Sicht. Beitrag in „Verkehrsmedizin". In: Wagner, H.-J. (Hrsg.). Springer-Verlag, Berlin Heidelberg, New York, Tokyo (1984)
2. Appel, H., Kramer, F.: Biomechanik und Kraftfahrzeugsicherheit. Script from the Institut für Fahrzeugtechnik, Technische Universität Berlin, Berlin (1987)
3. Burg, H., Moser, A.: Handbuch Verkehrsunfallrekonstruktion – Unfallaufnahme, Fahrdynamik, Simulation, 2. Aufl. Vieweg + Teubner|GWV Fachverlage GmbH, Wiesbaden (2009)
4. Collision Deformation Index – SAE J 224 MAR80. Report of the Automotive Safety Committee, approved Jan. 1971, completely revised by the Motor Vehicle Safety System Testing Committee March 1980. SAE Handbook, Vol. 4, Warrendale, PA (USA) (1986)
5. Compton, C. P.: The Use of Public Crash Data in Biomechanical Research. In: Nahum, A. M., Melvin, J. W. (Hrsg.) Accidental Injury – Biomechanics and Prevention. Springer-Verlag, Berlin, Heidelberg, New York, Tokyo (1993)
6. Danner, M., Appel, H., Schimkat, H.: Entwicklung kompatibler Fahrzeuge. Final report of the research project TV 7661, Bundesminister für Forschung und Technologie (1980)
7. HUK-Verband, Büro für KFZ-Technik: Fahrzeugsicherheit 90 – Analyse von PKW-Unfällen, Grundlage für künftige Forschungsarbeiten. München (1994)
8. Georgi, A.: GIDAS – Unfallforschung vor Ort an der TU Dresden. Lecture as part of the Dresden Safety Colloquium at the Hochschule für Technik und Wirtschaft (HTW), Dresden (2004)
9. Kramer, F.: Unfallrekonstruktion. Script for the course of the same name at the Hochschule für Technik und Wirtschaft (HTW), Dresden (2010/11)
10. Kramer, F., Kramer, M., Lüders, M., Herpich, T., Class, U.: Die Bedeutung von OoP-Situationen im Straßenverkehr hinsichtlich künftiger Auslegung von Insassenschutz-Systemen. VDI-Tagung „Innovativer Kfz-Insassen- und Partnerschutz" in Berlin (2003)

11. Kramer, F.: Analyse des Unfallgeschehens zur Ermittlung der Unfallkenngröße für Frontalkollisionen und der Verletzungsschwere sagittal belasteter PKW-Insassen. Forschungsbericht Nr. 325/88. Institut für Fahrzeugtechnik, Technische Universität Berlin, Berlin (1988)
12. Kramer, F.: Volkswirtschaftlicher Schaden aufgrund der Verletzungsfolgekosten. Unpublished working paper and evaluation program AUSWERT (Version 3.8 – applied July 11, 2012). Hochschule für Technik und Wirtschaft (HTW), Dresden (January 2008)
13. Kramer, F., Deter, T.: Zur Quantifizierung der Straßenverkehrssicherheit. Verkehrsunfall und Fahrzeugtechnik, Heft 1 (January 1992)
14. Kramer, F.: Schutzkriterien für den Fahrzeug-Insassen im Falle sagittaler Belastung. Dissertation an der Technischen Universität Berlin. Fortschritt-Berichte, VDI-Reihe 12. Bd 137 (1989)
15. Nagel, U.: Rechnergestützte Unfallrekonstruktion – Programmsystem auf der Basis der Stoß- und der Kraftrechnung. Diplomarbeit am Institut für Fahrzeugtechnik, Technische Universität Berlin (1991)
16. Praxenthaler, H., Wagner, H.: Verkehrsmedizin in Gegenwart und Zukunft. In: Wagner, H. (Hrsg.) Verkehrsmedizin. Springer, Berlin, Heidelberg, New York, Tokyo (1984)
17. Richter, B., Appel, H., Hoefs, R., Langwieder, K.: Entwicklung von PKW im Hinblick auf einen volkswirtschaftlich optimalen Insassenschutz. Final report of the research project TV 8036, Bundesminister für Forschung und Technologie (1984)
18. Schneider, W.: Verhalten des Menschen im Straßenverkehr als Risikofaktor und seine Beeinflussung. In: Wagner, H. (Hrsg.) Verkehrsmedizin. Springer-Verlag, Berlin, Heidelberg, New York, Tokyo (1984)
19. Schöneburg, R..: Integrale Sicherheit von Kraftfahrzeugen / Modul 2: Unfallforschung, Biomechanik und Sicherheitsanforderungen. Script of the joint course of the same name at the TU- and Hochschule für Technik und Wirtschaft (HTW), Dresden (2019)
20. Statistisches Bundesamt Wiesbaden (Hrsg.): Todesursachen in Deutschland 2010. Fachserie 12, Reihe 4. https://www.statistischebibliothek.de/mir/servlets/MCRFileNodeServlet/DEHeft_derivate_00010400/2120400107004.pdf (2010). Accessed 17 Nov 2023
21. Statistisches Bundesamt Wiesbaden (Hrsg.): Verkehrsunfälle 2011. Fachserie 8, Reihe 7. https://www.statistischebibliothek.de/mir/servlets/MCRFileNodeServlet/DEHeft_derivate_00011362/2080700117004.pdf (2011). Accessed 17 Nov 2023
22. Zeidler, F.: Die Analyse von Straßenverkehrsunfällen mit verletzten PKW-Insassen unter besonderer Berücksichtigung von versetzten Frontalkollisionen mit Abgleiten der Fahrzeuge. Dissertation at the Institut für Fahrzeugtechnik, Technische Universität Berlin. Verlag Information Ambs GmbH, Kippenheim (1982)

Biomechanics

3

Wolfram Hell, Florian Kramer and Rodolfo Schöneburg

While the **biomechanics** in active vehicle safety as well as in work science focuses on ergometric (work to be performed) and dynamometric (force to be applied) aspects and their influencing factors, biomechanics in the context of passive safety deals with the mechanical resilience of the living body or body parts with the aim of analyzing the injury mechanics of accident victims and providing objective criteria for distinguishing between reversible and irreversible injuries. Biomechanics thus refers to the study of the mechanical behavior of the living body or its components, i.e., static and dynamic states and processes are analytically described here [98]. Research in injury biomechanics has evolved from simple ad-hoc experiments to well-organized studies. The main goals of this research are:

1. Identification and explanation of injury mechanisms
2. Quantification of the mechanical response of body components to impacts
3. Determination of tolerance limits to impacts
4. Evaluation of safety devices and techniques for the evaluation of prevention systems [127]

W. Hell (✉)
Ludwig-Maximilians-Universität, München, Germany
e-mail: wolfram.hell@t-online.de

R. Schöneburg
RSC Safety Engineering, Hechingen, Germany

F. Kramer

R. Schöneburg (ed.), *Integral Vehicle Safety*,
https://doi.org/10.1007/978-3-658-48100-1_3

The following will illustrate the **injury mechanics** of organ and tissue injuries as well as fractures. This will involve the **anatomy of the human body** and its behavior under stress. Injury mechanics is relevant for understanding the **injury causation** in accidents and the prevention of accident-specific injuries through the use of safety measures. Understanding the resilience of tissue is also relevant in terms of the treatment and healing of injuries as well as the recovery of accident victims. Since the origin and extent of injuries can be traced back to physical stresses, mechanics is important, and the body's response to the physical load is rooted in biology. Thus, the determination of protection criteria that allow a statement about the strain of the body or individual body parts is essentially mechanics. Biomechanics also builds on knowledge from physics, mathematics, chemistry, biology, **sports physiology, neurophysiology** and anatomy. The design of vehicles and protection systems (e.g., restraint systems, helmets, guardrails) and their facilities to ensure only a permissible strain or to avoid an excessively high load is the competence and responsibility of safety engineers and requires interdisciplinary cooperation between various disciplines, so biomechanics is understood as an engineering science link between biology, medicine, and applied mechanics.

However, humans are not machines and already **Otto Messerer** noted in 1880 that age, constitution, and gender make significant differences in the elasticity and strength of bones [74], fracture mechanisms (torsion fractures, Messerer wedge) were extensively researched and illustrated by him.

Also, **John Paul Stapp** (Colonel USAF) was a pioneer in the study of the effects of **acceleration forces** on the human body. Particularly noteworthy are his **self-experiments** on **rocket sleds,** in which he subjected himself to great stresses and contributed to the safety of passengers in airplanes and the further development of **seat belts** with his results. A *Stapp Car Crash Conference* (traffic safety conference) named after him still takes place today. Stapp believed that the resilience of humans during his series of experiments had not yet been reached and was significantly higher than assumed. In 1954, he accelerated to 632 mph (approx. 1,017 km/h) in one run (facing the direction of travel) and came to a complete stop in 1.4 s. At the moment of maximum deceleration, the force of gravity was 46.2 times (i.e., 46.2 g), and for a period of 1.1 s it accounted for 25 g. This is the highest acceleration a human has voluntarily withstood so far. Throughout his research time, Stapp often announced that private **vehicles** needed to be made safer. He was determined to promote automobile accident research and the introduction of seat belts. His research on the biomechanical stress limits of humans was also applied in aerospace (backward-facing seating position during landing of space capsule, ejection seats for jet pilots).

Nils Bohlin developed the 3-point seat belt for cars in 1958, which was described by the European Patent Office as one of the 10 most important inventions of the modern era, saving millions of lives.

The human body and the basics of mechanics have not changed significantly, yet there are still uncertainties about certain injury mechanisms and tolerance limits, so there is a continuous need for research in this area.

3.1 Anatomy of the Human Body and Injury Mechanisms

In anatomy, the cartesian coordinate system in x, y and z direction is not used, but anatomical planes have been defined [45], which are shown in Fig. 3.1. The planes are referred to as median or sagittal plane, coronal plane, and transverse or horizontal plane. The directions are distinguished according to [86] in ventral (towards the belly) or anterior (front) or dorsal (back) or posterior (back), in lateral (sideways, outward) or medial (forming the center) and in superior (higher) or distal (further away from the torso) or inferior (lower) or proximal (close to the body).

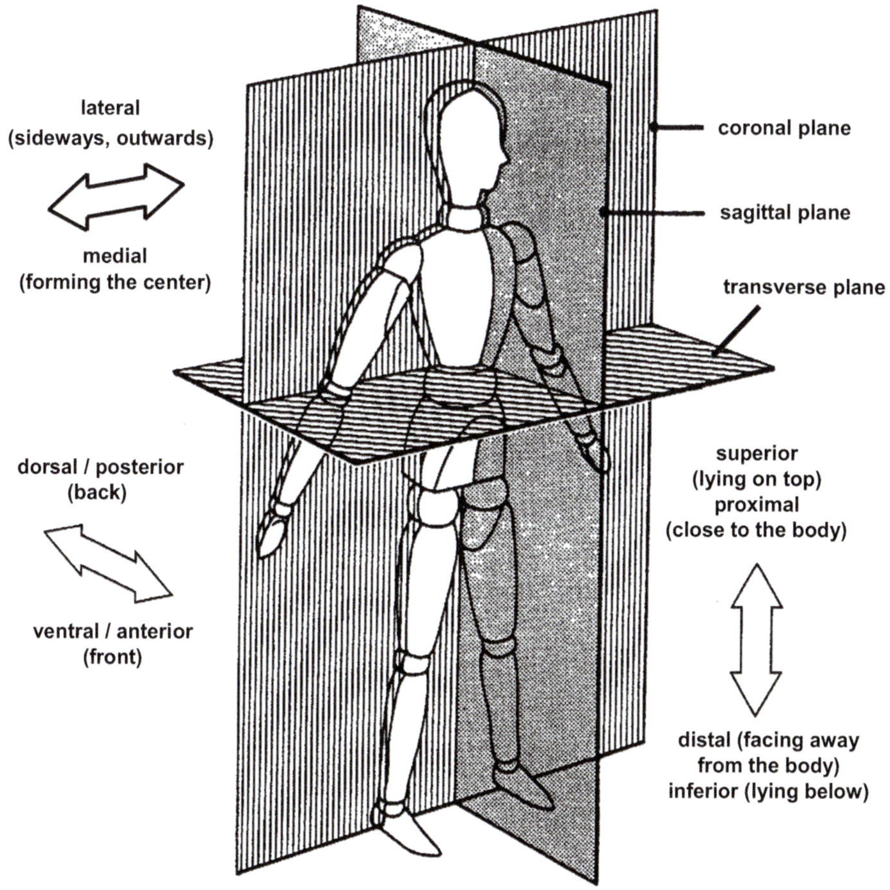

Fig. 3.1 Anatomical directions and planes (from [87])

3.1.1 The Head

The head is the most exposed and the most endangered part of the body in traffic accidents. Fig. 3.2 shows an example of the distribution of injuries to car occupants in frontal collisions; the high proportion of head and facial injuries compared to injuries to other body regions is striking. The high risk to the head and the resulting need for effective protection against injuries results from the fact that injury to the brain structure is currently irreversible, i.e., not curable.

In the anatomy of the head, we distinguish between the external organs, the soft and tissue parts, the bony skull, and the brain, which is embedded in the skull and forms the transition to the spinal cord with the midbrain.

The **scalp** is 5 to 7 mm thick and contains three layers of tissue: The hair-bearing skin consisting of the epidermis (vascular outer skin) and the dermis, which accommodates the hair follicles and is equipped with smooth muscle tissue (goosebumps, erects finer hairs). Below this is the subcutaneous tissue, in which clusters of fat cells, blood vessels, and nerves are embedded. And finally, there is the muscle tissue as the third location. Below this three-layered scalp lies a loose cellular tissue on a fibrous skin, which encloses the bone structure (periosteum). The thickness, firmness, and mobility of these outer three scalp layers can be understood as protective mechanisms in their interaction, just like the rounded skull contour. If a tensile force acts on the scalp, the three skin layers move together like a tissue.

The **skull** is the most complex structure of the human skeleton. Its bony structure is clearly formed and contains the brain, the eyes, the ears, the nose, and the teeth. The

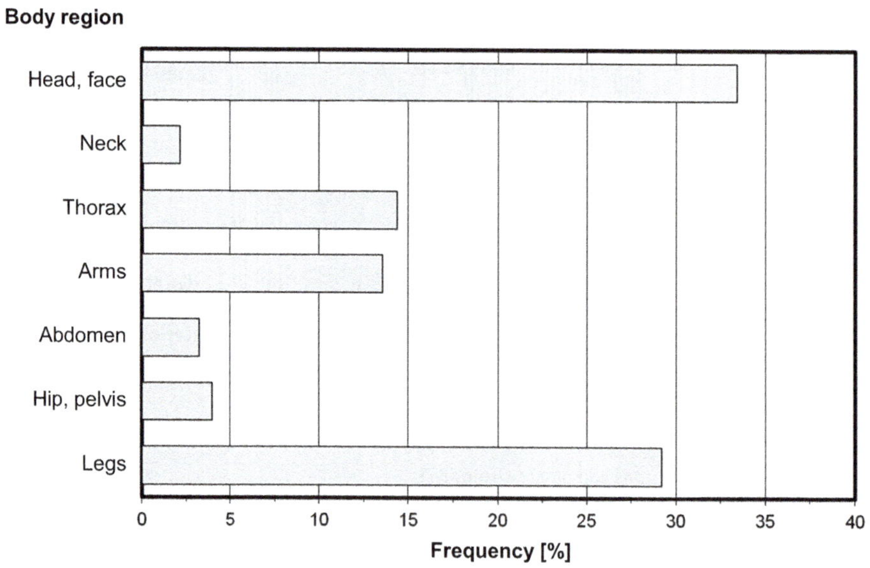

Fig. 3.2 Body part-specific injuries of car occupants in frontal collisions

thickness of the skull varies between 4 and 7 mm and thus provides a suitable receptacle and sufficient protection for the organs. The skull consists of 8 bone shells for accommodating the brain, 14 bones form the facial skull, and the total number of teeth varies and can reach up to 32. In addition to the facial skull (eye, nose, and mouth cavity), the skull cap is formed by the ethmoid bone, frontal bone, parietal bone, occipital bone, and the paired temporal bone and the sphenoid bone (Fig. 3.3). The interior of the skull cap is concave (arched) and has a relatively smooth surface. The skull base is an irregular bone plate with depressions and elevations as well as small holes (foramen) for blood vessels and nerves as well as the large occipital hole (foramen magnum) as the passage of the extended spinal cord.

Three protective membranes, known as **meninges and spinal cord**, surround and thus protect the **brain** and the spinal cord. The function of these membranes is to isolate the brain and spinal cord from the surrounding bones. The membranes consist of connective tissue and are simultaneously formed as part of the blood vessel walls and as nerve sheaths if they establish a connection between the brain and skull. There are three protective brain membranes: the outer tough layer (dura mater), the spider web skin (arachnoid), and the inner delicate layer (pia mater).

The dura mater (outer tough layer) is a resilient connective tissue membrane that surrounds the spinal cord and consists of two layers in the skull. The outer periosteal layer (also referred to as the cranial periosteum) lines the inner surface of the skull, while the inner meningeal layer encloses the brain. In most areas, these two layers are fused; however, they separate in certain regions to form venous sinuses that drain blood from the brain. Folds of the meningeal layer, such as the falx cerebri, extend deeply into the longitudinal fissure between the right and left cerebral hemispheres. Finally, the cerebellar tent supports the posterior brain hemisphere like a ledge.

The arachnoid is a connective tissue membrane that is laid over the grooves and convolutions of the brain and spinal cord. The outer surface of the arachnoid adheres to the

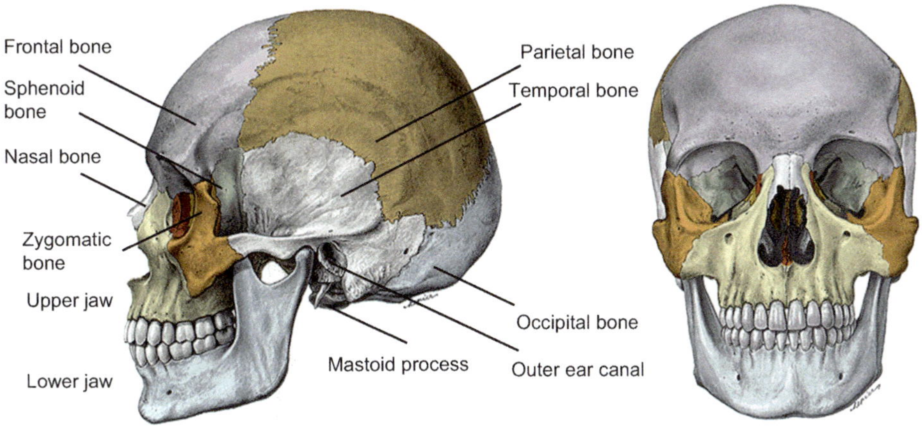

Fig. 3.3 Anatomical orientation image of the skull (according to [87])

hard brain skin, the dura mater, and limits the capillary subdural space from the inside; the inner surface is connected to the soft brain membrane, the pia mater, by a network of connective tissue trabeculae.

The inner delicate layer (pia mater) itself is a thin membrane of fine connective tissue, which is equipped with numerous small blood vessels. It lies close to the surface of the brain or the spinal cord. The space between the arachnoid and the soft brain skin, as well as the brain chambers (ventricles), are filled with a colorless fluid (cerebrospinal fluid) that supplies the brain with nutrients and dampens external shocks [68].

The **brain** is structurally and functionally divided into five parts (see Fig. 3.4): cerebrum, cerebellum, midbrain (mesencephalon), diencephalon, and medulla oblongata. Additionally, it has four brain ventricles, three membranes (meninges), two glands (pituitary and pineal gland), twelve pairs of cranial nerves, and the cranial arteries and veins. Its average length is approximately 165 mm and its greatest width is about 140 mm; depending on body size, its weight in men is approximately 1.36 kg and slightly less in women. The cerebrum accounts for about 7/8 of the weight and is divided into a right and a left hemisphere [68].

A detailed analysis of **head injuries** is difficult. Although the types of injuries (injuries and fractures) are largely known, their origin, especially the relation to injury type (mechanical strain), is still controversial and debatable. The reasons for this lie, on the one hand, in the lack of a clear correlation between injury type and nature (sufficiently complex mathematical simulation models are difficult to verify) and, on the other hand, in the problem of including the influence of age and the transferability to humans reliably in cadaver or animal experiments. Head injuries are differentiated into [60]

- Contusions, bruises, and lacerations and incisions,
- Fractures of the facial skull and skull fractures, which are strongly influenced by the age of the occupant, and
- Brain injuries, in which the mechanical strain essentially dominates the injury severity.

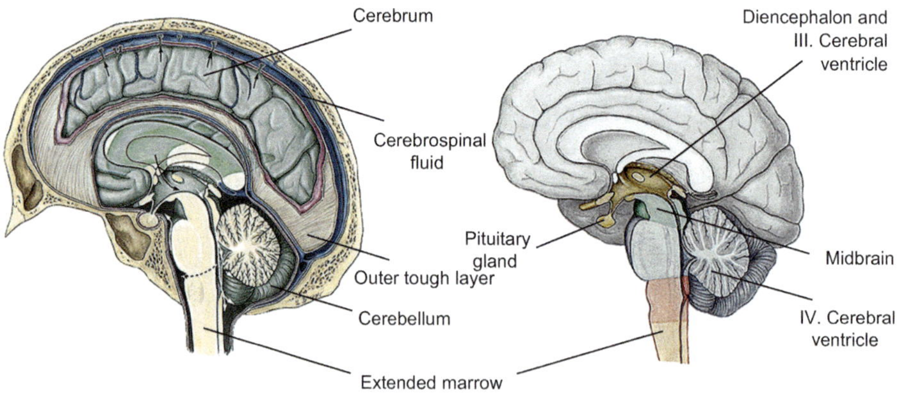

Fig. 3.4 Longitudinal section through the middle of the skull and the **brain** [87]

Skull fractures are understood to be fractures of the bony skull, which can be ruled out if there is no impact on "hard" structures (of the vehicle interior, but also of the vehicle exterior). The fractures can be differentiated into bursting and bending fractures. Bursting fractures occur with induced vibrations at approximately 360 Hz and large amplitudes, setting the skull into opposing vibrations (anti-resonance). Bending fractures are the result of loads with a higher frequency component of about 950 Hz; here lies the resonance frequency of the bony skull [42, 69, 101]. A clear correlation between skull fractures and brain injuries does not exist, yet various studies show that the occurrence of dangerous complications after initially mild brain injuries is often associated with skull fractures [48]. However, in [17] injured individuals with severe brain hemorrhages were examined, in which skull fractures were diagnosed in half of the cases, while no fractures could be detected in the other half of the patients. The reason for these uncertainties is that skull base fractures are very difficult to detect clinically, even after X-ray examinations.

Brain injuries are clinically divided into two major categories, namely diffuse and localized injuries. The diffuse injuries, which are found in about three out of four cases in vehicle and pedestrian accidents [31], include brain swelling, concussions (commotio, contusio, diffuse axonal injury DAI) and extensive injuries. Localized injuries are those such as traumatically induced bleeding between the dura and skull bone (epidural hematomas, Fig. 3.5), bleeding between the dura and arachnoid (subdural hematomas, Fig. 3.6) usually caused by tearing of the bridging veins, large bruises within the brain occurring in skull fractures (intracerebral hematomas, Fig. 3.7) and contusions, which are bruises and crush injuries of the brain that occur at the moment of impact of a blunt trauma by collision of adjacent cortical regions with the skull bone. The contusions (Fig. 3.8) are found both on the side facing the impact (coup effect) and on the side away from the impact (contre-coup effect) and are often accompanied by commotion syndromes

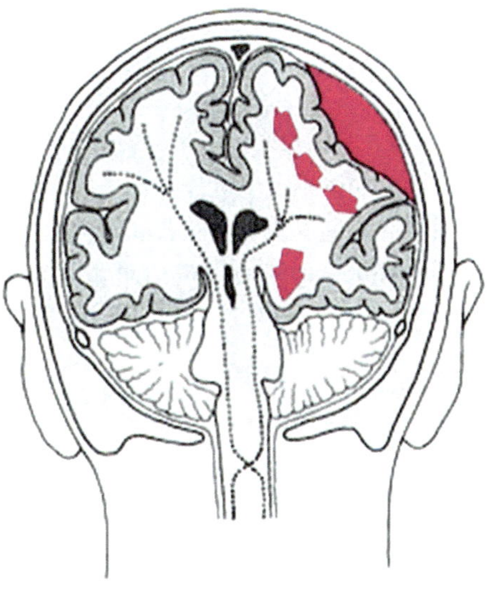

Fig. 3.5 Epidural hematoma: Bruise between the skull cap and the hard meninges (from [99, 100])

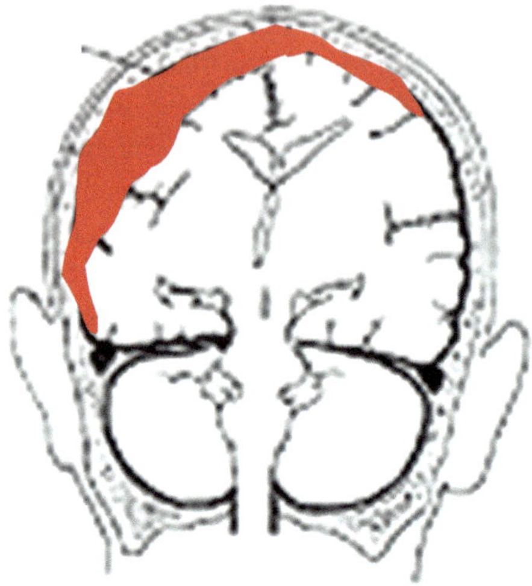

Fig. 3.6 Subdural hematoma: Bruise beneath the meninges (from [99, 100])

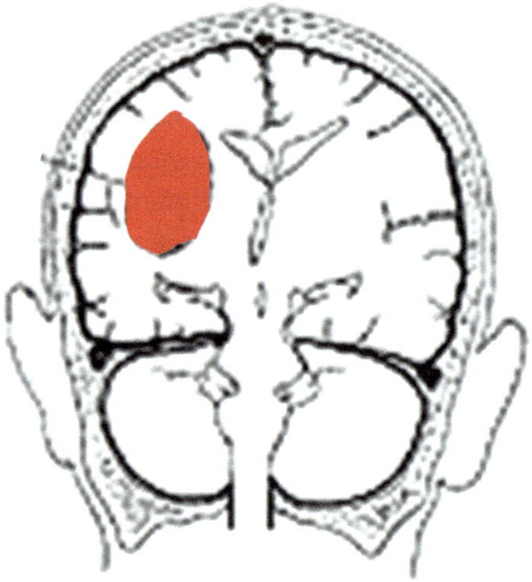

Fig. 3.7 Intracerebral hematoma: Bruise within the brain substance (from [99, 100])

(commotio cerebri: concussion), i.e., traumatically induced reversible damage to the brain without anatomically tangible substrate [86]. Subdural hematomas and extensive injuries are the most significant injuries in accident-related deaths [31].

3.1.2 The Spine

In this section, only the region of the **cervical spine (C-spine)** will be considered, as it exhibits a highly complex mechanism from a mechanical and structural aspect [120, 121]

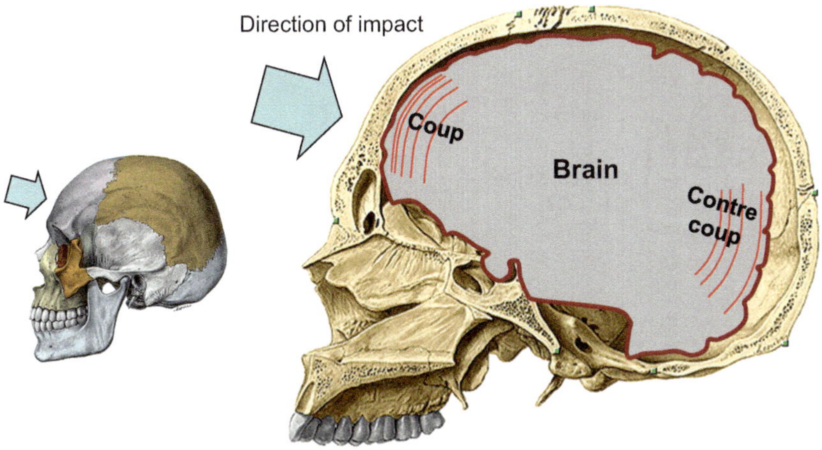

Fig. 3.8 Appearance of brain contusions (coup and contre-coup injuries), [87]

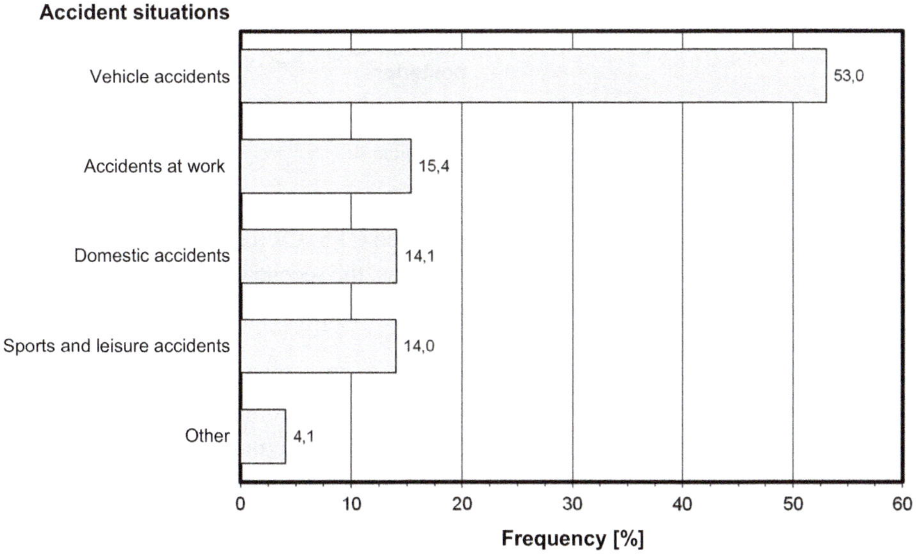

Fig. 3.9 Cervical spine injuries in different accident situations (from [66])

and is also dominant in traffic accidents compared to other causes of accidents according to [66] (Fig. 3.9). The cervical spine itself is the upper part of the human spine (columna vertebralis), which is divided into the following regions according to [86] (Fig. 3.10):

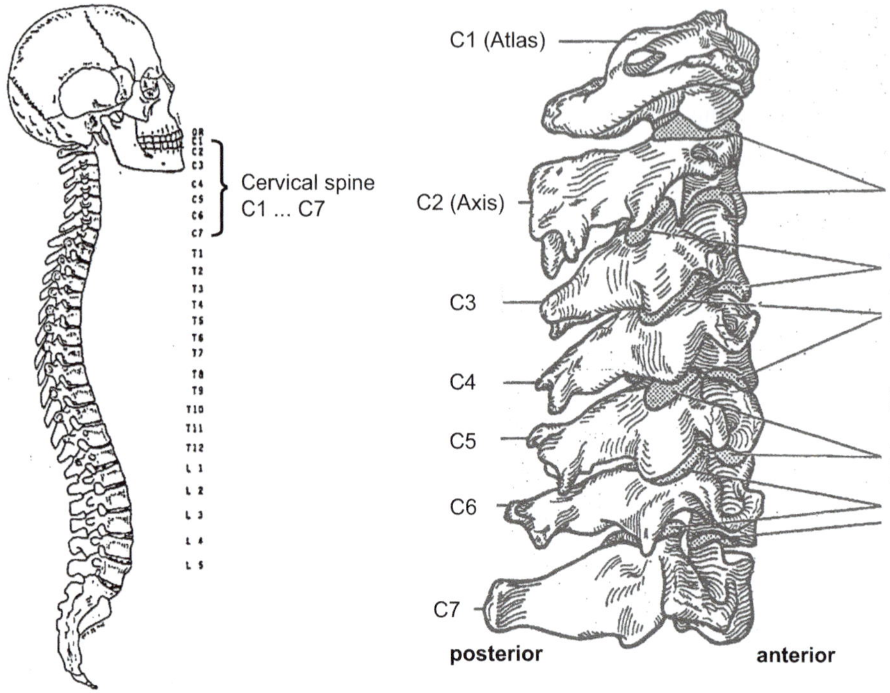

Fig. 3.10 Side view of the entire spine and the cervical spine

- 7 cervical vertebrae (vertebrae cervicales) labeled C1 to C7 form the cervical spine, of which the first cervical vertebra C1 is called atlas, the second C2 is called axis and the seventh C7 is called vertebrae prominens due to its palpable and visibly protruding spinous process,
- 12 thoracic vertebrae (vertebrae thoracicae) labeled T1 to T12 form the thoracic spine,
- 5 lumbar vertebrae (vertebrae lumbales) labeled L1 to L5 form the lumbar spine,
- 5 sacral vertebrae (vertebrae sacrales) fused to form the sacrum, and
- 4 to 5 rudimentary and often fused coccygeal vertebrae (vertebrae coccygeae).

The **region of the cervical spine** extends from the (imaginary) connecting line between the base of the skull and the lower jaw to the level of the first rib or the clavicle and has two basic functions: It carries and supports the head and allows its mobility, and it also serves as a conduit to connect structures of the head and face with those of the thorax and pelvis. In the anterior region of the neck, the supply organs are arranged. These are the smaller muscles of the front neck, the esophagus as a connection to the stomach, the larynx and the upper trachea leading to the lungs, the thyroid gland, and arteries and veins for blood supply to the head, brain, and face. In addition, important nerve strands run vertically through this region from the brain and spinal cord of the cervical spine. In the posterior part of the neck, the central area of the cervical vertebrae is located, whose lateral and posterior surfaces are surrounded by the main neck muscles (Fig. 3.11). In addition, the nerve strands that originate from the spinal cord of the cervical spine run in and between the lateral muscles into the posterior region of the neck.

The **mobility of the cervical spine** is limited by the joint connection, the normal extension of the ligaments that surround the vertebrae, and by the muscles that are attached to the **skull** and the cervical vertebrae. Nodding is a movement between the skull and the first vertebra C1 (atlas), all other movements of the head and neck, such as forward bending (flexion), backward extension (extension), tilting (inclination) and turning (rotation) of the head to one side, generally occur between each of the cervical vertebrae (Fig. 3.12). The only possible movement that can occur between the vertebrae C1 (atlas) and C2 (axis) is the shaking of the head (e.g., in a denial). The great mobility of the spine is given by the lateral joint connections. Neck movements, however, are limited due to the joint capsules that surround all joints of the cervical spine, the small amount of compression of the intervertebral discs, and the limited stretchability of the ligaments between the adjacent vertebrae (band between the spinous processes and the vertebral arches, neck band, as well as anterior and posterior longitudinal band). Excessive strains in any direction thus necessarily lead to either an overstrain of the ligaments, compression of the intervertebral discs or the bones of the cervical spine, occasionally associated with ligament tears and vertebral fractures [71].

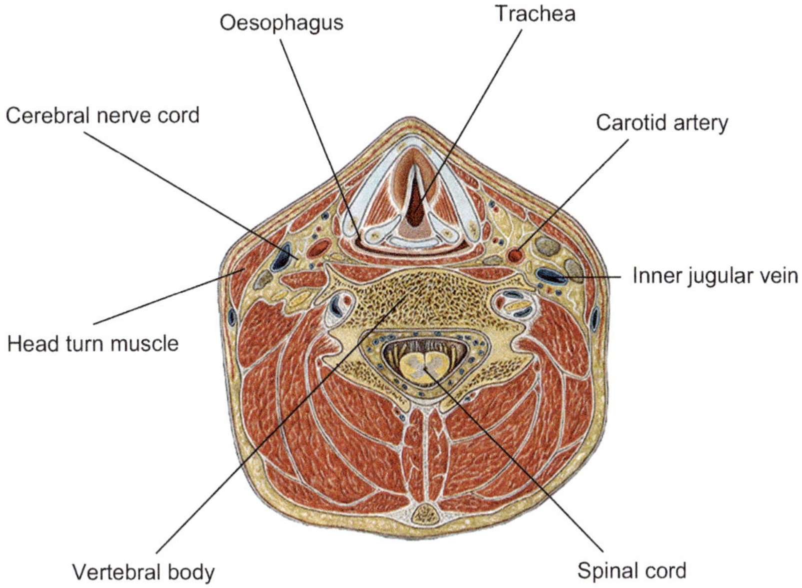

Fig. 3.11 Horizontal section through the neck (according to [87])

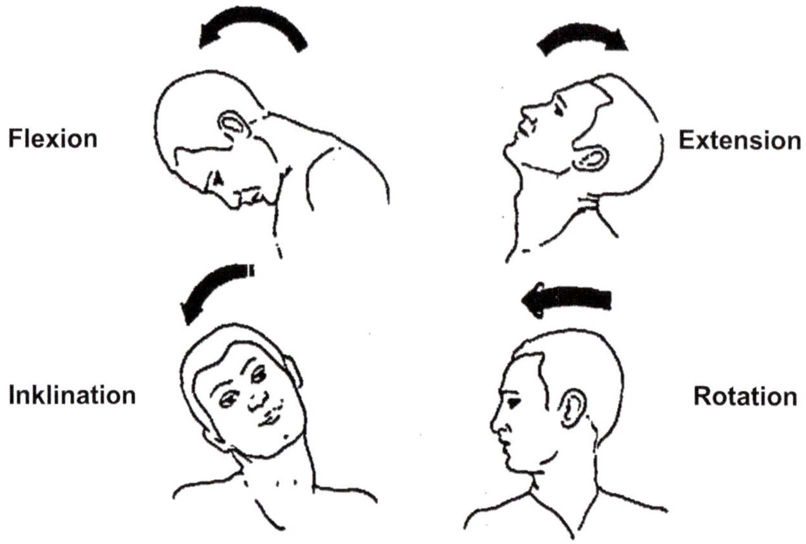

Fig. 3.12 The four general movements of the head and the cervical spine (from [71])

The **cervical nerves,** which originate from the spinal cord of the cervical spine, pass outward between the vertebral pedicles and immediately recombine after passing through. In this way, the composition of nerves forms a proper plexus, which reaches various body regions through restructuring and redistribution of individual nerve fibers.

The cervical plexus (plexus cervicalis, CI to CIV) supplies the skin of the neck, the sides of the face with strong branches to the muscles that are arranged next to and behind the cervical vertebrae (Fig. 3.13). The nerve strands from the lower cervical vertebrae (between C4 to C7 and between the thoracic vertebrae T1 and T2) form the extensive brachial plexus (plexus brachialis, CV to CVIII as well as TI) and branch out to the arm nerves [71].

The **classification of injuries** of the cervical spine is based on retrospective data evaluations of injured persons, movement studies using mathematical models, from video recordings, experiments with corpses or parts of corpses, and subsequent evaluations of various, already published injury circumstances [66]. Accordingly, the classification of injuries varies depending on the objective from a radiological, pathological, neurological, or mechanical perspective, often subjected to a certain simplification for the sake of presentation (or must be).

The following presents (in excerpts) the **injury mechanics** of the cervical spine as outlined in [71], where the dynamic aspect is in the foreground together with the clinical one. An excessive **forward movement of the head** relative to the cervical spine causes shear forces that cause a displacement between the atlas C1 and the occiput; the consequence for those affected is usually fatal. This injury type is often found in pedestrians

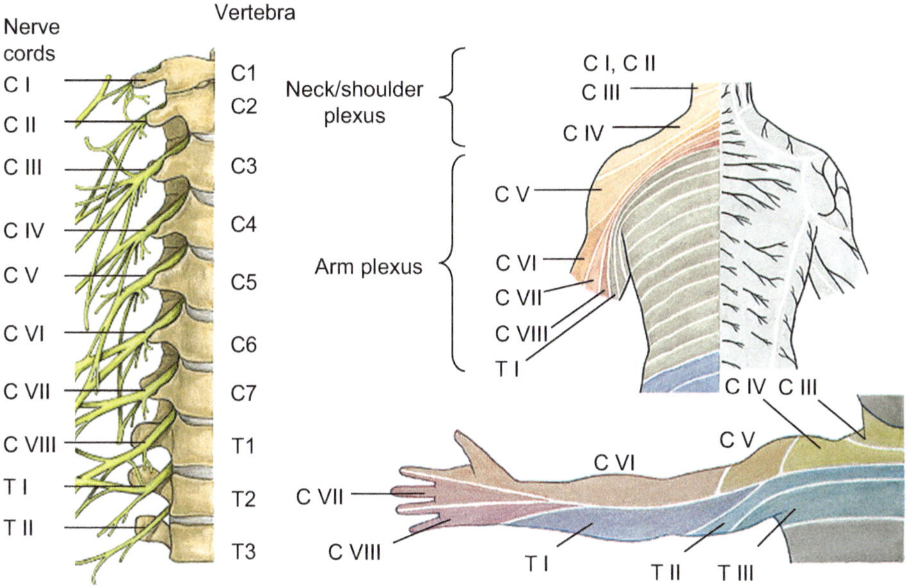

Fig. 3.13 The nerve strands of the cervical and brachial plexus and the distribution of the nerve segments in the neck, chest, and upper extremities (from [87])

who are hit by a vehicle. The integrity of the occiput/atlas joint is primarily maintained by ligament structures that tear in such injuries. The posterior arch of the vertebra C1 essentially breaks through compression between the arch and the pronounced spinous process of the vertebra C2. This fracture usually occurs at the depression, formed by the vertebral artery (arteria vertebralis), through which the annular bone has its thinnest point (Fig. 3.14a). In addition, in flexions of the cervical spine, splinter fractures occur because the strong band grown on the spinous processes tears away the narrow part of the atlas arch (C1). The bursting of the annular vertebra C1, caused by **axial load** from the occiput-joint head onto the overlying joint side of the vertebra C2, is referred to as the so-called Jefferson-fracture. The result is the classic fracture of the atlas into four segments (Fig. 3.14b).

The rotation between vertebrae C1 (atlas) and C2 (axis) takes place around the odontoid process (dens) of the axis, which protrudes into the front part of the atlas . **Dens fractures** are classified into three types: Type 1 is an oblique fracture through the end of the odontoid process and usually represents an avulsion fracture. Type 2 characterizes a fracture at the base of the odontoid process, i.e., at the vertebral body of the axis, and Type 3 finally describes a fracture that extends to the entire vertebral body of the axis. In the event of a facial impact, e.g., during a forward movement of the torso, the dens is slightly displaced backwards when a fracture occurs (Fig. 3.15).

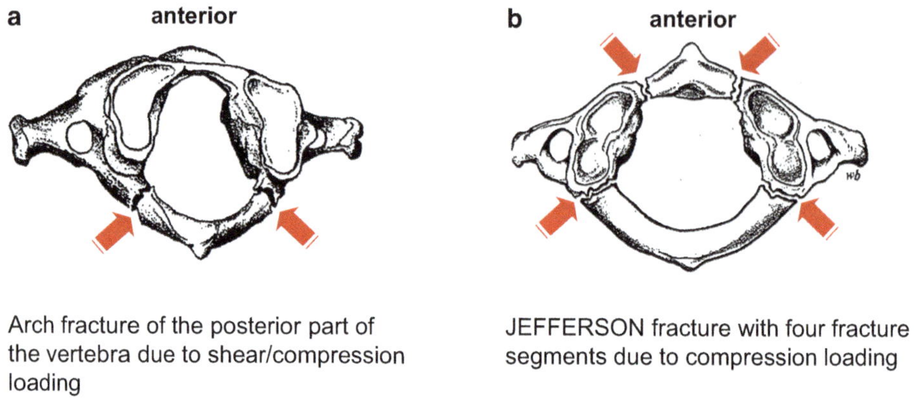

Fig. 3.14 **a**, **b** Typical fractures of the cervical vertebra C1 (atlas), according to [71]

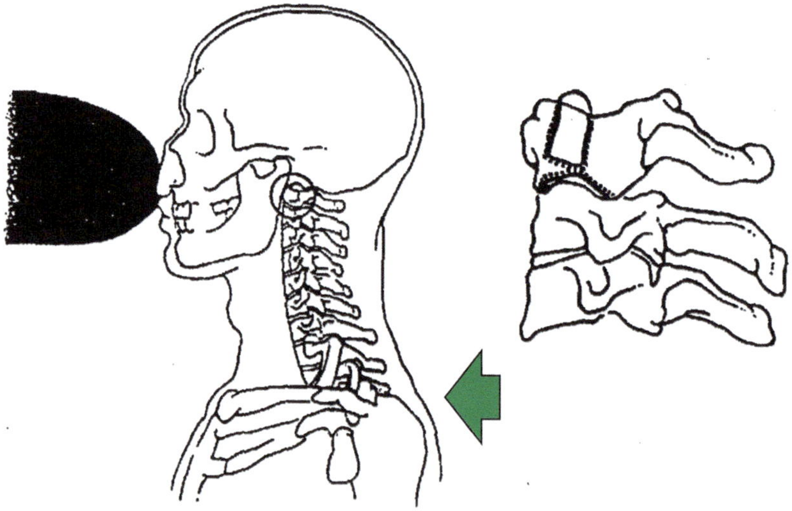

Fig. 3.15 Dens fracture with backward displacement due to a facial impact and simultaneous torso forward movement (according to [71])

Furthermore, a blow to the lower occiput or a fixation of the lower occiput during a backward torso movement induces a shearing and a forward-displaced fracture of the odontoid process (Fig. 3.16). The third mechanism occurs when the lower face collides with a structure, resulting in a diffraction of the head. If the torso continues to move forward, the neck below the atlas experiences an extension. The resulting bending moment causes the dens to break at its base, with or without forward displacement. Shear forces acting on the occiput or face can again displace the atlas on the axis forward or backward (Fig. 3.17). If the force vector acts eccentrically and thereby generates a moment around

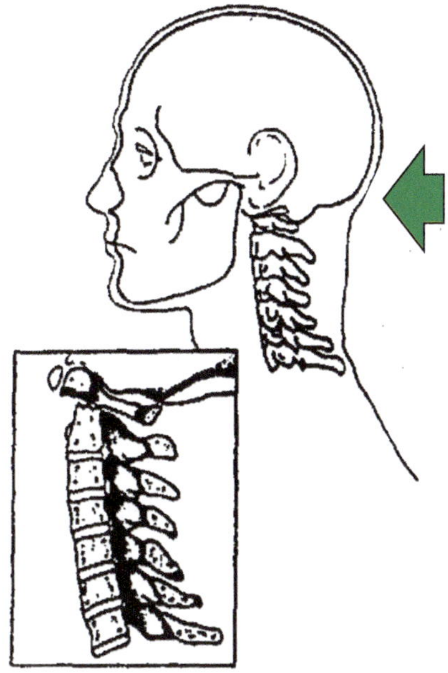

Fig. 3.16 Dens fracture with forward displacement due to a blow to the lower occiput or fixation of this area and simultaneous torso backward movement [71]

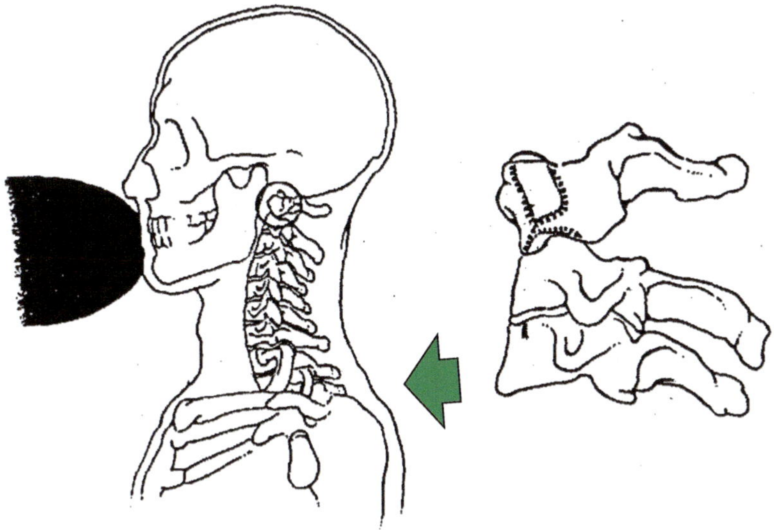

Fig. 3.17 Dens fracture at its base due to a bending moment during impact against the lower face and simultaneous torso forward movement (according to [71])

the vertical axis, a rotational subluxation (incomplete contortion, where the joint surfaces remain in contact) is conceivable.

The extension of the anterior vertebral elements due to an **extension of the neck**, where the posterior elements are subjected to compression, results in an interaction of two spinous processes and occasionally also with the posterior part of the vertebral arch. This usually results in a fracture of the C2 vertebra, where the posterior part of the vertebral body is separated from the anterior part (Fig. 3.18). The analogy of these injuries to those of judicially hanged led to the naming of this injury complex: "Hangman's fracture".

Occasionally, fractures of the C2 vertebra also occur in the joint area, which are then analogous to the more frequently occurring conditions in the lower spine. These injuries are only sometimes fatal and are often survived without neurological deficit. They usually occur through the fixation of the chin, while the body below the head drops downward (Fig. 3.19). This situation causes an extension combined with tensile load (tension).

Axial compression injuries occur when the head is restricted in movement, while at the same time the torso moves upward and aligns with the head and neck to form a straight line. This alignment causes a force to act along the longitudinal axis of the head, neck, and torso, leading to injury. If the vertical force is sufficiently large, this load results in a typical burst fracture of a vertebral body (Fig. 3.20). The origin of this fracture can be explained by the transmission of the compression force within the cervical spine: The intervertebral discs located between the vertebral bodies act as absorption elements during shocks, but due to their viscous behavior, they are essentially not compressible. With increasing force, the transparent cartilage plates, which encompass the adjacent surfaces of the vertebral bodies, bulge. The reason for this lies in the rugged bone structure of the vertebral bodies with their extensive vascular channels; with increasing load of the vertebral body, these blood vessels empty and the cartilage plate is pressed in. Consequently, there is initially a fracture of this cartilage plate. With a further increase in load, the intervertebral disc is pressed into the vertebral body, leading to the

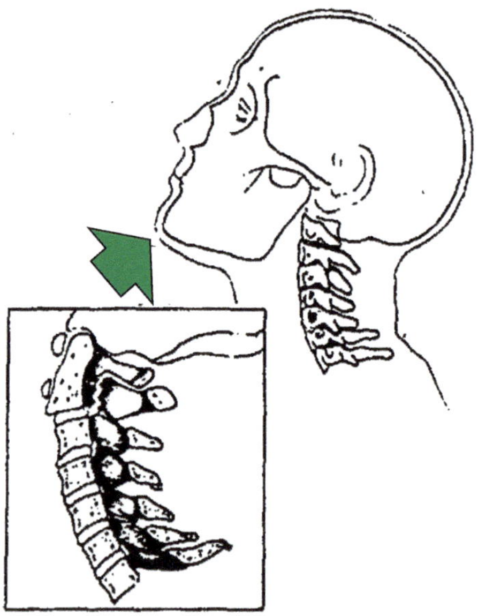

Fig. 3.18 Fracture of the C2 vertebra with separation of the posterior part of the vertebral body from the anterior (so-called "Hangman's fracture") due to an extension of the neck [71]

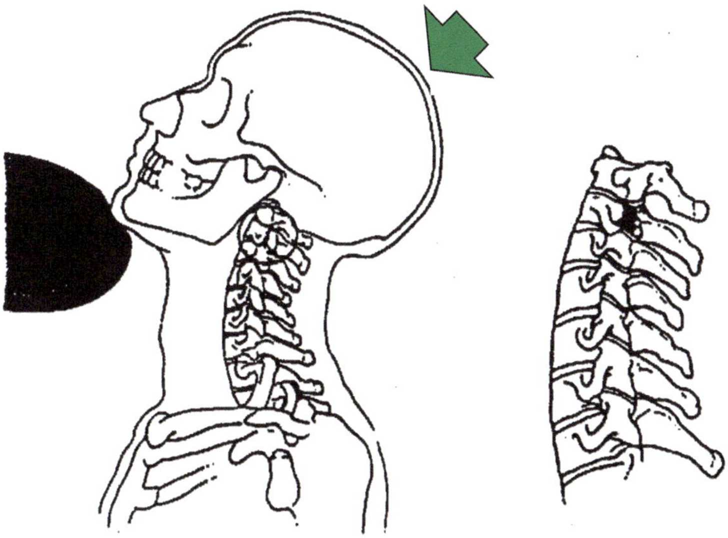

Fig. 3.19 Fracture of the C2 vertebra due to the fixation of the chin and simultaneous diving of the body due to a combination of extension and tensile load (tension), [71]

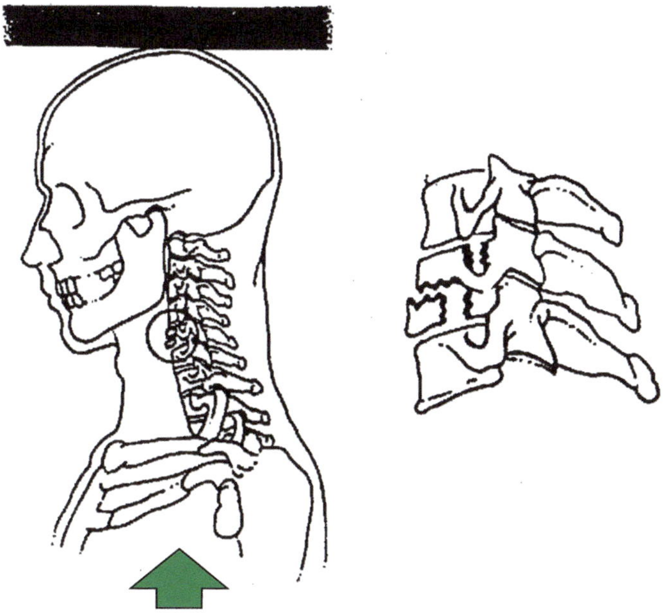

Fig. 3.20 Burst fracture of the C5 vertebra due to axial compression (according to [71])

typical burst fracture, which literally shatters the vertebral body. The rear part of the vertebral body usually shifts backwards towards the spinal cord, while the front fragment is pushed forward.

Flexion/compression injuries occur when a force attacks at the posterior vertex of the head, but not at the extreme rear or in the area of the occiput. A similar situation also arises with restricted movement of the head, while the torso continues to move and the head and upper body form an angle that induces a **flexion of the cervical spine**. If the force exceeds a certain level, a compression fracture of a vertebral body in the lower area of the cervical spine (C3 to C7) occurs in its typical wedge-shaped appearance (Fig. 3.21). Due to the compression of the vertebral bodies, the ligaments between the spinous processes of adjacent vertebrae (ligamentum interspinalia) are stretched, and occasionally a fracture of one or more spinous processes occurs (Fig. 3.22).

Injuries due to flexion/shear stresses occur due to forces that grab at the back of the head perpendicular to the longitudinal axis of the head; they also cause a flexion of the cervical spine. The shear stress that is effective in this case usually results in a ligament tear in the lower part of the cervical spine with simultaneous forward displacement of the vertebral bodies, whereby the lower joint surface of the vertebra slides over the upper one of the underlying vertebra. A gaping of the space between the adjacent spinous processes due to the ligament tear at the level of the dislocation is the result (Fig. 3.23). The relative displacement of the involved vertebrae can detach the anterior longitudinal ligament of the spine, which is fused to the anterior surfaces of the vertebral bodies and runs from the occiput to the sacrum, from the upper front side of the vertebral body;

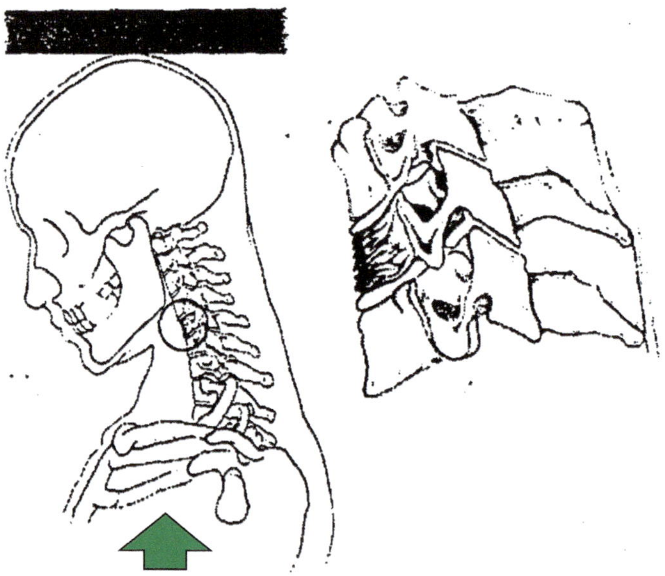

Fig. 3.21 Wedge-shaped fracture of the C5 vertebra due to a flexion/compression load [71]

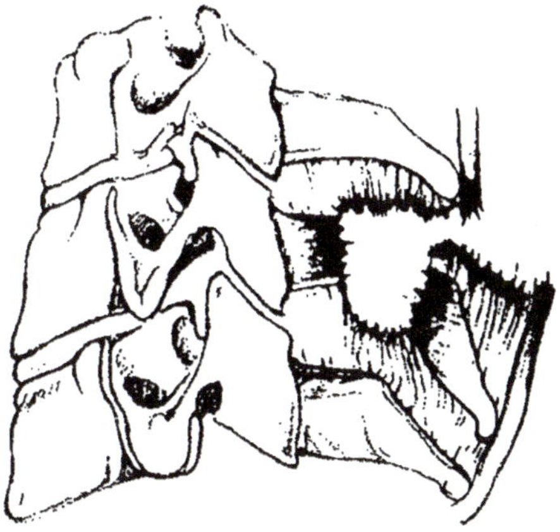

Fig. 3.22 Tear of the band between the spinous processes of adjacent vertebrae and fracture of the spinous process due to overstretching (from [71])

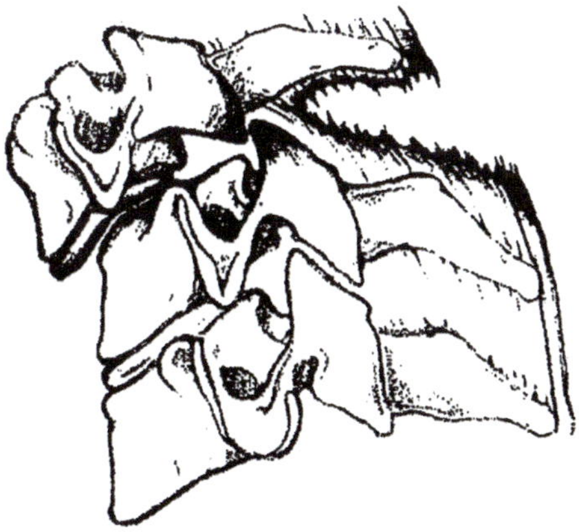

Fig. 3.23 Forward displacement and ligament tear between adjacent spinous processes without fractures due to flexion/shear load (from [71])

bone injuries usually do not occur. Also without significant bone fractures, the load is absorbed in cases where a force at the back of the head acts in a direction that still induces a flexion with simultaneous shear effect, without causing significant compression of the vertebral bodies (Fig. 3.24).

If the loads described so far occur in combination with a **rotation of the head/neck area**, they lead to significant ligament tears and detachments as well as to unilateral or alternating vertebral dislocations without significant bone injuries. However, the displacements already occur at a lower stress level than with forces that result in flexions and compressions. In investigations of accidents where neither the kinematics nor the injury pattern indicated a flexion/shear mechanism, but also in some cases of typical flexion/shear injuries, an additional tensile load of the neck with remarkable strains in the order of magnitude of the described dislocations was found due to the rotational movement, resulting in tears of the anterior and posterior longitudinal ligament, shattering of the intervertebral disc, and tearing of the joint capsules. The occurrence of these injuries is extremely uncertain, so a clear injury pattern cannot be assumed.

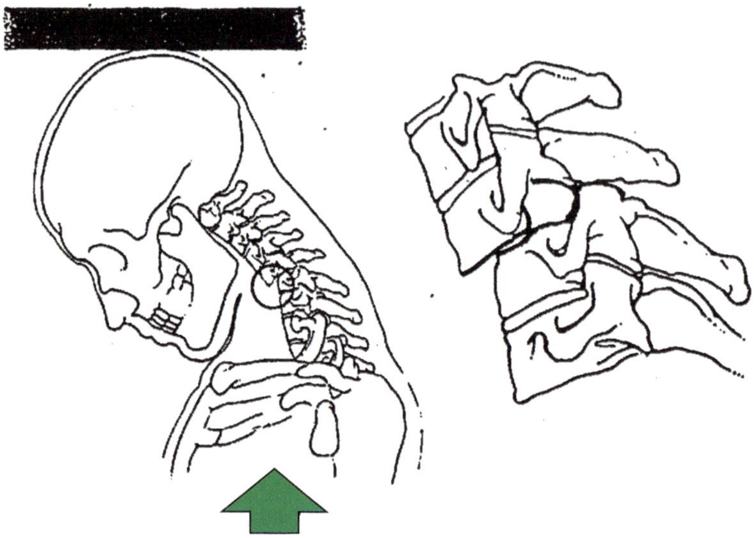

Fig. 3.24 Ventral dislocation without fracture of the C5 vertebra due to a flexion/shear load [71]

Injuries due to extension/compression loads occur with sufficiently high forces that act in the forehead area of the head or in cases where the head is hindered in its movement options and at the same time the torso essentially continues to move in the direction of the longitudinal axis. This load induces an **extension of the neck** and a corresponding load of the vertebrae. The posterior vertebral elements act as a pivot point, so that the occurring tensile forces can lead to the tearing of the anterior longitudinal ligament and to the rupture of the intervertebral discs. The backward displacement of the upper vertebrae causes a narrowing of the vertebral canal and thus a compression of the spinal cord either by the folding of the ligaments between adjacent vertebral arches or by a constriction between the rear lower edge of one vertebral body and the protruding edge of the underlying vertebra (Fig. 3.25).

An additional vertical compression induces fractures of the lower surface of the upper vertebra and allows a forward movement of the underlying vertebra. In this process, a small piece of bone from the lower front edge of the displaced vertebral body can be broken out through the torn anterior longitudinal ligament (Fig. 3.26). However, fractures of the spinous process and the vertebral arch can also occur, which tend to shift upwards (Fig. 3.27).

A lateral flexion/compression load is spoken of when the head is subjected to a lateral movement. The neck is thereby forced into a **lateral diffraction**, which results in a strain of the vertebral side area and can lead to fractures of the structure located on the compressed side (Fig. 3.28). If a rotation of the head occurs simultaneously, splinters can detach from the affected vertebral body.

Fig. 3.25 Impairment of the vertebral canal due to folded ligaments of adjacent vertebral arches during forced extension [71]

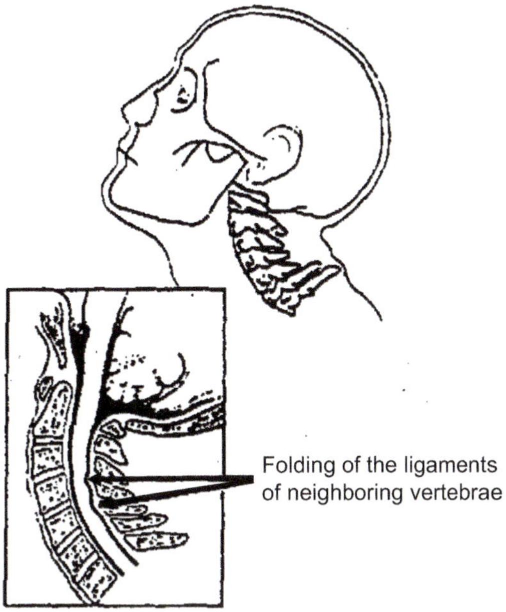

Fig. 3.26 Tear of the anterior longitudinal ligament and breakout of a fragment from the anterior underside of a vertebral body due to an extension/ compression load (from [71])

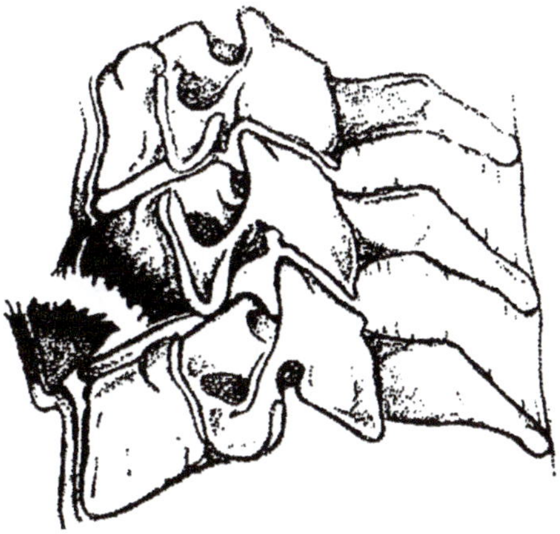

Direct injuries usually occur through a **punch to the back of the neck** delivered with sufficient force, especially in the lower area. The result can be a fracture of the spinous process of one or more vertebrae. With corresponding **lateral loads** against the neck, fractures of the transverse process of the vertebrae can be the result. Direct trauma to the neck is rarely associated with other bony injuries, without a considerable force being effective.

In rear-end collisions, cervical spine distortions often occur. These are usually a floating movement with an S-shape of the cervical spine and subsequent **extension**. The

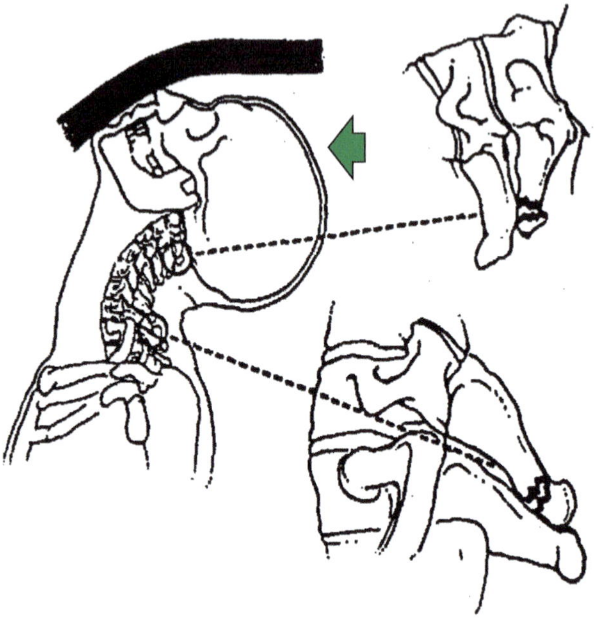

Fig. 3.27 Fractures of the spinous process of the vertebrae C1 and C7 due to an extension/compression load (according to [71])

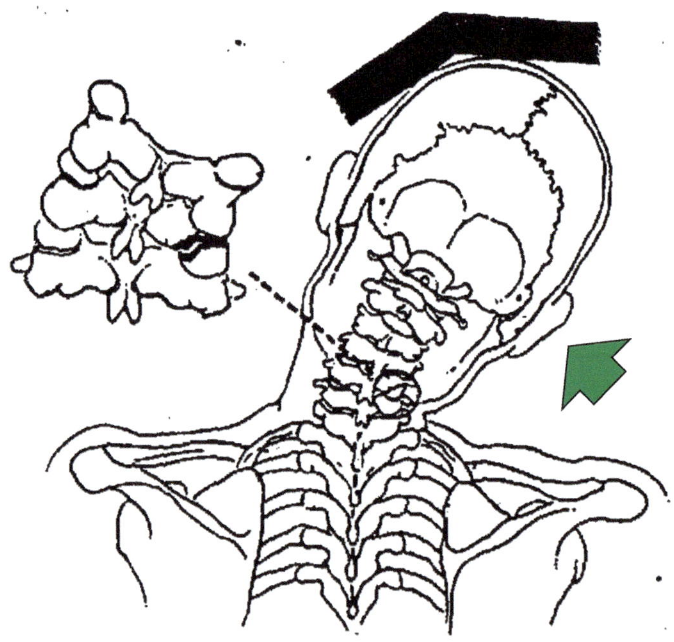

Fig. 3.28 Fracture of the lateral structure of the vertebra C5 due to a lateral flexion/compression load (according to [71])

secondary **flexion movement** (for example, when falling into the seat belt) is usually low in energy. Painful strains or sprains of the neck are not uncommon [107], the assessment of which requires special knowledge in the field of biomechanics [120, 122].

The majority of **soft tissue injuries** affect the neck muscles. The injuries previously observed in the front part of the neck due to missing headrests hardly occur anymore. In these cases, with a sufficiently high load, tears of the muscle fibers, bleeding in the esophagus and muscles, separation of the anterior longitudinal ligaments, and damage to the intervertebral discs occurred.

Nightingale et al. examined the flexibility and strength of 52 female motion segments in both flexion and extension [80]. In 2007, a similar study was conducted on 41 male cervical spines [81]. The study was the first to provide statistically significant tolerance data in pure bending (Table 3.1). They found that the upper cervical spine was stronger than the lower cervical spine. In both men and women, the failures of the lower cervical spine were complete joint tears (tear of all ligaments and the intervertebral disc) with some associated avulsion fractures of the vertebral bodies. The most common injuries of the upper cervical spine were dens fractures, which mainly occurred in extension, but were also observed in flexion.

3.1.3 The Thorax

The thorax, as the chest is called, consists of the bony chest and the organs underneath. It is bounded downwards by the diaphragm, a thin muscular partition between the chest and abdominal cavity. The chest wall is designed as a compliant cavity and protects the respiratory and the heart/vascular system.

Table 3.1 Average bending tolerance of the cervical spine (according to [81, 80])

	Upper cervical spine		Lower cervical spine	
	Moment Nm	Angle degree	Moment Nm	Angle degree
Flexion				
Man	39.0	58.7	20.5	14.0
Woman	23.7	56.2	17.4	19.3
Extension				
Man	49.5	42.4	17.1	15.1
Woman	43.3	50.2	21.2	20.5

The **chest** (Fig. 3.29) comprises twelve thoracic vertebrae (T1 to T12), twelve pairs of ribs, and the sternum. At the back of the spine, each of the twelve ribs is articulately connected to the corresponding thoracic vertebra. At the front of the chest, each of the top seven ribs (costae verae) is cartilaginously fused with the sternum, while the lower five ribs (costae spuriae) have no direct connection to the sternum. Ribs 8 to 10 form the rib arch and are connected to the seventh rib by a cartilage-like attachment; the last two ribs end freely [86]. The chest wall includes the bony ribs, cartilage, and muscles, which allow the chest to expand and contract during the breathing process. The muscles connect the ribs to each other and complete the compliant chest, which provides protection to the internal organs. In addition, blood vessels and nerve strands fill the space between the adjacent ribs [114].

The **lung** (pulmo) consists of two lung wings, with the left lung wing having two lobes (top and bottom) and the right wing having three lobes (top, middle, and bottom) (Fig. 3.30). The lungs are enveloped by a serous membrane (pleura visceralis); additional membranes (pleura parietalis) form the inner surface of the chest wall, enclose in the **mediastinal cavity** (mediastinum) the organs of the thoracic center and cover the diaphragm [86]. The heart, the large blood vessels leading to and from it, the mammary gland, the oesophagus, the lower section of the trachea, and the lymph nodes of the chest are located in the mediastinal cavity, which is bounded at the front by the sternum and at the back by the thoracic spine [14]. The trachea, a 10 to 12 cm long section of the respiratory tract, begins below the cricoid cartilage and ends with the branches into the right and left main bronchi at the level of the fourth thoracic vertebra. The main bronchi branch according to the lung lobes into three on the right and two on the left lobe bronchi, which in turn divide into the segment bronchi and finally open finely branched at

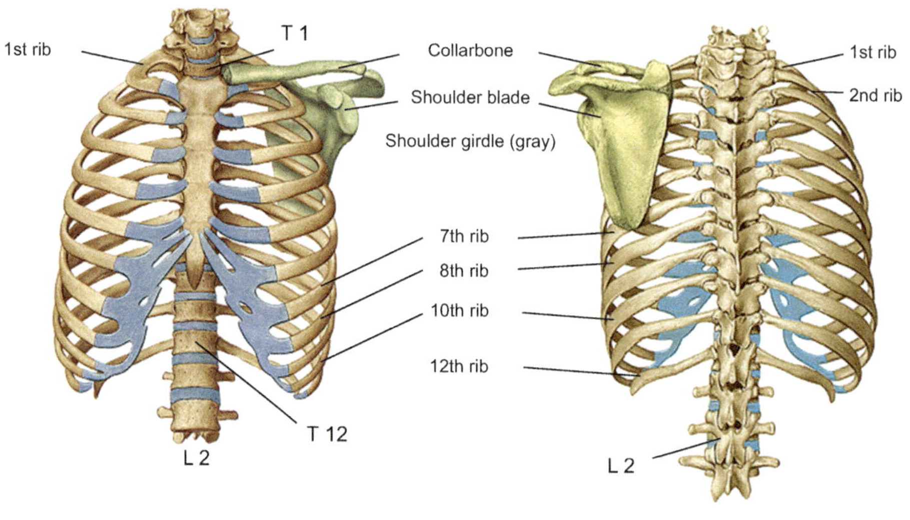

Fig. 3.29 Anatomical skeleton of the thorax (according to [87])

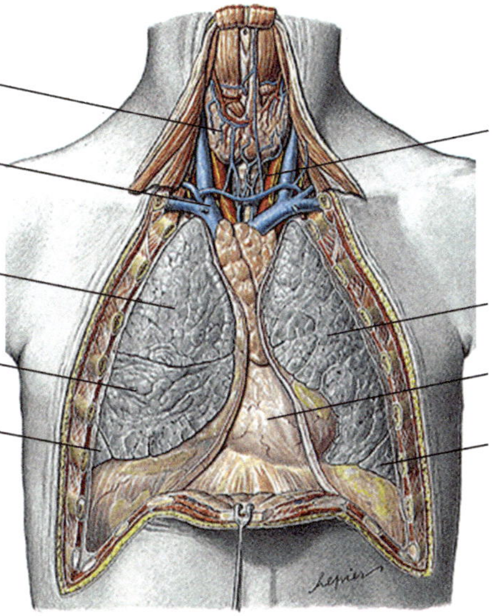

Fig. 3.30 Positional relationships in the chest (from ventral), according to [86, 87]

the alveoli (Fig. 3.31), where the oxygen/carbon dioxide exchange with the blood takes place [86, 87].

The **heart** (Latin: cor) is a muscular hollow organ located in the lower part of the chest in the middle mediastinum between the lungs and on the diaphragm. It is about the size of a fist and weighs 300 g in adult men and 250 g in women [14]. The heart is divided into four chambers, the left and right atrium and the left and right ventricle (ventriculus). The deoxygenated blood returning from the body (except the lungs) flows through the vertically running upper and lower vena cava (vena cava superior and inferior) and enters the right atrium. From the right chamber, the venous blood is pumped through the trunk of the pulmonary artery (truncus pulmonalis) and reaches the lung through the right and left pulmonary artery (arteria pulmonalis dextra and sinistra). There, the blood releases carbon dioxide to the breathing air in the alveoli, is enriched with oxygen and flows through the four pulmonary veins to the left atrium. From here, the arterial, i.e., oxygen-enriched blood, reaches the thick-walled left ventricle, from which it is pumped through the aorta to all body regions (except to the lungs) [86, 87].

Fig. 3.32 shows the front and back view of the heart muscle with its vessels. The heart and the arterial and venous coronary vessels are surrounded by a two-layered, fibrous connective and adipose tissue, the pericardium, which protects the heart muscle against overstretching. Between the two pericardial layers or sheets, there is only a thin film of serous fluid.

Injuries of the thorax, but also of the abdominal area, are less common in contact with vehicle interior parts compared to injuries of the head and extremities (see Fig. 3.2).

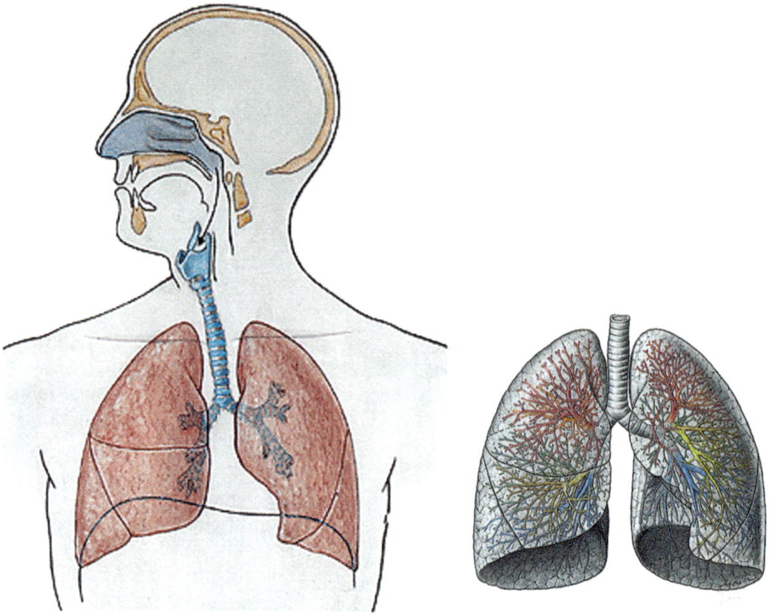

Fig. 3.31 Ramification of the trachea into the main bronchi (left illustration) and their branching (from [87])

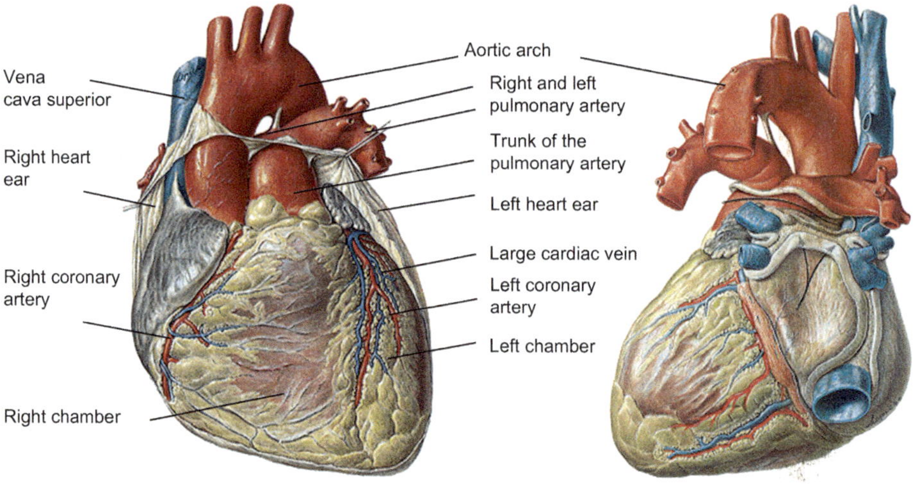

Fig. 3.32 Heart with opened pericardium (left illustration from ventral, right from dorsal), according to [87]

However, if only major injuries (AIS > 2) are considered, thoracic and abdominal injuries are the most common after head injuries. The driver's most frequent contact is with the steering wheel and its driver airbag, while the passenger is most often injured by the passenger airbag. As Fig. 3.33 shows, fractures of the chest are most prominent for both occupants, followed by lung injuries of the driver and injuries to the liver and heart. The liver and spleen are located below the diaphragm, i.e., outside the thorax, but are in the area of the lower lateral rib arch, where they can be injured by the steering wheel rim [14].

Nirula and Pintar [83] analyzed the databases of the US National Automotive Sampling System (NASS) from 1993 to 2001 and the databases of the Crash Injury Research and Engineering Network (CIREN) from 1996 to 2004. The incidence of severe thoracic injuries (AIS 3 and higher) in NASS and CIREN was 5.5% and 33% respectively. The steering wheel, the door panel, the armrest, and the seat were identified as contact points associated with an increased risk of severe chest injuries. The door panel and the armrest were repeatedly a common cause of severe injuries.

In a study on crashed motor vehicles in the United Kingdom, Morris et. al. [75] examined data on vehicle accident injuries to determine the relative injury risk of occupants of different age groups. For all occupants, the body region most frequently injured in a frontal impact was the chest. Older occupants were predictably at a higher risk of suffering chest injuries of maximum AIS 3+ (MAIS 3+). In frontal impacts, the vast majority of chest injuries were caused by the restraint system, while other vehicle interior components accounted for only 4% of injuries. A significant portion of middle-aged and older

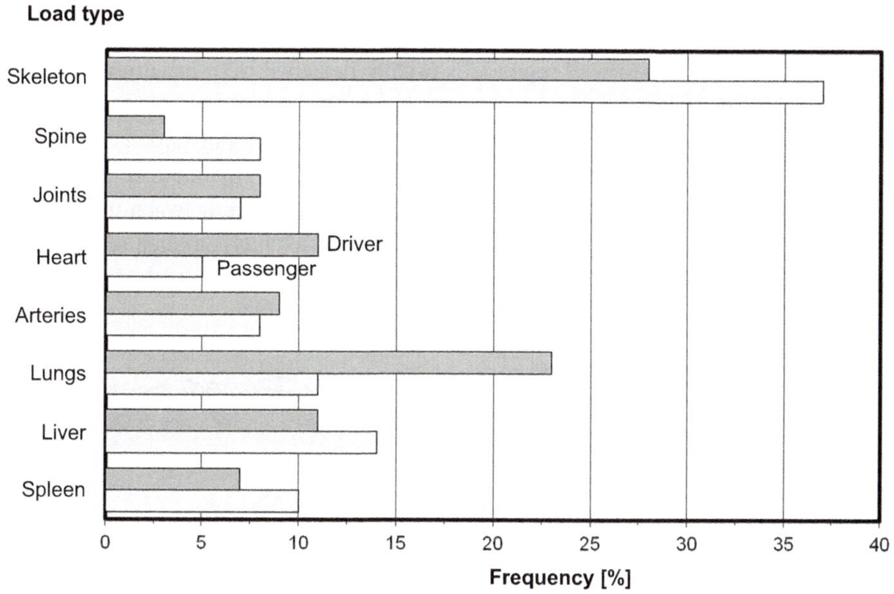

Fig. 3.33 Frequency of thoracic injuries (AIS > 2) for driver and passenger (from [14])

passengers were female. It was found that a seat belt pretensioner has a general effect in reducing the risk of MAIS 3+ chest injuries for all age groups. The anatomy of the thorax and chest significantly influences the injury pattern and the **injury mechanics** of the thorax, which are described in excerpts from [114]. Its lateral extension is greater than that in the anterior/posterior direction, so that a direct sagittal load generally leads to single fractures of several ribs, but not to the collapse of the chest wall. A lateral strain, as occurs in side collisions, on the other hand, leads to rib fractures with two or more fracture points on each rib (Fig. 3.34) and can result in thoracic instability. The reason for this is that in the lateral region the ribs lie close to the surface, have less protective muscle tissue, and have a stronger curvature.

Although **rib fractures** often occur in conjunction with severe chest injuries, the extent of bony injuries does not necessarily correspond to the overall severity of the injuries. Life-threatening injuries to chest organs can occur without rib or sternum fractures, especially in younger adults, whose ribcage is characterized by higher flexibility. However, when rib fractures occur, only a reduced proportion of the energy is transferred to the internal organs, thus reducing the probability of injury. Most bony injuries result from a thorax impact, which causes compression of the chest and deformation of the internal organs; this can be seen in the illustrations of Fig. 3.35 from a computational simulation with 30% thorax compression. The result can be a contusion of internal organs, i.e., a bruise due to blunt force. In the case of lateral impact, it can lead to fractures at several points of a single rib, but also multiple fractures, so-called serial rib fractures, are possible (Fig. 3.36). As a result there is a risk of instability of the chest wall and organ injuries due to the displacement of the fracture ends into the chest cavity.

Multiple fractures at different ribs cause a **thoracic instability**, which due to the undersupply of air to the affected lung leads to a significant impairment of respiratory activity. The unstable chest wall segment is usually detached from its surrounding tissue and thus allows it to be sucked in during inhalation. The remaining chest wall sinks in, instead of expanding; during exhalation, the segment is moved outward: This is referred to as paradoxical breathing [86]. Although the breath thrust is maintained, the flattening of the chest wall during inhalation reduces the negative pressure between the two layers of the pleura, so that the affected lung is no longer sufficiently supplied with breathing air; as a result, the blood has too little oxygen (bluish-red coloration of the lips).

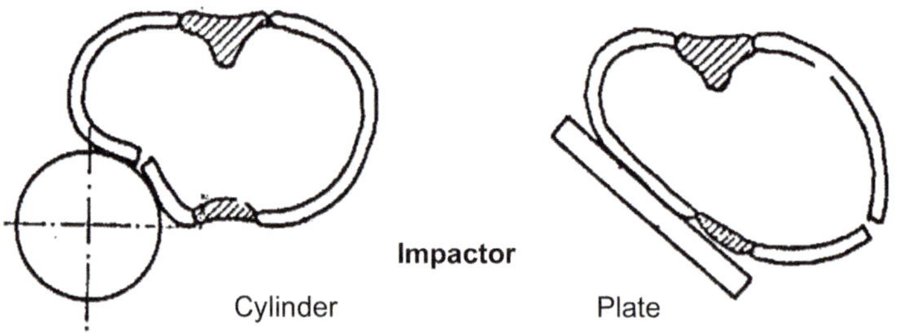

Fig. 3.34 Type of rib fracture in lateral impact with different impact bodies (from [54])

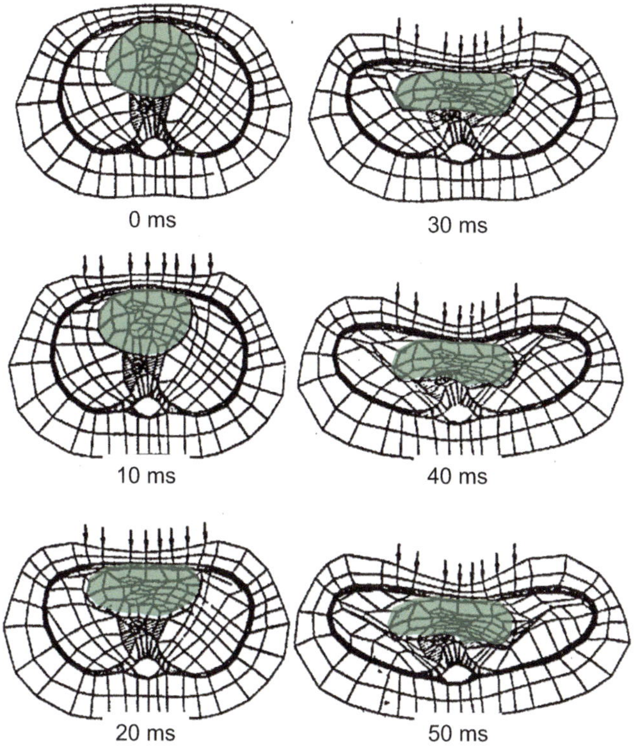

Fig. 3.35 Thorax compression and deformation of internal organs using computational simulation at 30% compression (according to [114])

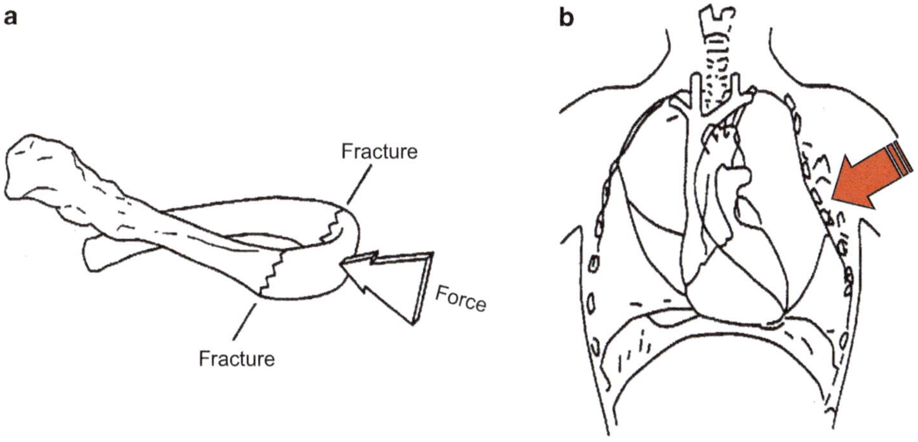

Fig. 3.36 Fracture of a rib at several points (left) and rib series fracture during lateral impact (according to [114])

The penetration of sharp fracture ends of the fractured ribs can lead to perforation and lacerations of the lung. Through an open wound, the pleura is exposed to atmospheric pressure, outside air enters into the pleural cavity (pleura) and the lung can collapse in on itself; this is referred to as **pneumothorax**. Superficial tears of the lung tissue lead to spontaneous adhesion through clotted blood, while deep tears, especially of main airways and blood vessels, usually result in significant blood loss. The blood enters the thoracic cavity and one speaks of a **hemothorax,** which can also lead to a lung collapse. The blood loss resulting from a chest wall injury reduces the circulating blood, as part of it flows into the chest cavity, and is usually more pronounced in connection with thoracic instabilities than with serial rib fractures. A bruise that leads to a hemothorax can also be caused by tears in the intercostal muscles and vessels, especially when the injury is close to the back of the chest wall, as strong blood vessels run in this area, which are connected to the main veins of the mediastinum.

The strain of blunt force on the chest usually leads to fractures of the sternum and ribs, so that the heart is compressed between the sternum and spine. Depending on the extent of the compression, **injuries to the heart** (crush injuries, aortic ruptures or tears in the heart muscle) can occur (Fig. 3.37). Tears that also affect the pericardium lead to a hemothorax and are fatal in most situations. However, if the pericardial tear is less extensive, the clotted blood can close the tear. In this case, or if the pericardium remains intact, blood collects between the heart and pericardium, resulting in a hematoma, the hemopericardium (pericardium: heart sac). The hemopericardium can result in cardiac tamponade, which compresses the vena cava under increasing pressure and leads to cardiac arrest.

In the case of blunt force, heart injuries with ruptures are possible, but they occur relatively rarely. Interestingly, they were mostly diagnosed as such in "unrecognized" injuries that led to death. **Cardiac contusion** (bruising) is the most common injury, but it is difficult to diagnose and can occasionally be obscured by the presence of other injuries. More severe injuries to the heart result from **lacerations of the heart muscle tissue**, i.e., tears in the tissue that occur on the outer or inner surface. However, a heart trauma can also cause functional effects, such as rhythm and heart action disorders, which can lead

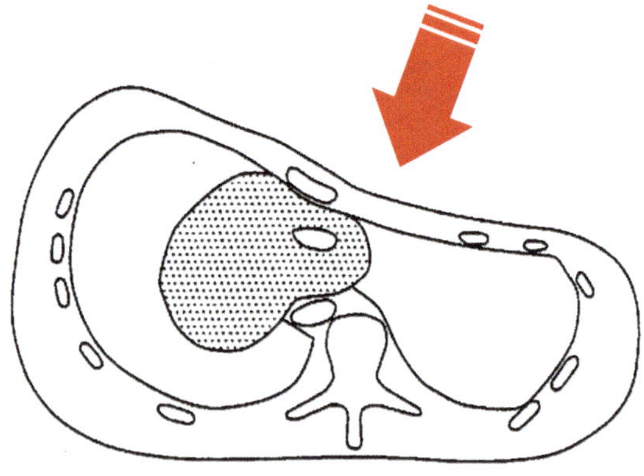

Fig. 3.37 Compression of the heart resulting in crush injuries, aortic ruptures or tears in the heart muscle (according to [114])

to death in severe impairment. A heart rhythm disorder is one of the most common signs of a heart muscle contusion.

Injuries to the aorta and the main blood vessels due to blunt force usually result in the tearing of the vessel wall. In many cases, the rupture is complete and thus fatal. In survivors of such injuries, it happens that the vessel wall laceration does not include the outermost layer of the wall, which is made up of elastic fibers, so that the blood largely remains within the vascular system, but the resulting expansion of the vessel wall can lead to death at a later time. If the blunt force acts on the right side of the chest and is directed from bottom to top, the resulting injuries are often accompanied by aortic lacerations (Fig. 3.38). The aortic tear usually affects the descending part of the vessel immediately behind the branch of the subclavian artery. This is the location of the transition between the relatively movable aortic arch and the descending section, which is fixed between the descending aorta and the thorax by the intercostal vessels. The "scooping up" of the heart during thoracic compression overextends the aorta on the front surface and can cause the vessels to tear off (Fig. 3.39).

Ruptures of the aorta and the blood vessels usually occur in conjunction with other thoracic injuries. A tear of the significant ascending vessels (carotid and subclavian arteries) can result from simultaneous chest compression with hyperextension of the neck and overstretching of these vessels. This injury situation is characterized by indentation of the upper chest and further injuries (Fig. 3.40).

The diaphragm (diaphragma) is a muscular partition between the thorax and the abdomen and can be injured by blunt force acting in the abdominal area (Fig. 3.41);

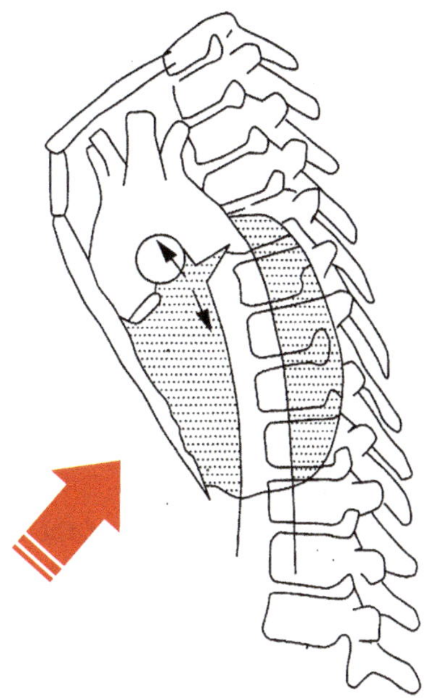

Fig. 3.38 Blunt force against the right side of the chest directed from bottom to top leads to aortic lacerations (according to [114])

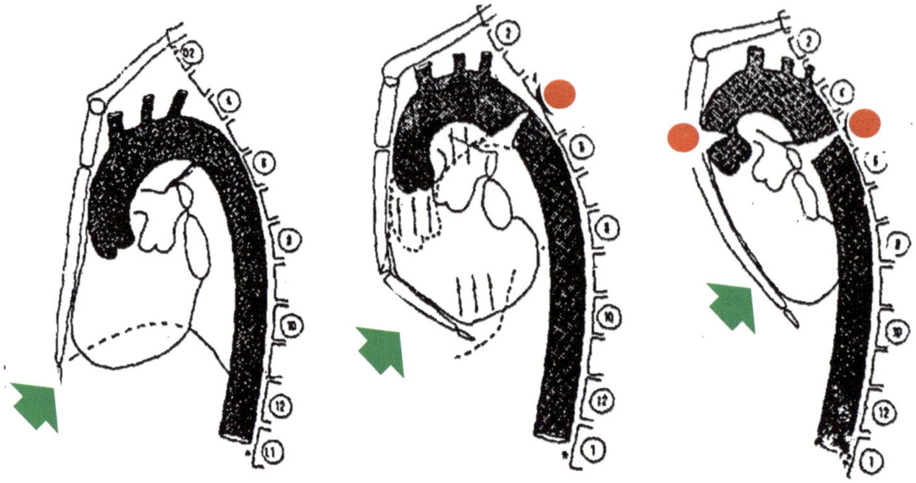

Fig. 3.39 "Scooping up" of the heart during thoracic compression leads to aortic and vessel ruptures (according to [114])

Fig. 3.40 Chest compression with hyperextension of the neck leads to vascular lacerations (according to [114])

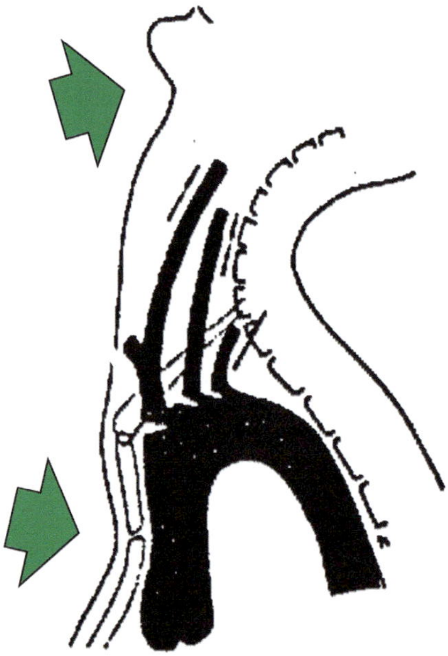

the resulting **laceration of the diaphragm** is essentially limited to the left area. Due to the negative pressure in the chest, the tear tends to remain open, so that parts of the abdominal organs can enter the chest through the opening (hernia). The consequence of the resulting pressure on the lung is a partial emptying of the affected lung section and a displacement of the heart to the opposite side.

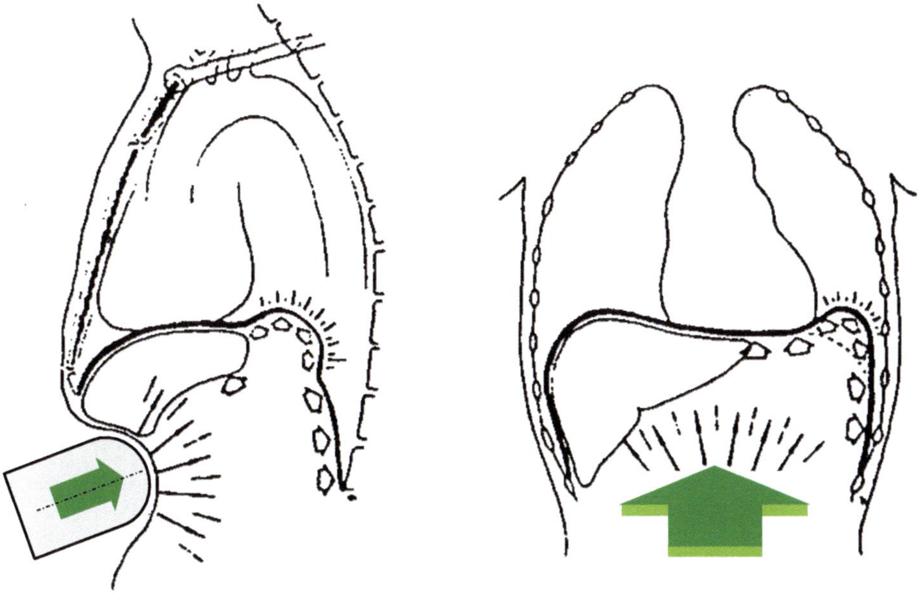

Fig. 3.41 Lacerations of the diaphragm due to blunt force against the abdomen (from [114])

The **esophagus** is protected by its location in the rear area of the mediastinum, but injuries can also occur here due to blunt force against the thorax.

3.1.4 The Abdomen and the Pelvis

The **abdomen**, also referred to as the belly or lower body, is the largest body cavity, but strictly speaking, it cannot be considered a "cavity" as the organs located in the abdominal region completely fill the entire space. The abdominal organs include two **main types of organs**, namely the compact and the hollow organs, which exhibit completely different behavior under various mechanical loads. The compact organs are liver (hepar), spleen (lien), pancreas, the kidneys (ren) with the adrenal glands (glandula suprarenalis) and in women the ovaries; the hollow organs include stomach (stomachus), large and small intestine (intestinum crassum and intestinum tenue), urinary bladder (vesica urinaria) and again only in women the uterus. The relative position of the organs is shown in Fig. 3.42. It shows that the liver occupies the largest space, surrounded by the lower, right-sided chest, while the stomach and spleen are located on the left side of the lower chest. The pancreas is diagonally located at the back of the abdomen. The **abdominal area** is bounded above by the diaphragm (diaphragma), which forms a dome over the viscera and extends high into the thorax to the level of the attachment of the fourth rib to the sternum. The abdominal area extends downwards to the bony pelvis with its muscles. The upper part of the abdomen is surrounded on all sides by the chest and at the back by

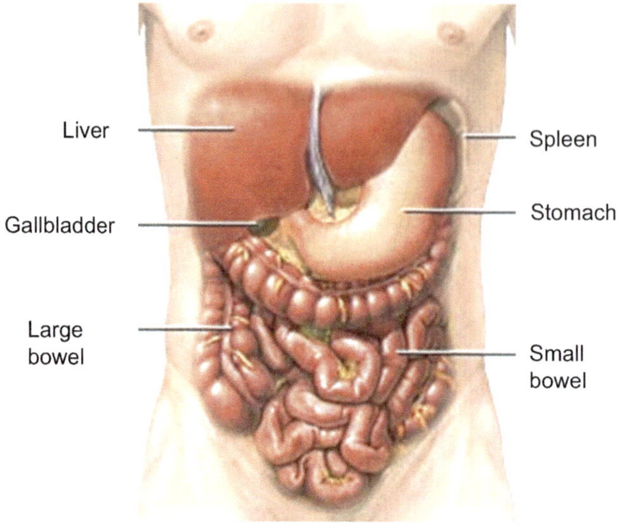

Fig. 3.42 The abdomen and its organs (according to [87])

the spine, which is usually not counted as part of the abdominal area, whereas the middle part is surrounded by muscles, but again at the back by the lumbar spine. The muscles extend to the lower part of the abdomen, which is bounded at the back by the sacrum and the coccyx and laterally by the iliac bone of the pelvis [93].

The **main blood vessels** of the abdominal area are the abdominal aorta (aorta abdominalis) and the inferior vena cava (vena cava inferior), which are located in the posterior area of the abdominal space right in front of the spine and from which branches lead to all parts of the abdomen. They reach the abdominal area through a separate opening in the diaphragm and divide at the level of the fifth lumbar vertebra (directly below the navel) into the left and right hip artery and vein (arteria and vena iliaca communis).

A characteristic in the distinction between the compact and the hollow organs lies in their density. The compact organs are equipped with vessels that contain fluids, and therefore have a higher density than the hollow organs. In addition, the individual organs are surrounded by a membrane coated with a thin film of fluid; moreover, some organs (e.g., liver, spleen, and intestines) are only loosely connected to their surroundings by folded ligaments, so they can be relatively easily moved against each other. This high mobility, but also the localization of the abdominal organs, has significant effects on the **injury mechanics**: Experiments on body parts in lying, sitting, or upright position yield completely different results because the abdominal organ under investigation can evade in different directions. In addition, the results depend on whether an organ is directly in front of the spine, where, for example, the risk of being crushed in a frontal collision is greater than for an organ located laterally next to the spine. Similar differences can be seen when comparing the two kidneys: The right kidney, due to the presence of the liver, usually lies somewhat lower than the left kidney and is more at risk due to the lack of protective effect of the lower ribs than the left kidney. This is basically true for all abdominal organs that are located in the area of the lower chest and enjoy some protection from the ribs. Although they are not as stiff as the upper ribs due to their only indirect attachment to the sternum, the lower ribs still offer a load-distributing surface and thus some resistance to deformation, especially in the case of blunt force from the side or from behind [93].

The **liver** (hepar) is the largest of the compact organs in the abdomen and has the highest mortality rate when subjected to blunt force [93]. Such injuries can result from a strain at low speed due to simple compressions against the spine or the rear abdominal wall, but viscous injuries can also occur, where high strain within the liver builds up fluid pressure leading to extensive tensile and shear stresses. In addition, the liver can be injured by relative movement against its surroundings at high accelerations; the injuries then usually occur at the attachment points and can be traced back to overstretching of the ligaments and blood vessels. Finally, liver tears can be caused by penetration of broken ribs when blunt force occurs.

Due to the heavily vascularized structure of the **spleen** (lien), traumatic organ destructions show a significantly high mortality risk. The two classic injury mechanisms of the spleen, as distinguished in [93], are an impact against the lower left area of the abdomen, where the spleen is directly hit, and an indirect trauma caused by a high acceleration of the body and the resulting large displacement of the spleen relative to its attachment point. In the first case, deep tears to complete shattering of the organ can often occur due to compression at low strains or due to viscous injuries at higher speeds. Injuries from indirect strain occur in the form of tears of the vascular stem or tears of the vascular and nerve strands at the spleen; the reason for this is seen in the overstretching of the attachment during the acceleration phase.

The **kidneys** (ren), located on both sides of the spine, are surrounded by muscles on the sides and in the rear area of the abdomen and in front by other abdominal organs. Based on this anatomical arrangement, comprehensive protection against blunt force can be assumed for the kidneys, only in the case of side impact of the abdomen, kidney injuries seem to be overrepresented. However, up to 90% of all kidney injuries without penetration are of minor severity, provided the tissue skin remains intact [93]. Kidney contusions (bruises) and minor tears below the enveloping membrane, often accompanied by hematomas, occur. Injuries to the vascular stem of the kidney also result from relative movement during the acceleration of the body, whereas injuries to the tissue skin are usually the result of a direct blow to the kidney.

Injuries to the **pancreas** and the **duodenum** are due to blunt force in about 25% of cases [93], often involving compression between vehicle interior parts and the spine. The injuries to the pancreas range from superficial contusions to complete severance and destruction of the organ. In the case of duodenum injuries, bruises and perforations occur with and without pancreatic injuries. However, contusions and lacerations can also occur due to contact with the spine, tears due to shearing during relative movements, or tears due to increasing internal pressure in the case of bilateral closure. Injuries to the **intestines** (intestinum) can essentially be traced back to the same injury mechanics as duodenum injuries with the corresponding injury patterns, with mortality in the event of blunt force highly correlated with the number and severity of accompanying injuries to other organs in the abdominal region [93].

The **urinary bladder** (vesica urinaria) is relatively well protected against blunt force due to its location within the bony pelvis; therefore, injuries to the urinary bladder without pelvic fractures rarely occur. However, if pelvic fractures do occur, they can lead to ruptures and lacerations caused by bony fragments.

In victims of motor vehicle accidents, injuries to the **diaphragm** (diaphragma) are common, with the number of ruptures being higher in side collisions than in frontal collisions. The injury mechanics of the diaphragm include shear stresses of the stretched membrane, tearing of the diaphragm from its attachments, and transmission of fluid pressure within the intestines.

Injuries to the large **blood vessels** in the abdominal area due to blunt force are relatively rare due to their protected location in the rear area in front of the spine. However, when injuries do occur, they can present in the form of contusions, lacerations, and severance, leading to bleeding, thrombosis (blood clot formation through coagulation), blood sac formations, or arteriovenous connections. Trauma to the abdominal aorta is highly likely to be fatal and can be caused by both direct and indirect stresses.

The relationship between the anatomical conditions and the injury mechanics in the abdominal region becomes clear when comparing the frequency of severe injuries (AIS > 3) that occur in **side collisions** for the driver in left-sided and for the passenger in right-sided impacts. The main contact points were, according to a US analysis [94], the interior contact surfaces of the vehicle (34%), the armrest (29%), and on the driver's side the steering system (9%) or on the passenger's side the dashboard with glove compartment (20%). The resulting organ injuries in the abdominal region are compared in terms of their frequency in Fig. 3.43. For example, if you compare the liver injuries in right-sided impacts with those in left-sided impacts, the nearly double frequency can be attributed to the anatomical arrangement of the liver.

This section would be incomplete without a brief presentation of abdominal injuries due to the use of the **seat belt**. In general, abdominal injuries can be understood as blunt force trauma resulting from the interaction of webbing and abdomen. They can be divided into injuries caused by submarining, i.e., the pelvis sliding under the lap belt and its displacement into the abdominal area, and injuries caused by malfunctions in seat belt use, such as the location of the seat belt webbing on the occupant, [79]. A careless location of the seat belt webbing on the body, referred to as a **location anomaly**, is the main cause of injuries to belted occupants in motor vehicle accidents [94], while most severe **submarining injuries** only occur at higher speed changes (> 50 km/h) [63]. It should not go unmentioned that while belted occupants have a higher risk of abdominal injuries

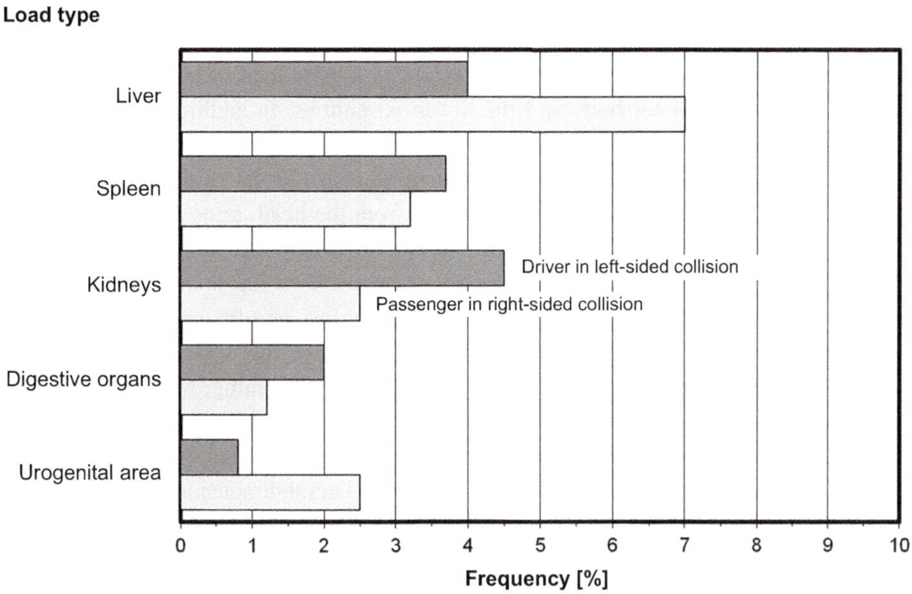

Fig. 3.43 Frequency of severe abdominal injuries (AIS > 3) in side collisions with left and right-sided impacts (according to [94])

of any severity, the risk of severe injuries is about half as high as for unbelted occupants (see Sect. 2.4.7).

Recent analyses by Yoganandan et al. [126] examined the frequency and severity of **abdominal injuries in frontal crashes** as well as impact-facing and impact-remote seating positions in side collisions using the NASS-CDS database for the years 1993–1998. For a combined dataset of belted and unbelted drivers and passengers, the distribution of organ injury frequency varied for each crash mode when comparing AIS 2+ and AIS 3+ injuries. The organs most frequently injured in frontal crashes were the liver (39%), spleen (29%), and digestive organs (11%) at AIS 2+ and liver (35%), spleen (28%), and arteries at AIS 3+.

In side crashes facing the impact, the spleen (49%), kidneys (20%), and liver (16%) were most frequently injured at AIS 2+, and the spleen (47%), diaphragm (17%), and liver (13%) were the most frequently injured organs at AIS 3+. The injury frequency in accidents on the side away from the impact was kidney (38%), liver (32%), and digestive tract (15%) for AIS 2+, and liver (50%), spleen (13%), and kidney (12%) for AIS 3+.

The spleen was the most frequently injured organ in left side impacts, while the kidneys and liver were more frequently injured in right side impacts. The kidneys are protected by fat encasements and their retroperitoneal location; therefore, kidney injuries tended to be less severe and made up a larger percentage of AIS 2+ injuries in side impacts. Similarly, injuries to the digestive tract are typically less severe on the scale of threat to life, and therefore these injuries made up a larger percentage when considering AIS 2+ abdominal injuries in frontal and side impacts.

The **pelvis** is a ring bone that can be considered a main load-bearing structure and is located between the upper body and the lower extremities. In addition, the pelvis, with its upwardly open, bowl-like shape, carries the abdominal organs and provides them with support and, at least in the lower area, protection against direct strain. Mechanically, it forms the only transmission path of the weight from the head, arms, and upper body to the ground. Therefore, the pelvic structure is also more massive than that of the head or thorax. The ring-shaped pelvis consists of four bones: Two hip bones (os coxae) form the lateral and anterior boundary, the sacrum (os sacrum), and the coccyx (os coccygis) the posterior closure. There are many differences between the male and female pelvis, but the most significant is seen in the shape of the pelvic opening, which is completely enclosed by a bony structure. Fig. 3.44 shows a male and a female pelvis, it is clearly visible that the opening in the female pelvis, namely the birth canal, is almost circular, whereas the opening in the male pelvis is wider in its lateral dimensions [50].

The **hip bone** (os coxae) is a large, relatively flat and irregularly shaped bony structure that encompasses the majority of the pelvic ring. It consists of three interlocking bones, the ilium (os ilium), the ischium (os ischii), and the pubic bone (os pubis). The dividing line between these runs in the area of the hip joint socket (acetabulum), as illustrated in Fig. 3.45. The **iliac bone** (os ilium) is divided into two areas, the obliquely ascending wing and the body, which forms the acetabulum of the hip joint. Most of the features and surface properties are of minor biomechanical significance and will therefore not be further elaborated. However, it seems necessary to discuss a specific point on the iliac crest (crista iliaca), namely the anterior superior iliac spine (spina iliace anterior superior). This is considered an anatomical anchor point, which is supposed to prevent the lap belt from sliding over the upper edge of the pelvis, the iliac crest, and triggering a submarining effect. The iliac bone is connected on its medial inner side by the sacrum

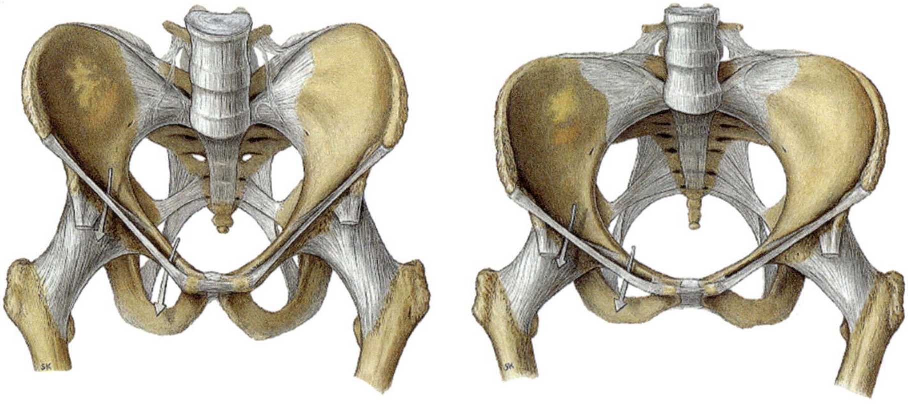

Fig. 3.44 Front view of a male (left illustration) and a female pelvis (from [87])

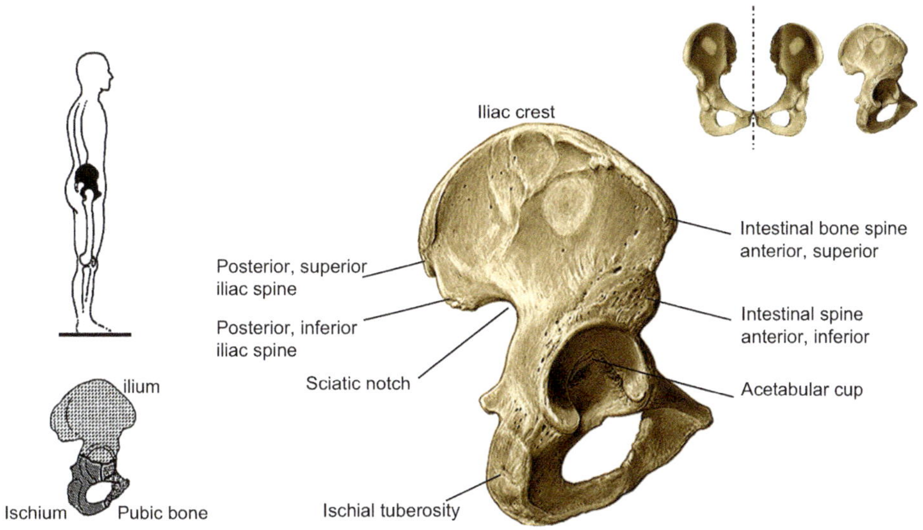

Fig. 3.45 The hip bone with iliac bone, ischium, and pubic bone [45, 87]

(os sacrum). This inner side is covered with a layer of cartilage, but the connection point only allows limited relative movement. The **ischium** (os ischii) forms the lower rear part of the hip bone and can be divided into body and branch. The ischial body includes the rear third of the joint socket. At its lowest point, it transitions into the ischial tuberosity, which provides support to the lower torso when sitting (see Figs. 3.45 and 3.46). The ischial branch is a thin, flattened part that is connected to the rear part of the pubic branch. The **pubic bone** (os pubis) itself is an irregularly shaped bone, which consists of a body and two branches, the upper and lower pubic branch. The pubic body forms the front third of the joint socket. The upper branch extends from the body to the median sagittal plane, where it unites with the branch of the opposite side. This connection point is the pubic symphysis (symphysis pubica), which forms a joint with limited mobility due to the cartilaginous disc filling the gap between the bone ends. The lower pubic branch forms the arch to the ischial tuberosity, which encloses a pronounced opening (foramen obturatum), and unites at the pubic symphysis with the branch of the opposite side [50].

The sacrum and the coccyx form the rear wall of the pelvic ring, with the **sacrum** (os sacrum) being a fusion of five vertebrae and having a triangular shape. The lateral surfaces form a stable joint with the pelvis and transfer the weight of the lower torso via the pelvis to the legs or the ischial tuberosities, depending on the position. The **coccyx** (os coccygis) is a rudimentary tail made up of three to five fused vertebrae, which, however, are not formed like normal vertebral bodies, for example, they lack the rear structure of the spinous processes; the last vertebra is merely a bony nodule. The joint between the

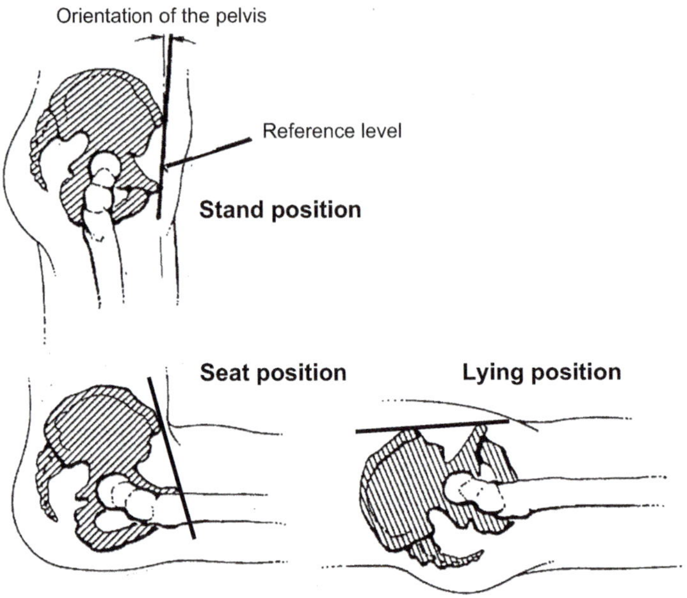

Fig. 3.46 Orientation of the human pelvis in standing, upright sitting, and lying position (from [45])

sacrum and coccyx has only limited mobility due to the thin, cartilaginous disc between them, which is stiffer than a normal intervertebral disc [50].

The **orientation of the pelvis** is difficult to quantify due to its irregular shape. It varies from person to person and shows the smallest differences when standing compared to the vertical axis, for example, compared to the sitting or lying position (Fig. 3.46). In [50], a reference plane is defined that appears as a straight line in side view, running through the front points of the anterior superior iliac spine and the pubic symphysis. These three orientation points can be easily palpated, so no X-rays are required to determine them. In standing individuals, the orientation of this reference plane is approximately 10° (see Fig. 3.46).

Since the **femur** (regio femoris) is connected to the pelvis through the hip joint and pelvis injuries often occur in this area, the anatomy of the proximal (upper) femur will be anticipated here. In detail, it consists of the joint head, the neck, and the femur region (see Fig. 3.49). The head has a spherical shape and forms the hip joint together with the joint socket. The surface of the joint head is covered with a transparent cartilage layer. Such joints maintain high mobility through the secretion of joint lubricant. Blood supply is provided via the femoral neck, the structure beneath the joint head. This neck has the shape of a truncated cone and connects the head with the rest of the femur; its axis forms an angle of approximately 125 degrees with the femoral shaft, in women the angle can be less than 90 degrees [50]. The upper region of the proximal femoral shaft is characterized by the trochanters, the outward-facing trochanter (greater trochanter, trochanter major) and the inward-facing trochanter (trochanter minor).

The following overview of the **injury mechanics of the pelvis** and the injury patterns that occur is excerpted from [50]; it includes both pelvic injuries and injuries to the hip joint and the proximal femur. The extension to the hip joint and the femur is necessary because their injuries often occur simultaneously with injuries to the pelvis.

Isolated pelvic injuries, which occur in motor vehicle accidents, can be classified into three areas: singular fractures of the pelvic ring, two or more fractures of the pelvic ring, and fractures of the sacrum and coccyx. If only one **simple pelvic ring fracture** occurs, there is no significant displacement of the broken segment. Examples of isolated fractures are unilateral fractures of the upper or lower pubic ramus, but also of both rami (ramus inferior ossis pubis); they do not require surgical intervention. Other isolated fractures include simple fractures of the ilium associated with slight separation or complete splitting of the pubic symphysis and displacement (subluxation) of the sacrum. These injuries result from lighter impact situations. In side collisions with an impact against the upper outer edge of the proximal femur, the greater trochanter, fractures of the pubic branches are common. Through **multiple fractures**, the pelvic ring becomes unstable, so they are usually associated with large displacements of the fragments. The two main types of injury are fractures of the pubis alone or fractures of the pubic bone in conjunction with fractures of the ilium. Multiple injuries of the pubis include two or more fractures of the branch and result in displacement of the pubic symphysis. They occur in pedestrians who are hit laterally by motor vehicles at the pelvis, generally on the side away from the impact. Although surgical interventions are not necessary, injuries to the urinary tract may occur, e.g., ruptures of the urethra, which require the creation of an artificial outlet at the bladder. The most common type of combined injury to the ilium and pubic bone segment involves the displacement of the pubic symphysis with simultaneous displacement of the sacroiliac joint. In addition, there are fractures of the ilium accompanied by displacement of the pubic symphysis or by fractures of both branches with simultaneous dislocation of the sacroiliac joint. Here too, injuries to the urinary tract and bladder are common. Such injuries, unlike the injury mechanics in the case of lateral strain mentioned above, are caused by a front-to-back load typical of frontal collisions. Extensive pelvic injuries can result from **fractures of the sacrum**, which usually lie in the area of load concentration, e.g., transversely through the foramina. Since these openings surrounded by bony arches carry the sacrum nerve roots, sacrum fractures can be associated with injuries to the nerves. A sufficiently high vertical acceleration can cause compressions and a loss of height of the sacrum. **Injuries to the coccyx** occur with vertical stresses in seating position and can result in contusions, fractures or dislocations of the coccyx. Although such injuries are not serious, they can be extremely painful. The majority of **tissue and vascular injuries** associated with pelvic fractures lead to bleeding, which is caused by large blood vessels on the pelvic wall and by the broken surface itself. The extent of blood loss can be very large and can only be stopped by tying off the internal iliac artery. Abdominal injuries usually occur with pelvic fractures, but they have already been discussed above.

Injuries to the hip joint occur either directly in the joint or in its immediate vicinity and affect one or more of the following structures: Acetabulum, femoral head (caput femoris), and proximal femur (regio femoris proximalis). In younger people, damage can occur to the proximal end of the femur, while in older people, injuries often occur at the hip due to bone fragility (spontaneous deformation) and decreasing elasticity. With external strain, hip injuries can be divided into two main groups: Traumatic hip dislocations and fractures of the femoral neck (see Fig. 3.47).

A **hip dislocation** occurs relatively frequently in a seated position due to a rearward strain along the femur, as in this position the hip joint is not held by any bony structure. Vehicle occupants who are exposed to a head-on collision and have their legs crossed expose themselves to this risk. If, on the other hand, the bending angle between the hip and femur is somewhat larger, comparable strain can cause **fractures of the posterior acetabular rim** along with dislocations of the hip joint. In addition, dislocations in the opposite direction, i.e., forward, are conceivable: for example, if the lower torso is held in the seat by the lap belt during a head-on collision, while the femur continues to move due to inertia, forward (anterior) dislocations of the hip joint are possible. With a lateral load on the hip, e.g., in side collisions, a central dislocation accompanied by **fractures of the acetabulum** can occur. If the occupant sitting on the impact side is hit by the hard structure of the door at the level of the greater trochanter at the end of the proximal femur, this can result in the femoral head piercing the thin wall of the acetabulum. Such injuries can also occur in a standing position, e.g., in pedestrian accidents. In backward dislocations of the hip joint, where fractures of the acetabulum rim occur, a **fracture of the femoral head** can also occur; there is a risk of long-term impairment

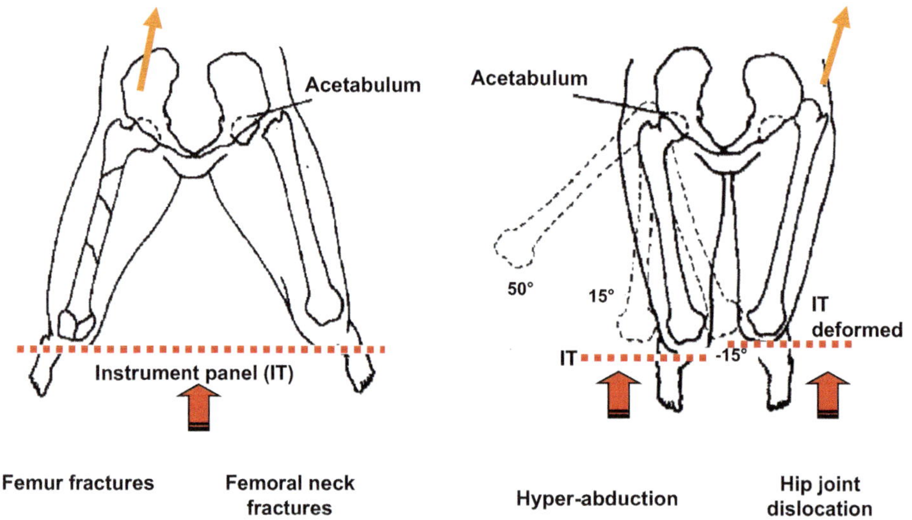

Fig. 3.47 Injury mechanisms of the pelvis, hip joint, and proximal femur in sagittally loaded occupants (according to [124])

due to a trauma-induced joint inflammation. Hip joint dislocations accompanied by **injuries to the femoral neck** are relatively rare and usually only occur in older people. The injury mechanics result from a high bending strain on the femoral neck, which can cause a fracture of the neck in the event of a dislocation of the hip joint. The consequence of such injuries is often bone decay at the femoral head due to insufficient blood supply. Dislocations with a **fracture of the femoral shaft** are also rare and their injury mechanics are largely unknown. However, the presence of a transverse fracture often suggests a bending of the shaft.

The isolated **fractures of the femoral neck** can be divided into two groups, within and outside the fibrous tissue that surrounds the hip joint. This hip joint capsule extends from the pelvis to the intertrochanteric crest between the rolling hills (Trochanter) on the proximal femur. If a fracture occurs within the capsule, it is usually a fracture at the lower edge of the joint head or a fracture of the femoral neck. Fractures outside the capsule can either occur between the trochanters or on them themselves.

An overview of **injury mechanisms** of the pelvis, hip joint, and proximal femur in sagittally loaded car occupants is shown in Fig. 3.47. The strain occurs through the introduction of knee/dashboard contact forces into the femur and hip area; it causes fractures of the femoral shaft, femoral neck, or hip joint region with sufficiently high accident severity. Depending on the size of the femoral spread, hyper-abductions and dislocations of the hip joint can also occur.

Even though pelvic fractures are usually only rated with AIS 2 or 3, according to criteria for **long-term injuries** (BG injury cost scale), the pelvic fracture is an injury with a frequent complication rate and long hospital stay duration, so that in the RESIKOMED study [90] pelvic injuries in car side collisions on the impacted side are by far the most common long-term injuries and consequently there is a great **need for prevention**.

3.1.5 The Extremities

With extremities, the outermost ends (of the body) are referred to, in humans these are the limbs - arms and legs.

The **arm** (brachium) consists of the upper arm, forearm, and hand (Fig. 3.48); the upper arm is extremely flexibly connected to the torso via the shoulder joint (articulatio humeri) through the shoulder blade (scapula) with the coracoid process (processus coracoideus) and the collarbone (clavicula). The bony basis of the upper arm is formed by the humerus, in the forearm it is the ulna and radius and in the hand the carpus, consisting of eight bones, the metacarpus, consisting of five bones, and the fingers (phalanges), consisting of two bones in the thumb and three bones in the other fingers. The combination of differently constructed joints (shoulder joint: ball-and-socket, elbow joint: pivot hinge, wrist: ellipsoid, and finger joints: hinge joints) and the opposability of the thumb gives the arm and hand their unique movement and functionality. Muscles coming from the trunk move the arm in the shoulder joint (forward and backward movement, adduction

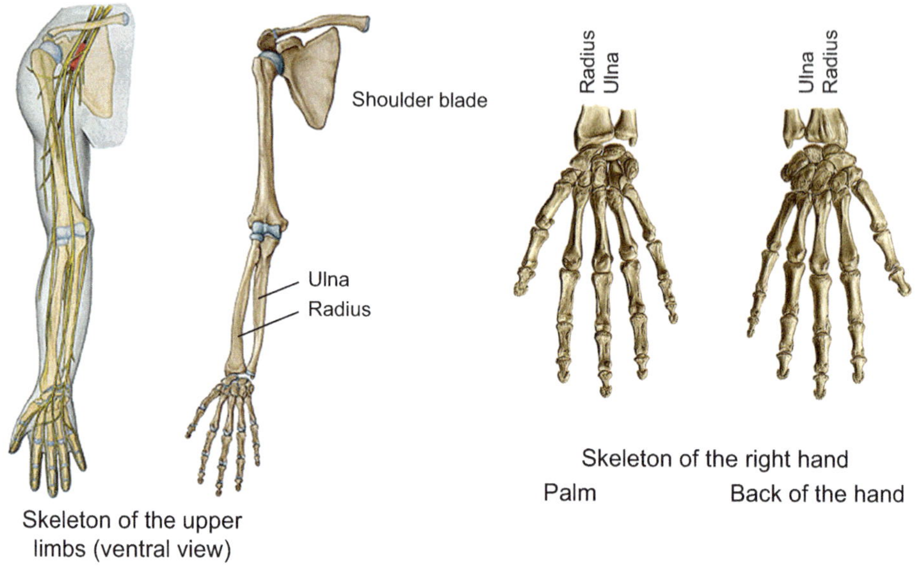

Fig. 3.48 Arm and hand skeleton (according to [87])

and abduction, rotational movement). The flexor (musculus biceps brachii) located at the front of the upper arm and the extensor (musculus triceps brachii) located at the back act on the shoulder and elbow joint. The muscles located on the forearm cause movements of the forearm (rotational movement: pronation and supination) and act on the wrist and the thumb and finger joints (diffraction and extension). The fine adjustment of the fingers and thumb is done by the short, small hand muscles. The major nerve trunks and blood vessels pass through the axilla to reach the arm. The brachial artery (arteria brachialis) is located on the inside in the biceps groove and branches in the area of the elbow bend. Two large branches are referred to as arteria radialis and arteria ulnaris according to the forearm bones. The pulsations of the arteria radialis are palpable at the transition to the hand. The deep arm veins accompany the arteries, the superficial ones form variable patterns and are often punctured for intravenous injections or blood sampling. The nerves of the arm emerge from the brachial plexus (plexus brachialis) with three large trunks; their main branches are the median nerve (nervus medianus), the radial nerve (nervus radialis), and the ulnar nerve (nervus ulnaris). The nervus radialis wraps around the humerus to the back and is very vulnerable here. Similarly vulnerable is the nervus ulnaris on the back surface of the elbow joint ("funny bone"), where it can be irritated by impact [11].

The skeletal structure of the human **leg** consists of a femur, two lower leg bones, and the bones of the foot. The femur (thigh bone) is the strongest tubular bone in the body (Fig. 3.49). Its shaft transitions into an angularly offset piece, the femoral neck, which carries the spherical joint surface of the femoral head; this, together with the acetabulum

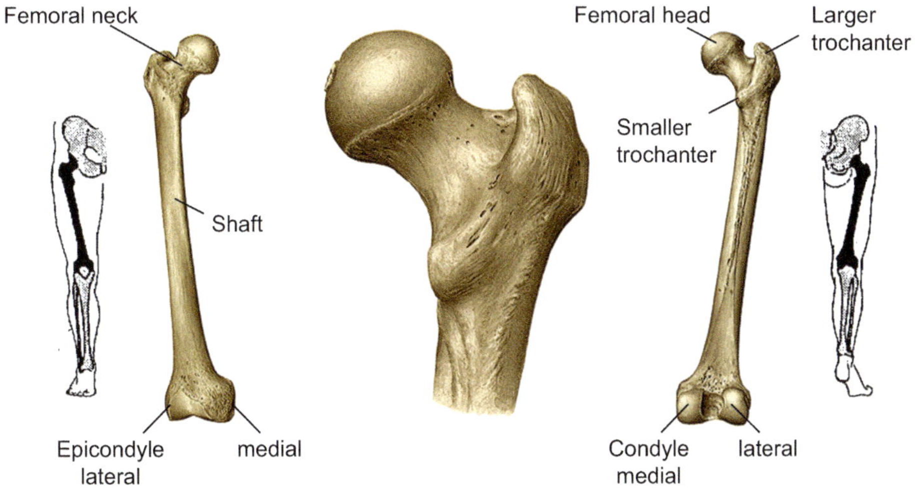

Fig. 3.49 Femur, front view (left) and rear view (according to [45, 87])

that surrounds it, forms the hip joint. The lower end widens into two powerful large condyles (rounded joint structures), whose cartilage-covered surfaces form the upper joint surface for the knee joint.

The bones of the lower leg (Fig. 3.50) are the inwardly located, stronger tibia (shin bone) that protrudes sharply under the skin of the front of the lower leg, and the slimmer, externally located fibula (calf bone). Both bones are connected by an interosseous membrane serving as an origin surface for muscles. The space between the joint knobs of the femur and the joint surfaces of the tibia is filled by semicircular, wedge-shaped cartilage discs (menisci) (Fig. 3.51). From the groove between the joint knobs, two intersecting, strong ligaments (cruciate ligaments), which tense in any position of the knee and thus secure it, extend forward and backward to the central edge of the tibia.

The kneecap (patella) is embedded in the joint capsule reinforced by ligaments. Depending on the force introduction, a very different injury risk for ligament ruptures can occur: As stated in [115], there is a low risk if the derivation of the knee contact force in the femur and pelvis only occurs via the kneecap (Fig. 3.52a). A significant displacement between the lower leg and the knee joint can result from a force applied to the tibial tuberosity (shinbone bump) (Fig. 3.52b), as even with low loads there is a high risk of ligament ruptures. This risk is reduced due to the smaller relative displacement in the knee joint when the contact force is introduced both into the patella and into the tibia (Fig. 3.52c). If the consideration is extended to the **lower extremities**, a knee contact with the lower instrument panel will occur as the pelvis displacement forward increases (Fig. 3.53). In addition, the intrusion in the footwell restricts the movement space, leading to a hyper-flexion of the ankle joint and due to the restraint of the knee joint on the deformed instrument panel, a pressure/bending strain of the lower leg with a high risk of fractures of the lower extremities.

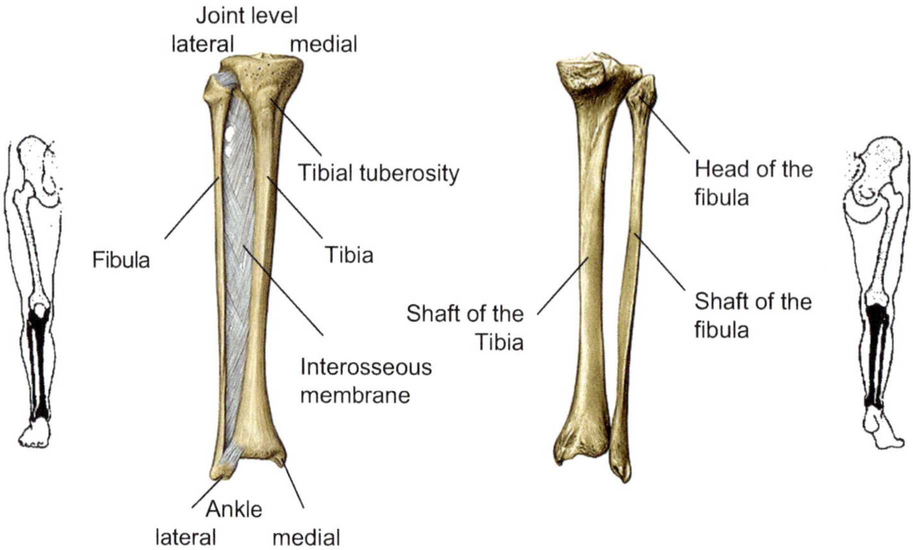

Fig. 3.50 Lower leg bones with tibia and fibula, front view (left) and rear view (according to [45, 87])

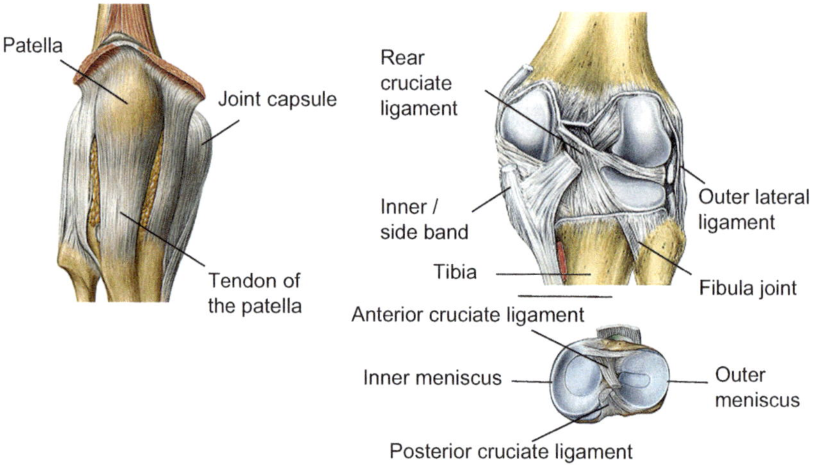

Fig. 3.51 Knee joint with patella and ligaments, front view (top left), rear view (top right) and top view (according to [87])

The lower ends of the shinbone and fibula form the two ankles on the outside and a joint surface for the upper ankle joint on the inside. The foot joints (articulationes pedis) include an upper ankle joint located between the two ankles (ankle joint, a hinge joint for raising and lowering the foot) and a lower ankle joint (for rotating foot movements). The foot (pes) connected to the lower leg through these joints consists of the tarsus with

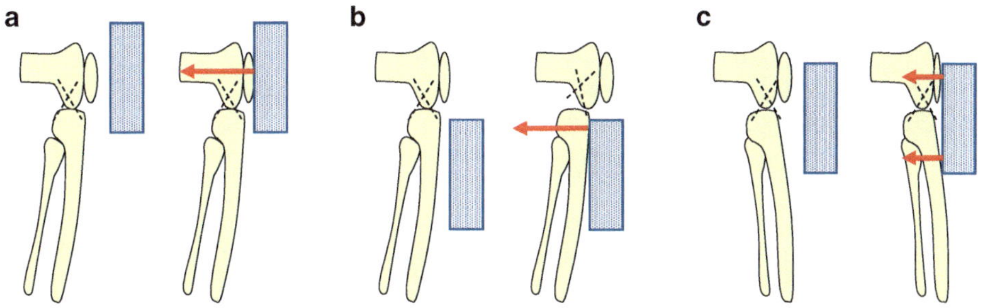

Fig. 3.52 Tendencies of displacement in the knee joint depending on the type of force introduction (from [115, 33])

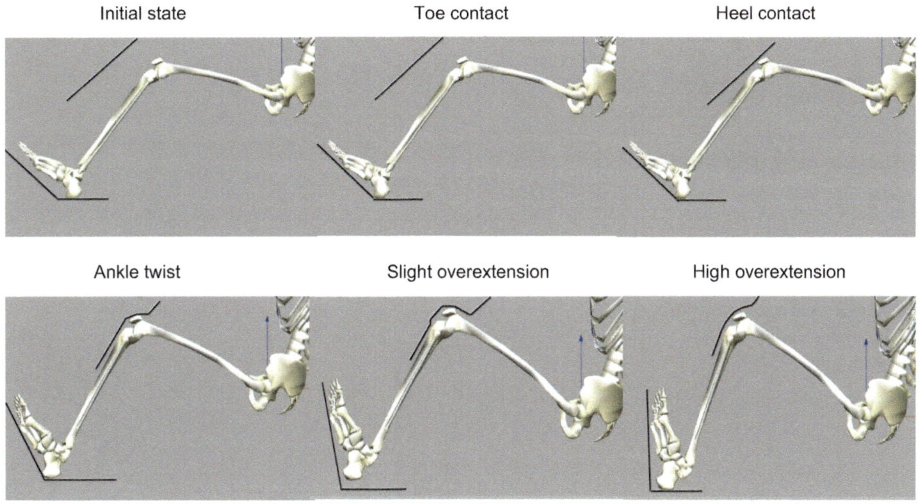

Fig. 3.53 Injury kinematics of the lower extremities in frontal collisions [88]

the tarsal bones (tarsalia), the metatarsus with the elongated, tightly connected metatarsal bones (metatarsalia) and the toes. The originally twelve skeletal elements of the tarsus are often reduced and fused; in humans, the foot skeleton consists of the bones of the five (two- or three-part) toes, the seven tarsal bones (heel bone, talus, navicular bone, cuboid bone and outer, middle and inner cuneiform bone) and the five metatarsal bones (Fig. 3.54). On the underside, a foot arch is formed, which is supported from the ground at three padded points (heel bone and the ends of the inner and outer metatarsal bones).

The individual **movement options** of the upper and lower leg, which are shown in Fig. 3.55, are caused by the muscles; according to [45] we distinguish flexion, extension and rotation as well as abduction (stretching) and adduction (pulling). The muscles that move the leg in the hip joint originate from the pelvis and the spine.

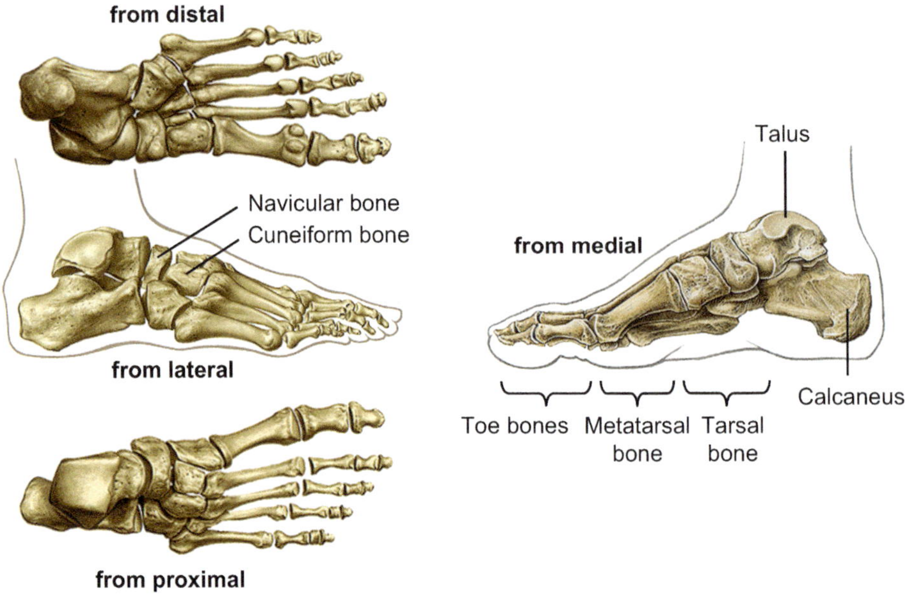

Fig. 3.54 Foot skeleton, seen from various views [87]

Fig. 3.55 Movement options of the upper and lower leg (according to [45])

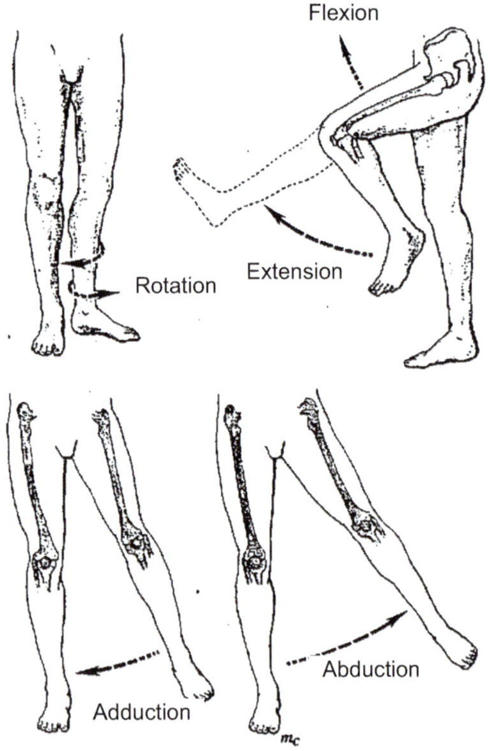

The hip flexors (for flexion) are located at the front, the hip extensors (for extension), especially the large gluteal muscle, are located at the back, the abductors are located on the outside and the adductors are located on the inside (Fig. 3.56). The extensors of the lower leg are located on the front side, while the knee flexors are located at the back. When the knee is bent, the tendons of the flexors can be felt as two strong cords on either side of the back of the knee (popliteal fossa) (Fig. 3.57). The strongest muscle on the lower leg is the triceps surae muscle, which consists of the two-headed gastrocnemius muscle and the soleus muscle; it attaches to the heel bone with the achilles tendon. The femoral artery (arteria femoralis), passes under the inguinal ligament into the thigh region and runs along the front of the thigh, then passes through a muscle canal in the last quarter of the thigh to the back into the popliteal fossa and divides into two main branches. The nerves for the skin and muscles of the leg originate from a nerve plexus emerging from the lumbar and sacral cord. The largest is the sciatic nerve (nervus ischiadicus); it exits the pelvis in the gluteal muscle region, runs along the back of the thigh to the popliteal fossa and divides there into two branches [11].

Injuries to the upper extremities of road users are relatively common, accounting for about 29% (for car occupants see Fig. 3.2), but due to their low injury severity and thus their low injury-related costs, they are of lesser importance [55]. In addition, the

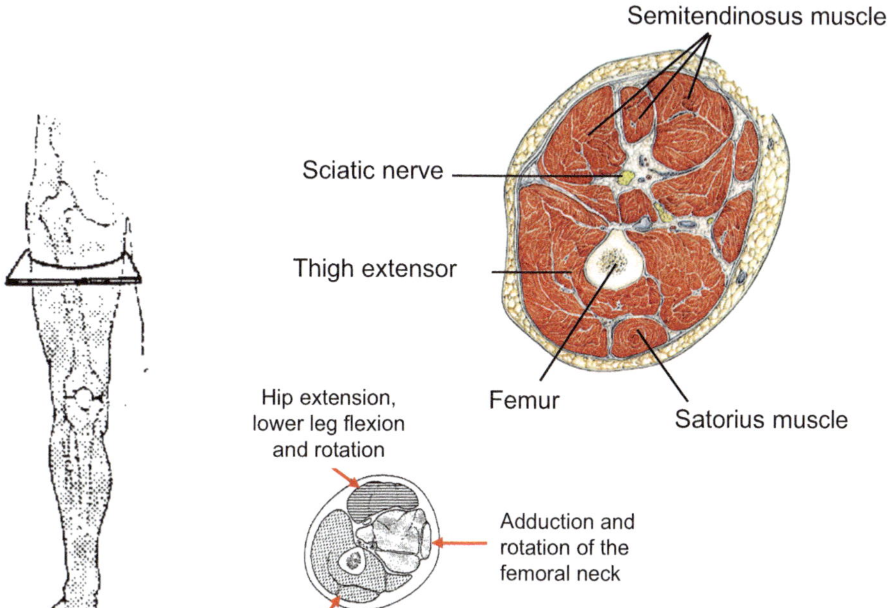

Fig. 3.56 Cross-section of the thigh with bones, muscles, blood vessels and nerves (according to [45, 87])

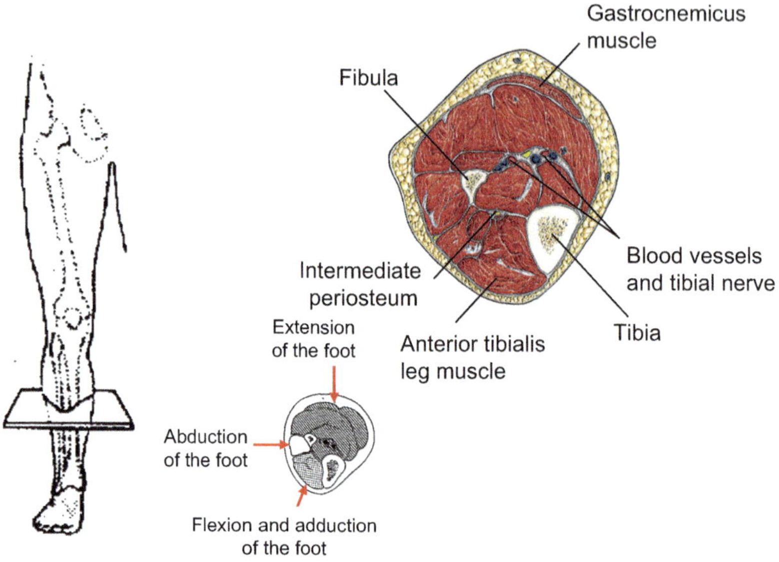

Fig. 3.57 Cross-section of the lower leg with bones, muscles, blood vessels and nerves (according to [45, 87])

anatomical structure of the arms, as the overview of the skeleton of the extremities in Table 3.2 shows, but also injury causation and patterns are largely comparable to the conditions of the legs, so that the following will mainly focus on the **injury mechanics of the lower extremities**.

Most bony components of the limbs are characterized by long, tubular bones, in which one can distinguish the middle piece (diaphysis), the proximal or distal end or joint piece (epiphysis), and the length growth zone (metaphysis) located between the

Table 3.2 Overview of the skeletal structure of the extremities

Arm	Leg
Upper arm bone	Femur bone
Not applicable	Patella
Ulna	Fibula
Radius	Tibia
8 Carpal bones	7 Tarsal bone
5 Metacarpal bones	5 Metatarsal bones
5 Fingers with 2 or 3 finger bones	5 Toes with 2 or 3 toe bones

middle and end pieces. These bone sections show considerable differences in their biomechanical properties due to their different types of bone tissue. The injury patterns of the middle pieces resemble the damage pattern of a correspondingly stressed cylindrical tube, while the injury patterns of the end pieces and the length growth zones are harder to systematize due to their variable bone structure. A **fracture** is a break in the bone structure; fractures with and without displacement of the bone fragments are distinguished. In a fracture without displacement, the fragments are still in their normal anatomical position, for example, when cracks occur in the bone, i.e., the bone is chipped. These fractures usually result from a **direct strain** with relatively low energy, while fractures with fragment displacements require higher energy. If fractures occur at some distance from the stress point, this is referred to as **indirect strain**; this injury type is relatively common in accident occurrence. A tensile load on the bone, for example, results in a fracture (transverse fracture) running perpendicular to the force. Avulsion fractures, in which a piece of bone is torn out by a load of the ligaments or tendons, usually also occur perpendicular to the direction of action of the force. A bending stress on a long tubular bone will cause a double fracture with a triangular fragment and a torsion will cause a spiral fracture. If a compression stress acts on a bone along its longitudinal axis, a longitudinal crack in the bone and a compression fracture within the joint will occur. Finally, there is the oblique fracture, which is caused by a combined strain from bending and longitudinal load or bending with torsion and longitudinal load [64]. In Fig. 3.58, the various fractures are summarized taking into account the respective strain.

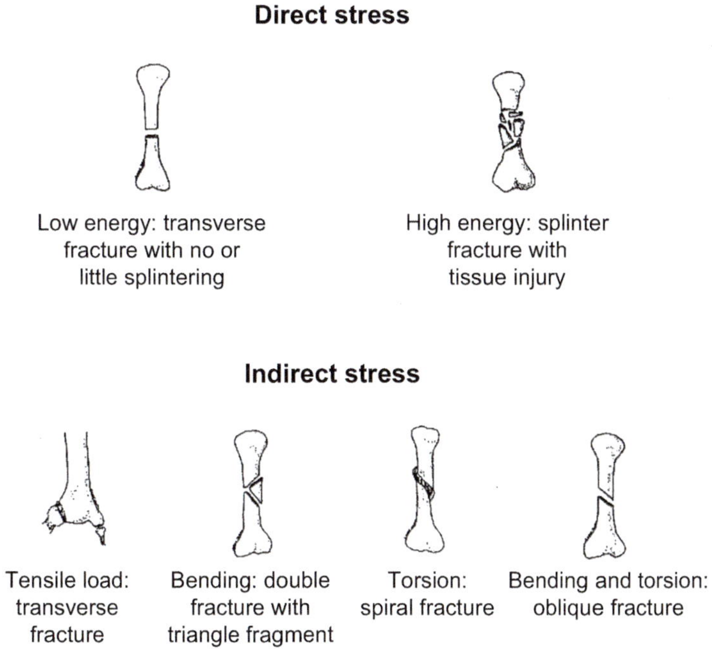

Fig. 3.58 Summary of bone fractures by injury type and type of stress (from [64])

Bone fractures can also be distinguished according to clinical criteria, namely into **open or closed fractures** [86]. A closed fracture is present when the skin and tissue in the immediate vicinity of the fracture remain uninjured; the bone is not exposed to any external contamination. An open fracture, on the other hand, is present when an object penetrates, injures the skin and tissue, and the underlying bone breaks on contact, or when the broken bone itself causes skin and tissue injuries; there is therefore a risk of infection. However, the absence of a wound does not necessarily mean that there is no damage to the skin and tissue, rather, even considerable injuries between the skin and the bone are possible without an open wound being present.

In the injury of limbs, displacements, so-called **dislocations**, can occur, in which the joint ends of the bones completely separate from each other; there is then no longer any contact between the joint surfaces. However, if the joint surfaces remain together, this is referred to as a subluxation. Similar to the above distinction between open and closed fracture, one can also distinguish open and closed dislocations in the case of dislocation.

Ligaments consist of tendons and fibrous tissue and serve to hold joints together and prevent abnormal joint movements; they are usually closely surrounded by the joint capsules. **Injuries of the ligaments** can be divided into three types: ligament tears with complete rupture of the tendons and tissue, strains, in which the ligaments are stretched beyond the elastic range but otherwise remain intact, and minor strains, i.e., a stretch within the elastic limits. Ligament tears and strains lead to instabilities of the joint and occasionally to degenerative arthritis (joint inflammation) due to abnormal joint movements during an atypical load.

In the case of **injuries to the nerves**, two types are distinguished: compression of the nerves, where the sheath remains intact, but the nerve fibers are interrupted. The nerves regenerate independently, starting from the cell body near or in the spine in the direction of the injury (about 1 mm per day). The most severe form of nerve injuries, however, is seen in the complete tear of the nerves and requires neurosurgical treatment. In general, nerve injuries often show defective healing, so prevention plays a major role here.

A remark on the significance of **injuries to the lower extremities** should conclude this subchapter: Injuries in this region are extremely rarely fatal, and then either in connection with injuries to other body regions or when complications occur. However, their treatment and the duration of the healing process, especially in the case of severe injuries, is longer than for correspondingly severe injuries to other parts of the body; permanent impairments cannot be ruled out. In Fig. 3.59, the values for the duration of hospital stay with different injury severity of the lower extremities are compared to the duration for the corresponding injury in other body regions from [46]. It shows that the duration of inpatient treatment for severe leg injuries (AIS > 2) is about twice as long as for comparably severe injuries to other parts of the body.

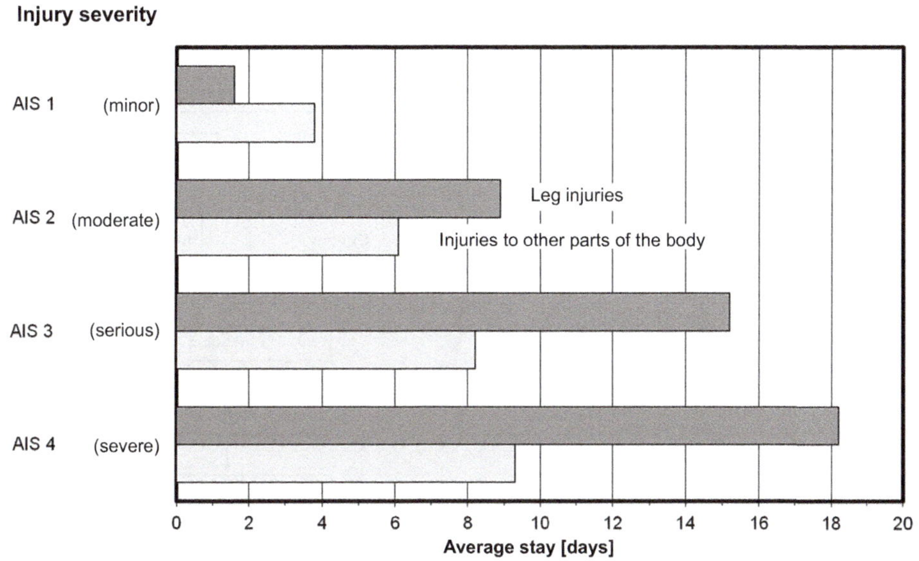

Fig. 3.59 Average hospital stay for injuries AIS > 2 of the lower extremities compared to injuries of other body regions (according to [46])

3.2 Injury Severity and its Monetization

The injury severity characterizes the size of physiological or structural changes in living humans (see Fig. 3.60) due to mechanical force. For example, a comminuted fracture of the femur or a second-degree burn is a characterization of the injury severity. However, it can only be fully estimated by a clinical diagnosis or by an autopsy of the deceased accident victim. In contrast, in most cases, a less exact estimate is made shortly after the accident, which is based neither on a clinical diagnosis nor on an autopsy of the victim. An acceptable, but currently still inaccurate estimate of the injury severity is occasionally made in the emergency room of a trauma clinic. However, this estimate is often insufficient, as for various reasons the symptoms of some injuries can only be recognized several days after the accident, such as instabilities of the knee joint due to a ligament tear or whiplash injuries due to an overstretching (hyperextension) of the neck.

3.2.1 Injury Severity

The degree of injury severity can usually only be determined by a comprehensive medical diagnosis and estimated by referring to the degree of life-threatening danger, the duration of hospital stay, the duration of illness, the risk of permanent physical disability, and other characteristics. Some classifications are only of historical significance,

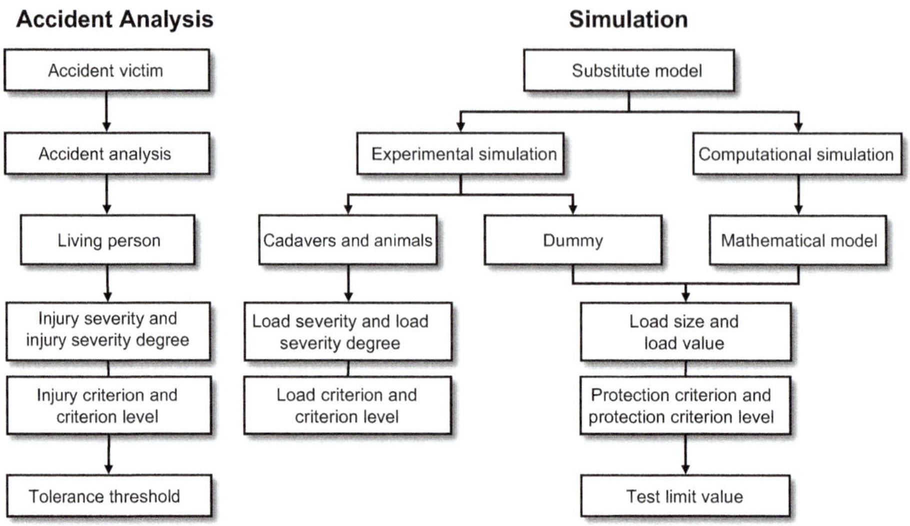

Fig. 3.60 Structure of accident analysis and simulation (according to [55])

and it is therefore not surprising that the assignment of injury severity varied from one application site to another, as one institution considered one characteristic in the injury assessment to be more important and relevant than the other. The result was that the same injuries were classified differently in different countries. Consequently, the so-called "abbreviated injury scale" AIS was developed, which originally dates back to 1959. This now internationally recognized injury severity scale, which takes into account the above-mentioned characteristics to a large extent, has undergone various modifications in recent decades; for example, an injury classified as AIS 3 in 1960 would today be rated only as AIS 2 in terms of severity. The AIS classification can be helpful when applied carefully, but it can also lead to confusion if applied incorrectly. In addition to the **AIS scale**, there are other scales that have been proposed for determining the degree of injury severity, especially in polytraumatized victims, but have not been established (e.g. [65, 15]).

To describe the frequency of injury severity degrees of a certain collective of accident victims, the probability of occurrence is represented as a distribution function of the severity degrees depending on the accident severity or the **load size**. This is necessary because different physiological characteristics of the affected accident victims, but also properties of the contact system and the vehicle body structure, have a scattering effect on the injury severity degrees under otherwise identical external conditions in the accident occurrence.

3.2.2 Injury Scaling According to AIS

Due to the significant problems in the quantitative assessment of injuries, numerous proposals for assessing injury severity have been published with the increase in road traffic accidents, approximately since the 1940s, of which the AIS scale, first published in 1971, has gained international acceptance [16]. The original form comprised a total of ten points and ranged from 0 to 9; it was primarily designed for use in traffic accidents (impact or blunt force). Over the past more than 50 years, the scale has been expanded to cover all conceivable injuries. In its current form [6], the scale now comprises six points and ranges from AIS 1 (minor) to AIS 6 (currently not survivable). The injury severity grades AIS 7, 8 and 9 are no longer used, as for example AIS 9 has included more than three fatal injuries. In the interim AIS revisions, each of these injuries was separately assigned to the severity level AIS 6. AIS 9 now means that the injury severity is unknown, while it was classified as AIS 99 in the original AIS coding. Based on a catalogue of injury descriptions divided according to seven body regions (namely head, neck, thorax, abdomen and pelvic contents, spine, extremities and bony pelvis) and the body surface, the injuries observed are classified according to the five criteria:

- Degree of life threat,
- Treatment duration,
- Permanent damage,
- Energy absorption and
- Frequency of an injury

to determine their severity. This is predominantly based on the degree of life threat, the other criteria are of secondary importance. The AIS scale appreciates the traumatological assessment of the injuries and does not consider the consequences of the injury such as the occurrence of death. This is at most separately coded and statistically represented by the **fatality rate** per AIS severity level. In Tab. 3.3, AIS injury severity grades are exemplarily summarized with some injury descriptions.

In the AIS manual, the most important injuries to be observed in practice are listed in an alphabetical list and evaluated in terms of their injury severity. This list should definitely be used when classifying injuries; it is an important tool for achieving a uniform, internationally comparable injury scale. By applying the identical scaling standards, developments in injury severity—for example as a result of constructive or legislative measures—can be better recognized, and the efficiency assessment of treatment methods finds a quantifiable basis through the introduction of a generally recognized scale.

Each individual injury can be assigned a severity level of the AIS scale, but it only considers each injury as if no further injuries had occurred. However, since many accident victims suffer more than just one injury, the severity level of the most severe individual injury is used as MAIS (maximum AIS) to assess the overall injury severity of a polytraumatized patient since the AIS revision in the 1980s.

Table 3.3 Injury severity according to AIS 2005, injury examples and fatality rate (according to [6])

AIS	Severity	Examples of injuries	Mortality rate [%]
0	uninjured		0.00
1	minor	Abrasions, cuts, sprains, bruises; 1st and 2nd degree burns up to 10% of the surface area	0.00
2	moderate	Extensive abrasion and contusion, extensive soft tissue injuries, slight concussion with amnesia; 2nd degree burns up to 15% of the surface area	0.07
3	serious: not life-threatening	Skull fracture without cerebrospinal fluid leakage, concussion with loss of consciousness, pneumothorax; 2nd degree burns up to 25 % of the surface area	2.91
4	severe: life-threatening, survival probable	Skull fracture with cerebrospinal fluid leakage, concussion with loss of consciousness for up to 24 hours, perforation of the chest; 2nd or 3rd degree burns up to 35% of the surface area	6.88
5	critical: survival uncertain	Skull fracture with brain stem hemorrhage, organ rupture or rupture; 3rd degree burns up to 90 % of the surface	32.32
6	maximum: rated as practically unsurvivable	Massive head contusion, brain stem laceration, skull base fracture, thoracic contusion, aortic rupture and transection; separation between thorax and pelvis	100.00
9	unknown		unknown

In the EU, injuries of severity level AIS 3+ are to be separately reported in the national statistics by 2030 in order to be able to evaluate traffic safety programs in more detail. According to the ETSC report PIN of 2022 [25], this has so far unfortunately only been achieved in a few countries, such as Sweden, Switzerland or Austria. In Germany, these severely injured—often also referred to as life-threateningly injured—are only estimated by the BASt based on the accident database GIDAS. According to this, of the approximately 60,000 seriously injured per year in Germany (as of 2019), about 15,000, or 25%, are severely injured with a severity level of AIS 3+.

For Germany, there is therefore a need for action to explicitly report AIS 3+ injured, as is done in Switzerland or Austria, in the traffic accident statistics and thus to reliably identify relevant prevention priorities.

3.2.3 Other Injury Scaling

The result and treatment of a patient with **multiple injuries** often differs compared to the injury condition of patients with only one injured body part. To better assess the mortality, i.e., mortality, of multiple injured accident victims, various methods were developed based on the AIS scale, two of which are to be presented below: The **ISS procedure** (ISS: injury severity score) proposed by Baker et al. in [7] and the **PODS value** (PODS: probability of death score) by Sommers according to [54].

The ISS procedure is used for coding multiple injuries. The ISS value (3.1) is the sum of the squares of the highest AIS values of the three (I to III) most severely injured body regions:

$$\mathrm{ISS} = \mathrm{AIS}_{\mathrm{I}}^2 + \mathrm{AIS}_{\mathrm{II}}^2 + \mathrm{AIS}_{\mathrm{III}}^2 \qquad (3.1)$$

The body regions are head, neck, thorax, abdomen and pelvic contents, spine, extremities and bony pelvis as well as body surface. In the calculation of the ISS value, none of these body regions is overrepresented. The maximum value can be at most 108, namely when all three considered body parts have an injury severity of AIS 6; if it occurs only in one body part, the ISS value is set to 75. If an injury with AIS 9 (unknown) is coded, the ISS value must not be calculated. The ISS value correlates highly with the mortality of accident victims.

The AIS coding is an anatomical injury scale, it does not take into account pre-existing conditions, nor the age of the patient or their physiological condition. With the **probability of death classification PODS** (probability of death score), an alternative to the ISS procedure was introduced in 1981. The PODS value can be calculated without (3.2) and with consideration of the age of the injured person (3.3):

$$\mathrm{PODS} = 2.2 \cdot (\text{highest AIS}) + 0.9 \cdot (\text{second highest AIS}) - 11.3 \qquad (3.2)$$

and taking into account the age

$$\mathrm{PODS_a} = 2.7\ (\text{highest AIS}) + 1.0\ (\text{second highest AIS}) + 0.6 \cdot (\text{age}) - 15.4 \qquad (3.3)$$

The PODS values are supposed to match better with actual deaths compared to the ISS values and allow a clearer interpretation of the individual values. Further calculations (injury statistics, benefit/cost analysis) should be carried out when the procedure is backed by a broader statistical basis [54].

The description of a close, functional relationship between mechanical impact and injury severity has not yet been achieved with any proposed scaling system. The main reason in the case of the AIS scale lies in the fundamental characteristic of this scaling system, namely in the number of evaluation criteria that lead to the classification of an injury. The AIS value is a summary of the above five criteria, so a close statistical correlation between the injury severity and any accident severity is hardly to be expected. In [55], various accident parameters were examined with regard to body part-specific injury severity grades. Correlation coefficients for pelvis injury severity grades using the speed change Δv (Eq. 2.2) and the specific accident performance SPUL (Eq. 2.3) were determined between 0.74 and 0.92. In addition to the AIS scaling characteristics, the influence of age and other anthropometric factors are responsible for the relatively low correlation between accident parameter and injury severity at other body parts.

3.2.4 Monetary Evaluation of Injury Severity

The AIS scaling system has proven to be an excellent unit of measurement for assessing injury severity in the past, but it is not helpful in predicting the consequences of injuries. When evaluating injuries, the question arose about the composition and measurability of the accident consequences (Fig. 3.61). The direct consequences of an accident include, in the broadest sense, personal injuries (physical and psychological) and property damage (to and in the vehicle and in the environment). The individual components of the accident consequences are therefore measured in different dimensions: monetarily, verbally or ordinally scaled. In the interest of a meaningful and understandable evaluation of the accident consequences, it is therefore absolutely necessary to find a uniform dimensioning of these consequences. The presentation of property damage can initially be omitted here, as they are to be dealt with in Chap. 4 (evaluation of vehicle safety); Here, it is therefore considered sufficient to focus only on injury-related costs in the following.

Since the late 1960s, various approaches to cost/benefit analysis of traffic safety measures have been published (e.g., in [40, 82]), with the work of Jäger and Lindenlaub [49] providing the user with a tool to evaluate the efficiency of such measures. Through modifications in the calculation [27, 56, 70], newer findings from an economic and medical perspective were taken into account, which led to the following improvements for a consistent monetary evaluation of injury severity [56]:

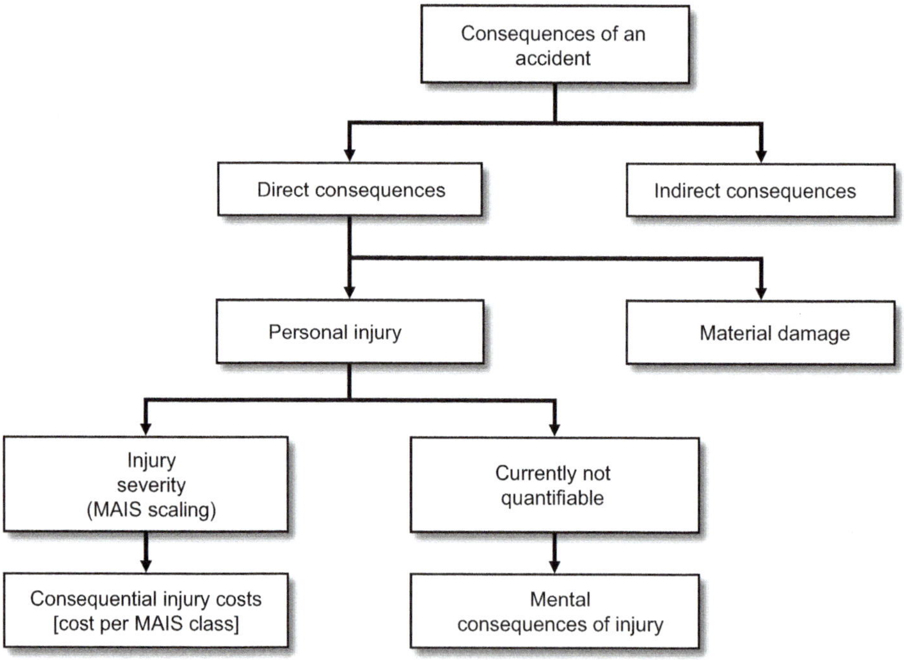

Fig. 3.61 Classification of accident consequences (according to [54])

- Revision of the yield value for children and adolescents,
- Introduction of the fatality rate,
- Replacement of the cost value by the yield value and
- Consideration of the latest economic situation by taking into account the increased unemployment and the growth of the so-called shadow economy.

The resulting **injury-related costs** per MAIS injury category are rounded and summarized for various years in Table 3.4 and shown for 2011 in Fig. 3.62. They have been successfully applied and published in various research projects for quantifying the consequences of injuries and road traffic safety (e.g., [91, 5, 57, 58]).

Table 3.4 Injury-related costs for various years (from [56, 59])

MAIS	Severity	Consequential injury costs [€] for the years				
		1980 according to [56]	**1990** according to [59]	**2000** according to [59]	**2010** according to [59]	**2011** according to [59]
0	uninjured	0	0	0	0	0
1	minor	3.600	5.900	8.500	10.000	11.000
2	moderate	18.000	30.000	44.000	53.000	54.000
3	serious	65.000	110.000	150.000	190.000	190.000
4	severe	150.000	240.000	350.000	420.000	430.000
5	critical	360.000	600.000	860.000	1.100.000	1.100.000
6	maximum	470.000	780.000	1.100.000	1.400.000	1.400.000

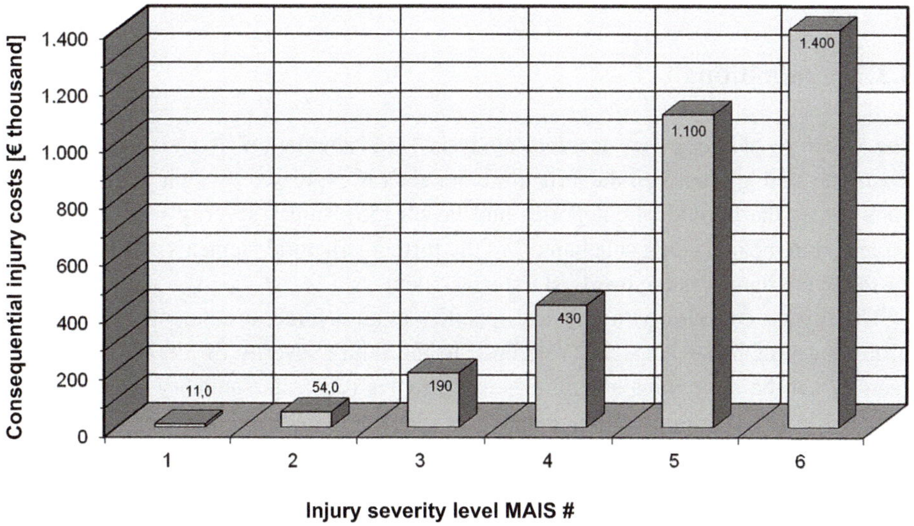

Fig. 3.62 Injury-related costs for the year 2011

The injury-related costs are defined for the overall injury severity grades (MAIS) and averaged over gender, constitution and age distribution of all road users. The application to individual severity grades of body-specific injuries is not without problems, as the differing expression of the mentioned criteria in the classification of injury severity, especially life threat, treatment duration and permanent damage, can lead to cost shifts among different road users (e.g., car occupants compared to riders of motorized two-wheelers).

3.3 Injury and Protection Criteria

Biomechanical investigations aim, among other things, to identify and quantitatively determine the factors responsible for the type and extent of injuries. Ideally, it would be possible to determine the functional relationship between the mechanical impact and the injury outcome. However, the quantitative description of the mechanical impact is not as straightforward as often assumed. For example, to assess the consequences of an injury in a vehicle/vehicle collision, it is not sufficient to consider only the collision speed; at least the deceleration of the center of gravity must also be captured in an accident-specific parameter. However, even the formulation of the accident parameters is not sufficient for describing the mechanical stress, as the properties of the contact structure can vary and can lead to different injury-causing forces. The following will therefore show which paths that have been and continue to be taken in the **biomechanics research** to find injury-relevant mechanical stress parameters and to determine their limit values. Some of these quantities are reflected in legal safety standards, while others have so far only been published and recommended for use in scientific papers and publications; they are in a stage of discussion. However, some of the terms used should first be explained.

3.3.1 Definitions

The structure of the areas 'accident analysis' and 'simulation' is shown in Fig. 3.60, where the analogy between the structural elements is expressed by their respective positions across the individual categories and layers [55]. Injury severity has already been explained in the preceding subchapter, so the further structural elements can be discussed for the living human being involved in the accident.

The **injury criterion** is a physical quantity (e.g., a force, a moment, or an acceleration) that establishes a specific reference to the injury severity of a body region. This quantity can be a combination of various variables (e.g., the summary of translational and rotational acceleration into a quantity for characterizing head injuries). For example, if the injury severity for skull/brain traumas, expressed by AIS, increases with accelerations, then the physical quantity "acceleration" is the injury criterion for the head. The relationship between the physical quantity and the corresponding injury severity of a certain body region should be as clear as possible and valid for all loads and all load cases. The **injury criteria level** (e.g., 80 g) denotes a certain value of the injury criterion in the living human being and is expressed by numerical values of physical parameters such as force, moment, acceleration, pressure, etc. If the value of the load is only large enough and its duration long enough so that a certain threshold is exceeded, defined changes occur with respect to the physiological or mechanical state of certain body parts. In some cases, it may be necessary to define several different thresholds for a load type, e.g., for the viscous criterion of the chest: one for compression and another for compression velocity.

Since different injury mechanisms are conceivable for the same body region, there are also various **tolerance thresholds**. When using the term tolerance threshold, it is therefore first necessary to refer to

- physical parameters or functions of a variable,
- type of living human (age, gender),
- body part,
- type of injury, and
- injury severity.

The tolerance threshold for a specific accident collective can only be determined statistically due to the previously mentioned range of variability. However, it does not mark the dividing line between life-threatening and non-life-threatening injuries, but rather indicates a certain criteria level at which a certain proportion of the injury distribution is present, which appears no longer tolerable or still tolerable (Fig. 3.63).

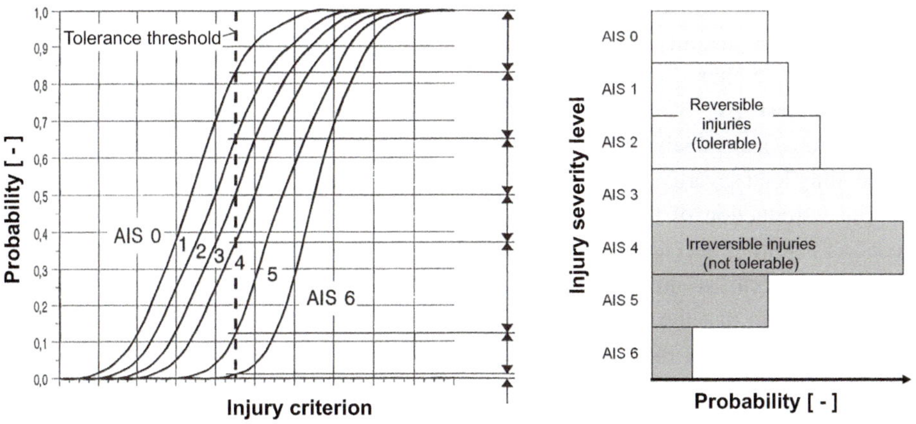

Fig. 3.63 Schematic distribution of injury probability for determining the tolerance threshold (according to [55])

The terms used so far refer exclusively to the **living human body** in accident occurrence. It is therefore only logical to also provide corresponding expressions for the substitute models used in the field of simulation (see Fig. 3.60).

Experiments on living humans are only conducted on a voluntary basis; however, volunteers are only subjected to stress in areas where injuries are definitely not expected. Therefore, physical and theoretical models serve as substitutes for the living human. In simulations with **physical models**, human corpses, anesthetized or killed animals are examined in the experiment. So-called dummies are used for safety tests, but also for basic research purposes. **Theoretical models** refer to electrical or electronic analogies for experimental simulation and, to a much greater extent, mathematical models. Depending on the question and research objective, physical and theoretical models represent either the entire human body or parts of it. The models used as a substitute for the living human in the simulation are specified by spring stiffness, damping characteristics, mass distribution, and moments of inertia and should, as far as possible, include the distribution of age, gender, and constitution of the accident collective represented by them.

The **load severity** characterizes, similar to the injury severity in the living human, the extent of the mechanical change in the form of a structural failure in the dead human body or animal body due to mechanical influences. When using dummies or theoretical models, the term "load severity" must be avoided, as the transmission behavior of the human (living or dead) body and that of the dummy or the mathematical model is different; for example, the fracture of "ribs" on the dummy does not provide any information about the injury severity on the chest of the accident victim. The **load severity degree** corresponds to the injury severity degree in the living human and can be similarly scaled. However, the use of AIS grades should be avoided or at least specially marked to prevent misunderstandings.

The **load criterion** is a physical quantity (e.g., a force, a moment, or an acceleration) that refers to the load severity of a specific body region of the dead human body or animal body. This quantity, comparable to the injury criterion in the living human body, can be a single characteristic value, but also a function of different variables. The **load criterion level** (e.g., 80 g) characterizes a certain value of the load criterion on the body part of a physical substitute model (cadaver or animal), which is sufficient to cause a corresponding load severity.

The introduction of the term **load size** on the dummy and on the occupant model (e.g., as resulting acceleration of the head) is necessary, as a structural destruction on the dummy occurs at much higher mechanical loads and does not occur at all in typical experiments, so that the destruction as a measure for a certain injury severity is unsuitable. The load size is the experimentally measured or computationally determined output quantity of the respective dummy or mathematical model used due to mechanical action. It corresponds to the injury severity in the living human. The commonly used term "dummy load" is misleading, as it not only characterized the measured load of the dummies used in the experiment, but was also used for the load calculated on the mathematically described occupant. The **load value** (e.g., 10,000 N for the femur longitudinal force) is the numerical expression of the load size.

Protection criteria are those load parameters that are measured on the dummy or calculated by the mathematical model and are in interrelation with the injury criteria of the living human body. They can also be composed of several load parameters if a single parameter is not sufficient for a comprehensive description of the injury mechanics. For example, in one of the later sections, a head protection criterion is shown that is composed of translational and rotational head acceleration. If the overall relationship between the individual injury criterion and the protection criterion is known and clear in its application, certain corresponding values can be found from both functions (Fig. 3.64). The relationship can be described using a closed mathematical function or statistically. Since the injury criteria level corresponds with the **protection criteria level**, multiple possible injury criteria result in an equal number of protection criteria.

The **test limit values** are fixed values of the protection criteria levels for defined test situations. They result from the compromise of the conflict of objectives between the need for protection, expressed by the tolerance threshold, and the technical and constructive possibilities. For example, the HIC (Head Injury Criterion) was developed as a protection criterion with a level of HIC = 1,000 for the head. This level was incorporated into European safety legislation. In contrast, the test limit value is significantly lower, at about HIC = 700–750, in order not to exceed the protection criteria level in acceptance tests, considering the range of variation in the test results.

3.3.2 Investigation Methods for Determining Protection Criteria

The load limits of humans can be determined by various investigation methods under simplified or real accident conditions. The investigation under simplified load situations offers well reproducible test conditions and thus a good safeguarding of the results for

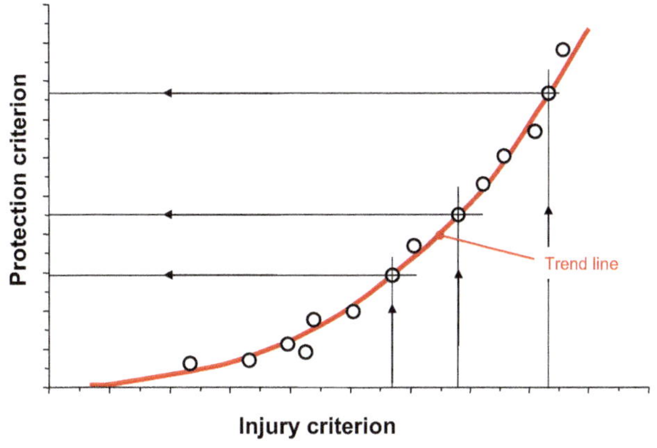

Fig. 3.64 Hypothetical relationship between injury and protection criterion (according to [55])

the investigated load case; however, the transmission of the results to the events in a real traffic accident is often difficult. The transition to more realistic test conditions significantly increases the number of parameters and thus makes it more difficult to secure results. Most of the load limits defined to date are based on statistically insufficiently secured results, as the number of variables is large, but the number of tests carried out is usually relatively small. The investigation methods can be distinguished according to the test objects used, namely tests with

- living humans (accident participants),
- volunteers,
- athletes, artists and dancers,
- human cadavers and parts of cadavers,
- animals, and
- anthropomorphic measuring devices (dummies).

In a real accident, the **living human** is exposed to loads that can lead to more or less severe injuries. In the case of injuries that are caused without or despite the use of restraint systems by contact with interior parts, the forces necessary to deform the contact zones can be used to determine injury criteria. The injuries of the occupant are precisely recorded and the forces necessary to deform the impact zones under the determined accident conditions are experimentally recorded. Assuming that the forces measured by measurement technology and those occurring in the accident are the same (depending on the quality of the experimental simulation of the accident), the injury can be assigned to the causing force. From a sufficiently large number of investigations, injury criteria and their limit values for different body regions can be derived. The resulting scatter range of the injury criteria levels determined in this way is caused by the diversity of the living accident victims, the "biological material", and by not completely or precisely known accident conditions.

Experiments with **volunteers** are problematic, as approaching the presumed injury threshold endangers a person's health. Therefore, in most volunteer experiments, the determined values form lower limits with a safe distance to the injury threshold. Nevertheless, the high risk involved in conducting such experiments should not be underestimated. Valuable insights into the influence of muscle forces on the sequence of movement can be gained from living humans when comparative experiments are conducted under similar conditions with corpses. Muscle tension in response to recognizing a dangerous situation can often mitigate the risk of injury in a real accident. Due to the influence of support and reaction forces, a similar movement sequence occurs in volunteer experiments as in an actual accident. However, the measured values from volunteer experiments cannot be used as injury criteria, as the strain must be below the hazard threshold, on the other hand, the test subjects are usually young men who are physically more resilient than the average person.

In experiments with **athletes, artists and dancers**, the fact is utilized that in many sports or in the practice of some professions, parts of the body are stressed to the limit of endurance. As a relatively new method of data collection, these stresses are measured (directly or from video recordings), for example, in boxers, football players, skiers, dancers, and artists. Similar to volunteer experiments, the stresses here are usually below the injury limit, and the test subjects are well-trained and therefore physically highly resilient.

The use of **cadavers** for experiments to determine load criteria has met with considerable ethical and legal concerns in Germany in the past; today, such experiments are only conducted to a very limited extent, as they encounter resistance and misunderstanding in the public and can only be carried out today with the approval of so-called ethics committees. However, the available cadavers are often unfavorably distributed in terms of age, as usually only those who have died of natural causes can be used. Younger deceased can only be selected with restrictions, as they often died in accidents and the bodies are usually pre-damaged as a result. As part of the individual case method applied in the "Cooperation Network Biomechanics" (KOB), specific, real accidents were recreated as accurately as possible in vehicle tests using both cadavers and dummies. The injuries of the respective accident victim were correlated with the severity of the load, i.e., with the "injuries" of the human cadaver and the load values of the dummy [62, 39]. Although the experiments aimed at realistic accident conditions, the actual accident occurrence remained unconsidered in its entirety even in these investigations.

However, the results obtained with the help of cadaver experiments show inadequacies that will be presented below: Bone strength decreases significantly with increasing age beyond the fourth decade of life. Seat belt experiments with complete cadavers show that the number of rib fractures increases with age and due to the increasing thoracic instability, internal organs such as the liver, spleen, lung, and diaphragm are also damaged. Joint diseases and bone loss, lack of support and reaction forces, lack of muscle tone, lung and vascular pressure, altered tissue elasticity, and usually lower body temperature during the experiment lead to deviations in resilience compared to the living. However, bruises, abrasions, cuts, fractures, and vascular ruptures are similar in fresh cadavers as in living accident victims.

While experiments with complete cadavers are usually used to investigate certain accident constellations or to reconstruct individual accident situations, experiments with **body parts** are conducted to determine load criteria for body parts. These body parts are usually examined quasi-statically using special devices, resulting in lower tolerable criterion levels compared to the dynamic load case. Influences of neighboring body parts can only be imperfectly taken into account in this examination method.

The load criterion levels obtained in experiments with complete cadavers or body parts are usually transferred to the living as injury criteria levels. This simplifying approach is based on the following consideration: If the load criterion level is set as the limit in the cadaver where the intolerable injury just occurs, then with the injury criteria level in the living human body due to the more favorable physical conditions at the same load, one is "on the safe side".

Experiments with **animals** provide the opportunity to stress the living organism to the point of destruction. Of course, animal experiments also require the positive vote of an ethics committee. In this way, it is particularly possible to determine the different results in living, dead unpreserved, and preserved test bodies and to use them for the transmission of results from corpse experiments to living humans.

In Fig. 3.65, an experimental setup is shown in which the injury mechanics of the cervical spine were investigated using anesthetized pigs [107]. However, it is very difficult to draw conclusions from the animal's load limits to those of humans. Although various animals (monkeys, pigs) are similar to humans in certain body regions, due to the different general body structure and the differing weight distribution, the load criteria levels from animal experiments cannot be directly transferred as injury criteria levels to humans.

For the experimental simulation of accidents, almost exclusively **anthropomorphic test devices,** so-called dummies, are used today, which are equipped in the head, chest, hip, and upper and lower legs with acceleration, displacement, force, and moment

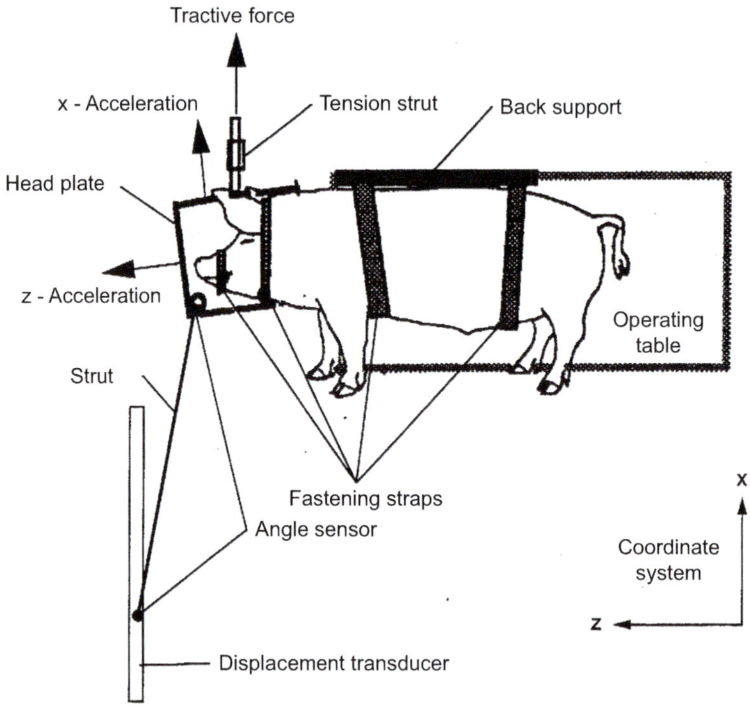

Fig. 3.65 Experimental setup for the investigation of traumas of the cervical spine using anesthetized pigs (from [107])

sensors. The development and improvement of these dummies has now, starting from a simple dummy (Sierra SAM 1949, see Table 8.4) led to ever more current generations of dummies (HYBRID I-III, THOR Dummy, WorldSID). The problem lies in the production of a dummy that is as similar as possible to humans and therefore complicated, but on the other hand, must have a sufficiently simple and reliable design to deliver reproducible results. In the construction of such measuring dummies, size, shape, mass distribution, kinematics, and joint characteristics of humans are replicated. The loads measured on the dummy during the crash test are compared with predetermined protection criteria levels, and depending on whether the measured values are exceeded or not, conclusions can be drawn about "death" or "survival". To determine the protection criteria, a method was used in which the injuries of accident victims were compared with the load values of dummies, which were determined under more or less simplified test conditions. In this process, individual values were compared with each other; the real accident occurrence was otherwise disregarded [55].

As early as 1978, the essential approaches to the method of direct coupling of results from accident analysis and computational simulation were first presented [43]. This method has subsequently been further developed, not least due to two research projects [91, 20], into the so-called **statistical biomechanics** [129] and incorporated into the results of the aforementioned research projects. Using the "method of equivalent accident characteristics" (EAC method), risk functions for all relevant body regions were developed in sagittal load in [55]. This method serves to combine injuries and loads and is based on the following elements (Fig. 3.66):

- Using the analysis of the data material from accidents, injury severity levels for all relevant body regions are provided,
- The computational occupant simulation allows the determination of load values on various, mathematically described body parts, and
- The accident characteristic SPUL (specific accident power) characterizes the severity of the frontal accident and forms the reference basis for the correlation between the results of the accident analysis and the computational occupant simulation.

The relationship shown in Fig. 3.66 shows the probability of head injuries, expressed by the injury severity AIS [6], depending on the equivalent accident characteristic. This accident characteristic indicates the accident severity (see Sect. 2.2.2); for frontal collisions, the specific accident power (SPUL, Eq. 2.3) was developed and used in [55]. The results from the computational occupant simulation are also shown in relation to the specific accident performance. The applied **GAMBIT value** for head load takes into account both the translational and the rotational head acceleration of the occupant model, it is explained in more detail in Sect. 3.3.4. Other load sizes for the head could also be the HIC or the $a_{3\,\text{ms}}$ value. For the same accident severity, a certain distribution of injury severity levels results from the accident data and a certain head load from the occupant

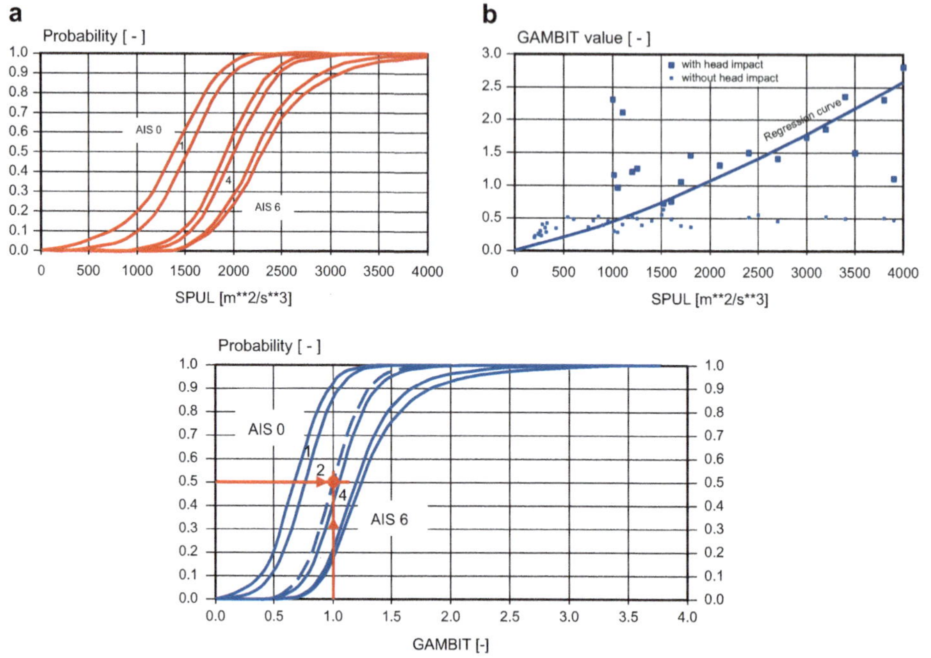

Fig. 3.66 Provision of the risk function for head injuries from results of the accident analysis (top left) and the computational occupant simulation (top right)

simulation, so that a correlation between injury probability and occupant load can finally be established by eliminating the accident severity. By dividing the injury severity levels—for the head and the extremities the boundary is between AIS 2 and 3, for the other body regions between AIS 3 and 4—a boundary curve is provided, with the help of which the probability for reversible and irreversible injuries can be shown. This boundary curve, which is shown dashed in Fig. 3.66, is the risk function; it indicates the risk of irreversible injuries depending on the respective load size. In Fig. 3.67, the risk functions are exemplarily shown for head and chest load in frontal collisions. From this, the protection criteria levels for HIC on the head and for $a_{3\ \mathrm{ms}}$-acceleration of the chest can be read off at a probability of $P = 0.5$.

In all the mentioned methods, attempts were made to include the living human as a victim of an accident. However, only in the method of statistical biomechanics was the real accident situation, at least as far as it seemed relevant for the car occupants, as well as the entire collective of car occupants taken into account. The primary goal, however, continues to be to ensure the conformity of the protection criteria and their levels to the living human. For this, further efforts must be made in the field of biomechanics, experimental and simulation technology to improve the conformity (**biofidelity**) between the dummies or models used on the one hand and the living human on the other. Significant steps in this direction can be seen in the provision of a new generation of dummies

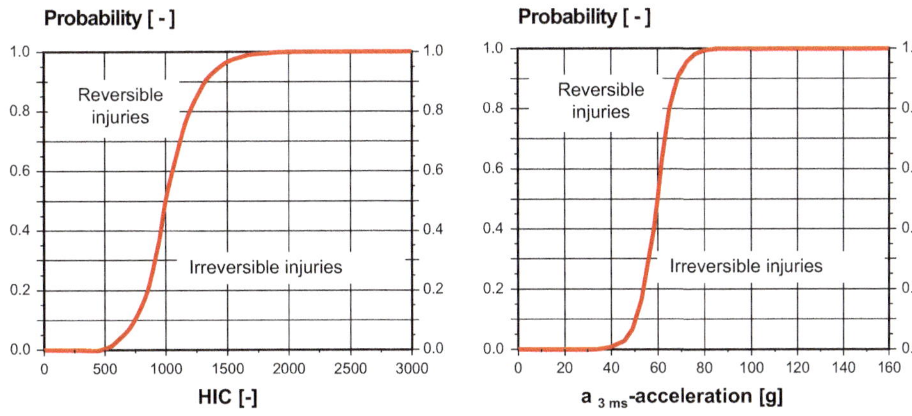

Fig. 3.67 Injury risk functions for head and chest load in frontal collisions

(THOR for sagittal strain and WorldSID for lateral strain, see Sect. 8.5.3) and improved occupant models.

3.3.3 Legally Defined Protection Criteria

The protection criteria levels defined in international safety legislation only guarantee a minimum level of passive safety, as they, like all such legal regulations, do not reflect the state of science, but are rather the result of the consensus achievable at the time of adoption between research, development (state of the art) and societal (often only national) demand.

The protection criteria contained in the current regulations and rules for passive safety can, almost without exception, be traced back to the regulations (FMVSS = Federal Motor Vehicle Safety Standard) issued by the US Department of Transportation, National Highway Traffic Safety Administration since the late 1960s, or the basic investigations that preceded them. These primarily related to occupants in frontal collisions, and were later extended to side and rear collisions.

In Table 3.5, the essential domestic and foreign **safety regulations** are assigned to the corresponding sub-areas of passive safety, see Chap. 4. The approval conditions required in the Road Traffic Licensing Regulations, Germany (StVZO) only describe general requirements. The guidelines of the United Nations (UN) and the regulations of the Economic Commission for Europe (ECE), which are largely identical in content and partly based on US regulations, focus on the examination of vehicle components. Nowadays, integral tests (experiments with complete vehicles) are carried out using dummies. The earliest regulations were issued in the USA in the form of a consumer information (rating) program by the traffic authority. They are continuously

Table 3.5 Safety legislations (2023) and hints (by underlining) on protection criteria

Counterparty protection	StVZO	UN/ECE	USA
Vehicle exterior shape	32	26	-
Wheel cover	36a	30	-
Wheel nuts, wheel caps	32	26	211
Fuel supply system	45, 46	34	301
Hood closure system	30	-	113
Bumper	32	42	215
Self-protection	**StVZO**	**UN/ECE**	**USA**
Roof indentation protection	30	29	216
Side door strength	-	95, 135	214
Door locks and hinges	35a	11, 32, 33	206
Steering column shift	-	12, 94, 137	204
Windshield mounting	-	43	212
Windshield intrusion	-	43	219
Dashboard impact	30	21, 32, 33	201, 208
Steering wheel impact	30	12, 94, 137	203
Sun visor impact	30	21	201
Seat, seat backrests	30, 35a	16, 17, 21, 44, 80	201, 207
Headrests	30	25, 80	202
Seat belts, occupant protection systems	35a	14, 16, 94, 95, 135, 137	208, 209, 210
Child restraint systems	35a	44	213
Flammability of interior parts	30	34	302
Safety helmets	-	22	-

supplemented and adapted to the presumed requirements. In contrast to the type approval process in Germany and in European foreign countries (but also in non-European countries that submit to UN regulations), approval in the USA and Canada is based on the self-certification process, the so-called **self-certification**. The regulations marked by underlining in Table 3.5 contain protection criteria levels that must not be exceeded under certain test conditions. These limit values are summarized in Tables 3.6 and 3.7 as examples for the most important laws on frontal impact and side impact and assigned to the individual body parts. The currently defined protection criteria and their limit values required in laws and consumer protection tests are subject to frequent adjustments. However, they can be very well taken from the annually published Safety Companion [96].

As a basis for the **protection criteria levels for the head** (see Table 3.6), the protection criteria derived from the resulting translational accelerations, which are measured at the center of gravity of the dummy's head, are used. These are the head injury criterion HIC (Eq. 3.4) and the acceleration value a_{3ms}, which lasts over a measured period of 3 (or 5 in the case of safety helmet testing) milliseconds.

Table 3.6 Frontal collisions passenger cars: protection criteria and levels according to current safety legislation [96]

Configuration		Rigid Barrier In-Position					Deformable Barrier In-Position		Out of Position			
Regulation		CMVSS 208 (old), ADR 69/00, FMVSS 208 (old)	FMVSS 208 CMVSS 208		UN R137		UN R94, ADR 73/00	FMVSS 208 CMVSS 208	FMVSS 208 CMVSS 208			
Dummy		Hybrid III	Hybrid III	Hybrid III	Hybrid III	Hybrid III	Hybrid III	Hybrid III	Hybrid III	Hybrid III	Hybrid III	CRABI
Size		50 % male	50 % male	5 % female	50 % male	5 % female	50 % male	5 % female	5 % female	6 year	3 year	1 year
Region	Criterion											
Head	HIC/HPC36 [-]	1000 (FMVSS, ADR)			1000	1000	1000					
	HIC_{15} [-]	700 (CMVSS)	700	700				700	700	700	570	390
	a_{3ms} [g]				80	80	80					
Neck	N_{ij} [-] (4 Values)		1.0	1.0				1.0	1.0	1.0	1.0	1.0
	$F_{x,shear}$ [kN]				3.1	2.7	3.1 @ 0 ms 1.5 @ 25-35 ms 1.1 @ ≥ 45 ms					
	$F_{z,tension}$ [kN]		4.17	2.62	3.3	2.9	3.3 @ 0 ms 2.9 @ 35 ms 1.1 @ ≥ 60 ms	2.62	2.07	1.49	1.13	0.78
	$F_{z,compr.}$ [kN]		4.0	2.52				2.52	2.52	1.82	1.38	0.96
	M_y [Nm]				57	57	57					
Chest	a_{3ms} [g]	60	60	60				60	60	60	55	50
	Deflection [mm]	76.2 (FMVSS, ADR) 50 (CMVSS)	63	52	42	34	42	52	52	40	34	30[1]
	VC [m/s]				1.0	1.0	1.0					
Femur	Axial Force [kN]	10	10	6.805	9.07	7	9.07 @ 0 ms 7.58 @ > 10 ms	6.805	6.8			
Knee	Displacement [mm]						15					
Tibia	TI [-]						1.3 (4 Values)					
	Axial $Force_{compr.}$ [kN]						8.0					

Safety knowledge by carhs

1 currently no measurement possible

Table 3.7 Side collisions passenger cars: protection criteria and levels according to current safety legislation [130]

Configuration		Barrier Side Impact			Pole Side Impact		
Regulation		FMVSS 214 CMVSS 214		UN R95 ADR 72/00	FMVSS 214 CMVSS 214		UN R135 GTR 14 ADR 85/00
Dummy		ES-2 re	SID IIs	ES-2	ES-2 re	SID IIs	WorldSID
Size		50 % male	5 % female	50 % male	50 % male	5 % female	50 % male
Region	Criterion						
Head	HIC/HPC36 [-]	1000	1000	1000	1000	1000	1000
Shoulder	$F_{lateral}$ [kN]						3.0
Chest	Deflection [mm]	44		42	44		55
	VC [m/s]			1.0			
Abdomen	Peak force [kN]	2.5		2.5	2.5		
	Deflection [mm]						65
Lower Spine	Acceleration [g]		82				75
Pelvis	PSPF [kN]	6.0	5.525	6.0	6.0	5.525	3.36

The threshold to be considered as the limit of the **HIC** is 1,000 according to European safety laws [23, 24] and 390–700 according to FMVSS 208 [22], where the formulation of the expression

$$\mathrm{HIC} = \max\left\{ \left[\frac{1}{t_2 - t_1} \int_{t_1}^{t_2} a(t)\mathrm{d}t \right]^{2.5} \cdot (t_2 - t_1) \right\} \tag{3.4}$$

with the units a [g] and $t_{1/2}$ [s] corresponds to the mathematical description of the WSU curve. The "WSU curve" stands for Wayne State University cerebral concussion tolerance curve [36, 37], which was determined as the boundary curve between life-threatening and non-life-threatening brain injuries (Fig. 3.68). The boundary conditions to be observed when using the HIC, such as the head's impact in the forehead area on structures, filter characteristics, and numerical treatment in the computer-aided data processing of the measured acceleration/time signal, among others, are predefined (summarized in [60]). In particular, the circumstance of the head impact was often disregarded in the experimental application and interpretation for the head injury risk. To avoid this inadequacy, the criteria HIC_{15} and HIC_{36} were introduced in the legal requirements; here, the HIC value according to the above formulation is considered in the HIC_{15} only in the interval of $t_2 - t_1 = 0.015$ s, i.e., within 15 ms, or in the HIC_{36} in the interval $t_2 - t_1 = 0.036$ s.

The 15 ms time interval represents the stiff, hard impact according to recent **biomechanics research**, while the 36 ms interval describes the "softer" head impact. The

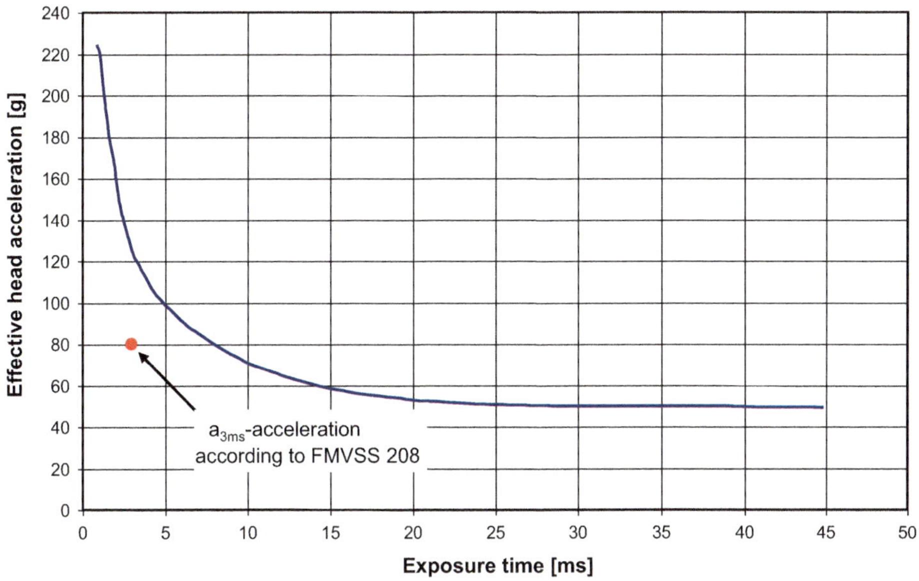

Fig. 3.68 WSU boundary curve (= Wayne State University cerebral concussion tolerance curve) and the 3 ms value of a=80 g

criteria levels HIC_{15} = 700 and HIC_{36} = 1,000 directly correlate with each other according to Eq. 3.4 [76]. It is therefore evident that the need for protection, but also the current development in defining the biomechanical protection criteria levels, has resulted in a consistent, justified reduction. The increase of the HIC value to 1,250 or even 1,500, initially planned for the ECE regulation for frontal collisions in the mid-1980s [23] and discussed in various publications [119, 28, 102], has thus become obsolete and must be seen more as motivated by safety policy than supported by new scientific findings.

The prescribed **head protection criterion HPC** in the European regulation UN-R 95 for side collisions [24] is calculated according to the same regulation (3.4) as the HIC criterion and has the same level, namely HPC = 1,000.

The HIC value, under otherwise identical conditions, is strongly dependent on the characteristics of the acceleration/time course. Fig. 3.69 shows various courses and approximation formulas [123] for calculating the HIC value, with the HIC values ranging from 423 (triangle pulse) to 712 (half-sine pulse) to 1,717 (rectangle pulse) for identical maximum acceleration and the same pulse duration.

The $a_{3\ \mathrm{ms}}$ value of 80 g can also be traced back to the WSU curve, where, however, the duration of the acceleration as well as the "effective acceleration" are only inadequately defined [60]. Using the results from Got [34], the limit value $a_{3\ \mathrm{ms}}$ = 80 g results in a peak value of about 125 g. This value corresponds to the 3 ms value of the WSU curve (marked in Fig. 3.68). The protection criteria level from the safety helmet regulation according to UN-R 22 prescribes a time window of 5 milliseconds, and a higher

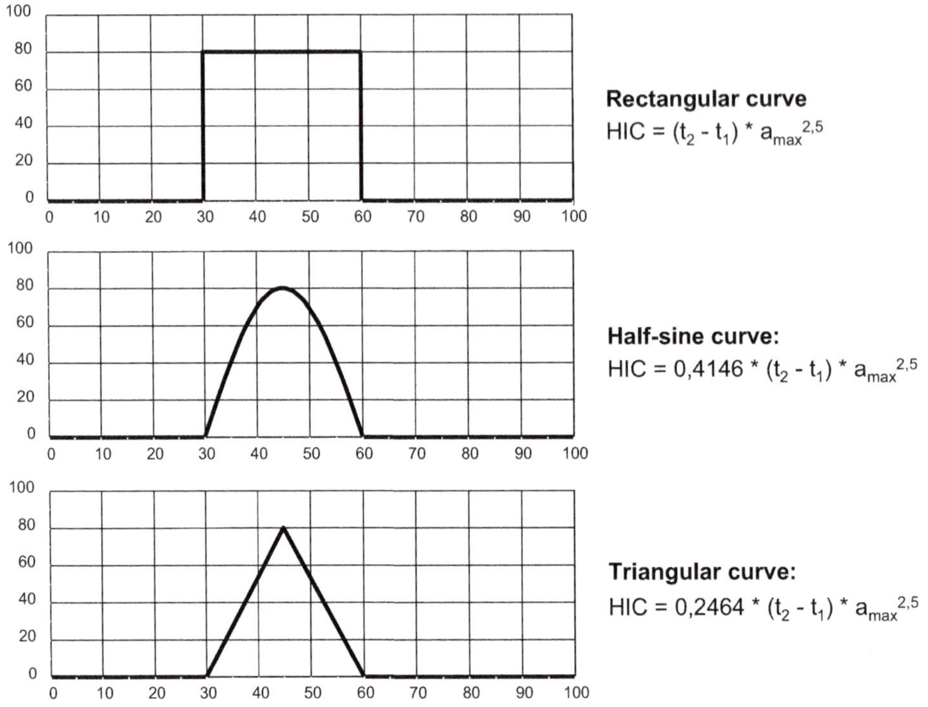

Fig. 3.69 Acceleration/time diagrams and approximation formula for HIC calculation (according to [123])

acceleration level, namely $a_{5\ \mathrm{ms}} = 150$ g, is permitted; falling below the maximum peak value of $a_{\max} = 300$ g is considered tolerable. Beier and Schuller [8] however, consider the prescribed time window to be superfluous, or at least questionable, as they view the peak acceleration, based on various studies [34, 103, 104], as a suitable "tolerance criterion" for injuries to the human brain.

The cause of injuries to the cervical spine are forces and moments, which serve as the basis for the **protection criteria level for the neck** (neck injury predictor); it can be categorized by four quadrants: Extension/tension, extension/compression, flexion/compression, and flexion/tension (see Fig. 3.70). The criteria for neck injuries thus result from the axial compressive and tensile forces at the head/neck transition and their duration of action. Similarly, the neck moment is derived from the bending moment around the transverse axis at the head/neck transition during extension/flexion movements.

However, there are two additional criteria for neck protection: the neck injury criterion **NIC** (according to [10]) and the normalized neck injury criterion N_{ij} (according to [22]). As a protection criterion, the NIC criterion is used to assess the injury risk of the **cervical spine**—typically in rear-end collision accidents with a Δv of 16-24 km/h—, and

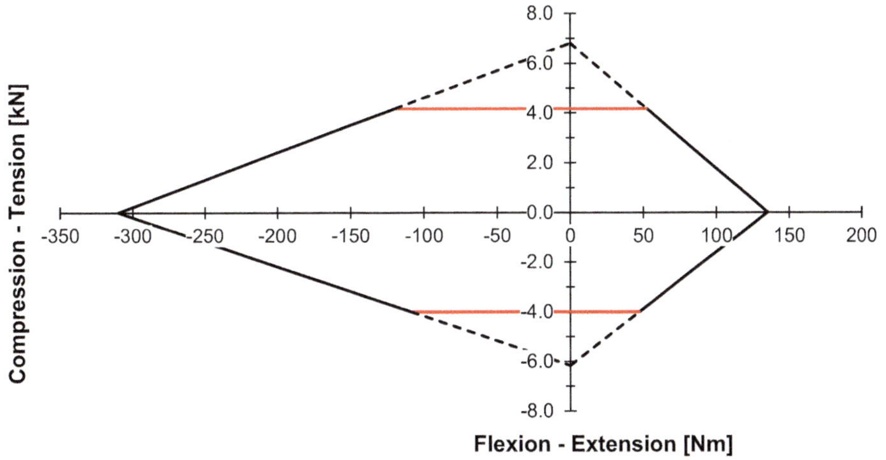

Fig. 3.70 Validity range for protection criteria for the cervical spine according to FMVSS 208 (example for the protection criteria level, HYBRID III 50% dummy)

this within the time span up to 150 ms after the start of the collision. The NIC is based on the hypothesis of a pressure wave in the spinal canal. The criterion is used to estimate the probability that a pressure wave caused by rapid bending leads to injuries. The protection criteria level is at $NIC_{max} = 15\ m^2/s^2$ and it is calculated from the relative acceleration between the head and the first thoracic vertebra and the integrated speed derived from it:

$$\mathrm{NIC}(t) = 0.2 \cdot a_{\mathrm{rel}}(t) + [(v_{\mathrm{rel}}(t)]^2 \tag{3.5}$$

$$\text{with } a_{\mathrm{rel}}(t) = a_x^{T1}(t) - a_x^{\mathrm{head}}(t) \text{ und} \tag{3.6}$$

$$v_{\mathrm{rel}}(t) = \int a_{\mathrm{rel}}(t)\mathrm{d}t \tag{3.7}$$

The N_{ij} criterion is based on the hypothesis that the interaction of axial neck forces (tension/compression) and neck moments (extension/flexion) leads to injuries of the cervical spine. The $\mathbf{N_{ij}}$ value is intended to estimate the risk of sustaining severe cervical spine injuries in **head-on collisions**. The value corresponds to the respective combination of forces (tension/compression) and moments (extension/flexion) NCF, NCE, NTF, NTE [9, 128]. In the normalized neck criterion N_{ij} and the critical values for $F_{z,\mathrm{crit}}$ and $M_{y,\mathrm{crit}}$, the occupant size (or the size of the dummy used) is taken into account, they are summarized in Table 3.8 along with other criteria and exemplified for the 50% dummy in Fig. 3.70. When using the 5% dummy, different criteria levels must be considered in the dynamic and static test. The forces and moments are determined at the same time at the

Table 3.8 Protection criteria and criteria levels according to FMVSS 208 depending on the dummy size and the test configuration

Safety standard	FMVSS 208 old	FMVSS 208 - Final rule					
Dummy size	Hill 50 %	Hill 50 %	Hill 5 %		6-year old	3-year old	12-month
Test constellation		in position dynamic	in position dynamic	OoP static	OoP static	OoP static	OoP static
Protection criteria	Protection criteria level						
Head HIC	1.000(36)	700(15)	700(15)	700(15)	700(15)	570(15)	390(15)
Neck criterion N_{ij}	—	1,0	1,0	1,0	1,0	1,0	1,0
Neck F_z critical [N] tensile force	—	6.806	4.287	3.880	2.800	2.120	1.460
Neck $F_{z\,crit}$ [N] compression force	—	6.160	3.880	3.880	2.800	2.120	1.460
Neck $M_{y\,critical}$ [Nm] flexion	—	310	155	155	93	68	43
Neck $M_{y\,critical}$ [Nm] extension	—	135	67	61	37	27	17
Neck $F_{z\,max}$ [N] tensile force	—	4.170	2.620	2.070	1.490	1.130	780
Neck $F_{z\,max}$ [N] compression force	—	4.000	2.520	2.520	1.820	1.380	960
Chest a_3 ms acceleration [g]	60	60	60	60	60	55	50
Chest indentation [mm]	76	63	52	52	40	34	not specified
Femur longitudinal force [kN]	10,0	10,0	6,8	6,8	not specified	not specified	not specified

head/neck transition. The calculation is based on measured forces and moments, which are related to specified critical values:

$$N_{ij}(t) = F_{Nz}(t) + M_{Ny}(t) \leq 1{,}0 \tag{3.8}$$

$$\text{with} \quad F_{Nz} = F_z / F_{z,\text{crit}} \quad \text{and} \tag{3.9}$$

$$M_{Ny}(t) = M_y / M_{y,\text{crit}} \tag{3.10}$$

The **protection criteria levels for the chest** (see Table 3.6) initially refer to the resulting acceleration measured at the center of gravity of the dummy's chest. The severity index SI

$$\mathrm{SI} = \int_0^t [a(t)]^{2.5} \mathrm{d}t \text{ with } a[g] \tag{3.11}$$

is the original description of the WSU curve formulated by Gadd [30] for the head, which was adopted as a protection criterion for the chest with the criteria level 1,000 [95].

In modification of the permissible $a_{3\,\mathrm{ms}}$ value for the head, the limit for the chest is reduced to 60 g. The use of the impact force in both European guidelines and regulations as well as in American legislation refers more to the unbelted occupant and can only be understood from the historical development of safety legislation, as the belted occupant only hits the steering wheel with the chest at high accident severity or improper seat belt use. With the specified maximum load force of 11,100 N, which is equally specified in the American safety standards and in the UN guidelines or ECE regulations, the achievable limit value for a test block mass of about 35 kg corresponds to an acceleration of approx. 32 g and is thus significantly lower than the 3 ms value.

The permissible protection criteria level for **child restraint systems** is set at 50 g in the UN-R 44 regulation and is therefore lower than for adult occupants. It applies to child dummies in the age stages of 9 months, 3, 6 and 10 years, while the US standards allow 50 g for the 12-month-old child dummy, 55 g for the three-year-old, and 60 g for the six-year-old child dummy and the adult dummy as criteria level for the time barrier of 3 milliseconds. The mentioned UN regulation for the examination of child restraint systems also includes a special feature in the form of the protection criterion for the longitudinal load on the spine: The limit for this is the vertically measured acceleration difference between "lower body and head" with a limit of $\Delta a_{3\,\mathrm{ms}} = 30$ g. However, the approach seems questionable due to the different measurement times.

In addition to the accelerations or the SI value derived from them and the impact force, a protection criterion in the form of the **relative compression path** between the sternum and spine has been prescribed in the US standards (FMVSS 208) since 1986 [22]. The test limit value of 50.8 millimeters (= 2 in) must not be exceeded for such loads that are caused by the impact on surfaces of any kind. Excluded from this is the contact between occupants and airbag systems, as a large-scale force introduction into the torso is assumed here; in these cases, the limit is 63 mm (= 2.5 in) for the HIII-50M and 52 mm for the HIII-5F dummy.

The **compression of the chest** is also prescribed in the UN-R 95 for lateral load [24], the maximum rib indentation of one of the three dummy ribs (EuroSID) relative to the spine must not exceed a value of RDC = 42 mm (RDC: rib deflection criterion). As an additional protection criterion, the formulated **viscous criterion VC** by Viano et al. in [116] is used in side collisions. It consists of the two physical quantities, the relative compression path $c\ (t)$ and the compression speed $v(t)$:

$$\mathrm{VC} = \max\{c(t) \cdot v(t)\} = \max\left\{\frac{s(t)}{\text{halve thorax} - \text{width}} \cdot \frac{\mathrm{d}[s(t)]}{\mathrm{d}t}\right\}. \tag{3.12}$$

In [117] results from cadaver tests are shown, in which only irreversible injuries with injury severity levels of AIS 4 and 5 occurred at VC > 1.0 m/s. This protection criterion was also established in the regulation for frontal collisions UN-R 94 [23] and is applied as a protection criterion for sagittal thorax loads. However, the calculation of the viscosity criterion VC is done in the form

$$\begin{aligned} \mathrm{VC} &= \max\{c(t)\cdot v(t)\} \\ &= \max\left\{\frac{D(t)}{0.229}\cdot\frac{8\cdot[D(t+1)-D(t-1)]-[D(t+2)-D(t-2)]}{12\cdot\Delta t}\right\} \end{aligned} \tag{3.13}$$

with $D(t)$: indentation [m] at time t,

Δt: time interval [s] between the individual measurements, it should be a maximum of $\Delta t_{max} = 1.25\cdot 10^{-4}$ s.

The injury index for lateral chest loads, **TTI** (thoracic trauma index), defined by Eppinger et al. in [26]

$$\mathrm{TTI} = 1.4\cdot \mathrm{age} + 1/2\cdot\left(\mathrm{RIB}_y + \mathrm{T12}_y\right)\cdot\frac{m}{m_{\mathrm{strd}}} \tag{3.14}$$

is based on an analysis of results from several sled test series using cadavers, conducted at the University of Heidelberg, where a significant influence of the age of the post-mortem test objects **(PMTO)** could be determined. The acceleration term describes the average of the maximum lateral acceleration values at the fourth, impact-side rib RIB_y and at the twelfth thoracic vertebra $\mathrm{T12}_y$. The acceleration term is scaled by the mass m of the test object relative to a standard mass of $m_{\mathrm{Strd}} = 75$ kg [14]. For experimental use on the side dummy SID, the above relationship was modified in [21] neglecting the influence of age in the form

$$\mathrm{TTI} = 1/2\cdot\left(a_{y\ max}^{\mathrm{rib}} + a_{y\ max}^{\mathrm{spine}}\right) \tag{3.15}$$

and used for comparison of different, laterally loaded dummies (US-SID—no longer in use—and Euro-SID). The maximum values of the lateral acceleration component measured at one of the ribs and at the lower spine (at the level of the T12 vertebra) are considered as a measure of the effective load on the chest in the lateral direction. Regardless of the load direction, the measurement of acceleration difference is not without problems, as the phase position of independent measurements generally affects the level of the absolute acceleration difference.

Although pelvic acceleration is usually measured in all safety-related dummy investigations, there is currently no legally prescribed **protection criterion for the pelvis/abdominal area** (see Table 3.6). Only visual inspections are prescribed and indications are given that the lap belt must "stay on the pelvis" during the respective investigation and the "belt position must not be shifted into the abdominal area" (according to FMVSS 209). When checking child restraint systems according to UN-R 44, modeling clay is applied to the abdominal area to check for seat belt displacement, which must not be damaged after the test. The UN-R 94 test regulations [23] do not contain any criteria for controlling the submarining effect.

For the lateral load, the introduction of side impact tests according to FMVSS 214 in 1990 took into account the lateral acceleration as a protection criterion; it has a test limit value of 130 g. In contrast, the European regulation UN-R 95 for side collisions [24] provides for maximum forces as protection criteria: the limit values are 2.5 kN for the abdominal force (APF: abdominal peak force) and 6.0 kN for the pelvic force (measured at the pubic symphysis (PSPF: pubic symphysis peak force) (see Table 3.7).

As a **protection criterion for the extremities** (see Table 3.6), the femur axial (longitudinal) force exists based on the legally defined requirements. This criterion can also be understood as a **submarining** criterion due to the close kinematic relationship with the pelvis. Adomeit states in [2] that knee injuries only occur with too large pelvic displacement values. Such large pelvic displacements can usually be seen as a result of the pelvic seat belt sliding up into the abdominal area, i.e., a submarining of the pelvis under the seat belt [55]. However, this relationship only applies in cases where the restraint system is effective without the use of a knee pad.

Due to varying forces between the femur and tibia, a displacement in the knee can occur (see Fig. 3.52), which poses a risk for knee injuries in the form of ligament ruptures. Therefore, the displacement between the upper and lower leg was introduced as a protection criterion for the knee joint and was set at a maximum value of 15 mm in the UN-R 94 regulation [23].

To assess the risk of injury to the tibia, two criteria were developed, which are stipulated in the UN-R 94 regulation for frontal collisions and are also required in Euro NCAP for side collisions. They assess the probability of a fracture of the tibia, which is reflected in long healing processes and correspondingly high injury-related costs. The tibia compression force criterion **TCFC** on the tibia evaluates the axial compressive force that is introduced between the lower and upper end of the lower leg; it must not exceed the protection criteria level of 8.0 kN. This value was determined by Yamada et al. in [125] as the maximum tolerable threshold based on experimental results. Since not only compressive forces but especially bending moments occur when in contact with the instrument panel, Mertz et al. developed the tibia index **TI** in [72], in which threshold values from combined bending and compression loads were determined on the basis of a regression analysis and found their way into the assessment of the risk of injury as a protection criterion. The tibia index is calculated according to Eq. 3.16 and must not exceed the level of 1.3—originally the value was at TI = 1.0:

$$\mathrm{TI} = \left|\frac{F_z}{F_{z,\mathrm{crit}}}\right| + \left|\frac{M_{\mathrm{res}}}{M_{\mathrm{res,crit}}}\right| \tag{3.16}$$

$$\text{with} \quad M_{\mathrm{res}} = \sqrt{M_x^2 + M_y^2} \text{ whereby:} \tag{3.17}$$

F_z: axial compressive force [kN] in z direction,
$F_{z,\mathrm{crit}}$: critical compressive force [kN] in z direction, see Table 3.9,

M_x: bending moment [Nm] around the x axis,
M_y: bending moment [Nm] around the y axis and
$M_{res,crit}$: critical bending moment [Nm], see Table 3.9.

It is noticeable that the protection criteria levels given in Table 3.9 for the tibia index show significantly higher values than the maximum tolerable axial compressive force of 8.0 kN according to the tibia compressive force criterion (TCFC) on the shinbone. If one follows the line shown in Fig. 3.71 (for the 50% dummy) for TI = 1.3, the maximum possible resulting bending moment $M_{res} = 225$ Nm is reached at increasing compressive force (even beyond 8.0 kN), it is only reduced to lower values from a force of $F_z > 10.8$ kN. With a maximum permissible force of $F_z = 35.9$ kN, which must not be exceeded for TI ≤ 1.3, the bending moment still has a value of $M_{res} = 67.5$ Nm. However, it can easily be seen from the representation in Fig. 3.71 that the protection criteria levels TCFC and TI, which must be complied with in parallel in the European regulation UN-R 94, appear contradictory, since the protection criterion TI = 1.3 for an axial compressive force F_{max} > 8.0 kN cannot be fully utilized. It would therefore only be logical to allow as a protection criterion only the area limited by the bending moment $M_{res,crit} = 225$ Nm and the axial compressive force TCFC = 8.0 kN (see the grey highlighted area in Fig. 3.71).

3.3.4 New Protection Criteria and Current Discussions

The protection criteria established in safety legislation have been subject to ongoing dispute since their establishment, during which the level of the various levels is questioned, or at least queried. In addition, however, the criteria themselves are also discussed in terms of their meaningfulness and alternative injury-relevant stress and protection criteria are proposed. The following explanations are intended to show the current state of discussion without claiming to be complete.

The qualitative development of biomechanical research can essentially be considered complete by the 1970s. However, due to significantly improved experimental and computational simulation capabilities in terms of test, measurement, evaluation and modeling, notable quantitative advances could be observed. The basic disagreement, however,

Table 3.9 Critical forces and moments for the calculation of the tibia index depending on the dummy size

Dummy	Crit. compressive force $F_{z,crit}$ [kN]	Crit. bending moment $M_{res,crit}$ [Nm]
HYBRID III, male 95 %	44,2	307
HYBRID III, male 50 %	35,9	225
HYBRID III, female 5 %	22,9	115

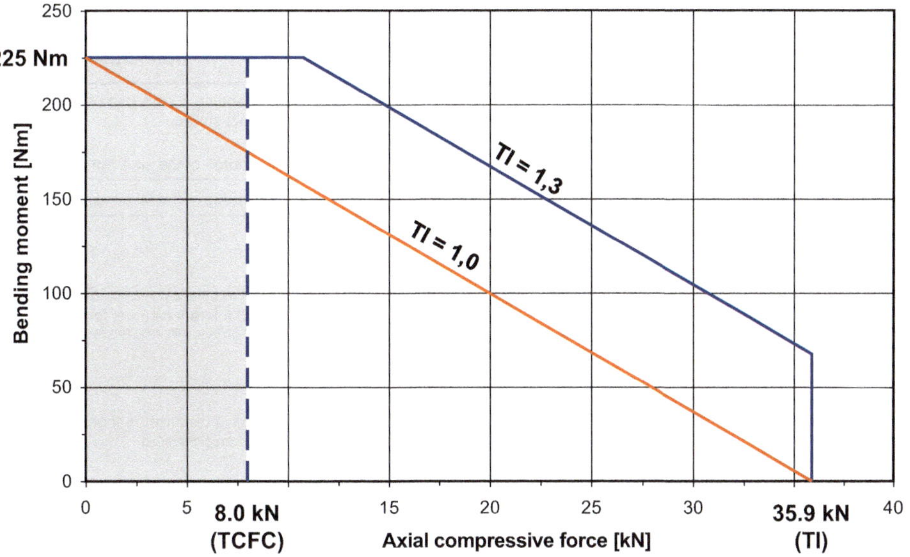

Fig. 3.71 Boundary lines of the tibia index TI, exemplified for the 50% dummy

which must be seen in the problem of the transferability of injury-inducing stresses on physical and theoretical models to the living human, still exists.

In the case of the **load of the human head** during an accident, two different manifestations are to be assumed: the translational and the rotational load. In the case of a head impact, the translational generally dominates, while the rotational load dominates in the case of an oblique or non-impact. It can also be tentatively stated that without head impact, no skull injuries occur. If contact occurs between the head and vehicle parts, brain injuries occur alone or in combination with skull injuries, depending on the intensity [60, 61].

Taking into account the types of stress and injury, the load and protection criteria can be distinguished. The criteria summarized in Table 3.10 are divided according to the types of injury: In the case of the bony skull, the contact force (usually measured is the **translational acceleration**, recently also a force equivalent [35]) leads to injuries, i.e. fractures. The apparent independence between force and acceleration is due to the evaluation of independent studies, which have been based on different experimental conditions [3]. The parameters impact surface (size, stiffness) and force introduction point on the head (forehead, nose, upper, lower jaw) are of fundamental influence; the load limits on the facial skull are significantly below the limits of the brain skull.

In the case of brain injuries, due to the biomechanical laws, both the translational and the rotational load are equally important. For the translational load, which can result in injuries both at the impact-facing (contusio cerebri at the frontal pole) and at the impact-opposite site (contre-coup effect), the accelerations and the derived, time-weighted criteria HIC (head injury criterion) and SI (severity index) are under discussion

Table 3.10 Protection and load criteria levels for the head (according to [55, 3])

Body region	Criterion	Level	Remarks
Skull cap and base	a_{max}	80 g	Depending on the size of the Impact surface
	F_{max}	930 … 9.000 N	Fractures of the skull bones
Facial skull	F_{max}	800 … 4.500 N	Fractures of the facial skull
Brain	a_{max}	100 … 300 g	
	$a(\Delta t)_{max}$	60 g	Limit acceleration according to the WSU tolerance curve; for Δt > 45 ms
	HIC	1.000	
	SI	1.000 … 1.500	
	x_{rel}	0,9 … 3,8 cm	Relative displacement between brain and cranium
	BrIC	0,7 … 1,3	Brain injury criterion for the assessment of injuries caused by rotational loads
	$\ddot{\varphi}_{max}$	1,8 … 3,0	Angular acceleration, depending on the value of the simultaneously effective translatory acceleration
	GAMBIT	0,2 … 3,0	Result from computational occupant simulation without and with head impact

as injury indicators; they are based on the WSU tolerance curve mentioned in the previous chapter. Starting from simple theoretical models—whose properties were calibrated using impedance measurements in animal experiments—various **relative displacements between the brain and skull** (= neurocranium: skullcap and base) were identified as tolerable in [67] and compared with the "classical" criteria. The criterion of "permissible displacement" (MSC = maximum strain criterion or NMSC = new mean strain criterion) was extended in all load directions by determining further mechanical characteristics with the help of animal experiments, in which the loads were applied from different directions. Thus, the displacement criterion can no longer only be applied for the sagittal, but also for the lateral impact [105].

The **rotational load** of the head usually causes extensive brain hemorrhages due to the inertia-induced relative motion between skull and brain. When a certain relative displacement is exceeded, the bridging veins tear [60]. In the basic works on load criteria in the form of angular accelerations (e.g., in [1, 32, 41, 44, 92, 111]), laboratory experiments were conducted on monkey heads, using test setups with and without receiving cuffs for fixation of the cervical spine [41, 92] or monkeys clamped in impactor setups were subjected to a targeted load [32, 111]. The transmission of the determined limit values to the human head was carried out by the approach of similarity mechanical relationships (Fig. 3.72). The relatively large range of the load criteria level given in Table 3.10 is due to the additional influence of various other parameters such as translational acceleration, angular velocity, and duration of action (see Fig. 3.73). On the other hand, different types of injuries (contusion or commotion syndrome) and tolerance thresholds were used in the studies.

To assess the risk of brain injury due to rotation of the head, a kinematically-based brain injury criterion (BrIC—brain injury criterion) was developed, Eq. 3.18). The criterion was developed using two different FE models of a human head or brain. It was found that the angular velocity correlated best with the cumulative strain damage measure (CSDM) and the first principal strain (maximum principal strain, MPS) [128]. Consequently, the **BrIC** is calculated by combining the angular velocities, related to the head around its three local axes compared to direction-dependent critical values. [108, 109].

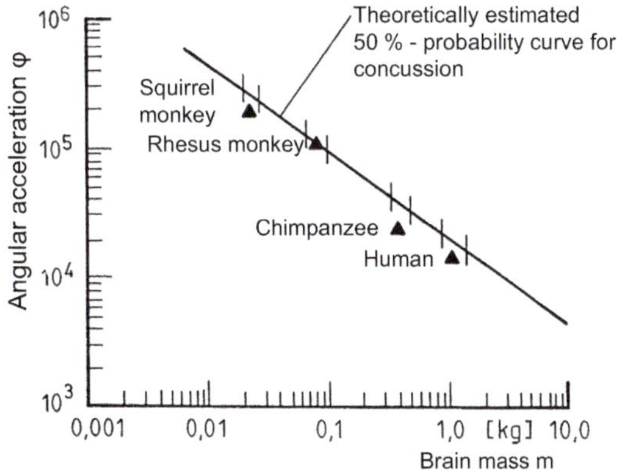

Fig. 3.72 Transmission of injury levels determined in monkeys for concussions to humans (according to [84])

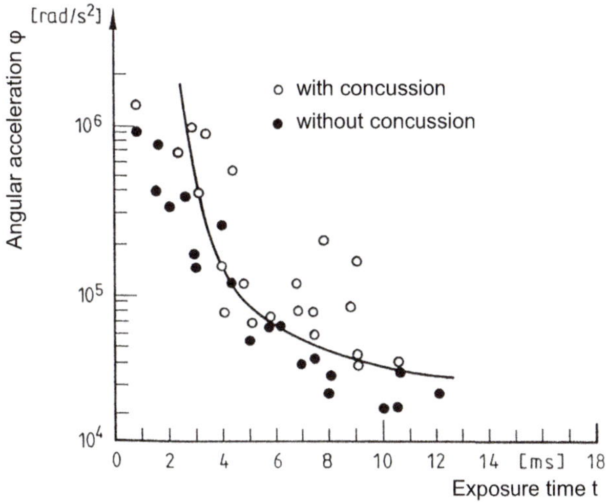

Fig. 3.73 Tolerance curve for rhesus monkeys—angular acceleration depending on the duration of action (according to [84])

$$\mathrm{BrIC} = \sqrt{\left(\frac{\omega_x}{\omega_{xC}}\right)^2 + \left(\frac{\omega_y}{\omega_{yC}}\right)^2 + \left(\frac{\omega_z}{\omega_{zC}}\right)^2} \tag{3.18}$$

$$P(AIS\ 3+) = 1 - e^{-\left(\frac{BrIC-0.523}{0.531}\right)^{1.8}}$$

$$P(AIS\ 4+) = 1 - e^{-\left(\frac{BrIC-0.523}{0.647}\right)^{1.8}}$$

$$\omega_{\mathrm{xC}} = 66.25\ \mathrm{rad/s};\ \omega_{\mathrm{yC}} = 56.45\ \mathrm{rad/s};\ \omega_{\mathrm{zC}} = 42.87\ \mathrm{rad/s}$$

Mild traumatic brain injuries often occur in vehicle accidents, i.e., injuries of severity AIS 1 or 2. With the introduction of the advanced THOR dummy, more measurement options are available compared to the Hybrid III dummy, serving as a tool for defining a brain injury criterion. The calculation of the criterion "diffuse axonal multi-axis generel evaluation" **(DAMAGE)** is defined in the technical report for the THOR dummy according to ISO/TR 19222 [47]. DAMAGE is based on the motion equations of a coupled second-order system with three degrees of freedom and predicts the maximum brain load based on the direction-dependent angular acceleration time courses of a head impact. The parameters for effective mass, stiffness, and damping were determined with simplified rotational impulses. The goal is to detect reversible brain injuries with the protection criterion DAMAGE [18, 29]. To find a physical equivalent for head injuries, the separate consideration of translational and rotational load is insufficient, as skull fractures associated with brain trauma can usually be attributed to the impact of blunt force (such as head impact) and thus the translational head acceleration appears sufficient , skull/ brain injuries, however, result from both translational and rotational relative movements. A possibility of joint evaluation of both types of stress was proposed by Newman in [77] with the acceleration model for the brain injury threshold (**GAMBIT**: generalized acceleration model for brain injury threshold):

$$\mathrm{GAMBIT} = \sqrt[k]{\left(\frac{a}{a_c}\right)^n + \left(\frac{\ddot{\phi}}{\ddot{\phi}_c}\right)^m}. \tag{3.19}$$

Here, a denotes the translational and $\ddot{\phi}$ the rotational acceleration, each related to their threshold values, indicated with c. Using statistical biomechanics, where head injuries from accident analysis and accelerations from computational occupant simulation were correlated, the unknown variables of the equation could be determined in [55] using the calculus of variations, so that they could be applied in the form

$$\mathrm{GAMBIT} = \sqrt[2.5]{\left(\frac{a}{250}\right)^{2.5} + \left(\frac{\ddot{\phi}}{25}\right)^{2.5}} \tag{3.20}$$

with the units of translational a [g] and the rotational acceleration $\ddot{\phi}$ [krad/s^2], (see Fig. 3.66). The **GAMBIT limit curve**, which separates reversible and irreversible injuries with a 50% probability ($P = 0.5$), has the value GAMBIT = 1.0; without head impact, values of GAMBIT ≤ 0.62 were determined. In Fig. 3.74, the limit curve GAMBIT = 1.0 and curves with constant GAMBIT values are shown depending on the translational and rotational head acceleration. The threshold values and exponents shown in Eq. 3.20 could essentially be confirmed by newer works (e.g., in [78]).

In various studies on the influence of load directions, it was found that the HIC value cannot be directly applied to all types of impacts. Therefore, a global head **protection criterion HIP** (head impact power) based on impact kinetics was developed [51], which can be used in particular in computational simulation, both in MBS (multi-body system) and FEM (finite element method) models. This allows criteria levels for different directions to be determined, taking into account selected failure criteria. The head impact power HIP can be calculated in the form

$$\begin{aligned}\mathrm{HIP} = {} & m \cdot a_x \int a_x dt + m \cdot a_y \int a_y dt + m \cdot a_z \int a_z dt \\ & + I_{xx} \cdot a_x \int a_x dt + I_{yy} \cdot a_y \int a_y dt + I_{zz} \cdot a_z \int a_x dt\end{aligned} \tag{3.21}$$

Here, the x-axis is defined in the posterior/anterior direction, the y-axis in the lateral direction, and the z-axis in the interior/superior direction; the unit corresponds to that of power [Nm/s = W]. A value of $\mathrm{HIP}_{max} \leq 4.3$ kW is currently being discussed as the protection criteria level.

Due to the high mobility and the complex mechanical structure of the **cervical spine (HWS) and neck**, a variety of injury patterns and corresponding diverse load criteria appear. Table 3.11 includes criteria for both HWS injuries and injuries due to loads directly applied to the neck. Criteria for surface and muscle injuries are not included, as these either have a low severity or occur in conjunction with head, HWS injuries or whiplash traumas [106].

In the presented load criteria, the first group in Table 3.11 can be traced back to the relative movement (anterior/posterior) between the head and chest. The different **load quantities** result from the respective test conditions and measurement-related factors; however, they are not independent of each other, as results from the computational occupant simulation show [55]. To measure head accelerations, shocks were applied to the thorax anterior and posterior without causing head impact. These shocks caused flexional or extensional **distortion.** In the studies by Burow [13], the flexion (or extension) angles of the cervical spine were determined photogrammetrically. Mertz et al. [73] determined in their laboratory tests the moment occurring at the head joint, which can be considered an excellent indicator for distortions. Compared to the bending moments determined in this way, considerably higher bending moments, namely 370 Nm at a 50% probability of irreversible HWS injuries, were determined with the help of a simplified HWS model at the neck/torso transition (C7/T1) [55]. The shear forces were indirectly determined in [13] under a simplified model assumption from the measured head

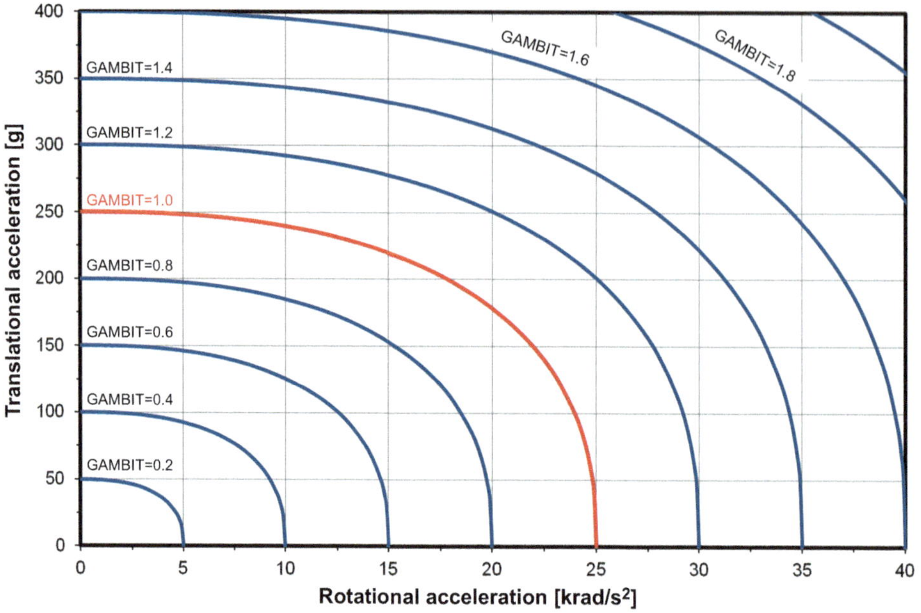

Fig. 3.74 GAMBIT curves for evaluating translational and rotational head acceleration

accelerations. They can also be derived from the bending moments occurring on the cervical spine.

Numerous investigations have been carried out to determine a load criterion in the form of tensile and compressive forces in the direction of the HWS axis, in which the force introduction was either directly into the neck or indirectly via the head (superior/inferior). The only work to determine a criteria level for longitudinal forces in dynamic impact, in which the load was measured on eleven cadavers, is described by Culver et al. in [19].

In the last group of Table 3.11, a criterion for neck injuries is discussed; the larynx is considered the least protected region of the neck. While the prepared laryngeal ring cartilage already breaks at low forces (approximately 150–250 N), the destruction of the larynx is not expected until about 500 N.

During a head-on collision, the forces of the (generally diagonally running) shoulder belt and the airbag act on the **chest** of the belted occupant. If the occupant is not belted, the chest is subjected to contact forces from the steering wheel, the airbag, the dashboard, or—in the case of rear passengers—from the front seat back. Regarding the frequency of individual injuries in the belted occupant, injury patterns can be identified that can be characterized by direct or indirect belt action: skin abrasions and contusions as well as fractures of the thoracic skeleton, especially of the sternum and ribs, are consequences of a direct action of the belt strap (Fig. 3.75). Indirect consequences include spinal injuries, both to the bony structure and to the muscles, and organ injuries due to

Table 3.11 Protection and load criteria levels for the neck (according to [55])

Body region	Criterion	Level	Remarks
Cervical spine	$a_{a\text{-}p\ max}$	30 ... 40 g	Flexion; anteriorly introduced into thorax Shock
	$a_{p\text{-}a\ max}$	15 ... 18 g	Extension; posterior joint inserted into thorax
	$\alpha_{F\ max}$	80 ... 100°	Flexion-bending angle of the cervical spine
	$\alpha_{E\ max}$	80 ... 90°	Extension- Bending spine of the cervical spine
	$M_{F\ max}$	190 Nm	Flexion, measured at the head joint
	$M_{E\ max}$	57 Nm	Extension, measured at the head joint
	M_{max}	370 Nm	Bending moment (flexion or extension), calculated at transition C7/T1
	F_{shear}	1.800 ... 2.600 N	Shear stress
	$F_{Z\ max}$	1.100 ... 2.600 N	Tensile load
	$F_{D\ max}$	3.600 ... 5.700 N	Pressure load
Larynx	F_B	178 ... 244 N	Rupture of the laryngeal ring cartilage with anteriorly applied load
	F_B	490 N	Collapse of the laryngeal structure with anteriorly applied load

the relative movement between the bony thorax and the internal organs of the chest or upper abdomen. Such injuries, such as aortic or myocardial ruptures, have decreased in frequency with the increase in seat belt use [97].

Chest injuries are highly dependent on the age (Fig. 3.76) and constitution of the accident victim; this explains the relatively wide range of load criteria levels given in Table 3.12 [52]. On the other hand, the movement behavior of the chest of the belted and airbag-protected occupant during the collision phase is more uniform and the injury mechanics are less complex than in the head, so fewer criteria seem sufficient here. The chest criteria are closely linked mechanical quantities such as resulting acceleration, load force, compression path and speed, and deformation energy of the chest as an integral of the path-dependent force. The viscosity criterion VC (viscous tolerance criterion), however, deviates from the fiction of the rigid thorax and takes into account not only the elasticity of the chest but also the interaction and displacement possibility of the chest organs [116]. This effect is also included in the thoracic trauma index TTI in the form of the arithmetic mean acceleration (see Eq. 3.15). The problem here, however, is that the displacement (and velocity) criterion VC for side collisions according to the European safety regulations and the TTI criterion based on accelerations according to the American standard contain different injury risks or suggest different safety measures with the aim of equal loads.

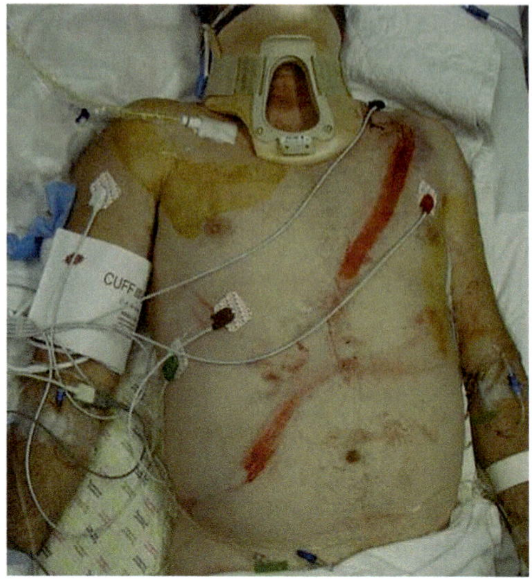

Fig. 3.75 Clearly visible seat belt bruise on the upper body of an accident victim (from [38])

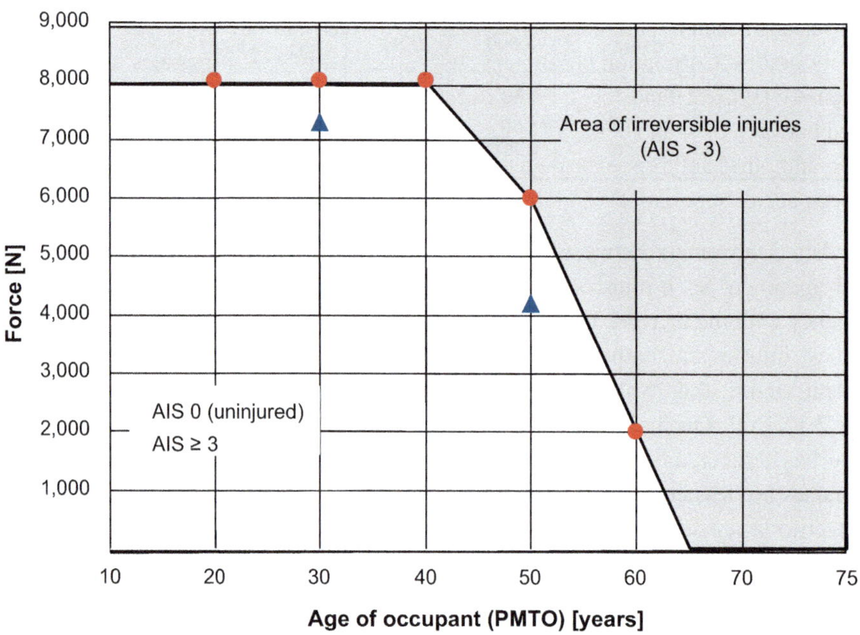

Fig. 3.76 Thoracic injuries in relation to shoulder belt forces depending on age (according to [52])

Two main causes are responsible for the **traumatization of the abdominal area and pelvis**: lap belt syndromes due to the so-called **submarining** effect and injuries to the pelvis/hip region due to high loads applied directly to the pelvis or through the femur [85]. The bony pelvic girdle can withstand very high loads without losing its stability. However, this requires a large-area force application (as is usually the case with the lap

Table 3.12 Protection and load criteria levels for thorax and spine (according to [55])

Body region	Criterion	Level	Remarks
Thorax	a_{max}	60 g	t > 3 milliseconds
	a_{max}	80 g	t < 3 milliseconds
	F_{max}	5,000 ... 9,080 N	With sagittal impactor loading
	F_{max}	4,750 N	Shoulder belt forces
	s_{max}	40 mm	Compression at 50% probability of irreversible injury from statistical biomechanics
	s_{max}	50 mm	Compression: 1986 in FMVSS 208 with 76.2 mm for airbag, otherwise 50.8 mm
	E_{pot}	350 Nm	Chest compression energy
	VC	1,00 m/s	Viscous Tolerance Criterion $[v\ .\ c]_{max}$; 39 cadaver tests, thoracic injuries AIS > 4
	TTI	120 ... 150 g	Chest traumatization index measured on dummies in vehicle side impacts
Spine	$a_{z\,max}$	± 45 g, Δt < 40 ms	Perpendicular to the backrest
	$a_{x\,max}$	± 16 g, Δt < 40 ms	In the direction of the backrest

belt) and a constant point of force application below the iliac crest (crista iliaca) during the collision phase. Rarely, in frontal collisions in conjunction with soft seat padding, unfavorable seat belt geometry, or handling errors in seat belt use, the pelvis submerges under the pelvic belt loop (hence submarining) or the pelvic belt slides up over the iliac crests. The forces that occur in this process act on the abdominal wall and lead to ruptures and lacerations of pelvic organs, thus **intra-abdominal injuries,** and—albeit only at relatively high accident severity—fractures of the lumbar spine [85]. The forces inducing abdominal injuries are significantly lower than the forces that can lead to injuries of the bony pelvis. Table 3.13 compiles criteria for abdominal injuries that reflect the current state of discussion. It should be noted that the pelvic belt force, its temporal change, and the pelvic rotation angle characterize less a tolerable load of the lower abdomen than the occurrence or avoidance of the **submarining effect** and resulting injuries. The direct measurement of the forces on the pelvis as an indicator for the sliding up of the belt loop is carried out using two competing methods; they are schematically represented in Fig. 3.77. The pair of APR (anterior-posterior rotation) force transducers mounted on the iliac crests can now be supplemented by the SWING sensors (= sensor iliac wing), which are also used on both sides and are embedded in the aluminum structure of the slotted pelvic scoops (Fig. 3.78). With the help of these strain gauge transducers, the moment acting on the pelvic bone is measured, which occurs in a characteristic way when the lap belt slides from the iliac crest (crista iliaca) into the abdominal area. This measurement option allows for the precise investigation of submarining effects [112].

On the other hand, Adomeit proposes in [2] kinematic quantities for assessing the submarining effect in terms of a yes/no criterion. This was consistently pursued with the help of statistical biomechanics in [55]: By combining accident analysis data and results from computational occupant simulation, a **prediction model for pelvis and abdominal injuries** (Fig. 3.79) could be developed, in which the **critical pelvic rotation angle** of $\alpha_{crit} = 20°$ was determined as a distinguishing feature between stable pelvic kinematics and the submarining effect. If the pelvic rotation angle is smaller than the critical angle ($\alpha_{max} \leq \alpha_{crit}$), a strain of the pelvic ring bone can be assumed, which with a 50%

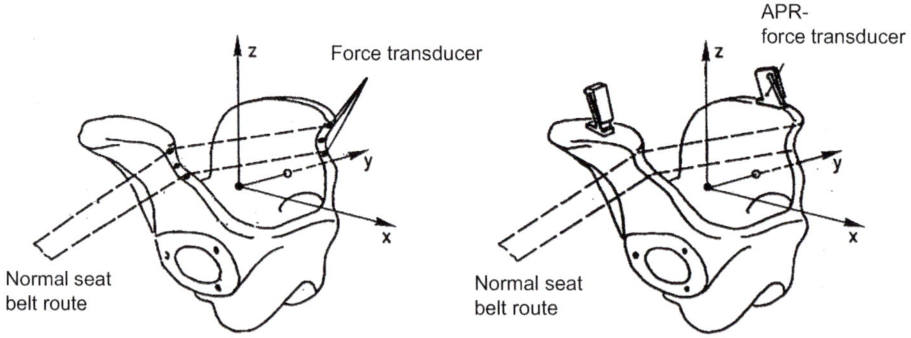

Fig. 3.77 Measurement methods for determining the submarining effect (from [55])

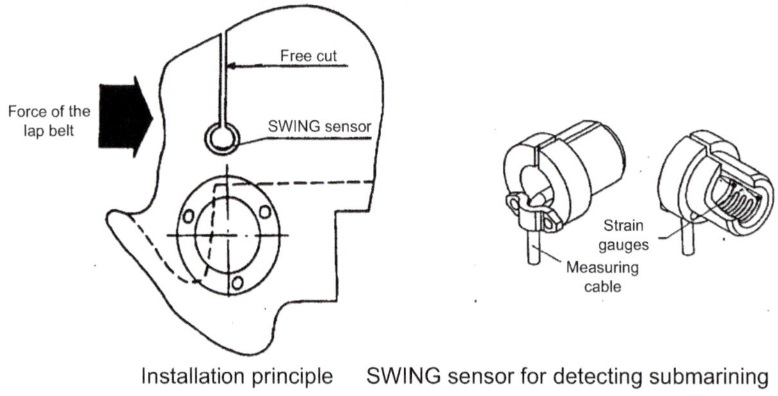

Fig. 3.78 SWING sensor and basic installation on the dummy pelvis of the 50% dummy, HYBRID III (from [112])

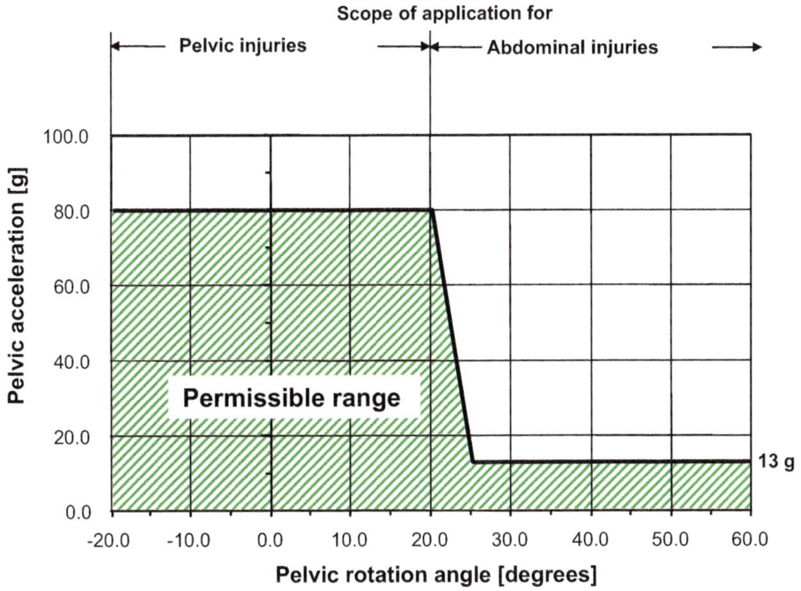

Fig. 3.79 Model for pelvis and abdominal injuries with a 50% probability of irreversible injuries

probability of irreversible (usually bony) injuries makes a resulting pelvic acceleration of a_{max} = 80 g appear tolerable. If the pelvic rotation angle is larger than the critical angle ($\alpha_{max} > \alpha_{crit}$), **submarining** must be assumed. The permissible tolerance threshold is then significantly lower, namely at a resulting pelvic acceleration of a_{max} = 13 g (see Table 3.13). The determination of the **pelvic rotation angle** is carried out by the double integration of the measurement signal of a rotation angle acceleration sensor mounted in the pelvis. The use and reliability of this relatively simple measurement method have now been demonstrated in numerous experiments [53].

As already established above, the pelvic ring bone is highly resilient, so the acceleration limit given in Table 3.13 is not only to be understood as a load criterion level for the bony pelvis itself, but also represents a level for internal injuries due to relative displacements and resulting organ ruptures. In the case of pelvic load due to a knee impact and the associated injury in the hip area, pelvic injuries can hardly be separated from injuries of the lower extremities. Thus, with a force introduction in the knee, it is unpredictable at which point in the stress chain kneecap—knee joint—femur—femoral neck—hip area an injury occurs. For fractures of the pelvis in the area of the hip joint socket, the higher load value of the criterion level given in Table 3.12 is rather assumed—to a large extent also age-dependent here.

The **upper extremities** most frequently experience a load during frontal collisions, which is introduced into the shoulder through the hands on the lower and upper arm by bracing or impacting the steering wheel or dashboard. The wide range given in

Table 3.13 Protection and load criteria levels for the abdominal area and the pelvis (according to [55])

Body region	Criterion	Level	Remarks
Abdominal area	F_{max}	500 ... 1,700 N	Load capacity of kidney and liver
	p_{max}	35 N/cm^2	Surface pressure
	$F_{BG\ max}$	800 N	Pelvic seat belt force, measured with APR transducer on the dummy pelvis
	ΔF_{SIP}	80 %	Change in the force measured on the SIP pelvis the force curve within a time of 5 ms
	α_{max}	40°	Pelvic rotation angle with vertical thoracic displacement of max. 40 mm
	α_{krit}	> 20°	Pelvic rotation angle as a sub-Criterion for abdominal injuries
	a_{max}	13 g	Result. pelvis acceleration with $\alpha > \alpha_{critical}$ (statistical biomechanics)
Lumbar spine relative to the pelvis	F_{max}	6.67 kN	Load limit between pelvis and lumbar spine
Pelvis	a_{max}	60 g	Correlation between accident injuries and simulation results
	a_{max}	80 g	Result. pelvis acceleration with $\alpha \leq \alpha_{critical}$ (statistical biomechanics)
Hip area	F_{max}	6,400 ... 12,500 N	Force application at the knee

Table 3.14 for such tolerable loads is explained by the significant influence of the parameters muscle tension, current arm bending angle, and bone fragility (due to the age of the occupant). In studies with volunteers, bracing forces of 3.2 kN were measured without any injuries [118], so the load criterion should be in the range above 3 kN. A shear load of the arm is rarely observed in a frontal impact. The criteria shown in Table 3.14 for this should therefore provide more insight into the resilience of the elbow and wrist. These criteria cannot be directly used as an equivalent for the load of the upper extremities when impacting interior parts. For this purpose, a tolerable **hand contact force** of $F_{max} = 3.8$ kN was determined using statistical biomechanics in [55] by correlating arm and hand injuries with calculated contact forces.

Compared to the studies on the resilience of the arms, there are significantly more results for the **lower extremities**. Thus, the scatter range of resilience for the femur is not only narrower, but also values can be given for the load duration as an influencing factor [4, 12]. The higher force level is supported by studies on dissected femurs and can be explained by the stronger development of the bone structure compared to the development of the upper extremities. The kneecap and joint lose their stability at a strain about

Table 3.14 Protection and load criteria levels for the extremities (cited from [55])

Body region	Criterion	Level	Remarks
Arm, total	F_{max}	150 ... 6,400 N	Supporting force
Elbow	$F_{shearer}$	180 ... 220 N	Shear stress on the elbow
Wrist	$F_{shearer}$	100 ... 180 N	Shear stress on the wrist
Hand	F_{max}	3,800 N	Hand contact force from statistical biomechanics
Leg or femur	F_{max}	8,000 ... 13,000 N	Strongly dependent on the duration of the load
Knee	F_{max}	5,000 ... 11,300 N	
Lower leg	$F_{x\,max}$	3,700 ... 4,300 N	Bending stress due to shear force
	$F_{y\,max}$	$\leq$ 16,000 N	Longitudinal load
Foot	F_{max}	$\leq$ 20,000 N	Volunteer tests
	F_{max}	10,000 N	Foot contact force from statistical biomechanics

15% lower than the femur length force [12, 113] and at a rather short load duration of less than 6 milliseconds [4]. The load criteria levels for the transverse load of the lower leg mainly result from studies on pedestrian accidents; however, they can be transferred to the vehicle occupant, for example, for the penetration of the bulkhead or the pedal assembly into the car footwell. Of greater importance, however, is likely to be the criterion for longitudinal load, as a high longitudinal force component can be observed in frontal collisions. Given the high force level, it should be noted that the values given in Table 3.14 are very short-term measurements. While the lower leg is capable of applying this force level, the imprinted forces often lead to destruction of the knee joint [110].

For the frequently occurring and protracted injuries to the foot and ankle, there are only a few study results. Forces of 20 kN were measured on volunteers without causing injuries [89]. No further details are given about the test conditions. In [55], a high correlation between leg and foot injuries on the one hand and the **foot contact force** on the other hand was demonstrated. With a 50% probability of irreversible injuries to the lower extremities, a level of $F_{max} = 10$ kN can be assumed for the foot contact force as a protection criterion (see Table 3.14).

The biomechanical limit values are constantly being further investigated and partially replaced by newer, better criteria.

References

1. Abel, J.M., Gennarelli, T.A., Segawa, H.: Incidence and severity of cerebral concussion in the rhesus monkey following sagittal plane angular Acceleration. 22. Stapp Car Crash Conference. Warrendale (1978)
2. Adomeit, D.: Neue Bewertungsgrößen für die Frontalaufprallsicherheit des gurtgesicherten Insassen. Dissertation. Technische Universität Berlin, Berlin (1980)
3. Allsop, D.L.: Skull and Facial Bone Trauma: Experimental Aspects. In: Nahum, A.M., Melvin, J.W. (Hrsg.) Accidental Injury – Biomechanics and Prevention. Springer-Verlag, New York (1993)
4. Appel, H., Färber, E., Heger, A.: Biomechanische Belastungsgrenzen. Forschungsbericht Nr. 194 im Auftrag der VOlkswagenwerk AG. Technische Universität Berlin, Berlin (1975)
5. Appel, H., Kramer, F., Glatz, W., Lutter, G., Baumann, J., Weller, M.: Quantifizierung der passiven Sicherheit für PKW-Insassen. Bericht zum Forschungsprojekt FP 8517/2 der Bundesanstalt für Straßenwesen. BASt (Hrsg.), Bergisch Gladbach (1991)
6. Association for the Advancement of Automotive Medicine: The Abbreviated Injury Scale – AIS, Rev. 2005. Association for the Advancement of Automotive Medicine, Des Plaine, IL (USA) (2005)
7. Baker, S.P., O'Neill, B., Haddon, W., Jr., Long, W.B.: The Injury Severity Score: A Method for Describing Patients with Multiple Injuries and Evaluation Emergency Care. J. Trauma **14**, 187–196 (1974)
8. Beier, G., Schuller, E.: Verletzungsmechanische Anmerkungen zum Schutzhelm. Jahrestagung 1986 der Deutschen Gesellschaft für Verkehrsmedizin e. V., Hannover (1986)
9. Boström, O., Svensson, M.Y., Aldman, B., Hansson, H.A., Häland, Y., Lövsund, P., Seeman, T., Suneson, A., Säljö, A., Örtengren, T.: A New Neck Injury Criterion Candidate – Based on Injury Findings in the Cervical Spinal Ganglia after Experimental Neck Extension Trauma. Proceedings of the International Conference on the Biomechanics of Impact IRCOBI, Dublin, Ireland (1996)
10. Boström, O., Bohmann, K., Håland, Y., Kullgren, A., Krafft, M.: New AIS1 Long-term Neck Injury Criteria Candidates based on real Frontal Crash Analysis. IRCOBI Conference. Montpellier (F) (2000)
11. Brockhaus Enzyklopädie in 24 Bänden, 19., völlig neu bearbeitete Aufl. F.A. Brockhaus GmbH, Mannheim (1987)
12. Brun-Cassan, F., Leung, Y. C., Tarriere, C., Fayon, A., Patel, A., Got, C., Hureau, J.: Determination of Knee-Femur-Pelvis Tolerance from the Simulation of Car Frontal Impacts. VII. IRCOBI Conference, Köln (1982)
13. Burow, K.: Zur Verletzungsmechanik der Halswirbelsäule. Dissertation. Technische Universität Berlin, Berlin (1974)
14. Cavanaugh, J.M.: The Biomechanics of Thoracic Trauma. In: Nahum, A.M., Melvin, J.W. (Hrsg.) Accidental injury -biomechanics and prevention. Springer-Verlag, New York (1993)
15. Committee on Medical Aspects of Automotive Safety: Rating the severity of tissue samage: II. The comprehensive scale. The Journal of the American Medical Association **220**, 717–720 (1972)
16. Committee on Medical Aspects of Automotive Safety: Rating the Severity of Tissue Damage: I. The Abbreviated Injury Scale. The Journal of the American Medical Association **215**, 277–280 (1971)
17. Cooper, P.R.: Skull Fracture and Traumatic Cerebrospinal Fluid Fistulas. In: Cooper, P.R. (Hrsg.) Head injury, S. 65–82. Williams and Wilkins, Baltimore/London (1982)
18. Craig et.al.: Injury ciiteria for the THOR 50th Male ATD. NHTSA (2020)

19. Culver, R., Bender, M., Melvin, J.W.: Mechanisms, tolerances and responses obtained under dynamic superior/inferior head impact – a pilot study. UM-HSRI-78–21. (Mai 1978)
20. Danner, M., Appel, H., Schimkat, H.: Entwicklung kompatibler Fahrzeuge. Abschlußbericht des Forschungsprojekts TV 7661 des Bundesministers für Forschung und Technologie (1980)
21. Decoo, A.C.M., Versmissen, E.G., Wismans, J.: Evaluation of European and USA Draft Regulations for Side-Impact Collisions. 11. International ESV Conference. Washington D.C. (USA) (1987)
22. Department of Transportation, National Highway Traffic Safety Admin. Docket No. 571.208, Occupant Crash Protection – Frontal Barrier Crash (2004)
23. ECE-R 94, 2003–02–13 Aufprallschutz: Frontalaufprall, Einheitliche Bedingungen für die Genehmigung der Kraftfahrzeuge hinsichtlich des Schutzes bei einem Frontalaufprall (Uniform Provisions Concerning the Approval of Vehicles with Regard to the Protection of the Occupants in the Event of a Frontal Collision) (2003)
24. ECE-R 95, 2005–02–21; Aufprallschutz: Seitenaufprall; Einheitliche Bedingungen für die Genehmigung der Kraftfahrzeuge hinsichtlich des Schutzes der Insassen bei einem Seitenaufprall (Uniform Provisions Concerning the Approval of Vehicles with Regard to the Protection of the Occupants in the Event of a Lateral Collision) (2005)
25. ETSC: Annual Road Safety Perfomance Index Report PIN. https://etsc.eu/wp-content/uploads/16-PIN-annual-report_FINAL_WEB_1506_2.pdf (2022)
26. Eppinger, R.H., Marcus, J.H., Morgan, R.M.: Development of Dummy and Injury Index for NHTSA's Thoracic Side Impact Protection Research Program, SAE technical paper series 840885, Government/Industry Meeting and Exposition. Washington D.C. (USA) (1984)
27. Fechner, H.: Ermittlung volkwirtschaftlicher Kosten von Verletzungsfolgen aus Straßenverkehrsunfällen. Diploma thesis, Technische Universität Berlin (1983)
28. Färber, E., Kramer, F.: On the Application of the HIC as Head Protection Criterion. International IRCOBI/AAAM Conference. Göteborg (S) (1985)
29. Gabler, L., Crandall, J., Panzer, M.: Development of a second-order system for rapid estimation of maximum brain strain. Annals of Biomedical Engineering, Vol. **47**, Issue **9**, Pages 1971–1981 (2019)
30. Gadd, C.W.: Use of the Weighted-Impuls Criteria for Estimating Injury Hasard. 10. Stapp Car Crash Conference (1966)
31. Gennarelli, T.A.: Mechanistic approach to head injuries: Clinical and experimental studies of the important types of injury. In: Ommaya, A.K. (Hrsg.) Head and Neck Injury Criteria: A Consensus Workshop. U.S. Department of Transportation, NHTSA, Washington, DC (1981)
32. Gennarelli, T.A., Thibault, L.E., Ommaya, A.K.: Pathophysiologic responses to rotational and tranlational accelerations of the head. 16. Stapp Car Crash Conference. Detroit (1972)
33. Glatter, J.: Nutzwert- und Risikobetrachtung eines Knieairbags aus Sicht der Unfallforschung. Diplomarbeit an der Hochschule für Technik und Wirtschaft (HTW), Dresden (2005)
34. Got, C., Patel, A., Fayon, A., Tarriere, C., Walfisch, G.: Results of experimental head impacts on cadavers: The various data obtained and their relation to some measured physical parameters. 22. Stapp Car Crash Conference (1978)
35. Grösch, L., Kassing, L., Katz, E., Stecher, J., Zeidler, F.: Die Beurteilung der Wirksamkeit von Airbag-Systemen mit Hilfe neuer Schutzkriterien. Automobil-Industrie **87**(2) (1987)
36. Gurdjian, E.S., Lissner, H.R., Latimer, F.R., Haddad, B.F., Webster, J.E.: Quantitative Determination of Acceleration and Intercranial Pressure in Experimental Head Injury. Neurology **3**, 417–423 (1953)
37. Gurdjian, E.S., Roberts, V.L., Thomas, L.M.: Tolerance Curves of Acceleration and Intercranial Pressure and Protective Index in Experimental Head Injury. Journal of Trauma, Vol. 6, Issue 5, Pages 600–604 (1966)

38. Gurt- bzw.: Prellmarken aus. http://notarztkurs.tripod.com/techrett.htm. Stand: Juli 2005
39. Heger, A., Appel, H.: Korrelationsmöglichkeiten von Unfallgeschehen und Versuch am Beispiel des Fußgängerunfalls. Jahrestagung 1982 der Deutschen Gesellschaft für Verkehrsmedizin e. V., Berlin (1982)
40. Helms, E.: Ökonomische Grundlagen zur Erfassung der Unfallkosten im Straßenverkehr. Dissertation, Bonn (1971)
41. Hirsch, A.E., Ommaya, A.K., Mahone, R.M.: Tolerance of Subhuman Primate Brain to Cerebral Concussion. Report 2876, Dep. of the Navy, Naval Ship Research and Development Center. Washington, D.C. (1968)
42. Hodgson, V.R., Brinn, J., Thomas, L.M., Greenberg, S.W.: Fracture Behavoir of the Skull Frontal Bone Against Cylindrical Surfaces. SAE Paper No. 700909, Proceeding of the 14th Stapp Car Crash Conference (1970)
43. Hofmann, J., Kramer, F., Dausend, K.: Side collisions, comparison of dummy loadings with injuries of real accidents. III. IRCOBI Conference. Lyon (F) (1978)
44. Holbourn, A.H.S.: Mechanics of Head Injuries. The Lancet. (Oct. 1943)
45. Huelke, D.F.: Anatomy of lower extremity – An overview. Biomechanics of impact trauma. Seminar in Orlando, FL (USA) (1990)
46. Huelke, D.F., O'Day, J., States, J.D.: Lower extremity injuries in automobile crashes. AAAM Conference: Crash performance standards and the biomechanics of impact: What are the relationships? Seminar in Orlando, FL (USA) (1990)
47. ISO TR 19222: Road vehicles — Injury risk curves for the THOR dummy;. Section 4.1. (2021)
48. Jennett, B.: Some medicolgal aspects of the management of acute head injury. BMJ **1**, 1383–1385 (1976)
49. Jäger, W., Lindenlaub, K.-H.: Nutzen/Kosten-Untersuchungen von Verkehrssicherheitsmaßnahmen. FAT-Schriftenreihe. Bd 5., Frankfurt/Main (1977)
50. King, A.I.: Injury to the Thoraco-Lumbar Spine and Pelvis. In: Nahum, A.M., Melvin, J.W. (Hrsg.) Accidental Injury -Biomechanics and Prevention. Springer-Verlag, New York (1993)
51. Kleiven S.: Influence of direction and duration of impacts to the human head evaluated using the Finite Element Method. International IRCOBI Conference (2005)
52. Kramer, F.: Ermittlung der maximal zulässigen Gurtkräfte an unterschiedlichen Gurtstraffern. Bericht Nr. SD2/9403–02 über Ergebnisse der Systemanalyse. TRW Repa, Alfdorf (1994)
53. Kramer, F.: Experimental determination of the pelvis rotational angle and its relation to a protection criterion for pelvic and abdominal injuries. International IRCOBI Conference. Lyon (France) (1994)
54. Kramer, F.: Passive Sicherheit/Biomechanik I/II. Lecture notes, Hochschule für Technik und Wirtschaft (HTW), Dresden (2012)
55. Kramer, F.: Schutzkriterien für den Fahrzeug-Insassen im Falle sagittaler Belastung. Dissertation, Technische Universität Berlin. Fortschritt-Berichte, VDI-Reihe 12. Bd 137 (1989)
56. Kramer, F., Melz, T.: Nutzwertbetrachtung bei der Entwicklung von Kraftfahrzeugen. Conference paper, VDI-Conference "Kraftfahrzeuge", Technische Universität Berlin, Berlin (1983)
57. Kramer, F., Deter, T.: Zur Quantifizierung der Straßenverkehrssicherheit. Verlag Information Ambs GmbH, Kippenheim. Verkehrsunfall und Fahrzeugtechnik **30**, (1992)
58. Kramer, F., Fruck, K., Bigi, D.: Sicherheitsindex – Eine Möglichkeit zur objektiven Bewertung von Insassenschutzsystemen. Conference „Rückhaltesysteme", Haus der Technik e. V. in Essen (1995)

59. Kramer, F.: Volkswirtschaftlicher Schaden aufgrund der Verletzungsfolgekosten. Nicht veröffentlichtes Arbeitspapier und Auswertungsprogramm AUSWERT (Version 3.8 – angewandt am 11.7.2012). Hochschule für Technik und Wirtschaft (HTW), Dresden (Januar 2008)
60. Kramer, F.: Über die Anwendung mechanisch-mathematischer Belastungskriterien als Äquivalent zu Kopfverletzungen. Bericht über die Tätigkeit als Gastwissenschaftler bei der Bundesanstalt für Straßenwesen (BASt). Research report Nr. 201/86. Technische Universität Berlin, Berlin (1985)
61. Kramer, F.: Analyse des Unfallgeschehens zur Ermittlung der Unfallkenngröße für Frontalkollisionen und der Verletzungsschwere sagittal belasteter PKW-Insassen. Research report Nr. 325/88. Technische Universität Berlin, Berlin (1988)
62. Lenz, K.-H., Appel, H., Cesari, D., Tarriere, C.: Joint Biomechanical Research Project (KOB). Final Report (1981)
63. Leung, Y.C., Tarriere, C., Lestrelin, D.: Submarining Injuries of 3pt. Belted Occupant in Frontal Collisions – Description, Mechanism and protection. SAE Paper No. 821158, Proceeding of the 26th Stapp Car Crash Conference (1982)
64. Levine, R.: Injury to the Extremities. In: Nahum, A.M., Melvin, J.W. (Hrsg.) Accidental injury – Biomechanics and prevention. Springer-Verlag, New York (1993)
65. Marsh, J.H.: Existing traffic accident injury causation – Data recording methods and proposal of an occupant injury classification scheme. Proceedings of the American Association for Automotive Medicine (1972)
66. McElhaney, J.H., Myers, B.S.: Biomechanical aspects of cervical trauma. In: Nahum, A.M., Melvin, J.W. (Hrsg.) Accidental injury – Biomechanics and prevention. Springer-Verlag, New York (1993)
67. McElhaney, J.H., Stalnaker, R.L., Roberts, V.L.: Biomechanical aspects of head injury. Proceeding of the Symposium on Human Impact Response. Held at the GM Research Laboratory, Warren, Michigan (USA) (1972)
68. Melvin, J.W., Lighthall, J.W., Ueno, K., Nahum, A.M., Melvin, J.W. (Hrsg.): Brain injury biomechanics in accidental injury – Biomechanics and prevention. Springer-Verlag, New York (1993)
69. Melvin, J.W., Evans, F.G.: A Strain Energy Approach to the Mechanics of Skull Fracture. SAE Paper No. 710871, Proceedings of the 15th Stapp Car Crash Conference (1971)
70. Melz, T.: Nutzen/Kosten-Untersuchung alternativer Konzeptionen zum Insassenschutz. Diploma thesis, Technische Universität Berlin, Berlin (1984)
71. Mendelsohn, R.A., Huelke, D.F.: Anatomy, Injury and Biomechanics of the Cervical Spine. AAAM Conference: Crash Performance Standards and the Biomechanics of Impact: What are the Relationships? Seminar in Orlando, FL (USA) (1990)
72. Mertz, H.J.: Anthropomorphic test devices. Accident injury in Biomechanics and Pervention, S. 66–84. Springer Verlag, New York (1993)
73. Mertz, H.J., Patrick, L.M.: Strength and Response of the Human Neck. 15. Stapp Car Crash Conference (1971)
74. Messerer, O.: Über Elasticität und Festigkeit der menschlichen Knochen. Cotta Verlag, Stuttgart (1880)
75. Morris, A., Welsh R.: Requirements fo the crash protection of older vehicle passengers. Ann Proc. Associc. Adv. Automotive Medicine Med **47**,165–180 (2003)
76. National Highway Traffic Safety Administration (NHTSA): Advanced airbag SNPRM preliminary economic assessment – injury criteria. http://www.nhtsa.dot.gov/cars/rules/rulings/AAirbagSNPRM/pea-III.n.html (2002)
77. Newman, J.A.: A generalized acceleration model for brain injury threshold (GAMBIT). International IRCOBI Conference. Zürich (1986)

78. Newman, J.A.: The biomechanics of head trauma and the development of the modern Helmet. How far have we really come? Bertil Aldman Memorial Lecture, IRCOBI Conference. Prague (CR) (2005)
79. Niederer, P., Walz, F., Zollinger, U.: Adverse effects of seat belts and causes of belt failures in severe car accidents in Switzerland during 1976. Proceedings 21st Stapp Car Crash Conference, New Orleans, LA (USA) (1977)
80. Nightingale, R.W., Chancey, V.C., Ottaviano, D., Luck, J.F., Tran, L.N., Prange, M.T., Myers, B.S.: Flexion and extension structural properties and strengths for male cervical spine segments. J. Biomech. **40**(3), 535–542 (2007)
81. Nightingale, R.W., Winkelstein, B.A., Knaub, K.E.: Myers, BS: Comperative bending strengths and structural properties of the upper and lower cervical spine. J. Biomech. **35**(6), 725–732 (2002)
82. Niklas, J.: Nutzen-Kosten-Analysen von Sicherheitsprogrammen im Bereich des Straßenverkehrs. Monograph series, Verband der Automobilindustrie e.V. (VDA). Bd 7. VDA, Frankfurt/Main (1970)
83. Nirula, R., Pintar, F.: Identification of vehicle components associated with severe thoracic injury in motor vehicle crashes: a CIREN and NASS analysis. Accid. Anal. And Prev. **40**(1), 137–141 (2008). https://doi.org/10.1016/j.aap2007.04.013
84. Ommaya, A.K., Yarnell, P., Hirsch, A.E., Harris, Z.H.: Scaling of experimental data on cerebral concussion in sub-human Primates to concussion threshold for Men. 11. Stapp Car Crash Conference. (1967)
85. Otte, D., Suren, E.G.: Gutachterliche und verkehrsmedizinische Aspekte zur Sitzposition und Gurtbenutzung in Unfallfahrzeugen. Annual coference 1984, Deutsche Gesellschaft für Verkehrsmedizin e. V. Köln (1984)
86. Pschyrembel, W.: Klinisches Wörterbuch, 255. Aufl. De Gruyter, Berlin, New York (1986)
87. Putz, R., Pabst, R.: Sobotta – Atlas der Anatomie des Menschen, 21. Aufl. Urban & Fischer, München (2002). CD-ROM-Service, Version 2
88. Ramsis Aircraft – Ergonomie Tool, Version: 3.8.16. Human Solutions. Kaiserslautern (2005)
89. Reidelbach, W., Zeidler, F.: Comparison of Injury Severity Assigned to Lower Extremity Skeletal Damages Versus Upper Body Lesions. 27. Annual Proceedings, American Association for Automotive Medicine (AAAM). San Antonio (USA) (1983)
90. Resikomed-Studie: Retrospektive Sicherheitsanalyse von Pkw Kollisionen mit Schwerverletzten. Institut für Fahrzeugsicherheit, München (1998)
91. Richter, B., Appel, H., Hoefs, R., Langwieder, K.: Entwicklung von PKW im Hinblick auf einen volkswirtschaftlich optimalen Insassenschutz. Abschlußbericht des vom Bundesminister für Forschung und Technologie (BMFT) geförderten Forschungsprojekts (TV 8036) (1984)
92. Rockoff, S.D., Ommaya, A.K.: Experimental head trauma cerebral angiographic observations in the early post-traumatic period. American Journal of Roentgenology, Vol. 91 (1964)
93. Rouhana, S.W.: Biomechanics of Abdominal Trauma. In: Nahum, A.M., Melvin, J.W. (Hrsg.) Accidental Injury-Biomechanics and Prevention. Springer-Verlag, New York (1993)
94. Rouhana, S.W., Foster, M.E.: Lateral Impact – an Analysis of the Statistics in the NCSS. SAE Paper No. 872203, Proceeding of the 29th Stapp Car Crash Conference (1985)
95. SAE J 885: Human tolerance to impact conditions as related to motor vehicle design. SAE Handbook, Vol IV (1994)
96. Safety Companion – Knowledge for Tomorrow's Automotive Engineering. Carhs Empowering Engineers. https://www.carhs.de/(2023). Zugegriffen: 18.Juni 2023
97. Schmidt, G., Kallieris, D., Barz, J., Mattern, R., Schulz, F., Schüler, F.: Belastbarkeitsgrenzen des angegurteten Fahrzeuginsassen bei der Frontalkollision. Monograph series, Forschungsvereinigung Automobiltechnik e.V.. Bd 15. FAT, Frankfurt/Main (1980)

98. Schmidtke, H.: Lehrbuch der Ergonomie. Bearbeitete und ergänzte Auflage. Carl Hanser Verlag, München Wien (1981)
99. Schmitt, K.-U., Niederer, P., Cronin, D., Muser, M., Walz, F.: Trauma-Biomechanik: Einführung in die Biomechanik von Verletzungen, 3. Aufl. Springer-Verlag, Berlin (2010)
100. Schmitt, K.-U., Niederer, P., Walz, F.: Trauma biomechanics – introduction to accident injury. Springer-Verlag, Berlin, Heidelberg, New York (2004)
101. Sellier, K.: Das Schädel-Hirn-Trauma. Rechtsmedizin **68**, 239–252 (1971). Springer-Verlag, Berlin, New York
102. Sievert, W.: Kenntnisstand der biomechanischen Grenzwerte. Innere Sicherheit im Kraftfahrzeug. Verlag TÜV Rheinland GmbH, Köln (1979)
103. Slobodnik, B.A.: SPH-4 Helmet Damage and Head Injury Correlation, S. 80–87. Fort Rucker, AL, United States Army Aeromedical Research Laboratory, USAARL Report No (1980)
104. Stalnaker, R.L., Melvin, J.W., Nusholtz, G.S., Alem, N.M., Benson, J.B.: Head Impact Response. 21. Stapp Car Crash Conference (1977)
105. Stalnaker, R.L., Lin, C.A., Guenther, D.A.: The Application of the New Mean Strain Criterion (NMSC). International IRCOBI/AAAM Conference. Göteborg (Schweden) (1985)
106. States, J.D.: Soft tissue injuries of the neck. The human neck – anatomy, injury mechanisms and biomechanics. Congress and Exposition SAE SP-438. Detroit (1979)
107. Svensson, M.Y.: Neck injuries in rear-end car collisions – sites and biomechanical causes of the injuries, test methods and preventive Measures. Dissertation, Chalmers University of Technology, Göteborg (S) (1993)
108. Takhounts, E.G.: Computational modeling and injury criteria for motor-vehicle crashes. Proceedings of the 59th Stapp Car Crash Conference. (2015)
109. Takhounts, E.G., Hasija, V., Craig, M.J.: BrIC and field brain injury risk. Proceedings of the 26th International Technical Conference for the Enhanced Safety of Vehicles (No 19–0154). https://www-esv.nhtsa.dot.gov/Proceedings/26/26ESV-000154.pdf (2019). Zugegriffen: 19. Juni 2023
110. Tarriere, C., Stcherbacheff, G., Duclos, P., Fayon, A.: The influence of the shape of the vehicle on the severeness of pedestrian injuries. International Congress of Automotive Safety (1974)
111. Unterharnscheidt, F., Higgins, L.S.: Traumatic lesions of brain and spinal cord due to non-deforming angular acceleration of the head. Texas Report on Biology and Medicine **27**(1) (1969)
112. Uriot, J., Page, M., Tarriere, C., Bendjellal, F., Loiseleux, F., Saloum, M., Fournier, P.: Measurement of submarining on HYBRID III 50 & 5 Percentile Dummies. 14. International ESV Conference. München (D) (1994)
113. Viano, D.C.: Femoral Impact Response and Fractures. V. IRCOBI Conference. Birmingham (1980)
114. Viano, D.C.: Chest: Anatomy, types and mechanisms of injury, Tolerance criteria and limits, and injury factors. AAAM Conference: Crash performance standards and the biomechanics of impact: What are the relationships? Seminar in Orlando, FL (USA) (1990)
115. Viano, D. C.: Bolster Impacts to the Knee and Tibia of Human Cadavers and an Anthropomorphic Dummy. SAE Paper No. 780896, Proceeding of the 22nd Stapp Car Crash Conference, Ann Arbor, MI (USA) (1970)
116. Viano, D.C., Lau, I.V.: Thoracic impact: A visous tolerance criterion. Proceedings at the X. Experimental Safety Vehicle Conference. Oxford (UK) (1985)
117. Viano, D.C.: Cause and control of automotive trauma. Bulletin of the New York Academy of Medicine **64**(5), (1988). New York

118. Wagner, R.: A 30 mph Front/Rear crash with human tėst persons. 23. Stapp Car Crash Conference. (1979)
119. Walfisch, G., Fayon, A., Tarriere, C., Chamouard, F., Guillon, F., Got, G., Patel, A., Hureau, J.: Human head tolerance to impact: Influence of the jerk (Rate of Onset of Linear Acceleration) on the occurrance of brain injuries. VI IRCOBI Conference. Salon de Provence (F) (1981)
120. Walz, F.: Biomechanische Aspekte der HWS-Verletzungen. Orthopäde **23**(4), 262–267 (1994)
121. Walz, F, Muser, M.: Biomechanical Aspects of Cervical Spine Injuries. SAE 950658 in SP-1077. SAE International Congress and Exhibition, Detroit, Michigan (USA) (1995)
122. Walz, F, Muser, M.: Biomechanical Assessment of Soft Tissue Cervical Spine Disorders and Expert Opinion in low Speed Collisions. Accident Analysis & Prevention, Vol. **32,** Issue 2, Pages 161-165 (2000)
123. Weissner, R.: Bewertung und Erprobung von Gurtsystemen beim Frontalstoß. Dissertation. Technische Universität Berlin, Berlin (1976)
124. Wykowski, E.: Use of Human lower Leg Model for Frontal Impact. EuroPam-User Meeting, Heidelberg (2001)
125. Yamada, H.: Mechanical Properties of Tendinous Tissue. In: Yamada, H. (Hrsg.) Strength of biological materials. Williams & Wilkins, Baltimore, MD (1970)
126. Yoganandan, N.: Pintar, FA, Genarelli, TA, Maltese MR: Patterns of abdominal injuries in frontal and side impacts. Annual Proceedings Association Advanced Automotive Medicine **44**, 17–36 (2000)
127. Yoganandan, N; Nahum, A.; Melvin, J.: Accidental injury biomechanics and prevention. ISBN 978–1–4939–1731–0. Springer Verlag, New York (2015)
128. Zeilinger, T.: Kinematik der Halswirbelsäule bei Frontal- und Heckkollisionen als Basis zur Validierung eines numerischen Menschmodells. Master thesis, TU Graz, 14–18 (2015)
129. Zobel, R.: Volkswirtschaftlich optimaler Insassenschutz – Ergebnisse der Simulationsrechnung. 11. BMFT conference „Kraftfahrzeuge und Straßenverkehr“. Münster (1984)
130. carhs.training: Protection Criteria for Side Impact (Legal Requirements). https://www.safety-wissen.com/object/B04/B04.S.9r7386936ry4lrnz6b47505gmbbuh63823036305/safetywissen. Accessed 22 June 2023

Evaluation of Vehicle Safety

4

Ralf Reuter and Florian Kramer

With the increase in road traffic due to the growing number of registered motor vehicles, the number of accidents also increased. Accordingly, legislative efforts began early to enact regulations for vehicles to improve traffic safety, without impairing mobility and traffic. To verify the safety measures, functional examinations are carried out on the one hand, on the other hand, the safety legislation for individual safety components or for the entirety of all safety devices represents a performance regulation, in which a distinction is made according to whether the legal requirements have been met or not. The evaluation, on the other hand, aims to distinguish vehicle safety not only according to yes/no criteria, but also to quantitatively capture the subcritical and supercritical area, so that, for example, in the context of investigations into passive safety, the occupant load values themselves represent the quality criterion and the level of the value correlates with the probability of the body-part-specific injury. This provides the consumer with information on the basis of which he can assess the *accident behavior*, in the context of investigations into active safety but also the *accident avoidance potential*, of a particular vehicle in comparison with another, within a vehicle class or even within the vehicle population.

R. Reuter (✉)
carhs.training GmbH, Alzenau, Germany
e-mail: ralf.reuter@carhs.de

Florian Kramer

R. Schöneburg (ed.), *Integral Vehicle Safety*,
https://doi.org/10.1007/978-3-658-48100-1_4

4.1 Quantification of Road Traffic Safety

It has already been shown elsewhere (see Fig. 1.1) that road traffic safety targets the human as a road user, the road vehicle, and the (traffic) environment. The accident-avoiding measures are assigned to the area of active safety and the accident consequence-reducing measures to passive safety. The following will now show to what extent safety has improved in the last six decades with an almost elevenfold increase in registered motor vehicles. The usual media indication of, for example, the percentage reduction in fatalities in accident occurrence compared to the previous year or the number of accident victims per registered vehicle or per kilometer "performance" seems insufficient, for two reasons [13]:

- The generalized statements do not allow any statement as to which of the two areas the improvement in safety is due to, and
- the damage that the national economy incurs annually due to injury-related costs can only be derived to a lesser extent from the fatally injured (2011: approx. 17.9% of the subsequent costs).

The number of registered motor vehicles increased about sevenfold in the period from 1960 to 2021, while the number of accidents only increased about twofold (Fig. 4.1). It should be noted that due to the reunification of Germany in 1990, the increase in registration numbers alone between 1990 and 2007—from 2008 without temporarily decommissioned vehicles—was approx. 20.7 million motor vehicles, which means an increase of 56.5%. The accident figures, on the other hand, initially increased from 1990 with 2.011 million to 1992 with 2.385 million accidents by 18.6% and then fortunately fell again to 2.361 million accidents by 2011; an increase of only 17.5% compared to 1990.

To separately display active and passive safety, initially the registered vehicles are put in relation to the vehicles involved in accidents. The safety index thus relates the number of vehicles not involved in an accident to one vehicle involved in an accident. It is presented as a quality criterion for active, or better primary safety [13], for the years 1953 to 2011 in Fig. 4.2. The so defined **index for active safety AkSIx** increases with a decreasing number of accident vehicles. Conversely, active safety increases with the stock (and to some extent the mileage) of registered vehicles with the same number of accident vehicles. By considering both the motor vehicle stock and the mileage, the measure of active safety is characterized by the probability of collision and the duration of exposure.

The **index for passive safety PaSIx** is characterized by the quotient of the number of all accidents and the injury-related costs for lightly, severely and fatally injured (for scale reasons in million €) [13]. In Fig. 4.3, the distribution of injury-related costs for the period 1953 to 2011 is shown and shows slightly increasing shares for lightly injured from about 22 to 30% and with a decreasing trend for severely injured from 54 to 52% and for fatally injured to finally 17.9%. By forming the quotient, the measure for passive safety increases with an increase in accidents with constant or decreasing subsequent

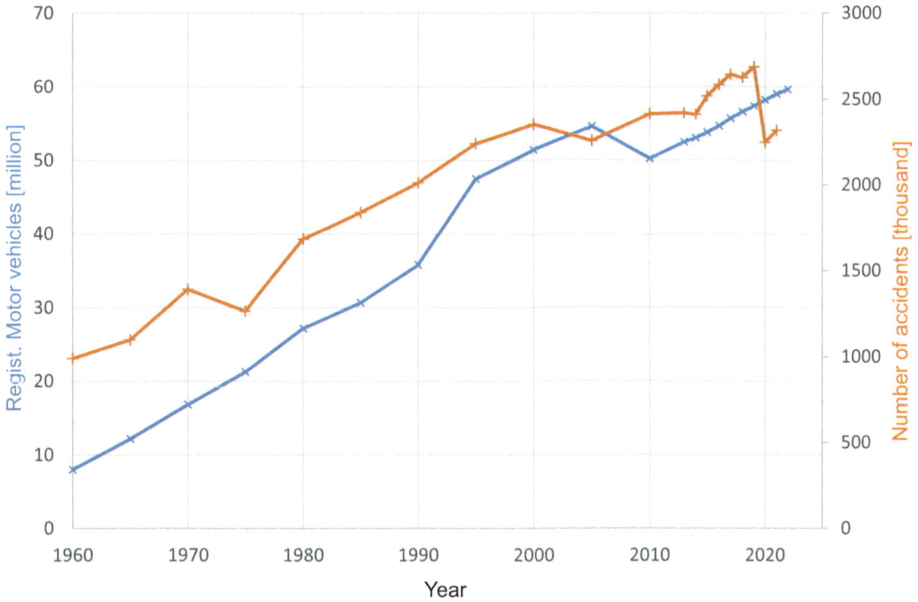

Fig. 4.1 Motor vehicle inventory [12] and number of accidents [24] in the period 1960 to 2021 (from 1991 including the new federal states and from 2008 without temporarily decommissioned vehicles)

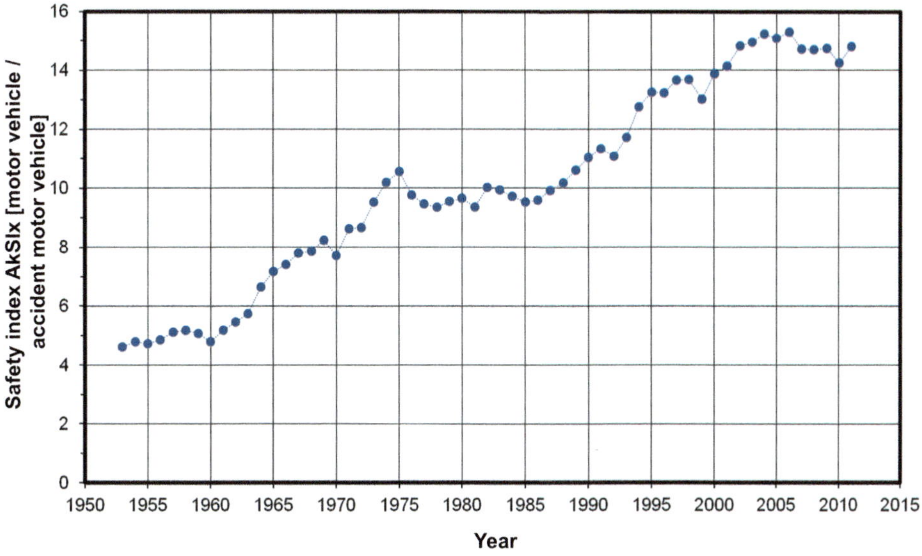

Fig. 4.2 Historical development of active safety in the period 1953 to 2011 (from 1991 including the new federal states)

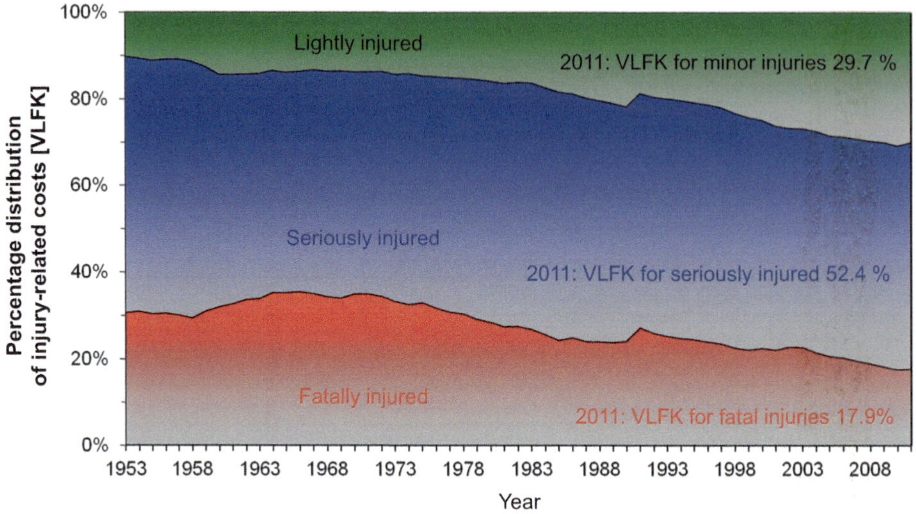

Fig. 4.3 Distribution of injury-related costs for lightly, severely and fatally injured in the period 1953 to 2011 (from 1991 including the new federal states)

costs; passive safety also increases with a constant number of accidents and decreasing injury-related costs.

In the historical representation of the measure for passive safety, which is shown in Fig. 4.4, a progressive increase in passive safety can be observed since the mid-1960s. The significant increase in the safety index in 1984 and 1985 can be attributed to the reinforcement of fines introduced in August 1984 for not wearing a seat belt.

A certain stagnation of passive safety has been observed since 1987, with a significant drop in the years 1991 to 1995, which must certainly be seen in connection with the reunification of Germany. There are numerous indications that suggest that the drop in passive safety is due to the transfer of a large number of used, defective cars to the new federal states in connection with a temporary excessive use. However, a recovery of safety can be observed in the subsequent period, which obviously results from the positive influence of the increasing proportion of cars equipped with front and side airbags in the total vehicle population.

The PaSIx index for passive safety shown here has already been used in a modified form: The safety figure shown in Fig. 2.8 was formed by the relation of the total number of injured and uninjured accident participants and the determined injury-related costs. Due to the different basis, the absolute values are not comparable, but by relating the number of accident vehicles or the number of accident participants to the specific injury-related costs, safety figures can be determined that aptly express the extent of safety.

Using the shown indices for active and passive safety, road traffic safety can be quantified; the metrics can be calculated from the accident numbers of the official federal statistics [23]. Due to their different dimensionality, they must not be converted into each

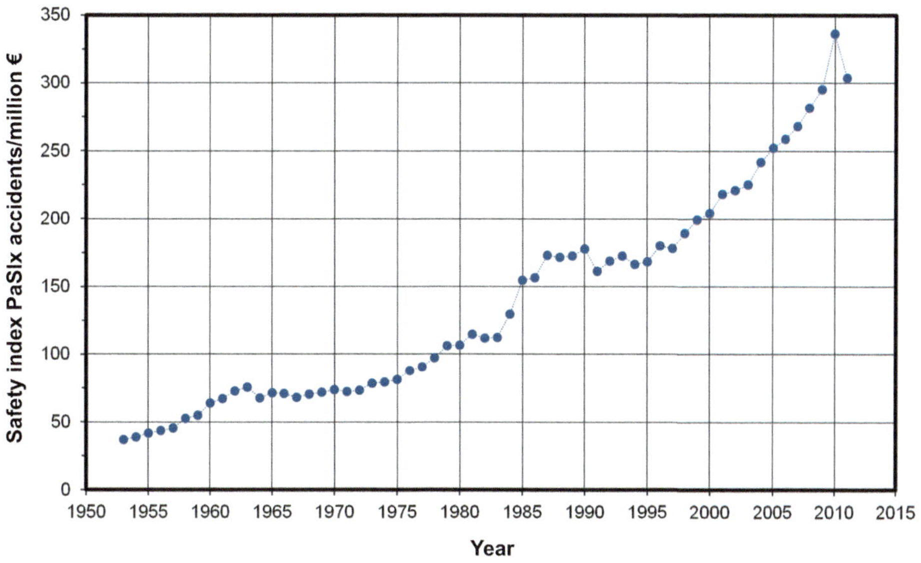

Fig. 4.4 Historical development of passive safety in the period 1953 to 2011 (from 1991 including the new federal states)

other, but they do allow a direct statement about the achieved improvement in the two safety areas and the potential for improvement contained therein. The increase in safety metrics shows quite different increases within a period of almost 60 years: for active safety more than three times (4.6 to 14.8), for passive safety more than eight times the value (36.6 to 304). These increases can only be directly attributed to individual legislative measures in a few cases (e.g., reinforcement of fines for not wearing seat belts in 1984), rather they are an expression of the overall effect of interlocking individual measures, initiated by the legislature, but also by car manufacturers, who met the safety needs of their customers. They have also been positively influenced by the publication of safety assessments from consumer protection tests (e.g., Euro NCAP).

4.2 Legislation

The legislative bodies worldwide reacted quite differently to the increase in accidents associated with the growing traffic volume, and the resulting injuries and fatalities, and formulated corresponding laws and standards with the aim of sustainably improving road traffic safety. With regard to vehicles, four approaches can essentially be distinguished:

- Development and construction of as safe as possible vehicles through design and performance regulations,

- Maintenance of the traffic-safe condition of vehicles by prescribing inspection intervals,
- Removal of trade barriers through uniform requirements regarding procurement, testing, and approval within the jurisdiction of member states, and
- Harmonization through the abandonment of national regulations in favor of international regulations and directives.

The legally defined regulations and criteria of safety legislation refer to requirements for the vehicle in everyday operation and in accidents. In addition, precisely specified regulations for the entire vehicle and certain vehicle components have been defined, which have grown to four to five times the number for individual states or groups of states over the past few decades [16].

4.2.1 Regulations in Germany and in Europe

The Road Traffic Approval Regulation (StVZO), which applies in Germany and was issued by the Federal Minister for Transport, Building and Urban Development, dates back to its history to 1937. The StVZO has been frequently revised and has increasingly lost its importance with regard to special national regulations for the type approval of vehicles; it has been successively transferred to other regulations. Since 2009, the EC Vehicle Approval Regulation (EC-FGV) also regulates the approval of vehicles in Germany with regard to vehicle safety based on UN regulations or EEC directives.

In the individual states of the European Union (EU), which was founded in 1957 as the European Economic Community (EEC), the construction and equipment regulations have developed differently, as have the requirements for the operation of motor vehicles in terms of dimensions, weights, and regular technical inspections. The aim of the EEC contract, which came into force in 1991, is therefore to see the removal of trade barriers and the implementation of a common transport policy, because the vehicle is on the one hand a commodity that is to be sold, but on the other hand, people and goods are to be transported safely with it [11]. To fulfill these tasks, regulations or directives are usually used for motor vehicles and trailers. Regulations have direct effect in the individual member states and replace existing national regulations with the community law contained in the regulation. Directives, on the other hand, do not have direct effect and therefore have to be ratified and implemented by the member states through a law or a national regulation.

The **EU (or EEC) directives** are based on optional application, i.e., instead of national regulations, those of the EU directives can be applied (optional harmonization). Currently, however, the EU is moving towards requiring member states to deny national operating permits if the vehicle does not comply with EU directives (total harmonization). In addition to the directives, decisions as well as recommendations and opinions are considered, although they are not binding. Due to the different regulations in all EU member states, the need for harmonization is obvious. The decisive factor for this is the creation of an **EU (or EEC) type approval,** according to which a vehicle is approved in each member state and this approval must and will be recognized in all other member states, without a new inspection having to be carried out in these. An automobile manufacturer therefore no longer has to obtain national approval in each of the currently 27 EU member states as in the past; in Germany, this is granted by the General Operating Permit issued by the German Federal Motor Transport Authority. Instead, a single EU type approval is sufficient, which is then recognized in the other states. For the issuance of a complete EU type approval of a vehicle of class M_1 according to Regulation (EU) 2018/858 as a prerequisite for the registration of a motor vehicle, 68 individual directives must be fulfilled, which cover the areas of active and passive safety as well as environmental protection.

In addition to national regulations and EU directives, the regulations of the ECE (Economic Commission for Europe), a sub-organization of the UN, are the third pillar of safety regulations for motor vehicles. These **UN regulations** are appendices to the 1958 agreement on the adoption of uniform conditions for the approval of equipment and parts of motor vehicles and on mutual recognition of approval. In addition to the now 27 EU member states, other non-European countries, such as Japan and Australia, are also parties to the agreement (Table 4.1).

Contracting parties that apply a UN regulation must approve vehicles and vehicle parts within the framework of the national operating permit or type approval procedure. The adoption of the regulations into national law is at the discretion of the governments, either optional (application parallel to existing national regulations at the choice of the automobile manufacturer) or mandatory (UN regulations replace national regulations). So far, 167 UN regulations exist (as of June 30, 2023), of which a large part (about two thirds) is applied in Germany [11].

Based on the so-called CARS-21 process, in which high-level delegations from governments, vehicle industry, and the EU commission have agreed on objectives and ways to strengthen the competitiveness of the European vehicle industry, the regulatory framework to be applied for type approval was also simplified [6]. By supplementing the Framework Directive 2007/46/EC on the type-approval procedure for vehicles [19] with Regulation (EC) No 661/2009, the so-called General Safety Regulation, numerous directives were replaced by UN regulations [27].

With the EU Regulation 2018/858 [25], the European type-approval procedure was revised and market surveillance was particularly strengthened. In addition, in 2020, the Regulation (EU) 2019/2144 [26], the so-called General Safety Regulation II came into force. This regulation introduces numerous new safety requirements and systems in stages, starting from 2022 to 2029. These include in the area of passive safety the introduction of an additional frontal impact (UN-R 137, 50 km/h impact on the rigid wall with 100% overlap), a pole-side impact (UN-R 135) and the expansion of the head impact area in the impactor tests for pedestrian protection. Furthermore, the General

Table 4.1 Signatory states of the 1958 agreement, UN regulations (as of March 2023)

ECE Symbol	*Country*	*ECE Symbol*	*Country*	*ECE Symbol*	*Country*
E 1	Germany	E 21	Portugal	E 43	Japan
E 2	France	E 22	Russian Federation	E 45	Australia
E 3	Italy	E 23	Greece	E 46	Ukraine
E 4	Netherlands	E 24	Ireland	E 47	South Africa
E 5	Sweden	E 25	Croatia	E 48	New Zealand
E 6	Belgium	E 26	Slovenia	E 49	Cyprus
E 7	Hungary	E 27	Slovakia	E 50	Malta
E 8	Czech Republic	E 28	Belarus	E 51	Republic of Korea
E 9	Spain	E 29	Estonia	E 52	Malaysia
E 10	Serbia	E 30	Republic of Moldova	E 53	Thailand
E 11	United Kingdom	E 31	Bosnia and Herzegovina	E 54	Albania
E 12	Austria	E 32	Latvia	E 55	Armenia
E 13	Luxembourg	E 34	Bulgaria	E 56	Montenegro
E 14	Switzerland	E 35	Kazakhstan	E 57	San Marino
E 16	Norway	E 36	Lithuania	E 58	Tunisia
E 17	Finland	E 37	Turkey	E 60	Georgia
E 18	Denmark	E 39	Azerbaijan	E 62	Egypt
E 19	Romania	E 40	North Macedonia	E 63	Nigeria
E 20	Poland	E 42	European Union	E 64	Pakistan
				E 65	Uganda

Safety Regulation II focuses on accident prevention systems, such as emergency brake assistants, lane keeping assistants or speed assistants.

4.2.2 Regulations in the USA and Other States

While in Europe a high degree of responsibility was imposed on the vehicle driver, so that the focus of legislative measures initially aimed at preventing accidents, those responsible in the USA believed that the road users, especially the vehicle drivers, can only be educated to a limited extent and therefore the occupants must be protected from the consequences of accidents by a suitable design of the vehicle. Only secondly, measures for accident prevention were implemented in the USA [20].

The increase in accidents involving serious and fatal injuries reached such proportions in the early 1960s that the American government, prompted by initiatives from powerful consumer organisations, was forced to enact drastic framework legislation. The Motor Vehicle Safety Act, decided in 1966, established detailed testing procedures for vehicle regulations regarding the behaviour of motor vehicles in everyday use and in accidents. These test regulations, which are described in detail under point 571 of the Automotive Safety Act and contain requirements and evaluation criteria, are formulated in the **Federal Motor Vehicle Safety Standards** (FMVSS) and refer for example to the vehicle interior (FMVSS 201 to 204 and 207), to the restraint systems and their components (FMVSS 208 to 210, 213 and 226), to the side structure (FMVSS 214) and to the integrity of the fuel system (FMVSS 301). The fulfillment of these regulations is not seen as a prerequisite for the admission of the motor vehicle to public road traffic, rather in the USA the **principle of self-certification** is applied by the automobile manufacturers, i.e. the manufacturers check the compliance with the standards independently and the vehicles are put into traffic, unlike in Europe, without an approval procedure. However, to ensure that the regulations are complied with, vehicles are specifically selected from the market at certain intervals by subordinate government agencies and checked on the basis of the safety standards.

In other states such as Australia (ADR), Japan, Korea (KMVSS), Canada (CMVSS) and Saudi Arabia (SAS), whose safety laws vary more or less strongly from the US safety standards, the fulfillment of the US or European regulations is often accepted as equivalent for road traffic approval [16].

The regulations in the so-called BRIC states have meanwhile gained great importance. The markets in Brazil, Russia, India and China have become some of the most important markets in the world for German vehicle manufacturers due to their large growth in terms of newly registered vehicles. Due to the sharply increasing vehicle stocks and the associated increase in traffic accidents, regulations on vehicle safety are also continuously being enacted here.

4.2.3 Summary of Regulations in Different Countries

Currently, numerous regulations are being revised and new ones are being prepared worldwide. The regulatory and directive framework applicable in Europe for type approval has been simplified and reduced by the framework directive.

As part of the so-called 1998 Agreement, regulations and test procedures are to be harmonized worldwide as far as possible through Global Technical Regulations (GTR); the signatory states are listed in Table 4.2. Currently, 23 GTRs are registered in the Global Register at the United Nations, four of which are assigned to the area of passive

Table 4.2 Signatory states of the 1998 agreement, Global Technical Regulations GTR (as of June 2022)

Signatory state	*Accession from*	*Signatory state*	*Accession from*
Australia	07.06.08	Republic of Moldova	17.03.07
Azerbaijan	14.06.02	Netherlands [1]	05.03.02
Belarus	03.03.15	New Zealand [2]	26.01.02
Canada	25.08.00	Nigeria	17.12.18
P.R. China	09.12.00	Norway	29.11.04
Cyprus	11.06.05	Romania	24.06.02
European Union	25.08.00	Russian Federation	25.08.00
Finland	07.08.01	San Marino	26.01.16
France	25.08.00	Slovakia	06.01.02
Germany	25.08.00	Slovenia	08.05.14
Hungary	21.08.01	South Africa	17.06.01
India	22.04.06	Spain	22.06.02
Italy	30.01.01	Sweden	01.02.03
Japan	25.08.00	Tajikistan	26.02.12
Kazakhstan	27.08.11	Turkey	01.09.01
Republic of Korea	01.01.01	Tunisia	01.01.08
Lithuania	25.07.06	United Kingdom of Great Britain and Northern Ireland	25.08.00
Luxembourg	15.11.05	United States of America	25.08.00
Malaysia	04.04.06	Uzbekistan	03.07.18

safety: GTR No. 1 on doors and locks, GTR No. 7 on headrests, GTR No. 9 on pedestrian protection and GTR No. 14 on pole-side impact.

The so-called "Blue Book" of the UNECE describes how the World Forum for Harmonization of Vehicle Regulations (WP 29) is structured and how international regulations come about according to the 1958 (UN regulations) and 1998 agreements (GTRs) [3].

The following Tables 4.3 and 4.4, without claiming to be exhaustive, provide an overview of the currently applicable legal regulations for the internal and external safety of motor vehicles in various countries and groups of states and describe the essential requirements and characteristics of these test regulations. Comparable legal requirements also exist in China (GB Standards) and Korea (KMVSS). These are not listed here due to the limited accessibility of the respective legal texts.

Due to the not insignificant differences between the UN regulations and the FMVSS regarding the impact tests for frontal and side collisions, these will be discussed in more detail. The differences relate to both the test setup and the test criteria to be met.

Table 4.3 Legal regulations and standards for internal safety in various countries

Country	Regulation	Requirements and characteristics for internal safety
Safety measures on and in the complete vehicle		
Europe	UN-R 94	Frontal impact on deformable 0° barrier with 40% overlap on the driver's side at 56 km/h; measurement on two HYBRID III-50M dummies driver and passenger side, criteria for head, neck, thorax, femur and knee; steering wheel displacement and rotation of the steering column. For high-voltage vehicles, verification of protection against electric shock after the crash test
	UN-R 95	Side impact with movable, deformable barrier (height 300 mm), at 50 km/h; measurement on an ES-2 dummy driver side, criteria for head, chest (VC criterion and indentation) and abdomen (internal, external and symphysis force). For high-voltage vehicles, verification of protection against electric shock after the crash test
	UN-R 135	Pole impact against rigid pole at 32 km/h at an angle of 75°. Measurements with a dummy (WorldSID-50M) on the driver's seat on the impact side. Criteria for head, shoulder, chest, abdomen, lumbar spine and pelvis
	UN-R 137	Frontal impact against rigid barrier 0° at 50 km/h. Measurements on two HYBRID III-50M dummies driver and HIII-5F passenger side, criteria for head, neck, chest and femur
USA/CDN	FMVSS 208	5 frontal impact configurations: 32–40 km/h against rigid barrier 0°, HYBRID III-5F driver and passenger side, unbelted 56 km/h against rigid barrier 0°, HYBRID III-5F driver and passenger side, belted 40 km/h against deformable barrier 0°, 40% overlap, HYBRID III-5F driver and passenger side, belted 32–40 km/h against rigid barrier −30°—+ 30°, HYBRID III-50M driver and passenger side, unbelted 56 km/h against rigid barrier 0°, HYBRID III-50M driver and passenger side, belted Criteria for head, neck, chest and femur
		Tests in so-called "Out of Position" situations Driver side with HYBRIS III-5F, passenger side with CRABI, HYBRID III 3 y/o and HYBRID III 6 y/o

(continued)

Table 4.3 (continued)

Country	Regulation	Requirements and characteristics for internal safety
USA/CDN	FMVSS 214	Side impact with movable, deformable barrier, moved sideways at 27°, at 54 km/h against driver's door of the stationary vehicle; measurements with two dummies (ES-2RE front, SID IIs rear) on driver's seat and in the rear on the impact side, criteria for head (HIC), chest (indentation or acceleration), abdomen and pelvis (force). Pole impact against rigid pole at up to 32 km/h at an angle of 75°. Measurements with a dummy (ES-2RE or SID IIs) on driver's seat on the impact side
Roof crush protection		
USA/CDN Saudi Arabia	FMVSS 216a SAS 266/267	Resistance to deformation of the front roof corners against static pressure of a test stamp. Convertible does not need to meet this requirement for USA/CDN
Side structure and side door strength		
Europe	UN-R 95	See under "Safety measures on and in the entire vehicle"
USA/CDN	FMVSS 214	See under "Safety measures on and in the entire vehicle"
USA/CDN Australia	FMVSS 214 ADR 29	Minimum requirements regarding minimum, initial, intermediate, and peak deformation resistance in static testing with cylindrical test stamp; alternative requirement in testing with built-in seats
Saudi Arabia	SAS 265/267	Like FMVSS 214a, but only peak deformation resistance (31.150 N) required
Door locks and hinges		
Europe	UN-R 11	Defined lock strength in longitudinal and transverse direction at half and full closing position; inertia analysis
USA/CDN	FMVSS 206	Hinge strength in longitudinal and transverse direction
Australia	ADR 2	
Japan	38	

(continued)

Table 4.3 (continued)

Country	Regulation	Requirements and characteristics for internal safety
Europe	UN-R 32 & 33	Maintenance of closure during frontal (48… 53 km/h) and rear impact (35… 38 km/h) against barrier or with rolling block; after impact, a defined number of doors must be able to be opened
Japan	25	General requirements for doors
International	GTR No. 1	Door locking and its retention
Steering column displacement		
Europe	UN-R 12	Max. permissible displacement (127 mm) at frontal impact (48… 53 km/h)
USA/CDN	FMVSS 204	
Australia	ADR 10 b	
Saudi Arabia	SAS 263/81 and 267/81	
Windshield Attachment		
USA/CDN	FMVSS 212	Minimum percentage (depending on the restraint system) of the windshield remaining in the frame after frontal impact (48… 53 km/h)
		Windshield penetration
USA/CDN	FMVSS 219	Integrity of a defined zone in front of the windshield after frontal impact (48… 53 km/h)
Dashboard Design and Impact		
Europe	UN-R 21	Occupant protection from sharp edges and protruding corners through minimum radii and energy-absorbing materials
Europe	UN-R 33	Defined minimum interior dimensions after frontal impact (48… 53 km/h)
USA/CDN Japan	FMVSS 201 34	Energy absorption capacity of the dashboard (head-form impact 24 km/h) and the armrests in the door panel; container lids in the interior must remain closed during longitudinal and lateral vehicle acceleration
Australia	ADR 21	Like FMVSS 201, but no requirements for armrests

(continued)

Table 4.3 (continued)

Country	Regulation	Requirements and characteristics for internal safety
USA/CDN	FMVSS 208	When using passive restraint systems: protection criteria with max. permissible load values on the head, chest and thighs after frontal impact (48… 53 km/h) with 0 to 30° barrier angle, side impact (32 km/h) and rollover (48… 53 km/h)
Japan	20	General requirements for elements in the vehicle interior
Steering Wheel Design and Impact		
Europe	UN -R 12	Max. permissible force on torso block after defined impact (24 km/h)
Australia	ADR 10 b	
Japan	27	
USA/CDN	FMVSS 203	Like UN-R 12; Exception: Vehicles with passive restraint systems (other than belt systems); they must comply with FMVSS 208
Sun Visor Design and Impact		
Europe	UN -R 21	Occupant protection from sharp edges and protruding corners through minimum radii and energy-absorbing materials
USA/CDN	FMVSS 201	Energy-absorbing structures; minimum radii for fastenings located in the area of the head shape
Australia	ADR 11	Like UN-R 21
Japan	40	Energy-absorbing structures, no sharp-edged internal parts (palm test); minimum radii for blind fastenings located in the area of the head shape
Seats and backrests		
Europe	UN-R 16	Design, strength, corrosion resistance, release and adjustment forces
	UN-R 17	Defined requirements for seat anchoring and headrests
Europe	UN-R 44	Dynamic and static testing of child restraint systems
	UN-R 21	Occupant protection from sharp edges and protruding corners through minimum radii and energy-absorbing materials

(continued)

Table 4.3 (continued)

Country	Regulation	Requirements and characteristics for internal safety
USA/CDN Japan	FMVSS 201 36	Energy absorption capacity of the front seat backs through static and dynamic tests
USA/CDN	FMVSS 207	Like EEC 74/408, but lower forces
Australia	ADR 3 a	
Japan	35	
	21	Minimum dimensions of the driver's seat and its positioning
	22	Minimum dimensions of other seats (except driver's seat)
Italy		For taxi vehicles, the rear seat width must be 450 mm per seat
Headrests		
Europe	UN-R 25	Strength through static and dynamic tests; shape, external design and geometric dimensions: minimum width 170 mm, minimum height above seat cover point 700 mm
USA/CDN Japan	FMVSS 202 a 32	Limitation of the insertion path, defined minimum impact area, geometric requirements for head to headrest distance. Dynamic test with Hybrid III possible
Australia	ADR 22 a	
Japan	32	Hardness test by palm test
International	GTR No. 7	Harmonized requirements for headrests. Dynamic test at the discretion of the contracting states with Hybrid III (like FMVSS 202a)
Seat belts, restraint systems		
Europe	UN-R 14	Location and strength of the belt anchorage points
	UN-R 44	Child restraint systems, static and dynamic tests
	UN-R 16	Design, strength by means of static testing, resistance to environmental influences, dimensions, release and adjustment forces, belt marking, abrasion resistance, installation and handling

(continued)

Table 4.3 (continued)

Country	Regulation	Requirements and characteristics for internal safety
	UN-R 129	Child restraint systems, static and dynamic tests, frontal, rear and side impact
	UN-145	ISOFIX anchorages and i-Size child seat positions
USA/CDN	FMVSS 208	Strength by means of dynamic testing; passive restraint systems; "Fasten seat belt" indicator light, warning buzzer; two-point belt for convertible allowed
USA/CDN Japan	FMVSS 209 31	Design, strength by means of static testing, resistance to environmental influences, dimensions, release and adjustment forces, belt marking, abrasion resistance
USA/CDN Japan	FMVSS 210 37	Location and strength of the belt anchorage points, in Canada (CDN) also location and strength of the belt anchorage points for child restraint systems (210.1). Static and dynamic tests on child restraint systems (in Canada 213)
USA/CDN	FMVSS 226	Protection of occupants from being ejected through the side window openings. Verification by means of impact test (guided impact from the vehicle interior) with head test body (ejection mitigation headform) at 16 and 20 km/h. Criterion: Head must not protrude more than 100 mm from the interior
Australia	ADR 4 d	Seat belt strength by means of static and dynamic testing, seat belt geometry, three-point seat belt for all outer seats
	ADR 5 b	Location and static strength of the belt anchorage points
	ADR 34 a	Location, static strength and accessibility of the anchorage points for child restraint systems
Japan	22.3	Three-point seat belts for front seats; strength requirements
Saudi Arabia	SAS 297/82	Warning device for unfastened driver's seat belt buckle
Flammability of interior materials		
Germany USA/CDN	StVZO 30 FMVSS 302	General requirements for occupant protection. Defined flame retardancy requirements for certain materials of the passenger compartment

(continued)

Table 4.4 Legal regulations and standards for external safety in various countries

Country	Regulation	Requirements and features for external safety
Vehicle exterior shape		
Europe	UN-R 26	Minimum radii for touchable edges between ground line and a height of 2 m; max. resistance forces for attached components
Europe	UN-R 127	Requirements for the protection potential before and in the event of a collision of pedestrians and other unprotected road users with a car. Component tests with head, leg and hip test bodies on the vehicle front in the area of the hood and bumper. From 2024 also in the area of the windshield
International	GTR No. 9	Harmonized requirements for pedestrian protection based on component tests
Wheel covers		
Germany	StVZO 36 a	Defined coverage areas (tread) to the front and rear
Wheel nuts, -hubcaps and -caps		
Germany	StVZO 32	General requirements for wheel attachment
Europe	UN-R 26	As EEC 74/483 for "vehicle exterior shape"
USA/CDN	FMVSS 211	No protruding parts
Fuel supply system		
Europe	UN-R 32, 33 and 34	General construction regulation; hydraulic pressure test of the fuel tank; requirements for plastic tanks; dynamic leak test; defined max. leakage after frontal impact (48 …53 km/h) and rear impact (35… 38 km/h) with pendulum or barrier
USA/CDN	FMVSS 301	At 0 ± 30° front (48 km/h), rear (80 km/h, 70% offset) and side impact (53 km/h) testing of the max. leakage/ min using a static vehicle rotation device

(continued)

Table 4.4 (continued)

Country	Regulation	Requirements and features for external safety
Canada	CMVSS 301.1	Requirements for vehicles that use LPG (liquified petroleum gas) as an energy source for propulsion: defined installation regulations, no gas leakage, no pressure drop allowed
	CMVSS 301.2	Requirements for vehicles that use CNG (compressed natural gas) as an energy source for propulsion: like CMVSS 301.1, additionally the fuel tank must withstand a pressure of approx. 20 bar
Japan	15	Filling nozzle at least 300 mm from the exhaust and 200 mm from electrical switches
	33	At 0° frontal impact (50 ± 2 km/h) testing of the max. leakage/min using a static vehicle rotation device
	131	(without static vehicle rotation device); plastic tanks only permissible under certain conditions
Saudi Arabia	SAS 264/267	Similar to FMVSS 301, but without static vehicle rotation device
Hood lock		
USA/CDN	FMVSS 113	Rear-hinged front hoods must have a second latch in the lock or a second lock
Bumper		
Europe	UN-R 42	Maintenance of traffic and operational safety after central longitudinal (4 km/h) and corner impact (2.5 km/h) using a test pendulum
USA	Part 581	Maintenance of traffic and operational safety after central longitudinal (4 km/h) and corner impact (2.5 km/h) using a test pendulum, as well as barrier impact (4 km/h) in the longitudinal direction; damage to the bumper and the fastening elements is permissible

(continued)

Table 4.4 (continued)

Country	Regulation	Requirements and features for external safety
CDN	CMVSS 215	Maintenance of traffic and operational safety after central longitudinal (8 km/h) and corner impact (4.8 km/h) using a test pendulum, as well as barrier impact (8 km/h) in the longitudinal direction; damage to the bumper, the fastening elements, and the car body is permissible
Saudi Arabia	SAS 273/82	Like UN-R 42, but no repairing or replacing of the bumper is allowed during the test

In the **crash tests for head-on collision** according to the UN-R 94 regulation, the impact test is carried out at a speed of $v = 56$ km/h against a deformable barrier with a driver-side overlap (offset deformable barrier—ODB) of 40%. Since 2022, vehicles in Europe have to comply with UN-R 137, which includes an impact test at a speed of $v = 50$ km/h against a rigid barrier with full overlap (full-width test). While the ODB test seats both front seats with HIII-50M dummies, the Full-width Test has a HIII-50M dummy as the driver and a HIII-5F dummy as the passenger. All dummies are belted.

In contrast, the US regulation FMVSS 208 also includes 2 crash tests with unbelted occupants. In total, this regulation includes 5 different crash test configurations with different occupants and different impact configurations. Figure 4.5 shows a comparative overview of the US and European regulations for frontal crash. The respective protection criteria and limit values are shown in Table 3.6-1.

With the introduction of the full-width test according to UN-R 137, Europe now also has a test that places higher demands on the restraint systems. This complements the ODB test, which primarily tests structural safety. In the USA, these two requirements are supplemented by the unbelted load cases, which place particularly high demands on the airbag, and the impact against the inclined barrier.

The **crash tests for lateral collision**, comparatively illustrated in Fig. 4.6, are characterized by the UN-R 95 regulation on the one hand by the right-angled impact of a sled with a deformable barrier on the stationary car to be tested at a speed of 50 km/h. According to FMVSS 214, however, an obliquely running trolley (crab walk barrier), also equipped with a deformable barrier, moves against the stationary car set at 27° oblique at a speed of 33.5 mph (~54 km/h). The oblique run is intended to simulate the self-motion of the struck car, which is stationary in the test, as observed in an accident.

Both in the USA and in Europe, in addition to the barrier side impact, a pole side impact (FMVSS 214/UN-R 135) is performed. In this case, the vehicle collides laterally at an angle of 75° with a stationary pole at a speed of 32 km/h. The European and

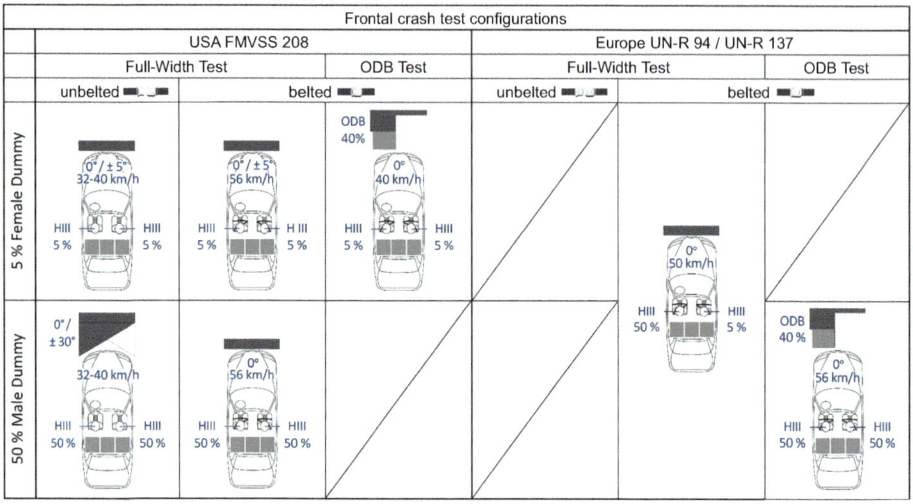

Fig. 4.5 Comparison of frontal crash test configurations in the USA and Europe

Fig. 4.6 Comparison of test configurations for side collisions for Europe (left) and for USA (right)

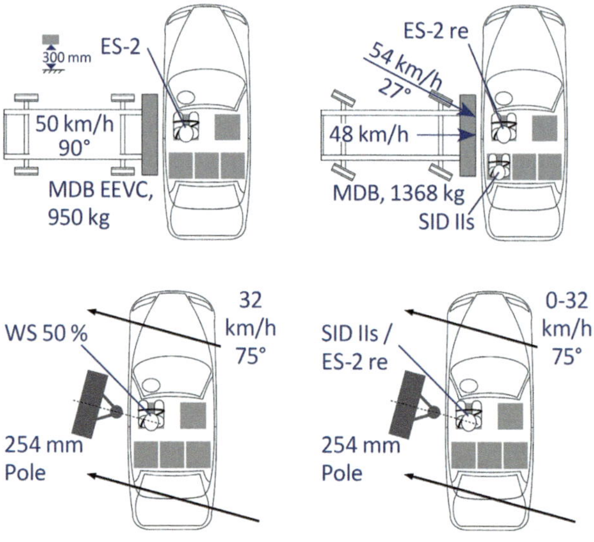

American regulations differ in terms of the dummies used. In Europe, the WorldSID-50M is used, while in the USA, the test must be passed with both the SID IIs and the ES-2 dummy. It should also be noted that in the USA, the pole side impact must be fulfilled in the entire speed range from 0 to 32 km/h, while the UN regulation only prescribes a test at 32 km/h. Therefore, in the USA, it may be necessary to perform additional tests in the medium speed range, just below the ignition threshold of the side airbags, to ensure and demonstrate compliance with the criteria. The protection criteria used in side collisions are described in detail in Table 3.6-2.

Test methods for pedestrian protection were developed in Europe as early as the 1980s. While one of the first proposals was based on conducting tests with dummies, it was quickly recognized that such a test method could, for example, only simulate an accident configuration of a pedestrian of a certain body size (anthropometry). However, the sequence of a pedestrian accident with a car is extremely complex and the potential impact areas are diverse, even considering different body sizes and impact configurations. Based on these findings, component tests were developed by the working groups WG (Working Group) 7, 10 and 17 of the EEVC (European Enhanced Vehicle Safety Committee) to examine the vehicle front in terms of its protection potential. Various test proposals were developed from these working groups. The last test proposal was submitted by the EEVC WG 17 in 1998 and updated again in 2002 [2]. It is largely reflected in the so-called Phase 2 of the directive 2003/102/EC, which will be described in more detail later. The newly developed impactors used are designed to simulate the impact load on the head, the pelvis and the lower extremities in accidents (Fig. 4.7). The test procedure for the bonnet area describes an angled head impact with a child or adult head impactor in free flight, the speed is 40 km/h. The head protection criterion HPC = 1,000 is used as the evaluation criterion [10], which is determined according to the calculation rule for the head protection criterion HIC (head injury criterion, see Eq. 3.4). The bonnet leading edge is checked at three points where the impactor hits depending on the geometry of the front of the car at a speed of 20 to 40 km/h. The direction of impact and

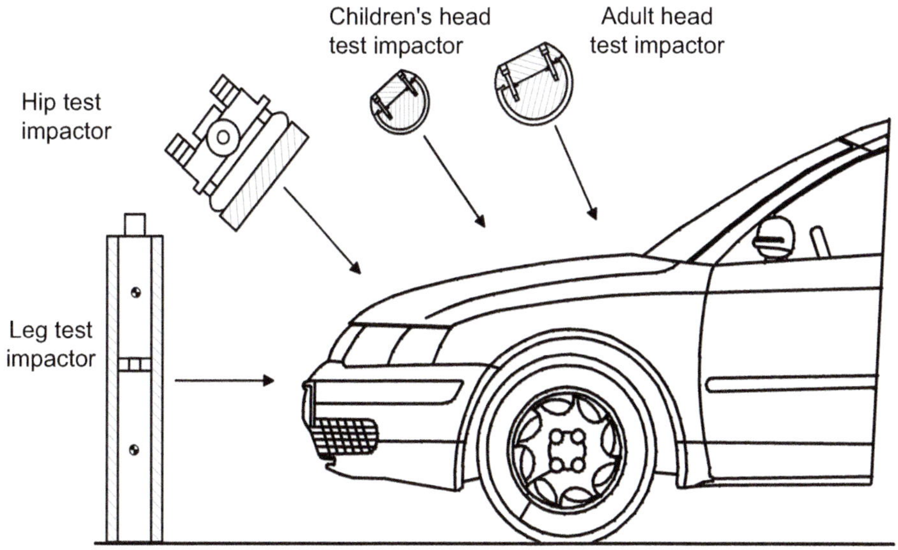

Fig. 4.7 Impactors for pedestrian protection according to EEVC WG 17

the mass of the impactor also depend on the front geometry and can be read from a diagram. Evaluation criteria are the shear force measured on the impactor ($F = 4$ kN) and the bending moment ($M = 220$ Nm). In the test procedure for bumpers, the measurement is also carried out at three points (middle and outside). The impactor used corresponds to the leg of an adult (50th percentile dummy), on which the loads on the knee and lower leg can be determined. It is catapulted in free flight at a speed of 40 km/h against the test points of the front bumper [16].

To improve the **safety of external, vulnerable road users,** e.g., cyclists in the event of a collision with a motor vehicle, the test requirements according to directive 2003/102/EC should be further tightened. This directive came into force at the end of December 2003 after lengthy and controversial discussions and provided for two phases. In Phase 1, all new vehicle models type-approved after October 1, 2005 had to meet the protection criteria shown in the left-hand illustration in Fig. 4.8 in component tests. Testing is carried out with a free-flying head impactor (mass $m = 3.5$ kg, diameter $d = 165$ mm, so-called child/adult headform) at an impact speed of 35 km/h on a defined area of the hood. Also, in tests with a leg impactor (the so-called TRL impactor) at 40 km/h in the area of the front bumper, the protection criteria for leg acceleration as well as for bending angles and shearing in the area of the knee joint must not be exceeded. For vehicles with higher-mounted bumpers (e.g., SUVs), the bumpers can also

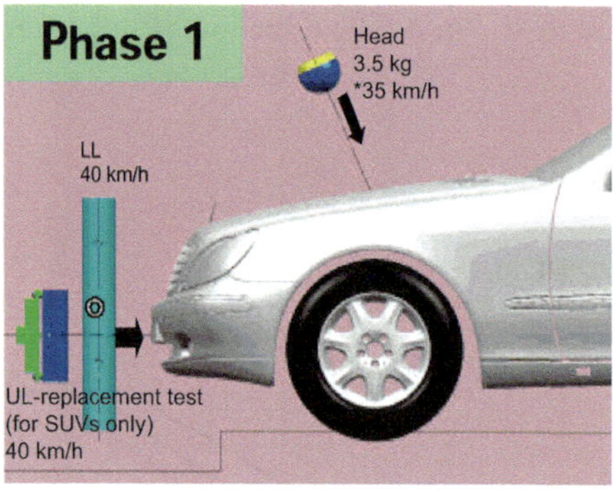

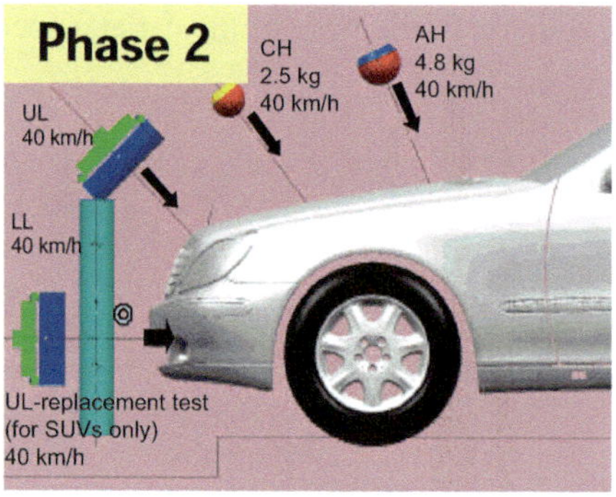

Phase 1:

Head: 3.5 kg, 35 km/h.
HIC 1,000 (2/3 of the area)
HIC 2,000 (1/3 of the area) 1,000 to 2,100 mm on canopy, 50°
(WSS: measure only, 35 km/h, 4.8 kg, 35°).

Leg (LL): 40 km/h; 200 g, 21°, 6 mm.

Hip (UL): measure on BLE only.

Hip (UL): SUV, bumper 7.5 kN.

Phase 2:

Head - child (CH), adult (AH): 2.5/4.8 kg, 40 km/h.
HIC 1,000 (entire surface) 1,000 to 1,500 mm or 1,500 to 2100 on hood, 50° or 65°.

Leg (LL): 40 km/h; 150 g, 21°, 6 mm.

Hip (UL): on BLE 5 kN, 300 Nm.

Hip (UL): SUV, bumper 5 kN, 300 Nm or corresponding measures!

Fig. 4.8 Phases 1 and 2 of the directive 2003/102/EC

be tested at 40 km/h with a guided upper legform impactor. An upper leg impact test in the area of the bonnet leading edge and tests in the windshield area were only carried out for monitoring purposes in Phase 1.

In Phase 2, following the directive 2003/102/EC, a significant increase in requirements, i.e., a reduction in the protection criteria level, was planned from 2010 or 2015 onwards (right-hand illustration in Fig. 4.8): These test requirements are based on the recommendations of the EEVC working group WG 17 described above. Article 5 of the EC directive stipulated a feasibility study in which the technical feasibility of the requirements proposed for Phase 2 was examined until mid-2004. These and other studies came to the essential conclusion that the requirements could only be technically implemented incompletely or only with very high, economically unjustifiable effort. Therefore, the EU commission developed a new proposal for a regulation that replaced Phase 2. This new regulation (EC) No 78/2009 was adopted by the European Parliament in January 2009 and came into force with publication in the Official Journal of the EU on February 4, 2009. The associated technical requirements are set out in Regulation (EC) No 631/2009 and were published in the Official Journal of the EU on July 25, 2009. In addition to requirements for passive safety, which are based on those of Phase 1 of the directive 2003/102/EC, components from the field of active safety, such as brake assistants, were also prescribed to improve the protection of pedestrians and other unprotected road users before and during a collision. Furthermore, the regulation provided that effective, collision-avoiding systems could be used, and the requirements for passive safety could be reduced or even eliminated in return.

With the General Safety Regulation II (GSR II) [26], regulations 78/2009 and 631/2009 were repealed. Instead of these regulations, UN-R 127 is now applied within the framework of European type approval. This prescribes the use of the new leg impactor Flex-PLI for the execution of leg impact tests. This is characterized by higher biofidelity, especially in the area of the knee. In addition, the GSR II envisages an extension of the head impact area up to a wrap around distance of 2,500 mm to improve the protection of cyclists who, due to their elevated seating position on the bicycle, can hit further back on the vehicle. Figure 4.9 shows the impact tests according to UN-R 127.

So far, only the legal requirements in and on the motor vehicle have been presented. However, it should not go unmentioned that there is also a legal regulation for **drivers of motorized two-wheelers** that is intended to ensure the protection of the head. The regulation UN-R 22 describes a procedure for testing the strength and shock absorption of protective helmets, in which, among other things, drop tests with helmeted dummy heads are carried out using a helmet test stand. The resulting head acceleration is measured as a criterion, which must not exceed the maximum value of a_{max} = 275 g or HIC_{max} = 2,880. Compared to the head protection criterion, for example according to FMVSS 208 ($a_{3\,ms}$ = 80 g), the criteria are considerably higher here, as a large-scale force distribution is ensured by the helmet and thus a lower injury risk can be assumed [7]. In addition, the brain injury criterion (BRIC) is limited to 0.78 and the maximum resulting rotational acceleration to 10,400 rad/s^2 when impacting on inclined surfaces.

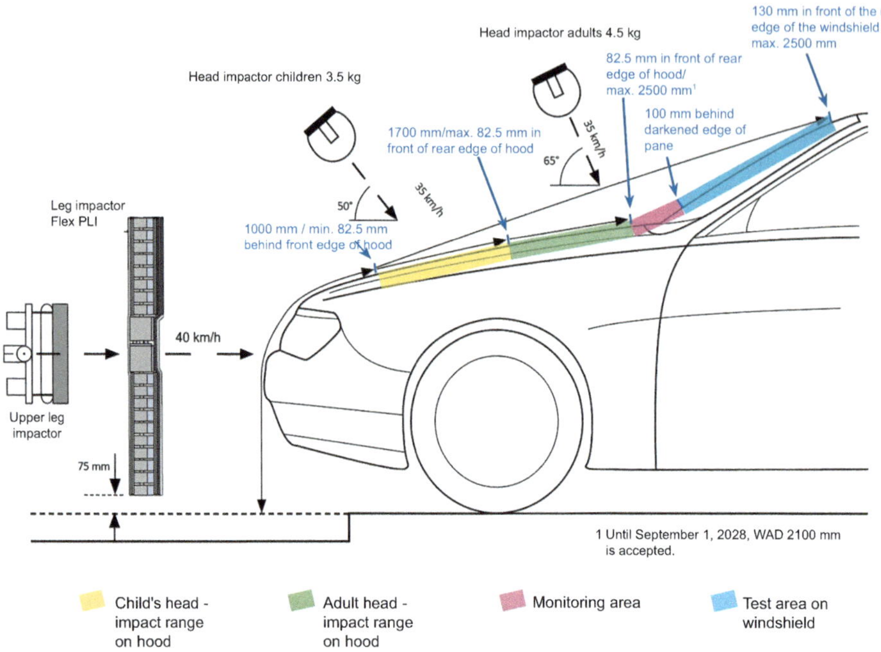

Fig. 4.9 Pedestrian protection impact tests according to UN-R 127

4.2.4 Future Regulations for Passive Safety

The improvement of passive safety of motor vehicles has taken a preferred place in vehicle development in recent decades, but it also continues to be the focus of interest of the professional audience and broad buyer groups. In addition to the further development of frontal impact, decisive impulses for increasing safety in side collisions, especially through the standard introduction of the side airbag for head and thorax area, have been added in recent years. This can also be read from the further development of legal regulations and standards. If one carefully observes the discussion in the relevant safety working groups and working groups, it can be assumed that further test regulations and evaluation criteria will be established to keep up with the state of technical development. The following will briefly discuss the expected development.

With the introduction of legal regulations for frontal and side collisions, minimum standards were established. As a result of technological progress and changing vehicle populations, e.g., increase in vehicle mass and wider distribution of newer vehicle concepts, such as off-road passenger cars (SUV: sports utility vehicle) and large limousines, but also the increasing stiffness of vehicles, the accident occurrence changes. Just as the widespread use of the front airbag reduced the number of severe and often fatal head and thorax injuries, injuries to the lower extremities have gained relative importance in

Head-on collisions	Description	Application
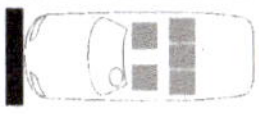	Rigid wall, full overlap	Law (e.g. U.S., UN) NCAP (e.g. Euro, U.S.)
	Deformable barrier, offset (40 %)	Law (e.g. U.S., UN) NCAP (e.g. IIHS, Japan)
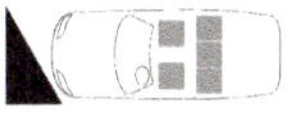	Rigid wall, 30° angle	Law (U.S.)
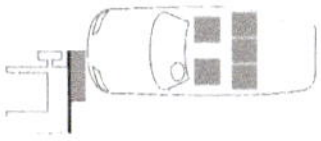	Mobile deformable barrier, offset (50 %)	NCAP (e.g. Euro, China)
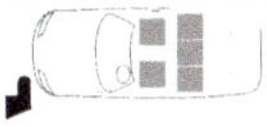	Rigid barrier, offset (25 %)	NCAP (IIHS, C-IASI)
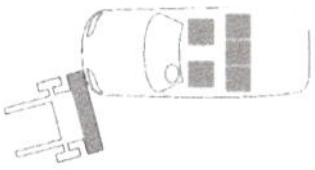	Mobile deformable barrier, 15° angle, offset (35 %)	Internal requirements Planned: NCAP (U.S.)
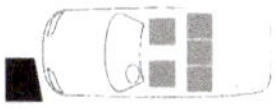	Rigid barrier, 10° angle, Offset (40 %)	Insurance crash (RCAR)
	Generic bumper, full overlap and offset (15 %)	Insurance crash (RCAR)
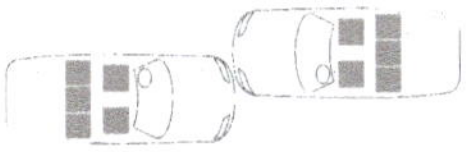	PASSENGER CAR	Internal requirements according to [29]
	TRUCK	Internal requirements according to [29]

Fig. 4.10 Test configurations for frontal collisions

frontal collisions today. Although these injuries are usually not life-threatening, they are associated with high economic follow-up costs and often a restriction of mobility for those affected.

A major challenge is the demographic change in many Western nations. With the increasingly aging population, the proportion of older vehicle occupants also increases. Older people are, among other things, less resilient from a biomechanical point of view.

Analyses of accident databases such as GIDAS have shown that older and smaller, predominantly female individuals are exposed to a significantly increased injury risk.

Thus, the legal regulations for frontal and side impact cannot be considered as completed to this day. A significant field are questions of compatibility, i.e., essentially, how different vehicles can optimally use each other's energy-absorbing structures in the event of a collision, for example, to reduce the consequences of the accident for the collision opponents. This applies to vehicles of the same categories, but especially to vehicles with significant differences in mass and size, up to questions of under-ride protection in commercial vehicles in the event of a frontal or rear impact by passenger cars. Both in Europe and in the USA, intensive work is being done on test procedures to assess the compatibility of vehicles. In consumer protection, Euro NCAP introduced a frontal crash test in 2020 with the so-called MPDB (mobile progressive deformable barrier) test, which evaluates both the self-protection and the partner protection of a vehicle. Based on this test procedure, a revision of the legal regulations is conceivable in the future.

Further revision is needed for the currently used deformable barriers in frontal and side impact, which still replicate the geometry and stiffness behavior of the vehicle front of earlier vehicle generations. The masses of the currently used side impact barriers also need to be adjusted to the conditions of the current average vehicle population. Here too, consumer protection tests have already done preliminary work and developed new barriers.

The EEVC WG 20 has worked on a static or geometric test procedure that aims to set minimum requirements for the design of seats and headrests to protect vehicle occupants from injuries in the area of the cervical spine. EEVC WG 20 also proposed a dynamic test procedure using the BioRID dummy and introduced it into the discussions during Phase 1 of GTR No. 7 on headrests. During the development of this phase, the NHTSA (National Highway Traffic Safety Administration) in its Final Rule FMVSS 202a also defined requirements for headrests that can be met by static tests (e.g., requirements for the minimum height of the headrest and the distance from the head to the headrest, the so-called back-set) and by a simulated rear impact (sled test with a Hybrid-III dummy). However, the dynamic procedure according to FMVSS 202a in particular still has deficiencies, which are due, among other things, to the use of the Hybrid-III dummy used, which has completely inadequate **biofidelity** (i.e., correspondence with the living human) for a rear impact and was never developed for such a load case. Nevertheless, this test was written into GTR No. 7 for the first phase after long, controversial discussions.

The EU commission has set itself the ambitious goal of significantly reducing the number of different directives and regulations and—as far as possible—simplifying the entire set of rules. This has already been largely achieved through the introduction of the General Safety Regulation. Parallel existing regulations are to be replaced by uniform regulations and also harmonized worldwide. An important role here is the development and introduction of Global Technical Regulations (GTR), which can be adopted by the signatory states on the basis of the "1998 Agreement". The first regulation (GTR No. 1) set requirements for door locks and door hinges of motor vehicles. Currently, 23 GTRs are registered with the United Nations (as of December 2022). GTR No. 7 (headrests) and GTR No. 9 (pedestrian protection) are currently being revised in a Phase 2. In this phase of GTR No. 7, the introduction of the BioRID dummy as a biofidelic tool for dynamic testing of seats is being discussed in detail. In Phase 2 of GTR No. 9, the main focus is on the implementation of a new leg test body, the Flex-PLI (flexible pedestrian legform impactor).

The great future challenge, however, is the increasing networking of active and passive safety systems, especially in connection with driver assistance systems (see Sects. 5.3 and 5.4). Systems already in the field, such as the PRE-SAFE® system from Mercedes-Benz, could then be combined with collision avoidance systems. However, this requires solving numerous technical and legal questions.

4.3 Safety Ratings

In addition to the legal requirements, there are more or less pronounced company-specific requirements in the automotive industry, which are often more numerous in terms of test arrangements and more extensive in terms of test criteria than the safety-legal test procedures. But also the test or rating programs carried out by state institutions (e.g., NHTSA with the US-NCAP) and consumer protection organizations (e.g., Euro NCAP) and insurance companies (e.g., IIHS) have found their way into the specifications of vehicle manufacturers. The test configurations shown in Figs. 4.10, 4.11, 4.12, 4.13 for frontal, side and rear collisions as well as rollovers serve as examples. The various test arrangements and the investigations based on them serve to ensure company-specific requirements. However, the findings from this are only indirectly accessible to the professional audience, for example through participation in safety committees, and can therefore only be used to a limited extent for a broadly based, universally accepted safety assessment.

In the experimentally-based evaluation, the results from safety tests are used as evaluation criteria and considered in relation to the protection criteria level [1]. This allows for an evaluation already in the development stage using prototypes or through computational simulation.

4.3.1 ADAC Test Procedures for Passive Safety of Cars

The General German Automobile Club (ADAC) began systematically testing cars for passive safety as early as the 1980s, thus taking a pioneering role in publishing such test results for consumer information and thus providing assistance in making purchasing

Side collisions	Description	Application
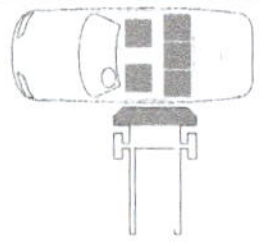	Mobile deformable barrier	Law (e.g. UN) NCAP (e.g. Euro, China)
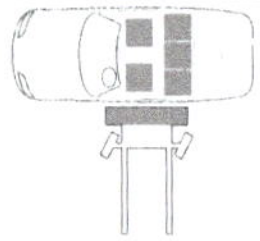	Mobile deformable crabbed barrier	Law (U.S.) NCAP (U.S.)
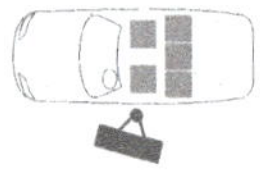	Rigid pole, 75° angle	Law (e.g. U.S., UN.) NCAP (e.g. Euro, U.S.)
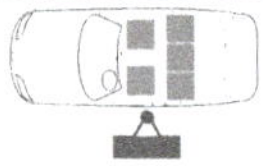	Rigid pole, 90° angle	NCAP (e.g. Latin)
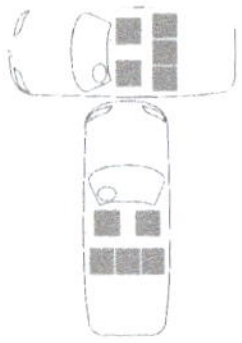	PASSENGER CAR	Internal requirements according to [29]

Fig. 4.11 Test configurations for side collisions

decisions. The ADAC test procedures have changed over time, have been further developed and adapted to technical progress. In addition to cars and child restraint systems, other vehicle categories, such as motorhomes, have also been and continue to be tested. Studies on vehicle compatibility were also carried out based on crash tests (e.g., off-road vehicles against compact class vehicles). In addition to frontal and side impact tests, rear-end collisions are also simulated and the protective potential for reducing the probability of injury in the area of the cervical spine is examined.

In the meantime, the ADAC regularly relies on the results of the European New Car Assessment Programme (Euro NCAP), of which it is a member and testing laboratory. In addition, the ADAC tests child seats annually using sled tests that simulate a frontal impact at 64 km/h and a side impact at 50 km/h. The requirements significantly exceed the legal requirements of UN-R 44 and UN-R 129.

In addition, the ADAC occasionally conducts vehicle crash tests in configurations not covered by the Euro NCAP rating, for example with varying dummy sizes, with loading or vehicle-vehicle collisions. The latter laid the foundation for the compatibility crash now introduced at Euro NCAP.

Rear-end collisions	Description	Application
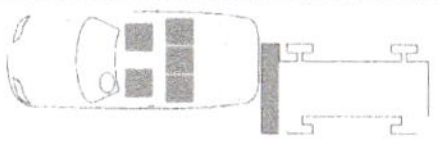	Mobile deformable barrier, offset (70 %)	Law (e.g. U.S.)
	Mobile rigid barrier, full coverage	Law (e.g. UN)
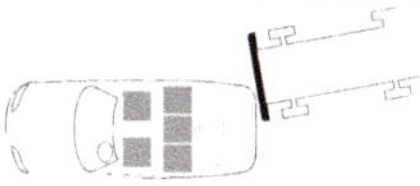	Mobile rigid barrier, 10° angle, offset (40 %)	Insurance crash (RCAR)
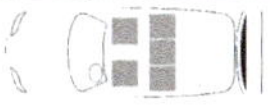	Generic bumper, full overlap and offset (15 %)	Insurance crash (RCAR)
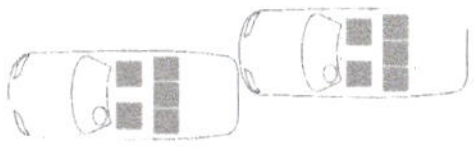	PASSENGER CAR	Internal requirements according to [29]

Fig. 4.12 Test configurations for rear collisions

Rollover accidents	Description	Application
	Roof crush test	Law (e.g. U.S.) NCAP (e.g. IIHS, C-IASI)
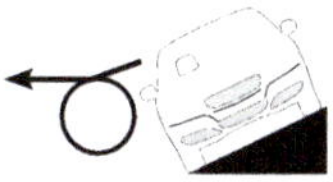	Rollover	Internal requirements according to [29]
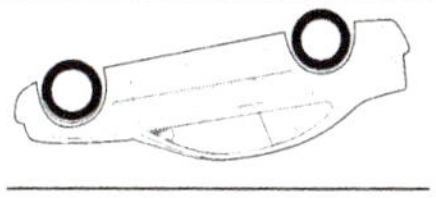	Roof drop test	Internal requirements according to [29]

Fig. 4.13 Test configurations for rollover accidents

4.3.2 New Car Assessment Programs (NCAP)—Vehicle Safety Rating Programs

In the last decades and especially in the last decade, various test programs have been established worldwide to investigate and publish the safety of new vehicles, thus providing consumers with a decision-making aid when buying a vehicle. These test procedures, known as the New Car Assessment Program (NCAP), essentially have in common that the legal requirements prescribed in the respective countries or regions serve as minimum requirements and vehicles that want to achieve a good rating must meet significantly higher and sometimes far exceeding and additional requirements. The first rating programs focused on accident protection, i.e., the passive safety of vehicles. Nowadays, consumer protection organizations consider the integral vehicle safety and also evaluate systems of active safety, such as the performance of automatic emergency braking systems or lane keeping systems. Such programs are carried out in the USA by the research institute of the NHTSA (National Highway Traffic Safety Administration), referred to as US-NCAP, in Japan by NASVA as so-called J-NCAP, in Australia as A-NCAP and in Europe by a consortium of 16 members as Euro NCAP. In recent years, the Latin-NCAP in Latin America, C-NCAP in China, K-NCAP in Korea, Asean NCAP in Malaysia, and TNCAP in Taiwan since 2022 have also been established. The oldest such test program is the subsequently described evaluation procedure US-NCAP.

As part of the "Decade of Action for Road Safety 2011–2020" [28] initiated by the World Health Organisation (WHO), the World Bank and the FIA Foundation and proclaimed by the United Nations in May 2011, a Global NCAP was also launched, which aims to establish and coordinate further NCAPs. Euro NCAP supports this global program in technical and organizational matters. Global NCAP has conducted consumer protection test procedures in markets where there are no established NCAP tests yet with SaferCarsForIndia and SaferCarsForAfrica. India has since announced the introduction of the so-called Bharat NCAP.

4.3.2.1 US-NCAP

The basis for the evaluation procedure New Car Assessment Program (NCAP) was created by the US Senate following a significant increase in accident and injury numbers through the enactment of a consumer-oriented legislative directive. This motor vehicle information and cost-saving act (Motor Vehicle Information and Cost Saving Act) from 1972 contains a call to the Department of Transportation (DOT) to comparatively examine newly offered vehicles on the US market from economic and safety-technical aspects and to make the results obtained accessible to the public. The declared goal was and is to prioritize the safety level over buyer interest and demand situation. However, initially, economic and energy policy issues were at the forefront, at least in the years of introduction. The growing proportion of smaller vehicles (1980: 40% share of compact and sub-compact cars in the vehicle population) at the beginning of the 1970s due to the emerging supply shortages and the restrictive regulatory measures (speed limit of 55 mph, determination of fleet consumption values) was disproportionately reflected in accident occurrence. Therefore, this vehicle class was initially subjected to a safety-technical examination:

- Occupant protection for driver and passenger in frontal and rear-end collisions,
- Integrity and stability of the windshield on the vehicle,
- Fire hazard after a collision and
- Comfort consideration at the occupant protection system.

The processing and additional use of insights from police records and insurance records were considered useful, but their limited informative value about currently introduced or even new vehicles was recognized. To compensate for this disadvantage, an experimental test- and evaluation procedure was established, which has been applied since the 1983 model year to the most common vehicles of domestic manufacturers and importers and affects an average of about 60 vehicle models per year. In defining the test conditions, a frontal impact test is based on a rigid, non-deformable barrier with an impact speed of 35 mph (≈56 km/h). In contrast to the impact test according to FMVSS 208, where tests are also carried out with unbelted occupants, the occupants (a HYBRID III-50M male on the driver's side and a HYBRID III-5F female on the passenger's side) are always belted in the NCAP test.

In addition to the frontal crash, the US-NCAP program includes a barrier-side impact with a 1,368 kg heavy barrier at an impact speed of 62 km/h. In this test, an ES-2RE dummy is placed on the driver's seat and a SID IIs dummy is placed on the shock side in the rear. This is supplemented by a pole-side impact at 32 km/h under 75°, in which a SID IIs dummy is used on the driver's seat. The rating is supplemented by a rollover evaluation, which is primarily based on the static stability factor (SSF), the ratio of track width to center of gravity height of the vehicle. The results of the individual tests are combined and published as an overall rating, with a maximum of 5 stars achievable.

A revision of the US-NCAP rating has been under discussion since 2015. In addition to a revision of the current crash tests, a number of new evaluations are to be introduced into the rating. These include pedestrian protection tests, the evaluation of a whole range of assistance systems, such as emergency brake assistants or lane keeping assistants, as well as the evaluation of the headlights. For political reasons, the revision of the rating has been repeatedly delayed. Most recently, a roadmap for the introduction of the new evaluations was published in 2022 [17].

In the area of passive safety, the planned new oblique frontal impact (see Fig. 4.14) with a mobile deformable barrier (OMDB = oblique moving deformable barrier) poses high demands on the car body due to the mass (2,486 kg) and the speed (90 km/h) of the barrier. At the same time, the test is challenging for the restraint systems due to the oblique impact and the use of the THOR dummy, which is significantly more flexible compared to the Hybrid III dummy, as they must prevent a strong head and upper body rotation in the direction of impact.

4.3.2.2 Euro NCAP

The European New Car Assessment Programme was initiated by a consortium that, in addition to some European government institutions (Germany, France, Great Britain,

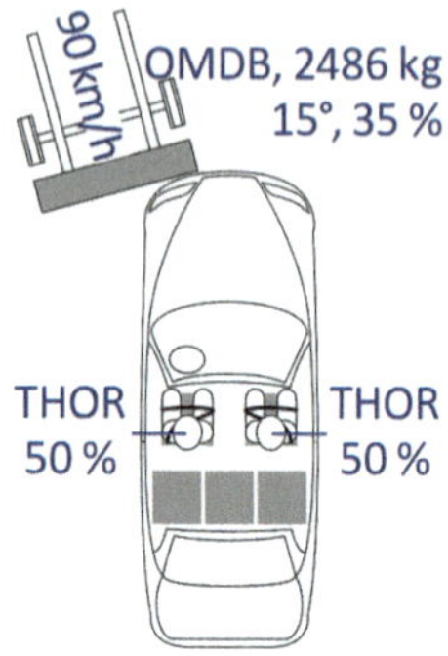

Fig. 4.14 Planned new frontal crash in US-NCAP—oblique moving deformable barrier

Catalonia, Luxembourg, Netherlands, Norway, Austria, Sweden), also includes automobile clubs (FIA Foundation, ADAC, ACI), the umbrella organization of European consumer protection organizations (ICRT), an expert organization (DEKRA), and representatives of the insurance industry (Thatcham, accident research of insurers) as members. The EU commission is not a member, but it politically supports Euro NCAP, as this test program can help achieve traffic policy goals in terms of increased road safety. The first results of the European New Car Assessment Programme were published in 1997.

The test program is continuously being developed. Currently (as of 2023), the rating is divided into 4 categories, the contents of which are described in Tables 4.5, 4.6, 4.7 and 4.8:

- Adult occupant protection
- Child occupant protection
- Protection of vulnerable road users (pedestrian and cyclist)
- Advanced driver assistance systems

For the individual categories, Euro NCAP determines the degree of fulfillment, i.e., the percentage of the maximum score of the respective category, for a vehicle to be evaluated. The degrees of fulfillment in the four categories form the basis for the overall rating with up to five stars. To achieve a certain number of stars, minimum fulfillment levels are required in the four categories, for example, from the year 2023 for 5 stars 80% in adult occupant protection, 80% in child protection, 70% in protection of vulnerable road users and 70% in the assistance systems. This "balancing" procedure is intended to ensure that well-rated vehicles achieve a high level of safety in all categories. Euro NCAP uses the determination of the minimum fulfillment levels also for the continuous adjustment of the requirements to the technical progress.

The **evaluation** in the test procedures for passive safety according to Euro NCAP is based on the body part-specific load values measured on the dummy. The legally prescribed limit values form a switch to zero points or to a red coloring of the relevant body region in the pictogram of the occupant. Between this red or zero-point area and

Table 4.5 Test procedures within the framework of adult occupant protection assessment in Euro NCAP (2023)

<table>
<tr><th colspan="3">Adult occupant protection</th></tr>
<tr><th colspan="2">Test procedure</th><th>Score</th></tr>
<tr><td>Frontal impact at 50 km/h against a 1,400 kg mobile deformable barrier (MPDB) with 50 km and 50% overlap on the driver's side. In addition to the occupant protection assessment, this test also evaluates compatibility</td><td>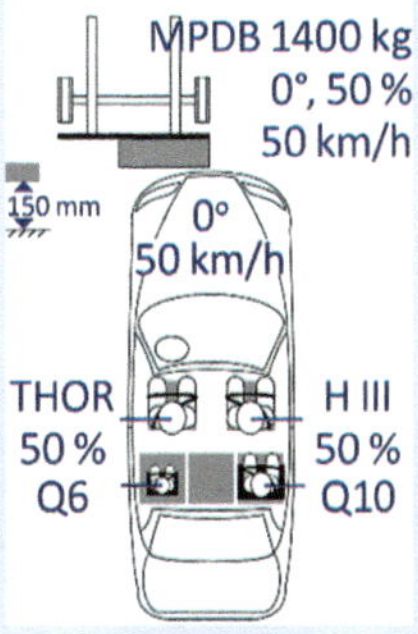
</td><td>8</td></tr>
<tr><td>Frontal impact at 50 km/h against a rigid wall</td><td>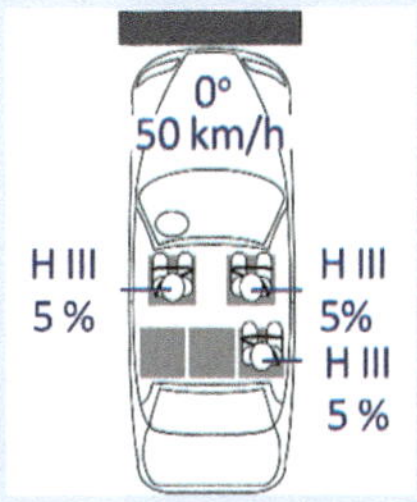
</td><td>8</td></tr>
<tr><td>Barrier side impact with a 1,400 kg mobile deformable barrier at 60 km/h</td><td>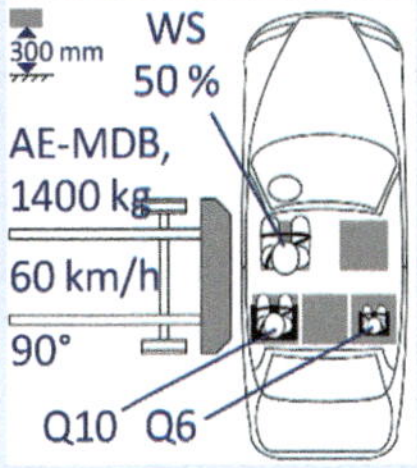
</td><td>6</td></tr>
<tr><td>Pole side impact at 75° against a rigid pole at 32 km/h</td><td>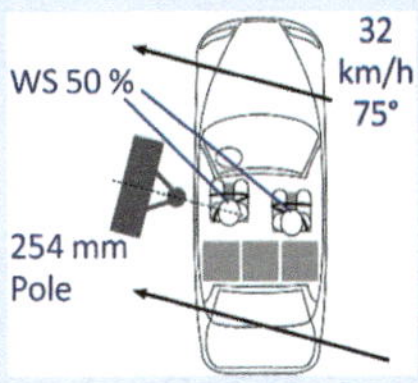
</td><td>6</td></tr>
<tr><td>Evaluation of the protection of occupants on the non-impact side during side impact using two sled tests with the pulses from barrier and pole side impact</td><td>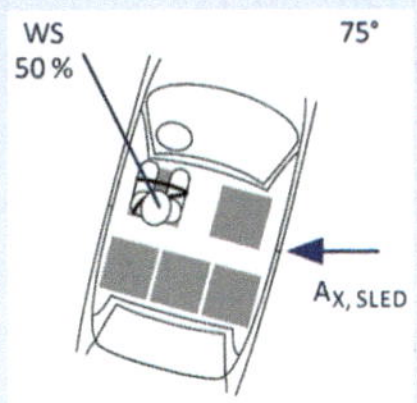
</td><td>4</td></tr>
</table>

(continued)

Table 4.5 (continued)

Adult occupant protection		
Test procedure		Score
Evaluation of the protection of the front occupants against whiplash injuries using sled tests with 2 generic pulses		3
Evaluation of the protection of the rear occupants against whiplash injuries using a geometry assessment of the rear seats and headrests		2
Evaluation of measures for safe rescue and extrication of occupants from the crashed vehicle	• Rescue sheet • Opening of doors & seatbelt buckles after accident • E-Call, multi-collision brake, protection when submerging in waters	4
Maximum total score		40

the lower limit values for the dummy load values to be assessed (green area, full score), a so-called "sliding scale" is used to linearly interpolate and determine the respective point evaluation. If multiple protection criteria are used for a body part, the worst value (worst scoring component) is included in the overall evaluation. An example of this is the area of the lower extremities, where, for example, the tibia index (cf. Eq. 3.16) and the lower leg force in the z-direction are determined. In addition, only the worst results of the driver or passenger side are included in the overall vehicle evaluation, e.g., for the frontal test.

In addition to the evaluations determined from the dummy load values, so-called modifiers—usually a point deduction—are applied, which are derived from the results of vehicle measurement, the evaluation of high-speed films, and the examination of the vehicles by a team of Euro NCAP inspectors. These modifiers are usually deducted from or assigned to the respective body part-specific evaluation (e.g., steering wheel displacement to the head, shortening of the passenger compartment to the chest, footwell and pedal intrusions to the lower extremities).

Different evaluation methods are used in the other test procedures. In some cases, points are awarded depending on performance, e.g., based on the achieved speed reduction in the emergency braking systems. In some cases, the evaluation is based on the pass/fail method, i.e., Euro NCAP either awards the full score for meeting the requirements or zero points for non-compliance, e.g., in the evaluation of lane assistance.

Euro NCAP continuously develops its test- and evaluation procedure and responds to both technical progress and real accident occurrence. Since 2009, Euro NCAP has adjusted the rating every 2 years. From 2026, Euro NCAP plans to switch to an adjustment interval of 3 years. At the same time, the consumer protection organization will fundamentally change the evaluation categories. The previous test procedures will now

Table 4.6 Test procedures within the scope of child occupant protection assessment in Euro NCAP (2023)

Child occupant protection		
Test procedure		Score
Frontal impact at 50 km/h against a 1,400 kg mobile deformable barrier (MPDB) with 50 km and 50% overlap on the driver's side	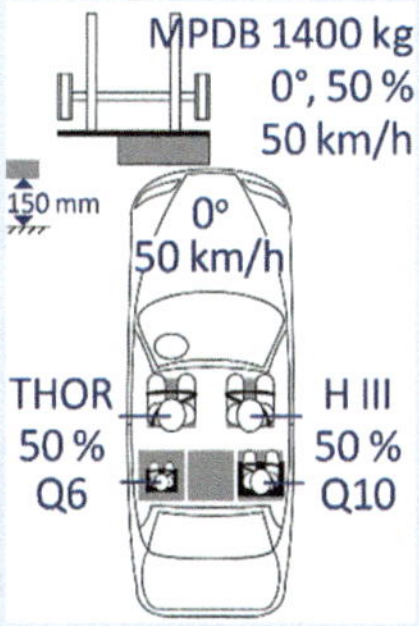	16
Barrier side impact with a 1,400 kg mobile deformable barrier at 60 km/h	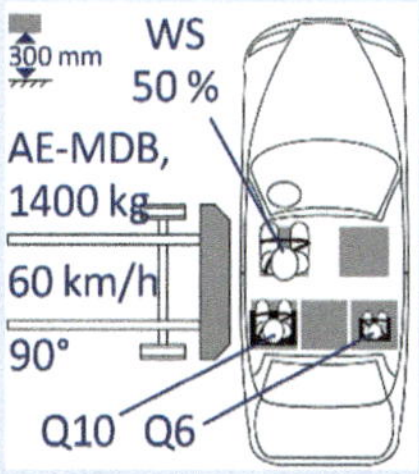	8
Child seat installation test	• Trouble-free and safe installation of child seats with and without ISOFIX	12
Vehicle-based assessment	• Availability of i-Size seating positions • Deactivation of the passenger airbag • Detection of children left in the vehicle (child presence detection—CPD)	13
Maximum total score		49

be categorized in phases. The phases are safe driving, accident prevention, crash protection and the post-crash phase, which is increasingly approaching an integral safety consideration, as explained in Sect. 1.3. The background to this change is in particular the intention to integrate assisted and automated driving functions into the rating in the future, which should then be evaluated in the safe driving phase [4].

In the area of accident protection, Euro NCAP is aiming for a more robust design of the restraint systems. For this purpose, in addition to the full scale crash tests, in future sled tests and computational simulations should evaluate a wider range of crash configurations in terms of impact speeds and, in particular, in terms of the occupants [5]. The Table 4.9 provides an overview of the future frontal impact tests.

Table 4.7 Test procedures in the context of assessing the protection of vulnerable road users in the Euro NCAP (2023)

Protection of vulnerable road users		
Test procedures		Score
Head impact with child and adult head	• 40 km/h • WAD 1000–WAD 2500 • Grid method with manufacturer's result forecast and verification tests by Euro NCAP	18
Leg impact with aPLI and upper legform	• aPLI impact at 40 km/h on vehicle front • Upper legform impact on WAD 775 with geometry-dependent speed	18
Emergency braking assistance for pedestrian protection	• 7 scenarios • Evaluation of AEB and FCW systems • Pedestrian crossing, along the roadway, behind the vehicle, in intersection situations • Tests during day and night	9
Emergency braking assistance for cyclist protection	• 7 scenarios • Evaluation of AEB and FCW systems • Cyclist crossing, along the roadway, in intersection situations, passing parked car (dooring) • Tests during day	9
Emergency braking assistance for protection of motorized two-wheelers	• 3 scenarios • Evaluation of AEB and FCW systems • Cyclist ahead, turning across path • Tests during day	6
Lateral guidance assistance for protection of motorized two-wheelers	• 2 scenarios • Cyclist overtaking, oncoming • Tests during day	3
Maximum total score		63

4.3.2.3 C-NCAP

The China Automotive Technology and Research Center (CATARC) has been conducting C-NCAP since 2006, which has continuously gained importance. While the requirements were initially quite moderate, the program is now at a level comparable to Euro NCAP. The requirements are also continuously developed and cover aspects of both passive and active safety. The following Tables 4.10, 4.11, 4.12 provide an overview of the test procedures carried out in 2023.

4.3.2.4 IIHS Rating

In the U.S. market, in addition to the US-NCAP evaluation procedure described above, the evaluation procedure of the Insurance Institute for Highway Safety (IIHS) is of great

Table 4.8 Test procedures in the context of the evaluation of safety assistance systems in the Euro NCAP (2023)

Safety assistance systems		
Test procedures		Score
Occupant status monitoring	• Seat belt warning • Monitoring of the driver's alertness and attention	3
Speed assistance	• Information about current speed limits, • Warning when exceeding the set maximum speed • Active prevention of exceeding the maximum speed	3
Lane assistance	• 12 scenarios • Lane keeping assistants (LKA) • Lane departure warning (LDW) • Blind spot assistant (BSM) • Information and documentation assistant lane keeping assistants (ELK) • Daytime tests	3
Emergency brake assistance car/car	• 7 scenarios • Evaluation of AEB and FCW systems • Car ahead, crossing, oncoming, turning oncoming • Daytime tests	9
Maximum total score		18

importance. Unlike the NCAP programs, the communication of results here is not via a star rating, but via a four-level grading scale Poor—Marginal—Acceptable—Good. In addition to these grades, the IIHS awards the "Top Safety Pick" or "Top Safety Pick +" distinction to vehicles that perform well in all tests.

The IIHS crash test procedures place very high demands in some areas. For example, the institute conducts the so-called small-overlap test [22], a frontal crash with 25% overlap, which often leads to deep intrusions into the passenger compartment and a pronounced rotation of the vehicle around the vertical axis. The latter results in a diagonal forward movement of the occupants, which often leads to them being inadequately restrained by the airbags.

The IIHS side impact test also places particularly high demands [21]. This is due to the high mass of the barrier vehicle, which has been 1.9 t since 2023 (compared to 1.4 t for the Euro NCAP). The resulting high impact energy leads to strong intrusions into the passenger compartment and requires massive reinforcements of the car body, especially in the area of the B-pillar.

Figure 4.15 provides an overview of the IIHS crash tests.

A test procedure strongly based on IIHS was also established in China under the name C-IASI (China Insurance Automotive Safety Index) since 2018. Unlike the IIHS

Table 4.9 Matrix of Euro NCAP frontal impact tests from 2026

	Hardware tests			
	Full scale crash tests		Sled tests	
Test	MPDB	FWDB	Robustness 1	Robustness 2
Scenario	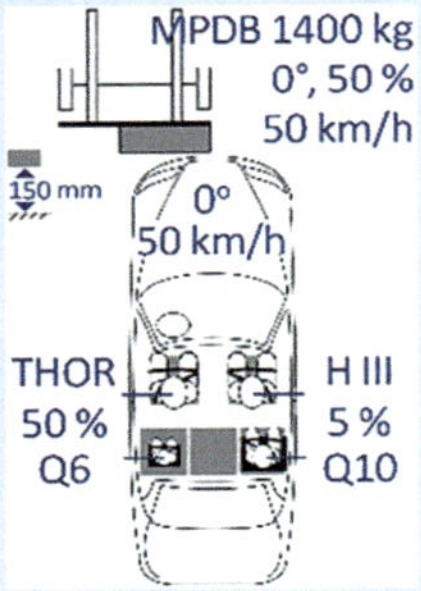	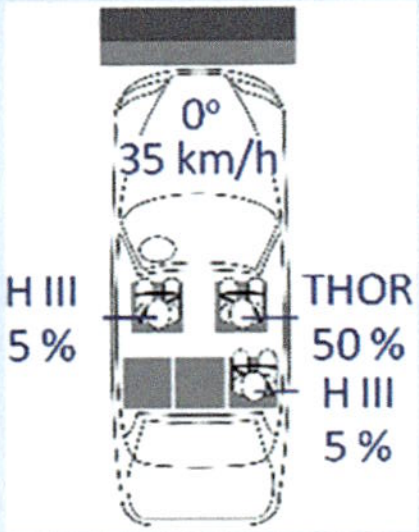	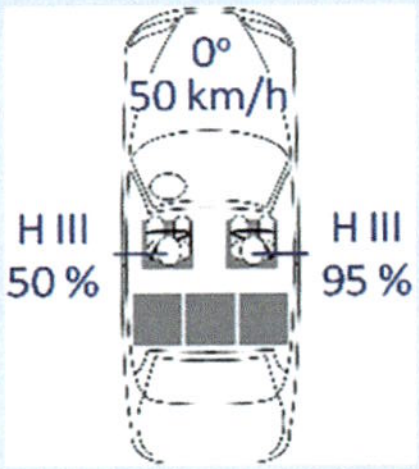	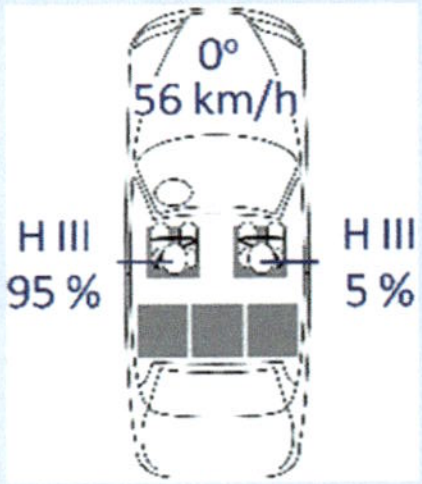
Speed	56 km/h	35 km/h	50 km/h	56 km/h
Driver	THOR 50% M	Hybrid III 5% F	Hybrid III 50% M	Hybrid III 95% M
Passenger	Hybrid III 5% F	THOR 50% M	Hybrid III 95% M	Hybrid III 5% F
2nd row driver's side	Q6			
2nd row passenger side	Q10	Hybrid III 5% F		

(continued)

Table 4.9 (continued)

	Virtual testing (computational simulation)			
Test	Robustness 1	Robustness 2	Robustness 3	Human body model
Scenario	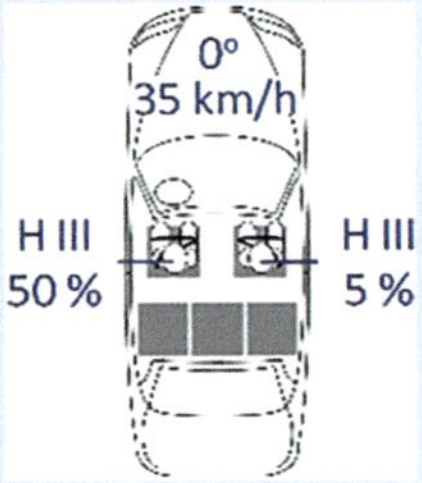	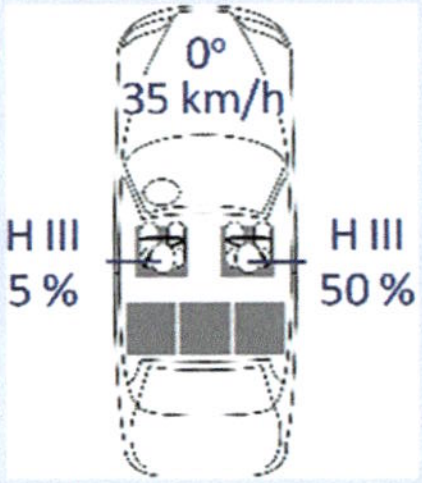	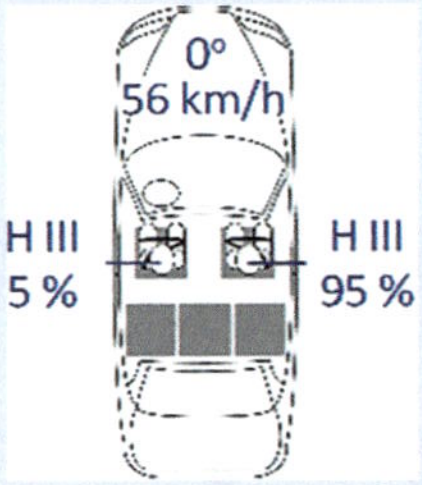	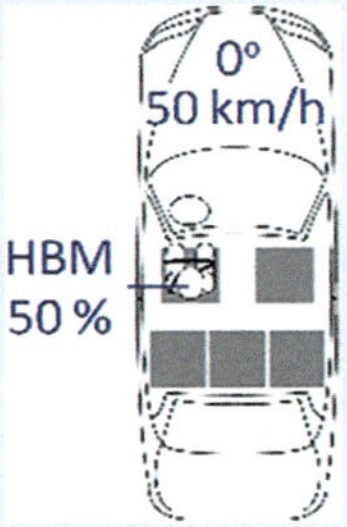
Speed	35 km/h	35 km/h	56 km/h	50 km/h
Driver	Hybrid III 50% M	Hybrid III 5% F	Hybrid III 5% F	Human body model medium size
Passenger	Hybrid III 5% F	Hybrid III 50% M	Hybrid III 95% M	-

Table 4.10 Occupant protection assessment in C-NCAP

Occupant protection		
Test procedure		Score
Frontal impact at 50 km/h against a 1,400 kg mobile deformable barrier (MPDB) with 50 km and 50% overlap on the driver's side. In addition to the occupant protection assessment, this test also evaluates the compatibility	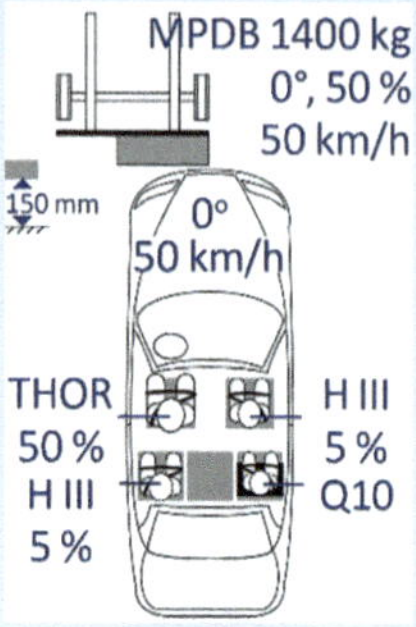	24
Frontal impact at 50 km/h against a rigid wall	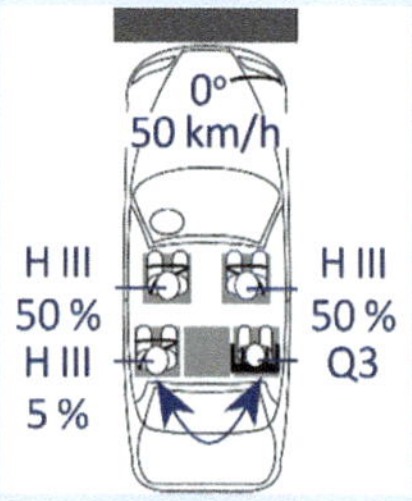	24
Barrier side impact with a 1,400 kg mobile deformable barrier at 60 km/h This test is only conducted for vehicles with combustion engines	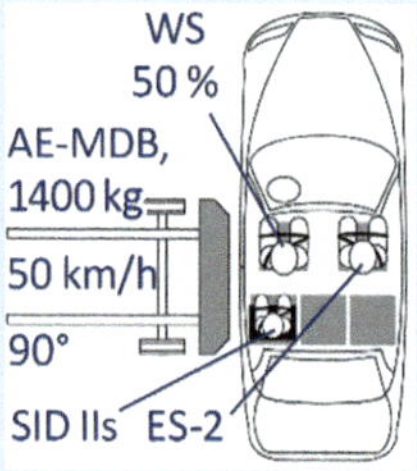	24
Pole side impact at 75° against a rigid pole at 32 km/h This test is only conducted for vehicles with electric or hybrid drive	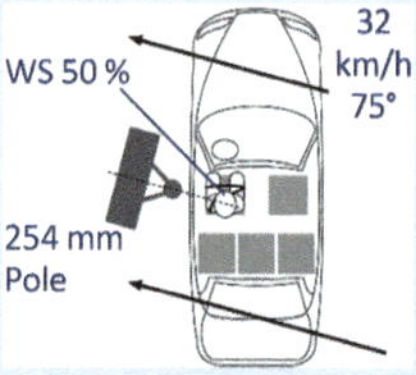	24
Assessment of the protection of front and rear occupants from whiplash injuries using sled tests with a generic pulse		7
Static child seat assessment	• Possibility of installing different types and sizes of child seats • Installation test	3

(continued)

Table 4.10 (continued)

Occupant protection		
Test procedure		Score
Protection of occupants from being ejected (ejection mitigation)	• Test according to U.S. regulation FMVSS 221, or • Pressure test for Curtain Airbag	2
Automatic emergency call	• E-Call	2
Maximum total score		86

Table 4.11 Pedestrian protection evaluation in C-NCAP

Pedestrian protection		
Test procedure		Score
Head impact with child and adult head	• 40 km/h • WAD 1000–WAD 2300	10
Leg impact with aPLI	• aPLI impact at 40 km/h on vehicle front	5
Maximum total score		15

rating, this also includes impact tests for pedestrian protection, based on the test procedures of Euro NCAP, as well as tests for evaluating the ease of repair of a vehicle.

4.4 Evaluation Based on Accident Statistics

In the evaluation of safety based on accident statistics, the assessment is made retrospectively using statistical analysis methods. The accident information evaluated is collected by the state (e.g., federal statistics) or by insurance companies over a sufficiently long period for statistical validation and stored in databases. The extent of statistically significant, usable accident data results from the frequency of accidents during the collection period [1]. The advantage of this evaluation is that the injury severity levels determined by trauma surgery of the accident participants can be correlated with the reconstructed accident severity and the relationship between the introduction of safety measures and preventable injuries can be statistically proven without the detour via physical or theoretical occupant models. The disadvantage, however, is the required time span between the introduction of safety measures and the retrospective determination of any safety improvements, as from the time of approval of improved motor vehicles for traffic, through the achievement of a noticeable proportion of these vehicles within the vehicle population to the time of the presence of a statistically significant number of accident

Table 4.12 Evaluation of active safety in C-NCAP

Active safety		
ADAS test procedure		Score
Occupant state monitoring	• Seat belt warning • Monitoring the driver's alertness and attention	3
Speed assistance	• Information about current speed limits, • Warning when exceeding the set maximum speed • Active prevention of exceeding the maximum speed	3
Lane assistance	• 2 scenarios • Lane keeping assistance (LKA) • Daytime tests	3
Emergency brake assistance car/car	• 2 scenarios • Evaluation of AEB and FCW systems • Car ahead • Daytime tests	11
Emergency brake assistance pedestrian, cyclist, motor scooter	• 5 scenarios • Evaluation of AEB and FCW systems • Pedestrian crossing, along the roadway, cyclist crossing, along the roadway, motor scooter crossing • Daytime tests	21
Human-machine interface of emergency brake systems	• Deactivation • Supplementary warning • Reversible belt tensioning	6
Electronic stability control (ESC)	• According to the Chinese regulation GB/T 30.677–2014 or GTR 8/UN-R 13H (R140)/ FMVSS 126	3
Optional ADAS systems	• Lane departure warning • Speed assistance • Blind spot warning	7
Maximum total score ADAS		56
Light testing procedure		**Score**
Low beam	• Illumination • Visibility of pedestrians	10
High beam	• Range • Visibility of pedestrians	
Bonus/Malus	• Adaptive light • Automatic switching on • Automatic headlight range control • Glare (malus)	
Maximum total score light		10

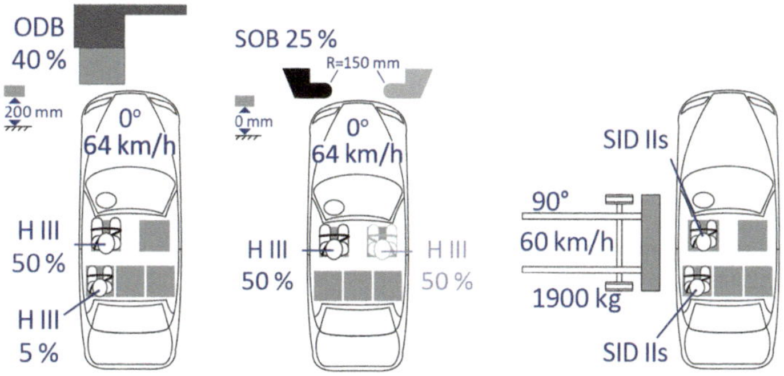

Fig. 4.15 IIHS crash tests: offset-crash, small overlap test (driver and passenger side), side impact barrier

vehicles, a certain amount of time is required. As a result, newer vehicle models are fundamentally exempt from such an evaluation.

4.4.1 Highway Loss Data Institute Report

The American Highway Loss Data Institute has been publishing comparative values for various vehicle types since the mid-1970s, which are based on the frequency and the resulting costs for personal and property damage. The basis for this evaluation are the analysis results of two insurance databases and police records. The qualitative assessment of individual vehicle types is made by relating the specific accident data to average values for the frequency of property damage and to the average ratio between personal and property damage [1].

4.4.2 FOLKSAM Report

The FOLKSAM Report is designed to alert consumers to safety-relevant differences between individual vehicle types and thus to increase the importance of vehicle safety in the purchasing decision. The safety related evaluation of specific vehicle types and models regularly published by the Swedish insurance group FOLKSAM INSURANCE is based on data from accidents of policyholders. The accident type, occupant-related sizes such as age and AIS-coded injuries, vehicle-related sizes such as mass ratio of the collision opponents and technical vehicle condition as well as information about the use and condition of the occupant protection systems are included in the data collection [1]. From this extensive accident data material, vehicle type-specific characteristics can be compared to the average values of all examined vehicles or the mean value of a certain vehicle category for evaluation. However, the risk statement for a vehicle type is not relativized with the registration numbers and includes all types of injuries regardless of their severity. In addition, the risk of fatal and serious injuries with consequential damage is validated against the average values through the definition of a safety index. Furthermore, injury distributions and patterns of the relevant body regions are evaluated.

4.4.3 Secondary Safety Rating System for Cars

This method developed in the UK is an analytical method in which the classification of safety-relevant vehicle components into different classes with good or bad technological solutions is carried out and related to a conception of the component to be evaluated that corresponds to the current state of knowledge. The relevance of the contact areas and the vehicle subsystems with regard to the injury risk for the occupants is derived from the accident occurrence and weighted taking into account injury-related cost rates. The assessment criteria include elements of self-protection (e.g., design of the door padding) as well as opponent protection (e.g., pedestrian-friendly design of the front hood). For each variable, up to 17 gradations are verbally categorized or defined by a matrix of individual criteria. The classification is carried out solely by visual inspections, taking into account constructive design features and material-technical assessment by an expert team. The overall evaluation of a vehicle results from the multiplication of all variables with the corresponding weighting factors and their summation [1].

4.5 Quantification of Passive Safety for Car Occupants and the Safety Criteria System SiKriS

The goal of the evaluation procedure, which was developed at the Technical University of Berlin as part of the long-term research project "quantification of passive safety for car occupants" funded by the Federal Highway Research Institute Germany (BASt), is to condense the load values measured on the dummy into a single safety index. The procedure can be located between the two extreme positions already described from a methodological point of view: On the one hand, the purely experimental evaluation by means of a frontal and a side impact test, as is carried out in NCAP procedures, and on the other hand, the evaluation by means of retrospective accident analysis, which is used in the vehicle safety rating by FOLKSAM. Compared to the most advanced procedure, the Euro NCAP, the procedure discussed here offers the advantages of not requiring component tests and being closely aligned with legally prescribed safety tests. However, an additional test, the so-called **compatibility crash test**, is required. The procedure includes three test configurations [1]:

- Frontal impact against a deformable 0° barrier at 50 km/h,
- Side impact at 90° with a movable, deformable barrier at 50 km/h, and
- Compatibility test, a 90° lateral collision between two identical vehicle models, also at 50 km/h.

For the applied evaluation algorithm, the relationships between injury severity and dummy loads on the one hand, and between body part-specific protection criteria and the corresponding degree of fulfillment on the other hand, were established. In addition, relevance factors from accident occurrence have been developed for weighting the partial results. Using this evaluation procedure, the load values determined by testing are normalized and evaluated. The degrees of fulfillment determined from this receive an accident-specific weighting through the relevance factors and can be combined into an overall safety index (SIX) via partial safety indices. This index provides information about the internal safety of cars with regard to self-protection and counterparty protection [1].

The special feature of the evaluation procedure briefly presented here lies in the consistent application of statistical biomechanics (see Sect. 3.3.2) and in the continuous evaluation possibility of load values associated with the help of body part-specific risk functions. In the meantime, the slightly modified evaluation algorithm under the term **safety criteria system SiKriS** has been incorporated into the exclusive evaluation of occupant protection systems using measured, but also computationally determined load values [14]. Taking into account suitable test arrangements and corresponding test results for external safety, the procedure could be extended and now allows the evaluation of the passive safety of cars not only with regard to internal, but also external safety [18].

Due to its universal applicability, the safety criteria system (SiKriS) will be explained in more detail below. With the help of this evaluation system, the safety index SIX (with $0 \leq \text{SIX} \leq 1$) is calculated for the evaluation of the measured or computationally determined load values of all relevant body parts, taking into account accident analysis, biomechanics research, and safety legislation (Fig. 4.16). The larger the SIX, the lower the injury risk or the higher the protective effect of the occupant or pedestrian protection system. The safety criteria system has now been extended to frontal and side collisions as well as pedestrian collisions [18]. Currently, the protection criteria shown in Table 4.13 can be applied, but the procedure allows for easy expansion, provided the risk function from biomechanics research is known, and can thus be extended to almost any accident configuration.

From the body part-specific risk function (see Fig. 3.67), which represents the probability of irreversible injuries depending on the respective load size, the safety level SG as a function of the normalized load value NBW is derived by inverting the function and relating the load size to the test limit value; the relationship is schematically represented in Fig. 4.17. For example, with a HIC = 700, the normalized load value NBW = 0.7 and the safety level SG ≈ 0.55.

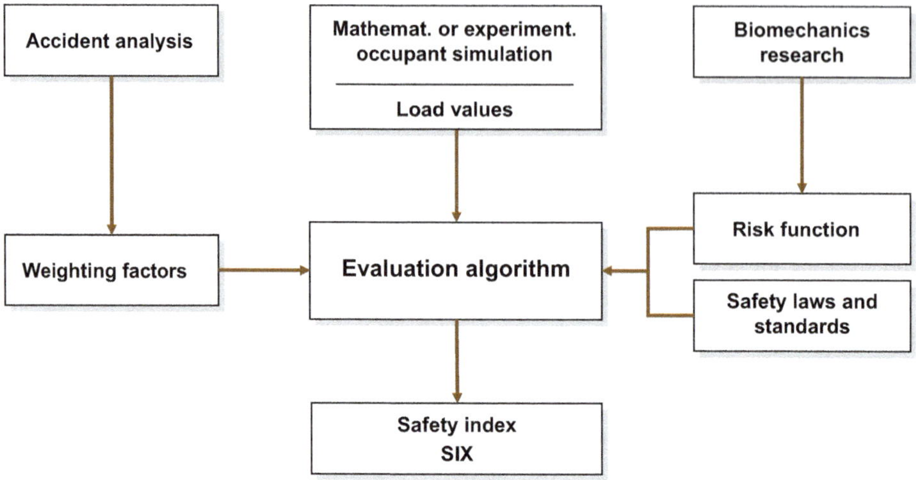

Fig. 4.16 Schematic of the SiKriS safety criteria system (from [14])

This safety level can be calculated for each body region, for which a risk function exists and for which load values (calculated or experimental) can be determined, using the evaluation algorithm according to the relationship:

$$\mathrm{SG}_j = f(\mathrm{NBW}_j), \quad \text{where} \quad -1 \leq \mathrm{SG}_j \leq +1 \text{ is,} \tag{4.1}$$

with

j: Current load size (e.g., HIC, neck bending moment, femur longitudinal force, etc.),
f: Evaluation function for the load size *j* and
NBW: Normalized load value; quotient of load value and protection criteria level

Using the weighting factors α from the relevance structure, which can be derived from the injury frequency and the injury-related costs of the accident data material taking into account the accident situation, the occupant seating position and the injured body part, the importance of the body part safety grades is relativized. They can thus be summarized for all body regions to the safety index SIX in the form

$$\mathrm{SIX} = \sum_{i=1}^{m} \alpha_i \cdot \mathrm{SG}_i, \text{where} \quad \sum \alpha_i = 1{,}0 \text{ is,} \tag{4.2}$$

with

i: Body region (e.g., head, neck, thorax, etc.) and
m: Number of currently considered body regions ($1 < m < n$; $n_{\max} = 7$).

Table 4.13 Protection criteria and protection criteria level at various body regions for frontal and side collisions as well as for accidents involving pedestrians

Body region	Protection criterion	Protection criteria level
Head	HIC36 or HPC	1,000
	HIC15	390… 700
	a3 ms	80 g
	a5 ms	150 g
	amax	300 g
	F	378 N
	GAMBIT	1
Neck	Nij	1
	NIC	15 m2/s2
	Fz max tension	4,170 N
	Fz max compression	4,000 N
	My max	57 Nm
Thorax	SI	1,000
	a3 ms	60 g
	sKompr	50.8 mm
	sKompr	50 mm
	RDC	42 mm
	Chest force	11,100 N
	VC	1.0 m/s
	TTI	85/90 g
Arms	Contact force max	3,800 N
Pelvis (α pelvis $\leq 20°$)	a3 ms	80 g
Pelvis, general	a3 ms	130 g
	Symphysis force max	6,000 N
	Transverse force max	2,500 N
Abdomen (α pelvis $> 20°$)	Resultant acceleration max (of pelvis)	13 g
Legs	Femur force max	7,580… 9,070 N
	Femur force max	10,000 N
	Tibia force max	8,000 N
	Knee displacement	15 mm
	TI	1.0 (formerly 1.3)
	Contact force max	10,000 N

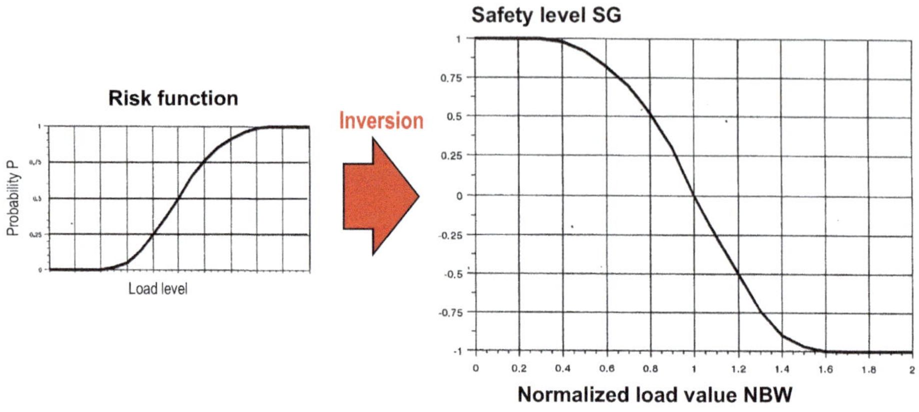

Fig. 4.17 Schematic representation of the transition from the risk to the evaluation function (from [14])

The safety index SIX focuses the experimentally or computationally determined load values on one value, taking into account biomechanical conditions and injury statistical aspects. However, no information is lost in reducing to a single value, as the individual results are retained. The safety criteria system SiKriS thus allows an objective evaluation of the effectiveness of protective measures and enables a reliable assessment of the internal and external safety of motor vehicles. The computer-aided evaluation procedure is almost arbitrarily modifiable, expandable and scalable.

In Fig. 4.18, the calculated safety index SIX is exemplarily shown as a parameter only for the head protection criterion HIC and the protection criterion for the thorax $a_{3\ \mathrm{ms}}$. The representation allows a clear assignment: for a safety index SIX > 0.6, none of the protection criteria levels for head and chest are exceeded, however, a target index of SIX = 0.8 should be achieved.

4.6 Injury-Related Costs and Material Damages

So far, the safety assessment was based solely on the basis of the occupant or pedestrian load and the resulting injury-related costs. The material damages resulting from traffic accidents, which are another type of direct accident consequences in addition to personal injuries (see Fig. 3.61), will be addressed in the following for the sake of completeness. According to the detailed accident figures of the federal statistics [23], it can be seen in the years from 1992 to 1999 that there was only a slight increase in the number of accidents and the injury-related costs (1.2 and 4.3% respectively). However, between

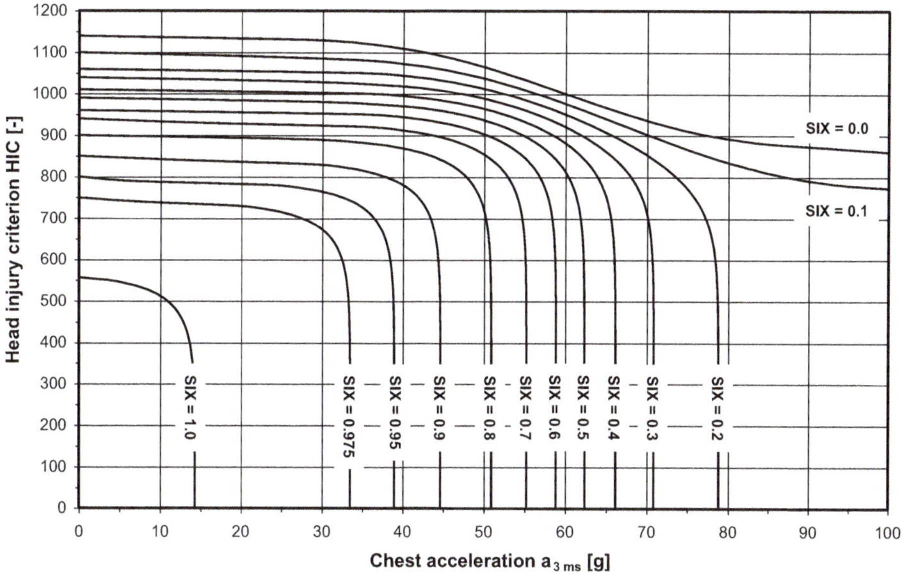

Fig. 4.18 Dependency between head and chest load with the safety index SIX as a parameter (from [18])

1999 and 2011, while the number of accidents decreased by about 2.2%, the injury-related costs decreased by about 17.7% during the same period. After the increase in both the number of accidents and the injury-related costs until 1999, the number of accidents stagnates, while the injury-related costs show an almost steady decrease (Fig. 4.19).

While the ratio of **accidents with material damage** to the total number of all police-reported accidents, i.e., accidents with and without personal injury, stabilized at a constant value of about 87% in the years 2000–2010 (Fig. 4.20), the absolute numbers of accidents with material damage only suggested an increase of about 3.3% since 1992 (1992: 1.989 million, 2011: 2.055 million accidents with material damage). Thereafter, the accidents with material damage increased in relation to the total number of accidents; however, the number of accidents with personal injury decreased by about 22.6% over the same period. It should be noted, however, that the accident figures are based on the police reporting system at the accident site. A lower effort in recording of **minor damages** distorts the overall picture, as damages up to 2,000,– € are disproportionately often classified. In addition, there are the accidents that are not reported to the police at all. To estimate the actual number of accidents occurring each year, it is useful to compare the accident rate, i.e., the number of accidents per 1,000 registered motor vehicles, which was approximately 44.6 in 2011, with the damage rate. According to the German Insurance Association (GDV), the number of damages reported to the motor vehicle

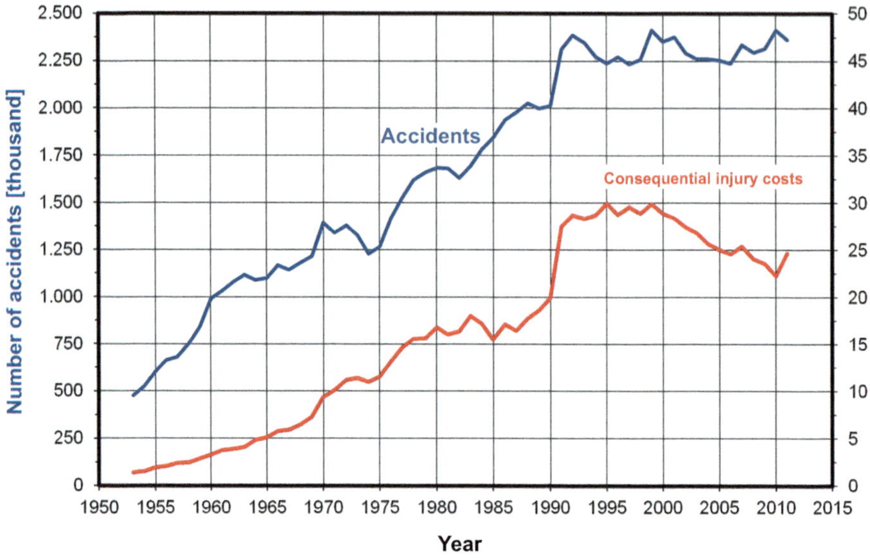

Fig. 4.19 Number of accidents and injury-related costs in the period 1953 to 2011 (from 1991 including the new federal states)

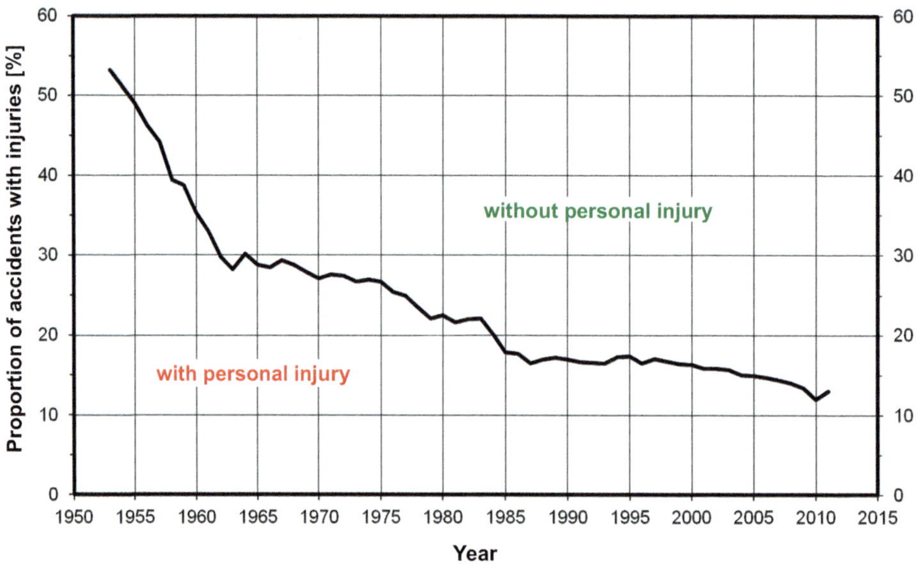

Fig. 4.20 Ratio of accidents with personal and material damage in the period 1953 to 2011 (from 1991 including the new federal states)

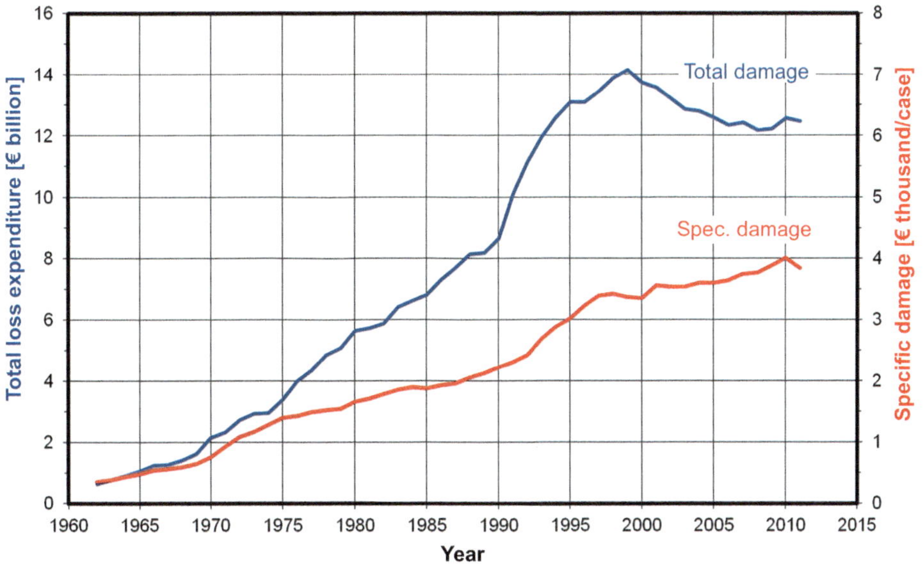

Fig. 4.21 Total damage expenses and specific damage from motor vehicle accident insurance (according to [9]) for road traffic accidents in the period 1962–2011

insurers in 2011 was 9,722 million [9]. This results in a damage rate of 78.9 damages per 1,000 insurance contracts, about 77% more than the accident rate. Thus, a total of about 4.2 million accidents in 2011 (compared to 2.361 million accidents according to federal statistics) can be realistically assumed.

The reported **damage expenses** [9] shown in Fig. 4.21 has been listed in the yearbooks of the German insurance industry since the early 1960s. It amounted to a total of about 12.452 billion € in 2011, which is a reduction of approximately 11.9% compared to the year 1999, when a decline was recorded for the first time, with 14.139 billion €. In contrast, the **specific damage,** i.e., the damage per regulated case, has increased by about 13.8% since 1999 to approximately 3,830 €/case in 2011. The specific damage exceeded the mark of 2,000 €/case in 1988 and that of 3,000 €/case in 1995.

Following an insurance-specific investigation of damage expenditure conducted in [15] (Fig. 4.22), the costs reimbursed by the insurance companies in 1994, amounting to a total of 12.58 billion €, were distributed approximately 20% for the settlement of personal injuries and 80% for the compensation of **material damage** [8]. It can be shown that the costs for personal injuries (medical treatment and compensation for pain and suffering) borne by the liability insurance amount to approximately 2.52 billion €, which is about a tenth of the actual consequential costs resulting from injuries, which amounted to approximately 24.6 billion € in 2011. The significant portion of these economic costs is obviously not covered by vehicle insurance, but is financed through other insurance carriers and private expenditures, particularly through societal expenditures (education

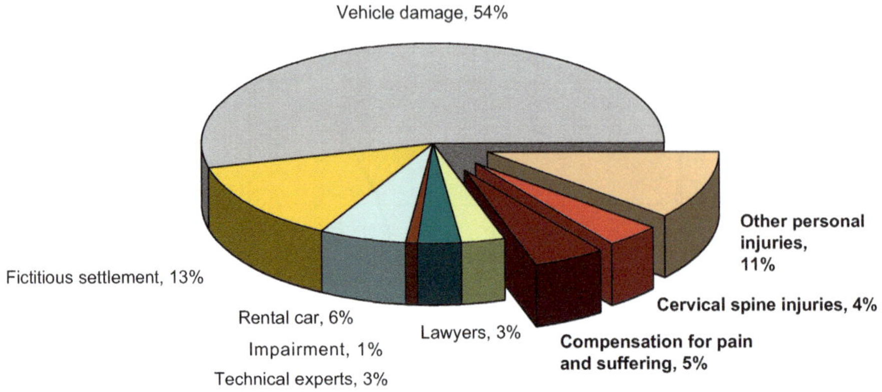

Fig. 4.22 Insurance damage expenses (in 1994: 12.58 billion €) for personal and material damages (according to [15])

costs, production loss, etc.). The material damages, which amount to approximately 10.1 billion €, consist of vehicle damage repairs (67%), incidental costs (17%), and fictitious settlements (16%), regardless of the actual damage repairs. Taking into account the dark figure due to unreported and uninsured accidents, it can be assumed that the direct consequences of accidents are divided in a ratio of approximately 2:1 into injury-related costs and material damages. Since the amount of damage in the repair of vehicle damages after so-called repair crashes plays a significant role in the classification of new vehicle models into the damage classes of comprehensive insurance, the question of the repairability of motor vehicles as a development criterion continues to gain importance.

References

1. Appel, H., Kramer, F., Glatz, W., Lutter, G. et al.: Quantifizierung der passiven Sicherheit für PKW-Insassen. Technische Universität Berlin. Report on the research project 8517/2, Bundesanstalt für Straßenwesen (Hrsg.). Bergisch Gladbach (1991)
2. EEVC Working Group 17 Report: Improved test methods to evaluate pedestrian protection afforded by passenger cars (December 1998 with September 2002 updates). http://www.eevc.org/wgpages/wg17/wg17index.htm
3. Economic Commission for Europe (Hrsg.).: World Forum for Harmonization of Vehicle Regulations (WP 29): How it Works—How to Join it, 4. Aufl. United Nations Publications, https://unece.org/sites/default/files/2022-07/ECE_TRANS_289_Rev.1_E_corrected.pdf, New York and Geneva (2022)
4. Euro NCAP: Euro NCAP Vision 2030—A safer future for mobilty. https://cdn.euroncap.com/media/74468/euro-ncap-roadmap-vision-2030.pdf. Accessed January 2, 2023 (2022)
5. Euro NCAP: Crash Protection Frontal Impact Protocol – Version 0.9, (2024), unter: https://www.euroncap.com/media/85407/euro-ncap-protocol-crash-protection-frontal-impact-v09.pdf (accessed February 10, 2024)

6. European Commission, Enterprise and Industry Directorate-General: CARS 21—A Competitive Automotive Regulatory System for the 21st Century (2006)
7. Florian, M.: Darstellung von Schutzmaßnahmen zur Verbesserung der Kraftfahrzeugsicherheit und deren Wirkprinzipien. Großer Beleg an der Technischen Universität Dresden, Dresden (1996)
8. Geier, S.: Neugewichtung bei den Schadensersatzleistungen für Personen- und Sachschäden? Deutscher Verkehrsgerichtstag in Goslar (1996)
9. Gesamtverband der deutschen Versicherungswirtschaft e. V. (GDV): Statistisches Taschenbuch der Versicherungswirtschaft. Berlin (2007)
10. Glaeser, K.-P.: Der Anprall des Kopfes auf die Fronthaube von PKW beim Fußgängerunfall – Entwicklung eines Prüfverfahrens. Berichte der Bundesanstalt für Straßenwesen, Fahrzeugtechnik, Heft F 14. Bergisch-Gladbach (1996)
11. Grupe, D.: Die Gesetz- und Verordnungsgebung auf dem Gebiet der Fahrzeugsicherheit in Deutschland und der Europäischen Union. 2. International Symposium on Sophisticated Car Occupant Safety Systems. Karlsruhe (1994)
12. Kraftfahrt-Bundesamt: Bestand an Kraftfahrzeugen und Kraftfahrzeuganhängern in den Jahren 1960 bis 2022 nach Fahrzeugklassen. https://www.kba.de/DE/Statistik/Fahrzeuge/Bestand/FahrzeugklassenAufbauarten/2022/b_fzkl_zeitreihen.html?nn=3524712&fromStatistic=3524712&yearFilter=2022&fromStatistic=3524712&yearFilter=2022. Accessed January 11, 2023 (2023)
13. Kramer, F., Deter, T.: Zur Quantifizierung der Straßenverkehrssicherheit. Verkehrsunfall und Fahrzeugtechnik. Bd. 1. Verlag Information Ambs GmbH, Kippenheim (1992)
14. Kramer, F., Fruck, K., Bigi, D.: Sicherheitsindex – eine Möglichkeit zur objektiven Bewertung von Insassenschutzsystemen. Contribution to the conference „Bedeutung der Rückhaltesysteme in der passiven Sicherheit", Haus der Technik. Essen (1995)
15. Kramer, F.: The Development of the KTI (Kraftfahrzeugtechnisches Institut)—Objectives, Tasks and Organisation as well as planned Projects. Conference The RCAR Meeting 1995. Rotherwick, UK (1995)
16. Lehmann, D.: Aktuelle und zukünftige Vorschriften zur Bewertung der passiven Sicherheit. 2. Internationales Symposium on Sophistcated Car Occupant Safety Systems. Karlsruhe (1994)
17. New Car Assessment Program, Request for comments (RFC), [Docket No. NHTSA-2021-0002], Federal Register/Vol. 87, No. 46/Wednesday, March 9 (2022)
18. Reimer, J., Werner, R.: Die Bewertung der passiven Sicherheit bei relevanten Kollisionen mit Hilfe des Sicherheitskriterien-Systems (SiKriS). Diploma thesis at Hochschule für Technik und Wirtschaft (HTW), Dresden (2005)
19. Richtlinie 2007/46/EG des Europäischen Parlaments und des Rates vom 5.9.2007 zur Schaffung eines Rahmens für die Genehmigung von Kraftfahrzeugen und Kraftfahrzeuganhängern sowie von Systemen, Bauteilen und selbstständigen technischen Einheiten für diese Fahrzeuge (Rahmenrichtlinie), Amtsblatt der europ7, L263/1-160 (2007)
20. Seiffert, U.: Fahrzeugsicherheit: Personenwagen. VDI-Verlag GmbH, Düsseldorf (1992)
21. Insurance Institute for Highway Safety: Side Impact Crashworthiness Evaluation 2.0 Crash Test Protocol—Version II. Ruckersville, USA (2022)
22. Insurance Institute for Highway Safety: Small Overlap Frontal Crashworthiness Evaluation Crash Test Protocol—Version VII. Ruckersville, USA (2021)
23. Statistisches Bundesamt Wiesbaden (Hrsg.): Verkehrsunfälle 2011. Fachserie 8, Reihe 7, Wiesbaden (2012)
24. Statistisches Bundesamt Wiesbaden (Hrsg.): Verkehrsunfälle und Verunglückte im Zeitvergleich (ab 1950). https://www.destatis.de/DE/Themen/Gesellschaft-Umwelt/Verkehrsunfaelle/Tabellen/liste-strassenverkehrsunfaelle.html#251628. Accessed January 11, 2023 (2023)

25. Verordnung (EU) 2018/858 des Europäischen Parlaments und des Rates vom 30. Mai 2018 über die Genehmigung und die Marktüberwachung von Kraftfahrzeugen und Kraftfahrzeuganhängern sowie von Systemen, Bauteilen und selbstständigen technischen Einheiten für diese Fahrzeuge, zur Änderung der Verordnungen (EG) Nr. 715/2007 und (EG) Nr. 595/2009 und zur Aufhebung der Richtlinie 2007/46/EG (Text von Bedeutung für den EWR.) (2018)
26. Verordnung (EU) 2019/2144 des Europäischen Parlaments und des Rates vom 27. November 2019 über die Typgenehmigung von Kraftfahrzeugen und Kraftfahrzeuganhängern sowie von Systemen, Bauteilen und selbstständigen technischen Einheiten für diese Fahrzeuge im Hinblick auf ihre allgemeine Sicherheit und den Schutz der Fahrzeuginsassen und von ungeschützten Verkehrsteilnehmern, zur Änderung der Verordnung (EU) 2018/858 des Europäischen Parlaments und des Rates und zur Aufhebung der Verordnungen (EG) Nr. 78/2009, (EG) Nr. 79/2009 und (EG) Nr. 661/2009 des Europäischen Parlaments und des Rates sowie der Verordnungen (EG) Nr. 631/2009, (EU) Nr. 406/2010, (EU) Nr. 672/2010, (EU) Nr. 1003/2010, (EU) Nr. 1005/2010, (EU) Nr. 1008/2010, (EU) Nr. 1009/2010, (EU) Nr. 19/2011, (EU) Nr. 109/2011, (EU) Nr. 458/2011, (EU) Nr. 65/2012, (EU) Nr. 130/2012, (EU) Nr. 347/2012, (EU) Nr. 351/2012, (EU) Nr. 1230/2012 und (EU) 2015/166 der Kommission (Text von Bedeutung für den EWR) (2019)
27. Verordnung (EG) No 661/2009 des Europäischen Parlaments und des Rates vom 13.7.2009 über die Typgenehmigung von Kraftfahrzeugen, Kraftfahrzeuganhängern und von Systemen, Bauteilen und selbstständigen technischen-Einheiten für diese Fahrzeuge hinsichtlich ihrer allgemeinen Sicherheit, Amtsblatt der europäischen Union, L200/1-24 (2009)
28. World Health Organization: Decade of action for road safety 2011–2020, Saving millions of lives. http://www.who.int/roadsafety/decade_of_action (2011)
29. Zeidler, F.: Die Bedeutung der Energy Equivalent Speed (EES) für die Unfallrekonstruktion und die Verletzungsmechanik. Brochure on a seminar held as part of an expert training. Mercedes-Benz, Sindelfingen (1992)

5 Accident Prevention and Severity Reduction

Adrian Zlocki and Lutz Eckstein

A key goal of vehicle development is to reduce the number of accidents to a minimum and to mitigate the consequences of accidents or, in the long term, to realize accident-free traffic ("Vision Zero", Sect. 1.4). The European Commission has set the goal of achieving this vision by 2050. To this end, the number of traffic fatalities is to be halved in the period from 2020 to 2030 and to reach a target value of approximately 11,400 people killed in road traffic across Europe by 2030. In 2021, the number of people killed in road traffic in Europe was approximately 19,900 [9].

Figure 5.1 shows the development of individual traffic, the police-recorded accidents, and the fatalities of the last decades based on 1975 in Germany. Due to the increasing complexity of the driving task, the number of accidents has steadily increased. Passive safety systems such as airbags or improved vehicle structures, legal regulations, but also infrastructural measures, such as road conditions and intersection concepts, have continuously reduced the number of people killed in these accidents. The introduction of active safety systems and advanced driver assistance systems (ADAS) in the last two decades also had a positive influence on the development of accident statistics.

Legal regulations, such as the helmet obligation for two-wheelers and the seat belt legislation in vehicles, have also been increasingly supplemented by the obligation of active safety systems in vehicles. For example, the legislator has made the equipment with emergency braking systems in commercial vehicles and buses mandatory more than a decade ago through European type-approval regulations. The technical requirements

A. Zlocki (✉)
fka GmbH, Aachen, Germany
e-mail: adrian.zlocki@fka.de

L. Eckstein
Aachen, Germany

R. Schöneburg (ed.), *Integral Vehicle Safety*,
https://doi.org/10.1007/978-3-658-48100-1_5

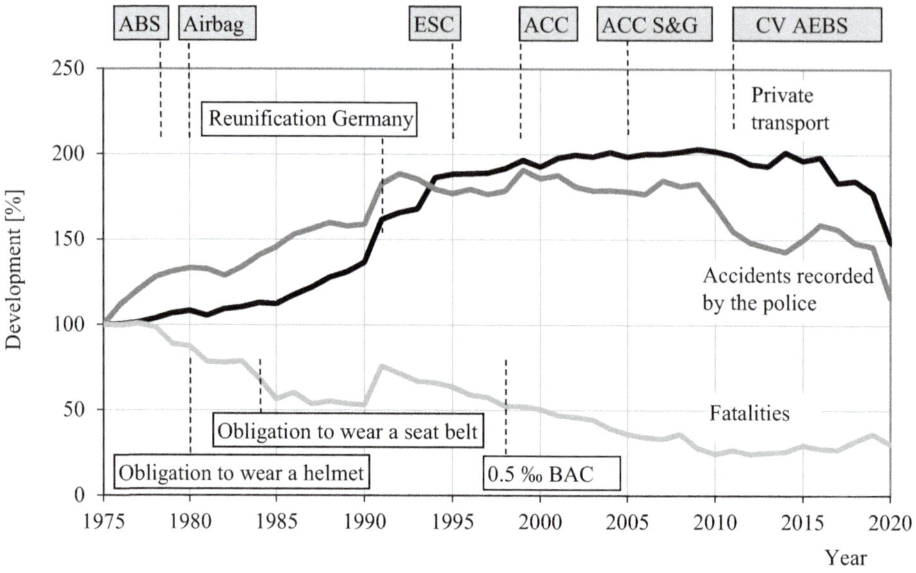

Fig. 5.1 Development of individual traffic, accidents, and fatalities in Germany (base 1975: 441.1 billion cars, 1.26 million accidents, and 14,870 fatalities) based on [16, 22]

for so-called AEBS (advanced emergency braking system) were specified identically in European Regulation No. 347/2012 and in UN-R 131. In a study to investigate the influence of emergency braking systems on the development of truck rear-end collisions on federal highways, the Federal Highway Research Institute Germany determined a total effect of up to 37% using the odds ratios methodology [23].

AEBS belongs to the class of active safety systems. Active safety describes all measures that prevent accidents or contribute to accident prevention. These include perceptual safety, operation and interaction safety, driver fitness (conditional safety), and particularly driving and handling safety. Active safety can be temporally divided into the phases of normal driving, accident prevention and pre-crash (see Fig. 5.2). Pre-crash refers to the phase in which preventive measures can be taken to reduce accident and injury severity. The integrative consideration of active and passive safety in the sense of an integral safety approach is based on the possibility of networking sensors and actuators of previously isolated systems with each other and thus generating new functions.

Systems for accident prevention and damage reduction can be classified based on their mode of operation using the three-level model of driving tasks (see Fig. 5.3). In the three-level model, the driving task is assigned to the three levels of navigation, path control, and stabilization [4]. On the navigation level, the driver decides on the driving route within an existing road network. During the drive itself, navigation extends to the perception of the necessary information to maintain the route and possibly adjust the driving route due to changed boundary conditions. On the path control level, the driver adjusts his driving style to the road course and surrounding traffic he perceives. The

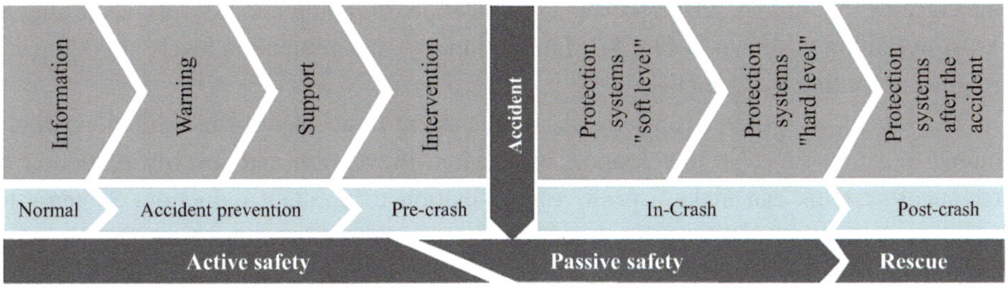

Fig. 5.2 Temporal sequence of a collision

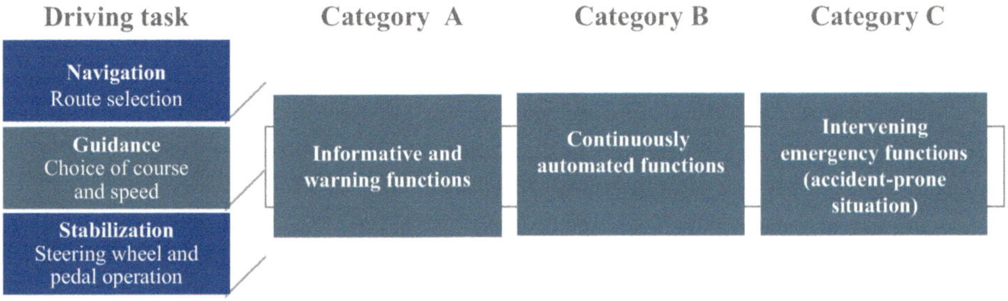

Fig. 5.3 Classification of systems on guidance level based on mode of operation [10]

stabilization level is characterized by the driver converting the chosen driving strategy into vehicle-side control variables, e.g., steering movement, accelerator pedal, brake, and gear position.

On the path control level, the vehicle task is again described and differentiated based on different modes of operation of systems into three categories. These are informative and warning functions, continuously automated functions, and emergency intervention functions in accident-prone situations. Functions for accident prevention and damage reduction are to be classified on the stabilization level and especially on the path control level, in addition to the driving dynamics control. The guidance level is divided into three further subcategories. Warning systems are found in category A, supporting systems in category B, and intervening systems in the pre-crash phase, especially in category C "emergency intervention functions" (see Fig. 5.3).

Category B of continuously operating automated driving functions is structured into automation levels. The classification of automation levels describes both the technical design of the system and the tasks remaining with the driver. The classification ranges from manual driving, where the driver permanently performs the longitudinal and lateral guidance, through "assisted", "partially automated", "highly automated" and "fully automated" to "driverless". The latter automation level corresponds to the "robot taxi", which performs all driving tasks at any time and under all environmental conditions manageable by human drivers. The classification of automation levels according to VDA definition is shown in Fig. 5.4. Internationally, the automation levels are defined in English according to SAE [14].

The human as a driver can only act with a delay in accident-prone situations due to human reaction time, so that support in accident prevention and severity reduction by technical systems can make a positive contribution to accident reduction. Systems for accident prevention and severity reduction become active in contrast to passive safety systems in time before an impending collision, see Sect. 1.3. The goal of these systems is to avoid the collision or to reduce the consequences of the collision. Especially in the pre-crash phase shortly before a collision, valuable measures can be initiated before the collision through appropriate sensor information about the environment, see Fig. 5.5. These measures consist in particular of reversible and an earlier triggering of irreversible restraint systems and technical systems to support the implementation of the driver's reaction.

The driver and the vehicle are the main factors in avoiding an impending collision. The physical infrastructure influences the course of the collision (road friction value), but so far has not been addressed by appropriate measures. Measures for the digital infrastructure and especially the vehicle-to-infrastructure communication V2I are being introduced. Manufacturers are equipping the first vehicle models with corresponding communication technologies.

The driver has the competence and responsibility for situation recognition and its assessment or situation evaluation. Once the situation has been correctly identified, the driver initiates measures to avoid the collision. Whether these are successful initially also depends on his abilities (reaction time, foot and hand forces etc.). The driver's reaction must be able to be implemented by the vehicle, i.e. there must be enough braking force to achieve the required braking distance or enough lateral force to evade. In this entire process of collision avoidance from situation recognition, through evaluation to the implementation of the reaction, the driver can be supported by technical systems or these can take over part of his tasks.

Conventional occupant protection systems are designed to recognize an accident during the impact, make decisions about the severity of the accident and initiate the

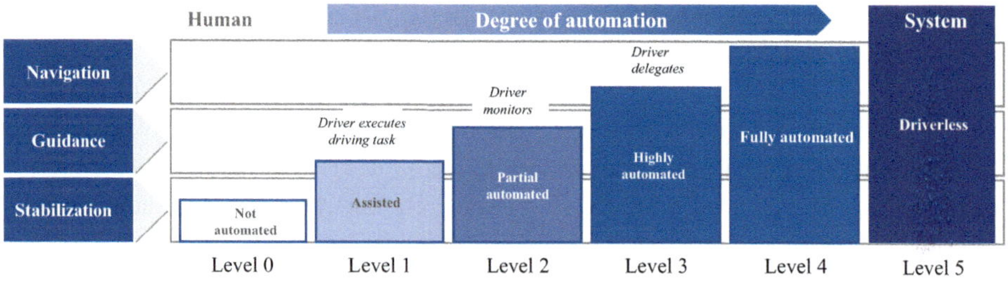

Fig. 5.4 Classification of automation levels according to VDA definition [7]

appropriate measures for personal protection, such as triggering the airbags. For these tasks, fractions of a second are usually available. Studies by Mercedes-Benz accident research of 371 accidents on the length of the pre-crash phase showed that in 59% of the collisions a longer pre-crash phase precedes, which already allows conclusions about the subsequent impact. These are available even before the collision occurs [12].

A pre-crash system essentially consists of the components perception, function and actuation. Perception is divided into environment sensors and vehicle condition sensors. The most important sensors are radar, LiDAR (laser scanner), infrared and image processing methods for environment detection as well as the vehicle condition sensors. The raw data captured by the sensors are then processed by the environment modeling and represented in a suitable environment model. Subsequently, the data relevant to the pre-crash system are extracted from the environment model and evaluated by the situation analysis. If a critical situation is detected, a suitable maneuver, such as braking or evasive maneuvers, is generated by the behavior generation and transmitted to the actuation. The basic structure of such systems is shown in Fig. 5.6.

The following describes this architecture for systems for accident prevention or accident consequence mitigation.

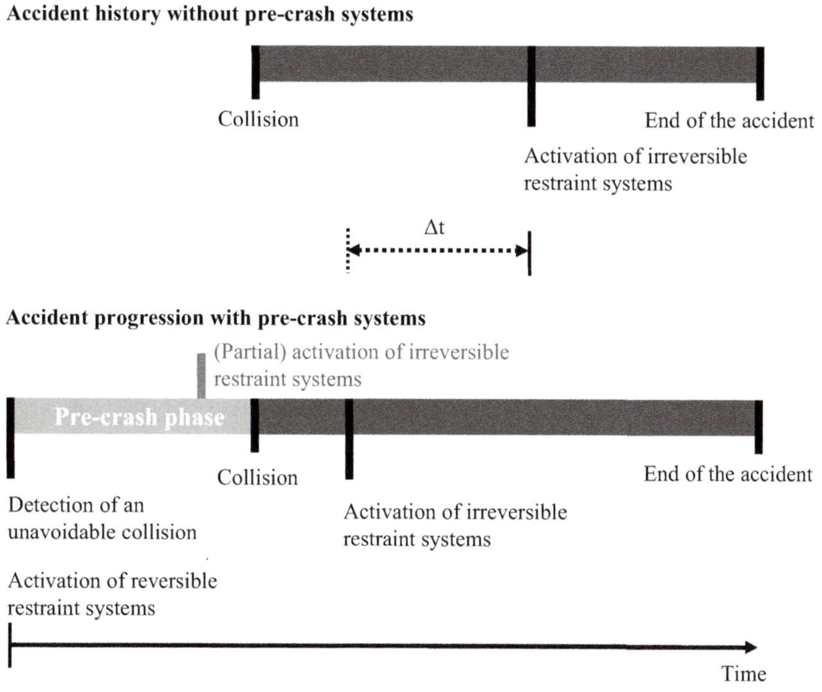

Fig. 5.5 Accident progression in the pre-crash phase and after the collision according to [3]

5.1 Perception of Critical Situations

Before a system for accident prevention or accident consequence mitigation can become active, the situation must first be recognized by the sensors installed in the vehicle. For this purpose, different sensors detect and monitor the vehicle condition and the vehicle environment. The raw data from the sensors are processed by algorithms and suitably represented as an environmental model. Subsequently, in the situation analysis, the collision risk, the severity of the collision, and possibly the exact time of collision are calculated based on different object and environment information, such as speed, distance, direction, and object classification. Appropriate metrics are used for this.

For environment detection, different distance sensors can generally be used:

- Radar sensors (both short-range and long-range radar)
- LiDAR sensors
- Cameras
- Ultrasonic sensors

For the sensor selection of a pre-crash system, among other things, the planned speed range must be taken into account, as the requirements for the sensor detection range can be derived directly from the speed range. Basically, the pre-crash application must recognize an impending danger source in time. Timely means at this point that the system must warn the driver at a time when the driver can still avoid the collision. This requires that there is enough distance to the source of danger so that the driver can react to the warning and perform evasive maneuvers (braking or steering). This results in the requirement for the sensor range to:

$$s_{sensor} > s_{reaction\ driver} + s_{evade}(v_{max\ system}) \quad (5.1)$$

The correct classification of a potential danger requires reliable detection first. This means that it must be ensured that an object is not mistakenly classified as dangerous ("false positive") or, conversely, a dangerous object is not recognized ("false negative"). Especially an intervention due to a "false positive" detection can unnecessarily lead to a

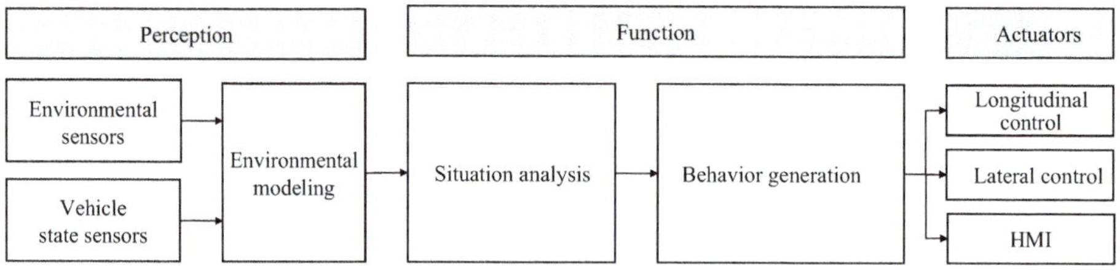

Fig. 5.6 Functional architecture of pre-crash systems based on [1]

dangerous situation in road traffic and should therefore be avoided. One way to ensure reliable detection is the data fusion of two or more sensors.

The data from powerful sensors, such as LiDAR and radar sensors as well as camera systems, are fused to utilize redundancies and ensure reliable environment detection. Data fusion plays an important role in the development of pre-crash systems. In Fig. 5.7, the results of data fusion for pre-crash relevant systems are shown.

Furthermore, data fusion can also be used for more accurate determination of the object position, as the weaknesses of individual sensors can be compensated. Thus, by combining a camera sensor (positive: lateral position determination; negative: distance determination) and a radar sensor (positive: distance measurement; negative: lateral position determination), a more accurate position of the detected object can be determined, see Fig. 5.8. This is particularly important with regard to the planning of evasion trajectories for pre-crash systems.

The situation analysis evaluates the fused sensor data and calculates the collision prediction based on the relevant object properties and the object classification, e.g., with the collision probability and the direction of impact relative to the own vehicle using metrics. Based on this data, the corresponding protection mechanisms can be activated by the actuator generation.

5.2 Metrics

To determine the potential danger in situation analysis, a measure to describe the hazard potential of an object must be found. Distance sensors generally provide the relative position and speed of the potential collision object in relation to the ego vehicle. However, this information alone does not yet describe how critical a situation is, and does not answer the question of whether a pre-crash system needs to intervene. Therefore, to describe the hazard potential of a situation, so-called metrics are used as a measure of the hazard potential.

Metrics, along with a selection condition (threshold), are necessary for the automatic identification of relevant, critical situations or scenarios. These are summarized under the term criticality metrics [11].

A distinction is made between deterministic and probabilistic metrics for the a priori assessment of criticality. Typical deterministic metrics are time-to-X metrics (TTX). The best-known representative is the time-to-collision (TCC). Other temporal metrics are the time-to-line-crossing (TLC), time-to-brake (TTB), time-to-steer (TTS) and time-to-kick-down (TTK), summarized as time-to-react (TTR). Other deterministic metrics use driving dynamic quantities for criticality estimation, such as the required deceleration (a_{req}), which determines the necessary braking deceleration for accident prevention. However, these metrics are only applicable if a collision is expected with the prediction used. No criticality is detected in predicted close passes. Probabilistic methods, on the other hand, are used primarily in robotics [11].

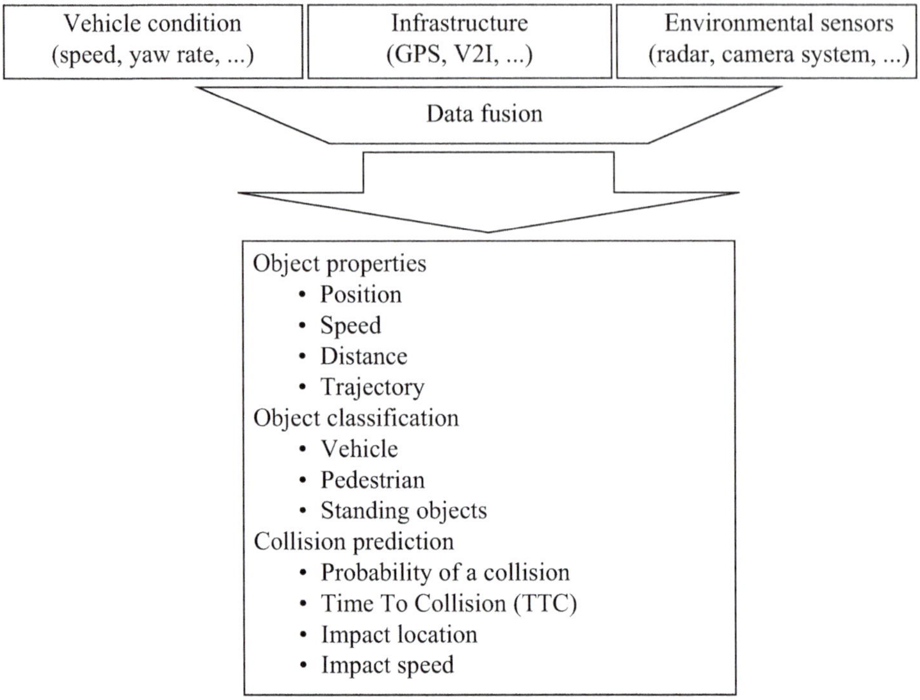

Fig. 5.7 Data fusion of different sensor systems

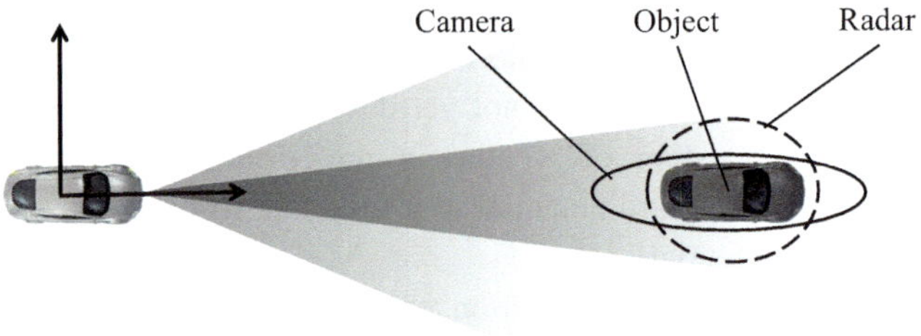

Fig. 5.8 Data fusion using the example of radar and camera sensors

In traffic research, the time-to-collision (TTC) has particularly established itself. This describes the time remaining until an accident occurs at a constant driving state (constant speed). It is calculated from the quotient of distance and relative speed between two vehicles. Positive TTC values result from an approach, negative values from a distance. For the assessment of traffic safety, the frequency of low and positive TTC values over a certain period and on a defined route section is relevant. All TTC values that are lower than about five seconds are considered dangerous. If TTC values are less than one second, an accident is usually unavoidable, as the driver's reaction time itself is about one second. However, such values can occur in reality without leading to an accident if the driver has already started to react. It should also be noted that the TTC can only be calculated if both vehicles are on a collision course (Fig. 5.9).

$$\mathrm{TTC} = -\frac{\mathrm{dx}}{dv} = -\frac{\mathrm{dx}}{v_1 - v_2} \tag{5.2}$$

If the vehicle is in the area of low TTCs shortly before a collision, so-called pre-crash systems become active. Pre-crash systems represent a combination of active and passive onboard systems. So far, vehicle safety has primarily pursued an approach to mitigate the consequences of accidents, in which sensors only provided information to existing safety systems for personal protection at the time of the accident. However, information from the so-called pre-crash phase shortly before the impact is helpful for an improved triggering strategy.

The time-to-lane-crossing (TLC) describes the duration until the lane marking is crossed if the current driving state is maintained. The TLC is calculated from the distance to the lane marking and the lateral speed of the vehicle, cf. Eq. 5.3.

$$\mathrm{TLC} = \frac{\mathrm{dy}}{\mathrm{v_y}} \tag{5.3}$$

Metrics are used in situation analysis to assess the dangers in critical situations and trigger pre-crash systems for accident prevention when thresholds are exceeded or not reached.

5.3 Systems for Accident Prevention

In principle, there are two ways to avoid an impending accident.

1. By timely warning the driver, his attention is drawn to the situation, thus giving him the opportunity to defuse the critical situation himself through an appropriate driving maneuver. During the driving maneuver, the driver can be supported by systems installed in the vehicle (e.g., brake assist).
2. A pre-crash system avoids the impending accident by intervening in the vehicle's behavior. Both lateral and longitudinal interventions are possible.

Different approaches are known for situation analysis, which are based on empirical studies or kinematic relationships and can use time-, distance-, and delay-based metrics.

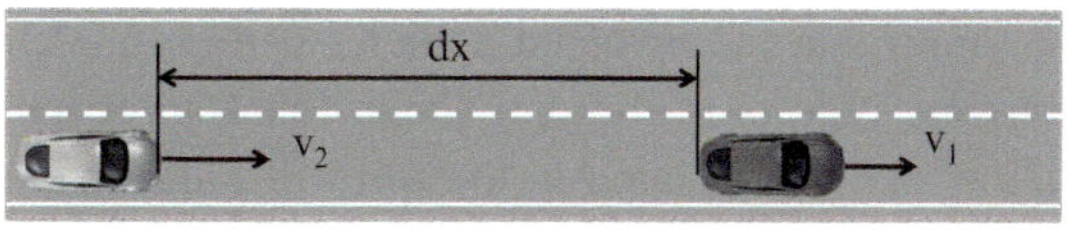

Fig. 5.9 Definition of the TTC

Warning systems (also called collision warning) are often used as pre-warning stages of an emergency braking system, i.e., before initiating an automatic emergency braking, a one- or multi-stage warning of the driver takes place, with the warning system using the same algorithms with different limits for the last escalation stage. The following describes warning systems for accident prevention and intervention in the pre-crash phase.

5.3.1 Warning Systems

In order for the driver to defuse a dangerous situation himself, his attention must be directed to the situation. This can be achieved by timely warning the driver.

Warning systems can be divided into warnings of critical conditions of the driver, warning of critical conditions of the vehicle, and warning of critical driving situations.

Critical conditions of the driver can be detected by so-called DMS (driver monitoring systems). These warn, for example, in case of fatigue.

Critical conditions of the vehicle include, for example, loss of pressure in the tires, which can be addressed by tire pressure monitoring systems that monitor the air pressure of the tires.

Critical driving situations can affect both longitudinal and lateral guidance. The lane departure warning is a driver warning that monitors the lateral guidance of the vehicle in the lane and warns when the vehicle threatens to leave the lane. The lane change assistant, on the other hand, usually monitors the lateral space behind the vehicle over a width of several lanes using radar sensors. If other road users enter the monitored area (especially in the area of the blind spot), the driver is warned as soon as he activates the turn signal, thus revealing his desire to change lanes. Blind spot detection systems warn the driver of dangerous lane changes when other road users approach from the side behind the vehicle. The systems use radar sensors or image processing to monitor the side and rear area of the vehicle. If a vehicle is in the neighboring lane in the driver's blind spot, information is initially provided. Similar to the lane change assistant, which monitors a larger area on the adjacent lanes, a warning is issued as soon as the driver activates the turn signal.

In longitudinal guidance, the distance warning can initially be given by a distance warning system that warns the driver of driving too close to the vehicle in front. In already critical situations, the so-called forward collision warning warns the driver of a collision with a vehicle in front.

The distance or collision warning uses the data from the distance sensors and is offered as an additional function to the ACC system. The algorithms used in warning systems can be divided into one- and multi-dimensional approaches. This classification assumes that the involved vehicles can be modeled as point-like, purely longitudinally moving objects in the one-dimensional approaches. The risk analysis of a traffic situation can be evaluated in the form of a distance-, deceleration-, or time-based metrics.

Accordingly, a warning is issued when the current distance is smaller than the warning limit distance calculated by the algorithm, the amount of required delay is greater than a predetermined reference value, or a warning time is less than a reference time.

Time-based metrics are often used for both distance and collision warning. These consist of the time remaining until the last possible moment at which the driver can initiate an accident-avoiding reaction ($t_{reaction}$) and the duration of the accident-avoiding reaction, e.g., emergency braking. This is referred to here as $t_{evasive\ maneuvers}$. To determine this time, both the vehicle speed and the maximum possible accelerations during the evasive maneuver must be taken into account.

$$t_{evasive\ maneuver} = \frac{v}{a_{max}} \tag{5.4}$$

For an emergency braking maneuver, the maximum possible deceleration a_{max} is given by:

$$a_{max} = g\,\mu \tag{5.5}$$

It also applies that the TTC—the remaining time until the accident—is composed of both temporal quantities.

$$TTC = t_{reaction} + t_{evasive\ maneuver} \tag{5.6}$$

For the determination of the criteria, the driver's reaction time can be taken into account. The reaction time is between 3 s (inattentive driver with slow reaction) and 0.7 s (extremely responsive driver). The reaction for an inattentive driver with a quick reaction is 1.3 s according to [15], see Fig. 5.10.

For warning the driver, there are a whole range of possible options. The warnings can be divided into visual (warning lights, warning symbols in the instrument cluster, warning symbols in the head-up display, etc.), acoustic (warning tones) and haptic warnings (e.g., vibrating steering wheel, accelerator pedal, brake pedal or seat belt, application of steering torque or braking). The different types of warning can of course also be combined.

However, the warning should always be understandable and intuitively comprehensible for the driver. It must be taken into account that the driver is in a critical situation and has only a short time to react. Therefore, certain criteria should be considered in the design.

For visual warnings, the following points should be considered [20]:

- Brightness/intensity of the warning,
- Visual acuity/spatial contrast sensitivity,
- Color specification,
- Optical representation (arrangement, similarity, continuity, resolution)
- As well as the position of the warning, which should be in the driver's field of vision.

For auditory warnings, the intensity of the warning, the position of the speakers, and the tone frequency (between 200 Hz and 6000 Hz) are decisive. Furthermore, possible ambient noises should also be considered in the design of the warning [20].

With the warning of the driver, measures can be taken in the vehicle at the same time to best support the driver in the event of possible evasive maneuvers. The technical measures can include, for example, sensitizing the chassis control systems for lateral evasion or the brake booster or brake assistant in the event of the driver braking. The sensitization of these systems can be achieved, among other things, by lowering trigger thresholds, pre-building brake pressure in the system, or by lightly applying the brake pads to the brake discs. All these measures have in common that they aim to reduce the dead time between driver reaction and vehicle reaction.

5.3.2 Support Systems

Basically, a distinction is made between systems for longitudinal and lateral guidance support. Supportive systems can be further divided into continuously supportive systems and those that support the driver in his action, such as the brake assistant. This increases the brake pressure depending on the brake pedal operation and possibly environmental information. Continuously supportive functions can be classified according to the automation levels according to SAE [14]. These range from driver assistance in longitudinal or lateral guidance to full automation of the vehicle. Up to and including level 2, the driver is fully involved in the driving task and is only supported by the system. From automation level 3, the system also takes over the entire driving task (see section "Automated Driving"). At this point, no further discussion will be made of continuously supportive systems, as they do contribute positively to accident prevention [13], but were not explicitly developed for collision avoidance, but to relieve the driver.

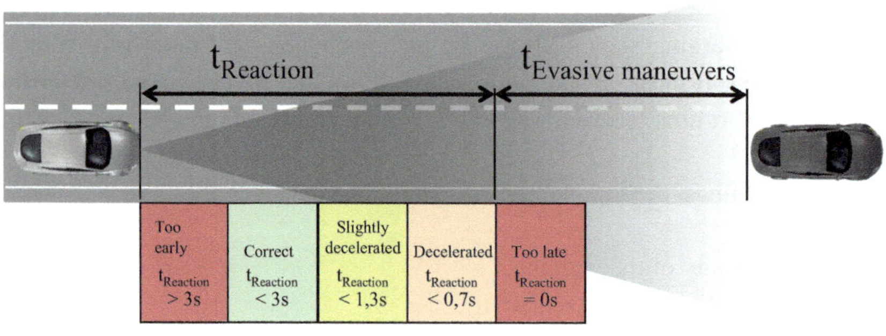

Fig. 5.10 ADAC criteria for a timely warning of the driver according to [15]

5.3.3 Intervening Systems

If a critical situation occurs in the pre-crash phase, measures for accident prevention can be initiated that intervene in the vehicle's driving dynamics. Ideally, a collision can be avoided, which is referred to as collision avoidance. From medium driving speeds, it is increasingly rare to avoid a collision, but it is at least mitigated by the interventions of the chassis systems, which is referred to as collision mitigation. In general, intervening systems can influence the longitudinal dynamics and or the lateral dynamics of the vehicle.

In the longitudinal direction, accident prevention is sought through an automatic emergency braking system that independently initiates a braking maneuver to prevent a collision. The braking takes place at a point in time before the crash when drivers are generally no longer in the location to avoid the accident.

For a stationary target vehicle, the minimum distance for collision avoidance through a braking maneuver can be estimated. The minimum longitudinal distance at which a braking maneuver is still possible is calculated from the following kinematics:

$$dx_{\min,brake} = \frac{v^2}{2\, a_{longitudinal,\max}} \tag{5.7}$$

with:

$dx_{min,brake}$: Minimum distance at which a collision-avoiding braking maneuver is possible [m]

v: Ego speed [m/s]

$a_{longitudinal,max}$: Maximum possible deceleration [m/s^2]

Accident prevention in the lateral direction works by means of an evasive maneuver automatically carried out by the system. For this, it is necessary that the system applies a yaw moment. This can be done by applying a steering moment or by braking individual wheels. Performing a lateral evasive maneuver places the highest demands on situation recognition, as it is not enough to just detect the impending collision. Rather, it must also be ensured during the evasive maneuver that there is no collision with a third road user. This means that the system must be in the location to recognize vehicles in adjacent lanes, if the currently driven lane is to be left during the intervention. Furthermore, infrastructural information (e.g., is the direction of travel in the adjacent lane the same) must also be taken into account. A lateral evasive maneuver is usually initiated at the latest possible point in time.

For a stationary target vehicle, the distance at which an evasive maneuver is still possible (see Fig. 5.11), assuming a circular evasive trajectory, can be determined in a simplified manner according to Eq. 5.8.

The minimum distance for an evasive maneuver is derived from the estimation of the radius of the evasion trajectory according to the lateral dynamics equation:

$$R = \frac{v^2}{a_{lateral,max}} \tag{5.8}$$

with:

R: Radius of the evasion trajectory [m]
v: Ego speed [m/s]
$a_{lateral,max}$: Maximum possible lateral acceleration [m/s²]

The Pythagorean theorem from the triangle between the vehicles and the center is expressed as:

$$\begin{aligned} &(R - B)^2 + dx^2 = R^2 \\ &\Rightarrow dx = \sqrt{R^2 - (R - B)^2} = \sqrt{2R \cdot B - B^2} \end{aligned} \tag{5.9}$$

Combining equations Eqs. 5.8 and 5.9 results in the minimum distance at which an evasive maneuver is still possible. If the longitudinal distance to the target vehicle is less, an evasive maneuver without collision is no longer feasible.

$$dx_{min,evade} = \sqrt{2\frac{v^2}{a_{lateral,max}} \cdot B - B^2} \tag{5.10}$$

In principle, an accident avoidance system can also combine both strategies for accident prevention. In this case, depending on the speed, it must be decided whether to react to the danger by means of an evasive maneuver or by means of a braking maneuver. The two equations Eqs. 5.10 and 5.7 provide a way to estimate the probability of an accident in the case of a stationary target vehicle.

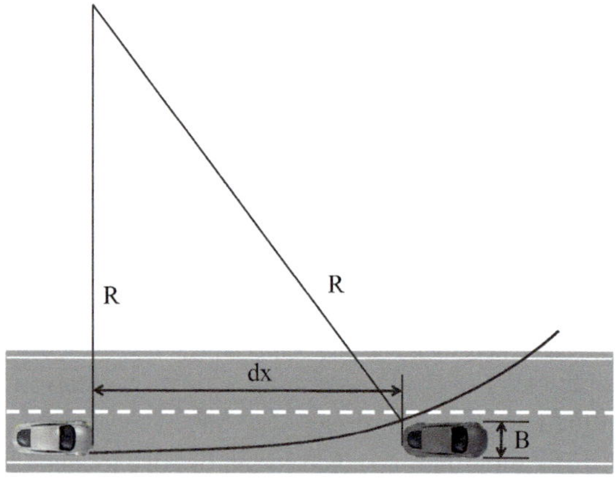

Fig. 5.11 Schematic diagram of an evasive maneuver

Figure 5.12 shows the minimum distances for collision avoidance by evasion and braking depending on the vehicle's own speed. The following assumptions were made for this diagram, which are based on real vehicle data:

- $a_{longitudinal,max} = 8\ m/s^2$
- $a_{lateral,max} = 6\ m/s^2$
- $B = 3\ m$

Although higher values for the maximum lateral acceleration and the maximum deceleration are achieved in reality, these cannot become effective immediately. The assumed values are considered as average values.

The diagram shows that at speeds less than 56 km/h, a potential collision can be more easily avoided by a braking maneuver than by an evasive maneuver. Braking maneuvers are more efficient for collision avoidance in this case. At higher speeds, evasion is more promising. Therefore, an autonomous collision avoidance system should only intervene at high speeds when evasion is no longer possible. This means that such a system could no longer avoid the collision by braking and should therefore also be able to steer autonomously. If the collision avoidance system is not able to perform an evasive maneuver, a collision is not avoidable, but the system can reduce the impact speed by the emergency braking maneuver and reduce the consequences of the accident.

In addition to stationary targets (e.g., ends of traffic jams), emergency braking systems are also designed for non-stationary targets. In this case, a movement of the vehicle in front is assumed. Now, whether a collision can be avoided depends on both the driving state of the following vehicle and the vehicle in front. The driving state of the vehicle in front can only be estimated until the time of the collision. For this estimate, a "worst-case" scenario is assumed, i.e., the vehicle in front accelerates (see Fig. 5.13). The predicted positions of the following and front vehicle are compared. In the case of an overlap, an emergency brake is triggered.

The course of vehicle deceleration for an emergency braking system for higher speeds is shown in Fig. 5.14. The warning cascade (initially visual and acoustic warning, then haptic warning) is intended to draw the driver's attention and motivate the driver's action in the next step. A speed reduction already occurs in the haptic warning phase, but it only reaches the full deceleration value of the vehicle in the pre-crash phase shortly before impact.

In the range of low speeds (see Fig. 5.12, speed less than 56 km/h), a purely longitudinally acting system can be used as a collision avoidance system. In this case, the system is activated at the moment when the driver can no longer prevent the collision by braking or steering, so that he is not overridden.

For speeds below 20 km/h, the distance at which an accident could be prevented by evasive maneuvers remains constant, as the vehicle limits the evasion trajectory through the turning circle. The turning circle radius is set at 5 m in the above example.

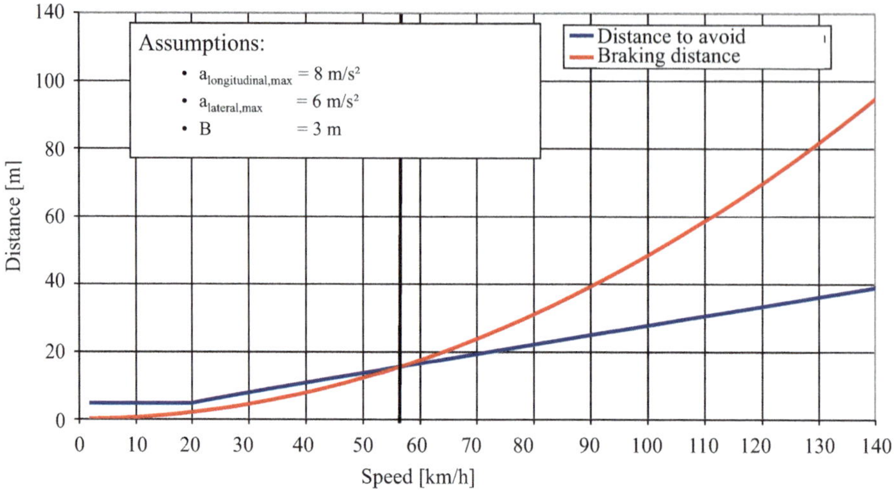

Fig. 5.12 Distances for accident prevention by evasion and braking depending on the vehicle speed

A system for the lower speed range up to 30 km/h was introduced in series in 2008 as the so-called "City Safety" system by Volvo Cars. The system's area of application is city and convoy traffic. The three-beam LiDAR sensor installed in the windshield detects other vehicles up to 10 m in front of the vehicle. The vehicle is braked at up to 0.7 g in the event of an impending rear-end collision. Thus, a collision can be avoided at a relative speed of up to 15 km/h. At relative speeds above 15 km/h, the collision is significantly mitigated. For this system, the warning phase for the visual/acoustic warning and the brake jerk from Fig. 5.14 is omitted due to the shortened time sequence.

Avoidance of accidents in which the opponents are in the blind spot or in an overtaking process will not be further discussed here. In principle, such accidents can be prevented following a warning by applying an opposing steering torque or by specifically braking individual wheels, similar to the lane keeping assistant.

5.4 Systems for Mitigating Accident Consequences

Mitigating accident consequences refers to measures that are initiated in the pre-crash phase with the aim of reducing the consequences of an accident. In contrast to the accident avoidance system, an accident usually occurs despite the system's intervention, as the systems only start braking when the driver no longer has any chance of preventing an impending collision. The transition between collision avoidance and collision mitigation is seamless, as many influencing parameters (friction value, behavior of others, etc.) affect the actual effect of such an intervention.

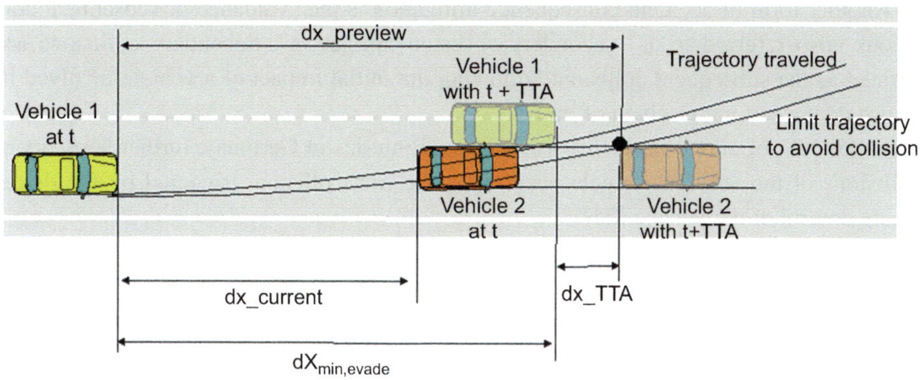

Fig. 5.13 Schematic diagram of an evasive maneuver for a non-stationary target object

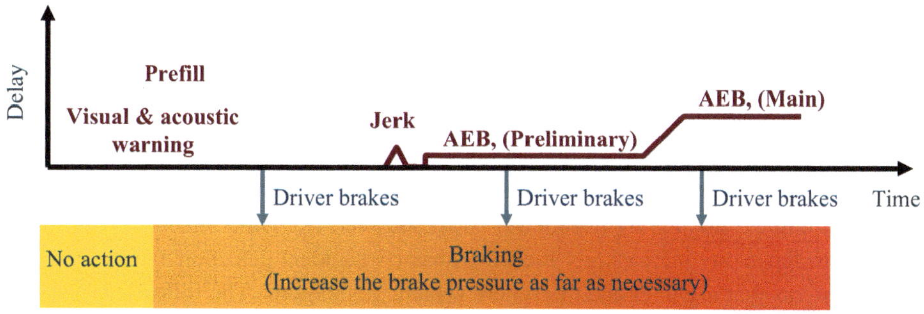

Fig. 5.14 Warning and braking cascade for collision avoidance systems in the higher speed range

In principle, there are two approaches to mitigating the consequences of accidents:

1. Reduction of speed. The reduction of speed aims to decrease the kinetic energy and thus also reduce the forces acting on the occupants during a crash. The speed is reduced by a brake intervention.
2. Optimization of the accident scenario. This involves attempting to improve the impact position or orientation through an intervention so that the vehicle's passive safety structures are optimally utilized. Optimizing the accident scenario often requires a lateral intervention. This intervention can be achieved through a steering actuator or by selectively braking individual wheels.

However, it should be noted that optimizing the accident scenario places high demands on the detection and decision logic as well as the actuator technology, as there is very little time left until the point of collision.

Another form of accident consequence mitigation is the avoidance of subsequent collisions, also referred to as "secondary collision mitigation". Secondary collisions are defined as the subsequent collisions following the initial impact of a vehicle involved in an accident event with a chain of impacts.

In 25–30% of all car accidents with personal injuries in Germany, further subsequent collisions of the accident vehicle occur after the first collision, triggered by the driver losing control of the vehicle [24].

The secondary collision mitigation system, also known as post-crash braking - or multi-collision braking system, uses the airbag sensors to detect the severity of the collision and triggers the secondary collision mitigation function from a certain threshold. The system automatically brakes the vehicle via the ESP control unit. By reducing the speed and the associated decrease in the vehicle's kinetic energy, further subsequent accidents are to be avoided or the severity of subsequent accidents reduced (see Fig. 5.15). The systems address subsequent collisions with other vehicles or infrastructure, as well as the impact of pedestrians and cyclists and the vehicle overturning.

5.4.1 Measures for Occupant Protection in the Pre-Crash Phase

The high demand of customers for safer vehicles promotes the development of measures for occupant protection in the pre-crash phase. The PRE-SAFE® system was first integrated into the S-Class by Daimler in 2002, see Sect. 1.3. Other manufacturers, especially from Japan, are increasingly focusing on early accident detection and the resulting safety systems that are integrated into new vehicle generations. The following examples are intended to demonstrate the range of functions of today's pre-crash systems.

The occupant protection system PRE-SAFE® detects a possible accident in advance and takes preventive protective measures. These include, for example, the rapid electrical seat belt pretensioning to reduce the seat belt slack, so that the driver and passenger are held in the best possible seating position before an impending collision and the airbags can work optimally at the time of impact. This measure is only activated if the seat belt buckle is fastened and the seat occupancy detection has recognized an occupant [17].

Figure 5.16 shows a comparison of the seat belt forces between seat belts with and without pretensioning over a period of 100 ms before and after the collision. In addition, a distinction is made between reversible and irreversible seat belt pretensioning by crash and pre-crash sensors. It can be seen that the levels of maximum seat belt forces for pre-tensioned systems are significantly lower compared to non-pretensioned systems. In addition to a seating position optimized for airbag systems, reversible seat belt pretensioning before the actual impact can also reduce the seat belt forces acting on the human body compared to a system without pretensioning. Furthermore, the temporal course of the seat belt forces of reversible pretensioning systems can be improved by the pretensioning and the seat belt force is applied to the body earlier [25].

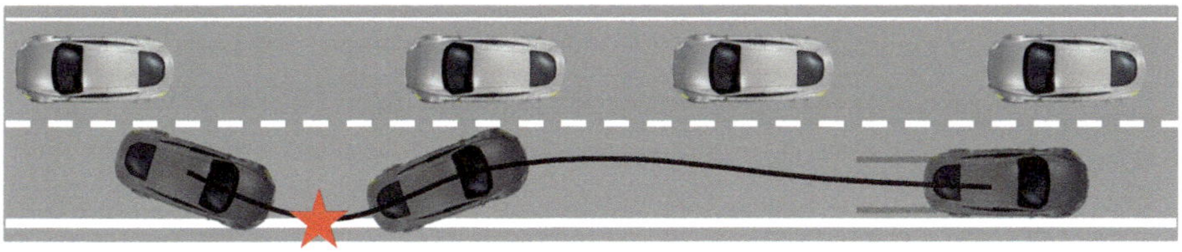

Fig. 5.15 Prevention of subsequent collisions using the secondary collision mitigation system

Reversible seat belt pretensioners, which can be pretensioned using electric motors, are already being used in pre-crash systems by different vehicle manufacturers today.

At the same time as the seat belt is tightened, the pre-crash systems can move the passenger seat and the electrically adjustable individual seats of the rear passengers into favorable positions and automatically close the sunroof and side windows when the vehicle is skidding. The driver's seat is not adjusted to avoid distracting the driver from the stressful driving situation. The components of the pre-crash systems are reversible. If the accident is prevented at the last moment, the preventive seat belt pretensioning automatically subsides and the occupants can return the seats and sunroof to their original positions. The systems are then immediately ready for use again [17].

Current production vehicles provide measures for the entire vehicle, such as preventive height adaptation, for example to increase the impact compatibility with a frontally colliding accident opponent.

5.4.2 Post-Crash Phase Systems

In addition to the possibilities in the pre-crash phase and during the accident to reduce the consequences of the accident, post-crash systems also allow the consequences of the accident to be reduced through optimized rescue management. The post-crash phase begins with the end of the in-crash phase and is therefore temporally to be classified behind the pre-crash systems, Figs. 1.7 and 5.2. The post-crash phase includes the run-out process of the vehicles and the subsequent rescue management.

If an accident has occurred, an emergency call must generally be made. The following are available for this purpose:

- Emergency call pillars,
- Mobile phones,
- Vehicle manufacturer related emergency call systems
- Witnesses or other participants.

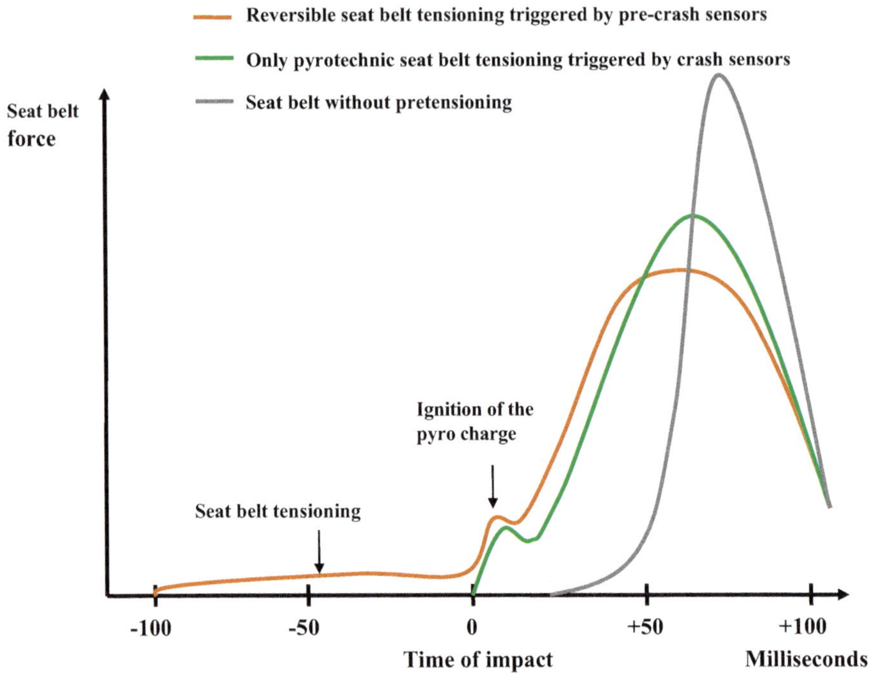

Fig. 5.16 Seat belt forces due to different seat belt systems according to [25]

If an accident occurs without witnesses, the occupants are left to their own devices. If people are no longer in the location due to serious injuries to send an emergency call, valuable time can pass until rescue measures are initiated.

Vehicle manufacturers have offered individual emergency call systems for rescue management after a collision for this case for many years. If an accident occurs, the systems send a signal to the manufacturer's own control center. The driver is connected to the center via the vehicle phone. If the driver cannot answer, the center immediately alerts the police and rescue service. The exact position data is known to the emergency call system thanks to satellite positioning using GPS. With some manufacturers, the emergency call is automatically made as soon as the airbag or other crash sensors are activated.

The disadvantage of these systems are the different concepts of the individual manufacturers. Each manufacturer represents its own proprietary system. For this reason, efforts were made at the European level to regulate emergency call systems under the name "eCall" uniformly by law. The EU Regulation 2015/758 regulates from 2015 the requirements for type approval for the introduction of the on-board eCall system based on the 112 emergency call in vehicles. Thus, eCall is mandatory to bring into the market for new vehicles in Europe [5].

In addition to the eCall system, many vehicles unlock the doors and disconnect the power line to the battery in the event of a collision, and in electric vehicles also the HV lines from the high-voltage battery.

In addition, it is conceivable that in addition to activating the rescue management, an accident site is also secured by vehicle-to-vehicle communication. In the event of a collision, the system recognizes the airbag deployment and transmits this information to the following vehicles. However, it is not only the information about the collision, but also about the location of this in relation to one's own vehicle that is important. The system must be able to determine whether the accident is on its own lane or, for example, at a nearby intersection. GPS systems are therefore necessary in both vehicles.

The so-called "Vehicle General Safety Regulation" GSR of 2019 requires accident data recorders, so-called event data recorders (EDR), to be integrated into passenger cars as a mandatory requirement. These record certain vehicle signals as soon as trigger thresholds are reached. This is intended to enable collisions to be clearly reconstructed afterwards in order to derive future measures for vehicle safety in the form of legal adjustments or function specifications from the knowledge [6].

5.5 Automated Driving

Driven by technological progress, particularly in terms of environmental detection and available computing power in the vehicle, various prototypes for automated driving behavior have been demonstrated to the general public. As early as the mid-1990s, the PROMETHEUS funding project produced automated driving research vehicles [21]. Another milestone for automated driving is the Grand Challenge of the US military research agency DARPA. The aim of this competition was to show that autonomous vehicles are capable of covering long distances in unstructured and unfortified terrain. In the first attempt in 2004, none of the 15 participating vehicles, which were developed at universities and through research collaborations, succeeded. The following year, four vehicles managed to autonomously cover a distance of 132 miles through the desert in a repeat attempt.

Unlike the Grand Challenge, the focus of the Urban Challenge 2007 was urban traffic. The participating vehicles had to cover a 60-mile course on the grounds of a military base. In this controlled field, vehicles driven by instructed drivers were on the move alongside the self-driving ones. The vehicles generally had the competence and responsibility to avoid other vehicles while observing the traffic rules valid in California. In addition to driving on the course on marked roads, these vehicles also had to act autonomously in zones without markings, for example, to park fully automatically in a parking lot. Eleven teams managed to complete the course within a given time frame with their automated vehicles. Compared to production vehicles of the time, the prototypes of the DARPA competitions were equipped with a multiple of computing power and an elaborate and expensive environment perception. The algorithms developed during these competitions, e.g., for trajectory planning, still form the basis for the further development of automated vehicles today.

In Germany, automated driving on the route between Mannheim and Pforzheim was demonstrated by Daimler in a modified S-Class in 2013. The route included urban and rural traffic, making the so-called Berta Benz Drive significantly more complex than other demonstrations at the time [2].

5.5.1 Structuring of Automated Driving

The topic of automated driving behavior can be divided into different levels, in which the driver, the vehicle, and the infrastructure are taken into account. In addition to purely technical aspects, such as the realization of driving behavior, the interaction between the system and the human (ergonomic and psychological level), legal as well as societal aspects need to be discussed and researched when introducing automated vehicles in order to develop viable technical solutions. The interaction between driver, vehicle, and environment forms the basis across all levels. Using the 5-level model [8], Fig. 5.17, topics or needs for action can be identified that must be researched before the market introduction of corresponding functions.

On a **societal level**, the acceptance of automated driving by different societal groups and the question of willingness to bear possible risks of automated driving need to be discussed. In addition to the presumed advantages of safer and more comfortable mobility, previously unknown types of accidents may occur as a result of the introduction of these functions, which stand in contrast to the advantages. Even considering macroeconomic aspects, such as the perspective of creating new industries and jobs, a more or less pronounced willingness emerges in a democratic society to adjust the legal framework in order to enable the development and market introduction of innovative technologies in the first place.

Of central importance is particularly the **regulatory level**. Here, vehicle-related technical regulation, but also laws regarding driver behavior and infrastructure-related laws are discussed. In Germany, many adjustments to the legal framework have been made in recent years, so that automated driving up to level 4 is legally possible.

On the **economic level**, suitable business models need to be established to justify the significant costs of development, validation, market introduction, and operation of automated vehicles, not only from the perspective of vehicle manufacturers and suppliers, but also, for example, municipalities, authorities, and service providers. Among other things, subscription models, fleet operations, car and ride sharing concepts are currently being tested. It is assumed that higher levels of automation will increasingly offer mobility as a service (MaaS), so that individual vehicle manufacturers will evolve into mobility providers.

The **level of human-machine interaction** (human factors layer) addresses human-related research questions, in the answering of which human cognition and psychology

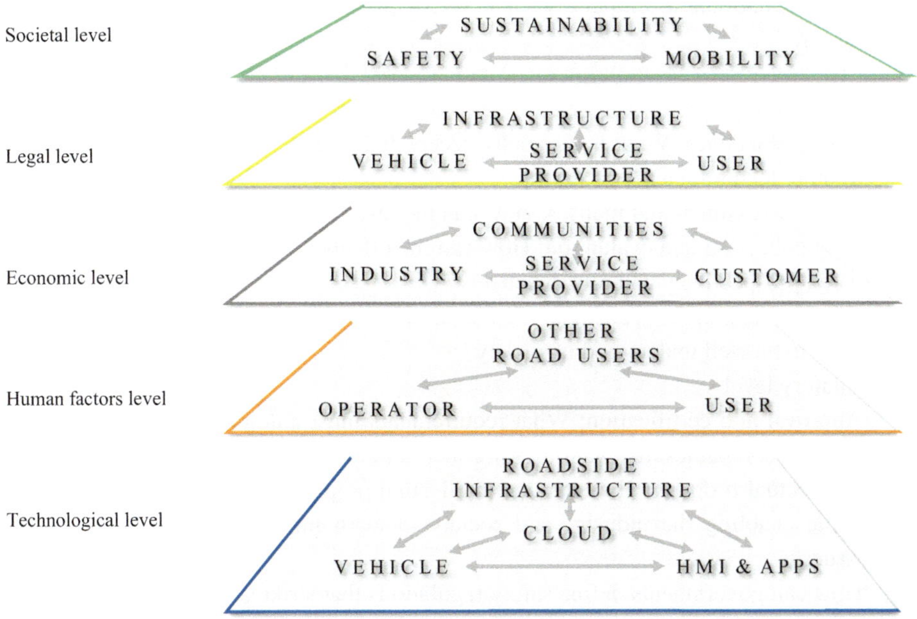

Fig. 5.17 5-level model for structuring automated driving according to [8]

play a role. Here, not only the driver or user is in focus, but also, for example, the effect of automated vehicles on other road users. Furthermore, the workplaces of people who are employed in a control center (legal technical supervision) must be designed in such a way that a "pilot" can accompany several vehicles from a distance.

The **technical level** is dedicated to the technical design of the interaction between driver, vehicle, and environment, from the sensory detection of the traffic environment to the assessment of the traffic situation, the planning and selection of a suitable trajectory of the vehicle, to its implementation via the coordinated actuation of vehicle systems such as steering system, drive, and brakes. In addition to the technical system itself, the strongly data-driven processes for the development, safeguarding, and updating of automated driving functions must also be designed so that the company wishing to introduce such a function into traffic can provide the necessary evidence to the licensing authorities. Since the sensors in the vehicle only have a limited range, which does not allow for comfortable control of the vehicle behavior in all driving situations, the networking of the vehicle with the environment or other vehicles and the "cloud" represents an important research or field of action.

The 5-level model is ideally suited to systematically structure central questions of automated driving, which in many cases have a direct relation to safety. A multitude of questions are currently being addressed and discussed in research and development projects at the national level, but also in an international context. Examples of individual questions are:

- Societal level
  - Safety objectives: What is a socially acceptable level of safety, also in comparison to other technologies?
  - Safety assessment and metrics: How can the level of safety be defined?
  - Data collection and evaluation: How can the effects of vehicle automation be evaluated prospectively and retrospectively?
  - Communication & acceptance: What information needs to be provided in what form to be well understood in society?
- Regulatory level
  - Approval and certification: What requirements must automated driving functions, but also companies that want to bring such functions into traffic, meet?
  - Intellectual property: How can the intellectual property of developers be protected while enabling the industry and society to learn important lessons from safety issues?
  - How can governments define safety regulations that strike a balance between innovation and protecting the public from unsafe, immature technologies, and how do they evolve over time?
- Economic level
  - Cost of development and safety assurance: How can data and scenarios be exchanged between organizations and countries?
  - Revenue sources: What additional income can be generated besides the revenue from the actual transport service to cover the costs?
- Interaction level
  - User-related interaction: What precautions must be taken so that a user can interact and drive safely with an automated vehicle, so that trust and a sufficient willingness to pay are established?
  - Technical supervision: How should the interaction in a control center be designed so that a "pilot" can accompany several vehicles?
  - Code of conduct for automated driving: How should automated vehicles interact with other road users?
  - Education: How can users and other road users be educated about the actual capabilities and limits of vehicle automation?
- Technical level
  - How should the environment detection be executed to meet the respective system requirements in terms of accuracy, latency, and fault tolerance?
  - How can the required computing power in the vehicle, but also in the backend and in control centers, be influenced and represented at manageable costs?
  - How can the actuators in the vehicle be designed redundantly so that the vehicle movement can be influenced sufficiently safely even in the event of a fault?

- How can the intelligence for automated driving be structured so that it can be made "capable of learning" through updates?
- How can a safety case be presented? How can existing methods be combined and what new methods are required?
- Securing of automated driving functions: How do the existing approaches complement each other and where are there gaps?
- Reference for safety consideration: How can established traffic safety statistics be used in the form of metrics? What potential do driver performance models have?
- Validity of models and simulations: How can models and simulations be improved to achieve sufficient validity? What procedures and criteria are required for the validation (and certification) of these instruments?
- Need for standardization: What is already taken into account and where is there additional need?

5.5.2 Functional Architecture of Automated Driving

The challenges of the technical level of automated driving behavior require interdisciplinary collaboration between vehicle engineering, computer science, control engineering, ergonomics, and psychology. The development of a powerful and modular E/E architecture for decoupling functions, software components, and hardware, as well as the early definition of clear interfaces, are the basis for successful function implementation and the agile creation and securing of function updates.

Known approaches divide the functional architecture of information processing into the areas of perception, planning, and action. This architecture has the disadvantage that the influence on the vehicle's longitudinal and lateral dynamics is fundamentally associated with the latency of the entire information processing.

Therefore, with the so-called A-model (see Fig. 5.18), a functional architecture is proposed in analogy to human information processing, which is divided into three levels according to Rasmussen [18]: the lowest level forms the skill-based behavior, which runs almost latency-free in the subconscious (e.g., reflexes), the middle level includes rule-based behavior (e.g., stopping at a red light), while the highest level represents the so-called knowledge-based behavior (e.g., route adjustment).

The information processing in the left, ascending leg of the 'A' initially extracts relevant features and objects from the sensor data and combines them into an environmental model. To make behavioral decisions, such as regarding a lane change, the automated driving function, like humans, must predict the future behavior of the relevant road users in the environment, which is naturally fraught with uncertainties. The actual behavior decision can be influenced by the control center in a driverless vehicle, up to the request for safe stopping, which is planned and implemented in detail by the respective vehicle.

The decided behavior is then planned out in the next step in the descending right leg of the 'A' in the form of concrete maneuvers and trajectories. The trajectory consists of the geometric path and the speed at which this path is traveled. To implement the trajectory, it is particularly useful to consider all chassis and drive systems integratively, as many systems can influence both the longitudinal and lateral dynamics of a vehicle.

Since the described information processing chain can take a time span of more than 100 to 200 ms, reflex-like behavior is a useful addition that can also come into play in the event of errors in higher layers of information processing.

Furthermore, it can be particularly useful for driverless vehicles to supplement the information processing in the vehicle with intelligence outside the vehicle. The operation of a fleet of driverless shuttle buses, for example, can be supported by environmental sensors at difficult-to-see intersections, which enable anticipatory and comfortable driving by completing the vehicle-based environment model with infrastructure-based information. In the UNICARagil project, it was also investigated to what extent driverless vehicles can gain spatially and temporally coded experiential knowledge from the chosen trajectories, which are then stored in a collective traffic memory and made available for all road users [26].

5.5.3 Safety Potentials of Automated Driving

Safety potentials and the potential societal benefits related to traffic safety due to increasing vehicle automation are a central issue for the introduction of automated driving. These are dependent on the market penetration of the systems. A methodological approach to address this issue was developed in [19].

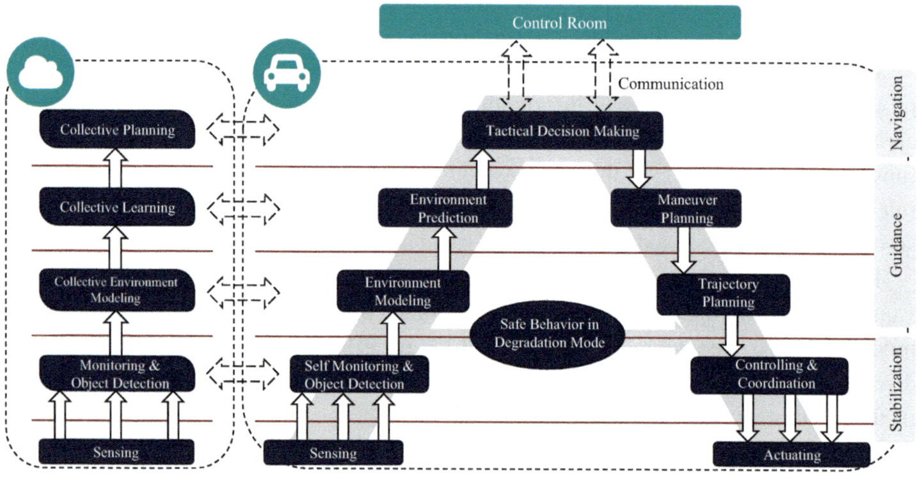

Fig. 5.18 A-model for the development of automated driving according to [26]

First, the fields of action of the respective driving functions are determined based on the function specification and the application domain. With these, the change in accident severity through accident resimulations and also the change in the frequency of occurrence of the scenarios are evaluated. Since automated vehicles, unlike active safety systems, work continuously, it is likely that certain accident scenarios (e.g., rear-end collision scenarios) will not be caused as frequently by automated driving functions. The change in the frequency of occurrence of different scenarios induced by automated vehicles is determined with a traffic simulation. The simulation results are scaled to the entire federal territory using an extrapolation methodology.

Five different automated driving functions of automation levels 3 and 4, see Fig. 5.4, were defined and examined. The individual automated driving functions to be considered are:

- Traffic jam chauffeur
- Highway chauffeur
- Commuter chauffeur
- Universal chauffeur
- Urban robot taxi

The application of the methodology to different automation functions results in the results shown in Fig. 5.19.

The automated driving functions can never prevent 100 % of the accidents in the domain, as accidents without car involvement are not addressed by the defined functions. A comparison of the number of potentially addressable accidents shows that urban operating driving functions can potentially address more accidents, as more accidents occur

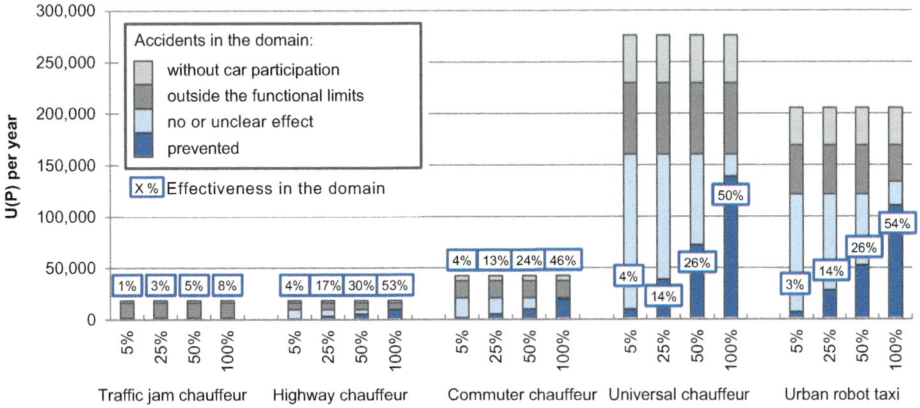

Fig. 5.19 Fields of action and effectiveness U(P) of the automated driving functions depending on market penetration [19]

on urban roads than interurban, especially more than on highways. When comparing the effectiveness, however, it turns out that these assume a comparable order of magnitude for all driving functions in the respective domain. The exception is the traffic jam chauffeur due to the low speed range. With a market penetration of 100 %, the driving functions have an effectiveness of approx. 46 % to 54 %, if they are related to all registered accidents with personal injury in the respective domain. It is estimated that a maximum of 83 % of all accidents can be avoided by vehicle automation of cars in the future. This assumes a fully automated driving function (level 5) that operates in all locations and on all road classes without speed restriction. This function would have to master all driving scenarios and have no functional restrictions and could also chauffeur people under the influence of alcohol or drugs. However, the series maturity of such a comprehensive fully automated driving function is not expected by experts in the next 20 years.

References

1. Bouzouraa, S., Reichel, M., Hofman, U., Siedesberger, K.-H., Siegel, A.: Grundlegende Architekturentscheidungen für hochautomatisierte Fahrerassistenzsysteme am Beispiel einer aktiven Gefahrenbremsung. 4. Tagung Fahrerassistenz, FTM, München (2010)
2. Dang, T. et. al.: Autonomes Fahren auf der historischen Bertha-Benz-Route. Technisches Messen. ISSN: 0170-575X, 0171-8096, 0340-4021, 0340-837X, 0365-7418, De Gruyter, Berlin (2015)
3. Domsch, C.: Vorausschauende passive Sicherheit am Beispiel des EU-Projektes CHAMELEON. 11. Aachen Kolloquium, Aachen (2002)
4. Donges, E.: Aspekte der aktiven Sicherheit bei der Führung von Personenkraftwagen. Automobil-Industrie **27**, 183–190 (1982)
5. EU-Verordnung 2015/758: des europäischen Parlaments und des Rates vom 29. April 2015 über Anforderungen für die Typgenehmigung zur Einführung des auf dem 112-Notruf basierenden bordeigenen eCall-Systems in Fahrzeugen und zur Änderung der Richtlinie 2007/46/EG, Brüssel (2015)
6. EU-Verordnung 2019/2144: des europäischen Parlaments und des Rates vom 27. November 2019 über die Typgenehmigung von Kraftfahrzeugen und Kraftfahrzeuganhängern sowie von Systemen, Bauteilen und selbstständigen technischen Einheiten für diese Fahrzeuge im Hinblick auf ihre allgemeine Sicherheit und den Schutz der Fahrzeuginsassen und von ungeschützten Verkehrsteilnehmern, Brüssel (2015)
7. Ebner, H.T.: Motivation und Handlungsbedarf für Automatisiertes Fahren. DVR-Kolloquium Automatisiertes Fahren. Bonn, 12,1013
8. Eckstein, L., Zlocki A.: Automated Driving – Concept and Evaluation. 23. Aachener Kolloquium, Aachen (2014)
9. Europäische Kommission: EU Road Safety Policy Framework 2021–2030 – Next steps towards "Vision Zero". Brüssel (2019)
10. Gasser, T., Seeck, A., Smith, B.W.: Rahmenbedingungen für die Fahrerassistenzentwicklung. Handbuch Fahrerassistenzsysteme, 3. Aufl. Springer, Wiesbaden (2015)
11. Junietz, P., Schneider, J., Winner, H.: Metrik zur Bewertung der Kritikalität von Verkehrssituationen und –szenarien. Uni-DAS Workshop Fahrerassistenz und automatisiertes Fahren, Walting (2017)

12. Justen, R., Baumann, K.-H., Schöneburg, R.: Pre-crash Erkennung, ein neuer Weg in der PKW-Sicherheit. VDI Bericht 1471. VDI Verlag GmbH, Düsseldorf (1999)
13. Kessler, C. et. al.: euroFOT – Final Report. Deliverable D11.3, euroFOT EU-Projekt, Grant agreement no.: 223945 (2012)
14. N.N.: Taxonomy and Definitions for Terms Related to On-Road Motor Vehicle Automated Driving Systems. SAE Standard J3016, USA (2018)
15. N.N.: Vergleichstest von Notbremsassistenten. Testbericht, ADAC Fahrzeugtest (2011)
16. N.N.: Verkehr in Zahlen 2022–2023, Kraftfahrt-Bundesamt, Flensburg (2022)
17. Paschek, L.: „Die Innovationen in der neuen Mercedes-Benz S-Klasse“, ATZonline, Springer Automotive Media, Wiesbaden (2004)
18. Rassmussen, J.: Skills, rules and knowledge; signals, signs and symbols, and other distinctions in human performance models. IEEE Trans. Syst. Man Cybern. SMC **13**(3), 257–266 (1983)
19. Rösener, C., Sauerbier, J., Zlocki, A., Eckstein, L., Hennecke, F., Kemper, D., Oeser, M.: „Potenzieller gesellschaftlicher Nutzen durch zunehmende Fahrzeugautomatisierung“, BASt-Bericht F 128 (2019)
20. Salvendy, G.: Handbook of human factors and ergonomics. Wiley, Hoboken (1997)
21. Siedersberger, K.H.: Komponenten zur automatischen Fahrzeugführung in sehenden (semi-) autonomen Fahrzeugen. Dissertation, Universität der Bundeswehr München, Institut für Systemdynamik und Flugmechanik, Neubiberg (2003)
22. Statistisches Bundesamt: Verkehrsunfälle 2023, Destatis. www.destatis.de (2023)
23. Straßgütl, F., Sander, D.: Einfluss von Notbremssystemen auf die Entwicklung von Lkw-Auffahrunfällen auf Bundesautobahnen. Berichte der Bundesanstalt für Straßenwesen Heft F 139 (2021)
24. Vitet, S., Schebdat, H.: Safety Benefit Evaluation of Secondary Collision Mitigation Braking. Paper Number 17–0269; 25th International Technical Conference on the Enhanced Safety of Vehicles (ESV) (2017)
25. Westerberg, L.: Future Safety Systems As a Challenge in an Changing Industry. Autoliv, Automotive News Congress (2002)
26. Woopen, T., et al.: UNICARagil – Disruptive Modular Architectures for Agile, Automated Vehicle Concepts. 27th Aachen Colloquium. Aachen, Germany (2018)

Occupant and Partner Protection

6

Frank Laakmann and Rodolfo Schöneburg

Occupant and partner protection aims to minimize or prevent the consequences of road accidents for occupants of different vehicle types, including passenger cars, commercial vehicles, and two-wheelers. Beyond vehicle occupants, this also includes road users outside the vehicle interior. These individuals are referred to as vulnerable road users (VRUs).

The development of protection concepts for vehicle occupants and VRUs is a challenging field of technology, as it requires the networking of different disciplines: The holistic view of traffic events, the analysis of accident root causes, the course of the accident, the possibilities for detecting an accident event and the development of strategies for intervention are examples of development tasks in occupant and partner protection (Fig. 6.1).

This chapter describes how protection concepts of integral safety for motor vehicles can be realized by linking the systems and components of passive safety with other domains. Examples of this include interventions by restraint systems already in the pre-crash phase or the fusion of sensors from different vehicle domains for differentiating intervention strategies and extended adaptability functions of restraint systems.

Close connections exist to *accident research* (Chap. 2), *biomechanics* (Chap. 3) and *accident detection* (Chap. 7). In addition, requirements for the protective effect of systems in vehicles and components and their verification by *legislation* and *safety ratings* must be taken into account in development and production (Chap. 4). Interfaces to active safety technologies for *accident prevention and mitigation*, described in Chap. 5, are addressed in several sections of this chapter.

F. Laakmann (✉)
ZF LIFETEC, Alfdorf, Germany
e-mail: occupantandpartnerprotection@outlook.com

R. Schöneburg (✉)
RSC Safety Engineering, Hechingen, Germany
e-mail: rsc@rschoeneburg.de

R. Schöneburg (ed.), *Integral Vehicle Safety*,
https://doi.org/10.1007/978-3-658-48100-1_6

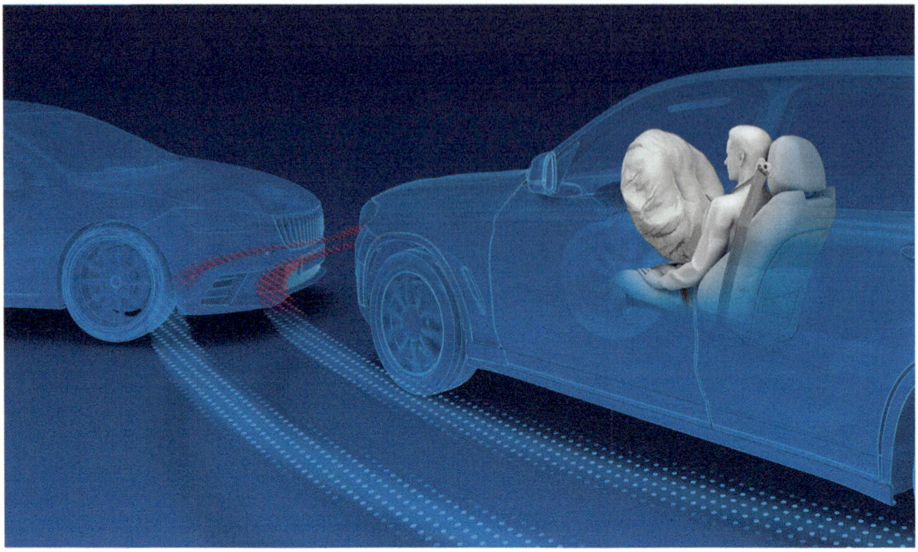

Fig. 6.1 Occupant and partner protection. (*Source*: ZF)

The structure of this chapter is shown in Fig. 6.2. The initial sections explain the basic concepts of passive safety (Sect. 6.1 and 6.2). The hierarchy of the subsystems of vehicle safety at the overall vehicle level and the classification of the systems of passive safety are described in Sect. 6.3.

The subsequent sections are divided according to the type of protection (*self-protection* and *partner protection*) as well as the considered vehicle segments: *passenger cars*, *commercial vehicles*as well as VRU and possibly their vehicles.

The protection mechanisms of a vehicle body structure and the passenger compartment of passenger cars are considered in Sect. 6.4. They also represent the system environment of the restraint systems of passive safety and thus influence the design of occupant protection concepts.

The core of this chapter is the introduction of *restraint system technologies*, especially of *seat belt systems* (Sect. 6.5) and *airbag systems* (Sect. 6.6). They contribute significantly to reducing the risk of injuriy—especially for vehicle occupants. They are supplemented by child restraint systems and technologies for occupant protection for further vehicle segments beyond passenger cars (Sect. 6.7).

Vulnerable road user and the passive safety technologies aimed at them are becoming increasingly important. On the one hand, the share of VRU out of all victims in traffic accidents is increasing due to advances in occupant protection. On the other hand, new

Fig. 6.2 Overview of topics and technology fields of the protection of vehicle occupants and vulnerable road users

challenges for traffic safety are arising from new mobility concepts and more diverse use of traffic spaces.

Section 6.8 includes measures for cars and commercial vehicles for partner protection of VRU as well as solutions for VRU vehicles. For VRU, solutions for self-protection through protective equipment and protective clothing are also presented.

Self-protection and partner protection in commercial vehicles are based on the same principles and technologies as in cars, but different requirements have to be considered and the system environment poses different challenges. Protective measures that are already in use are presented in Sect. 6.9. The vehicle segment of commercial vehicles includes both trucks and buses.

Passive safety also includes measures after the immediate accident event—the *post-crash phase* (phase: *after the accident,* Fig. 1.7). The *rescue chain* with its essential measures for the rescue and first aid of victims in traffic accidents is explained in Sect. 6.10.

6.1 Basics and Objectives

This first section is intended to explain basic terms of passive safety and to achieve a good understanding of their integration into vehicle safety. Often there is the perspective of a classic understanding with a focus on intervention in the in-crash phase as well as the extended perspective of integral safety.

6.1.1 Objectives of Passive Safety

The central objective of passive safety is the *reduction of accident consequences*. This includes those protective measures that help to avoid injuries or reduce the consequences of accidents for all persons involved in accidents—i.e., both *occupants* of vehicles and *vulnerable road users* such as pedestrian and cyclist. These objectives are also the subject of strategies and programs of the World Health Organization (WHO) and the United Nations (UN) to increase traffic safety [149].

The subfield of *occupant protection* is a significant technology field of passive safety, to which also *restraint systems* such as *seat belt systems* and *airbag systems* are assigned.

The focus of this publication is on the protection of road users and the goal of ensuring that they are not injured or less injured in accidents and that their lives are not endangered. Measures aimed at reducing property damage, such as repair techniques to reduce repair costs or improve insurance classification, are not covered here. Nevertheless, these are also important objectives in overall vehicle development and they are also influenced by the technologies presented here.

6.1.2 Self-Protection and Partner Protection

The consideration of the *direction of action* of passive safety measures leads to a distinction into two *types of protection* [61]. In Fig. 6.3, they are listed with the assignment of protection systems.

Self-protection serves to protect the occupants of one's own vehicle or, in the case of two-wheelers or similar vehicles, to protect the so-called riders. Systems and protective devices for self-protection include vehicle body and deformation structures, restraint systems, further systems in the compartment such as seating system, cockpit and protective

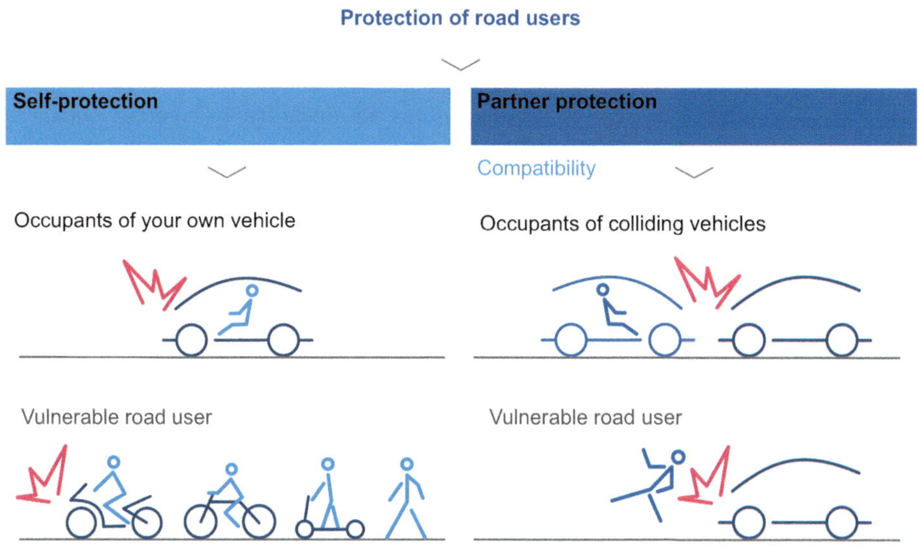

Fig. 6.3 Directions of action of passive safety systems: Self-protection and partner protection

devices such as child seats. For vulnerable road users such as two-wheeler riders, personal protective equipment is another measure for self-protection.

Partner protection serves to protect other road users outside of one's own vehicle through protective devices installed on one's own vehicle. This can refer to both occupants of other vehicles and vulnerable road users who collide with one's own vehicle. Vehicle body and deformation structures also function as systems and protective devices for partner protection. A measure in this area could be, for example, the optimization of the deformation behavior of the front structure on passenger cars or commercial vehicles. They are aimed at both self-protection and partner protection.

Partner protection also includes protection systems in the exterior of the vehicle. Examples of this are so-called pedestrian protection systems installed in passenger cars for the protection of VRU in case of frontal impact on one's own vehicle. Also, the goal of optimal effectiveness of protective devices of other vehicles counts towards partner protection. Examples of this are underride protection devices for commercial vehicles.

6.1.3 Direction of Action of Protective Measures

Direction of action of protective measures refers to whether they are more focused on *self-protection* or the protection of vehicle occupants *(internal safety)* or more on *partner protection* or the protection of occupants of other vehicles or vulnerable road users *(external safety)*.

Protective measures of *internal safety* are, for example, seat belt systems, airbags or padding in the interior design, as these measures act inwardly. In contrast, outwardly acting protective measures are the design of the outer surface as well as the shape and deformation behavior of the contact structure of vehicles. They are summarized under the term *external safety*.

In the 1990s, the compact car segment gained increasing importance again. While mid-size and larger passenger cars have a more or less sufficient deformation path, the design of smaller passenger cars (compact cars) has to compensate for a less capable front structure due to their small external dimensions. In addition, the passenger compartment is designed to be stiffer to ensure the survival space. This inevitably led to a relative decrease in external safety, which was noticeable by a significantly higher deceleration of the vehicle during the accident or the test. This reduction in external safety had to be compensated by internal safety. By foregoing a compliant, energy-absorbing deformation structure and using an improved occupant protection system, there was a shift in safety measures from outside to inside [64].

6.1.4 In-Crash Phase—Shorter than a Blink of an Eye

A traffic accident usually means a collision with an obstacle or another road user. The conversion of kinetic energy of the colliding objects takes place over a very short period of time. Therefore, in the in-crash phase from the moment of first contact t_0 until standstill, only a very short time span of usually less than 100 ms can be used.

A schematic comparison of the blink of an eye with an approximate duration of 150 to 200 ms and an accident event in the in-crash phase is shown in Fig. 6.4. The intervention of restraint systems begins just a few milliseconds after the impact. The in-crash phase is often significantly shorter than a blink of an eye.

6.1.5 Integral Protection Concepts

The measures of integral safety include those measures that are suitable for reducing accident risks, preventing accidents and minimizing accident consequences. The measures of active safety for accident prevention and severity reduction in the driving phase and in the pre-crash phase were already presented in Chap. 5.

An *exclusive* assignment of active safety to the phases *before* and passive safety to the phases *after* the collision *(in-crash-* and *post-crash phase)* is no longer purposeful. Rather, the close networking of *sensors, perception, cognition* and *actuators* of both technology domains in the pre-accident phase (Sect. 1.3) offers solutions for the

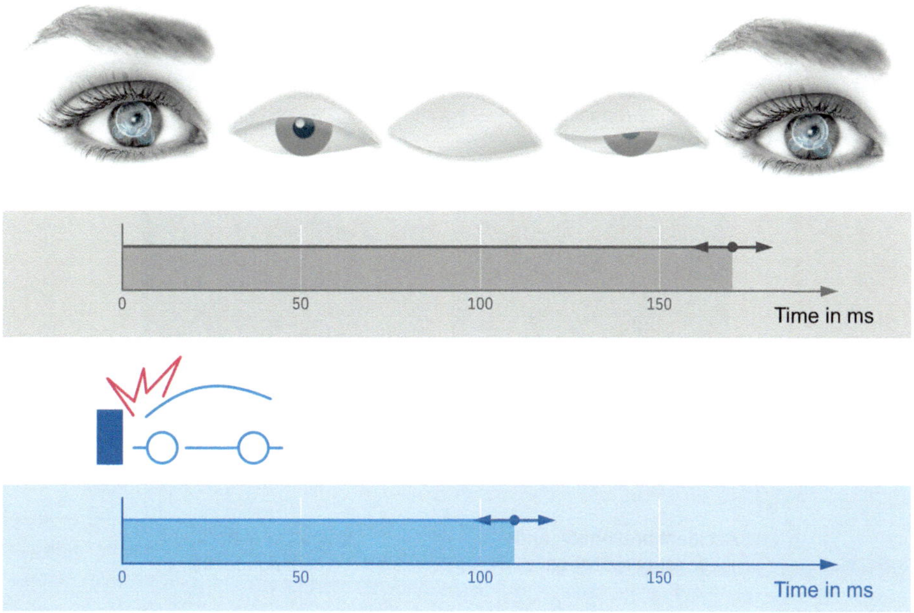

Fig. 6.4 Schematic comparison of the duration of a blink of an eye and a typical accident event as well as the intervention in the in-crash phase. (*Source*: ZF)

realization of protection functions of integral safety. Fig. 6.5 shows the development of passive safety technologies.

The *classic* focus of passive safety technologies in vehicles is on the mitigation of accident consequences and the measures aimed at this after the collision in the in-crash phase and the (immediate) post-crash phase. This original field of action has expanded in several areas to all phases before an accident event. The same applies to active safety technologies, which are integrated into functions of passive safety systems.

The subsequent measures in the post-crash phase along the stages of the *rescue chain,* such as securing the accident site and rescue services (Sect. 6.10), are rather processed in separate domains, such as emergency medicine. However, in the future, further networking of these areas of passive safety with those of active safety will lead to a further improvement in the mitigation of accident consequences for occupants and vulnerable road users.

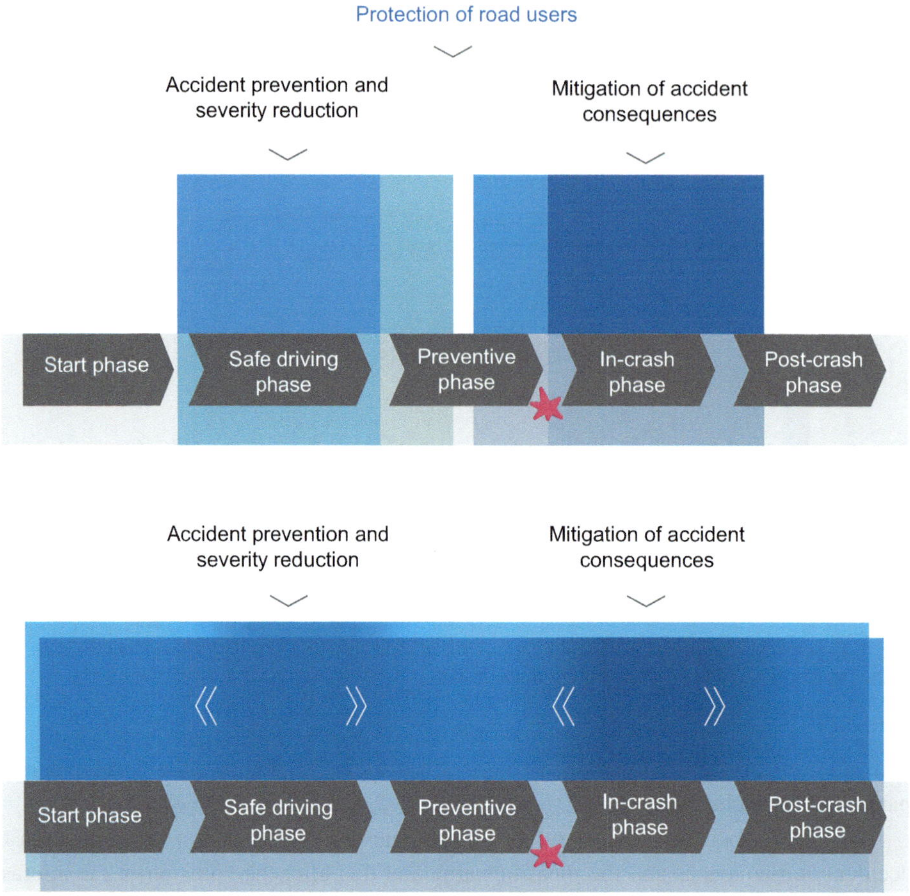

Fig. 6.5 Extended focus of active and passive safety

6.2 Protection Principles

Passive safety serves to *mitigate accident consequences* and injury risks for involved road users such as vehicle occupants and vulnerable road users. These goals and their integration into integral safety concepts were presented in the previous section (Sect. 6.1).

The collision of a vehicle with an obstacle, another vehicle, or another road user is an event in which high energies are converted. The basics for calculating energies and their conversion in collisions are derived in Sect. 6.2.1 with examples.

This is followed by an explanation of how occupants come to a stop from a certain speed after a collision, or how vulnerable road users—for example in a collision with a vehicle—are brought to the collision speed of the vehicle (Sect. 6.2.2). The necessary measures for this are based on *physical* and *biomechanical* principles and mechanisms, the basics of which were already explained in Chap. 2 and 3. In addition, there are legal and other requirements for protection systems, such as safety ratings or specific protection requirements for special use cases (Chap. 4).

In addition to considering energies, other calculation parameters are used for the calculation and design of occupant protection systems. The main ones among them are presented in Sect. 6.2.3 and 6.2.4. Further requirements and objectives in the design are shown in Sect. 6.2.5 and 6.2.6. The further design of the occupant protection systems of vehicles and the coordination of the individual subsystems are described afterwards (Sect. 6.3).

6.2.1 Energy Conversion in Collisions

The essential calculation parameter for the design of occupant protection concepts is the *energy* of the collision objects involved and their conversion during the accident. Their controlled transformation is the basis of occupant protection concepts and the reduction of injury risks for road users in an accident.

In physics, the contact of two bodies is referred to as an *impact* (also: *impulse*) when forces act between them in a very short time and thereby the state of motion of at least one of the bodies involved in the impact is greatly changed [44]. According to the *impact laws*, three essential constellations are distinguished in the collision of bodies:

- If the contact of the bodies does not result in permanent deformation, it is an *elastic impact.*
- On the other hand, if at least one of the bodies is permanently deformed, it is referred to as a *plastic impact* (also *inelastic impact*).
- The real conditions during the impact of bodies often contain both constellations. In the *real impact* (also: *partially elastic impact, actual impact*), the kinetic energy of the overall system is reduced by deformation work after the impact.

In classical mechanics, the kinetic energy ($[E_{\text{kin}}] = \text{J}$) of a body is calculated proportional to its mass and proportional to the square of its velocity:

$$E_{\text{kin}} = \frac{1}{2} m\, v^2 \tag{6.1}$$

For further calculations, the second *Newton's law* (principle of action) is used. It states that when a resultant force ($[F] = \text{N}$) acts on a body, it is accelerated in the direction of the force:

$$\vec{F} = m\, \vec{a} \tag{6.2}$$

This allows for the calculation of accelerations and forces acting on vehicles and occupants during a collision.

For two typical accident scenarios, the rigid wall impact and the collision of two vehicles, the kinetic energies and the achieved force levels for vehicles of different masses are determined.

6.2.1.1 Rigid Wall Impact

A first accident scenario is the *rigid wall impact* against a rigid obstacle. Here, ideally, the *complete conversion* of the *kinetic energy* E_{kin} into *mechanical work* ($[W_{def}] = J$) (deformation work). This means that the vehicle undergoes deformation and comes to a stop.

The following derivation is a *simplified* consideration of the real conditions during impact and a linear force-path characteristic of the front structure. Using the deformation force ($[F_{def}] = N$) at maximum deformation path ($[s_{def}] = m$) or the stiffness ($[c] = kNm^{-1}$), the deformation work is calculated as follows:

$$W_{def} = \frac{1}{2} F_{def}\, s_{def} = \frac{1}{2} c\, s_{def}^2 \tag{6.3}$$

The deformation work can also be determined in the *force-displacement diagram* (*F-s diagram*). The stiffness c refers as an idealized size to the deforming front carriage of the vehicle (deformation zone). For the stiffness, the deformation force and the maximum acceleration, the following equations result from equating and rearranging:

$$E_{kin} = W_{def} \Rightarrow \frac{1}{2} m\, v^2 = \frac{1}{2} c\, s_{def}^2 \tag{6.4}$$

$$c = \frac{mv^2}{s_{def}^2} \tag{6.5}$$

$$F_{def} = c\, s_{def} = \frac{mv^2}{s_{def}} \tag{6.6}$$

$$a_{def} = \frac{c\, s_{def}}{m} = \frac{F}{m} \tag{6.7}$$

From equation (6.7), it follows that for a given deformation path, the ratio of stiffness of the front carriage and the vehicle mass is crucial for the vehicle pulse and thus the load for vehicle occupants.

Calculation example: Rigid wall impact

Figure 6.6 shows a comparison of vehicles of different vehicle classes during the rigid wall impact (solid barrier) and the idealized parameters of vehicle deceleration as well as the force curve in the force-displacement diagram and the converted energy.

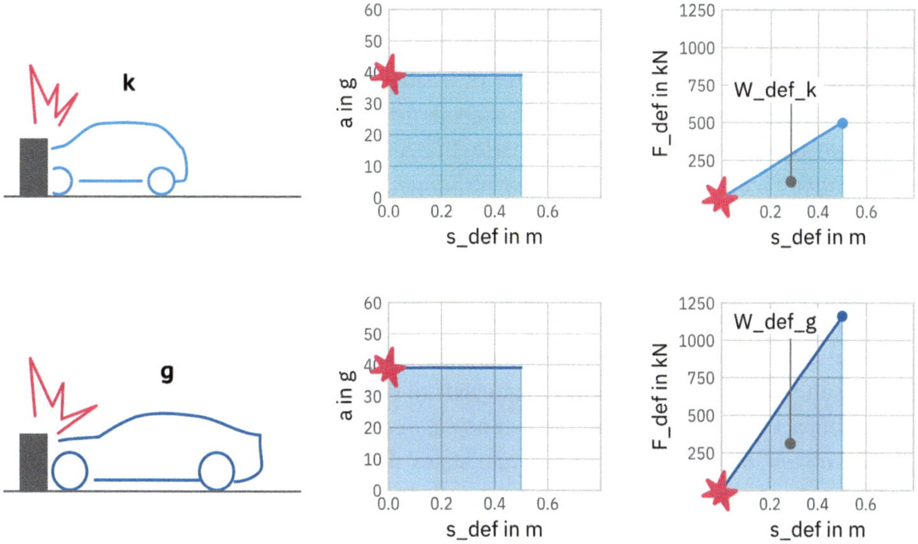

Fig. 6.6 Idealized examples of energy conversion during the rigid wall impact of vehicles of different masses: **k** small car, **g** large car

The parameters of the two vehicles are as follows:

- Small car (index k): Compact class, mass including payload: $m_k = 1{,}300$ kg, stiffness: $c_k = 1{,}003$ kNm^{-1}
- Large car (index g): Luxury class, mass including payload: $m_g = 3{,}000$ kg, stiffness: $c_g = 2{,}315$ kNm^{-1}

For impact speed and maximum deformation path of the front car, the same values are assumed for both vehicles:

- $v = 50$ kmh^{-1}, $s_{def} = 0{,}5$ m

The following values result for the kinetic energy, stiffness, deformation force and acceleration:

- Small car (index k): $E_{kin\ k} = 125$ kJ, $c_k = 1{,}003$ kNm^{-1}, $F_{def\ k} = 502$ kN
- Large car (index g): $E_{kin\ g} = 289$ kJ, $c_g = 2{,}315$ kNm–^{1}, $F_{def\ g} = 1{,}157$ kN

The deceleration for both vehicles is $a = 39$ g.

6.2.1.2 Car-to-Car Collision

The vast majority of accidents are not car-to-obstacle accidents, but car-to-car collisions. The following will analyze the collision of a small vehicle (car, index k) with a heavy vehicle (car, index g). The question arises about the respective energy absorption capacity of the two vehicles in this accident scenario. In reality, a collision of vehicles is a *partially elastic impact* with a small *coefficient of restitution,* which can initially also be calculated as an *ideal-plastic impact*. The following considerations are made to derive the conditions:

In an *ideal-plastic impact* (also: *fully-plastic impact*) of two bodies, two conditions always apply simultaneously—the *law of conservation of energy* and the *law of conservation of momentum*. They state that in a closed system—for example, consisting of two colliding vehicles—both energy and momentum are *conserved quantities*. Both conditions result in a system of equations that needs to be solved. The assumption of a straight central impact continues to apply, so that no *change in twist* needs to be considered [44].

The *law of conservation of momentum* states that the *momentum* in a closed system is also a conserved quantity. The momentum ($[p] = \text{kgms}^{-1}$) of a body is proportional to the mass and velocity of a body. Like velocity, it is a vector quantity:

$$\vec{p} = m\,\vec{v} \tag{6.8}$$

In classical mechanics, the kinetic energy ($[E_{kin}] = \text{J}$) of a body is calculated to be proportional to its mass and the square of its velocity:

$$m_\text{k}\,\vec{v}_\text{k} + m_\text{g}\,\vec{v}_\text{g} = \left(m_\text{k} + m_\text{g}\right)\vec{u}\,' \tag{6.9}$$

The *resulting speed* $[u']= \text{ms}^{-1}$) of the vehicles moving synchronously (due to ideal-plastic impact) after the collision is calculated from:

$$\vec{u}\,' = \frac{m_\text{k}\,\vec{v}_\text{k} + m_\text{g}\,\vec{v}_\text{g}}{m_\text{k} + m_\text{g}} \tag{6.10}$$

From the law of conservation of energy and the balances of the kinetic energies of both vehicles before and after the collision, the *deformation work* or the deformation energy is derived:

$$E_\text{kin k} + E_\text{kin g} = E'_\text{kin} + \Delta E \Rightarrow \Delta E = W'_\text{def} \tag{6.11}$$

$$W'_\text{def} = \Delta E = E_\text{kin k} + E_\text{kin g} - E'_\text{kin} \tag{6.12}$$

The *deformation energy* is divided between both vehicles:

$$W'_\text{def} = W_\text{def k} + W_\text{def g} \tag{6.13}$$

Since *action equals reaction* applies, the same *deformation force* $F_{\text{def}\,max}$ acts on both vehicles. Therefore, the following can be derived:

$$F_\text{def max} = F_\text{def k} = F_\text{def g} \tag{6.14}$$

If you solve (6.6) and insert it into (6.3), the result is:

$$W'_{\text{def}} = \frac{1}{2}\left(\frac{F^2_{\text{max}}}{c_{\text{k}}} + \frac{F^2_{\text{max}}}{c_{\text{g}}}\right) \tag{6.15}$$

Solving for $F_{def\ max}$ gives:

$$F_{\text{def max}} = \sqrt{\frac{2W'_{\text{def}}}{\frac{1}{c_{\text{k}}} + \frac{1}{c_{\text{g}}}}} \tag{6.16}$$

With the equations (6.6) and (6.7), the parameters of the accelerations and deformation paths of the two vehicles can be calculated.

With known stiffness c of the front carriage *(deformation zones)* of both vehicles, the further parameters for *deformation path* s_{def} and deceleration a_{def} can be calculated from this. In this second calculation example, the stiffness values from the first calculation example are adopted and used.

The *coefficient of restitution* ε is an indicator for the *rebound elasticity* of bodies. Under idealized conditions of the *ideal-plastic collision process,* in which the coefficient of restitution $\varepsilon = 0$ is assumed, the following applies:

$$\frac{m_{\text{k}}}{m_{\text{g}}} = \frac{c_{\text{k}}}{c_{\text{g}}} = \frac{s_{\text{def g}}}{s_{\text{def k}}} = \frac{\Delta v_{\text{g}}}{\Delta v_{\text{k}}} = \frac{a_{\text{g}}}{a_{\text{k}}} \tag{6.17}$$

As shown above, the same deformation force $F_{\text{def max}}$ acts on both vehicles, it follows that the energy absorption capacity is consumed in the smaller car and, on the other hand, a small portion of the available deformation energy is utilized in the larger car. The ratio of the absorbed deformation work, like that of the deformation paths, the changes in speed, and the decelerations, is inversely proportional to the ratio of the vehicle masses.

The derivation shows that the stiffness cof a front structure designed solely according to the criteria of the rigid wall impact test is proportional to the vehicle mass m. Furthermore, it is found that the smaller vehicle not only experiences a higher speed change Δv_{k} and thus a higher deceleration a_k, but also undergoes the larger deformation $s_{\text{def k}}$.

Calculation example: Car-to-car collision

Figure 6.7 shows the collision of vehicles of different vehicle classes and the idealized parameters of vehicle deceleration as well as the force course in the force-displacement diagram and the converted energy.

The parameters of the two vehicles, including the stiffness of the front end (*linear front-end pulse*), are taken from the previous example:

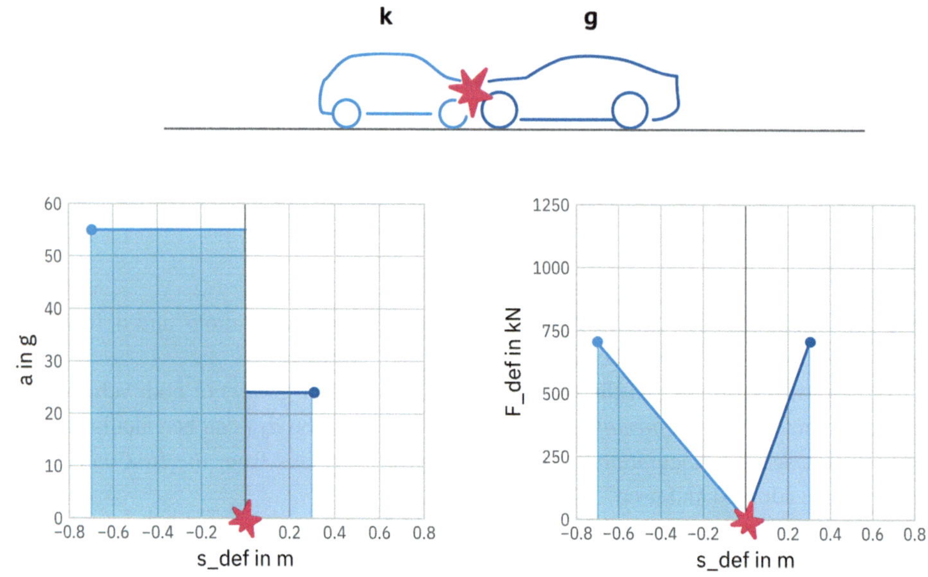

Fig. 6.7 Idealized examples of energy conversion in car/car collision of vehicles of different vehicle classes: **k** small car, **g** large car

- Small car (index k): Mass including payload: m_k= 1,300 kg, stiffness: c_k = 1,003 kN–^{1}
- Large car (index g): Mass including payload: m_g= 3,000 kg, stiffness: c_g = 2,315 kNm^{-1}

Both vehicles are moving towards each other at the same speed:

- $v_k = v_g = 50$ kmh–^{1}

From this, the speed of the synchronous movement of both vehicles after the impact is calculated:

- $\vec{u}' = -20$ kmh^{-1}

The difference in kinetic energy before and after the collision is equal to the sum of the deformation work on both vehicles: W'_{def}= 350 kJ. A deformation force of $F_{def} = 700$ kN acts on both vehicles.

The following values result for the vehicles:

- Small car (index k): $\Delta v_k = 70\ kmh{-}^1$, $s_{def\,k} = 700$ mm, $a_k = 550\ ms^{-2}$
- Large car (index g): $\Delta v_g = 30\ kmh{-}^1$, $s_{def\,g} = 300$ mm, $a_g = 240\ ms^{-2}$

The amounts of velocity changes Δv, deformation path s_{def} and deceleration a vary significantly.

Depending on the design goal, targeted measures can significantly reduce the injury risk of occupants, especially in smaller cars. In a research project of the 1980s, the following measures were identified [95]:

- Longer, softer deformation structure of the heavier vehicle
- Design of vehicles for a mass-dependent wall impact speed, i.e., heavier cars for a speed less than 50 km/h, so that softer structures can be realized for them
- Targeted positioning of heavy aggregates, such as the engine-transmission unit: Forward positioning in lighter cars, rearward positioning in heavier cars
- Disproportionate efforts for the restraint system in lighter vehicles

These measures are still fundamentally valid today. In particular, the aspect of compatibility has gained high importance in overall vehicle development and development of the vehicle body structure. New load cases have also been introduced that address these challenges in occupant protection (Chap. 4). The transformation of the drivetrain of vehicles and the associated higher weights of battery electric vehicles pose new challenges for passive safety, as the masses of colliding vehicles tend to continue to increase.

In reality, the transfer of momentum during vehicle collisions does not occur instantaneously, but is distributed over a certain period of time. This significantly determines the conditions for the vehicle's occupant protection system and possible protective measures. The further principles and the calculation sizes used for this are explained in the following sections.

6.2.2 Protective Functions in Collisions

The following illustration shows essential *protection concepts* of passive safety for occupants. It is formulated as requirements as well as measures that are used in the development and coordination of occupant protection systems. For a first overview and orientation, the following elements can be mentioned as essential measures:

- Vehicle body structure with deformation capability or other mechanisms for energy conversion (colloquially: *crumple zone)* and restraint systems
- Rigid passenger compartment (colloquially: *survival space*)
- Occupant deceleration and energy conversion (among other things through *restraint systems*)

The further, more detailed representation is roughly oriented towards the course of the load path—starting in the collision zone—towards the occupant as well as the temporal sequence of intervention measures starting with the impact. Suitable interventions already from the pre-crash phase are picked up further down in this section.

6.2.2.1 Introduction of Collision Forces into Vehicle Body Structure

The vehicle has an external structure with the ability to absorb and transmit forces and moments in a collision with an obstacle. Structures for self-protection have a high force level that is effective early on.

The vehicle body supports the targeted routing of the force flow, for example through high-strength cross, longitudinal and diagonal beams with high stiffness. But also rigid aggregates and main assemblies such as the drive unit or high-voltage battery can contribute to this.

The body structure enables the detection of the accident event by the crash sensors (Chap. 7), which is picked up again further down.

However, since requirements for partner protection, including those of vulnerable road users, must also be taken into account, there are conflicting objectives on the external vehicle body structure that must be taken into account in the design (Sect. 6.4 and 6.7).

6.2.2.2 Energy Conversion through Deformation of the Vehicle Body Structure

To reduce kinetic energy during a collision, it is converted into *deformation work* (also: *deformation energy*) and leads to permanent deformation of the vehicle body structure. For this purpose, the vehicle has defined areas with high deformation capacity for energy conversion and thus with relatively low *stiffness*. These are the *crumple zones*, which in an ideal behavior experience purely plastic deformation without *restitution (springback).*

In the front and rear area of the vehicle, crumple zones can usually be designed with a larger deformation path than in the side area. Since the forward displacement for the occupant is also dimensioned larger to the front than to the side, there are specific requirements for protection concepts in frontal impact and side impact.

The targeted control of deformation is achieved by special devices in the vehicle body. Examples are *crash boxes* (Sect. 6.4) or the targeted deformation of certain body areas (e.g. *longitudinal beam*). The vehicle deceleration should be as uniform as possible and without jumps.

6.2.2.3 Ensuring the Structural Integrity of the Passenger Compartment

A *significant* contribution to the protection of occupants is the preservation of the passenger compartment. This is a development goal opposite to the above-mentioned deformability of certain areas of the passenger compartment. The associated requirements are:

- **Preservation of the interior of the passenger compartment,** to prevent injuries from collapsing structures.
- **Prevention of collapsing or breaking components:** Similar to the vehicle body structure, the seat structure, among other things, must also be designed to be sufficiently rigid so that occupants do not suffer additional injuries. This also applies to other components in the interior or to assemblies that could cause intrusions into the interior.
- **Maintaining attachment points for restraint systems:** The rigid passenger compartment should support the preservation of the attachment points of restraint systems. This concerns, for example, the anchoring points of the three-point belt systems as well as the position and orientation of the airbag modules on the cross beam, on the seats or in the roof.
- **Maintaining forward displacement space for occupants** and **working space for restraint systems:** The rigid passenger compartment should maintain the forward displacement space for the occupants during the *restraint* as well as the working space for the restraint systems, especially the airbag modules.
- **Prevention of breaking components:** Occupants should not suffer injuries from breaking sharp-edged components in the interior. The rigid passenger compartment contributes to this. In addition, interior components such as the cockpit, steering wheel, seats and panels have appropriate contours and deformation properties.
- **Prevention of intrusions** into the passenger compartment by collapsing vehicle systems. An example of this is the conventional steering column, which can be displaced further into the interior in a frontal impact and can narrow the space in front of the driver with the steering wheel.

The rigid passenger compartment is achieved by a defined area with high stiffness or with zones of increasing strengths (Sect. 6.4).

The invention of the *rigid passenger compartment* in the 1950s by Bela Barényi at Daimler-Benz, together with the *crumple zone* (see above), was a breakthrough in passive safety.

6.2.2.4 Securing the Occupants at their Position

The next important measure is to protect the occupants from being ejected through the window openings, from unintentional opening of the doors or from openings in the roof and the associated high risk of injury from unprotected impact on surfaces.

In particular in the USA, occupant protection systems must meet special requirements against ejection in rollover accidents (*ejection mitigation*). For this purpose, specific versions of the curtain airbag are used (Sect. 6.6).

The essential measure is the securing of the occupants by the seat belt system, which locks the retractor in case of vehicle acceleration or rapidly extending webbing. Rigid seat structures contribute to this, as already described above. This is particularly true in the case of rear impact.

Securing the occupants at their position (also: *stabilization* of occupant posture) is also a contribution to prevent unwanted collisions of occupants with each other or, for example, the rear occupant colliding with the front seat with high risks of injury for both occupants. To expand the protection potential of vehicles for this aspect, the *center airbag* has been developed and already introduced in several vehicle models in recent years (Sect. 6.6).

6.2.2.5 Occupant Deceleration and Energy Conversion

Restraint combines several individual measures that make up a crucial part of occupant protection. The central measure here is the use of *restraint systems* with *seat belt* and *airbag systems*.

Basically, in the event of a collision, the velocity change Δv of the occupant must be achieved over a certain distance s. In a simplified view, he experiences a deceleration a (negative acceleration):

$$a = \frac{1}{2}\frac{\Delta v^2}{s} \tag{6.18}$$

Consequently, the crumple zone should be as large as possible. The distance s, which results from the deformation path of the vehicle and the displacement of the occupant relative to the vehicle, must be effectively utilized by the vehicle and the occupant protection system.

An earliest possible *coupling* is an ideal condition, so this is one of the motivations for using the pre-crash phase for reversible pretensioning and coupling of occupants by *active seat belt systems* (Sect. 6.5.10).

The *crash pulse* is the course of the vehicle deceleration (in a frontal impact) or the course of the vehicle acceleration (in the general case). A uniform course of the *crash pulse* and the occupant deceleration is a goal in the design of the vehicle body structure and the restraint systems.

The restraint includes the following individual measures in detail, which are described here for the case of a frontal impact:

- **Provision of the restraint system:** Restraint systems are either positioned on the occupant like the seat belt and can therefore be deployed very quickly. Or restraint systems like airbags need to be inflated first before they can support the restraint. This is also a reason for the interaction between seat belt and airbag systems mentioned above.

- **Early coupling of the occupant to the vehicle:** Early coupling makes the best possible use of the forward displacement necessary and available for the restraint (in frontal collisions). After the start of the collision (t_0), the occupant should experience as little or preferably no forward displacement without coupling. Typically, the ignition of the belt pretensioner starts about 8–12 ms after t_0. Without coupling of the occupant, he would hit the interior without deceleration after a forward displacement.
- **Utilization of the available forward displacement space:** After the coupling, a significant (!) objective of the design of occupant protection systems is to use the remaining available *forward displacement* to decelerate the occupants. Levers to achieve this goal include *adaptivity functions* in restraint systems and an airbag control unit, which can assess information from different sensors for accident type and specific requirements of occupants.
- **Optimization of force introduction:** When the occupant is restraint by the seat belt or supported by an airbag, loads act on the body. The design of the occupant protection system is such that forces act at those points that can best withstand these loads according to biomechanical criteria. For this purpose, biomechanical load values have been determined for the various body parts, which can be used for the technical design of the occupant protection system (Chap. 3).
- **Limitation of restraint forces:** Due to the force limitation in the coupled state and during the occurring forward displacement of the occupant, the deceleration of the occupant is less than that of the vehicle (Eq. 6.18). Overall, a reduction of the loads on the occupant is aimed for.
- **Influencing occupant kinematics:** The movement of the occupant during his forward displacement and during the restraint by restraint systems should occur in a defined manner, so that unwanted kinematics with high injury risks can be avoided. For the different body areas (including head, neck, thorax, pelvis, upper and lower extremities) there are different objectives for influencing the kinematics.

Vehicle collisions in reality are not purely plastic, but also have elastic components—caused by a variety of effects. This results in a springback, which can lead to a reversal of motion.

- **Reduction of rebound effects:** Energy storage during the collision—e.g. in vehicle body structure, interior components or even in the restraint system—results in springbacks due to the elastic properties of the components. In an idealized system, the *rebound* is completely eliminated. In reality, after the "*zero crossing*" (Fig. 6.8), there is often a reverse movement of the occupants. This backward movement between vehicle and occupant usually does not occur synchronously, as the occupant is still

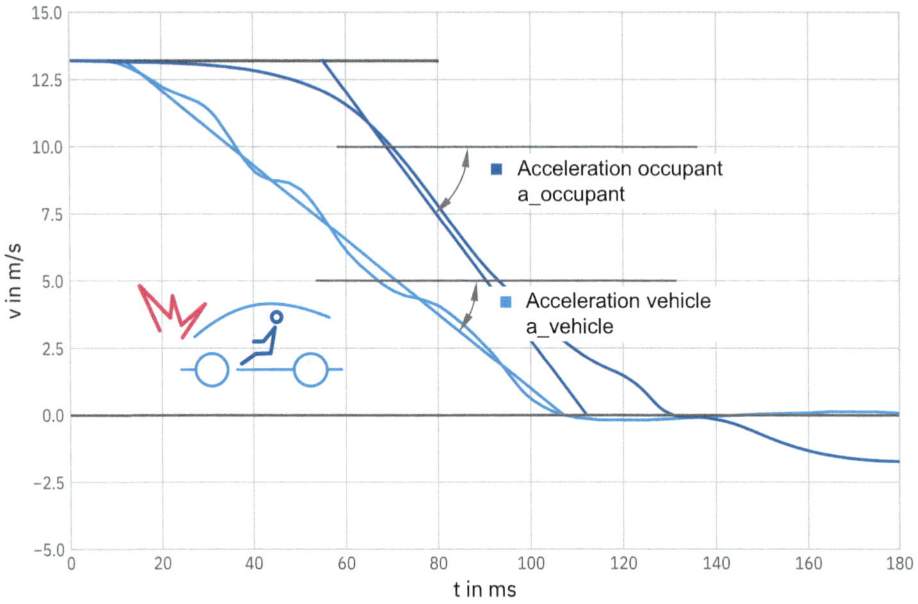

Fig. 6.8 Speeds of vehicle and occupant as a function of time and derivation of accelerations (v-t-diagram) [61]

moving forward while the vehicle is already in the *rebound* phase. This phase also carries injury risks, which can be reduced by specific protection systems (Sect. 6.7).

6.2.2.6 Safeguarding Critical Vehicle Systems through Deformation-Resistant Body Structures

An additional protective measure is the securing of systems and aggregates with high mass, such as the engine and transmission drive unit, against tearing away, so that injuries to occupants or other road users are prevented. This also applies to other sensitive systems, such as the fuel tank or high-voltage battery. If damaged, high injury risks for occupants would arise from subsequent effects such as fires.

6.2.3 Speed Alignment in the Event of a Collision

In a collision, the vehicle and the occupants inside are decelerated in a very short time. According to the second *Newton's law (principle of action)* (Eq. 6.2), a force F, which represents biomechanical loads, therefore acts on the occupants.

Depending on the question at hand, different perspectives can be taken when considering and modeling, each highlighting certain aspects of the accident event and the mechanisms of action of occupant protection systems in their real sequence. The following approaches are available:

- Force-displacement perspective: Representations and considerations as in the case examples shown above for the rigid wall impact and the vehicle-vehicle collision. This perspective rather represents the energies of the bodies involved in the collision and the energy conversion through the impact.
- Speed-time perspective: Representation and consideration of the processes over time. This highlights the course of forces and accelerations during the phases of the collision.

Both perspectives have their justification and are used for the development and coordination of occupant protection systems. However, challenges arise in practice when measuring energies. They are more difficult to determine in laboratory tests, which is why the speed-time perspective is chosen in many works.

The biomechanical load on occupants results from the adjustment of speed between the contact structure and the body. This *speed adjustment* will be considered in a simplified manner based on the relative speed between vehicle and occupants in a head-on collision.

The representation in Fig. 6.8 shows the speed-time courses of the vehicle and an occupant. It is clearly visible that the speed of the vehicle decreases rapidly after the start of the collision ($t = 0$), while the occupant initially continues to move at almost unchanged speed.

Only when the restraining forces from the occupant protection system become effective does the speed of the occupant reduce, until at a time of about $t = 140$ms he has the same speed as the vehicle. This phase after the zero crossing is referred to as the rest phase. After the zero crossing and the rest state, a rebound follows, which is characterized by a backward displacement of the occupant and negative speed ($t>$ 140 ms).

The drop in speed is a measure of the deceleration (negative acceleration), which is described here approximately by a straight line for the explanation of the basic processes.

By using restraint systems, the deceleration is initiated earlier and over a longer period of time compared to an unbraked impact on the contact structure. Contact with hard structures of the vehicle interior can be prevented, at least up to a certain impact speed, by using restraint systems.

The acceleration acting on the occupant can be reduced, among other things, by the following measures:

- Early-acting restraint systems, such as seat belt system with seat belt pretensioner, early-triggering airbag system, etc.
- Utilization or, if necessary, enlargement of the forward displacement path, for the driver, for example, by shifting the steering wheel away from the occupant.

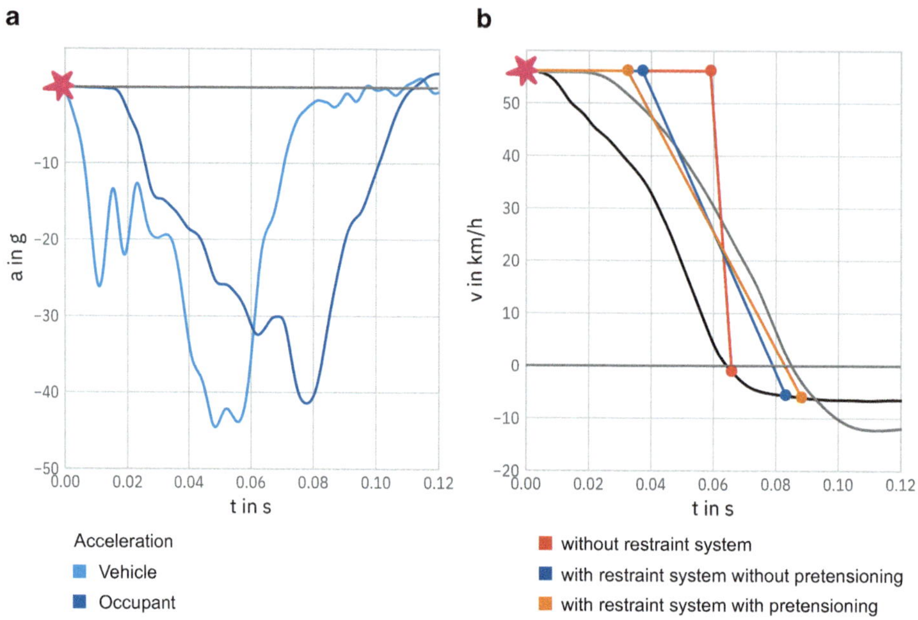

Fig. 6.9 Course of the **a** accelerations and the **b** speeds as a function of time. (*Source*: ZF)

These basic statements can also be transferred to other types of collisions and impacts.

Fig. 6.9a shows the course of accelerations of the vehicle and occupant during a collision. In Fig. 6.9b, the courses of the speed change of the occupant are shown in the following variants:

- No use of a restraint system and unbraked forward displacement of the occupant by approx. 300 mm. The occupant is subjected to an acceleration of $a_I = -239$ g (red).
- Use of a restraint system with occupant coupling without pretensioning. In this example, it means a forward displacement over a distance of 350 mm and an acceleration of the occupant of $a_I = -38$ g (blue).
- Use of a restraint system with occupant coupling with pretensioning. In this example, it results in an initial unbraked forward displacement of 25 mm and a forward displacement of 325 mm after coupling (orange).

6.2.4 Metrics for the Design of Occupant Protection Systems

6.2.4.1 Ride-Down Effect (RDE)

The *ride-down effect* is a parameter for assessing the coupling of the occupant during the collision process. It is formed from the relation of the deformation paths before and after the coupling. This parameter does not describe a single subsystem of the occupant protection system—such as the vehicle or the restraint system—but relates them to each other.

The calculation of the characteristic RDE results from the maximum deformation path of the vehicle $s_{\text{def max}}$ and the vehicle deformation path $s_{\text{def RD}}$ at the time $t = t_{\text{RD}}$, at which the coupling occurs and the restraint of the occupants begins. It is given in percent [3, 61]:

$$\text{RDE} = \frac{s_{\text{def max}} - s_{\text{def RD}}}{s_{\text{def max}}} \cdot 100 \tag{6.19}$$

An RDE value of 100 % means that the occupant is already decelerated from the point at which the vehicle also experiences a deceleration. An RDE value of 0 % characterizes a limit case in which the occupant is not subjected to any deceleration until the maximum dynamic vehicle deformation path is reached. The calculated RDE value is thus a necessary, but by no means a sufficient condition for an efficiently designed restraint system.

In practice, the calculation of the RDE value is done by using the temporal course of the resulting chest acceleration (Fig. 6.10). The 100 %-value corresponds to the (first) maximum of the chest acceleration (Fig. 6.10a). Using a straight line that intersects the 25 and 75 % point, the rising flank of the chest acceleration is approximated. The intersection of the line with the time axis represents the point in time t_{RD}, at which the seat belt retractor locks, the *belt slack* (Sect. 6.5) of the belt system is overcome and its effect occurs [3, 61].

Up to this point, the vehicle has covered the distance $s_{\text{def RD}}$ (Fig. 6.10b). The difference between the maximum dynamic deformation path $s_{\text{def max}}$ of the vehicle and the path covered ineffectively for the restraint effect $s_{\text{def RD}}$ describes the portion of the vehicle deceleration path that can be used for occupant deceleration. In this example, the ride-down effect calculates to a value of RDE = 24 %.

The later the point in time t_{RD} at which the restraint system becomes effective, the greater the already occurred forward displacement of the occupant (relative path between the occupant's chest and the vehicle) and the smaller the ride-down effect becomes. If the relative path is too large, the remaining forward displacement is no longer sufficient for adequate restraint, and contact between the chest and/or head of the occupant and parts of the vehicle interior can occur, which can lead to severe injuries.

In test series with a three-point automatic belt system according to UN-R 16 using the HYBRID-III 50M dummy, RDE values of 55 % were determined in the early 1990s, which characterized a good effectiveness of an occupant protection system [57].

From a series of tests, the RDE values are shown depending on the occupant's forward displacement path $s_{\text{Occupant } t=t_{\text{RD}}}$ at the time $t = t_{\text{RD}}$ (Fig. 6.11). It can be seen that with increasing forward displacement and later occurring time point t_{RD} the RDE value tends towards zero [3].

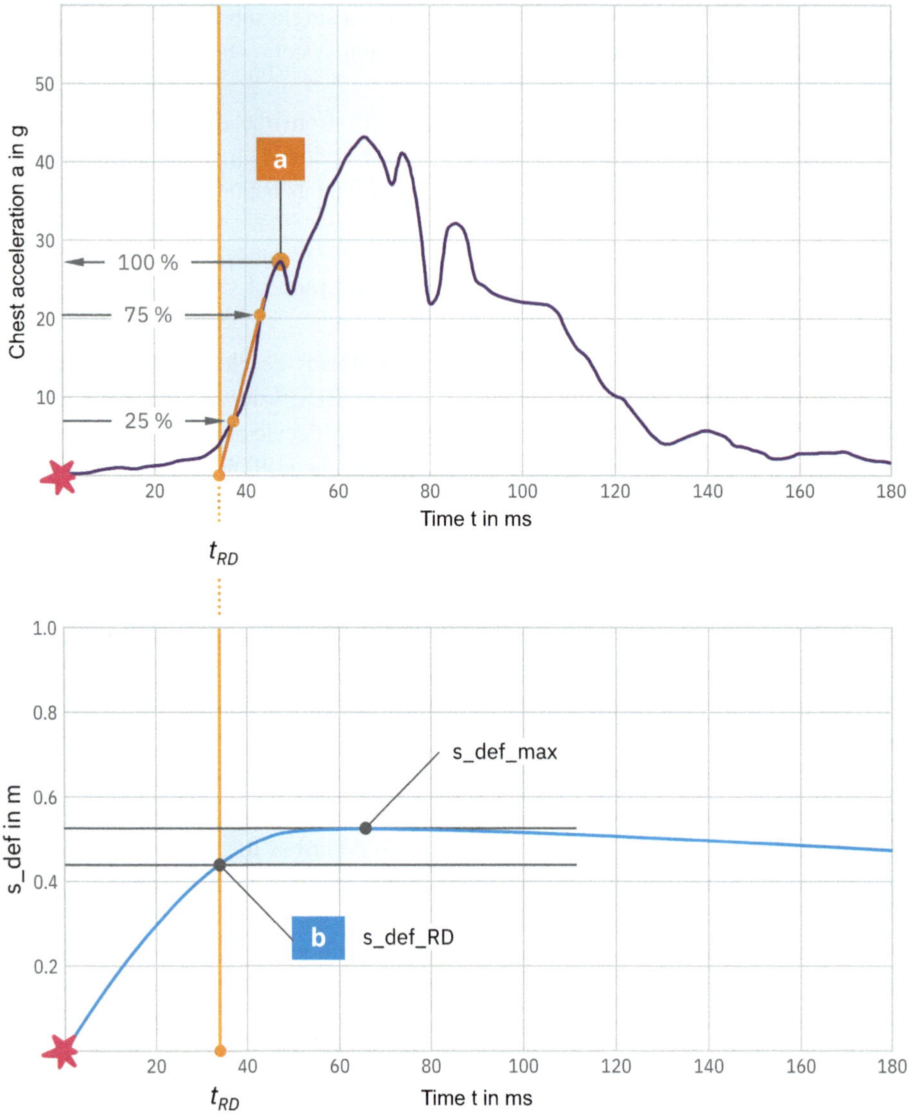

Fig. 6.10 Determination of the ride-down effect by analyzing the chest acceleration of the occupant and the deformation path of the vehicle (based on [3, 61])

6.2.4.2 Basic Metrics for Crash Pulse

Physically, the collision of a vehicle with an obstacle or another vehicle is a *real impact* (Sect. 6.2.1). Unlike an elastic impact, where the kinetic energy is fully preserved, large parts are converted into deformation and heat as a result of internal friction. The difference to a fully plastic impact lies in rebounds or restitutions.

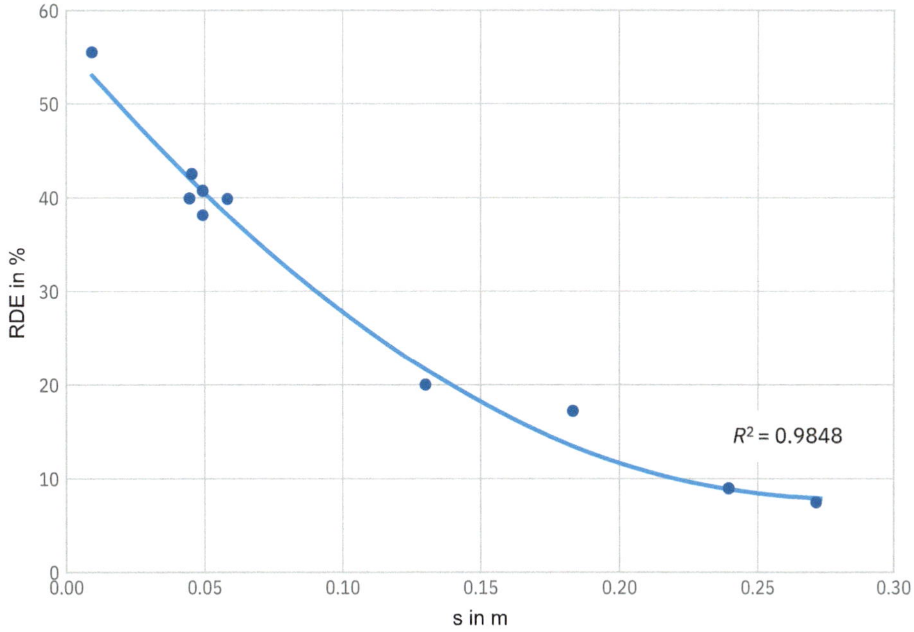

Fig. 6.11 Progression of the ride-down effect depending on the forward displacement of the occupant (based on [3, 61])

To describe and model real collision processes in the development of occupant protection systems, different *crash pulse criteria* were designed. They determine key figures that can be used, for example, for designs and comparisons of alternative system configurations.

Simple metrics for the crash pulse can be determined directly, for example:

- *Speed difference* of vehicles before and after the collision
- Maximum deceleration or *sliding mean*
- Average deceleration—often as the point in time of the speed zero crossing, corresponds to the average deceleration. The application only makes sense in the case of a rigid barrier.

Further key figures, such as the *occupant load criterion,* are derived from substitute models, which will be explained in more detail below.

6.2.4.3 Occupant Load Criterion (OLC)

The occupant load criterion (OLC) *(pulse criterion)* is based on a mechanical substitute model that approximates the loads on the occupant's chest during restraint [67]. A pulse criterion should describe the crash pulse independently of a restraint system.

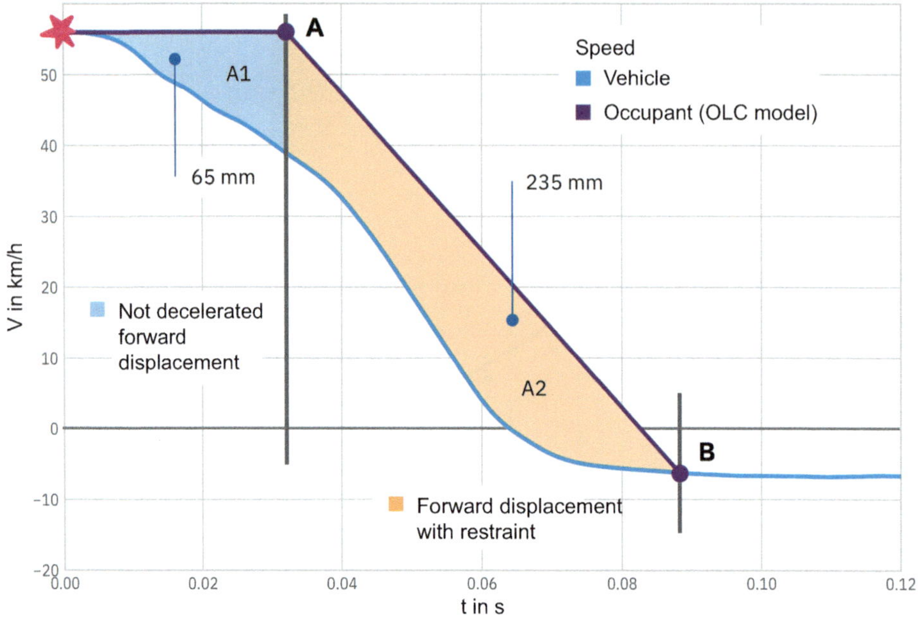

Fig. 6.12 Modeling of the crash pulse according to the occupant load criterion (OLC) (based on [67])

Fig. 6.12 continues the example from Fig. 6.9 and shows the same course of vehicle acceleration. The *OLC criterion* is now determined as follows: In the first phase of a frontal impact, the restraint forces are low and are ideally set to zero in the OLC model. The occupant moves without coupling at his initial speed $v_A = 56$ km/h, until a certain relative path between occupants and vehicle is reached ($s_A = 65$ mm). This distance is marked in Fig. 6.12 by the area **A1** between the vehicle and occupant speed curve. Subsequently, an optimal restraint is assumed, whereby the occupant is constantly decelerated using a further relative path to the vehicle of 235 mm (area **A2** in Fig. 6.12). The required deceleration that acts on the occupant in this area defines the *OLC value*.

6.2.4.4 Characteristic Shoulder Belt Force Level (CFL)

Validated simulation models for the dynamics of vehicle crashes are created and used for the design of a vehicle's safety system. They allow the safety system to be assessed depending on the load case using the *characteristic shoulder belt force level* (CFL) [78].

The effectiveness of a restraint system naturally depends on the operating conditions, i.e., the overall system of the vehicle (including the geometry of the belt system anchor points, seating position and inclination, crash sensors (accident detection), crashworthiness), the occupant (including weight, size, kinematics), the collision scenario (including delta speed, pre-crash dynamics due to AD/ADAS functions, impact vector and collision

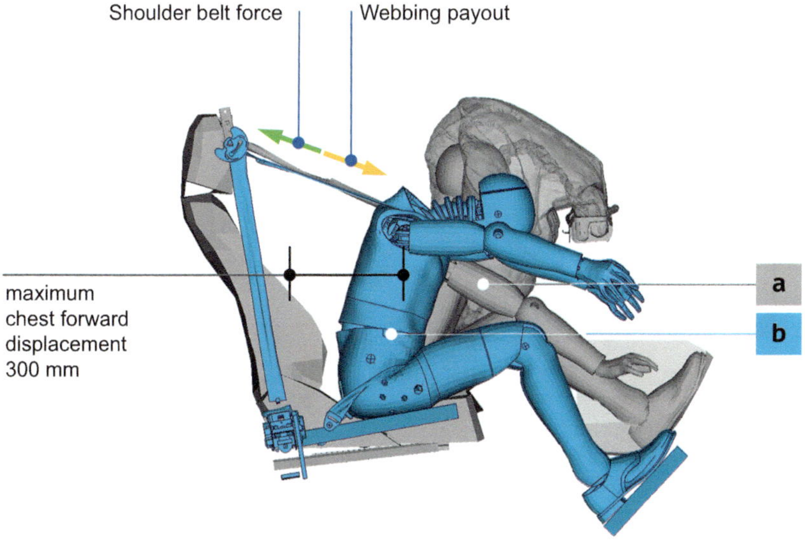

Fig. 6.13 Modeling to determine the characteristic shoulder belt force level (CFL) for evaluating the effectiveness of seat belt systems (based on [78])

location) as well as the possibly load-dependent activation strategy of the restraint system used.

The *characteristic shoulder belt force level* CFL is defined as follows: For an existing overall system including the restraint system (without airbag), a constant force limitation level at the retractor pretensioner is determined by iteration, so that the maximum chest forward displacement is exactly 300 mm. The associated shoulder belt force level corresponds to the evaluation size CFL (Fig. 6.13).

Fig. 6.13 shows the determination of the characteristic shoulder belt force level (CFL). Fig. 6.13a (gray) shows the maximum chest forward displacement in the overall system in combination with an airbag. Fig. 6.13b (blue) shows the maximum chest forward displacement of an occupant on a steel seat surface with generic foot support of 300 mm with a restraint through a seat belt and force limitation at CFL force level without the combination with an airbag.

Fig. 6.14 shows the progressions of chest forward displacement and shoulder belt force of the restraint in the overall system (gray) and of the seat belt system at CFL force level (blue). The curves for both systems coincide up to the coupling (also: *force closure*), the reaching of the first force limitation level (here at approx. 40 ms after t_0), almost exactly.

In the system without an airbag (Fig. 6.14b), the energy that is still present at this point is dissipated exclusively in the shoulder belt via the remaining forward

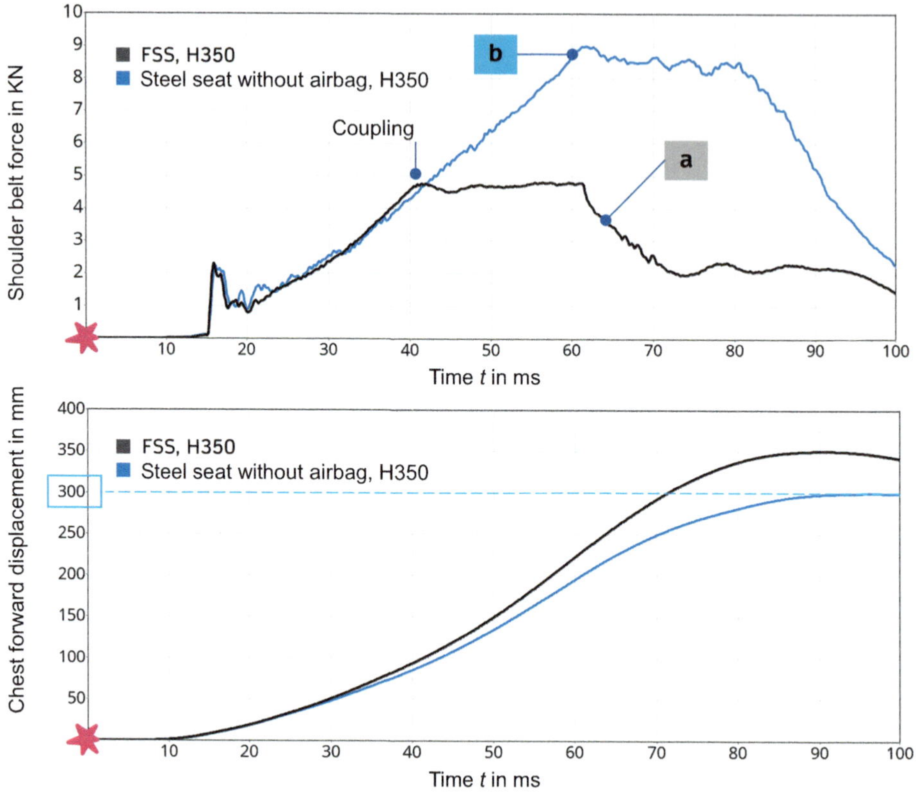

Fig. 6.14 Exemplary comparison of the progression of chest forward displacement and shoulder belt force: **a** Seat belt system in combination with airbag **b** Seat belt system with force limitation at CFL force level without combination with an airbag (based on [78])

displacement. Therefore, the level of required shoulder belt force (the less, the better) is a suitable measure for the restraint performance of the belt system.

It is clear that an effective coupling positively influences the virtual force level determined in this way. Because if a lot of energy or speed is already reduced during the coupling phase, less force is needed in the following force-limiting phase to still reach the same 300 mm chest forward displacement.

This method can be used in particular for comparing seat belt systems and for assessing activation strategies for reversible retractor pretensioners in the pre-crash phase. However, it also allows for comparative evaluation of crash pulses, crash sensors, ADAS dynamics, and vehicle environments with regard to occupant protection.

6.2.5 Compatibility Aspects in Collisions

Vehicle compatibility in accidents refers to the degree to which a vehicle's characteristics are harmonized with those of other vehicles or road users to reduce injury risk and crash severity in multi-vehicle or vehicle-to-vulnerable-user collisions. Key parameters include vehicle mass, structural stiffness, geometry, deformation patterns, and energy absorption capacity, all of which influence outcomes in such events. The following examples describe how compatibility is affected:

- *Vehicle masses:* There is a high probability that vehicles of different weights will collide with each other (see statistics—vehicle masses).
- *Stiffness:* Similarly, the stiffness of the colliding deformation structures varies, especially when not only collisions between two front parts, but also the different impact angles and impact points are taken into account.
- *Geometries:* The geometries of colliding vehicles vary, so that the respective contact point of a collision has a significant influence on the course of the collision.

Statistics—Vehicle masses
In 2022, the average unladen weight of new cars registered in Germany was about 1,700 kg. In 2014, this value was about 1,480 kg [115].

6.2.5.1 Vehicle Masses

In the case of accident severity of frontal collisions, a significant influence can be demonstrated, which can be attributed primarily to the *mass* of the involved vehicles, as opposed to car-to-obstacle collisions.

As a measure for the *accident severity*, the *speed change* Δv of the considered vehicle or the collision object can be used. It depends on the involved vehicle masses, the relative speed, and the *restitution coefficient* ε to characterize the elastic/plastic deformation behavior according to the *Newtonian impact laws* (Sect. 6.2.1). The larger the mass of the collision object in frontal collisions between two differently heavy cars, the smaller its speed change Δv will be.

Another parameter for characterizing the *accident severity* is the *vehicle deceleration* [2, 62], which depends heavily on the *deformation path*. With the simplifying equation of kinetic with deformation energy, it can be shown that the deceleration of a passenger compartment increases up to 100 g if the deformation structure is designed much stiffer and the penetration path is only 10–20 cm instead of 50–60 cm (Fig. 6.15). (1 g is equal to gravity, 100 g is 100 times gravity.)

Such hard structures, which allow a deformation of only 15 cm at a 50 km/h rigid wall impact, were investigated in the 1990s in connection with lightweight vehicles under the term *impact belt* [90].

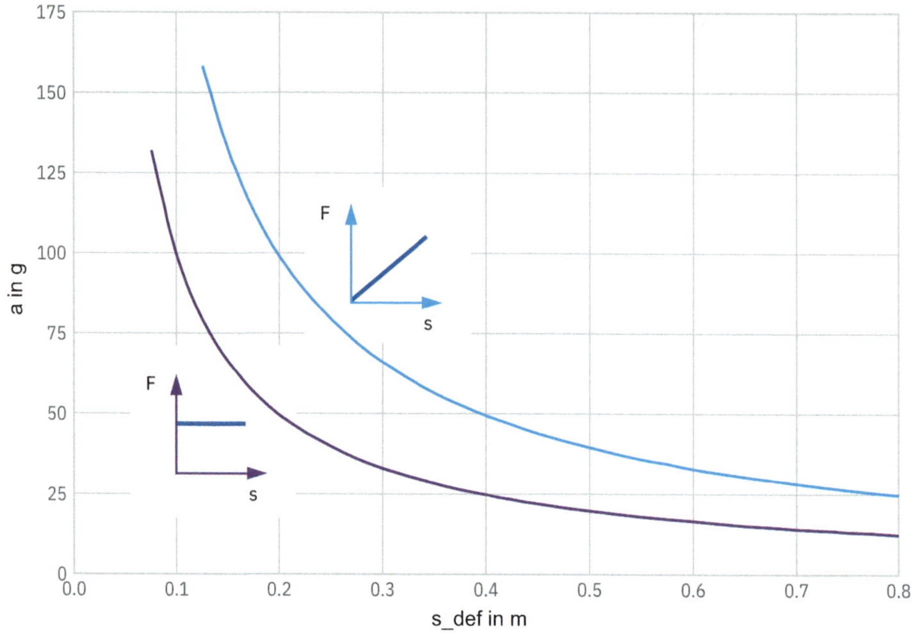

Fig. 6.15 Deceleration over deformation path during rigid wall impact (schematic energy consideration)

6.2.5.2 Stiffness of Deformation Structures

Another aspect of compatibility is seen in the adjustment of the stiffness ratios of the deformation structures. However, it must be taken into account that the design of the structural rigidity, for example, according to aspects of car-to-car collisions, also affects other accident (and test) configurations. Thus, the increase in stiffness in the front structure of smaller vehicles shows a positive effect, i.e., a lower injury risk in car/car collisions, but a higher injury potential in obstacle and side collisions. This phenomenon led to the consideration of the entire relevant accident event in a research project in the 1980s by setting up an accident scenario with three differently heavy cars in different accident configurations and simulating it mathematically [95]. Regarding the structural rigidity, the resulting constructive guidelines can be formulated as follows:

- Higher front stiffness in lighter cars
- Lower front stiffness in heavier cars
- Higher side stiffness in all cars

6.2.5.3 Vehicle Geometries

The third aspect of compatibility relates to the architecture of deformation structures, i.e., the geometric adaptation of structures in frontal and side collisions. Initially, the geometric conditions in car-to-truck collisions can be mentioned as unadapted circumstances in both frontal and rear and side collisions. As a remedy, the now legally required *front-* and *rear underride protection devices* on commercial vehicles are considered (Section 6.9).

After various initiatives to improve compatibility in collisions have been able to gather experience, this aspect has now been included in evaluation protocols for consumer ratings. Since 2020, Euro NCAP has been considering a *compatibility modifier* as part of a compatibility assessment [34] (Sect. 4.3.2.2).

6.2.5.4 Aspects of Compatibility through Integral Safety

With the development of concepts of integral safety, new functionalities will be added to occupant protection systems in the future to improve compatibility in collisions. Examples of this are:

- Functions for accident prevention and accident severity reduction such as emergency braking and evasive maneuvers
- Vehicle-vehicle communication to influence an impending collision event

Improving aspects of compatibility is a key approach to further reducing the consequences of accidents.

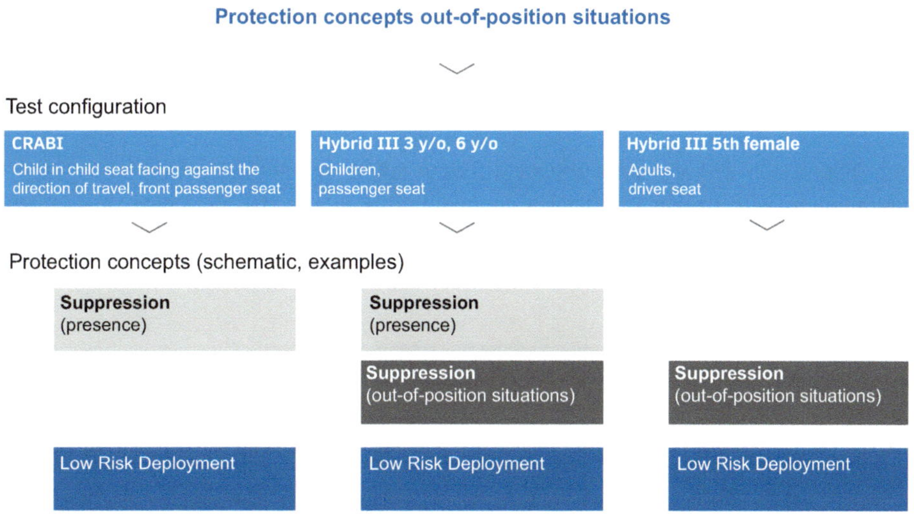

Fig. 6.16 Exemplary protection concepts for FMVSS 208 requirements for out-of-position situations

6.2.6 Out-of-Position Situations

In Chap. 4 it was already pointed out to *out-of-position* situations (OoP), for which requirements for restraint systems in frontal impact are formulated in the FMVSS 208 [141]. Their implementation is determined and evaluated in the development process (Chap. 8).

At the end of the 1990s, NHTSA studies made it clear that in addition to the proven general protection potential of airbags, there can be injury risks for small female occupants and children in out-of-position situations [9]. This led to the modification of FMVSS 208, which applies to protection systems in a frontal impact. For protection systems ina side impact (e.g., side airbag, curtain airbag), comparable evaluation criteria can be used within the framework of consumer protection ratings.

Figure 6.16 shows configurations of occupant protection system tests in frontal impact and the dummy types used for this on the corresponding seating positions, for example, a dummy (CRABI) in a child seat facing against the direction of travel on the passenger seat, further dummies (HYBRID III 3 year old and 6 year old child dummy) on the passenger seat and an adult (HYBRID III 5F) on the driver's seat.

In the design of the occupant protection system, the requirements can be implemented by following the protection concepts *suppression* (suppression of airbag deployment) or *low risk deployment* (deployment with low risk potential):

- Suppression (presence): If the presence of an occupant of the corresponding size or the child seat is detected, the airbag is deactivated.
- Suppression (out-of-position situation): If the out-of-position situation of an occupant of the corresponding size is detected, the airbag is deactivated.
- Low risk deployment: The performance characteristics of the airbag are designed in such a way that protection requirements can also be met when the airbag is deployed in out-of-position situations.

For suppression and the detection of different situations, suitable occupant sensors in combination with trigger electronics are required. For low risk deployment, a suitable coordination of the entire occupant protection system must be achieved when designing the airbag modules [63].

6.3 System Structure of Passive Safety Systems

This section provides a brief overview of some essential aspects of the design of occupant protection systems.

Occupant protection systems are complex systems in which many individual subsystems interact with each other and influence each other's effects. Based on the previously

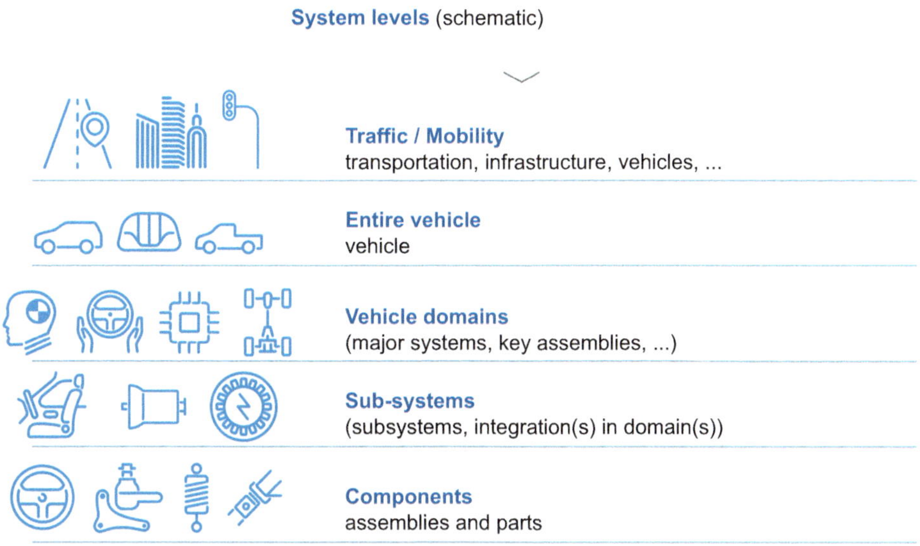

Fig. 6.17 Schematic classification of system levels in the development of occupant protection systems

presented protection principles of passive safety (Sect. 6.2) and supported by experiences with implemented solution concepts, typical system boundaries and system functionalities have emerged over time.

Those are often also the basis for the cooperation of the various partners in the value chains of the automotive industry. Subsystems from different technology domains of globally distributed partners come together to form coordinated occupant protection systems for specific vehicles or vehicle platforms.

6.3.1 System Levels from the Overall System to Components

An occupant protection system is hardly developed by a single developer or a development team alone. The respective focus of development activities is therefore often very strongly oriented towards individual tasks in a vehicle development project or their specific integration into the overall process (see also Chap. 8).

Figure 6.17 shows a schematic classification of system levels in the development of occupant protection systems. On an upper level, the traffic situation and the transport infrastructure represent a system environment for the motor vehicle. The prevailing conditions there are essential influencing factors for the vehicle during the journey, in the event of an impending collision and also during an accident.

At a higher system level, which encompasses the entire vehicle, it is necessary to define the occupant protection system of a vehicle, divide it into its subsystems within the vehicle and coordinate the interfaces with each other, so that by coordinating numerous target variables, the vehicle as a whole can be developed as efficiently as possible. The different technology domains need to be coordinated [69, 70]. In a first enumeration, this includes areas such as design, car body, interior, occupant protection, electrical/electronics (E/E), active safety, chassis, human machine interface (HMI), ergonomics, occupant protection, autonomous driving, connected car.

The design of the occupant protection system is divided on the underlying level for this reason already into various domains and subsystems. This particularly affects the following areas with associated main tasks:

- Vehicle body—design of the vehicle body structure and the passenger compartment, design and optimization of the behavior of the vehicle body structure in the event of a crash, design of attachment points for systems in the vehicle including the restraint systems
- Occupant protection and restraint systems—design and coordination of the restraint system and other systems in the vehicle with occupant protection functions, coordination of the core components of restraint systems such as seat belt systems and airbag modules with each other and in interaction with the other systems with occupant protection function
- Electrical/Electronics—development of sensors for crash detection, occupant sensing, trigger electronics for restraint systems (airbag control unit), interfaces to active safety systems (especially for integral safety functions), interior monitoring
- Interior—development and coordination of the systems in the vehicle interior such as seat system, cockpit, climate, entertainment, interior trim as well as the integration of restraint systems into the passenger compartment
- Compatibility and pedestrian protection: vehicle design and exterior design, overall vehicle and space coordination, environment sensors, pedestrian/VRU protection systems

At the sub-system level of occupant protection systems, it is about the design and development of the seat belt systems for the individual seating positions or the design of the airbag protection systems for protection in the event of frontal or side impact. For further systems—such as the steering system—, the occupant protection functions have to be designed and coordinated with the other components.

At the system level of occupant protection components, the technical requirements have to be implemented in individual components and assemblies, which are in themselves independent products with specific value chains. Examples of this are airbag modules, seat belt pretensioners or the trigger electronics as an airbag control unit.

This classification is intended to provide the reader with a basic scheme of typical task divisions in the development of occupant protection systems. A further representation of occupant protection systems and the essential subsystems can also be found in [97]. In the concrete situation, there will always be variants of this simplified representation.

6.3.2 Subsystems and System Boundaries

From one perspective, the subsystems of passive safety can be divided into sensors, signal processing, and actuators.

Classic sensors for passive safety are specific crash sensors that detect a collision within a few milliseconds.

In addition, sensors for the detection and classification of occupants have been developed and are now part of every occupant protection system. They contribute to the intervention strategy for the requirements of certain occupants in certain accident events, for example by triggering restraint systems.

As discussed in more detail in Chap. 5, the sensing of a critical driving situation or an impending collision is increasingly becoming part of the intervention strategies of integrated safety. The protective potential of vehicles will therefore continue to evolve, and the integration of signal or data processing from active safety sensors and classic passive safety sensors will increase.

Signal processing for passive safety systems is usually carried out by the airbag control unit (ACU), which initiate the activation of irreversible functions of restraint systems after an accident event. Functions of integrated safety can increasingly be carried out by both the airbag control unit of passive safety and systems of active safety or automated driving. They can activate functions in the actuators of passive safety from both sides.

From the system levels of an occupant protection system presented above, the main actors (and structures) of passive safety will be derived in more detail. They include, according to the focus chosen here, both protection systems for occupants and protection systems for vulnerable road users, serving both self-protection and partner protection (Fig. 6.18).

For occupant protection, the vehicle body or the vehicle body structure represents a first subsystem. It serves both self-protection and partner protection.

Various subsystems and components, which are summarized as restraint systems, follow: the components of seat belt systems, airbag modules, and other protection systems that are typically installed in vehicle systems assigned primarily to other domains (e.g., headrests as part of the seating system, child restraint systems).

There are also passive safety systems in motor vehicles specifically designed to protect vulnerable road users. Examples include pedestrian protection sensors, active front hoods, or airbags to protect vulnerable road users.

In addition, vulnerable road users, who are not protected by a passenger compartment, use personal protective equipment such as helmets and protective clothing.

6.3.3 Functional Sequence of Restraint Systems

The intervention of restraint systems in a collision event is characterized by the interaction of subsystems such as *sensors* and *airbag control unit*, as well as actuators like *seat belt systems* or *airbag modules*.

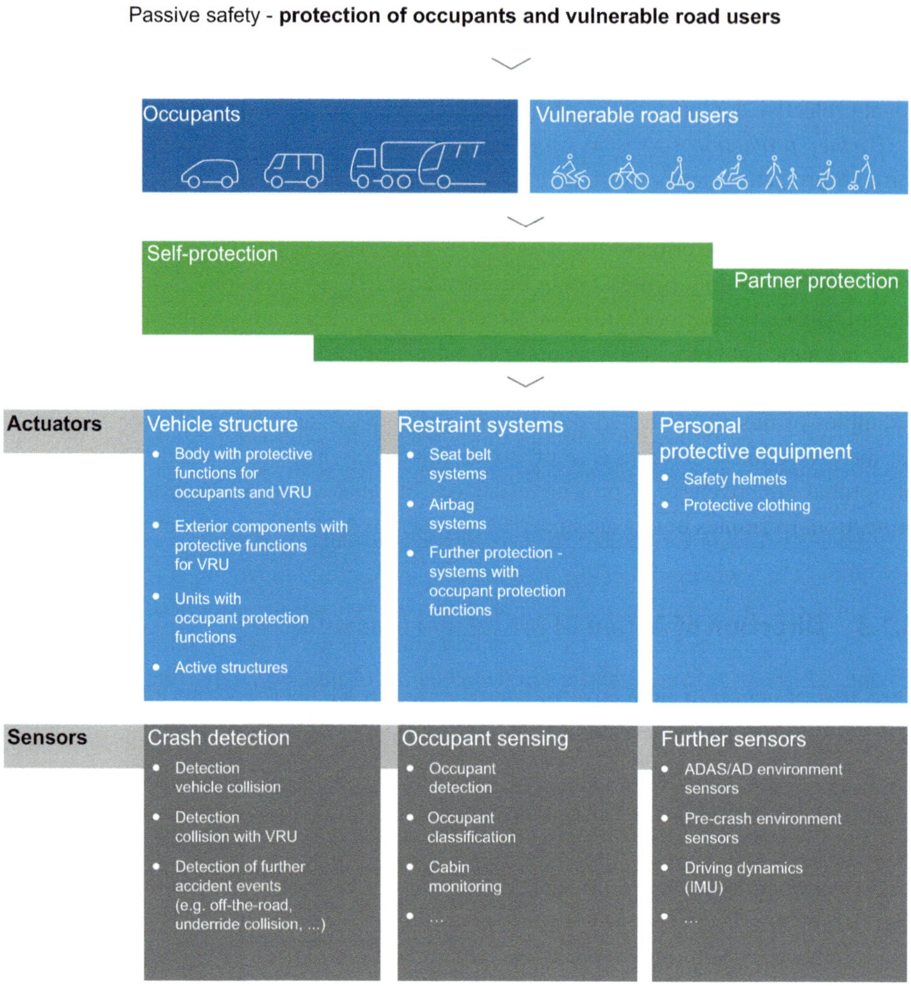

Fig. 6.18 Schematic classification of occupant protection subsystems

The crash sensors continuously sense various accelerations, as they occur in normal driving operation through shocks, accelerations and decelerations, as well as during repairs, parking bumps, extreme driving maneuvers and accidents, and transmit this as signals to a airbag control unit for diagnosis (Chap. 7).

Here, the data are compared with vehicle-specific and stored values. Additional sensors—such as the *occupant detection system* (ODS), the *occupant classification system* (OCS), the sensor technology of the safety belt buckles for the belt warning function (also: *seat belt reminder*, SBR)—can be included (Chap. 7).

Upon recognizing an accident endangering the occupants, specific restraint systems are usually activated according to a predetermined algorithm. For this purpose, the airbag control unit sends ignition impulses to the various pyrotechnic igniters in the *seat belt pretensioners* and *gas generators* of the *airbag modules* at specified times after the impact. They contain a pyrotechnic propellant charge, which is ignited by the igniter and burns off quickly. In the case of *seat belt retractor pretensioners*, the electrical current ignites *micro gas generators* (MGG) so that pretensioning for coupling the occupants to the vehicle takes place. In airbag modules, the propellant charge of the gas generator releases gas, which fills the folded airbags and they rapidly inflate.

To coordinate the various restraint systems with each other and in combination with other subsystems of the occupant protection system, a wide range of parameters can be adjusted in components as well as by the intervention strategy of the airbag control unit. For airbag modules, for example, the tuning of the airbag fabric (material, permeability, thread strength, fabric density, coating), seam leakage, airbag vents, adaptivity elements, airbag tethers, folding, shape of the airbag, inflator technology can be part of development processes.

6.4 Vehicle Body Structure and Passenger Compartment

The vehicle body structure is an essential subsystem of occupant protection and it represents an important part of the system environment for most other passive safety systems. In collisions, it is the immediate contact zone with an obstacle and influences the characteristics of the crash pulse acting on the occupants. Essentially, under the term *vehicle body structure* here, the aspects of occupant protection of the *vehicle body* are to be described, which in this sense also include doors, flaps or other components in the exterior.

As already explained in the principles of protection (Sect. 6.2), the ability to deform for energy conversion and the shape-retaining passenger compartment are essential measures for occupant protection. Therefore, this section deals with the *deformation structure* (also: *deformation zone,* colloquially: *crumple zone*) of the vehicle, which contributes significantly to *energy conversion* during collisions. In addition, the constructive implementation of *structural safety* (also: crash safety, *crash worthiness*) by the vehicle body is presented.

In addition to the car body, major systems and assemblies of large masses and/or high stiffness (e.g., engine, transmission, high-voltage drive battery) also influence the vehicle's protection potential. While in the past the drive train with combustion engine and transmission was the only variant to be considered, today electric vehicles with their electric motors, the high-voltage battery, fuel cell or hydrogen tanks set new conditions for the implementation of occupant protection concepts.

In addition to the car body and the key systems mentioned above, other systems of the vehicle, especially the seat system, the cockpit and the steering or steering column, also contribute to *structural safety* and help to reduce the risk of injury. Further aspects of vehicle body structure and occupant protection in special body variants (e.g., convertible) are pointed out in Sect. 6.7.

This section also addresses further aspects of partner protection. On the one hand, it is necessary to implement requirements for *compatibility* in the event of vehicle collisions and to protect the occupants of colliding vehicles. On the other hand, the requirements for the protection of *vulnerable road users* (VRU) have significantly increased. This will be discussed in more detail in another section (Sect. 6.8).

The vehicle body structure has—as already described in Sect. 6.2—the following two core tasks, which should be repeated here:

Energy conversion: Through defined and compliantly designed *deformation zones* (also: *crumple zones*) the vehicle body structure is intended to convert the kinetic energy of its own and possibly also the colliding vehicle in frontal, side and rear collisions. Here, kinetic energy is converted into deformation work and heat through internal friction.

Shape-retaining passenger compartment: On the other hand, the *passenger compartment* should be as resistant as possible to deformations in a collision. Collapsing vehicle structures or intrusions of components pose significant injury risks to occupants and can trap or crush them.

The shape-retaining vehicle body structure provides restraint systems such as seat belt systems and airbag modules with sufficient strength and geometric stability of anchoring points. It maintains defined action spaces for the deployment of airbags or the defined occupant displacement and kinematics during restraint.

For further tasks that the vehicle body structure has for occupant protection, reference is made to Sect. 6.2 and the explanations in [39].

Partner protection (Section 6.1.2 and 6.2.5): The vehicle body structure is also the contact zone with another vehicle in a collision with other road users, so requirements for compatibility and the protection of the occupants of the other colliding vehicles must be taken into account. In the event of a collision with vulnerable road users (Sect. 6.8), additional requirements for collisions with pedestrians and cyclists must be considered in the design of the contact structure and exterior.

Figure 6.19 shows the design of a modern vehicle body using high-strength steels and lightweight materials such as aluminium. In addition to the design in

- Steel shell construction and as a lightweight concept

For self-supporting vehicle bodies, further lightweight concepts are applied in passenger cars [31, 120]:

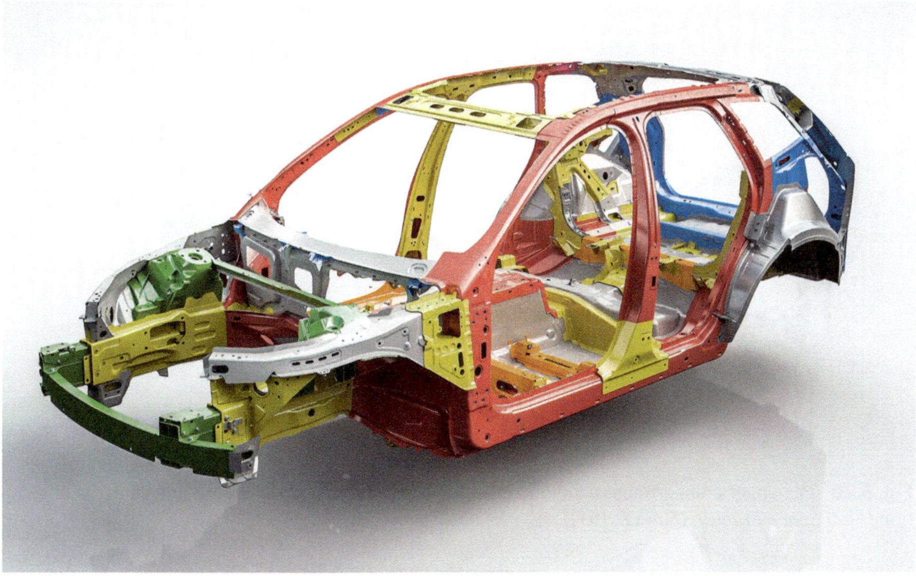

Fig. 6.19 Modern vehicle body structure with functions for self- and partner protection *(Volvo XC60)*. (*Source*: Volvo Cars)

- Full aluminium construction,
- Aluminium skeleton construction with profiles and formed parts,
- Fibre composite construction.

Figure 6.20 shows an example of a vehicle body in aluminium skeleton construction for reducing vehicle weight and also for reducing emissions during driving. Non-self-supporting vehicle bodies in ladder frame construction are becoming less and less common in passenger cars.

In frontal and rear impact, the front carriage and the rear vehicle are the main deformation zones of the vehicle body.

6.4.1 Front Carriage

The deformation structures on the vehicle serve to convert kinetic energy into deformation work (Sect. 6.1). This results in a force or acceleration curve that significantly influences the design of the occupant protection system.

The front deformation zone can be schematically divided into three zones (Fig. 6.21):

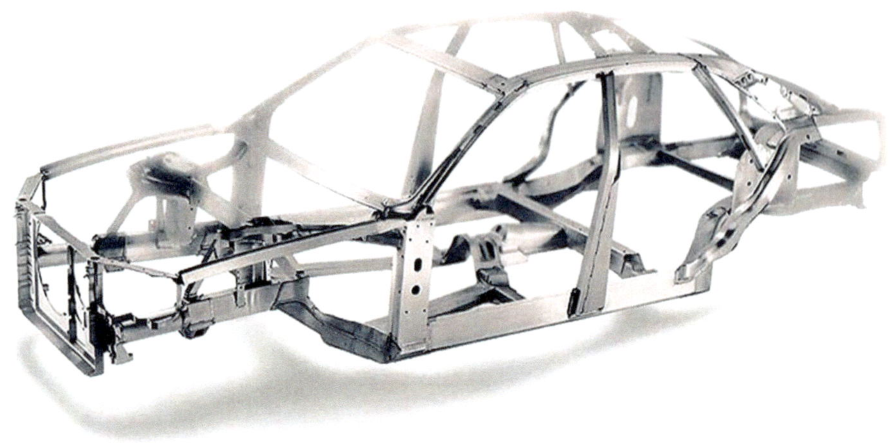

Fig. 6.20 Passenger compartment using the example of a vehicle body in aluminium skeleton construction *(Audi Space Frame)*. (*Source*: [61])

- Deformation zone for minor accidents and pedestrian/VRU protection zone
- Compatibility zone, which serves both self- and partner protection
- Self-protection zone of the rigid passenger compartment

The pedestrian protection zone contributes to partner protection against vulnerable road users (Sect. 6.8). The pedestrian protection zone should have a force level as low as possible when in contact with external road users, but this is limited downwards by a minimum strength for protection against minor collisions.

By appropriately designing this zone, accident consequential costs in the form of property damage at collision speeds up to approximately 15 km/h can be minimized *(minor accidents)*. The deformation forces can be specifically directed into a bumper or sleeve tube construction. The impact absorbers completely cushion minor bumps at very low speeds and show no damage. At collision speeds up to approximately 15 km/h, for example, replaceable sleeve tubes (also: *crash boxes)* absorb the impact energy.

This protection zone extends to the front edge of the front frame longitudinal members. However, since the vehicle must also have structures for self-protection with an early effective high force level, there are conflicting objectives to resolve here.

The area behind the compatibility zone serves both self and partner protection. As explained in Sect. 6.1, it should provide the best possible protection for the occupants of both vehicles in a collision of two differently weighted vehicles. When designing the deformation zone, collisions with differently weighted vehicles must be considered. An accident with a heavier car, for example, makes a hard front structure appear sensible, while an accident with a lighter car requires a softer structure. In other words, this zone must be designed softer for heavy vehicles than for lighter ones. However, since a car

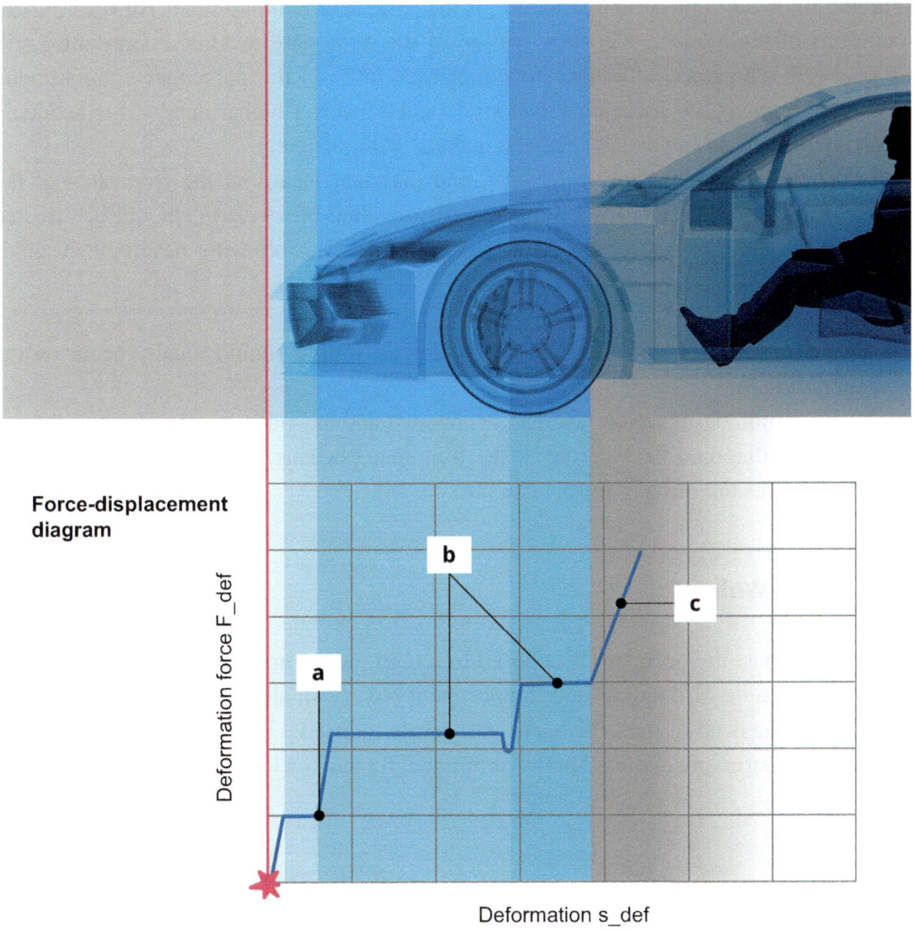

Fig. 6.21 Schematic representation of levels of deformation force along the deformation zones of a front structure in the force-displacement diagram: **a** Zone for protection in minor collisions and pedestrian/VRU protection, **b** Compatibility zone for partner and self-protection, **c** Self-protection zone of the rigid passenger compartment (based on [61])

can be involved in accidents with both light and heavy vehicles, the challenge is to find a compromise for a compatible force level. The compatibility zone ends in the area of the wheel suspension.

During a frontal impact, the front-end structure is designed to deform in a controlled manner. This aims to limit the deceleration of the passenger compartment to a relatively low level in order to protect the occupants. In contrast, the structure within the self-protection zone—located closer to the passenger compartment—should experience only minimal deformation. To ensure the stability of the passenger compartment, the stiffness in this zone is significantly increased. This is achieved through the controlled transfer of acting forces into the frame structure of the passenger compartment. However, under conditions of high crash severity, a portion of the remaining residual energy must still be absorbed. The force level therefore increases towards the passenger compartment. Although a rectangular force characteristic could convert a higher energy, the suddenly occurring high force can usually only be realized with difficulty.

By arranging longitudinal, transverse, and diagonal beams in the front area of the vehicle, forces in the offset frontal impact, the so-called *offset crash,* should be directed into the laterally arranged longitudinal beam, the sill. The following development goals are pursued:

- When a certain overlap, e.g., 30 %, is undercut, sliding should ideally occur on the diagonal beam.
- Up to a certain deformation, energy is converted and the vehicle is decelerated.
- The stable diagonal beam protects the foot space because intrusions are prevented, and the forces are directly introduced into the sill.

6.4.2 Side Wall

In side collisions, the occupant is exposed to a high risk of injury, as the available space between the inside of the door and the occupant is very small and the ability to deform compared to the front of the vehicle is significantly reduced. Figure 6.22 shows the structure of a body structure and load paths of initiated forces in a lateral collision.

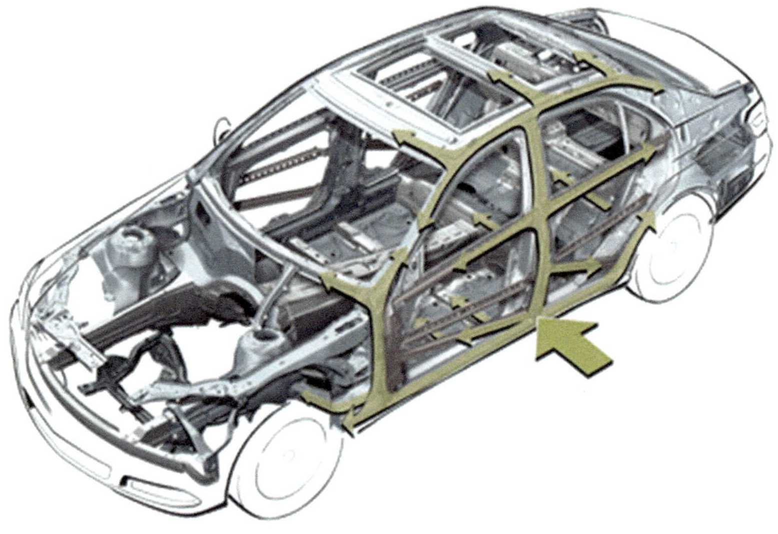

Fig. 6.22 Body structure and construction of the side wall for protection in side impact. (*Source*: MBpassion, [61])

In a lateral collision, the driver's door and the B-pillar can penetrate into the compartment of the vehicle. An unprotected occupant would be brought to a speed that approximates the penetration speed of the colliding vehicle.

To ensure strength in this area, stable door hinges and locks, large-volume reinforced sills and door frames, or additional sheet metal and tube constructions in the side areas contribute. Furthermore, the door frames are designed in such a way that additional support is provided on the door frame and the forces that occur are introduced into the stable door pillars, into the sill and into the roof frame.

The testing of structural safety in side impact is carried out with different load cases, such as for side and pole impact, which were already presented in Chap. 4.

6.4.3 Rear Vehicle

The deformation structure of the rear vehicle (Fig. 6.23) also consists of several zones:

The rearmost area is particularly at risk in smaller maneuvering and parking accidents. One goal here is to minimize repair costs. The first contact with the obstacle is made via the rear bumper. Collisions up to about 6 km/h can often occur without structural damage. Another measure is the design of the car body, in which individual deformed components can be easily replaced. In addition, hydraulic or pneumatic impact absorbers are also used.

The deformation structure of the rear vehicle is designed to be stiffer towards the passenger compartment in order to ensure structural safety in rear collisions at higher

Fig. 6.23 Deformed rear structure after a rear impact test. (*Source*: [61])

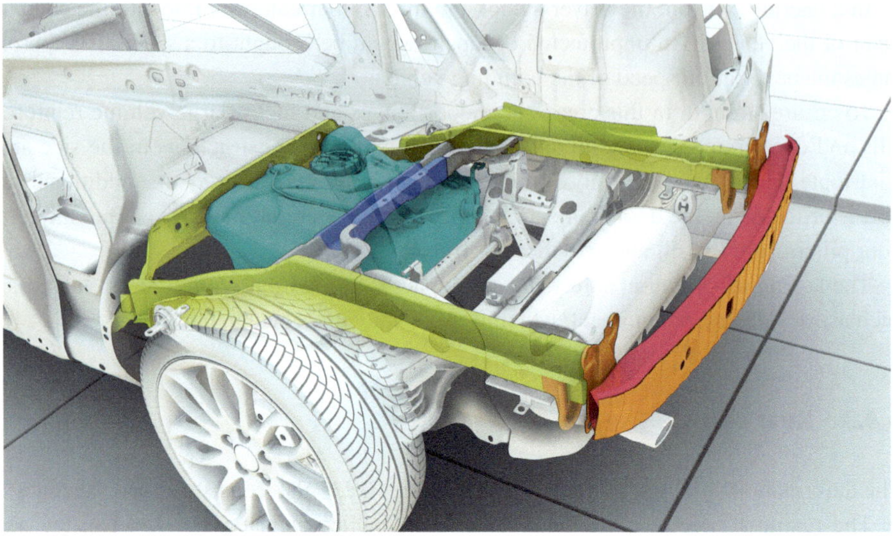

Fig. 6.24 Vehicle body structure with optimization of material qualities for defined stiffness and deformation ability for protection in rear impact. (*Source*: Volvo Cars)

speeds. Figure 6.24 shows the vehicle body structure of the rear vehicle with an optimization of different material qualities for defined stiffness and deformation ability to protect sensitive systems in the vehicle in the event of a rear impact. For vehicles with combustion engines, it is important to protect the fuel tank from damage. Therefore, installation sites are chosen that deform as little as possible in the event of an accident, for example under the rear seat bench and in front of or above the rear axle. This installation position offers additional advantages in terms of driving dynamics.

6.4.4 High-voltage Battery

The development of battery electric vehicles necessitates the integration of a *high-voltage battery*, which is characterized by high weight, large volume, and high protection needs against mechanical damage. These aspects have different effects on occupant protection.

Firstly, the vehicle mass of battery electric vehicles usually increases significantly compared to vehicles with combustion engines, as high-voltage drive batteries often weigh several hundred kilograms. Lithium batteries for motor vehicles currently still have a specific weight of about 150 to 200 Wh per kg. A 100 kWh battery for a car with a high range thus currently means a weight of more than 500 to 750 kg.

This influences the characteristics of the *crash pulse* and thus the tuning of the occupant protection system.

The technologies currently used for high-voltage batteries have high requirements for protection against mechanical impacts to reduce the risk of a *thermal runaway*. In this process—for example after a mechanical impact on the battery cells—charge transfers lead to a self-amplifying heat-producing process in which the cell overheats and the lithium contained in it can ignite.

To prevent this, the high-voltage battery packs are integrated into very rigid assemblies or into the vehicle body structure itself.

The effects of very rigid structures within the vehicle body on the occupant protection system must also be taken into account. In general, they have little deformation capacity for energy conversion. On the other hand, they can contribute to the structural integrity of the passenger compartment. When developing vehicle structures and occupant protection systems, these effects must be taken into account for both objectives.

6.4.5 Pressurized Gas Container

The installation of pressurized gas containers for vehicles with gas or hydrogen drive may increase in the coming years, so that such a body variant will often be part of vehicle programs. The large-volume pressurized gas containers with internal pressures up to 700 bar must be protected from damage in a collision to avoid associated risks, for example, due to a fire as a result of an accident. The first cars with hydrogen drive have already been tested in consumer protection ratings (Chap. 4).

Hydrogen drives may become more relevant for commercial vehicles or in the transporter segment for fleets (Sect. 6.9). Here, the installation is done in the underbody, in the roof, or in the area behind the driver's cab. So far, these systems have been integrated more into existing vehicle platforms. With these developments and vehicles, further experiences will have to be gathered in the future. See also the explanations on fire protection (Sect. 6.10).

6.5 Seat Belt Systems

The seat belt is the essential protective device for vehicle occupants and is rightly considered the *"number one lifesaver"*. Half of all car occupants who died in accidents in North America in 2021 (seat belt usage rate over 90 %) were not wearing seat belts [87, 88]. It is often counted among the most important inventions of the century [121, 89]. In 1985, the German Patent Office named it as one of eight inventions of the past 100 years with the greatest benefit to people [29].

This section presents seat belt systems and their components in use in cars as part of integral occupant protection concepts (Fig. 6.25). Seat belt systems, like airbags, are part of the restraint systems and are a significant product group of passive safety.

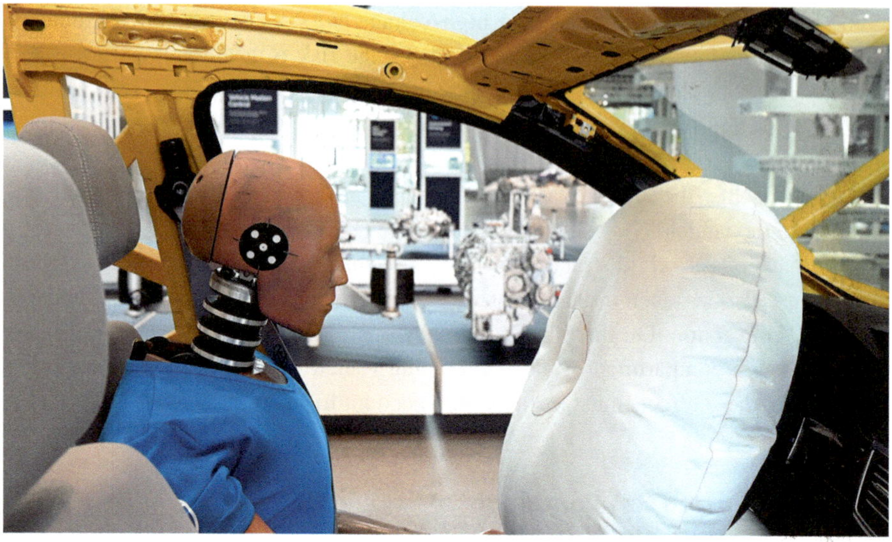

Fig. 6.25 Seat belt systems are fundamental for occupant protection. (*Source*: ZF)

The application of seat belt systems in commercial vehicles, especially in trucks and buses, is presented separately in Sect. 6.9.

6.5.1 The Beginnings of Seat Belts in Motor Vehicles

The spread of the seatbelt in motor vehicles began in 1958 with the registration of a *three-point seat belt system* by combining the pelvis and shoulder belt for patent by Nils Bohlin beeing an engineer at Volvo (first application: 1958, US patent US-3043625-A granted: 1962) [11]. Bohlin had previously worked on ejection seats for aircraft. Volvo then introduced the three-point belt as standard equipment in two vehicles in 1959. The patent was open for use by other vehicle manufacturers.

Starting from two anchor points on the B-pillar, the webbing runs diagonally across both the upper body and the pelvis (Fig. 6.26). At the transition from the shoulder to the lap belt, it was anchored to the vehicle center tunnel with a "fastening bracket" that could be "detached". Pelvic belts or diagonal belts were already known before, but they could not convince due to several disadvantages: Some were not an effective means against the impact of the head on the steering wheel and others could not prevent the occupant from slipping through (*submarining effect,* Sect. 3.1.4). In addition to the still existing injury risk, these belt systems were impractical and uncomfortable to use.

A first seat belt was installed in a vehicle in 1902. After that, belt systems were rather used in aviation. In the 1950s, an initiative by Vattenfall led to the development of the

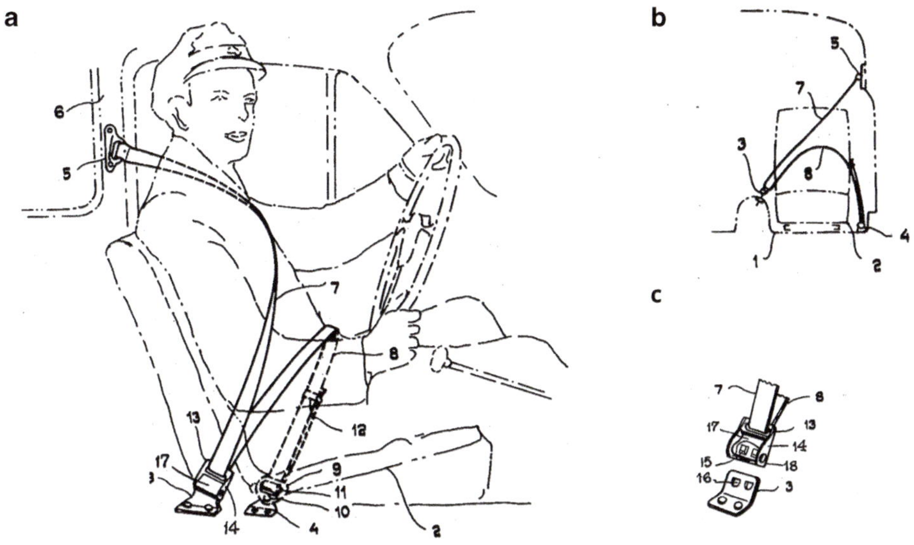

Fig. 6.26 Three-point seat belt—Illustrations for the patent specification: **a** Belt anchoring points, **b** Belt routing, **c** Belt buckle. (*Source*: [11])

Vattenfall system [58]. These and further developments led step by step to the further spread of seat belt systems.

Milestones

The list presents selected milestones in the development history, distribution, and legislation of seat belt systems:

- **1902** First **vehicle** equipped with a **safety belt:** Speed record car "Baker Torpedo", USA
- **1903** First **patents** for belt systems, both in France: Four-point belt system (Gustave-Désiré Leveau), Five-point belt system (Louis Renault)
- **1913** Use of safety belts in aircraft by the stunt pilot Célestin Pégoud in France
- **1956** Initiative by the company Vattenfall in Sweden for a **two-point belt system** *(System Vattenfall),* initially installed in company vehicles, manufactured under license by Lindblads Autoservice (today Autoliv)
- **1958** Patent application for the **three-point seat belt** by Volvo and Nils Bohlin in Sweden
- **1959** Market launch of the first standard **three-point seat belt system** by Volvo for the model PV544 *(system Volvo, Klippan belt)*

- **1963** Belt system with automatic seat belt retractor and locking function (automatic seat belt, *automatic belt*) (Britax or Excelsior Motor Company)
- **1966** Mandatory installation in new cars in the USA from 1968, "Federal Law Title 49 of the United States Code, Chapter 301, Motor Vehicle Safety Standard"
- **1970** Mandatory seat belts for outer seats behind the windshield, provided the vehicle is equipped with anchors
- **1973** Mandatory for an immobilizer in case of unbelted occupants (seat belt interlock, ignition interlock) in new vehicles; it was withdrawn in 1974 and led to the requirement for equipment of automatically applying belt systems (automatic belt) or driver and front passenger airbag in 1977
- **1974** In Germany, seat belt systems become mandatory on the front seats of new passenger cars
- **1976** Introduction of the **seat belt law** in Germany for the front seats of passenger cars, but initially without a fine
- **1979** Seat belt systems on the rear seats of passenger cars become mandatory in Germany for new vehicles
- **1979** Mandatory for three-point anchors and **three-point seat belts** for outer seats directly behind the windshield
- **1980** Market launch of **pyrotechnic seat belt retractor pretensioners** by Mercedes-Benz for the series 126 (S-Class) together with the market launch of the driver airbag as a supplemental restraint system
- **1984** Introduction of **fines** for violation of the seat belt law in passenger cars in Germany
- **1992** Seat belt systems become mandatory in Germany for new truck registrations
- **1997** Seat belts required for all seats in buses with a permissible total weight of more than 3.5 t in Germany
- **2002** Market launch of **active seat belt retractor pretensioners** by Mercedes-Benz for the series 220 (S-Class) (TRW, today ZF-LIFETEC)
- **2010, 2011, 2013** Market launches of **seat belt airbag systems** on rear seats by Ford in 2010 for the Explorer (Key Safety Systems), by Toyota in 2011 for the model Lexus LFA (Takata) and by Mercedes-Benz in 2013 for the series 222 (S-Class) (Autoliv) for rear seats

The early generation three-point belt systems were still *static belts,* i.e., they lacked an automatic belt retractor. It was introduced in 1963 as *automatic seat belt.*

First generations of belt systems could not fully exploit their protective potential and had disadvantages in some situations [91]. This was still used as a reason to argue very critically against the use of belt systems at the end of the 1970s, without sufficiently including the expected injuries in the event of non-use of the belt in these considerations. Nevertheless, this accelerated the improvement process and it remains to be noted that this individual measure had a very high efficiency in terms of improving vehicle safety.

6.5.2 The Long Journey to High Seat Belt Usage

The seat belt in the motor vehicle is a protection system that usually has to be applied by the vehicle occupant himself and can only then perform its protective function.

Its use by vehicle occupants has changed very positively since the 1970s, especially with the introduction of seat belt legislation. The *seat belt use rate* of occupants in Germany on front seats before the seat belt legislation in urban traffic at that time was only about 40%, but a level of over 98% has now been reached (Fig. 6.27). The seat belt use rate of occupants on rear rows of seats has improved even more significantly, which has also reached a very high level. The Federal Highway Research Institute Germany

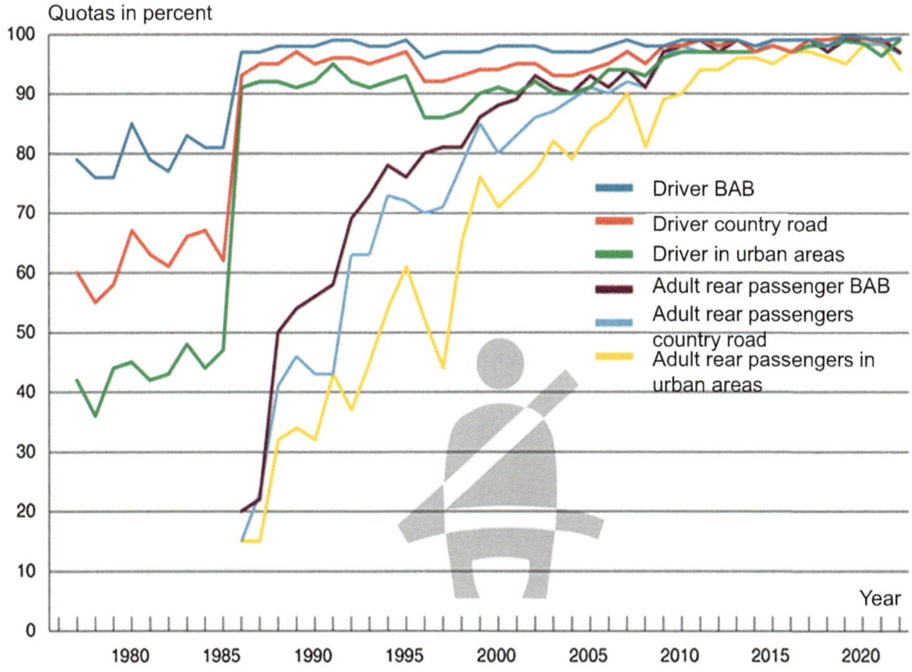

Fig. 6.27 Seat belt use rate of car occupants: Traffic observation 2021, adults, Germany, country roads and federal highways (BAB). (*Source*: Federal Highway Research Institute Germany, BASt, [38])

(BASt) conducts annual *traffic observations* among other things to determine the seat belt use rate, so that a continuous analysis of the behavior of vehicle occupants can be carried out over the years [38]. An average seat belt use rate of 98.8% (2021) was measured for all adult vehicle occupants. For children, a rate of 98.7% (2021) was determined for the use of child seats or seat belt.

In Chap. 4 the worldwide legal regulations in the regions for equipping motor vehicles with seat belt systems by vehicle manufacturers were already described. However, this did not immediately imply an obligation for the passengers to use the protection system. The implementation of the seat belt law in motor vehicles began in Europe in the 1970s. Fines were only imposed later, as the initial attempt was to motivate road users to use them through traffic education campaigns. The *seat belt reminder function* can be considered significant for improving the seat belt use rate, which triggers a clear warning to the driver or occupants if the seat belt is not fastened.

Wearing the seat belt is mandatory in many countries. However, in the USA, the only country in the world, tests of occupant protection systems with unbelted occupants are also carried out according to FMVSS 208 (Chap. 4).

6.5.3 Protection Principle of Seat Belt Systems

Seat belt systems are essential components of occupant protection systems. As a technology of passive safety, they serve to reduce injury risks in an accident. In the event of a vehicle colliding with an obstacle, the vehicle is abruptly decelerated. Due to the inertia of their masses, vehicle occupants would strike the steering wheel, the cockpit, or the seats at the driving speed of the vehicle if they were unprotected. An unhindered impact leads to high injury risks even at low speeds.

The protective principles of occupant protection systems and the objectives of seat belt systems have already been explained in Sect. 6.2. The principle of the protective effect of seat belt systems is to be clarified once again with the comparison in Fig. 6.28. It shows schematically simplified the speeds of the vehicle and the occupant in the vt-diagram as well as the deceleration acting on the occupant during the accident. Without restraint by the seat belt, the occupant would collide in the compartment and would therefore be exposed to high accelerations and high local contact forces over a short time (and short distance).

Seat belt systems perform protective functions in many accident scenarios: e.g. frontal impact, side impact, rear impact, rollover accident, and off-the-road accident. The protective effect of seat belt systems can be differentiated into the following essential sub-functions:

Stabilization of the occupant position: If the vehicle occupants are belted, they are stabilized to a certain extent in their seating position and the risk of being thrown out of the vehicle is reduced or can be prevented in many accident scenarios. This sub-function is

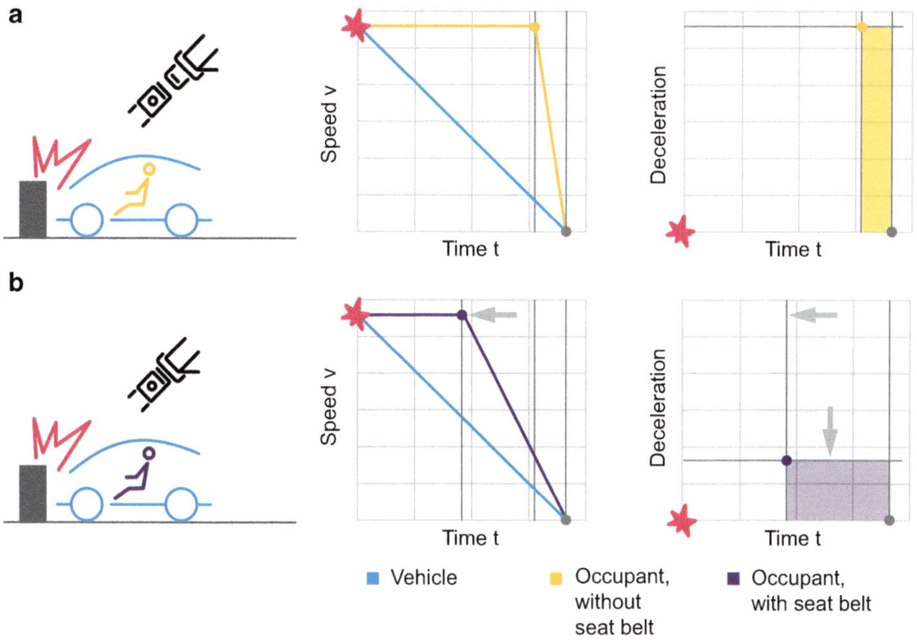

Fig. 6.28 Protective effect of the seat belt in the vt-diagram by comparing the restraint of **a** unbelted occupant and **b** belted occupant (schematic). (Based on [61])

ensured in static belts directly and in automatic belts by various *locking functions* of the *seat belt retractor*.

With the secure positioning of the occupants, the seat belt significantly contributes to the fact that other occupant protection systems (e.g. airbag systems, headrests, anti-whiplash systems etc.) can provide effective protection.

Seat belt systems also contribute to active safety during the first phase of the integral safety approach (phase: on the journey, Fig. 1.7) by stabilizing the occupant position. The driver is secured in his position by the seat belt, which supports the perception of his driving task. By adding a reversible pretensioning by active seat belt retractor pretensioners (ACR) to the driving maneuvers of driver assistance systems (ADAS), such as ADAS-AEB (*autonomous emergency braking*) or ADAS-ESA (*evasive steering assist, emerging steering assist*) this protective function can be further refined.

Coupling: The goal of the coupling of the occupants is the rapid build-up of restraint forces in the seat belt system, which are used to decelerate the occupants. This achieves an as early as possible start of the restraint and an unbraked *forward displacement* of the occupant can be avoided as far as possible. For coupling during the in-crash phase,

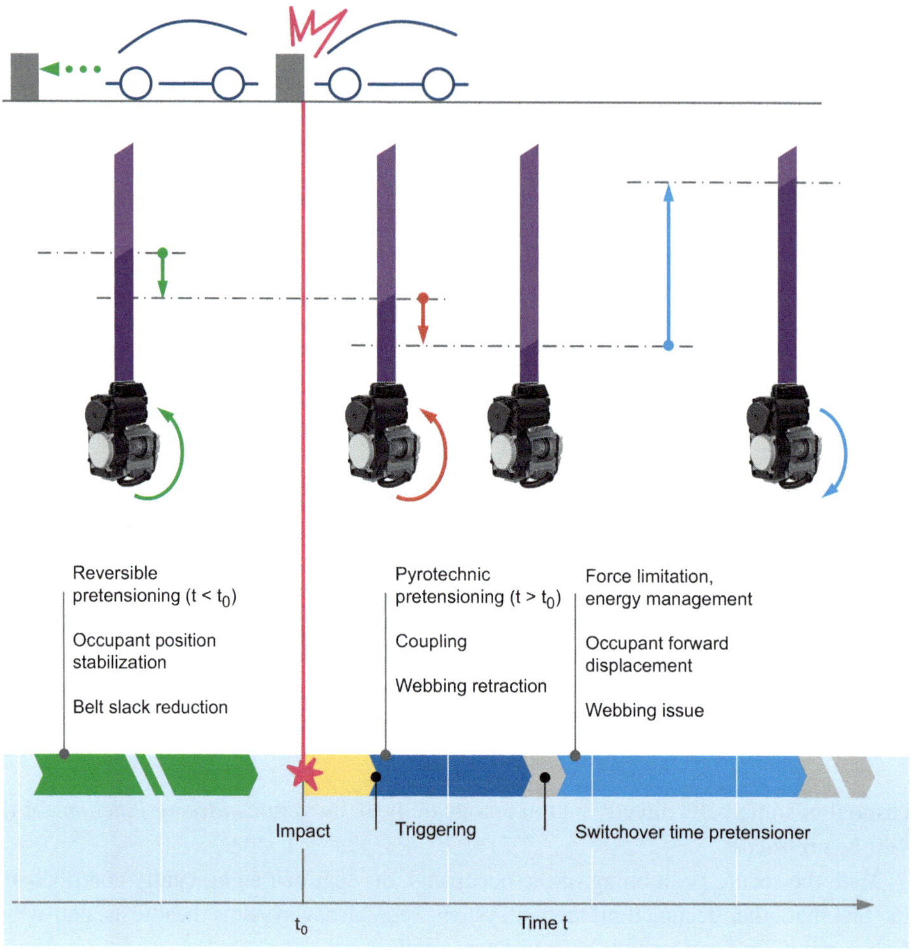

Fig. 6.29 Basic course of the belt movement in the three-point seat belt. (*Source*: ZF)

non-reversible pyrotechnic *seat-belt retractor pretensioners* (Sect. 6.5.9) and reversible *active seat belt retractor-pretensioners* (Sect. 6.5.10.1) are already used in the *pre-crash phase* (phase: *in case of danger,* Fig. 1.7) of the integral safety approach. The protective effect of the seat belt system is improved by using a larger proportion of the forward displacement path for occupant deceleration (speed adjustment, energy transformation), thereby reducing the load values (Fig. 6.29).

Force limitation in energy transformation: It contributes to the high protection potential that today's seat belt systems offer. In severe collisions with high speed of the own *(ego-vehicle)* or the colliding vehicle, the forces acting on the occupants would exceed a tolerable level and severe injuries, especially to internal organs, would be the result.

Therefore, seat belt systems have functions for *force limitation,* which were previously also implemented by defined tearing seams in the webbing (Sect. 6.5.8.1) or today usually by deformable *torsion bars* built into belt retractors. They allow the webbing to be dispensed with a defined belt force during the deceleration process in the shoulder and pelvic area, and the occupant can thus move further forward over the remaining forward displacement. The increased deceleration range reduces the restraining force acting on the occupant and the injury risk decreases.

Today, seat belt systems enable a defined control of the restraining force with their force limitation functions, so that a differentiated adjustment of the force limitation level to the respective occupant type or accident constellation can be implemented *(adaptivity).* The requirements for the level of the restraining force vary depending on the seating position. The systems available for this are explained in more detail in the following paragraph.

The protective functions of the seat belt system can be well assigned to the individual phases before, during, and after the collision (Fig. 6.30): Stabilization of the occupant position before the collision event. After the impact, the *in-crash phase* involves *non-reversible pretensioning* and *coupling* of the occupants, the *force limitation*, and the control of the force progression.

Figure 6.31 shows recordings of a system test for occupant protection systems and exemplarily the representation of the sequence of occupant kinematics in the pretensioning and force limitation phase, in which the webbing from the seat belt retractor is dispensed and a forward displacement of the occupant can be detected.

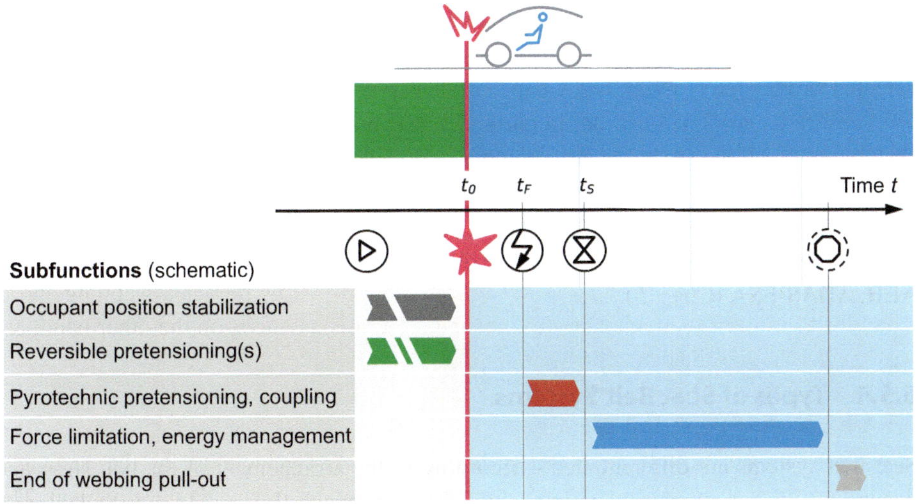

Fig. 6.30 Schematic representation of the protective functions of seat belt systems along the phases of a collision. (Based on [61])

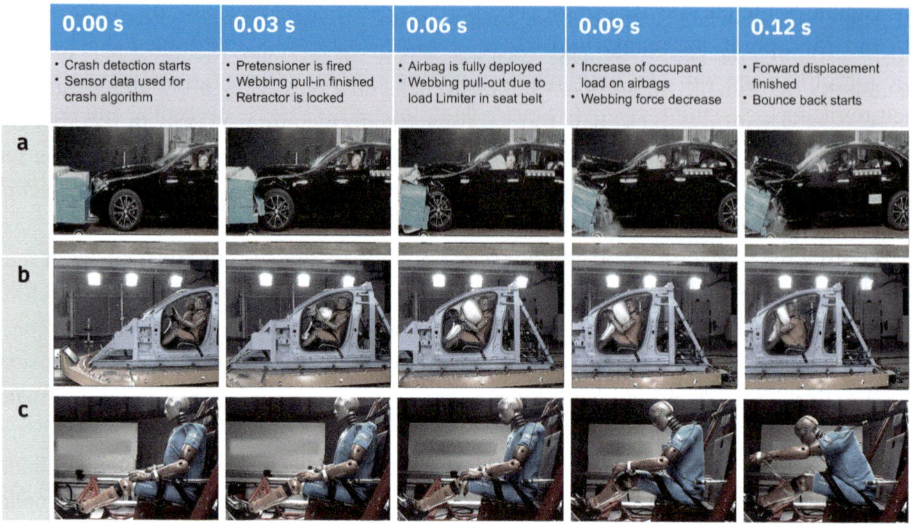

Fig. 6.31 System test for occupant protection systems and exemplary representation of occupant kinematics. (*Source*: ZF, based on [61])

An explanation of protective functions and intervention measures in the *in-crash phase* has already been given in Sect. 6.2. This particularly concerns the functions of belt retractors for *energy management* and the control of force levels. The advantages and extended protective functions of *active seat belt systems* are further elaborated in Sect. 6.5.10.

Concepts of integral safety aim at reducing the belt slack in the seat belt already before the collision, e.g. in critical driving situations or when a collision of the vehicle is detected as very likely (Sect. 6.2, Chap. 7). This serves to stabilize the occupant position or to ensure its nominal position. In particular, the early tight application of the seat belt ensures its favorable course at the shoulder and in the pelvis in the event of a collision [78]. This is achieved by the reversible pretensioning with an earlier start of the *coupling* by *active seat belt retractor pretensioners* (Sect. 6.5.10.1). Such reversible pretensioning can also be proactively combined with active safety driving maneuvers such as ADAS-AEB, ADAS-ESA (Chap. 5).

6.5.4 Types of Seat Belt Systems

Seat belt systems are differentiated—according to the arrangement of the belt strap and depending on the attachment points—into different types (Fig. 6.32): Lap belt, three-point belt, harness belts as four-, five- or six-point belts, and other special forms.

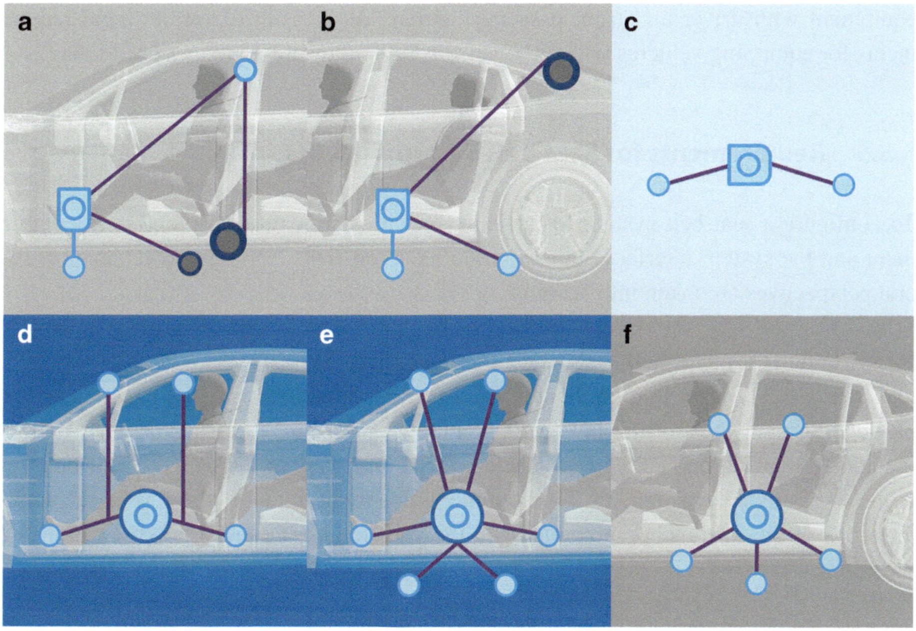

Fig. 6.32 Types of seat belt systems depending on belt anchoring points and belt routing: **a** Three-point belt with pillar loop (D-ring), **b** Three-point belt with direct pull, **c** Static lap belt, **d** Four-point belt system, **e** Six-point belt system, **f** Five-point belt system, e.g. also child seats. (*Source*: ZF)

Today, the *three-point belt system* with a retractor (also: automatic three-point belt, 3PGA) is the standard solution for seat belts in vehicles. The webbing runs in a first section from the upper anchoring point over the shoulder diagonally across the upper body to the pelvis. This area is the shoulder belt. The second section runs horizontally across the lap or the legs. This area is the *lap belt.*

Four-, *Five-* or *Six-point harness belt systems* are used in racing and safety requirements are specified for these applications, for example: FIA 8853/98 [40] or SFI 16.6 [100]. Simple *pelvic belts* are still used today, for example, in buses or airplanes. A lap belt can be designed with and without a seat belt retractor.

Auto-Belt Systems: In the 1980s and 1990s, *automatic belt systems* (also: automatic seat belt, *auto belt*) were used in the USA. They automatically applied the shoulder belt to the occupant when the door was closed. A mechanism moved the D-ring from the A-pillar to the B-pillar. They thus made manual buckling by inserting the belt tongue into the belt buckle by the occupant itself obsolete. A major disadvantage of these systems was that there was no lap belt portion and the femur had to be supported early on by other measures. Legal requirements demanded either such a belt system or the equipment with driver and front passenger airbag. As a result of further legal requirements for equipping vehicles with airbags from 1998, they are no longer used today.

6.5.5 Requirements for Seat Belt Systems in Cars

To configure a seat belt system for a seating position in a vehicle, the system environment and the system interfaces must first be recorded. This can only be achieved if several perspectives are taken into account.

Requirements for seat belt system protection functions
One perspective is the requirements for protection functions and their testing, which are legally regulated in different regions, among other things (see also Chap. 4). They include tests of components of seat belt systems and system tests of the seat belt systems in substitute environments, for example with a representation of the vehicle-specific positions of the belt anchorage points and the vehicle-specific crash pulse. Examples of these are:

- Testing of belt anchorage points in the vehicle body structure or in the seat system according to UN-R 14 [130]
- Tests of pelvis and chest preloading in sled tests according to UN-R 16 [131]

Another perspective for considering the system environment of seat belt systems is the vehicle's occupant protection system with its various subsystems. On the front seat rows, the interaction with the airbag systems as another restraint system must be considered, while this is usually not the case on the rear seats. Furthermore, interactions with the seating system, the steering column, or the vehicle body structure must also be considered—to name a few essential ones. Their interaction is investigated in computational and experimental system simulations. For this purpose, sled tests with vehicle-specific parameters are also carried out and the protective effect is tested and evaluated at the different system levels (Chap. 8). Examples of *system tests* are:

- System tests of seat belt systems according to UN-R 94 [132] (frontal impact on a deformable barrier with 40% overlap) and UN-R 137 [133] (frontal impact on a rigid barrier) are further tests for M1 class vehicles
- Tests according to FMVSS 208 for the USA [141]

Another perspective for considering the system environment is also the installation spaces for seat belt components and their physical integration into the vehicle. For each seating position, certain requirements and restrictions must be taken into account. In particular, the front and rear seat rows place different demands on seat belt systems. The optimization of ergonomics for the occupants in the vehicle interior is associated with the resolution of conflicting objectives.

The following will examine the influence of the position in the vehicle on the three-point belt system in more detail.

Requirements of different seating positions

The specific requirements for occupant protection systems for the respective seating positions in passenger cars lead to typical configurations of the seat belt systems. Seating positions in passenger cars are specified, among other things, by UN-R 16, so that requirements for the design of the belt system can already be derived from this [131]. Figure 6.33 shows the division of seating positions into typical seat rows in passenger cars:

Front seat row: This is where the driver's and passenger's seats are located, and in some cases also a middle seat. For passenger cars, adjustable single seats are typical in this position. The seat belt retractor is often housed in the B-pillar of the vehicle and guided into the interior via a D-ring or with an additional height adjuster. The other two anchorage points of the three-point belt, consisting of belt buckle and the so-called anchor bracket, are either attached to the car body or directly to the seat or seat rail. In the front seat row, a three-point belt system with a pyrotechnic seat belt retractor pretensioner and another anchor or buckle pretensioner is the most commonly installed belt system in conjunction with an airbag system.

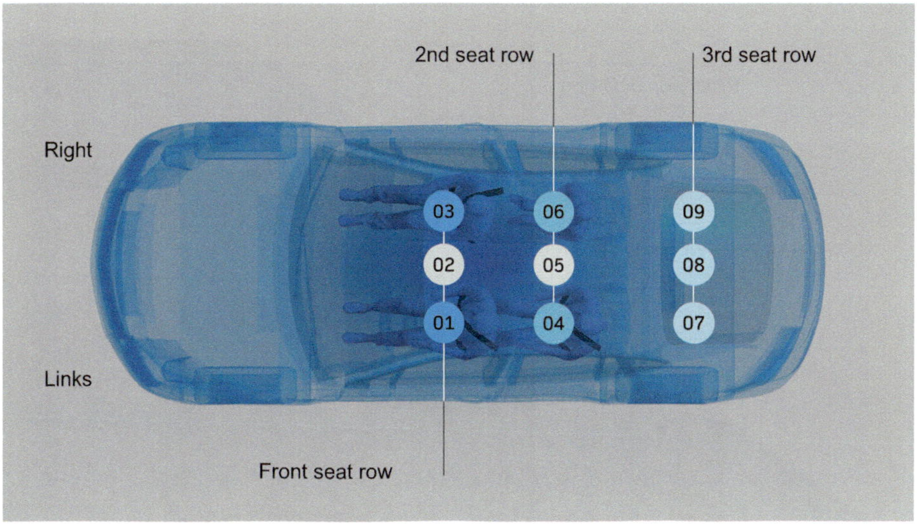

Fig. 6.33 Seating positions in passenger cars with requirements for seat belt systems. (*Source*: ZF)

Rear seat row: Typically, the rear seat row is equipped with a bench seat with three or sometimes only two seating positions. The belt routing also takes place via a pillar loop or through a direct pull from the C-pillar, if the seat belt retractor is mounted on the parcel shelf. Common configurations for the second row of seats are either a three-point belt system with a pyrotechnic seat belt retractor pretensioner or just a seat belt retractor.

The vehicle concept and the vehicle body structure define possible belt anchoring points of the three-point belt system. The possible installation space is also determined by further requirements of the interior design. Legal and other specific safety requirements for seat rows and seating positions must be taken into account for the configuration of the three-point belt systems. The combination of the three-point belt system with an airbag system also results in requirements for the pretensioning and the control of the restraining force by the seat belt retractor.

In Fig. 6.34, the essential components of a *three-point belt system* are listed, which can generally or optionally be installed:

- Webbing: The webbing runs from the retractor through the D-ring across the occupant's upper body.
- Seat belt tongue: The seat belt tongue guides the webbing in the transition from the shoulder belt to the pelvic belt area. Alternatively, a dynamic locking tongue can be installed.
- At the tongue position, a belt buckle, belt buckle pretensioner, or an active seat belt buckle (ACB, Active Control Buckle) will be installed. The anchoring of the belt buckle is either on the car body or on the seat.

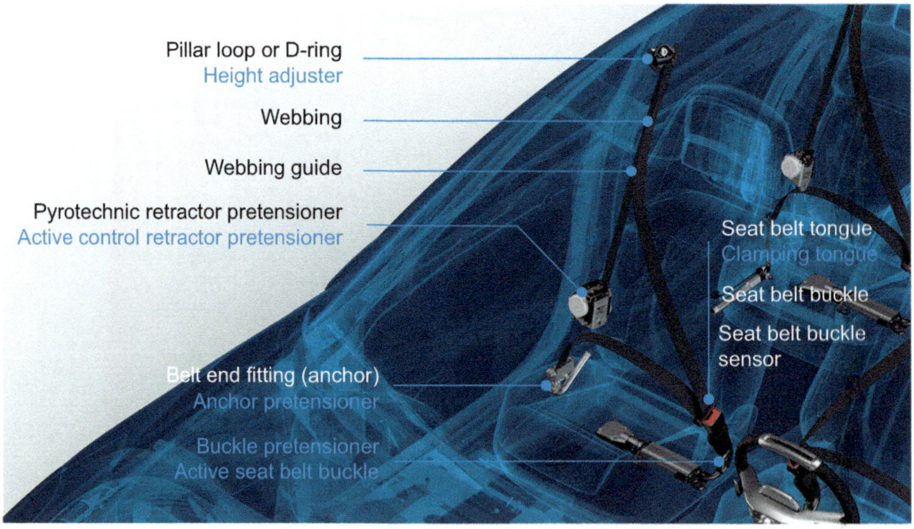

Fig. 6.34 Schematic representation of a three-point belt system for the front seating positions

- Anchor bracket: Either the webbing is anchored to a simple bracket again on the seat or the car body, or an anchor pretensioner is installed.
- Retractor, retractor pretensioner, active control retractor pretensioner: The storage of the belt webbing, many force limiting technologies as well as the locking functions are exemplary functions of the retractor systems. A detailed representation of the functions and types follows in Sect. 6.5.6.

The components outlined above form the basis for individual configurations of three-point seat belt systems, among other things, depending on the seating position, the vehicle, and specifications of the further occupant protection system. Examples of configurations of three-point seat belt systems for the front seat row are shown in Fig. 6.35a–d and for seating positions of the rear seat rows in Fig. 6.35e–f:

- Pyrotechnic seat belt retractor pretensioner for single pretensioning configuration
- Pyrotechnic seat belt retractor pretensioner and anchor pretensioner for dual pretensioning configuration
- Active control retractor pretensioner with seat belt buckle pretensioner among others for reversible pretensioning (pre-crash phase) and dual pretensioning configuration (in-crash phase)

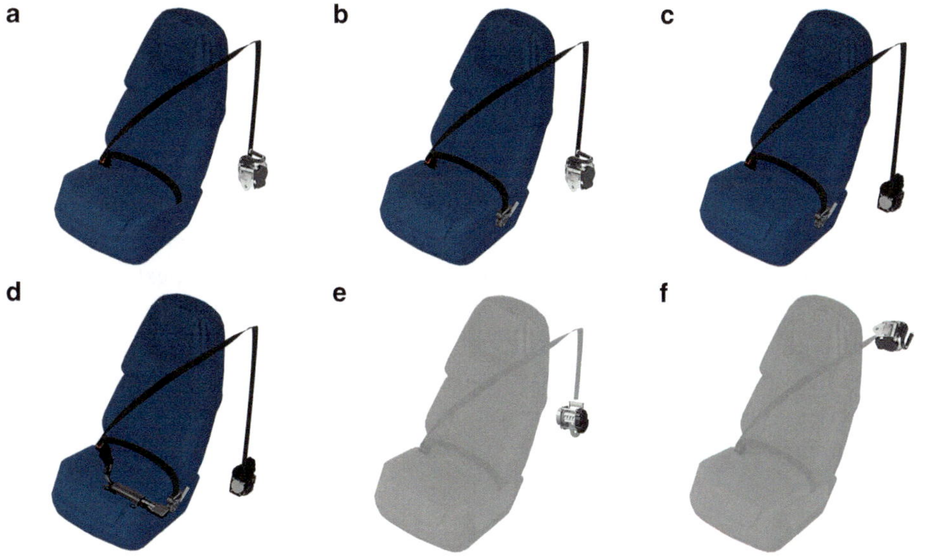

Fig. 6.35 Examples of three-point seat belt system configurations: **a**–**d**: for the front seat row, **e**–**f**: for rear seat rows (*Source*: ZF)

- Active control retractor pretensioner with seat belt buckle pretensioner among others for reversible pretensioning (pre-crash phase) and dual pretensioning configuration (in-crash phase) as well as active seat belt buckle
- Seat belt retractor without pretensioning
- Pyrotechnic seat belt retractor pretensioner for single pretensioning configuration

The belt routing starts at the retractor, which is usually installed in the lower area inside the B-pillar, and leads from there via the D-ring to the body of the occupant. Otherwise, in rear seat positions or with seat-integrated belt systems, three-point belt systems may avoid a D-ring leading to a *direct pullout* configuration. The webbing is then led directly from the seat belt retractor to the occupant, which influences the forces acting on the seat belt retractor. A lack of friction between webbing and D-ring and belt guides usually has a positive effect on the wearing comfort of the belt system.

Seat-integrated belt systems
The complete integration of the seat belt system into the seat can be an alternative to the conventional three-point belt system for vehicle applications with special requirements, if the belt anchorages are at least partially directly attached to the vehicle body structure. Examples are:

- Cars: Convertibles or coupés, where the B-pillar cannot be used for the upper belt anchorage point for a D-ring
- Minibuses: Middle seats or also removable seats
- Commercial vehicles: Seat belt systems that are fully integrated into suspension seats and are more comfortable compared to conventional arrangements

The implementation of new interior concepts and new functions for vehicle occupants is the subject of ongoing and future research, vehicle development, standardization and legislative processes [36, 59]. Figure 6.36 shows an example of a so-called *comfort seating position* (also: *non-nominal seating position*). It includes, for example, extended adjustment ranges for the backrest inclination and the longitudinal seat adjustment. The implementation of novel belt anchorage points for such a seating system and the integration of the seat belt components and necessary sensors into new installation positions are some of the tasks to be solved here. Here, seat-integrated belt systems can have advantages over conventional arrangements.

6.5.6 Basic Functions of Seat Belt Systems

This section presents the basic functions of seat belt systems as implemented in most three-point belt systems. Over time, certain assignments of functions to components have proven to be effective. However, the three-point seat belt system offers many possibilities

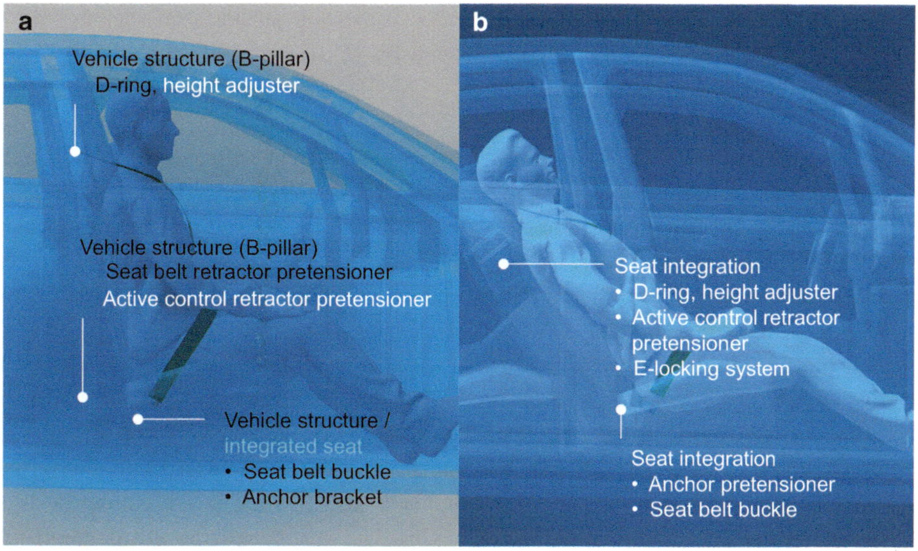

Fig. 6.36 Concept study of a three-point seat belt system for comfort seating positions. (*Source*: ZF)

to choose other assignments as well, which should not be overlooked, especially in the context of component new developments or when evaluating concept alternatives.

6.5.6.1 Belt Storage and Retraction

The *seat belt retractor* provides the occupants with freedom of movement during driving. The *webbing* wound on a *spool* is released against a slight spring force when the occupants move forward and is fully rewound when moving backward or when releasing the belt *(unbuckling)*. The *seat belt retractor* is a *belt storage.*

Even when the seat belt retractor is locked (Sect. 6.5.6.2), webbing can still be pulled out or dispensed, as the belt layers on the spool and the wound webbing can be compressed more tightly under higher load. This *belt slack* in the winding can, if not eliminated by a seat belt pretensioner, lead to a later coupling of the occupant. As a result, more forward displacement of the occupant is to be expected and the *ride-down effect* (Sect. 6.2.4.1) is negatively affected.

6.5.6.2 Locking Systems for Seat Belt Retractors

In automatic seat belt systems with seat belt retractor, the spool must be locked in the event of a collision so that no further webbing is dispensed *(locking function)*. By blocking against unrestricted belt webbing output, the occupants are restrained and protected against leaving the seating position or being thrown out of the vehicle (see also Sect. 6.2).

The spool is locked using two independent types of sensing and locking mechanisms. This redundancy is a legal requirement by UN-R 16 [131] or also GB 14.166 (CNCA-C11–04) [112]:

- **Sensing of belt webbing extraction:** This locking system with sensing of belt webbing extraction (*web sense,* WS) triggers when a certain belt webbing acceleration is exceeded. The spool and thus also the webbing are blocked and the occupant is restrained at his position (Fig. 6.37a). The triggering can already occur before the collision, for example, due to belt webbing extraction during dynamic driving maneuvers (e.g. braking, cornering) and in critical driving situations.
- **Sensing of vehicle accelerations:** This locking system is triggered by accelerations of the vehicle body (*vehicle sense*, VS). Usually, the mechanical sensor is an inert mass such as a ball *(ball sensor)* or a *pendulum.* When a certain acceleration (or deceleration) is reached, a locking mechanism blocks the spool and thus the further extraction of the belt webbing. The occupant is also secured in his seating position via this redundant locking system (Fig. 6.37b). The locking can, as already explained above, occur before the collision event.

In future vehicle projects, *mechatronic locking systems* will gain greater importance (Fig. 6.37c). Here, the mechanical sensor-actuator concept of sensing the vehicle deceleration—e.g. a *ball sensor*—is substituted by a mechatronic system. In the implementation of integral occupant protection concepts, the sensing of vehicle acceleration is done

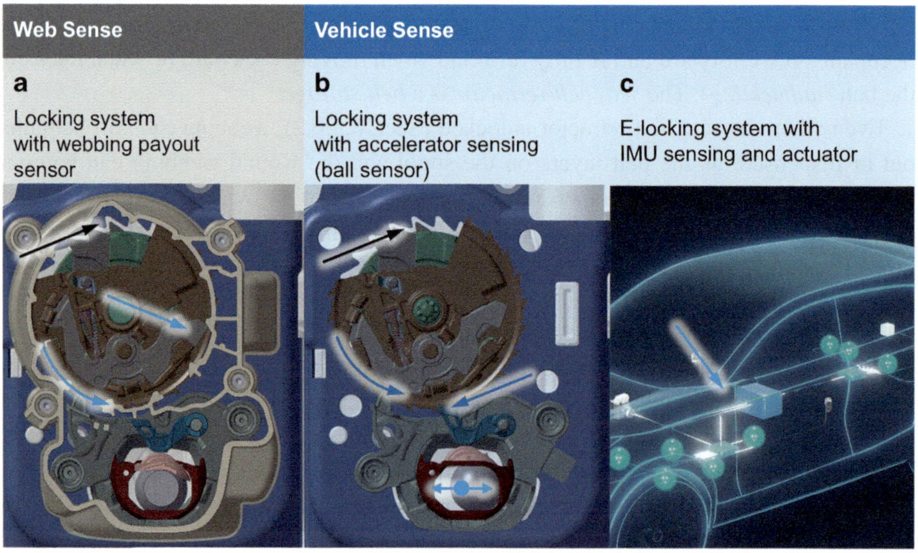

Fig. 6.37 Locking concepts for seat belt retractor: **a** *Web-sense* locking system, **b** *Vehicle-sense* locking system as ball sensor, **c** Mechatronic locking system *(e-locking)*. (*Source*: ZF)

by sensors of active safety such as the *inertial measurement unit* (IMU). The activation of the mechatronic locking system (*e-locking*) in the seat belt retractor is done via interfaces of the *airbag control unit* (Chap. 7).

This system offers advantages when installing seat belt retractor systems into the tiltable backrest of a seating system. This concept is an essential element for the realization of seat belt systems for *comfort seating positions*. Also, new customer functions to increase comfort or to improve the driving experience can be realized.

6.5.6.3 Occupant Coupling through Seat Belt Pretensioning

A development goal of occupant protection systems and seat belt systems is the early *coupling* of the occupant to the vehicle body structure, in order to positively influence the *ride-down effect* (Sect. 6.2.4.1). This is usually achieved by a *pyrotechnic pretensioning* after the impact. In Sect. 6.2, the protection principles of occupant protection systems were explained. It became clear that the maximum use of the forward displacement path is a significant objective in the development of occupant protection systems. For this purpose, the webbing should be tightened as quickly as possible after the impact. Ideally, the pretensioning occurs before the occupant's displacement process (relative movement in relation to the vehicle body structure) begins, so that as little forward displacement as possible without forces in the webbing, which slow down the occupant, is used.

By maximizing the use of the forward displacement path, energy conversion can take place over a longer distance, so that accelerations and thus the forces acting on the occupant can be reduced. The result should be a reduction in injury risks.

In Sect. 6.2, the comparison of the course of occupant acceleration without and with the use of a seat belt pretensioner was already shown. Early coupling with the help of a seat belt pretensioner should use a larger part of the forward displacement path for speed adjustment. This may require lower load limiter levels.

The advantages of *reversible pretensioning* before the impact by *active control retractor pretensioners* are discussed in Sect. 6.2 and 6.5.10.

6.5.6.4 Force Limitation

To limit the forces acting on the occupants during the restraint or during the energy conversion in the in-crash phase, concepts for *belt force limitation* (load limiting or energy management) can be installed.

They can be part of seat belt retractor pretensioners, but also of other components of the seat belt or occupant protection system. The webbing can also contribute to this.

After the coupling, the vehicle deceleration of the belt anchoring points connected to the vehicle is transferred to the occupant. If the belt forces generated in the process exceed the force limitation of the belt system, webbing is pulled out despite the blocking and leads—for example in a frontal impact—to a certain forward displacement of the occupant. Components for belt force limitation can be *torsion bars* in the *seat belt retractor*, *tear seams* on the webbing, deformation or friction elements with suitable energy conversion principles. Figure 6.38 shows the influence of belt force limitation on

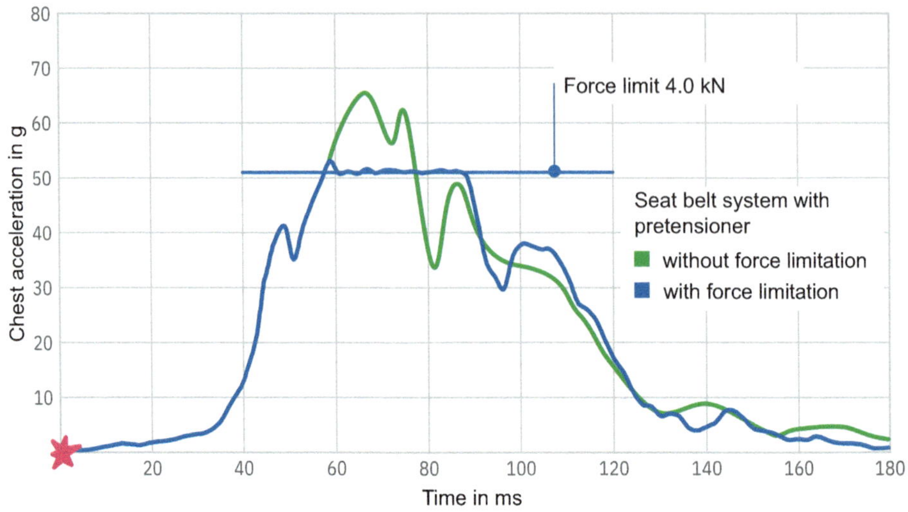

Fig. 6.38 Example of the influence of the belt force limiter on the dummy chest acceleration in a system test. (*Source*: [61])

the chest acceleration of a dummy. In a system setup with force limitation, it is reduced compared to the reference setup, but it lasts over a longer period of time. As a result, a reduction in injury risks is achieved through the lower chest load.

The possibilities for force limitation on the belt system are diverse, as they can basically be accommodated both at the belt attachment points (at the belt buckle and the retractor), at the D-ring (or at the height adjuster) or within the belt system (retractor, webbing). The respective position also determines the predominant body region that experiences a greater forward displacement or where a lower force level and thus a lower acceleration is effective.

In principle, for a certain occupant weight and a defined crash severity, there is a force level that would result in optimal restraint when using the maximum forward displacement path. The heavier an occupant and the higher the crash pulse, the higher the force level must be in order to reduce injury risks by colliding with structures in the interior.

Depending on the seating position, the protective effect is achieved through differently configured restraint systems—for example, the use of seat belts with or without combination with airbags. Therefore, different characteristics are common to influence the retraction force profile: Seat belt retractors usually have force limiting and energy management functions (*load limiting,* LL) such as *constant force limitation* (CLL) or depending on the restraint strategy sometimes also *progressive* (PLL) or *degressive force limitation* (DLL). These concepts often influence the level of restraint force through defined mechanical deformation processes. Figure 6.39 shows an overview of typical *energy management* functions for seat belt retractors and retractor pretensioners.

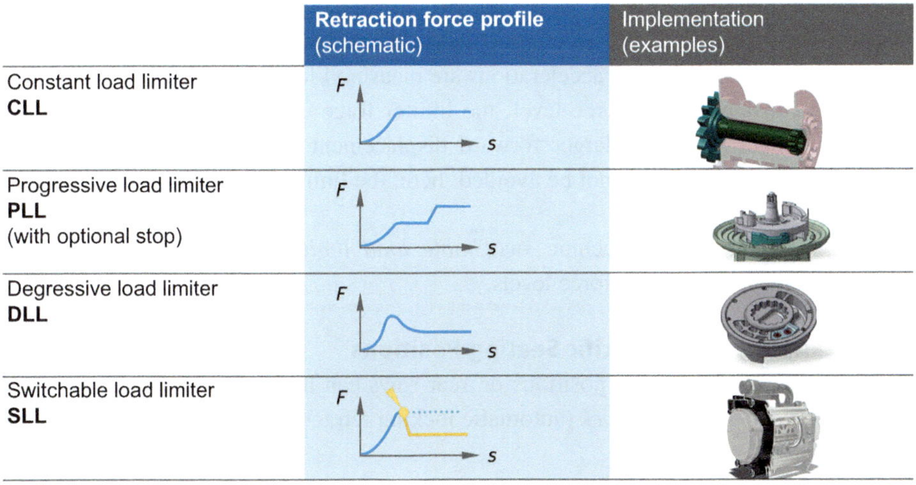

Fig. 6.39 Schematic representation of energy management functions for seat belt retractor and seat belt retractor pretensioners. (*Source*: ZF)

The constructive implementation of a constant force limitation is usually achieved by a *torsion bar,* which realizes both the force limitation and the belt webbing output through desired and defined plastic deformation. Torsion bars can be dimensioned for different force levels in different vehicle applications.

When tuning seat belt systems for entire vehicle programs, various vehicles can be grouped into groups with similar requirements for the force levels. The classification can be done according to vehicle mass and also according to crash pulse characteristics. For this purpose, the OLC (crash pulse criterion) derived in Sect. 6.2 can be used. If a complete vehicle simulation model for crash dynamics is available, as it is standardly co-developed during the development of a new vehicle, this tuning can also be optimized via the *characteristic shoulder belt force level* (CFL) with regard to load sensitivity [78].

6.5.6.5 Adaptivity Functions

In the development and tuning of occupant protection systems, an influence on the force limitation depending on the weight of the occupant or other parameters may be sought. This type of force limitation serves the implementation of *adaptivity concepts* (also: *individual safety*), which can be particularly desirable when seat belt systems and airbags interact.

For this purpose, *switchable force limiters,* which can assume one or more switching states, are used. They are controlled during the in-crash phase by the *airbag control unit* based on different parameters (Chap. 7). The occupant's weight (or size) is indirectly detected, for example via the seat and backrest adjustment, using sensors. The aim of such a design is to adapt the occupant load for lighter and heavier occupants to the generally lower values of the average occupant. Thus, compared to the average weight occupant, significantly higher accelerations are measured for lighter occupants due to the previously constant limiter force level; the limiter force could therefore be lower. The heavier occupant requires a larger forward displacement, so contact with the steering wheel or dashboard often cannot be avoided; here, the limiter force must therefore be set higher.

An example is the pyrotechnic *switchable load limiter* (SLL) with a pyrotechnic switching stage enabling two force levels.

6.5.6.6 Functions for Specific Seating Positions

Seat belt systems for seating positions on rear rows can have additional functions such as *stop function*, *child-proof lock* (automatic locking retractor, ALR) or *sensor block-out*:

ALR: For securing child seats, an additional function is integrated into the seat belt systems of the second row of seats, which is referred to as *child-proof lock* (*automatic locking retractor*, ALR). The function is activated after the webbing has been fully extended. The webbing can now only be shortened or rolled up and not extended. The function can be easily recognized by a typical clicking sound when the belt webbing is retracted. A child seat can thus be tightly secured on the rear seat bench. By opening the belt buckle, the webbing can be fully retracted and the function is deactivated. The ALR function is legally required in some regions, such as the USA, although the alternative ISOFIX system is now more commonly used for securing child seats (Sect. 6.7).

Sensor block-out: When folding backrests on the rear row of seats, deactivation of the locking mechanism may be necessary if otherwise the backrest could be prevented from being raised again. This is done with a *sensor block-out* function.

6.5.7 Further Development Goals of Seat Belt Systems

6.5.7.1 Ergonomics of Seat Belt Systems

The *ergonomics* of seat belt systems is a key development goal in the design of vehicle interiors [13]. The belt system should be easy to put on (including *accessibility*) and offer a good *wearing comfort* during the journey (including *restoring forces*). Indirectly, both requirements contribute to the *seat belt use rate* (Sect. 6.5.2) and thus crucially to occupant protection, because a belt that is not worn cannot provide any protective effect.

A frequently greatly underestimated aspect of the design of three-point seat belt systems is the design of the *belt routing*. The determination of *belt anchorage points* is usually derived from the overall vehicle concept, the geometry of the vehicle interior, the seat system, and the available installation spaces for seat belt retractors, seat belt buckles, and pretensioners. Key decisions are made in an early vehicle development phase. In this context, two development goals need to be considered:

Firstly, the **protective effect** of the occupant protection system as well as the seat belt system of the specific application for a vehicle model must be ensured. To ensure the restraining function, a closely fitting webbing is required in the shoulder belt area.

Secondly, as part of the interior design, the **handling** *(ergonomics)* of the seat belt system should be optimized and ergonomic criteria met. Here, the *accessibility* of the belt when buckling and unbuckling as well as the *belt routing* are the key aspects. The design of the D-ring also has a significant influence on the restoring forces and the *comfort perception* when rolling out and in the belt webbing during the journey as well as when buckling and unbuckling.

The handling of seat belt systems in a specific vehicle environment can be evaluated using simulation models (Fig. 6.40). In this context, the configuration of the three-point belt system in its system environment is optimized for different occupant proportions (*anthropometries*) and seating positions [13]. In detail, aspects such as the contact point and the belt release points at the shoulder or the accessibility when buckling and unbuckling are simulated.

In consumer surveys on vehicle quality, the experiences and problems of buyers with their vehicles are determined. An example are the studies of the market research

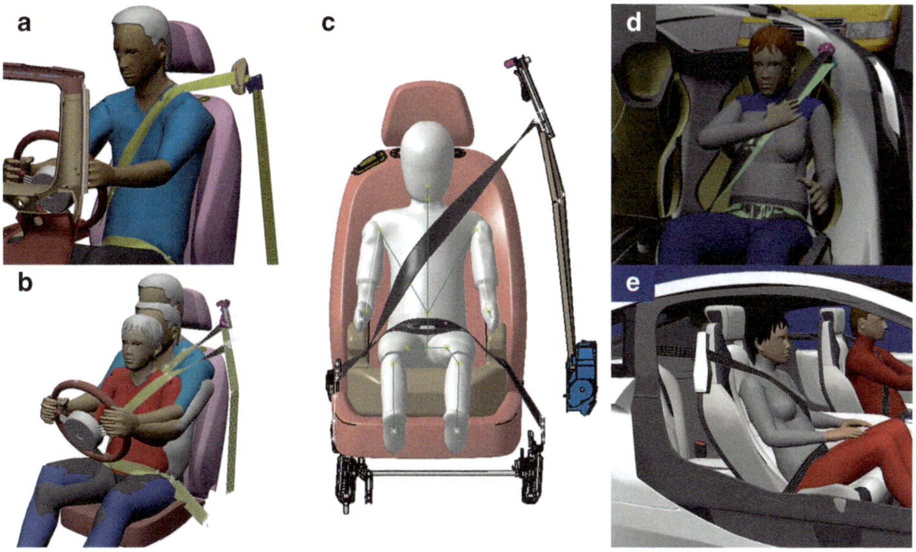

Fig. 6.40 Evaluation of the ergonomics of seat belt systems: **a** arrangement of seat belt system, **b** evaluation with test collective, **c** belt system when using child restraint systems, **d**–**e**: accessibility of the seat belt system *(RAMSIS Automotive)*. (*Source*: Human Solutions GmbH)

company J.D. Power: *Vehicle Dependability Study* (VDS), *U. S. Auto Performance, Execution and Layout Study* (APEAL) and *U. S. Initial Quality Study* (IQS). These capture impressions of the interior, the seats, and the use of seat belt systems, so they can be used for the development of future vehicle applications.

6.5.7.2 Minimization of Noise Disturbances caused by Seat Belt Retractors

Seat belt systems require secure connections to the vehicle body structure and are therefore directly attached to it by screw connections. During normal driving, vibrations of the vehicle body structure and the chassis—for example, caused by uneven road surfaces—are transmitted to the seat belt retractor and its components. These effects can cause rattling, clattering, humming, or knocking noises in the interior.

The analysis and optimization of this unwanted behavior is very complex, as the conditions at the different installation positions can vary greatly. First, synchronizations between involved vehicle domains (e.g., car body, occupant protection) need to be agreed upon in the vehicle development process at an early stage. At the connection points, the car body must be sufficiently rigid. Furthermore, sufficiently high natural frequencies must be achieved for the seat belt retractor in the local environment [43, 102].

Through NVH simulations (Fig. 6.41) and reference tests in the acoustics laboratory (Fig. 6.42), an optimal design of the car body and the connection points can be found among the involved development partners. As a result, noise disturbances are reduced and special insulation inside the interior is no longer necessary.

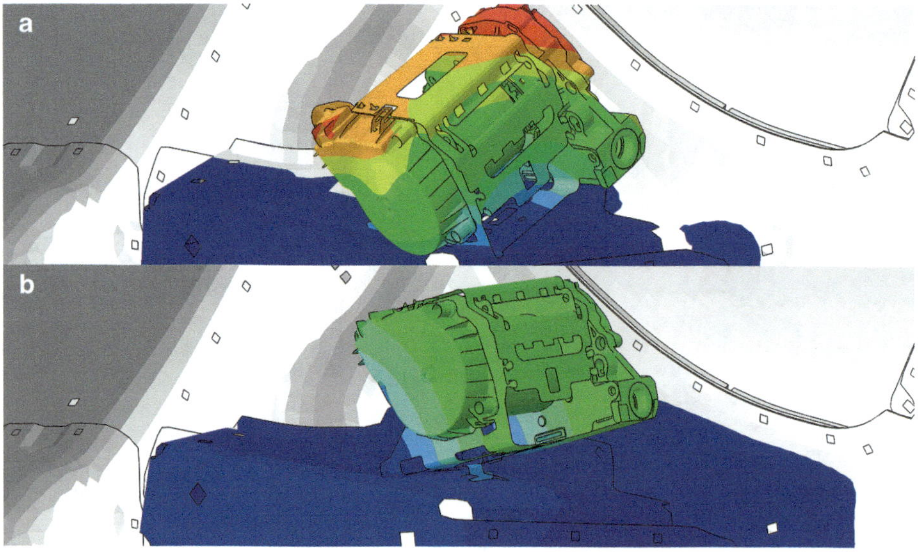

Fig. 6.41 NVH simulations for optimizing the interface between vehicle body and seat belt retractor. (*Source*: Virtual Vehicle, BMW Group, Autoliv, ZF, adapted from: [43])

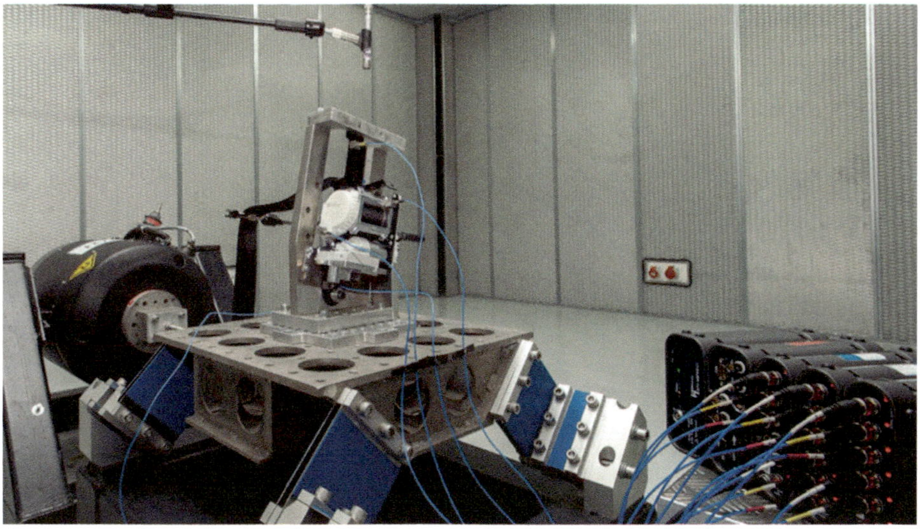

Fig. 6.42 NVH tests of seat belt retractor systems in the acoustics laboratory. (*Source*: ZF)

To eliminate further noise disturbances, alternative concepts for *locking systems* can be chosen (Sect. 6.5.9).

6.5.8 Webbing and Seat Belt Buckles as Core Components of Seat Belt Systems

This first of four sections describes basic technologies of seat belt systems, starting from the webbing, the seat belt tongue, the seat belt buckle, to webbing guiding components. The functions of seat belt systems previously presented, the constructive design of the components, and the implementation of specific development goals are discussed.

6.5.8.1 Webbing

The *webbing* consists of a *fabric* made of synthetic *yarn* such as high-strength polyester (PES or PET) and woven in different *belt constructions* as an endless strap. In the three-point belt system, it extends from the attachment point of the *anchor bracket* on the seat frame or also on the *vehicle body structure to* the *pelvic belt area* and from the *buckle* to the *shoulder belt area* and to the *D-ring*. It ends in the seat belt retractor, where it is attached to the spool, for example by means of a securing pin.

The webbing serves for force transmission and restraint during the course of a collision. Important in terms of restraint is also a good coupling with the occupant. Webbing should have a long lifespan and high haptic and optical quality even after long periods of

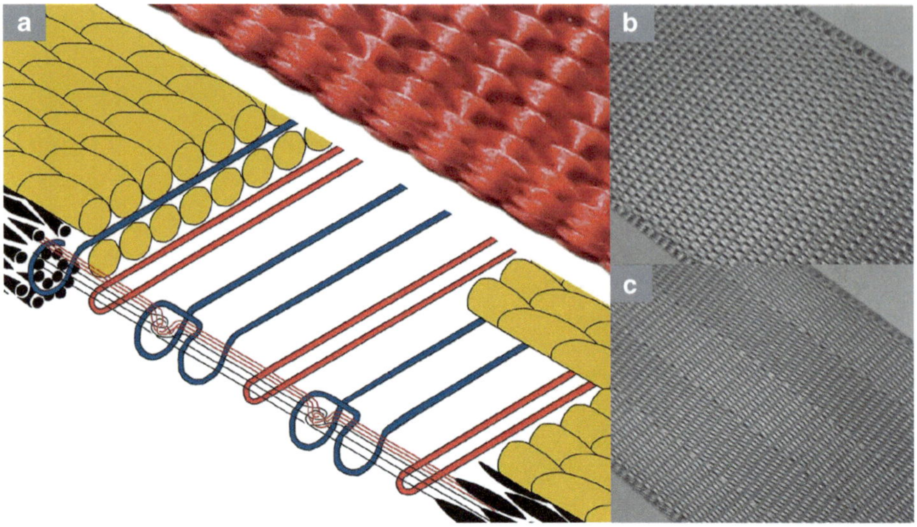

Fig. 6.43 Aspects of belt webbing constructions for achieving specific properties: **a** Example and schematic representation of a belt webbing construction. (*Source*: Berger GmbH & Co. Holding KG), Belt webbing design variants: **b** Optics without stripes, **c** Optics in five-stripe design (Sources: ZF)

use and high strain. Optical properties can be designed through belt webbing construction, binding technique, and weaving technique (Fig. 6.43).

Webbing is usually 46 to 48 mm wide and its length is up to 3,200 mm or more, depending on the design of the belt system. The elongation (reference elongation) of the belt webbing material is usually between 7 and 15 % at a certain load (reference load).

As *filament types*, mostly *untwisted* and *twisted multifilament* are used. To increase the transverse rigidity, also *monofilaments* can be used.

By incorporating dyes or soot particles, the belt webbing yarn can be dyed black, for example *(spinning black)*. However, to achieve optimal coordination with the color concept of the interior or to set accents in interior design, raw white fabric can be *piece-dyed* and fixed *(color fixation)*, making it available in several hundred colors. Further manufacturing steps are the thermofixation (adjustment of mechanical belt webbing properties) and the cutting of the webbing to its desired length.

The webbing is one of the few components in the vehicle that are in direct contact with the occupants, so comfort, ergonomics, well-being, and user experience play a major role. Therefore, additional functions in the webbing are an important field for innovations in the vehicle interior design.

An example of this is the heated seat belt webbing (heat belt) (Fig. 6.44). Immediately after starting the journey, it can heat up to 40 °C. This is intended to increase the energy efficiency of the vehicle and enhance the comfort perception of the occupants.

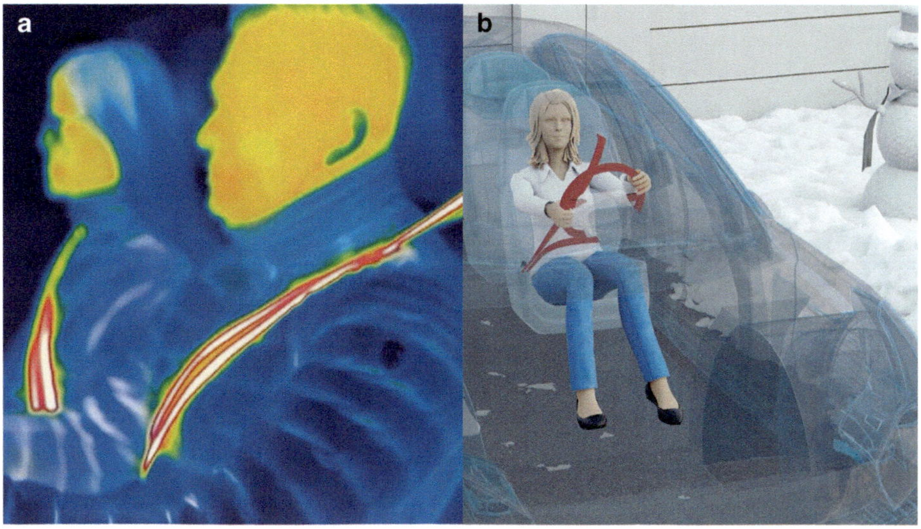

Fig. 6.44 **a** Heatable seat belt webbing, **b** In combination with steering wheel heating to increase the energy efficiency of the vehicle and the well-being of the occupants. (*Source*: ZF)

The more directly they allow this heat to reach the body, the better. Because if occupants, for example, do not wear thick winter jackets, the belt lies closer and directly on the body and can thus better fulfill its restraint function in the event of an accident.

6.5.8.2 Seat Belt Tongue and Dynamic Locking Tongue

When fastening, the closure of the belt system is achieved through the coupling between the *seat belt tongue* and the seat *belt buckle*. Tongues are used for most types of seat belt systems (Sect. 6.5.4). The following describes tongues for three-point belt systems in more detail.

During use, the webbing runs through the belt slot of the *seat belt tongue* (Fig. 6.45a) and at this position, the three-point belt divides into the shoulder belt and lap belt areas. Only in *two-retractor belt systems,* which represent a niche application, is the seat belt tongue firmly sewn to the webbing. This can result in an unusual, but also ergonomically favorable handling of the belt system when fastening.

In recent years, dynamic locking tongues (DLT) (Fig. 6.45b) have gained greater importance. They control the force exerted on the chest of the occupants, as by reducing the belt movement during the crash phase. Through such solution the parameters for the shoulder belt and the pelvic belt area can be influenced separately. A compact and lightweight design achieves a similar wearing comfort as with standard tongues.

Figure 6.46 shows images of the component manufacturing and assembly of belt buckles and tongues.

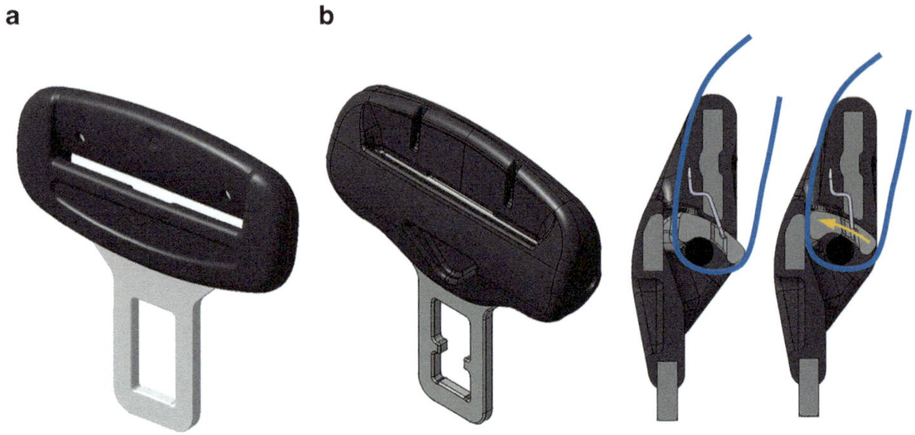

Fig. 6.45 Types of tongues: **a** Standard seat belt tongue and **b** Dynamic locking tongue. (*Source*: ZF)

Fig. 6.46 Component manufacturing and assembly of belt buckle and seat belt tongue. (*Source*: ZF)

Especially when installing belt systems on the rear seats, a mechanical coding of the seat belt tongue, for example through specific notches and the assignment of a corresponding belt buckle, can prevent confusion when fastening.

6.5.8.3 Seat Belt Buckle

Seat belt systems are closed by the coupling of *seat belt tongue* and seat *belt buckle* and opened again by pressing the red button on the seat belt buckle (Fig. 6.47).

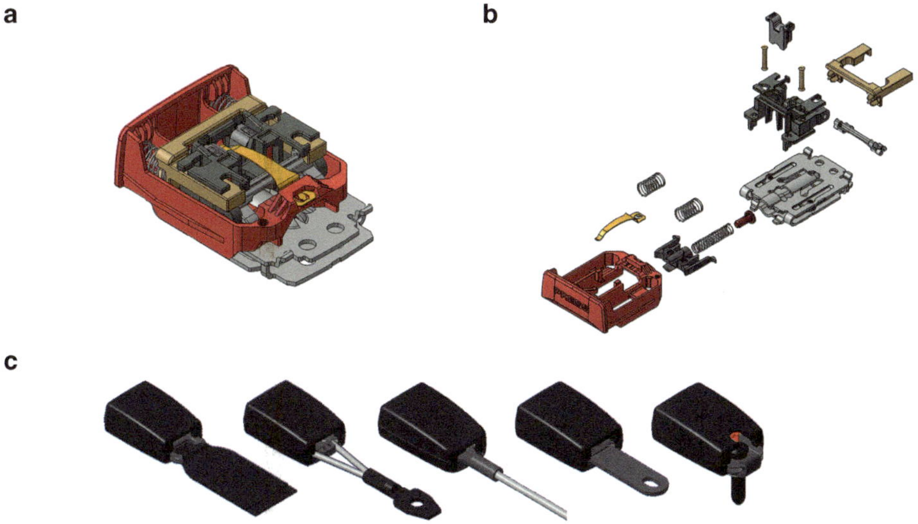

Fig. 6.47 Seat belt buckle *(RNS5):* **a** Buckle head, **b** Components latching cam buckle (RNS), **c** Fixation concepts for seat belt buckles—Webbing, cable, bracket, screwing (*Source*: ZF)

Among many available buckle variants, nowadays *latch lever* and *latching cam buckles* (RNS) are the most widespread. While the locking and unlocking mechanism as well as the housing are made of impact-resistant plastic, the load-bearing components (tongue, latch cam, and lock plate) are made of steel.

The anchoring of the buckle is done directly on the vehicle body structure or, in the case of longitudinally adjustable seats, directly on the seat rail. For this purpose, many different connection concepts are available (Fig. 6.47c): webbing, steel cable loop, steel cable with eyelets, bracket or screw connection.

In the interaction between the *airbag control unit* for passive safety (Chap. 7) and the sensors in the buckle, different vehicle functions can be realized and influenced. Examples are the *occupant sensing* (occupant detection system, ODS) or the *seat belt reminder function.*

The *seat belt reminder* (SBR) (also: *seat belt warning)* is the signaling to the driver whether the occupants are buckled up. This function is legally required in Europe (UN-R 16) and it is an important criterion in *consumer protection ratings* (Chap. 4). Therefore, buckles are usually equipped with sensors for the detection of a plugged seat belt tongue today. They are also referred to as *buckle switch* and work with different sensor technologies and interfaces towards the *airbag control unit* (airbag ECU, Chap. 7), from which the display in the cockpit and warning functions of the HMI is controlled. In various versions, the warning is given at the start of the journey or when unbuckling during the journey.

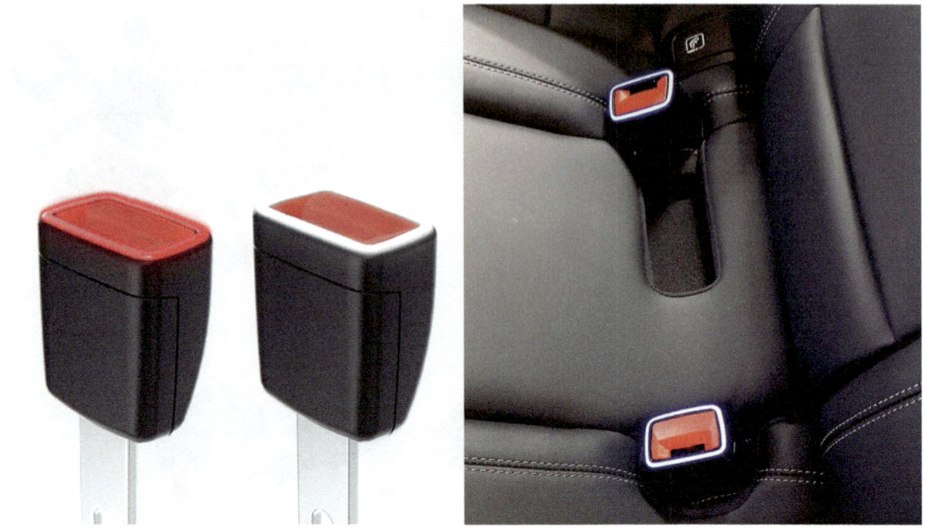

Fig. 6.48 Seat belt buckle illumination as an example of the connection of occupant protection, ergonomics, and design in the interior. (*Source*: ZF)

The sensing of the seat belt status is evaluated at individual seat positions for the decision to trigger airbags or seat belt pretensioners.

A good example of the close linkage of components of passive safety and occupant protection with aspects of interior design like *ergonomics, design* and HMI is the seat belt buckle illumination (Fig. 6.48). In the sense of integral safety, the protection potential for the occupants can be further increased. When entering the vehicle, the attention of the occupants is drawn to buckling up by the lighting and when unbuckling, the buckles can be found more quickly, especially on the rear seats. In the event of an accident, rescue workers can identify still buckled occupants by flashing buckles. The buckle illumination also contributes to the interior design and the positioning of safety functions in the vehicle.

6.5.8.4 Belt Routing

A frequently underestimated aspect of designing three-point seat belt systems is the design of the *belt routing*, which serves both the assurance of the occupant protection function and ergonomics (Sect. 6.5.7).

A first optimization field of the belt routing is the determination of *belt anchorage points*. They are already derived from the overall vehicle concept, the geometry of the vehicle interior, the seat system, and the available installation spaces for seat belt retractors, belt buckles, and belt pretensioners. Legal requirements such as UN-R 14 must be met for individual aspects of the positioning of anchorage points [130].

In the pelvic belt area, the requirements for positioning the buckle and the *effective belt anchorage point* (*H-point*) must be met. The location of the belt anchorage points also influences the accessibility when buckling and unbuckling.

Further development topics concern the optimization of the *retraction forces during driving* and the influences of *friction* on the webbing, which in turn affect the protection function and ergonomics.

Several seat belt components are available for guiding the belt webbing. The adjustment to differently sized occupants can be done with the *seat belt height adjuster*. The height adjuster is usually attached to the *B-pillar*. Attached to it is the *pillar loop* (also D-ring) through which the webbing is deflected from the shoulder to the retractor. In seat-integrated belt systems in cars as well as in commercial vehicles (Sect. 6.9.1.3), the *height adjuster* is located on the seat back frame.

Within the B-pillar, in seat-integrated systems within the seat backrest or also under the seat, further *webbing guides* may be necessary. It should be noted that each deflection affects the retraction and payout behavior of the belt webbing due to friction forces.

Ensuring the *accessibility* of the belt webbing and the belt routing in coupes and convertibles is not easy. Special protective devices for convertibles will be described in more detail later in Sect. 6.7. For seat belt systems, mechatronic seat *belt presenters* are available to improve the accessibility of the belt webbing when buckling up.

6.5.9 Seat Belt Retractor and Pyrotechnic Pretensioners

This section introduces the seat belt components with pyrotechnic pretensioning functions: the *seat belt retractor*, the pyrotechnic *seat belt retractor pretensioners* (R/P) as well as the *seat belt pretensioners* with the variants *buckle pretensioner* (BP) and *anchor pretensioner* (AP).

An overview of essential functions was already given in Sect. 6.5.6.

6.5.9.1 Seat Belt Retractor

The first *automatic belt system* with a seat belt retractor was introduced by Britax in 1963, four years after the introduction of the first *three-point belt system* in 1959. In contrast to the *static belt system*, manual adjustment of the belt length to the proportions of the occupant is not necessary. This has a positive effect on wearing comfort and the willingness to buckle up. In addition, the restraint is positively influenced, as the *belt slack* is automatically reduced.

The main components of the *seat belt retractor* are the *spool* for winding and storing the webbing, the *motor spring* to ensure a defined *retraction force* during use and when unbuckling, and the *retractor housing*. In addition, there are *locking systems* and other components for additional functions.

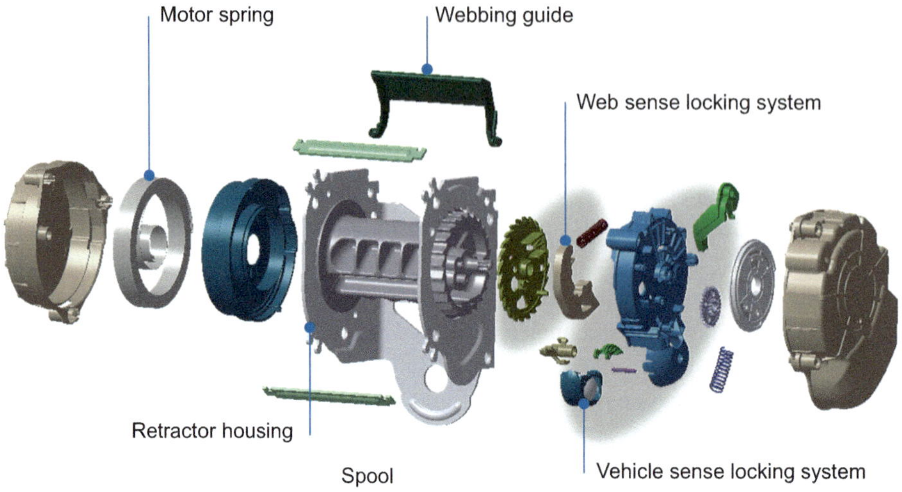

Fig. 6.49 Seat belt retractor *(floating spool retractor, FS1)*. (*Source*: ZF)

The components of a seat belt retractor are exemplarily shown in Fig. 6.49. It is a design with a floating spool. It is guided in the retractor housing. The motor spring keep the webbing continuously under tension, so that it positions as closely as possible to the occupant. While pyrotechnic seat belt retractors secure the seat belt with a lock pawl, here the spool itself slides into the retractor frame and locks. Seat belt retractors are built in variants for installation with a *D-ring* or for the *direct pullout* of webbing.

Today, in passenger cars, *seat belt retractors* are usually still used for seating positions in the rear or further rows of seats. *Seat belt retractors* are increasingly being replaced by *seat belt retractor pretensioners* with their expanded range of functions.

6.5.9.2 Pyrotechnic Seat Belt Retractor Pretensioners

The *seat belt retractor pretensioner* (R/P) consists in the extension of the seat belt retractor by a so-called pretensioning unit (Fig. 6.50). It can rapidly retract the webbing using a pyrotechnic drive in order to couple the occupant to the seat and thus to the vehicle as early as possible after a collision. The pretensioning begins before the occupant starts to move forward. The basics of this protection concept were explained in Sect. 6.2.

During the usual use of the three-point seat belt, a so-called *belt slack,* can occur, to which the following effects can contribute:

- Loosely fitting webbing, caused by thick clothing of the occupants
- System-related clearance in the three-point belt system, e.g. protruding seat belt buckles

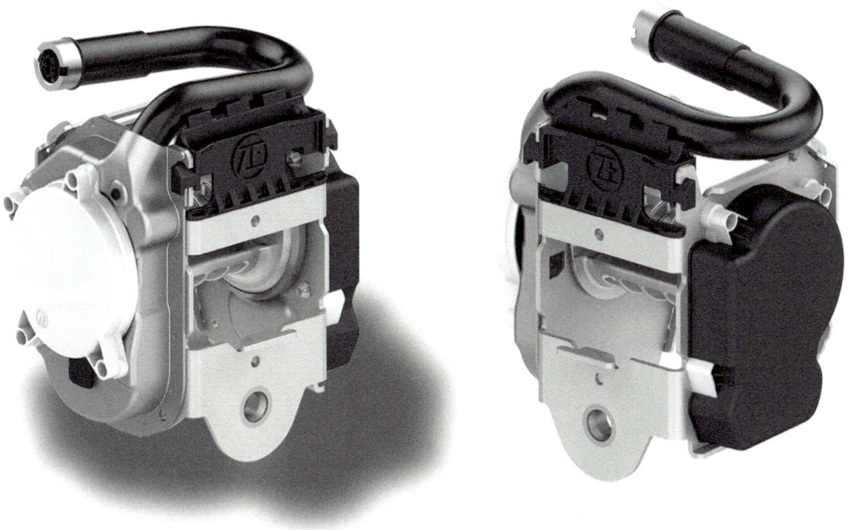

Fig. 6.50 Seat belt retractor pretensioner (snake-pretensioner, SPR6). (*Source*: ZF)

- Looseness and stretching of the belt webbing wound on the retractor spool (see above)
- Elasticity of the webbing

The function of the pretensioning unit reduces this *belt slack*. For pyrotechnic *seat belt retractor pretensioners*, a belt retraction of about 60–120 mm is typical depending on the conditions of use.

In the case of the pyrotechnic seat belt pretensioner, a micro gas generator (MGG) is activated by the airbag control unit after the collision is detected. It usually contains a propellant charge of *nitrocellulose* (NC). The gas produced during combustion exerts pressure on mechanisms to drive the *spool*. Various concepts have been developed for these mechanisms. A common design are *ball pretensioners*. Here, steel balls are located in a curved tube, from which the drive energy is positively transferred to a turbine and then to the spool of the belt retractor. A newer design is the *snake-pretensioner* (SPR), in which a plastic drive element is pressed into the pretensioner tube. This piston also drives a turbine and the associated spool of the retractor. The principle is shown in Fig. 6.51 [76].

For seat belt retractor pretensioners, two concepts are represented on the market, which differ in the flow of force from the pyrotechnic drive unit to the spool. In the so-called *spool pretensioner*, the force flow from the pyrotechnic drive unit goes directly into the spool. When transitioning from the phase of belt retraction to the phase of belt output and energy management, a slight drop in restraint force can often be noticeable. This effect is undesirable. In the concept of the *torsion bar pretensioner*, the force is

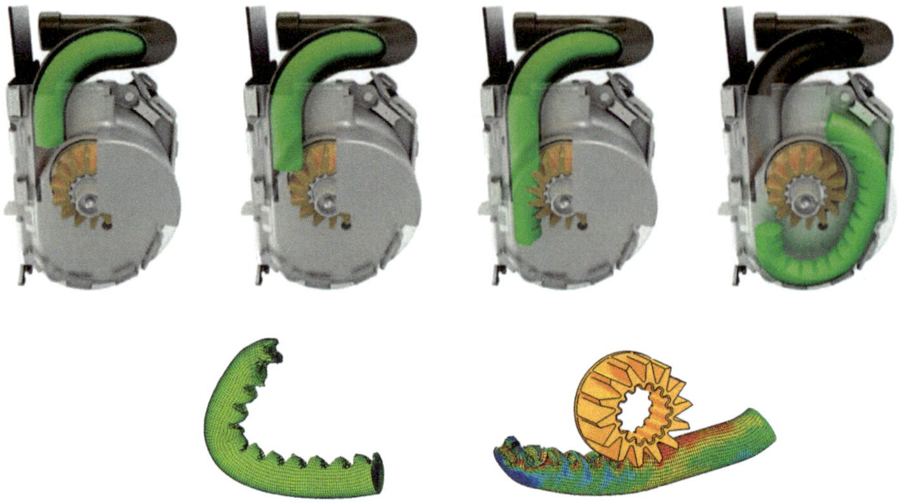

Fig. 6.51 Pyrotechnic pretensioner unit with polyamide plastic snake *(snake-pretensioner, SPR)*. (*Source*: ZF)

transmitted via the components of energy management (e.g., torsion bar) during the transition from the phase of belt retraction to the phase of belt output *(switching process)* without interruption, so that a more uniform force progression can be achieved.

6.5.9.3 Other Seat Belt Pretensioners

In addition to pretensioning by *seat belt retractor pretensioners*, *seat belt pretensioners* of various designs can also be used to tighten the belt, as the effect of pretensioning as a function of the occupant protection system is largely determined by the location of the pretensioning. The alternatives for positioning functions in the three-point seat belt system have already been pointed out in Sect. 6.5.5. Figure 6.52 shows possibilities and use of seat belt pretensioners for single and multiple pretensioning.

The pretensioning of the seat belt webbing primarily serves to reduce existing belt slack in the system and also to optimize the restraint and reduce occupant loads, particularly through:

- Reduction of belt slack even before the forward displacement of the occupant after the collision in the shoulder belt and lap belt area
- Restraint of the upper body by tightening the shoulder belt area
- Reduction of injury risks through submarining by tightening the lap belt

A *seat belt retractor pretensioner* acts on the shoulder belt area and causes a retraction of the belt webbing, which must be pulled through the *D-ring* and through the belt slot at the seat *belt tongue* (friction forces). It has the greatest effect on the restraint of

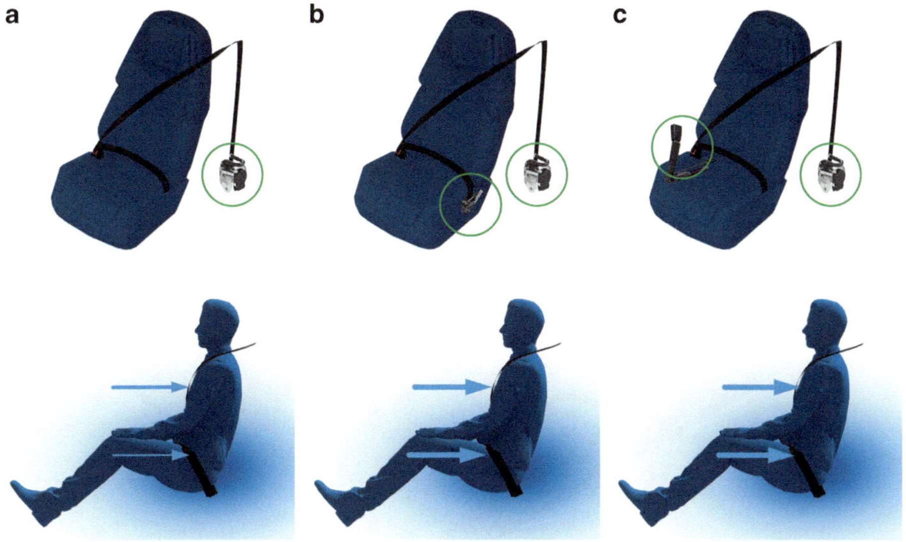

Fig. 6.52 Seat belt pretensioner for single and multiple pretensioning in the three-point seat belt: **a** Single pretensioning with buckle pretensioner and **b** Dual pretensioning with seat belt retractor pretensioner and anchor pretensioner or alternatively **c** Buckle pretensioner. (*Source*: ZF)

the upper body. The reduction of the *belt slack* in the lap belt area occurs with a time delay.

Compared to *single pretensioning*, *dual pretensioning* can also be chosen for a three-point seat belt. In the implementation, either the *buckle pretensioner* (BP) or the *anchor pretensioner* (AP) are used. Due to its central position, a *buckle pretensioner* simultaneously tightens the webbing in the shoulder and lap belt area, so this solution effectively achieves belt slack reduction and tightening. A challenge of this configuration is that the buckle pretensioner pulls the shoulder belt in the opposite direction to the retractor pretensioner. Therefore, this combination is rarely used.

Anchor pretensioners, on the other hand, act at the end of the lap belt and also cause a retraction of the belt webbing and belt slack reduction (Fig. 6.53). They are currently the most common configuration and are used together with a *seat belt retractor pretensioner*.

Seat belt pretensioners often have a pyrotechnic drive, in which a piston in the pretensioner tube is moved by a *micro inflator* (MGG) by approx. 80 to 160 mm. The piston is connected to the *webbing* or the *buckle* by means of a wire rope, so that the webbing can be retracted.

Anchor pretensioners are either integrated into the seat or they are attached to the vehicle body structure. The connection of the belt webbing is executed with different solutions such as mini buckles or seams. To increase the interior comfort and ergonomics in the footwell, space-saving concepts such as integration into seat structures are chosen.

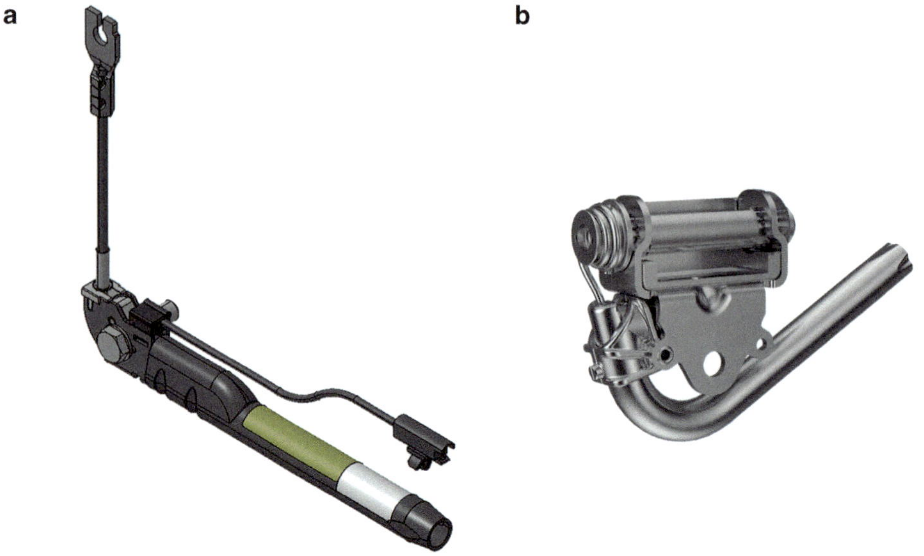

Fig. 6.53 Seat belt pretensioners and types: **a** Linear anchor pretensioner (*AP2*), **b** Rotary anchor pretensioner for alternative integration into the seat structure (*APR*). (*Source*: ZF)

At the same time, the requirements for the pretensioning performance increase, particularly due to new load cases and protection functions.

Concepts with *triple pretensioning* have also been developed, which combine *buckle* and *anchor pretensioner* together with a *retractor pretensioner*.

6.5.10 Active Control Seat Belt Systems and Mechatronic Systems

6.5.10.1 Active Control Retractor Pretensioner

The *active control retractor pretensioner* (ACR) (also: *reversible seat belt pretensioner, e-pretentioner)* is a protection system with a wide range of functions and integrations into several vehicle domains. In addition, it has a mechatronic drive unit that can perform reversible pretensioning and other functions in the pre-crash phase or even in the driving phase (Fig. 6.54).

The market launch of the *active control retractor pretensioner* took place in 2002 by Mercedes-Benz in the 220 series (S-Class) with a system from ZF (formerly: TRW). Since then, mechatronic seat belt systems have become widely spread across vehicle brands and segments, and they serve with their perceptible protection functions for product positioning and highlighting of vehicle safety features.

Fig. 6.54 Structure of an active control retractor pretensioner *(ACR8)*. (*Source*: ZF)

Typically, *active control retractor pretensioners* are part of the vehicle domain passive safety. In its structure, it has the functions of a *pyrotechnic seat belt retractor pretensioner* (Sect. 6.5.9) especially for the pyrotechnic pretensioning and *coupling* of the occupants in the *in-crash phase*. In a high proportion of cases, critical situations such as skidding, evasive maneuvers or emergency braking occur in the pre-crash phase, i.e., in the phase between normal driving operation and the accident event. This phase can last several seconds and can be used with an *active seat belt system* to initiate protective measures. Criteria for initiation are signals from active safety components such as brake assist, vehicle stability control, environment sensing. The early coupling of the occupants to the seat belt is a key protective function before the expected impact:

- Occupant protection function through early coupling: In normal driving operation, the functions of a belt retractor are initially fulfilled. When a critical driving situation is detected by sensing vehicle dynamics (accelerations due to driving operation) and chassis sensors or by the *environment sensors* (among others: environment cameras, environment radars), an airbag control unit detects a possible collision and activates the drive unit of the mechatronic belt retractor for a reversible pretensioning. This aims to reduce the *belt slack* and an early coupling of the occupants even before the impact [21].
  Depending on the configuration of the *belt retraction force,* the kinematics of the occupant's upper body can also be influenced in such a way that it assumes the *nominal position* as much as possible.

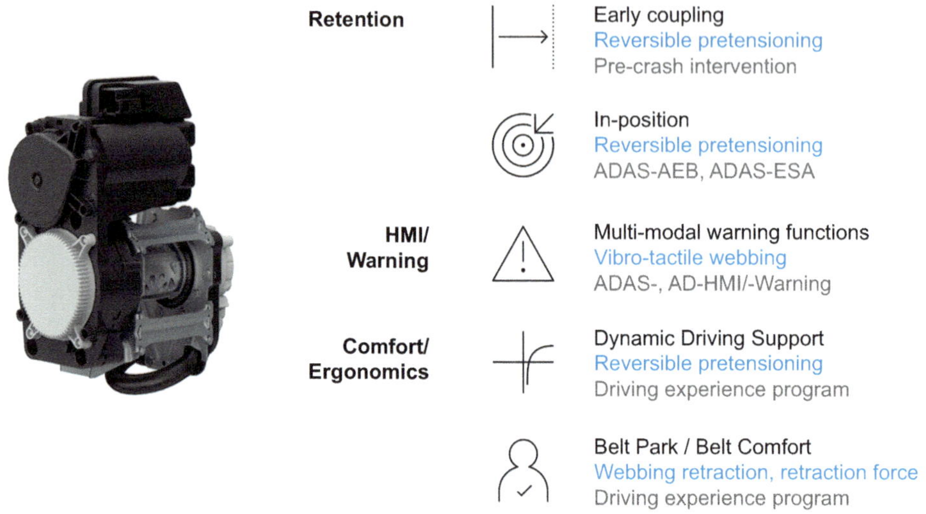

Fig. 6.55 Function spectrum of active control retractor pretensioners: protective functions pre-crash phase (restraint), HMI warning functions as part of AD/ADAS warning strategies, and comfort and ergonomic functions. (*Source*: ZF)

Meanwhile, with *active control retractor pretensioners*, further functions and usage scenarios (also: *use cases*) for systems of automated driving (advanced driver assist systems, ADAS, up to SAE level 2) or automated driving (AD, from SAE level 3) as well as for the *human machine interface* (HMI) are implemented. These are further examples of the implementation of integral safety concepts. The *active seat belt retractor pretensioner* thus serves, among other things, to realize the following additional vehicle functions (Fig. 6.55):

- **Early coupling in critical driving situations:** If the vehicle detects a high probability of an accident event in the *pre-crash phase* of the integral safety approach (phase: *in danger*, Fig. 1.7), a reversible pretensioning of the seat belt can reduce the *belt slack* to achieve as early a coupling of the occupants as possible. Depending on the force level, the aim may also be to position the occupants as close as possible to their nominal position.
- **Driver information, warning and alerting (HMI)** for ADAS/AD system functions (Fig. 6.56): An alert and warning of the driver assistance system (ADAS) or the *request to take over the driving task* (*take-over request*, TOR) by an AD system (e.g.: SAE-Level 3 or 4) can be signaled to the driver by a tactile signal and a pulsating webbing. This *vibro-tactile* signal can also be part of a *multi-modal* (e.g.: visual-tactile, visual-acoustic-tactile) *warning scenario* of the HMI system [71].

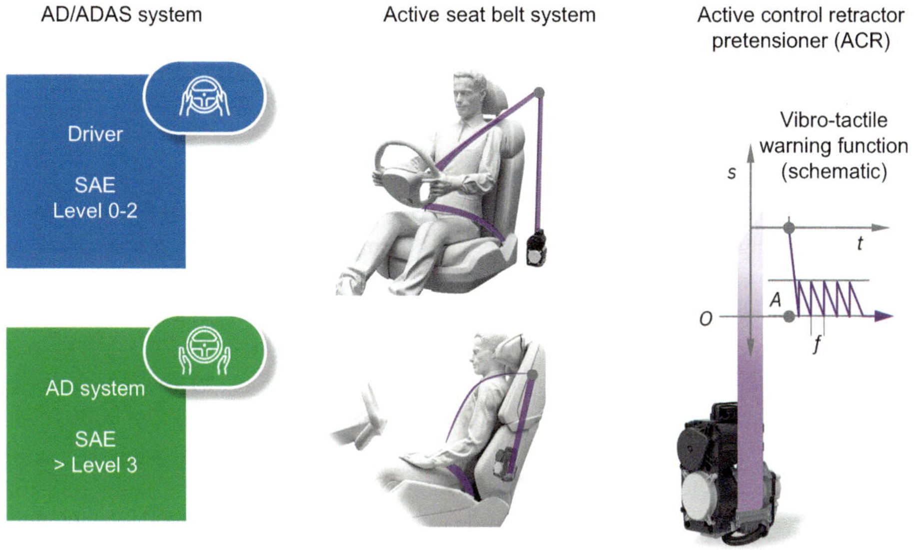

Fig. 6.56 ACR as part of multi-modal warning and alerting functions among others for functions of automated driving (AD, ADAS). (*Source*: ZF)

- **In-position AD/ADAS:** In AD/ADAS driving maneuvers initiated by the vehicle, an undesired influence on the occupant position in the longitudinal direction (e.g., by *autonomous emergency braking,* AEB) or lateral direction (e.g., by *evasive steering assist,* ESA) can occur. In the implementation of integral protection concepts, the occupants can be stabilized in their position by reversible pretensioning of the belt webbing when initiating driving maneuvers, before their positioning is negatively influenced by the dynamics of vehicle acceleration. This can also improve the driver's perception of the driving task. In addition, this function can also be combined with an *emergency stop assistance function* to achieve better occupant protection.
- **Ergonomics and operating comfort**: For the fastening and use of the seat belt system while driving, the webbing retraction force can be reduced in certain situations, thereby positively influencing the feeling when wearing the belt.
- **Driving experience:** The *active control retractor pretensioner* can also be used for special functions to enhance the driving experience. For this purpose, reversible pretensioning of the belt strap can be initiated, for example, before curve areas by selecting the driving mode, so that a stabilization of the occupant position anticipating the vehicle dynamics is achieved. In this way, the driving feeling can be adapted to customer expectations using a protective system already installed in the vehicle.

- **Reduction of belt slack at the start of the journey:** At the start of the journey, the belt slack can be reduced by *active control retractor pretensioner* by means of a certain belt strap retraction, so that in the event of an accident, the coupling can occur as early as possible. This function can also contribute to a positive experience with the vehicle's safety systems.

The *active control retractor pretensioner* (ACR) consists of a mechanical seat belt retractor with a locking system, a pyrotechnic pretensioning unit, and a mechatronic drive unit with an electric motor and an integrated control unit, as well as a clutch that separates the mechatronic drive unit from the mechanical seat belt retractor during normal driving operation and establishes power transmission when triggered (Fig. 6.54). The triggering occurs either via the *airbag control unit* of passive safety (*ECU*), via the driver assistance system, or via a central safety domain controller even before the impact. The system usually requires about 80 to 120 ms for the reversible pretensioning. If a collision occurs, the pyrotechnic seat belt pretensioner is activated. If there is no collision, the *active control retractor pretensioner* releases the reversibly retracted webbing and is ready for a similar situation again. Pyrotechnic belt pretensioners, on the other hand, are only triggered *after* the detection of an impact.

6.5.10.2 Active Seat Belt Buckle

The ergonomics of vehicle interiors and the perception of vehicle use by the occupants (customer experience) are becoming increasingly important in vehicle development. The focus on the interior for vehicle differentiation has been observed in numerous examples in recent history.

Aspects of ergonomics in the design of seat belt systems and their integration into vehicle interiors were already discussed in Sect. 6.5.7. The *accessibility* and *reachability* of the three-point belt system at the upper D-ring position as well as at the buckle position below influence the process of buckling and unbuckling. Furthermore, the position of the seat belt buckle can influence the wearing comfort during driving.

Another design aspect of seat belt systems is the location of the *effective belt anchorage point*, where specific requirements must be considered (Sect. 6.5.5). This often requires a specific location for the seat belt buckle, which can conflict with other objectives such as package, ergonomics, interior design.

The *active control buckle* (ACB) (also: *active buckle lifter, ABL)* allows the temporary elevation of the buckle position for buckling and unbuckling situations through a mechatronic drive unit, thus significantly improving the accessibility of the buckle [20]. By combining with sensors in the interior, different comfort functions with a perceptible customer experience can be implemented. This active highlighting of the seat belt also aims to further increase the attractiveness of this protection system and improve the seat belt usage rate (Fig. 6.57 and 6.58).

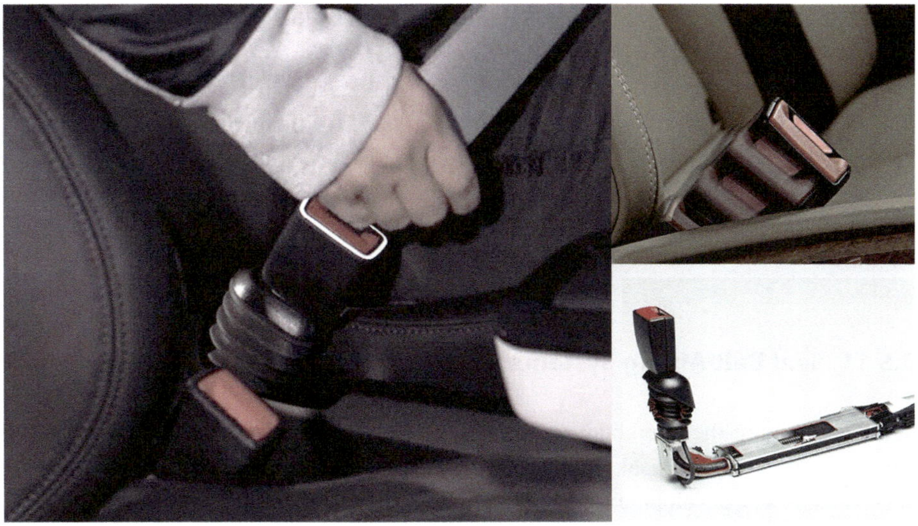

Fig. 6.57 Ergonomic handling of seat belt systems through active seat belt buckles. (*Source*: ZF)

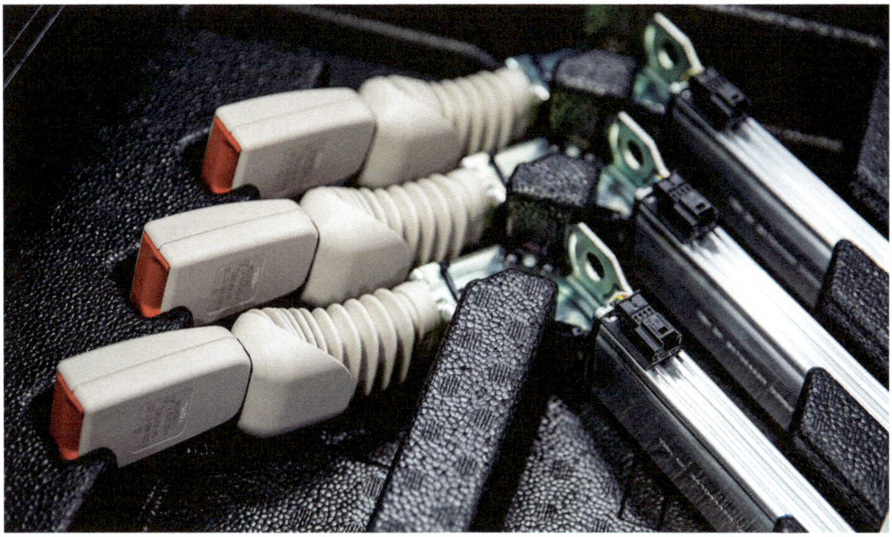

Fig. 6.58 Final assembly of mechatronic seat belt systems. (*Source*: ZF)

Active control buckle systems can be integrated into both front seats and standard rear seats as well as rear seats with extended *comfort seating positions*.

In an extended version, *active control buckle* systems and their mechatronic drive unit can also implement *reversible pretensioning functions* for the pre-crash phase or even in dynamic driving situations. By initiating the coupling earlier and reducing *belt slack*, the protection potential of the seat belt system can be extended. See also the presentation of vehicle functions for the *active control retractor pretensioner* (ACR) in the previous section.

6.5.11 Seat Belt Airbag Systems

Seat belts require the interaction of the belt webbing with the bodies of the occupants, which can lead to local peak loads in impact zones, particularly in the chest area of the occupants. A key development objective in the overall design of the occupant protection system is to minimize *chest deflection*.

The use of *load limiters* and *adaptive* functions for seat belt retractors contributes to this goal. Another approach to further minimize these loads is to reduce the *surface pressure* by enlarging and optimizing the impact zone. Early experiments with a simple widening of the belt webbing did not lead to the desired improvement, as it rolls up at the edges. Thus, the impact zone of the belt webbing remains largely unchanged.

Several concepts have since emerged for combining seat belt systems with airbag concepts. This results in a *seat belt airbag system* (Fig. 6.59). In the event of the belt system

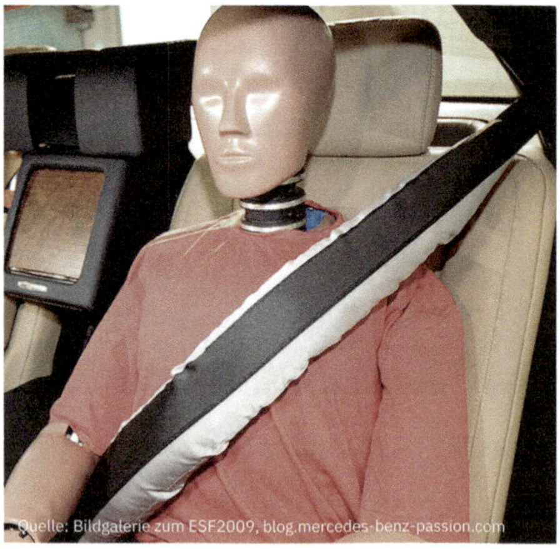

Fig. 6.59 Seat belt airbag system *(Mercedes-Benz, ESF2009)*. (*Source*: [61, 81])

being triggered in an accident, the *airbag* integrated into or attached to the webbing is deployed, enlarging the impact zone and reducing the surface pressure in the chest area. At the same time, the coupling between the seat belt and the occupant is improved by the deployment of the airbag, so that the effect of a seat belt pretensioner can be simulated. In addition, a suitable design of the airbag can provide support for the head and neck, which a conventional seat belt system cannot provide in the same way.

A component of the *seat belt airbag system* is an *inflator,* which is either mounted on the anchor bracket or coupled by a special seat belt buckle. The *three-point seat belt system* must be modified at the retractor, at the anchor bracket or also at the belt buckle.

So far, only individual vehicle models have been equipped with *seat belt airbag systems*. Examples of market launches are the *inflatable rear seat belt* by Key Safety Systems through Ford in 2010 for the *Explorer,* the *Belt Bag* by Autoliv through Mercedes-Benz in 2013 for the series 222 *(S-Class)* and the *Airbelt* by Takata through Toyota in 2011 for the model *Lexus LFA* [107, 16].

The implementations aim at occupants of the rear seat row and protection in frontal impact, because unlike the front seat row, there are usually no airbag systems installed in the rear seat. In some cases, the reduction of lateral head movement in a side impact or improved protective effect in frontal impact at higher speeds is also targeted.

A seat belt airbag system creates a different user experience. This can contribute to an increased sense of safety and an appreciation of protective devices, so that the motivation to wear the seat belt can be increased.

However, the technical implementation of a system solution and ensuring uncomplicated use for all age groups, especially for children, is a challenge.

6.6 Airbag Systems and Gas Generators

Airbag systems extend the protective effect of *restraint systems* and are now practically used in every vehicle as part of the occupant protection system for specific protection functions. Airbag systems are a significant product segment within passive safety.

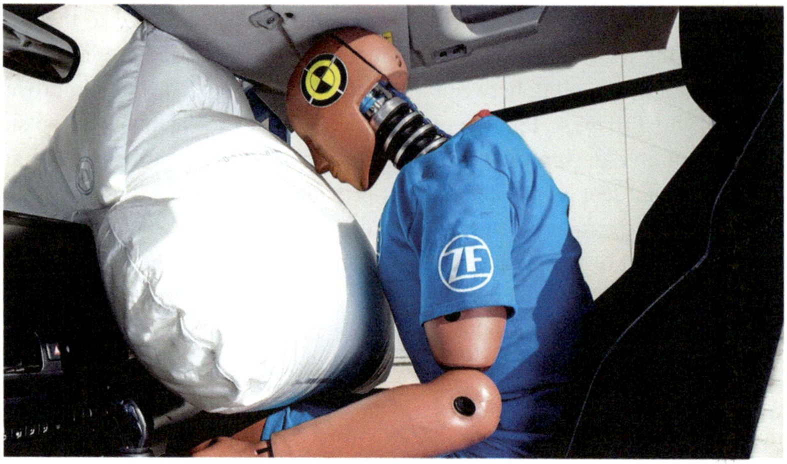

Fig. 6.60 Airbag module for vehicle interiors with modern design concepts and high differentiation requirements. (*Source*: ZF)

For airbag systems, terms such as *air sack, air cushion, bellows, impact cushion, crash cushion* or *safety cushion* are used. In the technical field, terms such as *inflatable restraint systems* (IRS) or *supplemental restraint systems* (SRS) are used.

Fig. 6.60 shows a modern front passenger airbag module with a large freedom for designing an attractive vehicle interior. Today, airbag systems, together with the deformation structures of the vehicle, the seat belt systems, and other protection systems in the vehicle, form a carefully coordinated occupant protection system.

This section presents airbag systems and their components for use in passenger cars as part of integral occupant protection concepts. First, the structure of protection systems with airbags, the structure of airbag modules, and general components are shown. In further sections, the currently available *airbag modules* for occupant protection in frontal impact (Sect. 6.6.7) and side impact (Sect. 6.6.8) are described. In a separate section Sect. 6.6.9, the technologies of *gas generators* as an essential component of airbags are presented.

6.6.1 Development Steps to Airbag Systems for Motor Vehicles

The idea of using an automatically inflating air cushion as impact protection for car occupants was already discussed in the 1960s. Initial preparatory work and patents date back to the 1950s: Walter Linderer (patent specification No. 896312, application: 1951, Germany) [73] and John Hetrick (US patent No. 2,649,311 application: 1952, USA) [49]. Figure 6.61 already shows many concepts for *airbag modules* still used today.

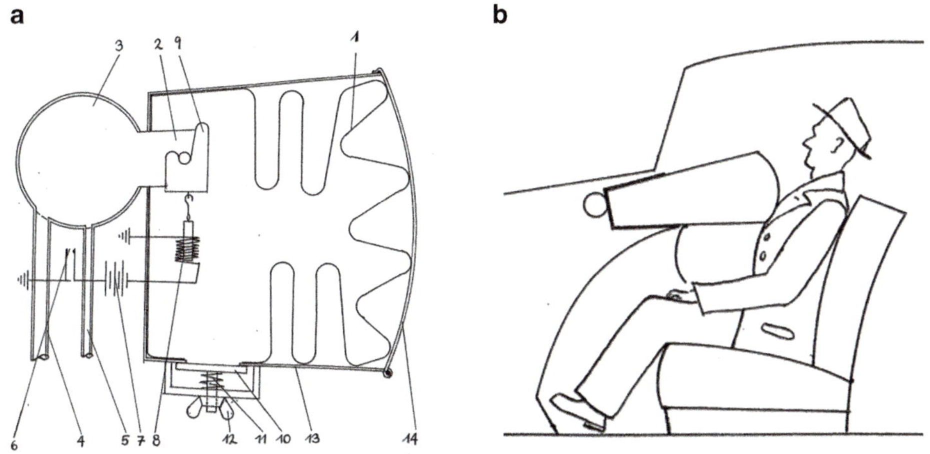

Fig. 6.61 Illustrations for the patent specification "Device for protecting persons in vehicles against injuries in collisions": **a** Example of the device, **b** Protective effect (Source: [73])

However, essential developments for further subsystems (e.g., crash sensing and electronic controls) were still missing for series-ready application in motor vehicles.

An early system for detecting a collision was developed in 1967 by Allen K. Breed (ball-in-tube sensor) [122]. It enabled the deployment of airbags in less than 30 ms—fast enough for use in frontal impact.

The first commercially available airbags were offered from 1973 as a *substituting restraint system* by General Motors. Under the name *air cushion restraint system* (ACRS), both driver and front passenger airbags as well as pelvic belts were installed [84]. However, a shoulder belt and the otherwise legally required *interlock system* (ignition interlock), which was supposed to prevent the vehicle from starting without fastened seat belts, were missing.

Further development work on the use of airbags in motor vehicles was started in 1966 at Mercedes-Benz. The particular challenges were the implementation of a short inflation time of the airbag, the high weight and the large construction volume of the high-pressure gas cylinders used at the time, but above all the reliability in operation. In 1971, a first patent application (patent specification no. DE 21 52 902 C2) was made and a breakthrough was achieved by transferring the principle of a solid propellant gas generator from the military sector [103].

The market launch of airbags as a *supplemental restraint system* (SRS) was carried out in 1980 by Mercedes-Benz with the series 126 (S-Class). The vehicle was equipped with a driver airbag and seat belt pretensioners as a coordinated restraint system [82]. Part of the system were also a crash sensor and an airbag control unit, which had been registered for patent by Bosch in 1976 as a *triggering apparatus for an inflation process in a bag* [96].

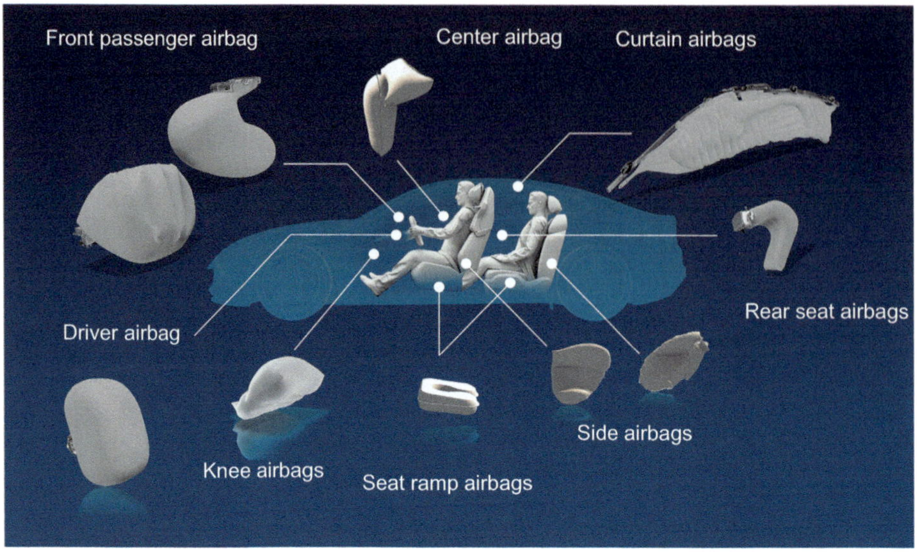

Fig. 6.62 Overview of airbag modules for occupant protection in passenger cars. (*Source*: ZF)

Figure 6.62 presents the essential airbag modules. Airbag modules for motor vehicles are often assigned according to their orientation to the essential load cases in frontal and side impact.

For protection during frontal impact, available types for occupants of the front row of seats are: the *driver airbag* and *front passenger airbag, knee airbag* and *seat ramp airbag*. For protection during frontal impact, airbags are also available for occupants in the rear row of seats, the *rear seat airbag* and seat belt airbag system, which is described in more detail in Sect. 6.5.

For occupant protection during **side impact**, the various types of *side airbags* for occupants in the front and rear rows of seats as well as a *curtain airbag* are available, which usually extends typically from the A-pillar along the entire compartment to the rear rows of seats. Both protection systems are installed on both the left and right side of the vehicle. The latest airbag for protection during side impact is the *center airbag* between the two front occupants.

Airbag systems for the protection of *vulnerable road users* (VRU, including pedestrians and cyclists), which are housed in the vehicle exterior, are described in Sect. 6.8.

Milestones

The following list highlights selected milestones in the development history, distribution, and legislation of airbag systems:

- **1951** Patent application "Device for protecting persons in vehicles from injuries in collisions" by Walter Linderer, Germany, Patent No. 896.312
- **1952** Patent application "Safety cushion assembly for automotive vehicles" by John Hetrick, USA, US Patent No. 2,649,311
- **1967** Invention of a crash sensor and triggering device *(ball-in-tube sensor)* for airbags by Allen K. Breed, USA
- **1971** Patent application "Impact protection device for the occupant of a motor vehicle" by Daimler-Benz, Patent No. DE 2152902 C2
- **1973** Market launch of **airbags in cars** in the USA as an alternative restraint system including **driver airbag** and front passenger airbag as well as lap belt by General Motors (Air Cushion Restraint System, ACRS), this equipment option was discontinued as early as 1976
- **1976** Patent application for airbag sensor technology: "Triggering apparatus for an inflation process in a bag" by Bosch
- **1980** Market launch of a restraint system with **driver airbag** as a supplemental restraint system by Mercedes-Benz with the series 126, S-Class
- **1987** Market launch of a restraint system with **front passenger airbag** as a supplemental restraint system by Mercedes-Benz
- **1993** Market launch of a first **rear seat airbag** for rear seat passengers on the passenger side by Nissan in the luxury series JHG50, President
- **1994** Market launch of **side airbags** by Volvo for the model 850
- **1996** Market launch of the **knee airbag** by Hyundai with the KIA Sportage for the USA and in 2003 by Toyota with the Avensis for Europe
- **1997, 1998** Market launches of side protection systems (by BMW for the 5 and 7 series as well as the **curtain airbag** by Volvo for the model S. 80
- **1998** Legal requirement in the USA for equipping cars with airbag systems for the passengers of the front row
- **2002** Market launch of **seat ramp airbags** (anti-submarining airbag) with metal air cushion by Renault for the three-door model Mégane
- **2009, 2013** Market launch of **seat belt integrated airbags** for extended protection of occupants in rear seat positions by Ford with the Explorer series and by Mercedes-Benz with the series 220, S-Class
- **2016** Market launch of an **airbag system** with reversible **pre-crash activation** for the protection of occupants in the front seats in side impacts by Mercedes-Benz with the series 213, E-Class
- **2020** Market launch of a **rear seat airbag system** for the protection of occupants in the rear row in frontal impact by Mercedes-Benz with the series 223, S-Class

6.6.2 Equipping Vehicles with Airbags

The use of airbag systems in motor vehicles followed the spread of seat belt systems and did not proceed uniformly in different regions of the world. In most countries, airbag systems are now supplementary protective devices of occupant protection systems and are used in combination with seat belt systems. Figure 6.63 provides a generalized overview of the equipment of vehicles with the essential airbag modules currently available for passenger cars for the markets China, Europe, India, Japan, South Korea, North America, South America.

Despite increasing convergence, differences in legal requirements persist to this day, so that system configuration, technical functions, and test scopes differ in detail in different parts of the world (Chap. 4).

These different approaches have had effects on the sizes of airbags, on the performance of gas generators, and on the corresponding airbag technologies.

Markets / **Airbag modules**	North America	Europe	China	Japan, South Korea	South America	India
Frontal impact						
Driver airbag	■	■	■	■	■	■
Front passenger airbag	■	■	■	■	■	◘
Knee airbag	◘	⭘	○	○	○	·
Front seat ramp airbag	◇	◇	◇	·	·	·
Rear seat ramp airbag	◇	◇	◇	·	·	·
Rear seat airbag	◇	◇	·	·	·	·
Side impact						
Front side airbag	■	■	◘	◘	◘	○
Rear side airbag	⭘	⭘	·	·	·	·
Curtain airbag	■	■	◘	◘	⭘	⭘
Front center airbag	○	⭘	○	○	·	·

Equipment rates (indicative): ■ 90-100 %, ◘ 70-90 %, ⭘ 20-50 %, ○ 5-20 %, ◇ individual projects, • not specified

Fig. 6.63 Equipment of current passenger cars with airbag modules by world region, data basis: 2023. (*Source*: ZF)

6.6.3 Functions and Protection Mechanisms of Airbags

The goal of airbag systems is to reduce the risk of injury to vehicle occupants in the event of a collision with a vehicle or obstacle. Airbag systems contribute to energy absorption and thus help to reduce the loads acting on the occupants during a collision.

Airbags can provide the best protection in combination with seat belt systems. The sometimes differing regulations for occupant protection systems in the USA have already been pointed out (Chap. 4). In frontal impact, seat belt systems often account for a larger share of energy conversion than airbag systems. Seat belts help maintain occupant position and control body motion, ensuring that airbags deploy and function effectively within their protective zones.

The concepts of occupant protection systems were already explained in Sect. 6.2. Here, the protective functions of airbag systems in an accident are further detailed. They include the following essential sub-functions:

Energy absorption: Airbag systems are part of a coordinated occupant protection system and contribute to the absorption of kinetic energy. This is intended to reduce the forces acting on the occupants during the accident, thereby minimizing the risk of injury.

Force introduction: One of the advantages of airbag systems is that they can have a high degree of self-adaptivity. The underlying physical principle of action

$$F = p\, A_{\text{eff}} \tag{6.20}$$

shows that with increased force application F by the occupant, the internal pressure p of the airbag is increased given an effective contact area A_{eff}. The ability to introduce forces over larger contact areas to occupants generally reduces injury risks due to lower local loads.

Occupant coupling: The targeted support of certain body areas in specific phases of the accident helps to reduce the loads acting on the occupants. In combination with seat belt systems, the contact time and protection areas of airbag systems can be specifically optimized. During a frontal impact, the coupling to an airbag system occurs at a later point in time compared to a seat belt system.

Targeted influence on occupant kinematics: Airbag systems can also be used to positively influence the kinematics of the occupants during restraint. In a frontal impact, airbag systems primarily influence the kinematics of the head and, in the case of a knee airbag, the legs and pelvis. Certain side airbag concepts aim to initiate a targeted push and influence the occupant's position inward before intrusions on the side wall begins.

Force limitation and adaptivity: The effect of airbag systems can be adapted to specific parameters of the accident or the occupants due to sensors in the vehicle. This allows the timing of deployment, the size of the airbag, the internal airbag pressure, or the restraining forces to be adjusted to further reduce loads on occupants. This is also intended to avoid excessive restraint on occupants, for example, in the case of smaller occupants or accidents with low impact energy.

Protection from impacts: The protective effect of an airbag also aims to prevent a direct impact on stiffer contact surfaces within the passenger compartment, such as the cockpit, windows, interior trim, or vehicle structures.

The protective functions of airbag systems unfold over time. For an overview of the basic protection concepts of occupant restraint systems refer to Sect. 6.2. The timing of an airbag deployment for the driver and passenger side in a frontal impact can be described schematically as follows: The deployment of the airbags occurs approximately 15 ms after the collision. At approximately 50 to 60 ms, the airbags are deployed and inflated. By specifically controlling the occupant kinematics using seat belt systems, the pelvic area of the occupants is restrained and the upper bodies and heads dive into the air cushion. A further reduction in the kinetic energy of the occupants is achieved with appropriately sized airbag vents or with the specifically coordinated, air-permeable fabric of the airbag.

At approximately 120 to 150 ms after the impact, the relative speed of the occupant to the car body is reduced and the airbags collapse. It should be avoided as much as possible that the occupant is thrown back into the seat back after the forward displacement (*rebound*). This is a significant development and tuning goal in the development of occupant protection systems.

6.6.4 Airbags as Part of the Occupant Protection System

The *airbag system* includes the subsystems *sensors*, *airbag control unit* (*airbag ECU*) and the *airbag modules* as actuators of the passive occupant protection system. This system provides the overall function consisting of sensing, detection of a collision or an accident situation, triggering and control of the restraint systems, intervention and restraint in the further accident occurrence. The protective effect of airbags occurs as a supporting restraint system together with seat belts and other protective devices. The vehicle interior and its subsystems such as cockpit, seats, interior trim, glazing, and vehicle body structure represent the immediate system environment of airbag modules as vehicle components. For the coordination of the protective effect within the passive occupant protection system, other subsystems such as vehicle body structure, deformation structure, steering, etc. must be considered.

Despite regionally differing requirements in the USA (FMVSS 208 [141] for load cases with unbelted occupants, Chap. 4), airbag systems are primarily used as supplementary restraint systems in combination with seat belt systems.

6.6.5 Structure of Airbag Modules

Airbag modules consist of the core components *inflator (gas generator)* and *airbag cushion* and usually also include a *housing* or a *cover* as an interface to the interior as well as the *mounting elements.*

The essential components of an airbag module are designed in the context of a preliminary development and are appropriately coordinated as a *modular system.* In the development process for a specific vehicle model, this allows for rapid adaptation to the specific requirements and operating conditions. The development scopes for a new airbag type or a new generation require a multiple of the time available for an individual vehicle application. The following list provides a general overview of the essential components for airbag modules.

Yarn for airbag fabric

For the *cushions* of *airbag modules, plastic fibers* (yarn) are processed into special *airbag fabric (textile).* The fibers for the airbag fabric are produced from polyamide 6 (PA 6), polyamide 6.6 (PA 6.6) or *polyethylene terephthalate* (PET). The strength of the fabric is influenced by the fineness of the yarn, given the same material.

Airbag fabric

In a weaving process, the airbag fabric is produced as sheet material from the plastic fibers. This intermediate product is referred to as flat fabric. The further processing of the airbag fabric takes place with the cutting and sewing of the airbag, before it is then completed into an airbag module.

In an alternative manufacturing process, the airbag fabric is produced in a single manufacturing step as an airbag cushion. These processes and also the airbags produced with them are referred to as *one piece woven (OPW).* The process is described in more detail below.

The requirements for airbag fabric are diverse and demanding.

Airbag fabric must be capable of withstanding high mechanical loads, so that the *tear resistance* of the fabric is a crucial design parameter. The strength of the fabric is influenced by the fineness of the yarn, given the same material.

The fabric must have a high *heat resistance*, as the inflowing gas can reach temperatures of several hundred degrees Celsius depending on the type of gas generator.

The function of the airbag must be ensured both at high and low ambient temperatures. This is accompanied by high demands on the product's lifespan.

With the *permeability* of the airbag fabric, the mode of operation of the airbag module can be influenced. It must be produced very uniformly.

The *weight* of the fabric should be as low as possible in order to have to use as little material as possible and to contribute to vehicle weight reductions.

The *processability* of the fabric is also a crucial criterion. This concerns the cutting, sewing and folding, which enable manual and automated folding processes.

Airbag coating

Airbag fabrics are, depending on the requirements for the airbag module and combination with gas generators, either *uncoated* or *coated*. Coating materials are, for example, *polychloroprene (synthetic rubber)* or *silicone*.

These coatings support sustained deployment throughout the crash event, which is particularly relevant for curtain airbags in rollover accidents, or they protect the fabric around the inflation opening near the gas generator.

Airbag folding

The operation of the airbag module is significantly influenced by the concepts and technologies used for the *folding*. The way the airbag is folded can specifically control the deployment process.

For folding airbags, different concepts and manufacturing processes are used. The most commonly used fold is a combination of *zigzag-* and *roll-fold*. For each airbag module type, there is a selection of preferred options.

For the curtain airbag, the traditional *leporello fold*, in which the airbag is folded over the generator, and an *accordion fold (also: Rabofsky folding)*, in which the airbag is located laterally on the generator, are distinguished.

For some time now, the *thermo-fixation* has been available as a subsequent process step for airbags. Here, the folding of the airbag and its outer contour are permanently fixed by temperature treatment, so that a closed airbag housing is not required anymore (for example, with certain knee airbags and curtain airbags). Through simultaneous compression, the volume of the folded airbag can be significantly reduced (for example, with compact driver airbag modules). The fixation has no influence on the airbag deployment.

Airbag functional elements

Tethers are attached inside or outside the airbag and can be used for very different purposes. They can reduce the deployment length and thus prevent contact between the airbag and occupants during the deployment process. They can also be used to influence the gas flow. Tethers can be specifically cut by pyrotechnic actuators such as *tether activation units* (TAU) during the course of the accident, so that the airbag deployment can be influenced.

Airbag's ability to convert energy can be significantly determined by the dimensioning of *outflow openings* (for coated fabrics) or the *air permeability* of the fabric. Through their defined characteristics, gas escapes from the airbag, allowing the occupant to immerse into the airbag.

Outflow openings can be specifically opened or closed by pyrotechnic actuators such as *tether activation units* (TAU) during the course of the accident by the airbag control unit depending on the design.

Tear seams are another alternative for controlling the retention behavior of the airbag.

Airbag housing

The *housing* of an airbag module is often both the connecting unit between the inflator and the airbag and between the airbag module and the vehicle. The inflator is attached and positioned relative to the diffusor of the airbag cushion with a mounting flange. Deep-drawn steel sheets, folded sheet metal structures, or plastics are used as housing materials.

In the past, all airbag modules required a housing. However, for weight reasons, a separate housing is increasingly being avoided through a higher integration with surrounding components is being sought. The inflator and the airbag cushion are then, for example, directly attached to the dashboard or the roof structure. Airbag housings can also be made of fabric with fastening elements or as a combination of fabric with a solid frame (fabric housing).

The secure connection of the airbag module to the vehicle on the vehicle body structure or another main assembly such as the dashboard or seat structure is established by screw connections.

Airbag cover

The *airbag cover* can be part of the airbag module if this function is not taken over by an adjacent component such as the dashboard, the seat, or the headliner. The cover can be a visible component in the interior for the occupants, for example in the area of the driver's airbag. As such, it must visually integrate into the overall image of the steering wheel or the dashboard. Otherwise, airbag modules are concealed, such as in the knee airbag.

Fabric- or metal-reinforced *polyurethanes* have established themselves as the material of choice. Composite materials are also used, in which case the inner cover consists of a *thermoplastic* material that is coated with a decorative surface made of *vinyl*.

In all cases, it must meet defined requirements in terms of tearing and opening behavior. This requires a high level of attention between the manufacturers of the airbag modules and those of the adjacent interior components during vehicle projects. The high variance of the materials used and constructions, as well as continuous component maturity level advancements during development processes and vehicle launches, may require adjustments.

Diffusor

To protect the airbag fabric from damage by inflowing hot gases, a special reinforcement may be attached to the inflation opening of the airbag. It consists of solid material or additional fabric layers with extended temperature resistance.

Fastening elements

Screw connections are chosen for the installation of airbag modules in vehicles due to the high forces during ignition.

Special clips for quick and secure assembly are used as fastening elements, especially in curtain modules.

6.6.6 System Environment of Airbag Modules and Installation Spaces

As **physical components**, airbag modules are *components* in the vehicle interior with interfaces to other interior main assemblies or the vehicle body structure. From a **functional perspective**, airbag modules are *actuators* within the occupant protection system. Therefore, airbag modules have numerous system interfaces that can be derived either from the installation space or the function.

The most important established *installation spaces* (also: *package zones*) for the integration of airbag modules into the vehicle interior are shown in Fig. 6.64. An installation space is particularly suitable when the airbag can unfold against the expected direction of movement of the occupant and thus unwanted interactions during deployment are avoided.

However, due to cramped or competing installation space situations, it may be necessary to develop concepts for alternative installation spaces. In the future, alternative installation spaces for airbag modules will play a significant role in the search for attractive interior concepts. An example of this is the integration of the front passenger airbag module into the roof to provide the passenger with additional movement space, a better visual appeal, or additional customer functions through displays (Sect. 6.6.7.2). The specific aspects of the package of the airbag modules will be discussed separately in the following sections.

Airbag modules / Mounting location	Steering wheel/ steering	Cockpit	Seating system	Roof	Other
Frontal impact					
Driver airbag	■	★	·	★	·
Front passenger airbag	·	■	·	O	·
Knee airbag	·	■	·	·	·
Front seat ramp airbag	·	·	■	·	·
Seat ramp airbag	·	·	◇	·	·
Rear seat airbag	·	·	■	★	·
Side impact					
Front side airbag	·	·	■	·	◇
Rear side airbag	·	·	O	·	■
Curtain airbag	·	·	·	■	O
Front center airbag	·	·	■	·	·

■ priority installation site O alternative installation site,
◇ individual projects, ★ future option, · not specified

Fig. 6.64 Options for the installation location (package) of airbag modules in the vehicle interior (schematic). (*Source*: ZF)

6.6.7 Airbag Modules for Occupant Protection in a Frontal Impact

Occupant protection in frontal impacts was the first application field for airbag systems. The focus of protection with airbags in frontal impact is still on the occupants in the front row of seats today.

The development of protection concepts for frontal impact is carried out in close coordination between the involved subsystems such as vehicle body structure, steering system, cockpit, interior, seats, glazing, occupant protection systems, and the sensors for accident detection (Chap. 7). The integrity of the passenger compartment during the collision is a crucial element for the effectiveness of airbags (Sect. 6.4).

Currently, the high variance of powertrain options across many vehicle models poses a significant challenge for the tuning of occupant protection concepts. On the one hand, the vehicle weight varies more than before, and for example, battery storage has a significant influence on the *crash pulse*. In addition, both novel vehicle structures and major systems such as the engine, transmission, and battery storage change the energy absorption behavior of the vehicle and influence the occupant protection system.

In many regions, airbag systems for occupant protection during frontal impact (Fig. 6.65) are designed in combination with seat belt systems, although different legal requirements apply in the USA (Chap. 4).

Beyond the classic concepts of passive safety, strategies of integral safety are also used for occupant protection during frontal impact by synchronizing active and passive

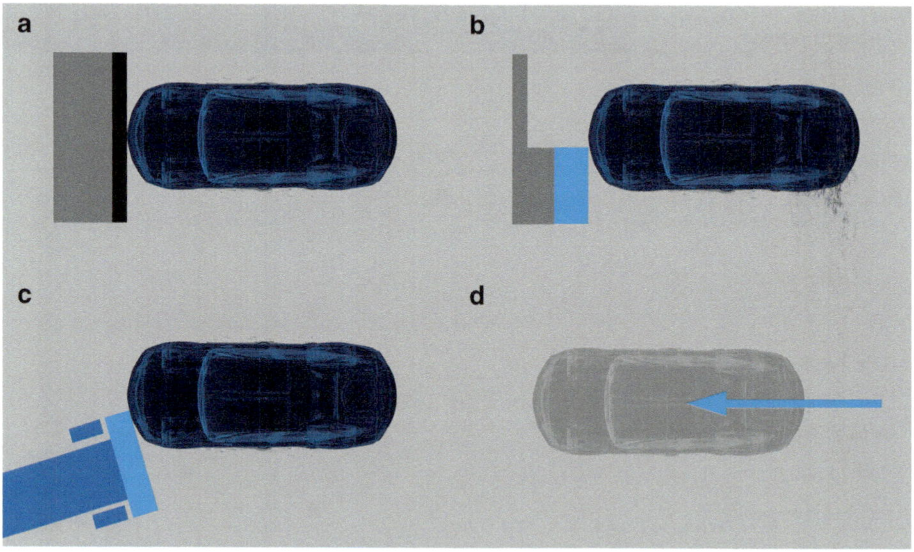

Fig. 6.65 Main focuses with test requirements for occupant protection during frontal impact: **a** Barrier tests, full overlap, **b** Barrier tests, partial overlap, **c** Barrier tests, dynamic barrier with oblique impact, and **d** Rear impact (*Source*: ZF)

Fig. 6.66 Arrangement of the airbag modules for the protection of occupants in the front and rear rows during frontal impact. (*Source*: ZF)

safety systems. So far, non-reversible actuators—like airbag systems—have not been used to intervene during the pre-crash phase of frontal collisions However, active control retractor pretensioners (Sect. 6.5) in combination with airbags aim at optimizing the occupant position before the collision and thus reducing injury risks.

The purpose of airbag systems for protection during frontal impact is to reduce injury risks through energy absorption and to prevent a direct impact on contact surfaces in the interior (Fig. 6.66).

In the following sections, the essential airbag systems aimed at occupant protection during frontal impact are described in more detail: driver airbag, front passenger airbag, knee airbag and seat ramp airbag. Only recently have there been increased activities to also bring airbag systems into use for occupants in the rear row. For this, the rear seat airbag is introduced below.

The most important legal requirements for testing occupant protection systems with airbag systems for frontal impact protection are specified, among others, in the following standards (Chap. 4):

- UN-R 94 (occupant protection frontal impact) [134] and UN-R 137 (frontal impact with focus on restraint systems) [135]
- FMVSS 201 (occupant protection in interior impact) (from 1967) [142] and FMVSS 208 (occupant crash protection) (from 1971) [141] for the USA, and
- GB/T 20913–2007 [113] (offset frontal collision) for China

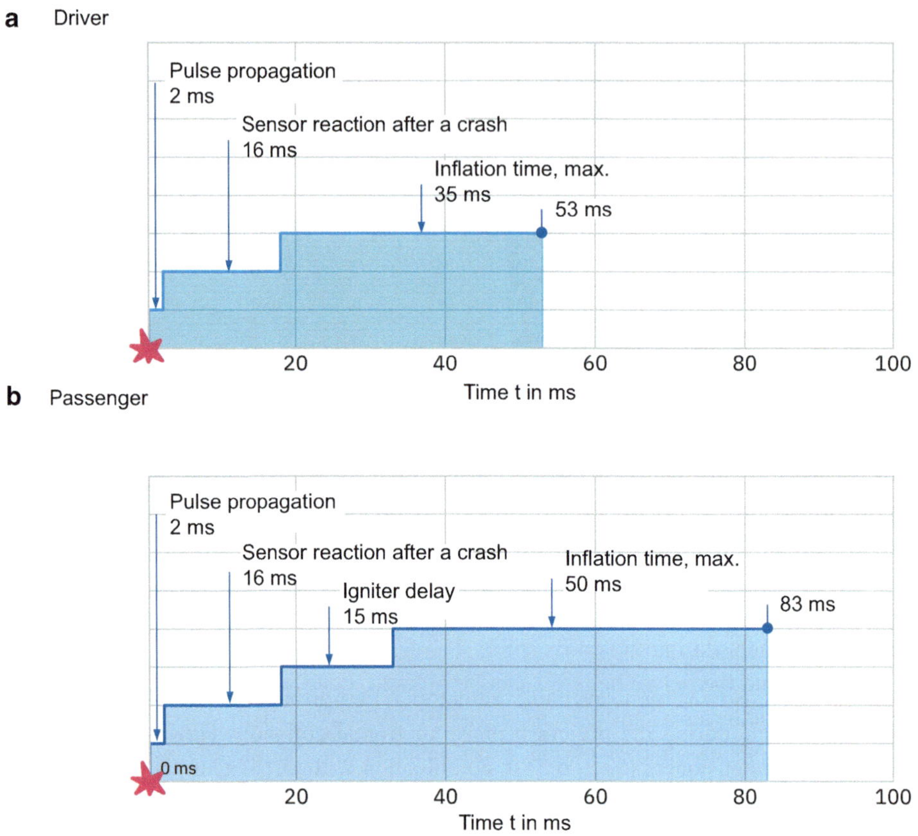

Fig. 6.67 Schematic representation of the timing of airbag ignition and deployment in frontal impact: **a** Driver and **b** Passenger. (*Source*: [61])

In recent years, the developments in frontal crash protection have been characterized by the introduction and expansion of test configurations with partial overlap and oblique impact. In some regions, the speed level was raised. Efforts were also made to increase the compatibility between different vehicle types.

Figure 6.67 shows, as an *example*, the timing of the airbag deployment for the driver's and passenger's side. It should first be noted that the seat belt system restrains the lower part of the occupant, allowing the upper torso to engage the airbag in a controlled manner following its defined deployment. The precisely calculated outflow openings or the specially developed air-permeable fabric of the airbag cause an energy reduction of the occupant's movement and prevent the body hitting the cushion from being thrown back into the seat back. Approximately 120 ms after the impact, the relative speed of the vehicle occupants to the car body is reduced, and the airbags collapse. The sequence of occupant movement in a 55 km/h barrier test is exemplarily shown in Fig. 6.68 for the driver and in Fig. 6.69 for the passenger.

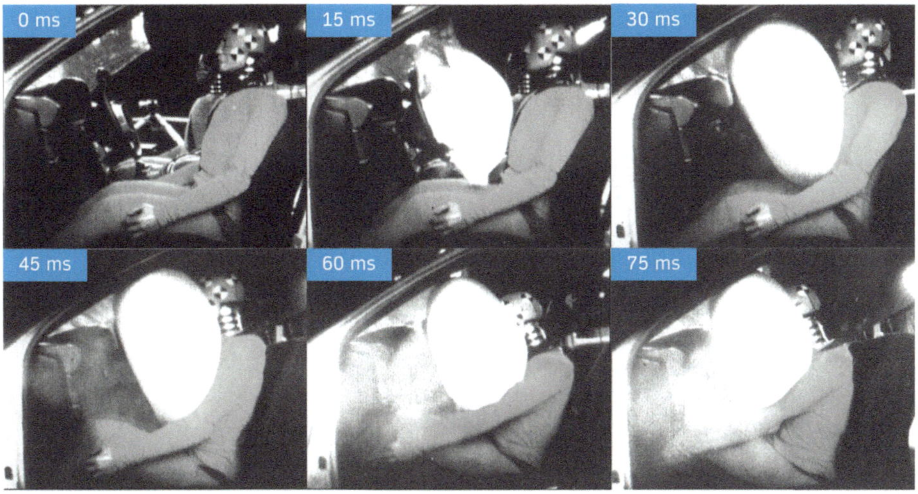

Fig. 6.68 Schematic representation of the timing of airbag ignition and deployment in frontal impact (driver). (*Source*: [61])

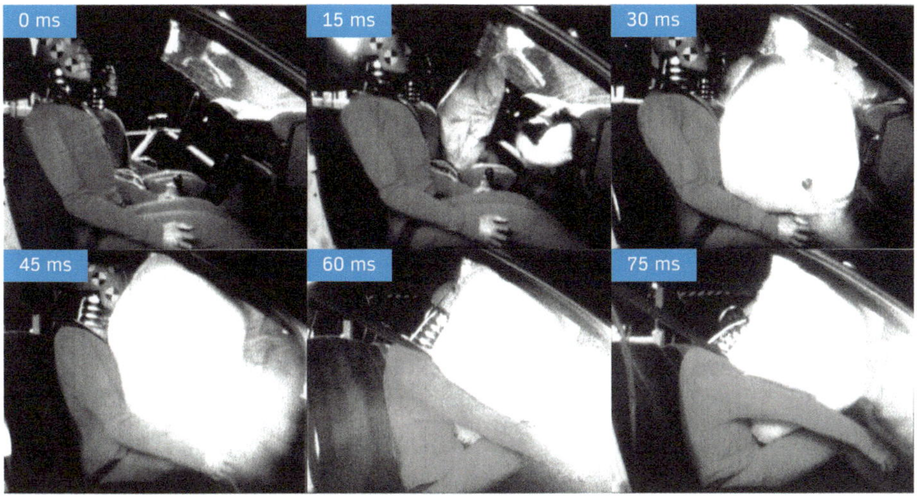

Fig. 6.69 Schematic representation of the timing of airbag ignition and deployment in frontal impact (passenger). (*Source*: [61])

6.6.7.1 Driver Airbag Module

The *driver airbag* (DAB) was the first airbag system (Sect. 6.6.1) for passenger cars and the first legally required airbag modules for cars. The "Intermodal Surface Transportation Efficiency Act" of 1991 regulated the equipment of cars with airbags for the passengers of the front row for vehicles in the USA from 1998 by law. Today, virtually all cars worldwide are equipped with driver airbags (Fig. 6.70).

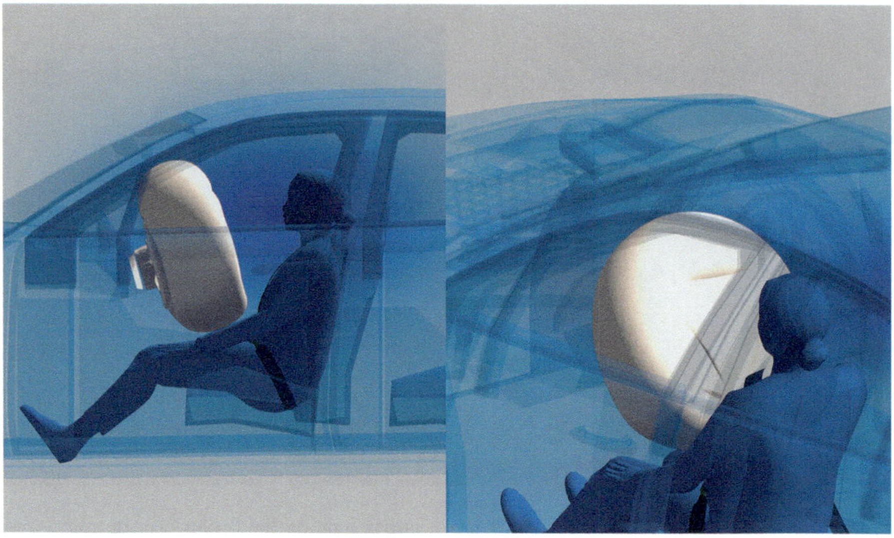

Fig. 6.70 Arrangement of a driver airbag module *(DAB dual contour)*. (*Source*: ZF)

Before the use of three-point seat belts and the use of driver airbags, there were significant injury risks from the driver's impact on the steering wheel, the steering column, the cockpit, and also on the windshield (Chap. 2).

The driver airbag contributes to reducing injury risks of occupants in the driver's position, particularly in frontal accidents. The driver airbag is primarily for the protection of the head, face, neck, and chest. Today, the driver airbag is part of a coordinated restraint concept, consisting of a seat belt system, steering wheel system, steering system, seat system, and for certain load cases also the side impact protection system. However, exceptions for the USA, which were already explained above, must be taken into account (Chap. 4).

In a frontal impact, the pretensioning and coupling of the occupant by the seat belt system occurs first. This is followed by the phase of occupant forward displacement and force limitation by the seat belt. Meanwhile, the ignition impulse to the inflator of the driver airbag occurs, the airbag unfolds in front of the occupant and fills the space between him and the steering wheel.

During the *restraint* by the airbag, forces are introduced into the body over a large area. Influencing the head and neck kinematics and avoiding impact is a key objective. Injury risks due to high head accelerations and neck loads should be reduced. The above-mentioned self-adaptivity of airbags supports this.

Steering wheel and driver airbag module represent a subsystem that is integrated as an independent product into different vehicle domains: chassis, steering, operating concept (HMI), cockpit, interior, automated driving (ADAS/AD), and vehicle safety. The occupant protection function is a significant subfunction of *steering wheel systems*. The driver airbag is significantly determined by the steering wheel system in the system environment, as typically the steering wheel and driver airbag module form a complete system (Fig. 6.71). Typically, the driver airbag module is integrated centrally in the steering wheel and is therefore placed directly in front of the driver. In addition to its primary function of protecting occupants, the horn function is usually activated via the driver airbag module by closing an electrical contact. Depending on manufacturer-specific designs and the partial or possible integration of operating or display elements, the complexity of the driver airbag module can be increased.

Today's driver airbag modules, see explosion diagram, are mechanically locked into a steering wheel or alternatively screwed in. By triggering an electrical signal at the pyrotechnic gas generator, hot gas continuously flows through an inlet opening into the driver airbag until it has reached its final shape or size after approximately 30 ms. At the beginning, predetermined breaking points, so-called tear lines, of the driver-visible and manufacturer logo-equipped driver airbag cap made of thermoplastic materials (TPE) are broken or torn open. The steering wheel rim supports the driver airbag when the occupant "dives" into the airbag. Rotationally symmetrical and circular airbag shapes are usually used, which have a volume of approximately 50 to 65 liters depending on the target market. The depth of the inflated airbag is controlled by tethers, which connect the front and back of the airbag. To ensure controlled outflow of the hot gas, at least one outflow opening is provided on the side of the airbag facing away from the driver.

The operating temperature of a driver airbag module is between −35° and +85 °C. In addition to meeting legal, customer-specific and consumer protection crash requirements,

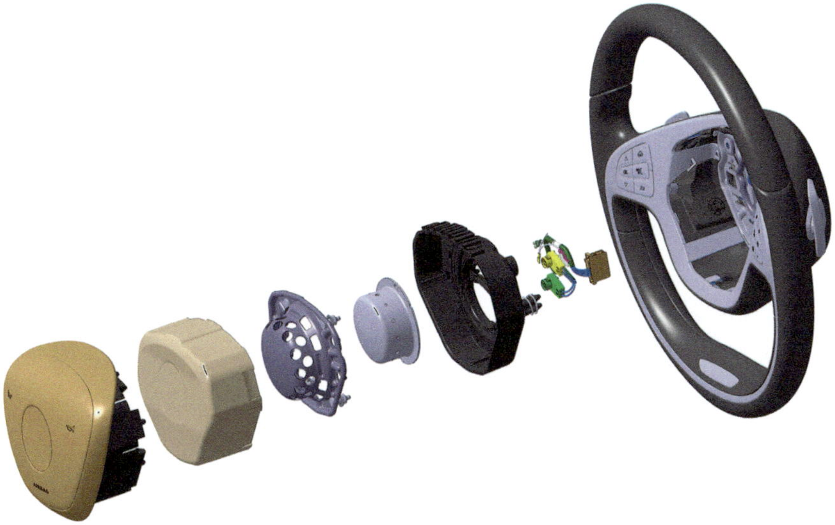

Fig. 6.71 System structure of the steering wheel system and integration with the driver airbag module. (*Source*: ZF)

for the US market, for example, out-of-position (OoP)-load cases according to FMVSS 208 must be met (Sect. 6.2).

In the event of a crash, forces are introduced into the body—especially on the head and chest. This absorbs the kinetic energy of the occupant and slows down the occupant's movement. The occupant hitting the steering wheel or interior components and overstretching the neck should be avoided.

Both, 1-stage and 2-stage gas generators are used in the driver airbag. While in the 1-stage inflator the combustion of the propellant takes place in a single stage, in the 2-stage inflator the propellant is controlled with a delay, typically after 5 ms, but this can also vary. 2-stage gas generators are particularly needed for the US-relevant load cases or requirements and for special OEM requirements.

Depending on the vehicle environment, for the U.S.-NCAP load case of the *5th percentile woman*, the integration of adaptivity functions on the driver airbag module may be necessary, as the energy input of an occupant into the inflated airbag differs significantly for a small and light person compared to a large and heavy person. Airbag adaptivities can be activated by igniting a pyrotechnic actuator (tether activation unit, TAU) to allow the gas in the airbag to flow out through another opening. The driver's classification is evaluated by the occupant sensor system of the airbag control unit (Chap. 7).

The integration of additional functions and the trend to make driver airbag modules increasingly compact requires the sewn-together airbag to be thermally fixed after the folding process in order to minimize the necessary installation space (Fig. 6.72).

The design of future driver airbag concepts will be determined, among other things, by innovative concepts for the vehicle interior and the design of automated driving. For example, it is conceivable that the driver will assume a comfort seating position in the

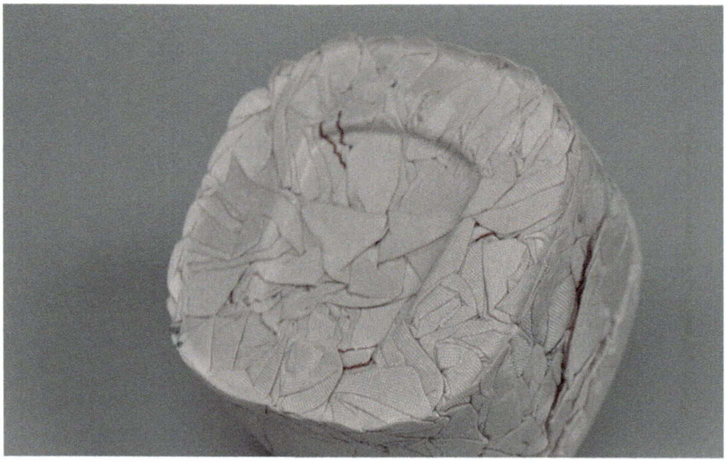

Fig. 6.72 Compact airbag modules through *thermo folding* technology. (*Source*: ZF)

future. To adequately protect the occupant in such a seating position, additional airbag volume can be released via another ignition unit. Installing the driver airbag for new interior and cockpit concepts in other package zones such as the headliner is also a viable option.

Occupant protection systems for protection in frontal impacts also include the steering system. It also has the ability to convert energy under load and can therefore extend the forward displacement of the driver. A further description of the interfaces between driver airbag and steering system in the coordination of occupant protection functions can be found in Sect. 6.7.

6.6.7.2 Front Passenger Airbag Module

The market introduction of *passenger airbags* (PAB) took place in 1973 in the USA by General Motors as an alternative restraint system under the name *air cushion restraint system* (ACRS). It included driver and front passenger airbag as well as pelvic belts (Sect. 6.6.1). However, this equipment option was already withdrawn from the market in 1976. With the market introduction of the passenger airbag as a supplemental restraint system by Mercedes-Benz in 1987, it then established itself as a protection system worldwide. The equipment rate for passenger airbags is now around 98 % across all regions. In the USA, passenger airbags have been legally required since 1998.

The front passenger airbag (Fig. 6.73) particularly contributes in combination with seat belt systems to reducing the risk of injury to occupants in the passenger position in a frontal impact. It is particularly aimed at protecting the head, face, neck, and chest. To

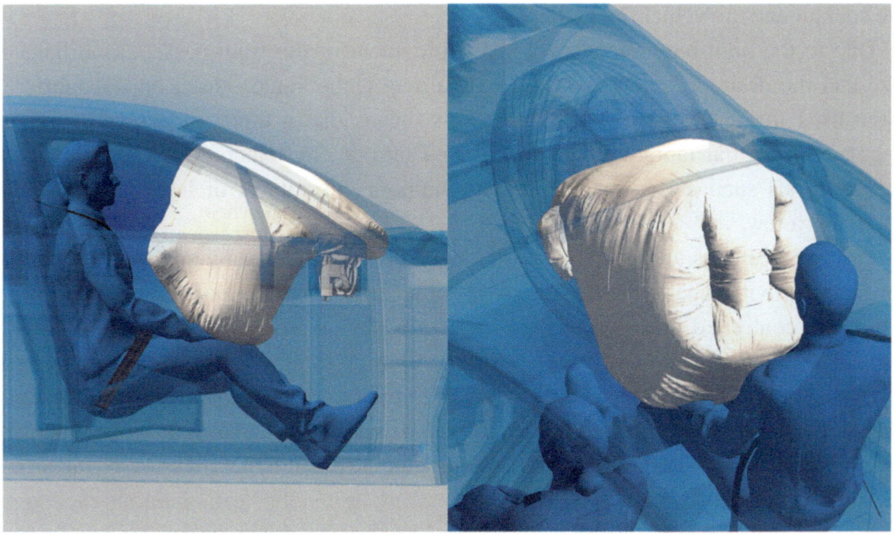

Fig. 6.73 Front passenger airbag module with dual contour technology. (*Source*: ZF)

improve occupant kinematics and to reduce the injury severity of the lower extremities, the knee airbag can be installed additionally.

Development activities have recently focused on the further development of requirements in frontal accidents. With the introduction of load cases such as the fixed barrier with *small overlap* and the *oblique frontal impact*, new challenges arose for the front passenger airbag. The protective effect had to be extended in a way that allows occupants to be supported even when hitting the airbag at an angle. In particular, the requirements for head protection and the relevant load values (Chap. 3) also increased. Development goals are the avoidance of head rotations or angular velocities and angular accelerations.

After the ignition impulse to the inflator and its ignition, the gas flowing out of the inflator unfolds the front passenger airbag in front of the occupant. The space between the occupant and the instrument panel is filled with airbag volume.

With the seat belt system, the phase of pretensioning and coupling is followed by the phase of force limitation with the upper body diving forward into the cushion of the passenger airbag.

For restraint, forces are introduced into the body over a large area. In this process, the kinetic energy of the occupant is absorbed and the movement of the occupant is slowed down.

Contact of the occupant with the contours of the interior should be avoided as much as possible.

The system environment for the front passenger airbag is significantly different than for the driver airbag. The installation location within the cockpit is further away from the occupant, so a much larger effective area needs to be covered. In addition, the deployment of the airbag is challenging due to the size of the airbag and the interactions with the cockpit and the vehicle's glazing.

Due to the high requirements for the deployment of the front passenger airbag, the size and the desired contour, so-called 3D airbag technologies based on flat fabric are often used in the design of the airbag. This means that a three-dimensional structure is created from flat fabric cuts by sewing and introducing tethers. It takes over various sub-functions such as occupant coupling, support and control of head kinematics, to name just a few.

Passenger airbags are equipped as single- or dual-stage airbag modules and also with further adaptivity functions such as actively controlled tethers or controlled airbag vents, which are controlled by a pyrotechnic actuator (*tether activation unit*, TAU). Similar to the *driver airbag*, the front passenger airbag often takes into account occupant classification, so elements for controlling adaptivity to the occupant are provided. For this purpose, dual-stage gas generators, tethers as well as passive or active *vent technologies* are used. To influence the *adaptivity*, for example, the airbag internal pressure or the size of the airbag can be influenced.

For the front passenger airbag module—like for the driver airbag—either *hot gas inflators* or also *hybrid inflators* are used. The airbag volume of a passenger airbag is usually 90 to 120 liters. For passenger airbags for the USA, different requirements apply (Chap. 4). They have an airbag volume of usually 110 to 135 liters. The weight of a module is about 1,100 to 2,500 g, with *fabric housing* technologies offering significant reduction potentials.

The front passenger airbag is positioned in the upper area of the cockpit in front of the passenger, so that the airbag can deploy in front of the occupant when the airbag is activated. To do this, the airbag pierces the surface of the cockpit module.

Typically, the front passenger airbag is mounted within the *cockpit* on the instrument panel or the *cross beam*. Therefore, the interface to the cockpit is of particular importance in the development process for the opening behavior and deployment of the airbag. The front passenger airbag often uses the *windshield* as a supporting surface during deployment. The coordination with the *glazing system* (Sect. 6.7) is therefore also an important aspect of vehicle integration.

The design of vehicle interiors is currently evolving by creating more free space for passengers and integrating larger displays and entertainment functions. Therefore, the aim is to use the large area of the cockpit in front of the passenger for this purpose, so that new mounting locations and concepts for the deployment of the front passenger airbag must be found.

One possibility is to accommodate the front passenger airbag in the roof structure, so that the cockpit area is not restricted by a front passenger airbag module (Fig. 6.74). This concept is available as *front passenger airbag* with bag-in-roof technology [18].

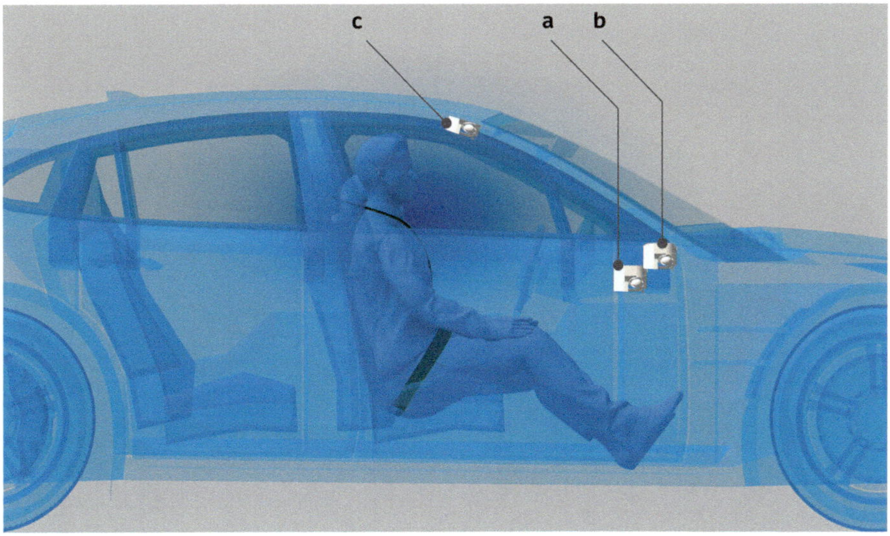

Fig. 6.74 Alternative package concepts for front passenger airbag module: **a** Cockpit integration, **b** Slim cockpit integration, **c** Front passenger airbag, bag-in-roof technology. (*Source*: ZF)

The installation of an infant carrier or a child seat on the passenger seat requires a possibility to deactivate the front passenger airbag using a sensor for the child seat or a manual switch. For user guidance, a safety label is mandated within the passenger compartment (Sect. 6.7).

6.6.7.3 Knee Airbag Module

To protect the lower extremities, *knee airbag modules* (KnAB) have been used as part of restraint systems since the mid 1990s. The first application was done in 1996 by Hyundai for the KIA Sportage marketed in the USA and in 2003 by Toyota for the *Avensis* in Europe.

The knee airbag provides its protective function in combination with other restraint systems such as the driver airbag or front passenger airbag and also the seat belt system. Due to the specific legal requirements in the USA by the FMVSS 208 [141] (*unbelted load case*), the knee airbag was originally used for load cases with unbelted occupants.

The protective function of the knee airbag is a targeted influence on the occupant kinematics as well as the prevention of injuries due to hard contact with the instrument panel.

The early coupling to the knee airbag allows the occupant to participate in the vehicle deceleration early on. This stabilizes the lower extremities, initiates the upper body rotation, and the energy conversion can be improved by a targeted influence on the forward displacement (Fig. 6.75).

A knee airbag can significantly influence occupant kinematics and thus also the load values, Fig. 6.76. Not infrequently, this can improve loads in the chest area and prevent head contact with the windshield. In the case of belted occupants, it aims for a positive influence on injury risks due to *submarining* and relieves the lap belt.

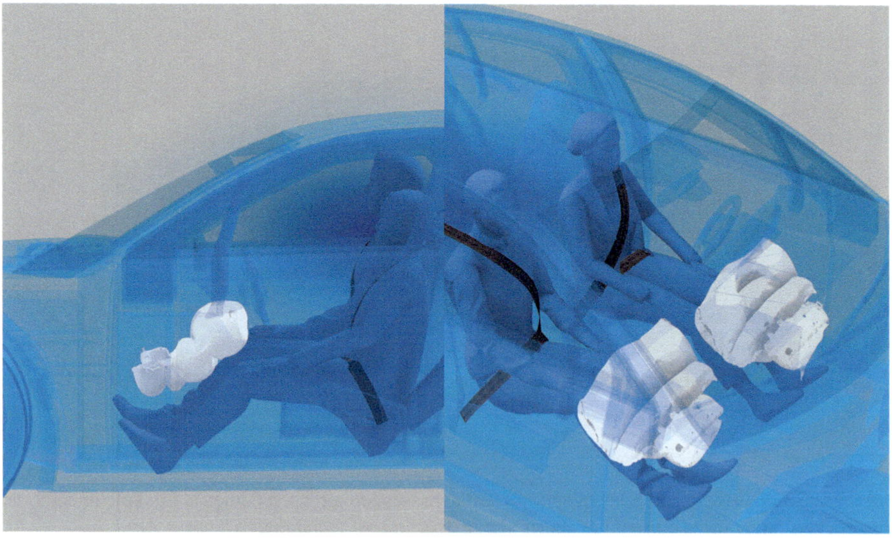

Fig. 6.75 Knee airbag modules for driver and passenger as low-mounted modules. (*Source*: ZF)

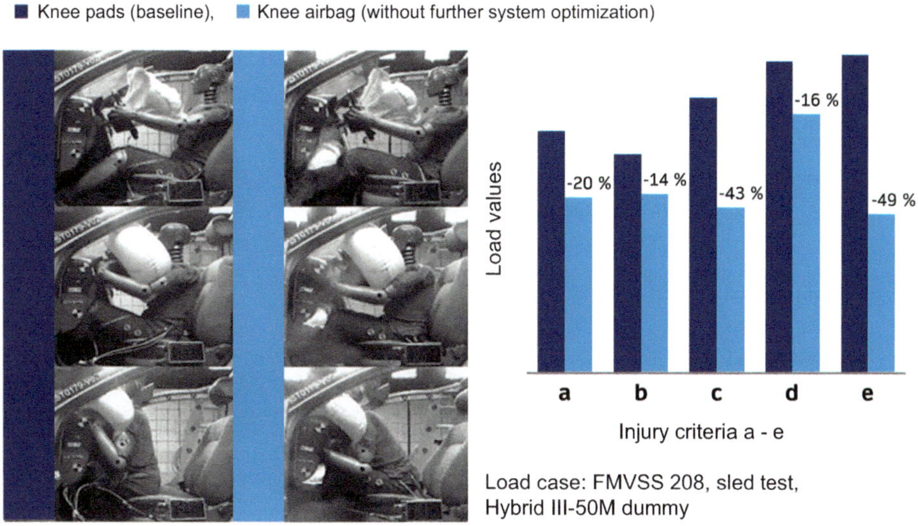

Fig. 6.76 Protective function of the knee airbag in system comparison with knee padding, representation of the reduced occupant load values: **a** Femur force, left, **b** Femur force, right, **c** Pelvis displacement, **d** Res. chest acceleration, **e** Head acceleration HIC—reduced values due to prevented head impact. (*Source*: ZF)

For the knee airbag, two different basic types result from the specific installation location in the cockpit. Either the knee airbag module as a mid-mounted module (MMM) with Class-A surface or as a low-mounted module (LMM) is concealed below the cockpit (Fig. 6.75). The airbag cushion first unfolds downwards and then upwards between the cockpit and the lower extremities. Knee airbags are primarily used on the driver side, but are also used on the passenger side to supplement the restraint system.

A challenge with the knee airbag is the rapid deployment of the air cushion in the narrowly measured space between the cockpit and the lower extremities, before the occupant begins to move forward.

Housings for knee airbags are made of metal and plastic or the module is encased solely by fabric, after the folded airbag has been compressed and thermally fixed (fabric housing technology). The weight of a module can thus be reduced from about 1,100 g to under 700 g (Fig. 6.77).

For knee airbags hybrid inflators are often used, as this represents the best compromise in terms of gas temperature and performance. The airbag volume of a knee airbag is usually 14 to 16 liters, but can also vary further.

6.6.7.4 Seat Ramp Airbag Module

For specific applications, the technology of *seat ramp airbags* (also: *anti-submarining airbag, anti-sliding airbag, seat cusion airbag*) has been developed, in which airbag modules are integrated into the seat cushion of the front or rear seats (Fig. 6.78). The

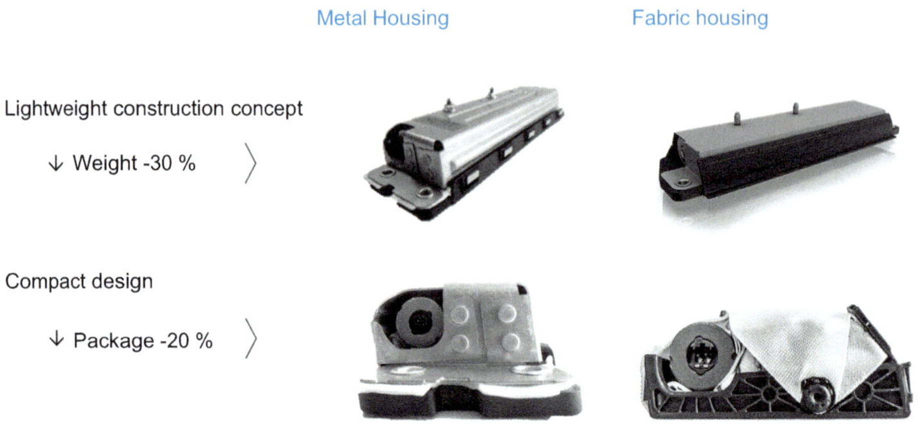

Fig. 6.77 Advances in reducing weight and space requirements through fabric housing technology in the knee airbag module. (*Source*: ZF)

first application of this type of airbag was made in 2002 by Renault for the three-door model Mégane.

The application of this type of airbag is currently limited to individual applications. However, an increasing importance is assumed for future interior concepts, in which greater degrees of freedom of occupant positions are to be enabled.

The use of this airbag is intended to specifically influence occupant kinematics in frontal impact. Depending on the orientation of the seat ramp airbag, the following development goals are pursued: Reduction of injury risks through *submarining,* as in these cases there are major injury risks among others for internal organs in the abdominal and pelvic area, support in the targeted restraint of the pelvis with positive effects for the occupant kinematics, reduction of spinal loads for future desired flexible seating positions (also: *comfort seating positions*).

As an integral part of the occupant protection system, it should be prevented that the occupant slides under the lap belt (known as *submarining*). This risk is particularly increased with a more reclined seating position. Secondly, the effect of the seat belt in restraint should be maintained. And thirdly, the restraint of the pelvic area should support a rotational movement of the upper body.

The construction of a seat ramp airbag module is influenced by the tight installation space in the lower seat structure and the high requirements for the ergonomics of the seat. The seat ramp airbag is mounted unfolded between the seat structure (seat pan) and seat cushion. As a rule, neither a cover nor an additional housing is required for this airbag system.

Furthermore, possible interactions with the different comfort functions installed in the seat must be taken into account. The use of 3D airbag technologies can achieve a

Fig. 6.78 Seat ramp airbag modules for front seat and rear seat applications *(3D-SRAB)*. (*Source*: ZF)

flexible airbag design, with which special areas for functional elements in the seat can be spared and adapted, without affecting the basic shape and function of the seat ramp airbag module.

In the seat ramp airbag hybrid inflators are used, which are characterized by a smooth rising performance curve and long delivery duration.

In addition, there are design variants in which the cushion is made from welded sheet metal cutouts. When triggered, this structure is inflated with an inflator and thus influences the contour of the seat cushion.

6.6.7.5 Rear Seat Airbag Module

To this day—with few exceptions—only seat belt systems are used to restrain occupants on the rear seats in the event of frontal impacts. The first development steps towards concepts beyond this with *rear seat airbags* (RSAB) resulted in the first market launch of an airbag for rear seat passengers on the passenger side by Nissan with the luxury series JHG50, *President* [92].

Figure 6.79 shows a rear seat airbag module with self-adaptive characteristics *(self conforming rear seat airbag, SCaRAB)*. As part of a program funded by the NHTSA to investigate protection potentials for rear seat passengers, occupant protection concepts and concepts for airbag modules that are adapted to the specific requirements for this application were tested and evaluated [51, 52].

The series application of a rear seat airbag was carried out in 2020 by Mercedes-Benz for the series 223 (S-Class) [83]. This airbag has a truss-like construction and is highly suitable for OoP situations (Sect. 6.2.6), which can occur especially on the rear seat [105].

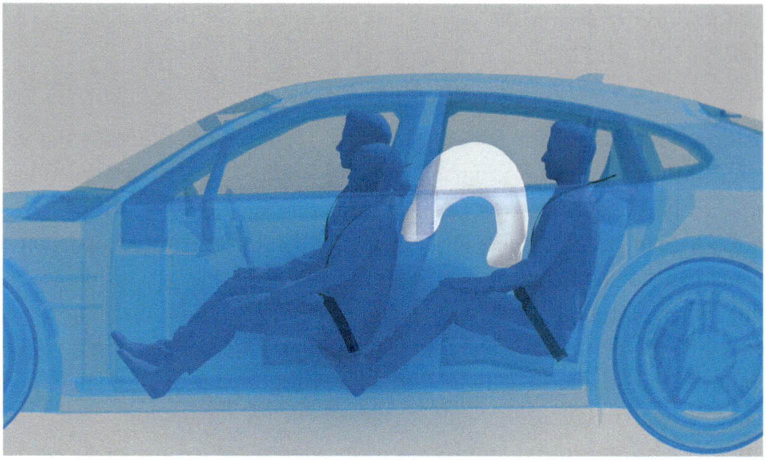

Fig. 6.79 Concept of a rear seat airbag module with self-adaptivity functions *(SCaRAB)*. (*Source*: ZF)

The purpose of a rear seat airbag module is to protect occupants on the rear seat row in the event of a frontal impact. The energy absorption takes place in conjunction with the seat belt system, by supporting the head, neck, and torso. In addition, the risk of hitting the backrest or headrest of the front seat should be reduced.

The larger contact area of an airbag compared to a seat belt is intended to reduce local loads, particularly on the head.

In principle, there are several options for the package of an airbag for rear seat passengers. At this point, the integration of such a module into the backrest of the front seat will be discussed, which offers particular advantages in the deployment of the airbag.

However, the system environment is challenging due to a high variability of interface parameters. The position of the airbag module varies due to the longitudinal, height, and backrest inclination adjustment. Furthermore, occupants differ in their size, mass, and specific protection requirements often more than in the front seat positions. The installation of child restraint systems must also be taken into account when designing the airbag system.

Due to the cramped installation space in the backrest of the front seat, pyrotechnic gas generators offer advantages for this airbag. To meet the requirements mentioned above, alternative design variants are available, which are chosen depending on the development direction and requirement profile. The airbag design can be similar to the front passenger airbagmodule, taking into account the smaller installation space and the support surfaces on the rear seat (Fig. 6.80a). Alternatively, the airbag can be designed similar to a knee airbag with reduced airbag volume (Fig. 6.80b). Or the airbag should have a pronounced self-adaptivity, so that the design can be based on the *SCaRAB* concept. The concept is adapted to the installation space and should self-adapt to different seat and backrest positions.

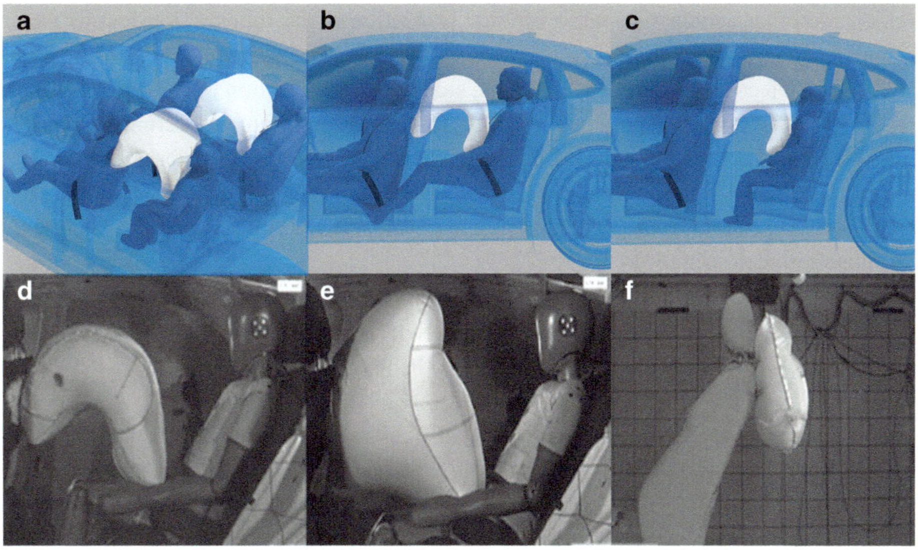

Fig. 6.80 **a–d** Rear seat airbag technology with high self-adaptivity *(SCaRAB),* **e–f** Design variants of rear seat airbags. (*Source*: ZF)

6.6.8 Airbag Modules for Occupant Protection in a Side Impact

Occupant protection in side impact presents a major challenge in vehicle construction, as the seating positions of the occupants are very close to the inner contour of the vehicle interior and possible collision areas. Unlike in frontal impact, where the front carriage usually offers a large deformation zone, little space is available.

Therefore, the development of side protection concepts is closely coordinated between the involved subsystems such as vehicle body structure, door modules, interior, seats and occupant protection systems as well as the sensors for accident detection (Chap. 7). Side protection is a focus of airbag technology. The following section presents the essential airbag modules for this application field.

Beyond the classic concepts of passive safety, concepts of integral safety are already being used for occupant protection in side impact by networking systems of active and passive safety. The starting point is the early detection of a possible collision before the actual contact and a coordinated intervention strategy with reversible or non-reversible actuators. Examples of this are occupant protection and assistance systems such as pre-crash side airbags or vehicles with chassis systems for pre-crash level adjustment.

At present and in the near future, development work on side protection is strongly focused on the changes brought about by battery electric vehicles and the integration of battery storage. This can change deformation behavior and load profiles, so that further developments are also necessary in the airbag systems.

The purpose of airbag systems in side protection is to reduce the risk of injury in a side impact with an obstacle or when another vehicle collides into the side of one's own vehicle. Another accident scenario for side protection is the rollover accident as a standalone event or as a result of a collision. Here, side protection aims to prevent the partial or complete ejection of occupants from the side windows (ejection mitigation).

The most important legal requirements for testing occupant protection systems for side impact protection are, among others, defined in the following standards (Chap. 4):

- UN-R 95 (protection of occupants in a side impact) [132] and UN-R 135 (pole side impact protection) [136]
- FMVSS 214 (side impact protection, adoption: 1990, phasing in: 1994–97) [143] and FMVSS 226 (ejection mitigation) [144] for the USA
- GB 20.071–2006 (lateral collision) [114] for China

Tests for side protection are carried out with movable deforming barriers (MDB) and with dynamic *pole tests*, Fig. 6.81. The requirements of the ejection mitigation are tested with *impactor tests*.

In recent years, the developments in side protection have been characterized by the introduction and expansion of test configurations in various regions as well as the further refinement of used ATD (anthropomorphic test devices, crash test dummies), see Sect. 8.5. With the introduction of the WorldSID dummy (*worldwide harmonized side impact dummy*), a higher *biofidelity*, but also a further global standardization in the development of occupant protection systems is aimed for.

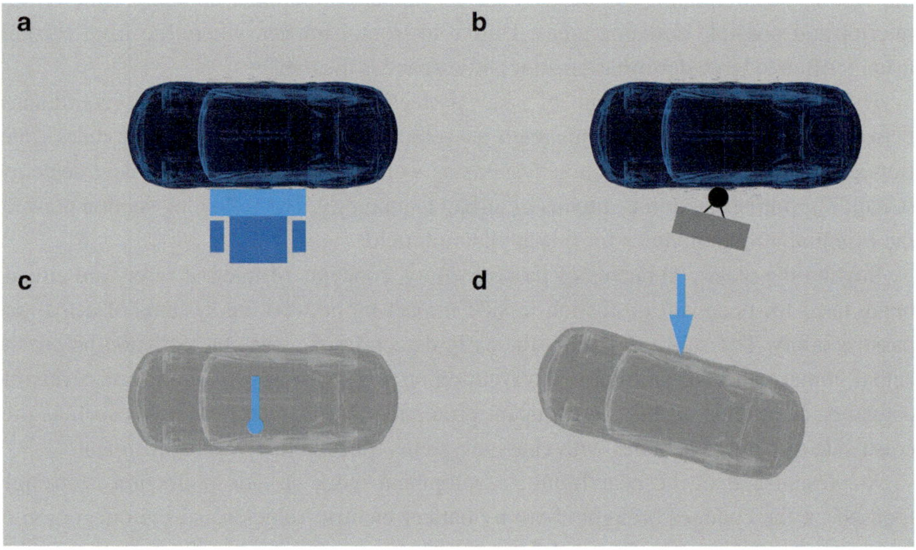

Fig. 6.81 Main focuses of occupant protection in side impact: **a** Barrier tests, **b** Pole tests, **c** Impactor tests, **d** Far-side impact tests. (*Source*: ZF)

Fig. 6.82 Arrangement of the airbag modules for protection in side impact for the occupants on the front and rear row of seats: **a** Side airbag, **b** Curtain airbag, **c** Center airbag. (*Source*: ZF)

The temporal behavior in a side impact differs significantly from that of a frontal impact. The collision must be recognized very quickly and safely after contact with an obstacle at time t_0 (the further time specifications refer to this), so that already at approx. t = 5 to 8 ms the ignition impulse is sent. Already after approx. t = 12 to 18 ms the airbag is deployed and inflated. In the following 30 to 40 ms, the side airbag delivers its protective effect and is deliberately deflated via outflow openings in the airbag (Sect. 6.6.8.1).

In the following sections, the specific airbag systems aimed at side protection are described in more detail: the side airbag and the curtain airbag for both sides of the vehicle as well as front and rear seat passengers and the center airbag on the front row of seats (Fig. 6.82).

6.6.8.1 Side Airbag

Efforts to improve occupant protection in side impacts have been focused since the early 1990s on the provision of *side airbags* (SAB). After successful experiences with driver and front passenger airbag systems could be gathered in the 1980s, this was the next significant development stage of airbag systems in motor vehicles.

The first market introduction of a backrest-integrated side airbag module took place in 1994 by Volvo for the 850 model [147].

This airbag is intended to reduce injury risks in side impacts and can include protection of the chest area, hip area, shoulder area, and head (head protection only in special forms such as head-thorax side airbag). The side airbag is often used in combination

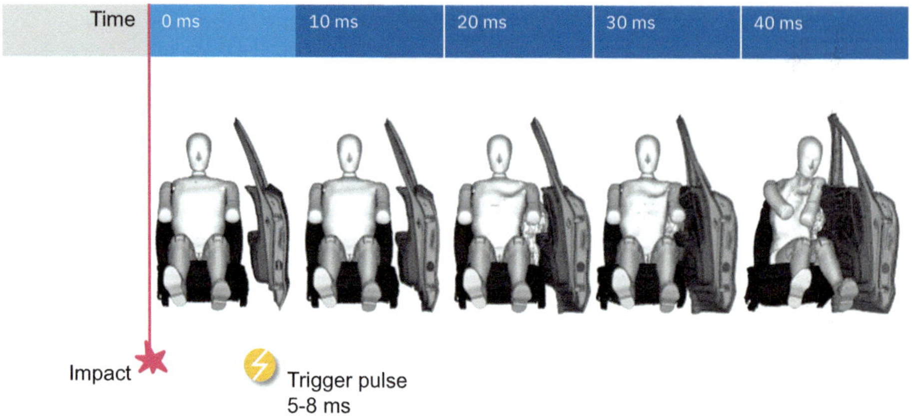

Fig. 6.83 Exemplary timing of the deployment of a side airbag. (*Source*: ZF)

with a *curtain airbag*. This is currently the most commonly chosen configuration. The types of side airbags are further explained below.

The side airbag is intended to provide protection against injuries in a very challenging environment. In the event of a lateral collision, it promptly unfolds into the free space between the occupant and the side structure, thus avoiding direct contact of the body regions with interior parts, such as the inside of the door and the B-pillar trim. In the case of a head-thorax side airbag, contact with roof frame trims is also avoided.

The time for the air cushion to unfold is very short. The ignition of the gas generator already occurs approx. $t = 5$ to 8 ms after the collision and its effect unfolds after about $t = 12$ to 18 ms. Due to the geometric conditions, both the intrusions into the interior and the displacement of the occupant relative to the inside of the passenger compartment begin very shortly after the collision. Furthermore, the possible interaction between the air cushion and the occupant, and especially his extremities (upper arm), must be taken into account. The timing for the airbag deployment, the application to the occupant, and the vehicle configuration during side impact are shown schematically in Fig. 6.83.

Today, various versions of side airbags with different protection functions are available:

- Thorax side airbag
- Thorax-pelvis side airbag
- Shoulder-thorax-pelvis side airbag
- Head-thorax-pelvis side airbag
- Head-thorax side airbag

The airbag has so far been sewn from flat fabric cuts into a flat airbag. Due to the increasing requirements of further load cases for side protection and the introduction of newer ATD generations (crash test dummies) such as the WorldSID dummy, more complex airbags with 3D technologies are now also being used. These constructions sometimes have two or more chambers, so that the restraint for the above-mentioned body regions can be influenced separately.

In the case of the side airbag module, due to the high requirements for deployment speed, *pyro-* or *hot gas generators* are often used. In a single-chamber airbag, pressures of approximately 100 to 120 kPa are typically achieved with a total volume of approximately 17 liters. The weight of a side airbag is approximately 400 g.

Front side airbags are installed on the frame of the backrest of the front seats. Either the seat cover is stretched over it, so that the module is not visible from the outside, or the side airbag has its own cover visible to the interior. Side airbags are also available for many vehicle types for the passengers in the rear seat row. They are then installed in the outer area of the rear backrest. Here, the interaction between the side airbag and possibly installed child restraint systems must be taken into account (Fig. 6.84).

6.6.8.2 Curtain Airbag Module

The *curtain airbag* (CAB) is another essential airbag system for occupant protection in side impact. This airbag is also referred to as *head airbag* or *side-curtain airbag, window bag, curtain shield airbag, inflatable curtain, side tubular bag*. It covers a large area above the windows along the interior and has distinctive profiles that indicate the individual impact areas (Fig. 6.85).

Developments for this type of airbag began in the 1990s. The first market launch was in 1997 by BMW for the 5 and 7 series as the *head protection system* or the *inflatable tubular structure* by Zodiac [153]. This was quickly followed by other concepts that were very similar to the products installed today, including the inflatable curtain airbag by Volvo for the Volvo S80 in 1998 [147].

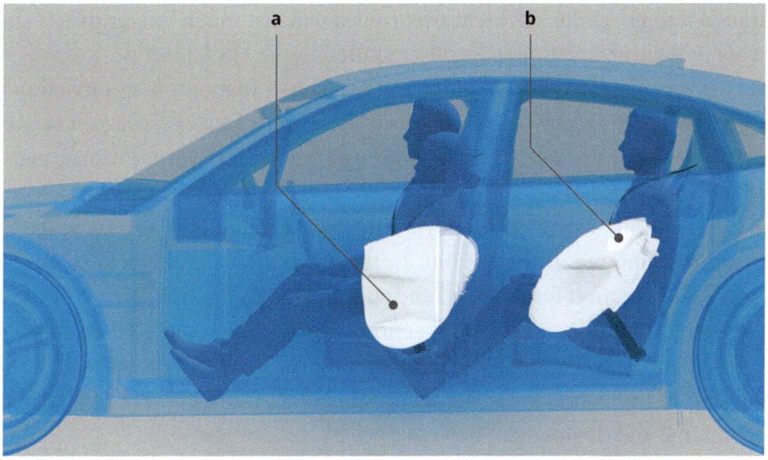

Fig. 6.84 Integration of side airbags for the front and rear seat row in: **a** Backrest of the front seat, **b** Side paneling of the rear seat system. (*Source*: ZF)

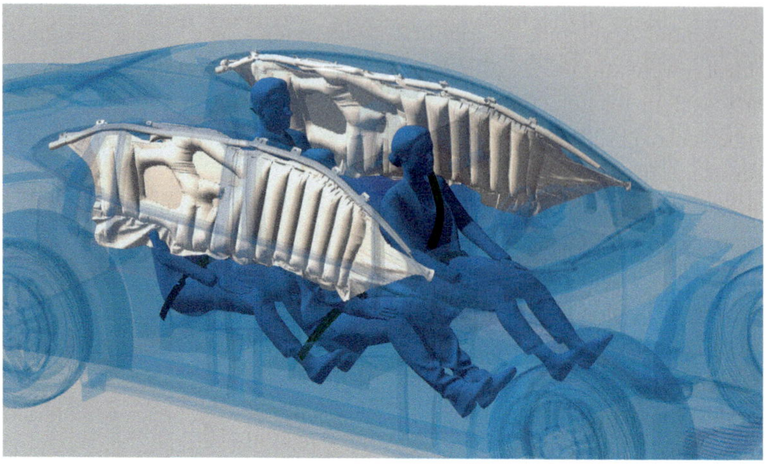

Fig. 6.85 Curtain airbag modules for the driver and passenger side. (*Source*: ZF)

The *curtain airbag* is practically always used in side protection in combination with a *side airbag*—and the seat belt.

The protective function of this airbag is aimed at side impact by another vehicle or side impact on an obstacle. The protective effect of the curtain airbag consists of lower head loads during side impact for the front and usually also for the rear occupants. It provides energy absorption during the impact of the occupant, increases the contact area for force introduction, and reduces accelerations by extending the time of force introduction. This design is also referred to as *curtain airbag 1st impact.*

In addition, curtain airbags can be specifically designed for protection against ejection in the event of a rollover accident. These variants have an extended standing time of the inflated airbag, as this accident occurrence can last much longer than a single first impact. This protective function is legally required in the USA [144].

Curtain airbags can perform an additional protective function in conjunction with the driver airbag or the front passenger airbag in the event of an *oblique frontal impact*. In this accident scenario, the risk of diving through between the front airbag and the side curtain airbag can be reduced.

Usually, the curtain airbag unfolds from the roof frame down to below the door sill, where it can be supported at the bottom. In this way, it covers the area of the side windows to provide protection for the head and shoulder area (Fig. 6.85).

In the construction of the cushion of a curtain airbag, the required effective zone must be covered. When the airbag is deployed and inflated, a rather narrow structure with an efficient design of protection areas for occupant contact is sought. This can also optimize the airbag volume and deployment behavior.

To achieve these design goals, two essential fabric technologies are used for curtain airbags. The airbag can be sewn from *fabric cutouts* (flat fabric), from which the inside

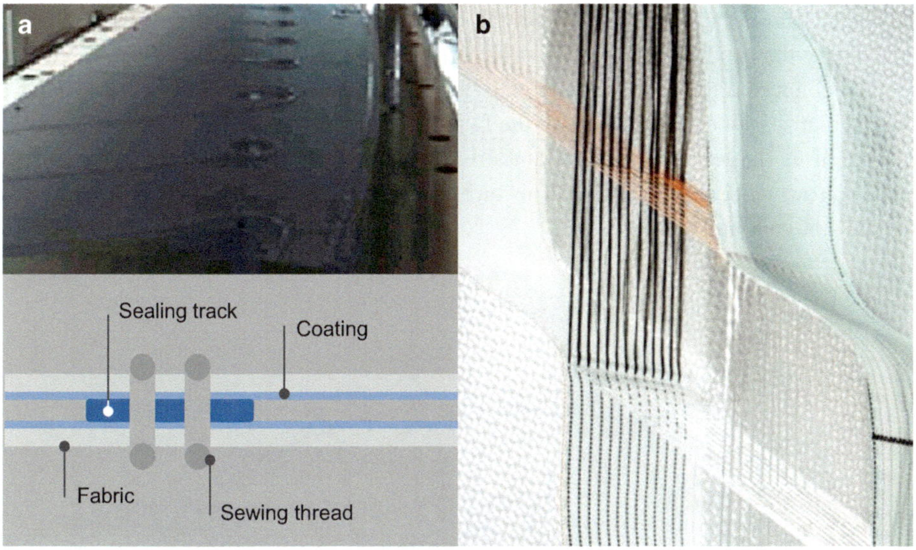

Fig. 6.86 Airbag and fabric technologies for the curtain airbag: **a** Technology for seam sealing for rollover curtain airbag modules with high pressure holding time requirements **b** *X-tether* technology for one-piece woven airbags. (*Source*: ZF)

and outside of the airbag are formed. Additional seams or tethers are incorporated over the surface to shape the protection areas.

A completely different approach is to weave the airbag in one piece in the desired contour and also with all the support threads required within the airbag in one piece. This process and also the airbags produced in this way are referred to as one-piece woven (OPW). The choice of the respective concept often depends on the specific requirements of the respective vehicle applications.

For sealing for long pressure retention in rollover protection, a seam sealing with silicone is carried out (Fig. 6.86a). For the folding of curtain airbags, primarily the roll-fold and less the *Z-fold* is used.

The *one piece woven* manufacturing process enables another technology for airbag design. With the direct weaving of stabilizing tethers in a crossed arrangement (X-tether technology), a targeted profiling of the airbag can be achieved. In this way, certain protection areas can be incorporated into the airbag cushion (Fig. 6.86b).

To reduce the required installation space in the roof lining, a thermal fixation of the folded airbag can also be carried out, so that a flat cross-section can be realized (*thermo-fixed roll folding*).

In the Curtain airbag module, due to the large airbag volume, hybrid inflators are used in a slim design. The airbag volume of a curtain airbag is usually 25 to 70 liters, but can also reach up to 120 liters. The weight of a module is about 1,000 to 2,500 g. The curtain airbag is either encased in a fabric and can gain additional stability and contouring for optimal integration in the roof area through thermo-fixation. In addition, the airbag can be held in shape by adhesive tapes with adapted tear properties at intervals of approx. 80 to 120 mm.

The curtain airbags are installed in the interior in the area of the roof frame and are covered by the headliner (Fig. 6.87). The inflator is directly screwed and the air cushion is often connected to the roof frame with clips. They usually extend from the A-pillar to the C-pillar—or in minibuses also to the D-pillar.

One of the largest available curtain airbags is installed in a transporter with 15 seats and five rows of seats (Fig. 6.88). This airbag has several gas generators and is approximately 4.50 m long [8].

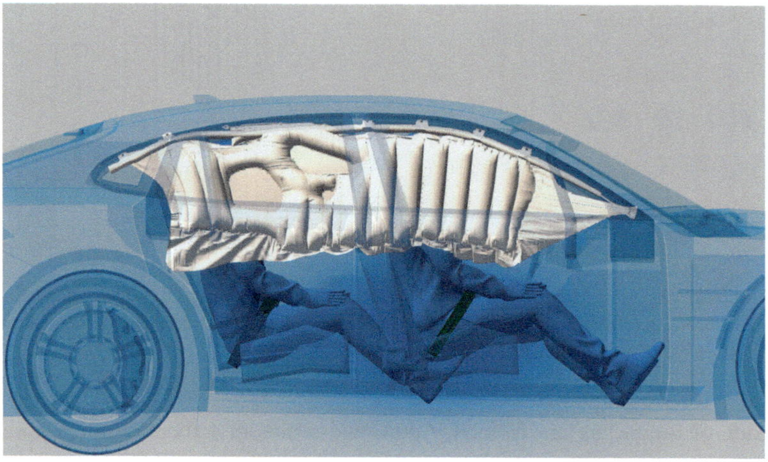

Fig. 6.87 Integration of the curtain airbag modules into the roof frame. (*Source*: ZF)

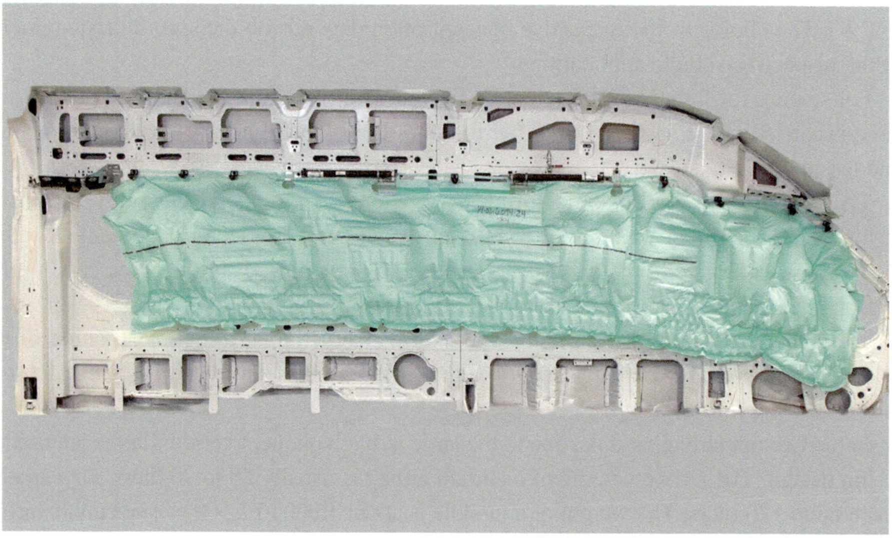

Fig. 6.88 Curtain airbag for van with five rows of seats (Ford Transit). (*Source*: ZF)

In convertibles without a fixed roof structure, curtain airbag variants are used that are installed in the upper door area and unfold from bottom to top.

As an alternative to the curtain airbag in combination with a side airbag, a head-thorax-*side airbag* can also be used under certain requirements (Sect. 6.6.8.1).

6.6.8.3 Center Airbag Module

In the recent past, new requirements for side impact protection led to the development of front *center airbag modules* (CeAB) (also: *middle airbag modules*). Depending on the orientation of the concept or the specific requirements, they are also known as *interaction bag* and as *far-side airbag.*

The protective function of this airbag is initially aimed at accident scenarios in which a side impact occurs on the side of the vehicle opposite the driver (*far side*). Cases were observed where, for example, occupants on the driver's side, despite wearing a three-point belt system, collided with the B-pillar on the passenger side, resulting in high injury risks.

Findings from accident research before the introduction of these airbags showed that in the USA up to 30 % of all fatalities and in Germany almost 30 % of all seriously injured were the result of such accident scenarios. This led to the introduction of evaluation criteria by Euro NCAP in 2020, which often require a center airbag to be achieved. Many vehicle models are now equipped with front-center airbags as standard (Fig. 6.89).

The primary protective function of the center airbag is to stabilize the upper body and head of the occupants so that they remain in their seating position. Another protective function of the center airbag aims to prevent contact between two occupants in the front seating positions. The inflated airbag is intended to prevent direct contact between

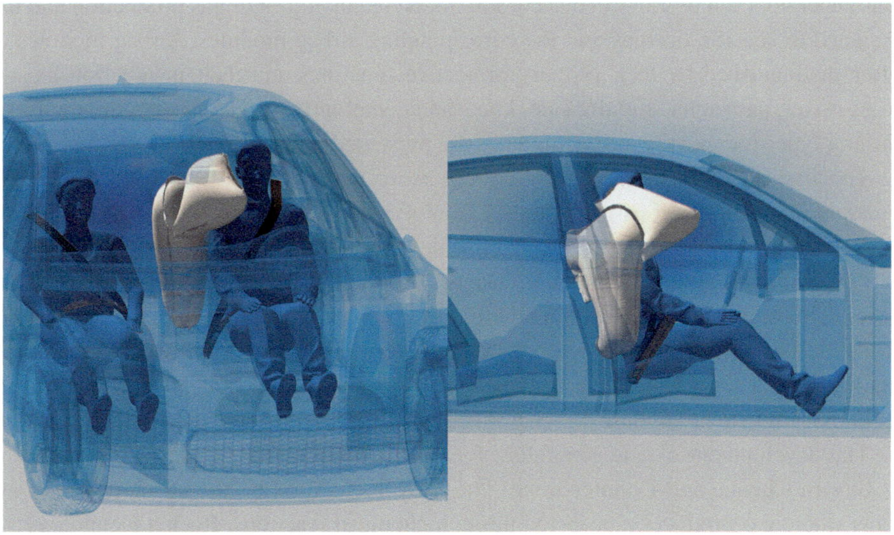

Fig. 6.89 Center airbag for the protection of front occupants in lateral collision. (*Source*: ZF)

occupants on both the impact-facing and impact-avoiding sides of the vehicle, providing protection for the head, neck, shoulder, and chest. Different scenarios and modeling can be found in [56].

In the center airbag module, *hybrid gas generators* are used. The airbag is sewn together from flat fabric pieces or woven in one piece. The airbag is made up of one or more chambers to optimize the support function and the airbag volume. The airbag can be stabilized by tethers during deployment.

The center airbag is mounted towards the vehicle center on the frame of the driver's seat backrest and integrated into the seat. In some cases, a center airbag is installed in both the driver's and passenger's seat. The airbag could be supported by the storage box of the center tunnel or is self-stabilizing where it is not available. This is particularly challenging in compact small cars or battery-electric vehicles. This airbag concept is currently hardly used in the rear seat, with few exceptions.

6.6.9 Gas Generators

Gas generators (GG) (also: *inflators*) are, in addition to the *airbag cushions*, the other essential components of *airbag modules*. They serve to generate gas, which in a fraction of a second causes a deployment of the airbag module as well as deployment and filling of the cushion with a defined amount of gas. The gas generators, their gas generation technology, their performance, and their gas generation characteristics are closely matched to the airbag modules and their purposes.

6.6.9.1 The Evolution of Gas Generator Design Concepts

Gas generators can be classified according to the technologies of gas production and then assigned to specific designs and the corresponding airbag modules. Airbag modules are often distinguished by their primary protective functions and their installation location (e.g., driver, passenger, curtain, side, knee, and center airbag). The most common uses of inflator types for specific airbag modules are presented in Sect. 6.6.9.4.

The first attempts to implement airbags were made in the 1950s and 1960s with *pressure gas containers* as gas storage. They were also the basis of the work of Walter Linderer *(compressed air container)* and John Hetrick *(air accumulator)* (Sect. 6.6.1).

For the pressurized gas container designs, experiments were conducted with different gases. The first vehicle applications with airbag systems (General Motors) in the 1970s still relied on this concept (Sect. 6.6.1). However, the system was withdrawn from the market after a short time. Due to their size, high weight, and maintenance effort, this technology had to be discarded at that time.

The development of gas generators for a series application began in the late 1960s at Bayern Chemie and Daimler-Benz. These works focused on the development of systems where gas production occurred through chemical reactions. Rocket technology was

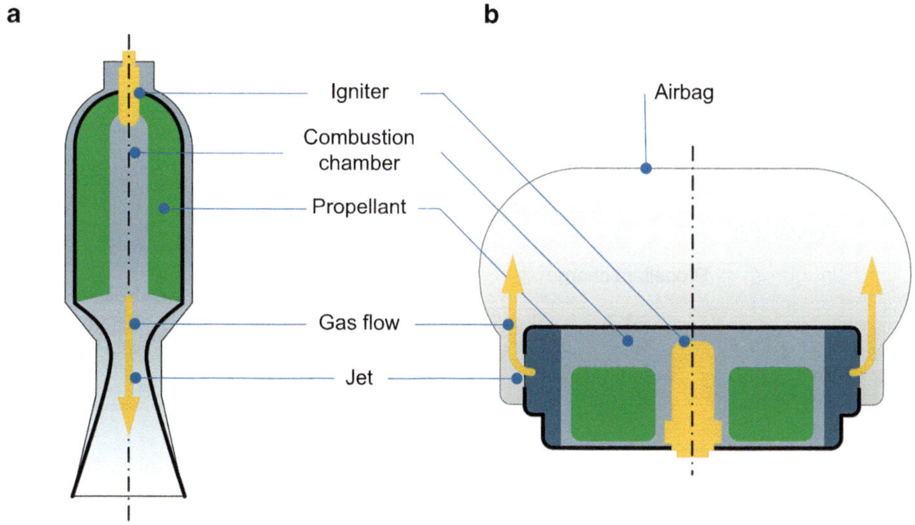

Fig. 6.90 Schematic representation of the transmission of technologies from **a** Rocket motor to **b** Pyrotechnic gas generators. (*Source*: [103])

applied and transferred to gas generators for airbag modules (Fig. 6.90). The goal here, as elsewhere, was to generate gas in a short time by burning pyrotechnic mixtures [103].

The market introduction based on this technology then took place in 1980 by Daimler-Benz with the series 126 (S-Class) with a *driver airbag module*. The inflator used a propellant based on *sodium azide,* which was installed in the form of tablets and ensured safe function over the lifespan. The first front *passenger airbag module* followed in 1987.

The following will show the further development steps to alternative inflator basic types. Further technical explanations of the concepts of gas generation and the inflator basic types can also be found in Sect. 6.6.9.5.

The gas generators of the first generation are among the *pyrotechnic* (also: pyrochemical) *gas generators*. They still represent the most common technology for gas generators today. Parallel to azide-free propellant recipes, the idea of the pressure gas storage was also revisited in the 1990s.

The combined use of pressure chambers and pyrotechnics in *hybrid gas generators* combined the advantages of both technologies. Figure 6.91 shows a comparison of *pyrotechnic* and *hybrid gas generators*. The high temperature of the chemical reactions is used in this type to heat the pressure gas, which cools down significantly due to expansion. Such gas generators are characterized by efficiency and comparatively low gas temperatures, which makes them particularly suitable for use in *passenger* and *curtain/head airbag modules* (both for *first-impact* and also *rollover airbag modules*). Technologies for hybrid gas generators were developed or brought to production readiness in the

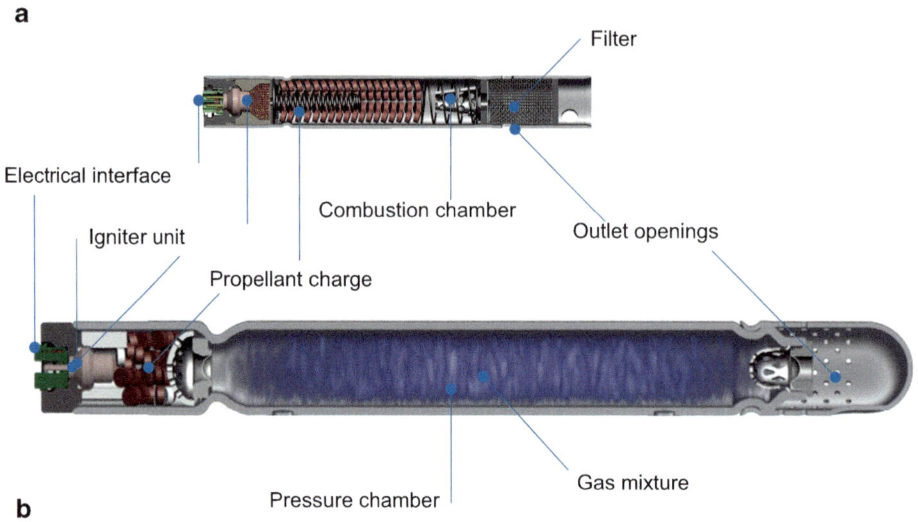

Fig. 6.91 Exemplary comparison of the structure of **a** Pyrotechnic gas generator *(SPI2 EVO)* and b Hybrid gas generator *(SHI2)*. (*Source*: ZF)

second half of the 1990s by companies such as ZF (then: Temic, TRW), Autoliv (then also: OEA), ARC, Daicel and Joyson (then: Takata).

Cold gas generators as pure pressure gas systems have been on the market as a niche product since the beginning of the 2000s. The advantage of this type of generator is based on the effect that the pressure gas cools down significantly during expansion. An airbag inflated in this way then increases its internal pressure over several seconds solely by heating the filling gas through the warmer airbag material and the friction heat during deployment. For use in *rollover curtain airbag modules*, this is a very advantageous characteristic, because this type of module must achieve airbag pressure holding times up to 6 s. The disadvantages of this technology are a larger space requirement for the pressure vessel and a higher component price, which is largely due to the use of the noble gas *helium*. This has the advantage that it cools down significantly less during the expansion of the gas from the pressure vessel to ambient pressure than other noble gases, such as argon. This allows smaller pressure vessels and enables use in vehicles. However, it is a rare raw material, which is also obtained with great energy expenditure and therefore expensive at only a few sources worldwide.

Improved *sealing systems* in the design of *airbags* and *airbag fabric* reduce the need for cold gas generators. They are increasingly being replaced by *hybrid gas generators* and adapted airbag modules.

Other special forms of gas generators on the market are types that generate their energy by burning hydrogen with oxygen. Such *hydrogen hot gas generators* produce pure steam during combustion and are therefore low in pollutants.

In addition to the gas generators themselves, individual components are also worth highlighting. Thus, igniters are also used in combination with other propellant charges or mechanical devices in occupant protection systems for vehicles, for example:

- Micro gas generators (MGG), which are used in seat belt retractor pretensioners (Sect. 6.5)
- Pyrotechnic tether activation units (TAU) for control of adaptivity for airbag cushions (Sect. 6.6)
- *Pyro-switches* (also: *pyro-fuses*) for the interruption of high-voltage circuits of battery electric vehicles upon detection of an accident
- *Pyrotechnic hood lifters* for the protection of vulnerable road users (Sect. 6.8)

In addition to the automotive field, the range of pyrotechnic protection devices for cyclists or athletes is gradually expanding.

6.6.9.2 Requirements for Gas Generators

Gas generators must meet a variety of specific requirements. This applies to both the design of the inflator components and the selection of raw materials, components, and manufacturing processes of the propellants and the gas generator. The following explanations address some essential requirements for gas generators.

When developing such pyrotechnic systems, the utmost care must be ensured in testing and validation of materials and processes. After all, the function for each individual inflator must be precisely guaranteed under all climatic conditions, even after many years on the road, down to the millisecond at the correct speed. The resulting gas must not cause any health problems.

When a gas generator is triggered, a—mostly chemical—reaction is initiated to release gas, with the aim of filling an *airbag*. It inflates within milliseconds and offers protection from contact with hard vehicle structures with its large and yielding impact surface. Depending on the application (airbag module) and inflator technology, the *gas quantity,* the *gas delivery speed* (also: *mass flow*) and the *gas temperatures* differ (Fig. 6.92).

It should be noted that for a *front passenger airbag* significantly larger gas volumes are needed than for example for a *side airbag*. However, since the function duration is very short for this module, it is sufficient to use a smaller gas quantity at a higher temperature. For a *curtain airbag*, on the other hand, the protection of the occupants should be ensured even in the event of a vehicle rollover. Therefore, a low gas temperature is maintained so that the air cushion remains inflated for the duration of the accident event *(airbag hold-open time)* for protection.

After an accident, the safety of the rescue forces must be ensured (Sect. 6.10). Non-deployed airbags (as in a frontal impact, side airbags are usually not activated) must not

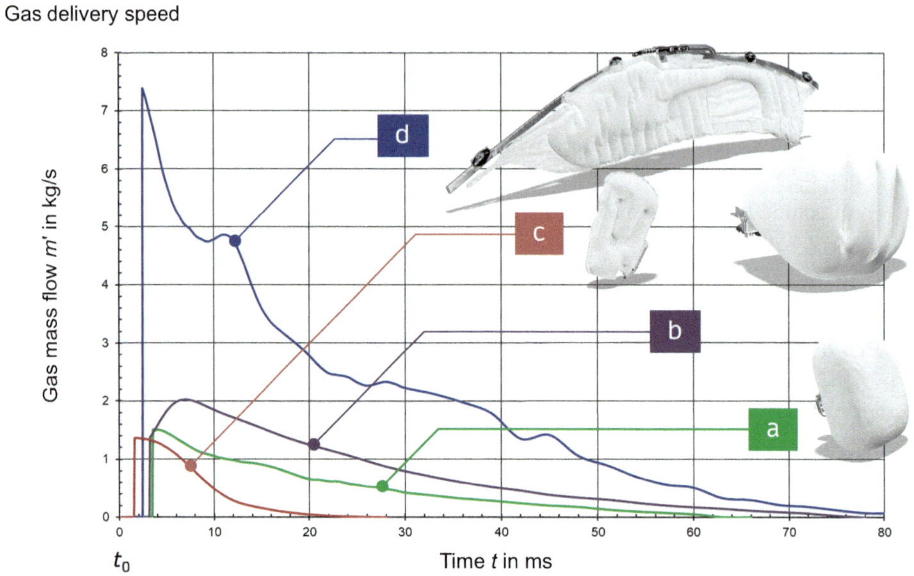

Fig. 6.92 Different requirements for gas generators using the example of gas delivery speed for various airbag application fields: **a** Driver airbag, **b** Front passenger airbag, **c** Side airbag, **d** Curtain airbag. (*Source*: ZF)

pose a danger or release harmful substances even during rescue measures (e.g., when rescue shears are used by rescue forces).

The safety during transport of gas generators is also regulated by law. If a fire occurs during transport, the pyrotechnic propellant must not pose a danger to the environment or the rescue forces.

Examples of requirements and standards for gas generator development are:

- DIN EN ISO 14451–5:2013–08 [53]
- USCAR 24 [99]

In addition to regionally valid requirements and standards, requirements from automobile manufacturers usually also have to be taken into account.

6.6.9.3 Inflator Components

This section describes the essential components of gas generators. Figure 6.93 shows the basic components of gas generators using the example of a single-stage pyrotechnic round gas generator. Depending on the inflator type, they are used or are partially obsolete.

Electrical interface

The *electrical interface* between the trigger electronics (Chap. 7) and the *igniter unit* and thus of the gas generator is usually a *two-pole plug* (squib connector) with a standardized

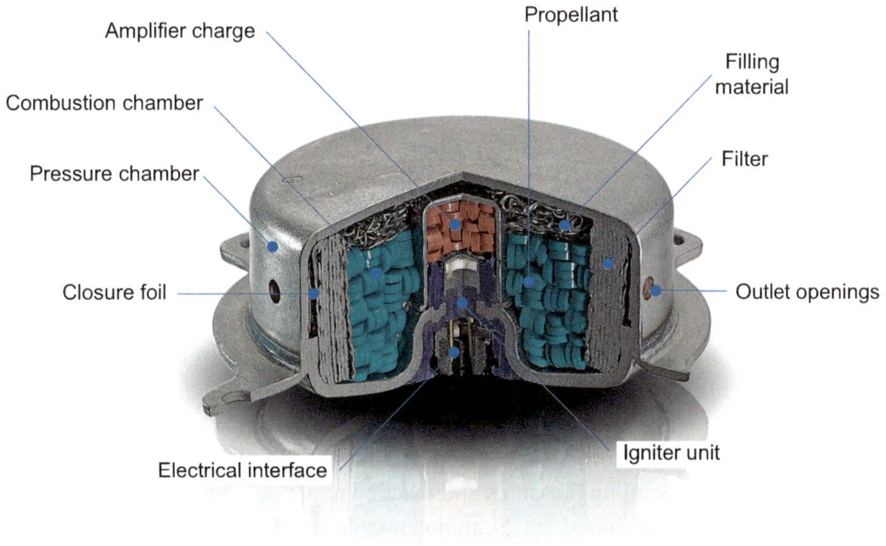

Fig. 6.93 Components of a gas generator using the example of a pyrotechnic round gas generator. (*Source*: ZF)

contour, which ensures that e.g. during the assembly of a multi-stage airbag, no connections can be swapped.

Igniter unit

To initiate a gas generator, a pyrotechnic igniter is activated in all systems used today. The entire ignition process usually takes place within less than 2 ms.

The *igniter unit* consists of the actual *igniter* and its surrounding components. The igniter consists of the *igniter body,* the two *contact pins,* the *bridge wire,* the *pyrotechnic charge* and the *igniter cap*. The *pyrotechnic charge* consists of a *metal-* or *metal hydride*-powder and an oxidizer, usually an alkali perchlorate. The igniter unit is tightly connected to the inflator housing and part of the combustion chamber.

Booster charge

The igniter unit alone is usually not powerful enough to ignite a stable combustion of the tablet-shaped propellant. Therefore, as an intermediate step, a booster charge is ignited. It consists of a granulate or small *tablets* of a pyrotechnic mixture, which is easy to ignite and burns at high speed. This produces hot particles and hot gases, which then enter the actual *combustion chamber* and ignite the propellant.

Propellant

The propellant used in a gas generator typically consists of a mixture of fuel, an oxidizer, and additional *additives*, which serve to optimize combustion characteristics and improve its processability in manufacturing.

$$H_a\ C_b\ N_cO_d + \text{oxidizer} \rightarrow \frac{1}{2}a\ H_2O + b\ CO_2 + \frac{1}{2}c\ N_2 + \text{slag} \tag{6.21}$$

In the first vehicle projects with airbags in the 1980s, gas generators were built with propellant based on on *sodium azide* (NaN_3). These gas sets had the advantage of burning stably and reproducibly at relatively low temperatures and operating pressures. However, they were difficult to handle and dispose of the materials, as sodium azide itself is classified as highly toxic and can easily form explosive compounds with various materials. Therefore, in the 1990s, research was conducted for alternative propellants and they were brought into series production.

All propellants for gas generators have in common that they do not "explode", but instead burn in a *controlled manne*r at the surface of the propellant.

The recipes commonly used today are often based on *guanidine nitrate* (GuNi) and *nitroguanidine* (NiGu). They are characterized by a higher gas yield and chemical stability, but require higher operating pressures and reaction temperatures.

In addition, propellants based on 5-aminotetrazole and *nitrocellulose* were tested. Such react already at relatively low operating pressures, but place high demands on the design of the gas generators.

Propellants based on on *ammonium nitrate* were also developed and used, but they underwent physical changes to the propellant bodies due to initially unknown long-term effects, as a result of which the combustion became uncontrollable. This was followed by a comprehensive recall, in which approximately 67 million airbags were recalled by the end of 2022 [119, 86].

The most common shape in which the propellant is used is the *tablet* (Fig. 6.94). For this purpose, a mixture of the raw materials is ground into a fine, homogeneous powder and then pressed into tablets.

Fig. 6.94 Shapes of propellant tablets for airbag inflators (*Source*: ZF)

In other cases, the raw materials are mixed with a liquid and wet ground. This doughy mass can now be shaped into propellant bodies(also: *molded body*) in a variety of geometries through an extrusion process—not dissimilar to the production of pasta. The shape and surface can influence the performance characteristics of the propellant. With *tablets* the surface decreases significantly ("degressive combustion"), with *molded bodies* with inner holes it can remain constant ("neutral") or with *multi-hole bodies* even increase ("progressive").

Combustion chamber

The combustion chamber accommodates the propellant. It is sealed by the igniter unit and the closure foil. The combination of the burning propellant surface, the size of the outlets, and the opening pressure of the *closure foil (inflator seal)* defines the performance characteristics of the gas generator. While the combustion chamber and the filter chamber used to be different components of a pyrotechnic generator, today the entire *housing* represents the combustion chamber.

Filling material

To compensate for normal manufacturing tolerances and to ensure the safe positioning of propellant tablets, elastic components, *filling bodies (filling pad),* are installed in the combustion chamber.

This ensures that the propellant does not suffer mechanical damage due to normal vibrations during driving, which could alter the performance characteristics of the airbag.

Pressure chamber

The crucial component of *hybrid-* and cold gas gas generators is the *pressure chamber* (Fig. 6.91b). It is sealed by the *igniter unit* and the *outflow membrane* (also: *exit membrane*). The filling gas (Ar, He, H_2, O_2 or mixtures) is under a pressure of 240–600 bar. Each pressure chamber is individually tested for leak-tightness after filling and sealing to ensure the safe function of the gas generator for its lifetime (Fig. 6.95).

Depending on the design, a propellant tank may also be located within the pressure vessel. The combustion of the propellant then heats the pressurized gas in the event of an activation until the outflow membrane bursts and releases the gas to fill the airbag.

Filter

To ensure thrust neutrality during transport, the generated gas is blown out through many small, opposing openings from the inflator. Before passing these outlet openings, the gas flows through porous, mostly metallic structures, which are referred to as *filters (also: filter packages)*. They filter out *slag particles* and the gas cools down significantly on the material of the filter.

Fig. 6.95 Gas generator manufacturing. (*Source*: ZF)

	1	2	3	4	5
Features	Type	Fuel	Form factor/ installation	Adaptivity	Electrical connection
Options	Pyrotechnic	Pyrotechnic, solid propellant *(GuNi, NiGu, ...)*	Round *(toroidal, disk type)*	1-stage	Plug *pin-type*
	Hybrid	No fuel	Pipe *(tubular)*	2-stage	alternatively
	Cold gas *(no propellant)*	Fuel, gaseous *(hydrogen,...)*	Aspirator	multilevel	
	Propellant, gaseous *(hydrogen)*		Alternatively *(module-integrated)*		
	Other		Gas lance *(inflator outside module)*		

Fig. 6.96 Essential and further features of the designs of gas generators. (*Source*: ZF)

6.6.9.4 Designs of Gas Generators for Airbag Modules

The primary protective functions and the installation location result in essential requirements for airbag modules. The selection of generator technology, the variation of the form factor, and the selection of the essential components result in a portfolio of different inflator series. Fig. 6.96 summarizes the essential features and their manifestations of currently available inflator variants.

Due to a broad portfolio of gas generators, it is necessary to make the optimal selection for a specific airbag module. The other components of the airbag module are then matched to the characteristics of the gas generator. Over time, standard concepts have emerged that represent an optimal solution across many vehicle applications.

In Fig. 6.97, the most common alternatives for the selection of inflator variants for specific airbag modules are listed in a generalized configuration scheme. These preferences show the current state of application. Due to specific requirements (e.g., new load cases or new space requirements in interior design), alternative technical solutions and configurations may also arise in individual vehicle projects.

6.6.9.5 Basic Types of Inflators

The development work on alternative technologies for gas generation and basic types of inflators, as well as the essential inflator components, were already described in Sect. 6.6.9.1. In Sect. 6.6.9.3, the main aspects of gas generation and the technical differences of the inflator types that are currently used in airbag modules in motor vehicles are discussed.

Design	Gas generation technology			Form factor	
Airbag modules	**Pyro**	**Hybrid**	**Cold gas**	**Round** *Toroidal*	**Tubular** *Tube*
Frontal impact					
Driver airbag	■	○	·	■	·
Front passenger airbag	■	◘	·	■	◘
Front passenger airbag bag-in-roof	·	■	·	·	■
Knee airbag	○	■	·	·	■
Seat ramp airbag	■	·	·	·	■
Rear seat airbag	·	·	·	·	·
Side impact					
Side airbag·front	■	○	·	·	■
Side airbag·rear	■	·	·	·	■
Curtain airbag·1st Impact	○	■	○	·	■
Curtain airbag·roll-over	·	■	◘	·	■
Center airbag·front	·	■	○	·	■

Configuration scheme (generalized): ■ more frequent option, ◘ further option, ○ less frequent option, · no specification

Fig. 6.97 Airbag modules and the configuration of gas generators. (*Source*: ZF)

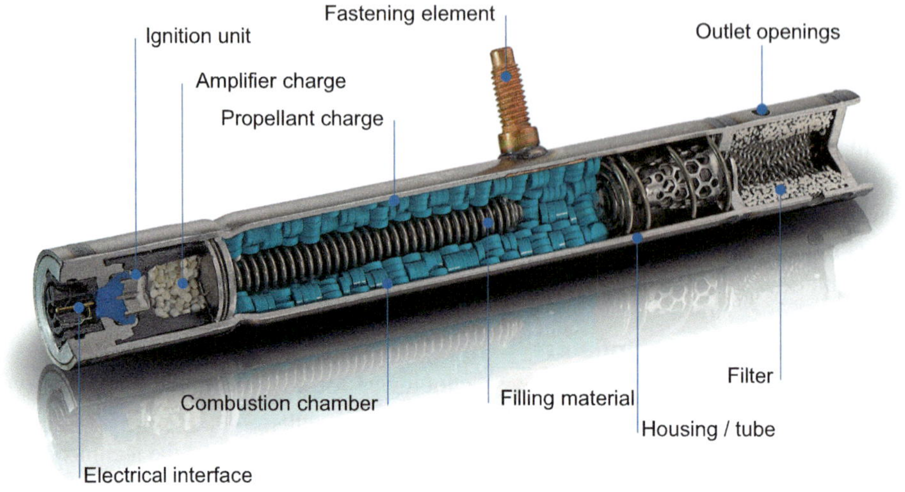

Fig. 6.98 Schematic structure of a single-stage pyrotechnic tube gas generator (tube hot gas generator *SPI2 EVO*). (*Source*: ZF)

Pyrotechnic Gas Generators

Pyrotechnic gas generators are produced as tubular gas generators (Fig. 6.98) or round gas generators (Fig. 6.93 and 6.99). The principle of operation is the same for both designs.

In an electric *igniter*, a mixture of a propellant (often a *zircon metal powder*) and an oxidizing agent (usually *potassium perchlorate*) is ignited by a glow wire within less than 1 ms. The electric igniter ignites a *booster charge* and the *propellant*. For a uniform, reproducible, and controllable ignition, the amount of this booster charge must be dimensioned in such a way that the main propellant is pressurized and the hot gases and particles effectively ignite the surface of the *propellant form* (usually *tablets*). They then generate gas during combustion, which flows into the airbag and inflates it. Before this, the combustion gases flow through *filter packages*, which reduce the temperature from approx. 1,800 K to approx. 800–900 K.

Figure 6.99 shows a dual stage pyrotechnic round inflator, which can adapt the restraint characteristics of the airbag to the size or position of the occupant or the accident severity by changing the ignition times of the two stages in both driver and front passenger airbag modules. The systems for *occupant detection* and the *crash sensors* provide relevant parameters, which are evaluated by the *airbag control unit* (Chap. 7), which then carry out the ignition of the stages.

Hybrid Gas Generators

In *hybrid gas generators*, the *propellant* is ignited directly by the *igniter*.

Depending on the inflator concept, a pressure wave (shock wave) is initiated, which runs through the gas in the pressure chamber and opens the *exit membrane*. All hybrid

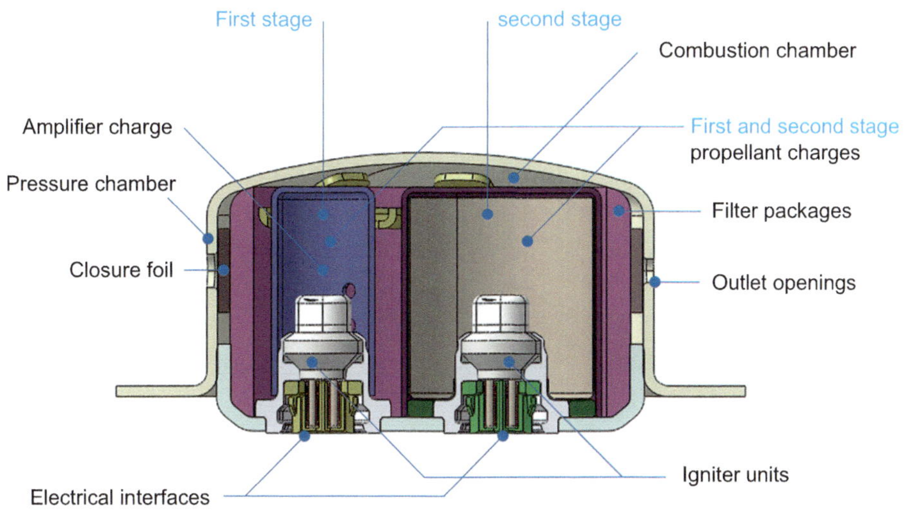

Fig. 6.99 Schematic structure of a dual stage pyrotechnic round gas generator *(DI10)*. (*Source*: ZF)

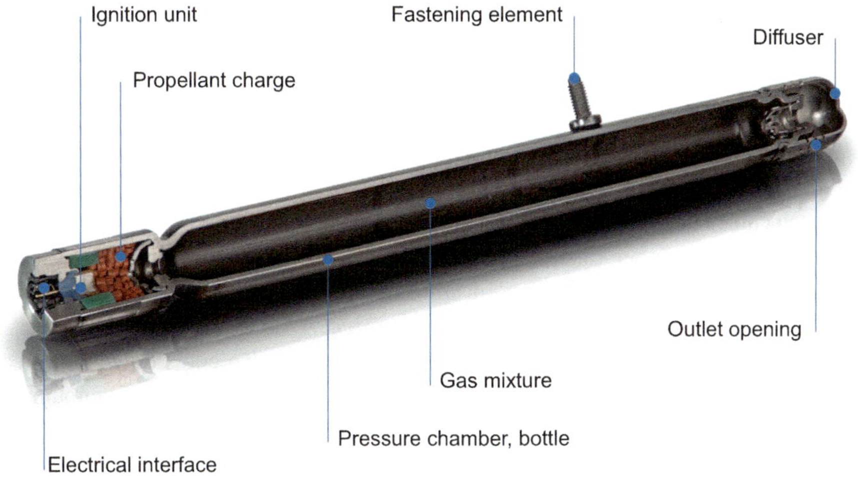

Fig. 6.100 Schematic structure of a hybrid gas generator with shock wave opening *(SHI2)*. (*Source*: ZF)

side impact protection systems have this *shock wave opening* in common, as here the inflator must deliver gas to the airbag in less than 2 ms. This enables the extremely short inflation times of the side impact protection systems. Figure 6.100 shows the schematic structure.

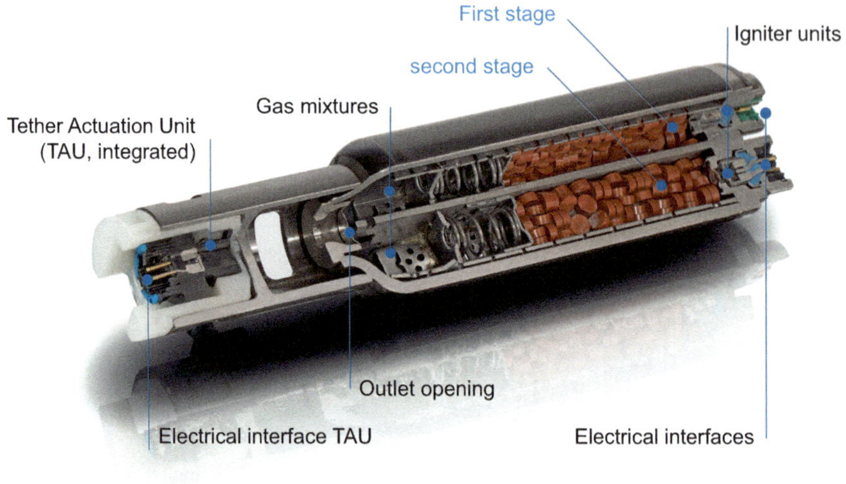

Fig. 6.101 Schematic structure of a dual stage hybrid gas generator *(PHI4 EVO)*. (*Source*: ZF)

Other hybrid gas generators operate according to the *overpressure principle*, in which the burning propellant heats the gas and opens the exit membrane when the opening pressure is reached. Fig. 6.101 shows the structure of a dual stage hybrid gas generator.

Cold Gas Generators

Cold gas generators (compressed gas inflator) are *pressurized gas container* systems, which are tightly sealed with a *membrane* (Fig. 6.102). Upon triggering, this membrane is opened by a pyrotechnic charge, usually a strong airbag igniter, which is fixed in close proximity.

The pressure chamber (also: pressurized gas container, gas storage, bottle) usually contains *helium* (He) with pressures of up to 600 bar. After opening the membrane, the pressurized gas flows out and fills the *airbag cushion*.

During this outflow, the gas expands very quickly but also cools down significantly. Therefore, airbags with cold gas inflators require a considerably higher amount of gas for inflation compared to hybrid or pyrotechnic systems, which generate gas at a higher temperature.

The use of *argon, nitrogen* or *air* as filling gas is not advantageous, as these gases cool down to undesired partial liquefaction when decompressing from 600 bar to ambient pressure.

Hydrogen Gas Generators

In the search for inflator technologies that produce as few particles and as low-pollutant gas as possible, the use of hydrogen is a possible alternative. In these gas generators, hydrogen and oxygen are burned. The result is hot steam, which can be used to inflate the airbag. The combustion is free of slag and low in pollutants. Two technologies have been developed for series production with hydrogen-powered gas generators:

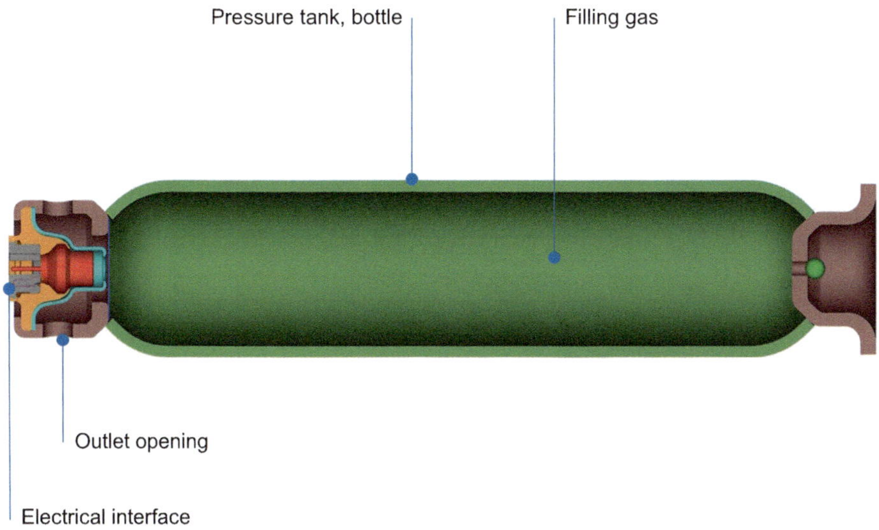

Fig. 6.102 Schematic structure of a cold gas generator *(ROI)*. (*Source*: ZF)

In the *single-bottle concept*, a mixture of hydrogen, oxygen, and argon is stored together in one container. Upon activation, it is ignited with a pyrotechnic *igniter*, causing the gases within the gas generator to react with each other. The gas emerging from the *bottle* is a mixture of steam, oxygen, and argon. The *HGI* inflator is an example of this type and is shown in Fig. 6.103.

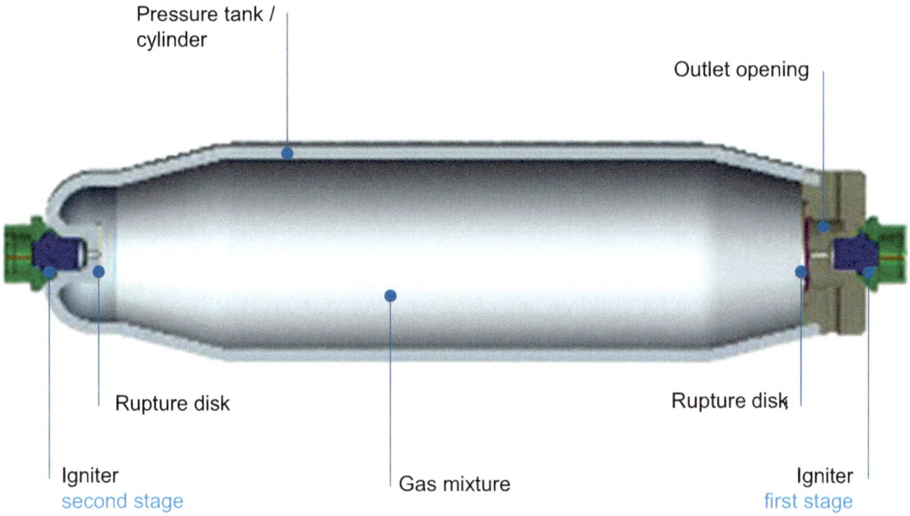

Fig. 6.103 Schematic structure of a single-chamber hydrogen inflator *(HGI)*. (*Source*: ZF)

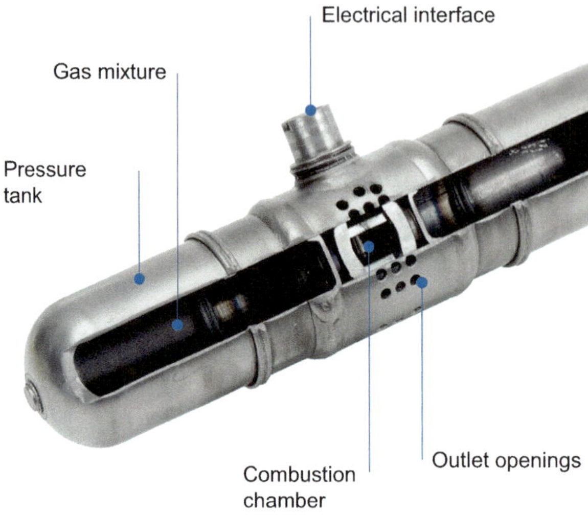

Fig. 6.104 Two-chamber hydrogen inflator *(APG, APGS)*. (*Source*: Autoliv)

In the *two-bottle concept*, one pressurized gas chamber holds a *hydrogen-argon* mixture, while the second contains an *oxygen-argon* mixture. An igniter opens the two containers simultaneously and ignites the emerging mixed gas in a central *combustion chamber*. The resulting hot steam then fills the airbag together with the non-reacting gas components [50]. An example of this type is shown in Fig. 6.104.

6.7 Child Restraint Systems and other Protection Systems for Occupants

In this section, *additional systems* with *protection functions for occupants* are presented. In addition to the vehicle body structure and the classic restraint systems such as seat belt and airbag systems, they also offer important sub-functions to reduce injury risks. The presentation follows along individual focus topics that have a special significance for occupant protection:

- Occupant protection for children and *child restraint systems* (CRS)
- Protection systems against *cervical spine distortion* (also: HWS syndrome)
- Steering systems with occupant protection functions
- Protection systems in convertible vehicles

These and other systems are located in the immediate system environment of restraint systems and interact with them. During vehicle development, various coordination processes take place to synchronize the systems and optimize their protective effect (Chap. 8).

6.7.1 Occupant Protection for Children and Child Restraint Systems

Child occupant protection is a special field of work. A child's body is still growing. In addition, a child's head is larger and heavier in relation to the total body weight than in adults [17, 74] (Fig. 6.105), which presents a particular injury risk. An occupant protection system designed only for adults would not be suitable for younger passengers. Geometric proportions suitable for children in the vehicle (e.g., anchorage points for seat belts and belt routing) differ significantly from adults, so usage instructions must be observed. When designing occupant protection systems for adults, the goal is that they do not pose an additional injury risk for children.

Legal and other requirements for child occupant protection are directed at child restraint systems. They also include requirements for vehicle manufacturers regarding the effect of occupant protection systems on children in a collision, as well as requirements for handling when installing child restraint systems in vehicles and for their manufacture. In addition, regulations apply for their use while driving.

In its early stages, the development of occupant protection concepts was focused on adults. However, assessment programs for new vehicles (*New Car Assessment Program,*

Fig. 6.105 Body proportions of occupants and their development from baby to adult. (based on Volvo Cars, [17])

NCAP) now contain specific requirements for child occupant protection (Chap. 4). Automobile clubs and other organizations regularly test and evaluate child restraint systems. At Euro NCAP, criteria in three areas are relevant [35]:

- Child restraint systems for installation in vehicles *(crash performance)*
- Vehicle safety functions *(safety features)*
- Installation of child restraint systems in vehicles *(CRS installation check)*

The special requirements for child occupant protection will be highlighted in the following section for different accident scenarios. The associated measures and systems will then be detailed further along the three areas listed.

The increased safety requirements in the evaluation of child safety through consumer ratings (for example, newly included dummies Q6 and Q10 from 2016 by Euro NCAP) were already pointed out in Chap. 4.

6.7.1.1 Challenges in Child Occupant Protection

Occupant protection for children, like for adults, should be ensured for different accident scenarios such as frontal, side, rear impact, rollover, and off-the-road accidents. Furthermore, the occupant protection concept for children needs to be integrated into the protection concept for adults.

For seat belt systems protecting children, a proper belt routing is particularly relevant. Seat belt systems must, as with adults, fulfill the function that the children remain as much as possible in their position and are not thrown out of the vehicle. Child restraint systems raise the seating position of children, thus enabling an adequate belt routing. In particular, child seats for older children may have additional guides for the webbing of the vehicle seat belt. Child restraint systems for younger children usually have an integrated belt system.

In the statistically most common situation of a frontal impact, due to the mentioned biomechanical conditions in children, loads on the head and neck should be avoided as much as possible. Therefore, a positioning against the direction of travel appears advantageous, provided the space conditions and the size of the child allow it. For larger and then forward-facing children, *Belt pretensioners* and adjusted *load limiters* contribute to reducing loads and reducing the risk of *submarining*.

In side impact situations, the different geometric conditions are also relevant. Both *side airbags* (SAB) and *curtain airbags* (CAB) primarily fulfill protective functions for adults. Therefore, child seats or booster seats first ensure that children are positioned higher to side and head airbags. Furthermore, child restraint systems can have additional protective elements against side impact. If several occupants sit next to each other at the same time, this also reduces risks for direct contact of the occupants in the event of an impact.

Child restraint systems only offer a protective function if they can be securely installed in the vehicle. While the systems used to be attached solely with the three-point seat belt, additional devices such as the *top thether* and especially the ISOFIX system are available today. Concepts for anchoring child restraint systems are described further below.

One aspect in terms of integral safety is the interaction with the child *during* the journey. This is often a motivation for placing the child seat on the front passenger seat. However, the risk of driver distraction may rise so that the traffic situation can no longer be observed, the risk of an accident increases significantly. If the view through the rearview mirror to the children in the back row is obscured, additional mirrors can be mounted, for example, on the rear headrest. Also, *interior monitoring systems* can be used for this purpose.

Due to the extensive safety equipment of child restraint systems, they have a significant volume and weight. This makes installation in the vehicle or daily transition between different vehicles difficult. Therefore, there is a permanent conflict of goals with child restraint systems, as integrated child seats have not been established for reasons such as seating comfort. A general goal is a high interoperability between the vehicles, their child seat detection systems and the child restraint systems, however, the compatibility between the sensors in the vehicle and the requirements for the child seat must be observed (Chap. 7) or airbag deactivations must be done manually.

6.7.1.2 Child Restraint Systems

It has been shown that child restraint systems are limited to a narrow age or weight range, as the requirements change rapidly with age and growth. Therefore, various classification systems have been defined for child restraint systems. They are based on the *weight* of the child in the UN-R 44 standard and groups from 0 to 3 are defined [137]. In the UN-R 129 standard, which has been in parallel use for several years, a classification according to *body size with weight limit* was made in three phases (Phase 1 to 3) [138].

Child restraint systems for installation in vehicles available on the market essentially comprise the following product categories (Fig. 6.106) (individual deviations possible with specific seats, as manufacturers decide on the permissible body sizes):

- *Infant tubs* and *infant carriers:* Suitable for an age of 0 to 18 months, a size of 40 to 85 cm, and a weight up to 13 kg. According to the i-Size standard babies should be transported exclusively with their backs facing the direction of travel until the 15th month of life to reduce loads in a frontal impact, which is the most common.
- *Toddler car seats*: Intended for an age between 9 months and 4 years, for a size of 60–105 cm, and a weight between 9 and 18 kg. These child seats have an integrated belt system.
- *Child seats:* for children suitable for an age between 3 and 12 years, for a size of 100–150 cm, and a weight between 15 and 36 kg. These child seats are usually used in combination with the vehicle's seat belt system.
- *Booster seats*: These products are only suitable for a size over 1.25 m and a weight from 22 kg. They only have a seat surface, but no backrest, side padding, and headrest. Although they are also available with ISOFIX system, they are often only fixed

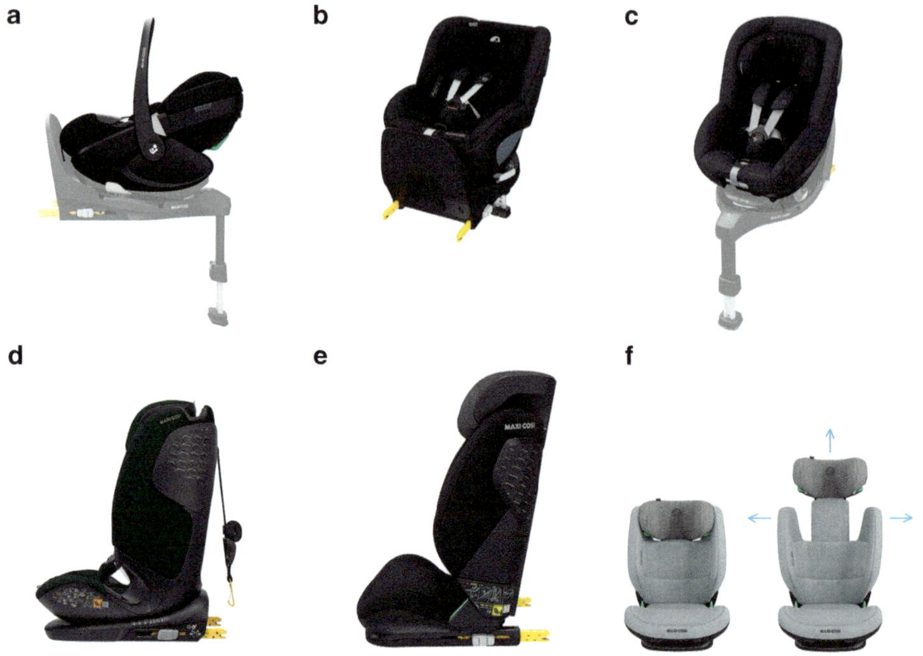

Fig. 6.106 Child restraint systems for different age, size, and weight classes. (*Source*: Dorel Juvenile—Maxi-Cosi)

with the lap belt. Despite their lower protection potential, they are recommended for short trips or even for children who are younger than 12 years and smaller than 150 cm with a weight of over 36 kg and thus still fall under the statutory child seat requirement.

Since the purchase of child restraint systems is associated with considerable costs, certain child seat types are also approved for a broader range of age, size, or weight classes.

Child seats must allow children a certain freedom of movement, even if they are equipped with a seat belt. This is especially true for longer journeys. Child seats for toddlers with their own harness system offer a high level of safety. However, the restraint of the thorax can cause a high load on the cervical spine due to an excessively large relative movement between the head and torso. When using the vehicle's seat belt system for child seats, it is important to ensure that the three-point belt fits optimally on the body, i.e., the lap belt must engage low on the pelvis and the shoulder belt must run diagonally across the chest. For this purpose, child seats, unlike pure booster seats, have a separate webbing guide (Fig. 6.106).

In a child restraint system, the children are either secured by their own harness system built into the seat (Sect. 6.5), or the vehicle's belt system is used. The former concept is often used for babies and toddlers and the latter more for child seats. When buckling up,

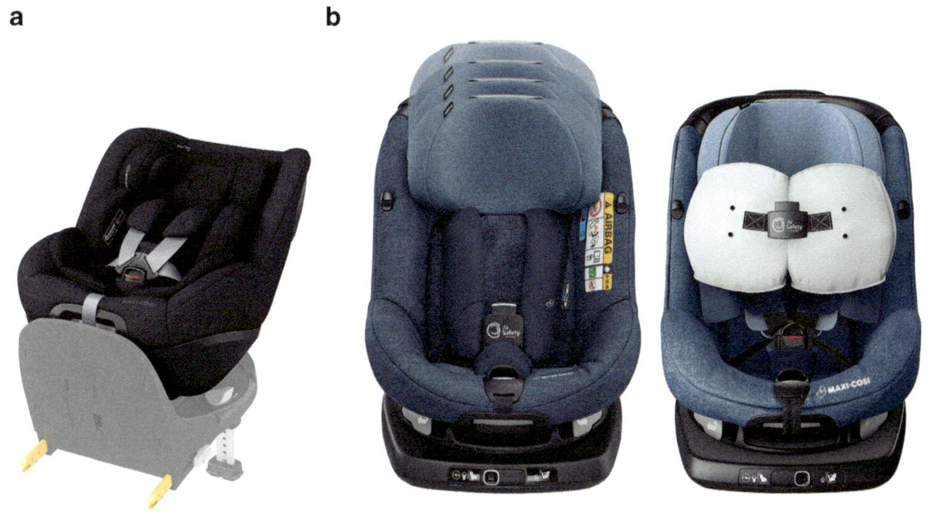

Fig. 6.107 Reduction of load values in frontal impact by **a** rear-facing child restraint systems and **b** child seat-integrated airbag system. (*Source*: Dorel Juvenile—Maxi-Cosi)

it is important to pay attention to a tight and correct webbing routing, otherwise either the restraint function cannot be ensured or injury risks arise from the belt cutting into body parts. Careful buckling helps to reduce injury risks from submarining (Sect. 6.5).

Child restraint systems that are mounted against the direction of travel are *(rear-facing child restraint systems),* offer advantages in frontal collisions (Fig. 6.107a), as the body loads are introduced over a large area and evenly distributed, and there are no body parts that are highly stressed due to relative movements compared to others. *Infant carriers,* that comply with i-Size, prescribe this rearward orientation exclusively until the 15th month of life.

Another approach is the additional installation of an airbag system in a child seat, which contains autonomous crash sensors and airbag trigger sensors in addition to the two airbag modules (Fig. 6.107b). Tightly installed with vehicle anchorages of the ISOFIX system (Sect. 6.7.1.3) mounted on the vehicle body structure, the collision can be detected and the airbags can be deployed before there is a forward displacement of the child. The loads on the chest and neck are reduced as the head can be supported on the airbags.

For protection in side impact events, side cushions and headrests with side guidance are installed in the child restraint systems to allow the torso and head to be supported. In this case, the interaction with the vehicle's side impact protection system such as side airbag or head airbag may be important.

Integrated child seat systems are permanently housed in the backrest or in the seat cushion of the rear seat assembly (Fig. 6.108 and 6.109). Such folding seats are practical because they are always available, space-saving and usually easy to use.

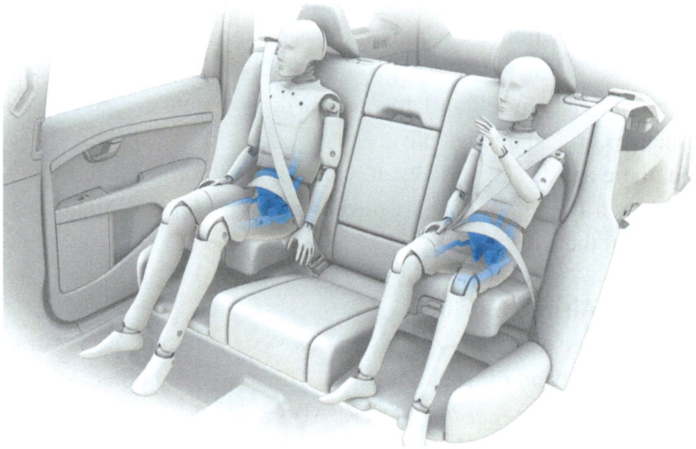

Fig. 6.108 Vehicle integration of child restraint systems. (*Source*: Volvo Cars)

Fig. 6.109 Seat-integrated child restraint system. (*Source*: Volvo Cars)

Nevertheless, these solutions have so far also had limitations, as integrated child seats are usually only approved for children from 15 kg. On the seating positions equipped with integrated child seats, adults tend to perceive comfort losses and it means additional weight for the vehicle manufacturer. Therefore, this equipment is only available as special equipment for individual vehicles and requires an individual purchase decision.

6.7.1.3 Vehicle Equipment for the Occupant Protection of Children

Child restraint systems only provide a protective function if they are securely installed in the vehicle. The challenge for vehicle manufacturers and manufacturers of child restraint systems lies in the variety of vehicles, child seats and different customer needs.

While child seats used to be secured solely by the three-point belt, specific concepts for child seat mounts such as the ISOFIX system have now become established. They simplify the otherwise often difficult handling, which can result in injury risks if not used properly.

Securing through the seat belt system: The securing of child restraint systems with the seat belt can be supported by the automatic locking retractor (ALR) of the seat belt retractor (Sect. 6.5). This is a requirement in certain regions. Here, the webbing is fully extended once, so that the locking function against belt webbing output is mechanically activated. The webbing can now lock the child seat for the journey.

An additional anchoring point is offered with the top tether. This anchoring provides significant protection against tipping of the child seat in a collision for both forward and rearward facing child seats, and particularly reduces the injury risk to the head. The top tether is stretched between the child seat and an anchoring point on the rear seat assembly or behind it on the vehicle body structure.

ISOFIX child seat mounting: A significant step was achieved in 1999 with the establishment of the ISOFIX system [54]. The standard defines the interface for both vehicles and child restraint systems and is regulated by law in many regions worldwide. Regional variations for the system are *universal child safety seat system* (UCSSS), *lower universal anchorage system* (LUAS), *lower anchors and tethers for children* (LATCH, USA) or *Canfix* (Canada). The regulations UN-R 14, UN-R 44 and UN-R 129 also contain regulations for the ISOFIX system [130, 137, 138].

ehicle anchorages for the ISOFIX system have been mandatory at the two outer seat positions of the rear seat row in Europe since 2014. Some vehicles have ISOFIX vehicle anchorages at further seat positions. The child seats are equipped with a quick release with signaling of the locking status. This means that the child restraint system is firmly connected to the vehicle at two points. A third anchoring point can be realized either via a support down to the vehicle floor or a *Top Tether*.

A secondary function of the ISOFIX system is the secure anchoring of the seat when no child is in it. This prevents injury risks in a collision from a child seat flying through the passenger compartment in the event of a crash.

Disadvantages of the ISOFIX system are the higher weight and often higher purchase price of the child restraint system.

Installation checkup: The assessment of vehicle safety equipment by NCAP includes an installation checkup of child restraint systems. Essential criteria are the geometric conditions for installing child seats, the length of the seat belt, the belt buckle position, access to ISOFIX vehicle anchorages, and the stability of the child restraint system installed in the vehicle [35]. For this purpose, several different child restraint systems available on the market were selected in advance, which are part of the tests. Suitable seating positions for the installation of child restraint systems must be marked.

Airbag deactivation: When child seats are placed on the front passenger seat, appropriate precautions must be taken to ensure that the triggering of the *passenger airbag* does not create an additional injury risk for the child [42]. The deactivation of the *passenger airbag* is carried out according to the vehicle's operating instructions. Typically, when installing rear-facing child seats, the airbag must be manually deactivated by a switch on the cockpit or automatically using sensors in the seat surface.

If alternative sensors such as camera-based interior monitoring systems were to be used in the future, injury risks due to incorrect operation could be reduced.

Disadvantages of previous automatic *airbag deactivation systems* are the often required specific approvals of child restraint systems by the vehicle manufacturers. The associated lists are stored in the vehicle documentation.

The results of a previous study on child restraint systems noted that usability and the prevention of potential misuse are relevant conditions for all further developments. The use of child restraint systems must remain understandable for consumers [14].

Occupant protection for children and child restraint systems

Numerous organizations are dedicated to the field of occupant protection for children, child restraint systems, and other topics related to the protection of children in road traffic. The following overview is a selection of further information and sources:

- ADAC—www.adac.de—Child safety
- ÖAMTC—www.oeamtc.at—Child safety
- ADAC Foundation—stiftung.adac.de—Traffic Heroes
- Euro NCAP—www.euroncap.com—Occupant protection for children—Expert Group Child Safety
- Federal Highway and Transport Research Institute (BASt), Germany—www.bast.de—Child safety
- Deutsche Verkehrswacht (DVW)—deutsche-verkehrswacht.de—Safety for Children
- UN ECE Vehicle Regulations—unece.org/transport/vehicle-regulations—World Forum for Harmonization of Vehicle Regulations (WP.29)—Working Party on Passive Safety (GRSP)—Informal Working Group (IWG) Child Restraints Systems (CRS)
- Accident research of the Insurers (UDV)—www.udv.de—Child safety
- ISO—www.iso.org—ISO Standard 13216—ISOFIX child seats for cars
- SAE International—www.sae.org—Children's Restraint Systems Committee
- IIHS—www.iihs.org—Child Safety
- NHTSA—www.nhtsa.gov—Child Safety

- Transport Canada—tc.canada.ca—Child Car Seat Safety
- Children's Hospital of Philadelphia—www.chop.edu—Center for Injury Research and Prevention—Child Safety

6.7.2 Protective Measures against Cervical Spine Distortion

This section will discuss the challenges of *cervical spine injuries* (CSD or whiplash), which are a particular requirement and a crucial occupant protection function of seating systems. Whiplash (also known as whip-lash phenomenon, cervical spine acceleration trauma, cervical spine distortion) can occur due to unwanted accelerations of the head during an accident occurrence.

The headrests attached to the seat back are a protective measure to prevent cervical spine injuries both when the upper body is thrown back in the event of a frontal impact (*rebound effect,* Sect. 6.2) as well as in the event of a rear-end collision with another vehicle.

Cervical spine injuries are responsible for a large proportion of accident-related costs (see also Sect. 4.6). They are also a common cause of impaired well-being after an accident. Due to the high costs of settling pain and suffering claims by insurance companies, the question of the constructive design and geometric arrangement of headrests gained importance in the 1990s.

Adjustable headrests: Studies in the 1990s found that manually adjustable headrests pose a risk as users often do not adjust the headrest correctly or do any adjustment at all [61]. It was also found that, in a large number of variable headrests, the locking mechanism did not withstand a load from the head, thus posing an additional injury risk.

From the evaluation of the geometric conditions in the head-headrest relation, it follows that headrests can only be considered safe if the vertical distance from the top edge of the headrest to the top of the head is 0 to 60 mm and the horizontal distance between the back of the head and the front edge of the headrest is 0 to 70 mm [94]. Manually or electrically adjustable headrests are often used to fulfill positioning requirements.

Headrests with protection against cervical spine distortion: Figure 6.110 shows a solution of headrests with *anti-whiplash* function. The first *active headrest* (also: *crash-active headrest*) was installed on front seats from 1997 as *Saab Active Head Restraint* (SAHR) by Lear as a supplier for Saab. In an accident, the occupant is pressed against the backrest due to the initiated rear-end impulse and thereby operates the mechanism. As a result, the headrest moves forward in fractions of a second. However, this mechanical system has since been replaced by other concepts.

In alternative systems, instead of mechanical activation, the headrest is activated electronically by the airbag control unit in the event of a collision. If a collision with

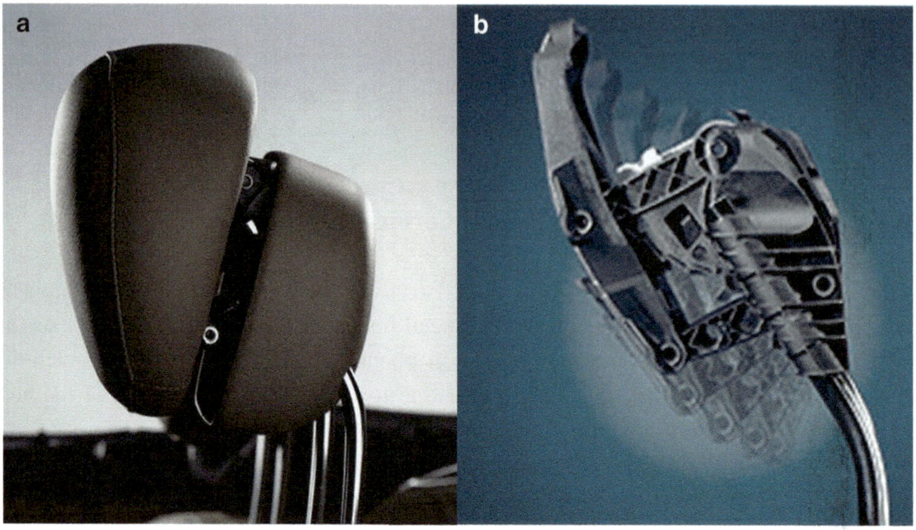

Fig. 6.110 Example of a headrest with anti-whiplash function. (*Source*: Mercedes-Benz, atzonline.de, [61])

a certain impact severity is detected by the crash sensors, the headrest is activated and pushed both forward and upward (Fig. 6.110). In such a system, the headrest can be pushed forward by about 40 mm and upward by 30 mm in less than 50 ms using pre-tensioned springs (Kramer 2013). This is intended to ensure early support of the head and prevent the risk of a cervical spine distortion in the event of a rear-end collision. Other concepts use a pyrotechnic system to activate the headrest.

Seat-Backrest with Protective Function against Cervical Spine Distortion

To reduce the risk of cervical spine injuries, especially in rear-end collisions, Volvo introduced a system for resetting the backrest known as the *whiplash protection system* (WHIPS) (from 1998). The technology includes a stencil gear in the backrest adjustment, into which a deformation element is integrated. In the event of a severe rear-end collision, a pin sheares off and then the deformation element is deformed with energy expenditure. As a result, the backrest moves backwards, then performs a rotational movement of up to 25°, and ensures a stable holding function in the event of an even stronger load. Fig. 6.111 shows the structure of this system and the kinematics of the occupant.

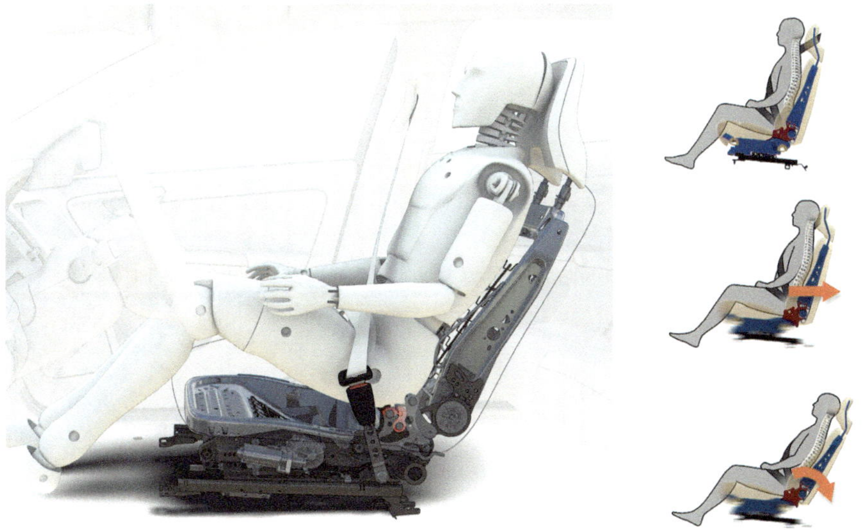

Fig. 6.111 Schematic representation of the *whiplash protection system (WHIPS)* for influencing occupant kinematics in rear-end collision. (*Source*: Volvo Cars)

6.7.3 Mechanical System for Frontal Impact Protection

From 1986, Audi relied on the system *procon-ten* (*programmed contraction and tension*), which was intensively advertised as a vehicle safety function. It thus became known to the general public and is mentioned here to illustrate stages of development for safety systems in the past.

An obvious solution to reduce the injury risk in *frontal impact* was to pull the steering wheel forward out of the contact area. The repositioning of the engine in a collision tightens cables to pull away the steering column and the steering wheel. This expands the possible *forward displacement space* in front of the driver. At the same time, the webbing of the seat belt was tightened for better *coupling* and *restraint* of the occupants on the front seats [61].

The mechanical system was then overtaken by the rapid developments in airbags and seat belt pretensioners.

6.7.4 Steering Systems with Occupant Protection Functions

Steering systems provide important occupant protection functions to reduce injury risks in vehicle collisions. The prevention of intrusions into the passenger compartment, the preservation of the forward displacement space for the occupant, and contributions to energy conversion are sub-functions of holistic occupant protection concepts (Sect. 6.2),

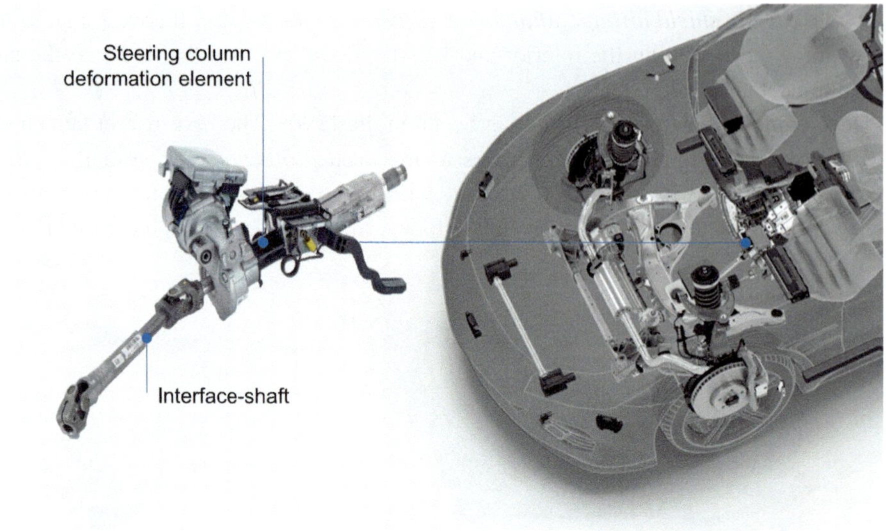

Fig. 6.112 Protection functions of the steering system *(ZF CD-EPS)* through deformable steering column. (*Source*: ZF)

which are realized by various components of steering systems. In addition, injury risks from components or controls of the steering system inside the passenger compartment should be avoided. Figure 6.112 shows their arrangement as an example in a *column-drive*-steering system.

In the event of deformations of the front-end in a frontal impact, injury risks due to penetration of the steering column and the attached mounted steering wheel into the passenger compartment are prevented by a suitable design of the *intermediate steering shaft*, for example, by allowing it to slide into each other.

Furthermore, in Fig. 6.112 the *deformation elements* of a *steering column* are shown. In case of a load from the deployed airbag by the driver and force introduction via the steering wheel and the steering column, they contribute to energy conversion through deformation. This results in a compression of the steering column by typically 80 to 100 mm, so that the *forward displacement* for the occupant can be extended [22]. As a result, loads and therefore injury risks for occupants can be reduced.

In conjunction with the seat belt system and the driver airbag module, *deformation elements* in the *steering column* contribute to *energy conversion*. For this purpose, a coordination of the restraint function between the steering system and the steering wheel system with the driver airbag module as well as the seat belt and seat system is required. In the vehicle development process, several engineering departments need to be synchronized for this purpose (Chap. 8).

As an extension of the adaptivity of the occupant protection system or an adaptation to requirements of certain markets, switchable *deformation elements* can be used.

This allows the protection function to be adapted, for example, to the mass or size of the occupant [66].

Also in *steer-by-wire steering systems* (SbW), these protection functions are taken into account. With deformation elements in the *hand wheel actuator,* energy conversion as part of the restraint concept is also enabled.

6.7.5 Protection Systems for Convertibles

Vehicles with an open-top passenger compartment, *convertibles* (cabriolets) pose a particular challenge for occupant protection, as there are increased injury risks in several accident scenarios compared to vehicles with a conventional body structure. Special protective measures and systems for convertibles are intended to counteract these disadvantages.

Currently, the market offering of convertibles has further decreased compared to the 1990s and 2000s. A study on US traffic accident statistics for the period 2014–18 has shown that differences in accident frequencies as well as fatalities between convertibles and other cars—in contrast to the proportion of ejected occupants—are not significant [46]. It is suspected that the structural disadvantages of convertibles are offset by more favorable weather or traffic conditions during the typical usage times of convertibles.

Compared to the conventional body structure version of the vehicle, additional measures for protection in case of rollover as well as frontal, side, and rear impact are used. In addition, in convertibles, due to different attachment points and installation spaces, adjustments to the restraint systems such as seat belts and airbags may be necessary. These challenges will be further explained.

Challenges in securing survival space in collision: Firstly, the vehicle body structure in the area of the floor group is designed to be stronger than in sedans, so that sufficient stiffness can be achieved in the event of a frontal or rear collision. In the case of a lateral collision, the lack of support from the roof structure must be compensated in the floor group of the convertible. The steel roofs that have been frequently applied in the meantime hardly contributed to increased vehicle stiffness.

Challenges in securing survival space in rollover accidents: In the event of a rollover, the survival space must be particularly secured both at the front in the area of the A-pillar and in the rear. The front area is significantly reinforced by an adapted *A-pillar* and the *windshield frame*. Here, a collapse of the A-pillar must be prevented in the event of a rollover.

In the rear, *roll bars* provide additional protection. In the past, they were often rigidly installed and they covered the entire vehicle width. However, they did not meet the demands on the design of the vehicle silhouette and the customer experience. Therefore, roll bars that are triggered in the event of a collision or rollover are often used in the rear today (Fig. 6.113). In the resting state, they are hidden in the car body and form a closed surface with the convertible top box cover.

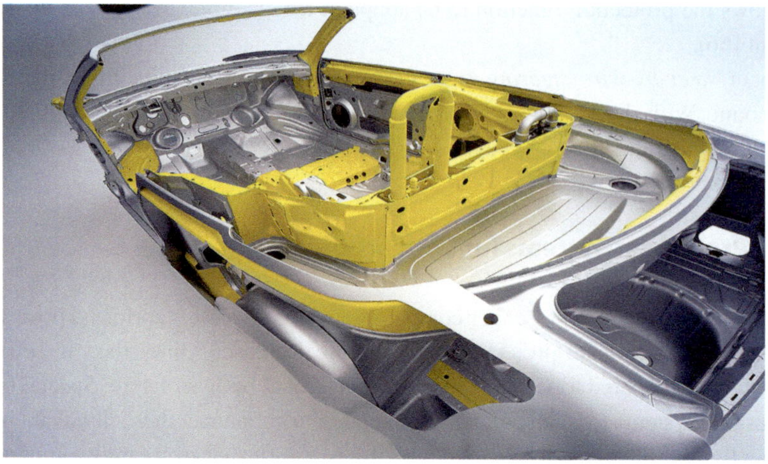

Fig. 6.113 Protection systems for convertibles: Rear headrests with protective function in case of rollover. (*Source*: Volvo Cars)

In addition, solutions are in use where the rear headrests simultaneously take on the function of the roll bar and help to ensure the survival space.

In vehicle tests, however, it is often pointed out that the protective measures on both the front seats (very flat and very far retracted windshield) and the rear seats (short roll bars) do not appear sufficient, especially for tall occupants [61].

Challenges for the seat belt system: Due to the missing B-pillar above the window line, the usual installation location for the D-ring (pillar loop) of the three-point seat belt system or the height adjuster is omitted. The seat belt is often led further back and at a lower position, which is disadvantageous especially for larger occupants.

Challenges for the airbag system: A replacement must be found for the usual installation location of the *curtain airbags* (CAB) in the roof frame. Therefore, the installation of airbags for protection in the event of a side impact and rollover often takes place in the area of the door sill. The deployment and unfolding of the CAB occur from bottom to top, so that the stability of the deployed airbag must be particularly taken into account.

6.7.6 Protective Effects of Glazing

Among the systems relevant for occupant protection, one would hardly think of the glazing installed in the vehicle. Nevertheless, the windshield, rear window, and side windows are safety-relevant systems and the glass for installation in the vehicle has special properties. The glazing also has a significant influence on the function of airbag systems. The proportion of the glazed vehicle surface has increased significantly in recent years.

The glazing primarily provides protection against external influences while offering a flawless view from the interior to the environment (transparency). It protects from environmental influences such as rain and wind while driving, as well as from sunlight (UV protection, IR protection), insulates against heat and cold, and reduces wind noise and traffic noise (noise protection). It serves as theft protection, is a design element through coloration, and is combined with other facilities such as antennas, heating elements, lights, or environmental sensors. Windshield glazing systems are often part of the design of the vehicle body structure today. They are firmly bonded with the vehicle body structure, thus increasing its stiffness.

6.7.6.1 Protective Functions of Glazing Systems

The glazing is part of the vehicle safety concept and offers several protective functions in the event of an accident:

Support surfaces for the deployment of airbags: The glazing can be used as a support surface for airbags during their deployment (Sect. 6.6). This ensures a safe positioning of the airbag and increased retention effect for the occupants. At the same time, however, it must be ensured that the glazing is not already stressed to the point of breaking during the deployment of the airbag. As measures for weight reduction also decrease the load capacity of the glazing, this remains a constant field of work in the development of occupant protection systems.

Safety against ejection of occupants: The goal of seat belts and airbag systems is the restraint of the occupants at their seating position. Here, their protective functions can work in combination. In rare cases, specific constellations may cause the occupants to be at risk of being ejected from the vehicle. The glazing as a mechanical barrier can prevent this in such a case. Glass and panoramic roofs are becoming increasingly important in this regard.

Protective function against the penetration of external objects: Should there be stone chipping from vehicles ahead or a collision with wild animals or other objects (which, for example, were thrown down from bridges) during the drive, the glazing offers the occupants significant protection against penetration into the interior. Resistance to stone chipping is a key development goal of glazing systems.

Protective function for vulnerable road users (VRU): In the event of a collision of VRUs such as pedestrians, cyclists, or motorcyclists with a vehicle, the glazing can contribute to energy absorption. Load values upon impact are demonstrated in the development of glazing systems through tests (Sect. 6.8).

Criteria for glazing systems in vehicles are defined, for example, in the following standards:

- GTR No. 6 Safety Glazing Materials for Motor Vehicles and Motor Vehicle Equipment [139]
- UN-R 43 Safety Glazing Materials [140]
- FMVSS No. 205 Glazing Materials [141]

Conventional glass led to a high potential for injury from cuts or eye injuries in the past due to the absence of seat belt and airbag systems, causing the occupants to collide with the window during an accident and injure themselves on the broken glass. The installation of both restraint systems and *safety glass* has significantly reduced these risks.

6.7.6.2 Safety Glass

To reduce the risk of cuts and eye injuries, glass for installation in motor vehicles is hardened. This makes it firstly more resistant and secondly, it shatters into small pieces in the event of a strong impact, which do not cause such injuries. Safety glass was already patented in 1929, and its use in motor vehicles began in the 1930s.

Based on the type of construction, a distinction is made between *single-pane safety glass* (ESG), which is only used in today's vehicle construction for side windows and partially for small sunroofs, and *laminated safety glass* (VSG), which can fulfill additional functions (Fig. 6.114).

Glazing made of *single-pane safety glass* (also: *tempered glass, toughened glass*) consists of a single-layer pre-stressed glass. Rapid cooling creates a tensile stress in the core

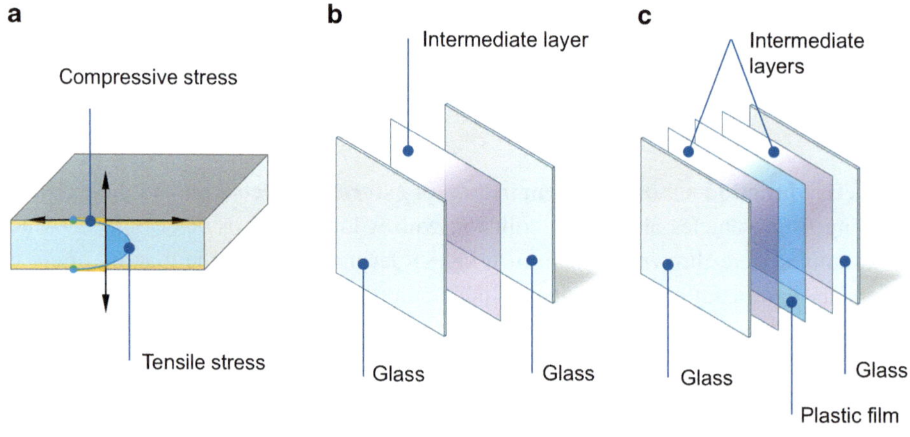

Fig. 6.114 Construction of safety glass **a** Single-pane safety glass (ESG) **b, c** Laminated safety glass (VSG). (Based on [61])

of the pane and a compressive stress on the surface. If the safety glass is highly stressed locally, the internal tension created by heat treatment is lost and the pane disintegrates into small blunt-edged crumbs. The greater the pre-stress of the pane, the smaller the crumbs after the break. Typical material thicknesses are 3–5 mm and the material is still commonly used today for side or rear windows and small sunroofs. Single-pane safety glass can also be provided with an additional plastic film on the inside.

Glazing made of *laminated safety glass* usually consists of a composite of at least three layers: glass—interlayer—glass. This technology was already developed by Edouard Benedictus in 1909. In vehicle construction, the *interlayer* is a film made of *polyvinyl butyral* (PVB). Since *laminated safety glass* does not use tempered glass like ESG, its breaking strength is significantly lower due to the lack of pre-stress. However, a higher energy expenditure is required to break through a VSG pane, which increases burglary protection in addition to the protective function in accidents.

Laminated safety glass is legally required for windshields and is also used for rear and side windows as well as roof glazing.

With these panes, the first cracks in the glass occur even at low loads. The crack pattern after an impact is star-shaped or spider-web-like and is essentially limited to the immediate vicinity of the load point. After the glass breaks, further energy is absorbed by the elasticity of the film. Only when there is a large bulge do cracks appear in the film and the VSG pane can be broken through (Fig. 6.115). The strength properties of the film are dependent on moisture and temperature, it tears more easily when wet than when dry. Its energy absorption capacity is highest at 20 to 23 °C. Outside this range, the VSG pane has a lower impact resistance.

The safety glass can be provided with an additional plastic film on the inside, so that even in the event of a strong impact, the splintering effect is reduced by the inner glass pane. This *laminated safety glass* still has load-bearing capacity and thus protective effect even when partially damaged. Other advantages include higher impact resistance, better behavior in terms of abrasion (wear) and scratching, and better resistance especially when exposed to combined effects of temperature, humidity, and solar radiation as well as temperature changes. The four-layer structure fulfills the following sub-functions:

- Outer glass pane: mechanical stability and protection against abrasion
- Interlayer: Energy absorption and elasticity
- Inner glass layer: Protection of the interlayer
- Inner protective film: Protection against cuts

Figure 6.114c shows a five-layer structure that possesses additional properties for protection against heating by the sun's infrared radiation. This is increasingly relevant for electric vehicles to increase their range. The integrity of the glazing system in interaction with airbag systems must be ensured in the vehicle's development process (Sect. 6.6).

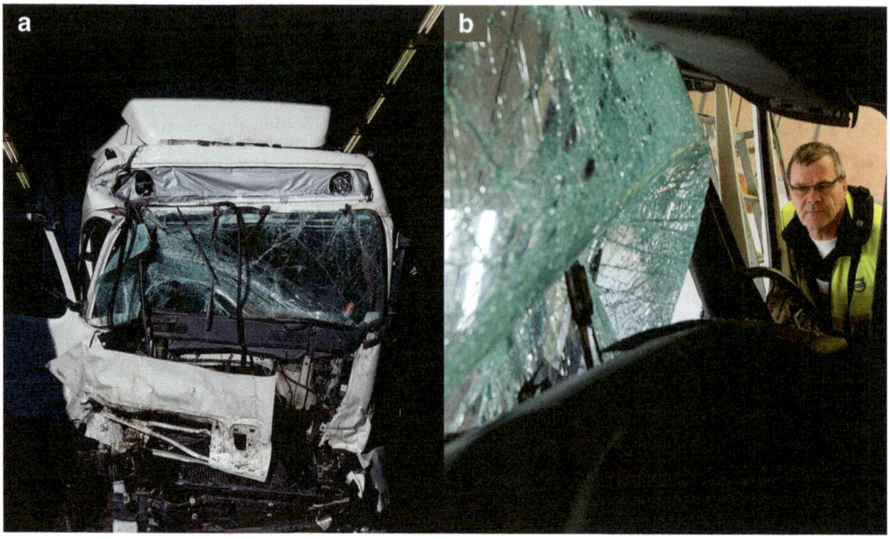

Fig. 6.115 Fracture patterns of safety glass in accident analysis. (*Source*: Volvo Trucks)

In addition to the material glass, fixed windows are also made exclusively of plastic (e.g., *polycarbonates,* PC). They are used in side windows, rear windows, or roof systems. The advantages over tempered safety glass are weight reduction and increased burglary protection.

6.7.6.3 System Interface between Glazing and Airbag Systems

Technical requirements for windshields or front windows are higher than the requirements for all other windows in the vehicle. The front passenger airbag, which is now used in almost every vehicle, can only fulfill its function by supporting a windshield firmly attached to the vehicle (usually glued). The vehicle deceleration caused by an accident accelerates the occupant towards the windshield and instrument panel. The occupant dives into the deployed airbag through this forward displacement and pushes it towards the windshield and instrument panel. At the same time, the airbag cushion is compressed by this.

The airbag is supposed to absorb the load initiated by the occupant. To do this, the airbag cushion must support itself on several parts of the vehicle interior, such as the windshield and the instrument panel. Thus, the airbag can absorb the force introduced by the occupant, absorb it, and effectively decelerate the occupant [1] (Sect. 6.6.6).

6.8 Protection of Vulnerable Road Users

The term vulnerable road user (VRU) includes several groups of road users. Unlike vehicle occupants, they are not protected by a closed vehicle interior (passenger compartment) and are therefore exposed to a particular injury risk (Fig. 6.116). VRUs are particularly pedestrians or cyclists, but also other groups with specific characteristics, which are further explained below.

This section describes both measures for cars as partner protection and the various measures as self-protection of vulnerable road users. Individual aspects of the protection of VRUs in commercial vehicles are also addressed in Sect. 6.9.

The following groups are to be summarized here under *vulnerable road user* (if they are not occupants of vehicles):

- Pedestrians
- Children
- Elderly road users
- People with disabilities
- Road users with scooters, e-scooters, kick scooters, kickboards, skateboards, roller skates, inline skates
- Wheelchair user (WCU)

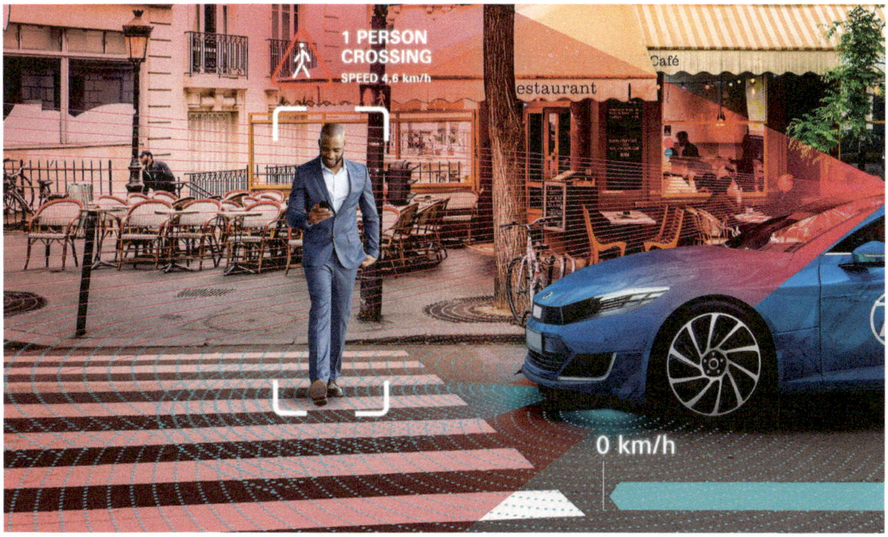

Fig. 6.116 Protection of vulnerable road users as partner protection by cars and commercial vehicles and as self-protection for micromobility. (*Source*: ZF)

- Cyclists, also with pedal assistance up to 25 km/h by auxiliary motor (*pedelec*), and drivers of *e-bikes* (fast pedelecs or *speed-pedelecs*) with pedal assistance up to 45 km/h by auxiliary motor

Road users on *motorcycles* or *motorized two-wheelers* (*powered two-wheeler*, PTW):

- Drivers of motorcycles, including mopeds, mopeds, small motorcycles, motor scooters, light motorcycles, motorcycles and also two-wheelers with electric motor without necessary pedal assistance

A growing focus in traffic safety is the protection of another group:

- Drivers of *micro electric vehicles* (eKF) (also *personal mobility devices, personal light electric vehicles*) of the *micromobility* (also: last-mile-mobility), e.g. *e-scooter*

Whether the protective measures listed further below also have potential for drivers of light vehicles or light electric vehicles of vehicle classes L6e or L7e depends on the specific vehicle type and the accident situation. Helmet obligation, insurance obligation and license plate obligation apply to individual vehicle segments. Minimum age, driver's license or test certificate may be a prerequisite for driving the respective vehicle.

6.8.1 Accident Statistics for Vulnerable Road Users

A significant part of all victims of traffic accidents are the group of *vulnerable road users*. Due to advances in the safety for vehicle occupants—especially through systems of integral vehicle safety in motor vehicles—the relative share of VRUs in the total number of victims of traffic accidents has significantly increased. According to the *Global Status Report on Road Safety 2018* of the World Health Organization (WHO), more than half of all victims of road accidents are VRUs [150].

The picture is similar in Europe. Figure 6.117 shows the so-called *collision matrix,* which counts traffic fatalities after collisions with other road users or vehicle types (EU 27, data base: 2021, except IE (2017), MT and SE (2019), CZ, EE, EL, CY, LV, SK (2020)). The group of *vulnerable road users* has a share of 46 % compared to 54 % for the share of killed vehicle occupants and other, unassigned traffic fatalities [30]. For VRU, collisions of pedestrians, motorcyclists and cyclists with cars as well as own accidents of motorcyclists are the most common accident constellations.

A significant step towards improved protection of VRU came about after consultations by the European Commission to improve pedestrian protection by establishing new test procedures for vehicles from 1998. With the aim of preventing serious

Collision matrix

	Traffic fatalities ... / ... in a Collision with...	Pedestrian	Cyclist	Light motorcycle	Motorcycle	PASSENGER CAR	Truck < 3.5 t	Truck > 3.5 t	Bus	Other	Single accidents	Total
VRU	Pedestrian	0	29	16	115	2.328	416	391	97	162	-	3.554
VRU	Cyclist	7	45	6	26	838	183	199	30	77	426	1.837
VRU	Light motorcyclist	0	1	7	6	232	42	27	5	20	175	515
VRU	Motorcyclist	10	8	8	91	1.386	231	207	13	85	1.197	3.236
Occupants	PASSENGER CAR	18	6	4	21	2.504	625	1.392	115	298	3.900	8.883
Occupants	Truck < 3.5 t	1	0	1	0	124	62	250	10	32	262	742
Occupants	Truck > 3.5 t	1	1	0	0	40	11	192	2	13	153	413
Occupants	Bus	3	0	0	0	6	7	12	4	14	74	120
	Other	2	3	2	4	169	34	52	4	27	300	597
	Total	42	93	44	263	7.627	1.611	2.722	280	728	6.487	19.897

Fig. 6.117 Traffic fatalities by involved collision partners (European Union, EU 27, data base 2021, except IE 2017, SE and MT 2019, CZ, CY, EE, EL, LV and SK 2020—based on [30])

injuries to unprotected road users (including pedestrians) in collisions with cars, the Directive 2003/102/EC on pedestrian protection subsequently came into force in 2003 (EU-Directive 2003/102/EG, 2003), see also Sect. 4.2.3.

Despite efforts to protect traffic accident victims, analyses show very different developments depending on the type of road user (Fig. 6.118). While in the European Union (EU 27), the number of killed vehicle occupants could be reduced by over 30% between 2010 and 2021, the developments among VRU are very different. The number of killed pedestrians decreased by 40%, that of motorcyclists by 28%, but that of cyclists by only 7%. The influences of the pandemic on traffic events must be taken into account in the developments of the years 2020 and 2021 [30].

In 2001, the European Commission established a program for road traffic safety for the years 2002–2010 [37]. However, its goal of halving the number of traffic accident victims could not be achieved. The graphic shown in Fig. 6.118 illustrates that more efforts are still required for the group of VRU to get closer to this goal in the future.

Approaches to traffic safety for the protection of VRU are not discussed in detail here. In addition to vehicle-side measures, these include [149]:

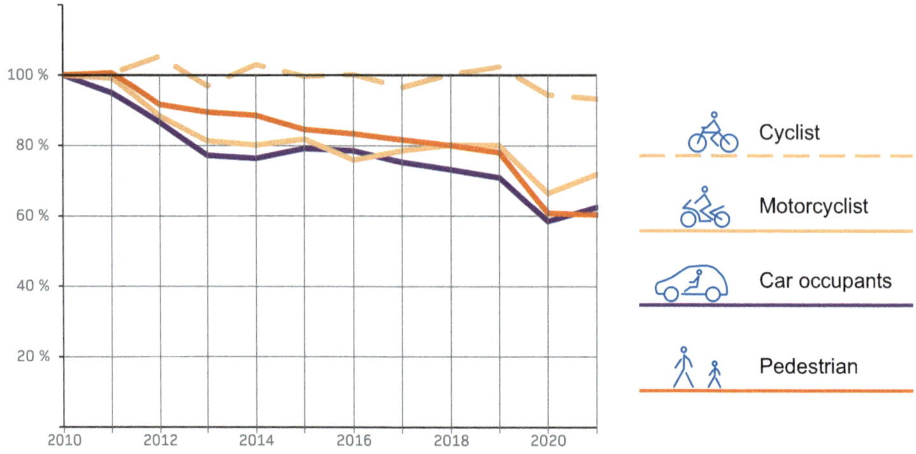

Fig. 6.118 Traffic safety statistics of the European Union (EU 27), traffic fatalities by road users, changes 2010–21 (data base 2021, except IE 2018, SE 2019, MT 2019, CY, LV and SK 2020). Source: [30]

- Traffic space design and transport infrastructure
- Identification and elimination of accident hotspots (e.g., junction areas, roundabouts) and areas with recurring accident constellations [93]
- Separation of traffic routes for pedestrians, PTW, cars
- Traffic education
- Driving license requirement
- Information campaigns

Vulnerable Road User

Numerous organizations are dedicated to the field of Vulnerable Road User and the relevant technologies for self and partner protection. The following overview is a selection of further information and sources:

- Federal Highway and Transport Research Institute (BASt), Germany—www.bast.de—Pedestrian and cyclist protection
- Deutsche Verkehrswacht (DVW)—deutsche-verkehrswacht.de—Protection of e-scooter, bicycle, elderly
- Euro NCAP—www.euroncap.com—Vulnerable Road User (VRU) Protection
- European Commission—Mobility and Transport—transport.ec.europa.eu—Vulnerable Road User
- IIHS—www.iihs.org—Motorcycles /Pedestrians and bicyclists
- NHTSA—www.nhtsa.gov—Motorcycle Safety /Bicycle Safety /Pedestrian Safety
- NTSB—www.ntsb.gov—Most Wanted List

- UN ECE Vehicle Regulations—unece.org/transport/vehicle-regulations—World Forum for Harmonization of Vehicle Regulations (WP.29)—Working Party on Passive Safety (GRSP)—Pedestrian Safety (PS), Protective Helmets (PH)
- German Insurers Accident Research (UDV)—www.udv.de—Protection of pedestrians /Protection of cyclists
- Vulnerable Road Users Safety Consortium (VRUSC)—vrusc.sae-itc.org

An overview of bicycle helmet requirements can be found at:

- Wikipedia—wikipedia.org—Bicycle helmet

6.8.2 Approaches for Self-Protection and Partner Protection of Vulnerable Road Users

Like occupant protection, the protection of vulnerable road users is a key objective of the integral safety of motor vehicles. This is also a field of work with increasing importance, as VRUs make up a significant proportion of the victims of traffic accidents with injuries or fatal outcomes.

The measures for the protection of VRUs in the event of a collision with other vehicles or even in accidents without external influence cover the following areas:

Protection of VRUs in motor vehicles (as *partner protection*) (Sect. 6.8.3)

- Vehicle design (including vehicle front geometry, deformation behavior of the front structure, component design and positioning)
- Protection systems with activation after collision detection (Chap. 7)
- Active safety systems (Chap. 5)
  For motor vehicles, there are safety requirements in many regions worldwide (including legal requirements, consumer rating) for the protection of VRUs (Chap. 4). This also applies to measures of vehicle design, as far as it serves to reduce the risk of injury to pedestrians or cyclists.

Protective equipment VRU (*self-protection*) (Sect. 6.8.4)

- Safety helmet, especially for cyclist or motorcyclist
- Protective clothing
- Protective clothing with activation of integrated protection systems after collision detection
  Even if there is not always a requirement to wear a helmet or protective clothing, their use is recommended to reduce the risk of injury when using vehicles by VRUs.

Vehicle design and **vehicle-based protection systems** for self-protection in **two-wheelers** (Sect. 6.8.5)

- Vehicle design
- Protection systems with activation after collision detection
- Active safety systems (Chap. 5)

  Since motorcycles, bicycles or e-scooters have different conditions compared to cars when installing safety systems, many such concepts have so far only reached prototype status and are rarely encountered in practice.

6.8.3 Protection of Vulnerable Road Users through Vehicle Design

In many regions worldwide, there are now safety requirements for motor vehicles (legal requirements, consumer ratings) enforcing the protection of VRUs. The focus here is on measures for protection in frontal impact between pedestrians and cyclists with cars. This technology field is often also referred to as pedestrian protection.

Clarification of Terms

In the literature, a distinction is sometimes made between measures of *passive pedestrian protection* and measures of *active pedestrian protection*.

Passive measures here include the design of the vehicle geometry, location of deformation zones, package (arrangement) of vehicle components, e.g., underneath the front flap.

Active pedestrian protection measures are understood to be systems with collision detection and triggering in the in-crash phase. These include, among others, windshield airbags and pyrotechnic or mechanical hood lifters.

Since these system components can usually be assigned to passive safety technologies, this quickly leads to a confusion of terms. For this reason, the use of these terms in this sense will be avoided in the following.

The focus of the further explanations in this section is on the measures of vehicle and body design as well as other protection systems to reduce injury risks that are to be expected in the event of a collision. They are assigned to passive safety.

Systems and functions of active safety such as *emergency brake assistants* (BAS), *autonomous emergency braking* (AEB) and *forward collision warning* (FCW), which particularly serve to protect VRU, are described in Chap. 5.

6.8.3.1 Requirements for the Protection of VRU in Passenger Cars

A significant step towards improved protection of VRU in passenger cars was the consideration of specific requirements and tests in the Euro NCAP protocol from 1997 onwards. Initially, these focused on pedestrians. Currently, the protection of VRU is a separate focus of vehicle evaluation at Euro NCAP. Tests for assessing the risk of injury currently include:

- *Head impact*
- *Upper leg area impact*
- *Lower leg area impact*

In addition, there are tests for autonomous emergency braking for pedestrians and cyclists [32, 33]. The establishment of the evaluation scopes for the protection of VRU was pioneering for other NCAP organizations. Criteria for pedestrian protection can now also be found in other NCAP programs (Chap. 4).

An important role is played by the essential criteria for determining the relevant *test areas* for the *head impact* in the front area of the vehicle (Chap. 4). They are defined by the *bonnet side reference line* on the sides and at the front and rear by the *bonnet leading edge reference line* and the *bonnet rear reference line*. A *grid* is defined across the entire relevant test area, which determines the positions for both calculation and possible tests.

Another parameter for the design and evaluation of pedestrian protection measures on the vehicle is the *wrap around distance* (WAD) *(wrap length)*. The measure runs vertically upwards from the standing surface reference plane on the ground over the front hood top or the front protection system and can then be continued over the windshield and the roof. Wrap lengths of 900 mm (WAD900), 1000 mm (WAD1000), 1700 mm (WAD1700) and 2100 mm (WAD2100) are relevant measures. In addition, further reference points and lines are defined for the evaluations, such as *bumper reference line(s), bumper corner(s), internal bumper reference line* (IBRL).

The evaluation procedure at Euro NCAP is now a combination of calculated load values for the *grid* by the vehicle manufacturer and random tests of individual grid points with test bodies by a test laboratory for the verification of the calculation.

For the *head impact* and the *upper leg area impact*, the following test bodies for laboratory tests including instrumentation are defined (Chap. 4):

- *Headform*: Free-flying impactor with measurement of translational accelerations in the center of gravity in three axes for the derivation of the HIC. There is no measurement of rotational accelerations and no representation of the neck.
- *Upper Legform*: Free-flying impactor

For the *lower leg area impact*, test bodies for measuring biomechanical loads are defined. Examples include:

- *Advanced pedestrian legform impacter (aPLI)* and *flexible pedestrian legform impactor* (FlexPLI) as a precursor. These free-flying impactors represent the leg without the ankle joint.

6.8.3.2 Vehicle Design and Constructive Measures for Passenger Cars to Protect Vulnerable Road Users

The introduction of test criteria for the protection of VRU has shaped the exterior design of passenger cars. The most efficient way to reduce the risk of injury to VRU, and especially to pedestrians and cyclists in a collision, is to optimize the design of potential collision areas. For the vehicle front end, this includes the *vehicle front* and *bumpers, front hood, fenders, front spoiler, lights* as well as *windshield, windshield root, roof edge* and *A-pillar*. However, this also includes vehicle components located beneath the car body exterior, provided they could influence the deformation in a collision with VRU, such as rigid, non-deformable components like the battery or shock absorbers. They can no longer be positioned directly beneath the car body exterior.

In addition to providing sufficiently large deformation paths in the contact area, care must also be taken to ensure that the force transmission can be initiated over a large area. Contact points must therefore not have any protruding edges. For these reasons, *bumpers* cover larger areas and are integrated into the vehicle front. They are pulled further down to achieve as flat a surface as possible. The injury risk to VRU can be reduced by using energy-absorbing front structures, deformable impact surfaces and compliant structures, e.g. by deformable radiator grille and headlight mounts, or the use of energy-absorbing materials [68].

Passive structural measures and approaches to meeting VRU protection requirements through the targeted design of individual vehicle components (see also Fig. 6.119):

- Optimization of front flap and fenders through material selection and dimensioning. It should be noted that lightweight materials such as CFRP or GFRP pose a major challenge for VRU protection measures, as they have different failure mechanisms and often less deformation capacity for energy absorption compared to steel.
- Joint design between front hood and fender
- Optimizations through defined breaking mounts or mounts with defined kinematics to allow diving—for example with main headlights, front hood hinges, locks and buffers
- Optimization of the windshield by extending the deformation path in the area of the instrument panel, by failing frame structure or by package adjustments at the front hood and windshield root
- Lowering of components in the engine compartment or adjustment of the geometry to extend the deformation path, such as battery, water tank, wiper system or radiator

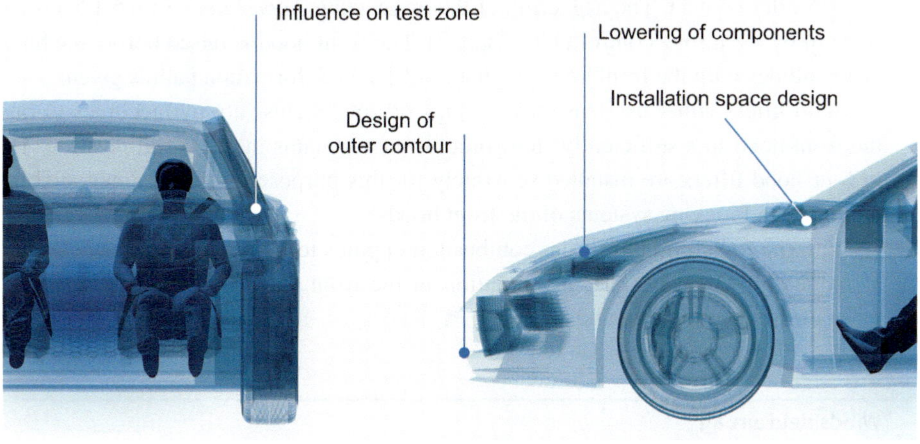

Fig. 6.119 Measures for protection of VRU at the vehicle front. (*Source*: ZF, based on [68])

- Targeted influence on the relevant test zone by drawing in edges and beads and designing the vehicle geometry

Optimizations of the A-pillar and the roof edge are hardly possible due to the very compact, but also rigid structures through design measures. In this area, improvements are often only achievable through additional protection systems.

For battery electric vehicles, the package situation may be less challenging due to smaller engines or other drivetrain design than for combustion engine vehicles. On the other hand, a shorter front end can pose new challenges for the protection of VRU.

6.8.3.3 Protection Systems with Activation after Collision

If the required protective effect is not achieved through constructive measures in the design of the vehicle front end or through other design objectives, such as vehicle design, additional protection systems can be applied. They consist of specific sensors for detecting collisions with VRU and actuators such as front hood lifters or external airbags and serve to extend the deformation path or cover certain exterior skin areas to reduce the load values for VRU.

Active hood

When designing the front end of a vehicle, if areas with high load values remain for *head impact* or *upper leg impact*, an *active hood* (*also: active bonnet)* can provide the possible deformation path. It reduces the risk that the pedestrian or cyclist colliding with the vehicle will come into contact with rigid vehicle components during the impact on the front hood and the subsequent deformation.

With this protection system, the collision of the VRU is detected by sensors in the bumper after contact. The triggering of the *pyrotechnic hood lifter* (Fig. 6.120) is carried out by the airbag control unit (Chap. 7). The front hood is raised before the hip or head collides with the front hood, so that a sufficient deformation path is given.

Hood lifters either use tensioned springs or now exclusively pyrotechnics to raise the front hood in a sufficiently short time. It then remains in the raised position. Two to four hood lifters are installed separately for this purpose or will be integrated into the hinge and locking systems of the front hoods.

With the transformation from combustion engines to battery electric vehicles with a frequently simplified package situation in the front carriage, a decreasing importance of these systems can be expected. ◀

Windshield airbag

A concept with different priorities is the external *windshield airbag* (also: *pedestrian airbag*). This external airbag is intended to cover both contact surfaces with high injury risk and low deformation ability, and as an additional option, also raise the front hood (Fig. 6.121). Contrary to the term 'windshield airbag', this airbag concept primarily serves to cover the window frame, i.e., A-pillars, window root, and front roof frame. The concept aims to protect not only pedestrians colliding with the vehicle but also cyclists.

The windshield airbag is an effective concept to cover A-pillars and possibly also the roof edge. However, challenges for the concept are weight, space requirements, and environmental influences in the exterior. Furthermore, after the airbag has been deployed, the concept must ensure the driver's view of the road, which can be

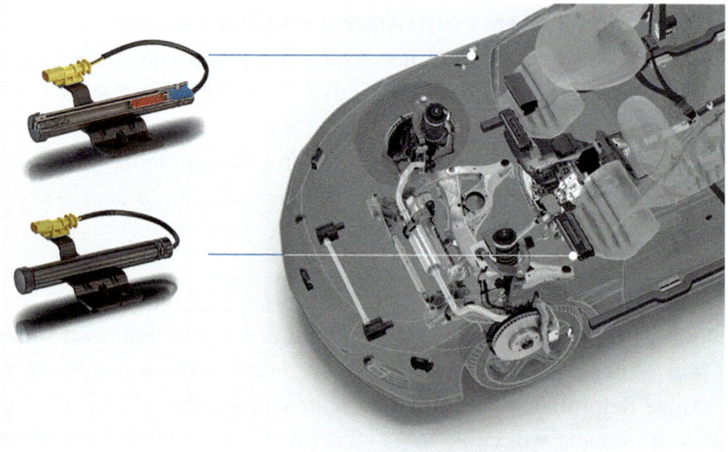

Fig. 6.120 Active hood through pyrotechnic hood lifters. (*Source*: ZF)

Fig. 6.121 Windshield airbag for covering window frames and A-pillars and for raising the front hood. (*Source*: Autoliv)

achieved through airbag design or additional devices. The large number of possible accident scenarios and kinematics of the VRU also pose great challenges in assessing the protective effect of the system. ◀

6.8.4 Protective Equipment for Self-Protection of VRU

As described in the introduction to this section, vulnerable road users do not have a closed passenger compartment, so they are exposed to significant injury risks in a collision. Vulnerable road users largely lack a protective vehicle body structure with energy absorption capacity. The accommodation and use of additional protection systems as in passenger cars are challenging and find little application (Sect. 6.8.5).

Therefore, wearing a personal protective equipment is often the only measure to reduce injury risks in a collision. This section presents especially protective clothing and *safety helmets* for cyclists as essential protective equipment for vulnerable road users. For motorcyclists and their passengers, there has been a largely helmet obligation in Germany since 1976 or 1980 respectively (StVZO §21a).

6.8.4.1 Head Protection

Head protection is now a very established measure for the group of cyclists among the vulnerable road users. The helmet obligation for motorcyclists, but also the increasing acceptance of bicycle helmets, are a clear sign of an increasing need for safety. The development of safety technologies for safety helmets has contributed significantly to their protective effect in different accident scenarios. This section describes exemplary challenges in head protection in typical accident situations and protective technologies developed for safety helmets that can be used by motorcyclists, cyclists, but also in other application areas.

Conventional safety helmets provide damping of linear head accelerations due to an impact. The helmet absorbs part of the energy of the impact and enlarges the area of force introduction. Figure 6.122 illustrates the reduction of head load values in simulated impact.

The impact of the two-wheeler rider often does not occur at a right angle to the obstacle, so rotational forces are introduced into the safety helmet. A rotational movement is created, which causes injury risks. The loads acting on the head can subsequently lead to severe concussions, among other things.

The MIPS system (*multi-directional impact protection*) supplements safety helmets with additional elements that allow a rotational movement of the helmet relative to the head. This aims to reduce the rotational forces acting on the head [47, 48, 12]. This system is integrated into helmets for motorcyclists and cyclists, but also into safety helmets for athletes (Fig. 6.123).

6.8.4.2 Protective Helmets for Motorcyclists

To reduce their risk of injury, riders of motorized two-wheelers (*powered two wheeler*, PTW) are required to wear *protective helmets*. Protective helmets provide effective

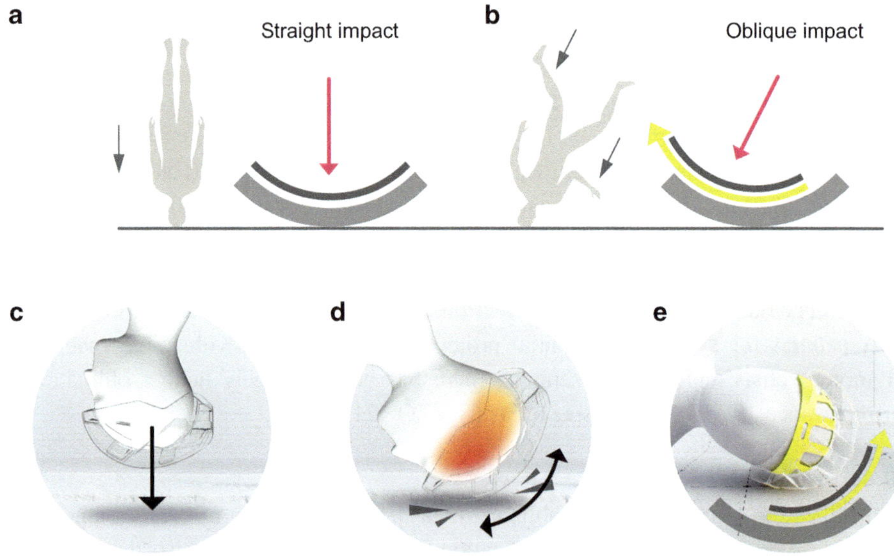

Fig. 6.122 Comparison of straight and oblique impact in accidents. (*Source*: MIPS)

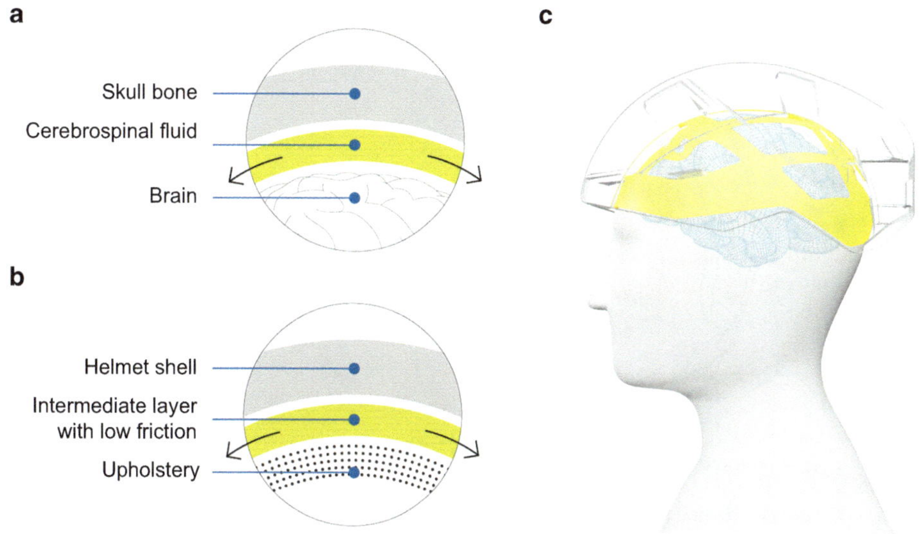

Fig. 6.123 Protection technology for reducing loads during rotational movement: **a** Protection principle brain, **b, c** Protection concept *MIPS* system. (*Source*: MIPS)

protection from impacts—whether in a collision with a vehicle or in the subsequent fall onto the road.

For motorcycles, *open protective helmets* (also *jet helmet*) and *full-face helmets* (also *full protective helmet*) are available. Jet helmets are open in the face area and sometimes without a visor. Full-face helmets offer higher protection against injuries in an accident due to their closed and thus more stable design with a chin bar. Hybrid forms are *modular helmets* or *flip-up helmets*.

A *full-face helmet* for motorcyclists (Fig. 6.124) is externally characterized by the stable, resistant *outer shell (helmet shell)*, which protects the helmet wearer from impact and cut injuries. Occurring impact forces are evenly distributed over a larger area. The outer shell also allows sliding on the asphalt, thus reducing the risk of injury due to compression of the neck.

The protective and comfort padding, usually made of polystyrene foam, prevents the undamped transmission of impact energy to the head. In addition, the *padding* of the protective helmet ensures the secure fit of the helmet and the highest possible wearing comfort. The neck padding has the competence and responsibility to protect the sensitive cervical vertebrae from injuries. The lower part of the face is covered by the chin bar, which is either firmly connected to the helmet or hinged to the helmet and can be flipped up.

The *visor* finally is a transparent shield placed in front of the face in the area of eyes and nose, which keeps insects and dust particles away and allows a correct and distortion-free view. In case of an impact, the safety helmet should remain on the head, for

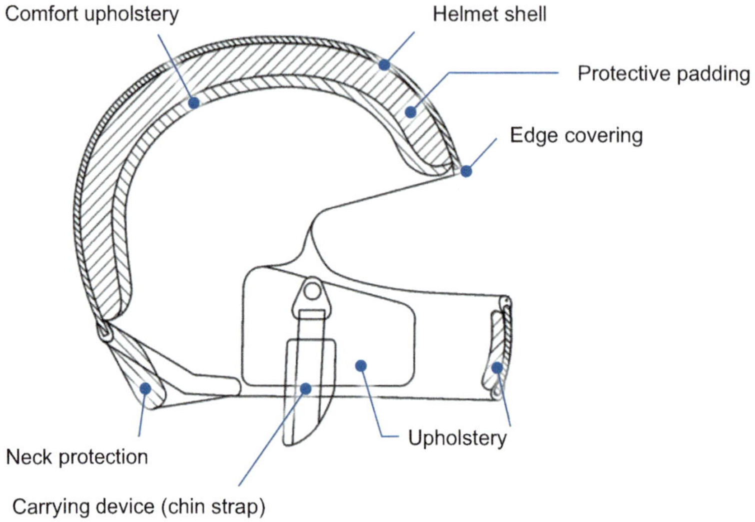

Fig. 6.124 Schematic structure of a motorcycle protective helmet. (Based on UN-R 22, [61])

this the *wearing system (chin strap, helmet lock)* must have a sufficiently high strength. A loss of the protective helmet during the accident must be prevented as far as possible. However, the device must also be able to be released safely and quickly in case of rescue.

Thermosetting plastics are used for helmet production. The GFK or fiberglass helmets produced in the lay-up process and subsequent hot pressing are insensitive to weather influences and solvents, but they tend to have a higher weight. If fiberglass material is replaced by carbon fiber material, the helmet is lighter, but usually also more expensive. The shock-absorbing property of the helmet interior can decrease with prolonged use and it can be sensitive to solvents or operating materials.

Motorcycle helmets with airbag systems: Airbag systems for head protection have been available for cyclists for some time, which are described further below. It was therefore obvious to also approach the integration of an airbag concept in motorcycle helmets in order to extend the protective effect of existing head protection systems. To further increase the damping capacity of protective helmets, especially for linear loads compared to conventional helmets, further concepts for various areas of application have emerged, which are currently in the prototype status [4, 5].

6.8.4.3 Head Protection for Cyclists

For bicycle helmets, the same key functions apply as for protective helmets for motorcyclists: They dampen impacts to the head and thus reduce the risk of injury in a collision with vehicles or a fall onto the ground. However, for cyclists, the requirements for energy absorption are significantly lower due to the lower speeds, the helmets are considerably lighter, and they often only cover the top of the head and the back of the head.

For cyclists, there are strong recommendations, but with few exceptions, no obligations to wear a helmet or protective clothing. Currently, a general helmet obligation exists, for example, in Albania and Malta, outside of closed localities in Slovakia and Spain, and for children of different ages in other European countries [148].

In Germany, there is no general helmet obligation for cyclists. However, the rate of helmet use has continuously increased over the past years and rose - measured across all age groups - from 15 % (2015) to 26 % (2020). Children up to 16 years stand out with an above-average wearing rate of over 75 %. The rate is lowest in the age group 17–21 [38] (Fig. 6.125).

Bicycle Helmets

Bicycle helmets as *micro-shell protective helmets* consist of two layers, each with specific protective functions, and are the most common variant. The hard outer shell is made of polycarbonate (PC), ABS (acrylonitrile-butadiene-styrene copolymer), fiberglass (GFK), or carbon fiber (CFK). This layer protects the underlying inner shell during use, but also ensures that the head slides on the ground during a fall, thus reducing injury risks due to neck strain. The inner shell is usually made of EPS (expanded polystyrene). Large ventilation openings are typical for bicycle helmets to

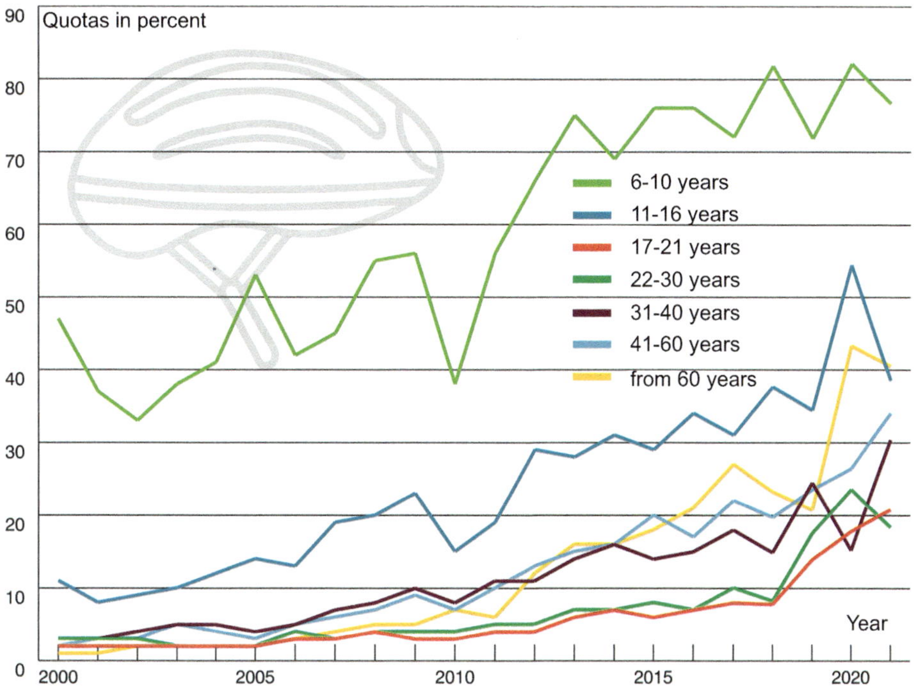

Fig. 6.125 Helmet wearing rate by age groups in Germany. (*Source*: Federal Highway Research Institute Germany, BASt, [38])

increase wearing comfort. As with motorcycle helmets, an adjustable *wearing system,* consisting of a *chin strap* and *helmet lock,* adjusts to the head size and secures against loss during a fall. Additional protective technologies such as the MIPS system (see above) can also be installed to further reduce injury risks.

Hard shell helmets, nutcase helmets, or *full-face helmets* are other variants.

In Europe, bicycle helmets are considered as personal protective equipment and fall under the European Directive 89/686/EEC. Requirements for materials, construction, or protective properties are defined in DIN EN 1078 and DIN EN 1080 [24, 25]. ◀

Head Protection for Cyclists with Airbag Systems

Special solutions are *bicycle airbags* (also: *airbag helmets, airbag-based head protection systems for cyclists*), which are activated upon detection of a fall by integrated and autonomous sensors. Thus, simplifying the description, either a bicycle helmet can be combined with an airbag system or the airbag can be used as the sole head protection (Fig. 6.126).

The latter variant has been offered for several years as an alternative to conventional bicycle helmets. This airbag system is placed around the neck like a neck brace before starting the ride, and the integrated trigger sensors and electronics are activated. They trigger the airbag as soon as the movement and acceleration patterns characteristic of a cyclist's fall are detected. The airbag unfolds along the neck and head and has a duration that covers typical falls. After the fall, the non-reversible system must be reprocessed by the manufacturer.

In product-related tests, this airbag for cyclists shows a reduction of head loads (HIC) compared to conventional bicycle helmets [65, 118]. Due to the concept, the challenges of head protection by an airbag lie in the sensing of an impending head impact, in the unfolding of the airbag before the impact, and in the duration for possible subsequent collisions. Certain accident scenarios can therefore either not be detected or the airbag cannot be deployed until the impact. The protective potential of the system was analyzed in further studies using simulations, component and system tests [151].

An alternative concept aims at combining the advantages of a helmet with a solid structure and the energy absorption capacity of an airbag. A prototype for this concept consists of a helmet, in which a textile structure—more air tubes than an airbag—is deployed between the inner and outer shell, thus improving the damping capacity of the protective helmet. However, an integrated system for triggering will also be required for this helmet concept. ◀

6.8.4.4 Protective Clothing for Cyclists

In addition to the safety helmet for the head of the two-wheeler rider, the *protective clothing* is another effective measure to reduce the risk of injury. It serves the cyclist with its protective effect in the event of an accident, it can increase his visibility, it provides weather protection in rain or sun exposure and contributes to good aerodynamics.

Fig. 6.126 Cyclist airbag system. (*Source*: Hövding)

It should offer a high wearing comfort and—like other clothing—follow a good design. However, there is usually no obligation to wear protective clothing.

Traditionally, it is made of leather or is now made in a combination of textiles in several layers with additional protectors. The outer layer must be very abrasion-resistant. Body parts at risk of falling, such as knees, elbows, shoulders, and back, are padded with *protectors* that absorb energy and improve abrasion behavior. They consist of one or more layers of energy-absorbing soft foam and can be provided with a tough outer shell. The use of external hard shells serves to distribute the impact energy over a large area, thus reducing the load when hitting sharp-edged obstacles.

Sturdy and back-of-the-hand padded *motorcycle gloves* and *motorcycle boots* complement the effective protective clothing.

For special areas of application and injury risks, further protective clothing has been developed. Often, motivation to develop new solutions comes from racing, or there is a transmission of concepts from other areas such as extreme sports.

Protective clothing with airbag systems: In cars, airbags have long been of greater importance as another group of protection systems alongside seat belt systems.

Motorcycles do not have an energy-absorbing passenger compartment and hardly any support surfaces for airbags can be used. The diversity of accident scenarios and the limited space conditions pose further challenges.

In order to make airbag technology usable for motorcyclists, numerous solutions for airbag systems integrated into protective clothing have been developed. These solutions are either offered as *airbag vests* that are worn under or over the other protective clothing, or the airbag system is integrated into the protective jacket or suit (Fig. 6.127).

The sensors, trigger electronics, and power supply are now fully integrated into the protective clothing, so that the system usually operates independently of the motorcycle.

Neck protectors: Additional protection of the neck and spine area by a *neck protector* (also: *neck brace)* is mainly used in racing. In a collision, the head can be overstretched forward, backward, or to the side, or the cervical spine (CWS) can be compressed by force on the helmet. The aim is to prevent or at least reduce the risk of injuries in the areas of the neck, spinal cord, and collarbone in severe collisions. Figure 6.128 shows an example of a system where the neck protector wraps around the rider's neck, shoulder, and chest like a neck brace, without excessively restricting the movement of the head.

Backpacks with airbag systems: Another solution consists of backpacks with integrated airbag systems. Here too, the airbag is intended to provide additional protection for the neck, shoulder, collarbones, and chest area in addition to the bicycle helmet.

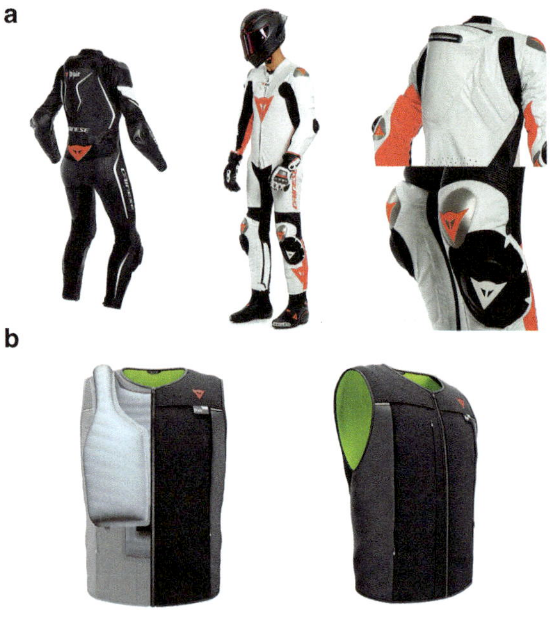

Fig. 6.127 Protective clothing with integrated airbag systems for motorcyclists. (*Source*: Dainese)

Fig. 6.128 Neck protection system for cyclists. (*Source*: Leatt)

Since wearing the backpack is required, these concepts are more aimed at injury risks in specific use cases (e.g., commuting, courier rides).

6.8.5 Self-Protection for Vulnerable Road Users through Vehicle Design and Vehicle-Based Protection Systems

The possibilities for design protection measures for motorcycles for self-protection are much more limited compared to a car. This section presents some concepts for the protection of drivers and passengers of *powered two-wheelers* (PTW). For motorcycles, bicycles, or e-scooters, compared to cars, there are different challenges to solve when installing safety systems. This affects both the crash sensors and airbag control unit as well as potential actuators such as airbags. Protection systems often rely on technologies that have already been established in cars, but they need to be adapted to the different conditions. Furthermore, different test and approval conditions apply to motorcycles.

Also, the accident scenarios are very diverse, so not the *one* system can provide comprehensive protection in a collision. The application and distribution of protection systems in PTWs are determined by the protection potential to be achieved, but also by criteria such as robustness, weight, everyday suitability, and cost. In addition, there is the safety need and the subjective risk assessment of the cyclists. Therefore, solutions often only have concept status and are only sporadically found in practice. Vehicle-based protection systems with triggering in the event of a collision, as presented below, are so far a niche application. Some are introduced below.

6.8.5.1 Vehicle Design and Structural Design in Motorcycles

Even the vehicle design concept of a motorcycle takes into account protection concepts for the *riders* (driver and passenger) as well as the motorcycle for different accident scenarios.

The collision matrix in Fig. 6.117 illustrates that about a third of traffic fatalities among motorcyclists are due to *single-vehicle accidents*. The analysis of single-vehicle accidents as well as experimental and computational simulations made it clear that *sliding* and *stopper devices* (also: *fall pads*, *crash pads*) represent a simple measure with a protective effect. The fallen and sliding motorcycle can be influenced in its movement behavior relative to the driver in such a way that the driver can detach from the machine [60]. This detachment is achieved by three contact points that were integrated into the full fairing on both sides of the motorcycle: two stoppers at the front and rear with the greatest possible distance from the wheel contact point with a non-slip surface and a slider near the footrests made of appropriate plastic material. They ensure that the motorcycle, regardless of whether the front or rear wheel slips, is forced into a kinematics that allows separation of the driver and machine. Moreover, the appropriately designed motorcycle side prevents the driver from being trapped between the motorcycle and the road surface.

6.8.5.2 Vehicle-Based Airbag Systems in Motorcycles

Currently, about half of the traffic fatalities among motorcyclists are due to collisions with cars (Fig. 6.117). A common accident scenario between motorcycles and cars are *intersection accidents*. Here, the injury risk for motorcyclists is high if the head collides with the rigid roof frame of the car due to the geometric conditions. Even at low speeds, the head accelerations then exceed tolerable load values even when wearing protective helmets.

One way to reduce the injury risk can be seen in the use of a *motorcycle airbag*, which can work in various ways:

- The load on the motorcyclist when hitting the car is reduced by the inflated airbag.
- The kinematics of the motorcyclist's body are influenced in such a way that the risk of a frontal collision with the car is reduced.
- These protective mechanisms can be realized through the installation position on the motorcycle and the shape of the airbag. For example, tethers help stabilize the location of the airbag, as there are hardly any support surfaces that can be used on the motorcycle. However, the conditions of a collision between a motorcycle and a car are very different, so there is still a high injury risk due to increased neck forces, neck bending moments, and head and chest accelerations.
- Injury risks from a further impact on the ground can hardly be reduced by a system installed on the vehicle side, so personal protective clothing still plays a major role here.

- The use of an airbag on a motorcycle, like in a car, requires a complete system, consisting of crash sensors, airbag control unit (Chap. 7) and the airbag module.

6.8.5.3 Airbag Concepts for E-Scooters

The increasing use of alternative vehicle concepts leads to new challenges in protecting road users. As already described above, several factors contribute to the suitability of protection systems in everyday life. Furthermore, it is always necessary to integrate a complete system (sensors, trigger electronics, actuator) into a vehicle. Figure 6.129 shows concepts of airbag modules with the integration into *E-scooters* or *scooters* especially for protection in case of lateral collision with a vehicle [6, 7].

The focus of further developments will be the protection potential in the most relevant accident scenarios as well as the best coordination of further development goals such as weight, installation space, robustness, operational readiness, and costs—to name just a few. Further applications of airbags in PTWs are to be expected in the future.

6.9 Occupant and Partner Protection in Commercial Vehicles

The protection mechanisms of passive safety and occupant protection systems were described in the previous section for the application field of passenger cars. When used in the segment of *commercial vehicles* (CV), however, dedicated requirements must be considered, so that this section provides a separate presentation of the protection systems

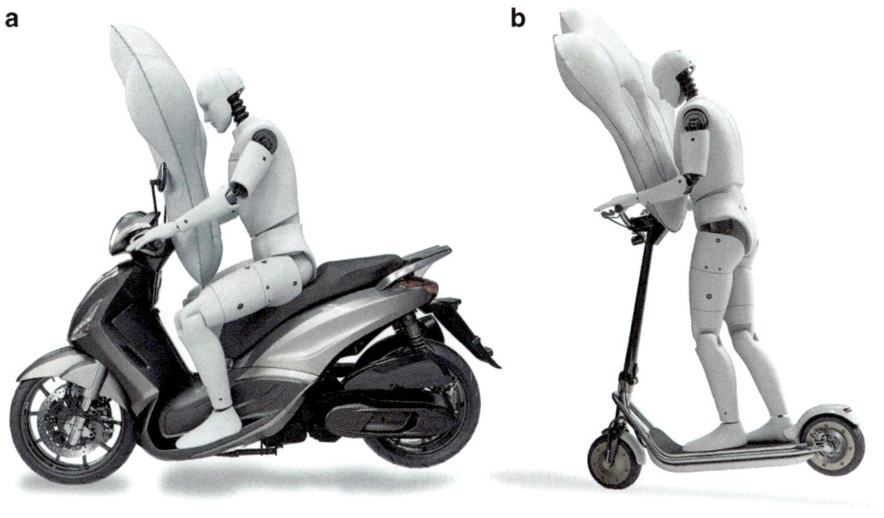

Fig. 6.129 Airbag concepts for two-wheelers: **a** Motor scooter, **b** E-scooter (electric micro vehicles). (*Source*: Autoliv)

in use. Due to the heterogeneity within the vehicle segment and the different operating conditions, the accident constellations and thus also the available protection concepts may differ.

This section will subsequently show the measures for *self-protection* of occupants in *trucks* (CV) and in *buses*. Vehicle occupants in CVs are both the driver and other occupants in the driver's cab. In the case of buses and autonomous vehicles, the aim is to protect the passengers in the passenger compartment (Fig. 6.130).

In accidents involving commercial vehicles and other road users such as cars, motorized two-wheelers, bicycles, or pedestrians, the size and mass ratios differ significantly. The higher injury risk is therefore clearly on the side of the road users colliding with CVs. This section introduces various aspects of partner protection in CVs, which aims at the protection of occupants of other vehicles and also at measures for the protection of vulnerable road users.

For CVs, very differentiated requirements for occupant and partner protection concepts have to be taken into account. One reason for this are the different classifications of vehicle segments and classes in individual regions and countries, where specific regulations may apply. Another reason is the use in very different road environments, such as public roads, cordoned off or even separated traffic routes, and also off-road, so that exceptions or deviating regulations have to be applied.

Fig. 6.130 Truck vehicle crash test *(Volvo FH)*. (*Source*: Volvo Trucks)

6.9.1 Occupant Protection in Trucks

6.9.1.1 Occupant Protection through Safe Vehicle Structures

The *driver's cab* (also *driver's cabin*) of a truck is adapted to the respective use and serves the driver as a workplace as well as a living space or it provides space for additional passengers. In Europe, due to the length limitation of commercial vehicles, almost exclusively *forward control trucks*are used. The *cabin* is positioned over the engine in these vehicles. The vehicles do not have a front end and therefore do not have the geometries common in passenger cars with comparatively larger deformation paths.

To increase strength and provide the required energy absorption capacity, longitudinal beams and sills are installed in the driver's cab. Stiffened front pillars, profiles in the rear wall, and reinforcements in the roof area also promote the stability of the driver's cab (Fig. 6.131). The windows glued into the structural profiles increase the stiffness and reduce the risk of occupants being thrown out during a rollover.

Since the 1960s, pendulum impact tests have been conducted to test the protection of occupants in the driver's cab of commercial vehicles (Fig. 6.132). This test is part of the vehicle approval process and is anchored in UN-R 29 [124]. In several tests, the front area, the front pillars (A-pillars), and the strength of the roof are examined, and the integrity of the driver's cab must be demonstrated.

6.9.1.2 Constructive Design of the Interior

The reduction of injury risks for the occupants and other aspects of traffic safety have also gained increasing importance in the design of the interiors of driver's cabins for commercial vehicles. Figure 6.133 shows an example of the driver's cabin of a commercial vehicle. Today, rounded surfaces and energy-absorbing constructions are common design principles for interior components. Soft, yielding, and tear-resistant surfaces

Fig. 6.131 Truck driver's cabin with stiffening structural components *(Volvo FH)*. (*Source*: Volvo Trucks)

Fig. 6.132 Pendulum test of a truck cabin, Volvo FH. (*Source*: Volvo Trucks)

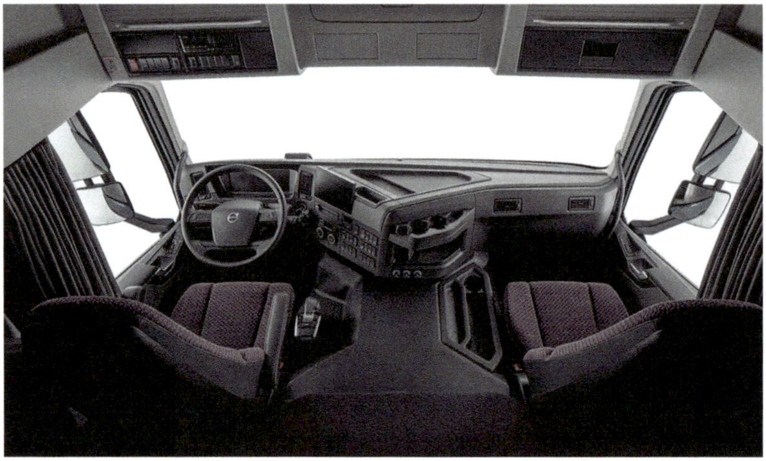

Fig. 6.133 Interior design of the driver's cab in trucks. (*Source*: Volvo Trucks)

prevent cuts during impact. An instrument panel, whose harder components are installed deeper, acts as a knee protector and absorbs the energy of the occupants.

Injury risks from the steering column and the steering wheel are reduced by constructive measures: A flat position of the steering wheel aims to ensure that in accident situations, the chest area, the higher load-bearing part of the upper body (Chap. 3), and not the head/neck area of the driver is affected. In addition, the steering wheel can also convert energy through its deformation ability.

6.9.1.3 Seat Belt Systems

In Germany, the requirement for seat belts has been in place since 1992 for trucks (StVZO 21, 21a) and for all passengers transported in the cabin [108, 109]. According to a study in Germany, heavy-duty trucks now standardly feature oscillating seats with integrated seat belt systems for the driver's seat, while on the passenger side, the seat belt is often attached to the cabin [79, 80]. Drivers reach weekly driving times of up to 56 hours—in addition, they spend more time in the cabin during breaks and rest periods. The improvement of the ergonomics of driver/passenger seats and the expansion of comfort functions also have implications for seat belt systems.

For truck-installed seat belt systems, the *seat-integrated* (seat-fixed) three-point belt system has significant advantages over the anchoring of the D-ring on the cabin wall. The air suspension systems integrated in the seat base lead to noticeable relative movements during the journey. The wearing comfort of the seat belt is significantly improved if all belt anchoring points are directly on the seat (Fig. 6.134).

In a truck, the driver usually sits upright, so the seat belt routing is favorable. A submarining (Chap. 3) and resulting pelvic and abdominal injuries are not very likely.

The equipment of seat belt systems in trucks increasingly includes seat belt retractor pretensioners, which improve the coupling of the occupant to the seat in the event of a collision.

The *active seat belt retractor pretensioner* (Fig. 6.135) also offers the truck driver additional safety functions. In the implementation of integral occupant protection

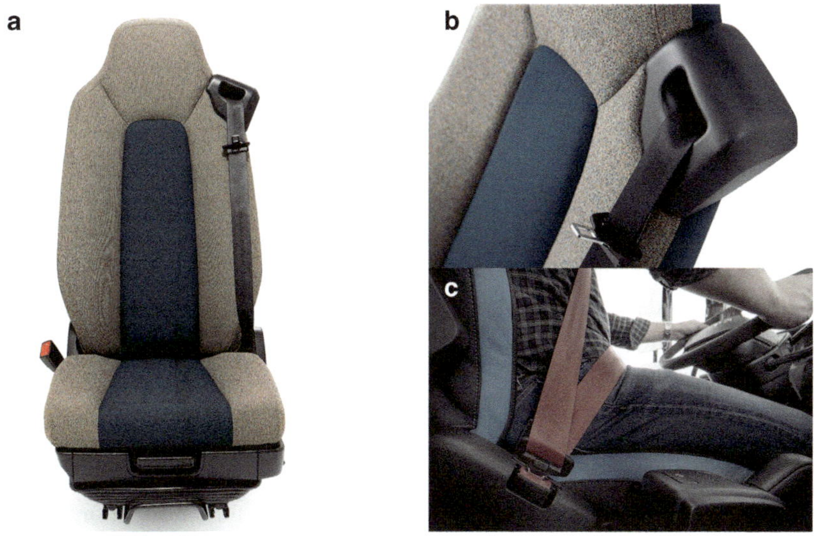

Fig. 6.134 Truck integral seat with seat-integrated three-point seat belt. (*Source*: Volvo Trucks)

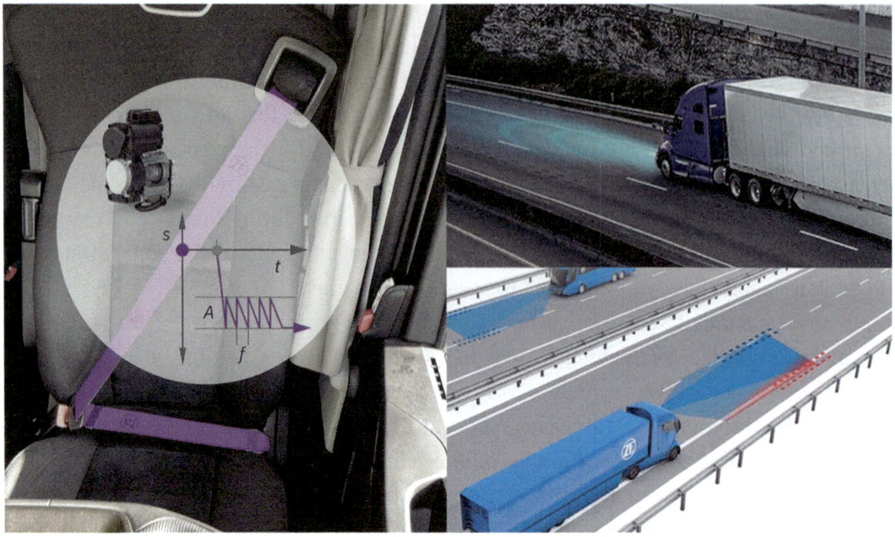

Fig. 6.135 Active seat belt retractor pretensioner for the implementation of multi-modal warning strategies in connection with driver assistance systems. (*Source*: ZF)

concepts, the *driver assistance system* (ADAS) of the vehicle can be supplemented by tactile alarm and warning functions through the seat belt.

6.9.1.4 Airbag Systems in Trucks

Since the 1980s, airbag applications for frontal and side impact in cars have become established, followed by products for trucks. Current studies also show that rear-end collisions as frontal impact in heavy trucks are the most common accident scenario [79, 80].

The use of airbag systems in trucks began in 1995 with a truck *driver airbag* for frontal impact in the Volvo FH [145]. This was also associated with the first application of a sensor system for accident detection in trucks. Equipping trucks with driver airbag systems is not yet common practice and is not required by law.

While seat belt systems provide effective and basic occupant protection functions for many different accident scenarios, airbag systems are more specifically tailored to certain accident scenarios. The driver airbag thus marked the beginning of a development towards further reduction of injury risks for truck occupants through airbag systems.

A subsequent application of airbag technology in trucks was the knee airbag. This airbag also aims to reduce injury risks in head-on collisions, which are caused by the exposed seating position of the driver and possible deformations of the driver's cab. The kinematics of the vehicle occupant during a frontal impact for a restraint in combination with seat belt and driver airbag is also improved.

Another frequently occurring accident scenario in trucks is the rollover accident. The most common cases can be traced back to the following causes: the truck leaving the road on a straight stretch and subsequent rollover on a slope, the truck tipping over during tight cornering or at high speed on descents, the truck rolling over after evasive maneuvers to prevent a head-on collision with an obstacle or vehicle.

A side airbag module for rollover accidents is intended to contribute to the reduction of injury risks by ejection from the vehicle and by energy absorption during the impact of the occupant [145, 146, 101] (Fig. 6.136). For installation in trucks, airbags attached to the cabin structure are available today, as well as seat-integrated systems. The latter are particularly aimed at installation in special vehicles or for certain fleets.

6.9.2 Occupant Protection in Buses and Coaches

The protective measures for occupants apply not only to trucks but also to *buses* and coaches. Here, in addition to the occupant protection for the driver, that for the passengers must also be considered. The following therefore discusses some specific protective devices for passengers in coaches, school buses, city buses, or mini buses.

6.9.2.1 Safe Vehicle Structures in Buses and Coaches

As with cars and trucks, the primary concern is to ensure the *structural integrity* of the passenger compartment so that in the event of accidents, and particularly in rollover accidents, a collaps of the vehicle body is avoided. Rollover accidents with buses, in which the roof structure fails in particular, have dramatic consequences for the passengers.

Therefore, for buses, reinforced roof structures and roll bars integrated into the vehicle body are implemented. Two basic concepts are used in bus bodywork. Traditionally,

Fig. 6.136 Side airbag module for use in trucks. (*Source*: Scania)

truck chassis were used as the basis for construction. This concept is widely used in the *ladder frame* design. The self-supporting car body is constructed differently, its concept is known as *integral* design (also: *shell*, *lattice frame* design). Kässbohrer developed a tubular frame for the first time in the early 1950s, which was able to successfully establish itself on the market. Low weight with high stiffness of the car body are still advantages of this concept. Today, hybrid forms such as *semi-integral* design or *composite* design are also found [77].

The requirements for the strength of bus structures in *rollover tests*, *load tests*, and *vehicle approval* are defined in UN-R 66 [125].

6.9.2.2 Seats and Seat Belts

In addition to the vehicle body, *seats* with integrated safety features are another essential element of passive safety for passengers in buses. Depending on the area of application of the bus, whether as a city bus, intercity bus, or coach, there are standing or seating options for passengers, the latter being the option that offers more safety. The requirements for the construction and anchoring of seats in buses are formulated in UN-R 80 regulation [126].

In modern buses, comfort features such as folding tables, handles, and cup holders should ideally not be mounted in potential impact zones of the head and chest, or they should be break-resistant and designed with sufficient deformation capacity [45]. The failure of the seat and backrest structure in a collision must be avoided through appropriate design. Through the design of seating systems for buses, reductions in occupant load can be achieved with *energy-absorbing backrests* to catch the upper body and support the head-neck region.

Figure 6.137 shows the *seating system* of a coach with its integrated safety features.

An effective safety measure in buses is also the use of seat belt systems. For coaches, the *equipment requirement* for seat belts at all seats has been in place since 1999. For exposed seats—these are seats in front of which there is no further row of seats—three-point belts are to be used, for the remaining seats two-point belts, i.e., lap belts, are to be used. For city buses in local transport, where standing places are also permitted, there are exceptions from the *seat belt legislation* [108–110].

For school buses seating systems are available that are particularly adapted to children, as they have a soft and homogeneous surface on the backrest. The seats are equipped with three-point seat belt systems as well as integrated child seats with five-point seat belt systems.

Potentials for the use of airbag systems in buses have been investigated, however, the concepts developed for this vehicle segment have not yet found series application.

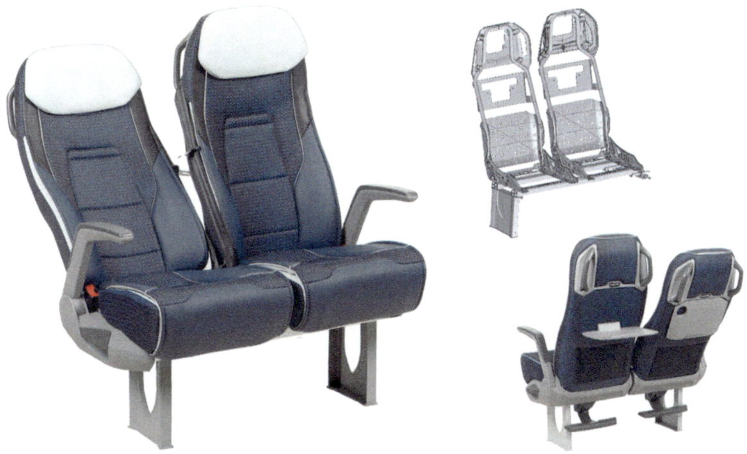

Fig. 6.137 Seating system for coaches with safety devices. (*Source*: Vogelsitze)

6.9.2.3 Additional Facilities for Passenger Protection

Passenger handles: On city buses, passengers often only occupy the seats for a short time and often stand in the aisle. Therefore, special attention must be paid to the *passenger handles* and *hand rails*: Accessibility must meet the *anthropometric* requirements (Chap. 3) and during cornering maneuvers as well as during braking and acceleration processes, they must be quickly and safely accessible for passengers. In addition, corners and edges should be softened by rounding and padding to provide favorable impact geometries in accident situations (Fig. 6.138).

To improve passenger stability when standing, the *floor coverings* are non-slip and firmly attached to the floor.

For strollers, but also for the transport of walkers, bicycles or e-scooters, city buses in particular can be equipped with *lashing devices* and special parking spaces. For the transport of wheelchair users, *wheelchair restraint systems* can be installed.

Further potential danger for passengers repeatedly arises from *luggage racks*, as luggage can fall from open storage compartments during cornering and other driving maneuvers and can endanger passengers. Lockable compartments, as in air travel, are however not common.

Escape routes for evacuation: Buses must be designed for the rapid evacuation of passengers in case of an accident. For this purpose, there are regulations for the number and size of *escape routes* and *emergency exits,* i.e., openings such as doors, windows, and roof hatches [111].

An improvement in escape options can be achieved if emergency release devices coult be triggered by first responders from the outside.

Fig. 6.138 Interior design in buses and passenger handles. (*Source*: MAN)

However, the glazing is utilized to increasingly contribute to the structural rigidity of the vehicle. However, this requires devices for opening the windows in the event of evacuation. An example of this are *detonating cords*, as used in armored limousines. They separate the window from the vehicle and should be triggerable from both inside and outside. However, protective measures against misuse must also be provided for them.

6.9.3 Partner Protection through Commercial Vehicle Design

In collisions between commercial vehicles and cars, cyclists, and pedestrians, the injury risk is clearly on the side of the other road users. Commercial vehicles, due to their mass, the low energy absorption capacity of their vehicle body structure, and the inhomogeneous vehicle surfaces, hardly meet the requirements for compatibility in collisions (Sect. 6.2).

In order to achieve a reduction in the injury risks of other road users, measures for partner protection in commercial vehicles are mandatory, which are assigned to passive safety and will be described in more detail below.

Further protective functions and potentials can be implemented with approaches of integral safety through the system-wide integration of technologies of active and passive safety. One example of this is the improvement of the triggering mechanism of restraint systems in under-ride accidents through the use of vehicle-environment sensors (Chap. 7).

6.9.3.1 Front Underride Protection Device for Commercial Vehicles

The front underride protection device (FUPD) is designed to protect the occupants of other vehicles by preventing the front of the commercial vehicle from being underrun and effectively utilizing the deformation structures of both vehicles. Without an underride protection device, the geometric incompatibility between commercial vehicles and passenger cars would lead to pronounced intrusions on the passenger car and a high injury risk for the car occupants. In frontal collisions between passenger cars and commercial vehicles equipped with a front underride protection device, the rapid underrun is largely prevented. The deceleration experienced by the passenger car is now supported in a controlled manner by the deformation structures of the commercial vehicle. In conjunction with the passenger car restraint system, this is intended to achieve overall lower occupant loads.

Such underride protection device installations have been legally required in Europe since 2003 for new vehicles with a permissible total mass of 3.5 to 12 tons and for those over 12 tons (vehicle categories N2 or N3) in Directive UN-R 93 [127]. Exceptions only apply to off-road vehicles, to trucks whose intended use is not compatible with the provisions for a front underride protection device, and to vehicles with a permissible total mass of 7.5 tons and a ground clearance of no more than 400 mm. Figure 6.139 shows an example of a vehicle crash test of an underride protection device installation. The test criteria for front underride protection devices are maximum permissible deformations and minimum resistance times, which must be met after being subjected to a defined force. Front underride protection devices have a maximum distance to the road surface of 400 mm at the lower edge. They have a profile height of at least 100 mm for N2 category vehicles or 120 mm for N3 category trucks and are sufficiently resistant to forces that act essentially parallel to the longitudinal axis of the vehicle.

6.9.3.2 Rear Underride Protection Device for Commercial Vehicles and Trailers

The rear underride protection device (RUPD) was the first underride protection device for trucks and trailers with a permissible total weight of 3.5 t. It was introduced as early as 1975 with UN-R 58 [128]. The rear underride protection device is designed to provide effective protection against underrunning for cars colliding with the rear of a truck or trailer. Without a rear underride protection device, due to the geometric incompatibility, the car penetrates very far under the truck or trailer, leading to pronounced intrusions of the car (Fig. 6.140). As a result, the front carriage of the car contributes little to energy absorption, and there are cases where no detection of a collision triggers the restraint systems. All these factors contribute to a high injury risk for the car occupants.

Analogous to the front underride protection device, the rear protective device is designed for forces that predominantly act parallel to the vehicle's longitudinal axis. Rear underrun protection systems have a maximum ground clearance of 550 mm in the unloaded state. The height of the cross-beam is at least 100 mm. Compared to the front

Fig. 6.139 Crash test of a truck underride protection device installation during side impact. (*Source*: Volvo Trucks)

Fig. 6.140 Car rear-end collision with a truck without a rear underride protection device. (*Source*: [61, 72])

protective device, the rear underride protection device has significantly more exceptions for specific vehicles.

Investigations into rear underride protection devices from the 1990s revealed improvement opportunities. If the rear underride protection device is installed too high, the deformation structure of the car cannot support itself on it [72]. The possible braking dive of the car and the resulting pitch of the car front exacerbate the situation. It has been proven that in such collisions, the rear of the truck was underrun by cars despite the protective device installed [106]. The rigid structures of the commercial vehicle penetrated deeply into the passenger compartment during underrunning, resulting in a serious injury risk for the front passengers.

6.9.3.3 Side Underride Protection Device for Commercial Vehicles and Trailers

With the *side underride protection device* (*lateral protection devices*, LPD) the main aim is to prevent that pedestrian, cyclist and motorcyclist from getting under the truck from the side (Fig. 6.141). They act more as deflectors and less as energy-absorbing deformation structures. The equipment requirement according to directive UN-R 73 applies to trucks of categories N2 and N3 as well as to trailer vehicles (classes 03 or 04), provided that the lateral body lower edge in the unloaded state has a height of more than 550 mm [129].

A side underride protection device should not increase the overall width of the vehicle. They should be installed as flush as possible to the outer contour of the vehicle and their outer surface should be smooth and as continuous as possible, without leaving open gaps. Materials often used are aluminum profiles or structures out of plastics. The test requirements stipulate that when a static force is applied at any point on the protective device, a certain deformation must not be exceeded, whereby the proof can also be provided by calculation.

With the introduction of side underride protection, the protection of pedestrians and cyclists was clearly in the foreground, as a static test force of only 1 kN was considered sufficient. These constructions are therefore not suitable for reliably preventing underriding in the event of a side impact by motorcycles or cars on the truck, trailer or semi-trailer.

Fig. 6.141 Side underride protection device for trucks, semi-trailers and trailers with additional optimization of aerodynamics. (*Source*: ZF)

6.10 Post-Crash Phase and Rescue Chain

The post-crash phase (Phase: *after the accident*, Fig. 1.7) is part of passive safety. The objectives after an accident event are to reduce the consequences of the accident and to avoid further damage to occupants and other road users as a result of the accident.

Chapter 2 already covered the essential accident events and their analysis. In Chap. 5 and 7 technologies of the *connected car* for networking with rescue services are presented.

This section addresses specific aspects of the rescue chain and shows links between occupant protection technologies and the approach of rescue forces after accidents. It is undisputed that the success of rescue measures depends on their timely initiation. A contribution to this is the collection, transmission and networking of available information. This is a focus of technologies of integral safety in the *post-crash phase.*

In a further section, measures for *fire protection* are exemplified. Due to the transformation of the automotive industry towards new drive technologies, this technology field has again attracted greater attention.

6.10.1 The Rescue Chain

This section describes the *rescue chain* following a traffic accident with the aim of providing an overview of the process and the rescue concepts and technologies used, and to identify starting points for further improvements in the future. It also provides references to vehicle construction and concepts of integral safety.

The *rescue chain* (also: chain of survival, life-saving chain) is the process that begins with the *emergency event* and ends with the *clinical care* of emergency patients. The rescue chain is divided into three phases and six process steps (Fig. 6.142): Process steps 1 to 3 are referred to as *immediate life-saving measures*: **1:** Securing, self-protection, **2:** Emergency call, **3:** First aid by first responders at the accident site. Process steps 4 and 5 are referred to as *extended life-saving measures*: **4:** First medical aid and **5:** Elementary diagnostics and therapy by the rescue service. This also includes the transport of injured persons to a clinic. The rescue chain ends with *defined measures* and the process step **6:** Clinical care by specialists in a clinic [152].

In the *rescue services*, public and private organizations such as *rescue services* work together across disciplines. They provide medical aid for emergency patients until their admission to a clinic *(professional preclinical medical aid)*. The care of emergency patients after traffic accidents is part of the rescue system. Rescue service organizations are sometimes specialized in individual types of operations, e.g. ground-based rescue, air rescue.

In Germany, medical aid for emergency patients is a public competence and responsibility and is regulated by rescue service laws at the state level. In the care of emergency patients, several professional fields work together with medical (including emergency

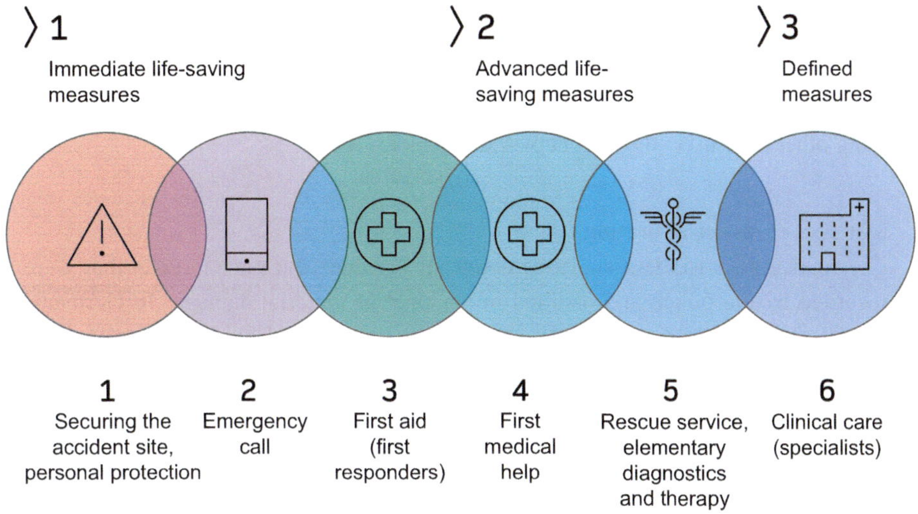

Fig. 6.142 Stages of the rescue chain. (Based on [152])

doctor, senior emergency doctor, rescue service management) and non-medical education (including emergency paramedic, rescue paramedic, rescue assistant).

Two basic concepts can be considered in the organization of a rescue system, which have actually established themselves in their manifestations: In many European countries, the principle of *medical care at the accident site* has been established for a long time (stay-and-play approach) [41]. It means that the emergency doctor comes to the patient at the accident site and works there with the non-medical staff. In contrast, in the Anglo-American area, the concept of *rapid transport* of emergency patients to the clinic is widespread (*load-and-go* or *scoop-and-run* approach) [152]. Therefore, job profiles, training and activities of the rescue forces can differ depending on the concept applied in each region.

However, these basic concepts and the associated treatment concepts such as ALS (*advanced level support)* or BLS *(basic level support)* for the stay-and-play approach or load-and-go approach continue to evolve. An example of this is the *go-and-treat* approach for the care of patients with *polytrauma* [10].

This section also aims to highlight the importance of performance and quality indicators as well as the necessary information technologies. They are an essential tool for identifying potential improvements in rescue services. Measurements are taken along the entire rescue chain. Performance and quality indicators are recorded for process times in emergency call handling, dispatch quality of the rescue control center, diagnostics, victim care, and transport to the clinic [75].

In addition to the advances in medical emergency care, there are significant technical developments that have allowed rescue services to evolve. Examples include air rescue, a central emergency call system, networking of control centers, automatic emergency call systems in vehicles (eCall), equipping rescue workers with emergency medical technology, providing information at the accident site, digitizing communication of rescue workers at the accident site, traffic control technology, and informing flowing traffic and warning other road users through networked vehicles (Chap. 5).

Statistics of rescue operations

The road traffic accident statistics count all accidents on public roads that were recorded by the police and resulted in personal or material damage. In Germany, a total of 2.3 million road traffic accidents were recorded by the police in 2021 [116]. A total of 259,000 accidents with personal injury were counted. This resulted in 268,000 minor injuries, 55,000 serious injuries, and 2,562 fatalities.

In the performance analysis of rescue services in Germany, 2.9 million rescue operations with and 4.3 million without emergency doctor deployment were counted (the current recording period is 2016/17). The number of rescue operations following traffic accidents was 177,000, accounting for only 2 % of the total volume compared to 19 % in 1977 [104].

The air rescue services in Germany carry out over 80,000 operations together [15, 117]. The proportion of air rescue operations with rescue helicopters due to traffic accidents is only about 11 % [98].

In recent years, the number of employees in rescue services in Germany has increased significantly: in 2021 there were 85,000 employees [28]. This includes the areas of emergency rescue and patient transport, the often voluntary workers are not included in this count.

In the future, technical solutions for data networking and provision will contribute to further improvements in rescue services. The equipping of vehicles with environment sensors and interior sensors complements established sensors for collision detection, vehicle dynamics, and occupant sensing and classification (Chap. 7). The more accurately the rescue control center can assess the accident situation at the time of the alarm, the more targeted it can dispatch the rescue services. A reordering due to misjudgment can lead to significant delays in emergency care.

After an accident, the targeted information of rescue workers is therefore a key goal of integral safety concepts. Through the enormous expansion of the communication capabilities of all participants in the rescue chain and linking of data through digitization technologies, further improvements can be achieved. Examples of this development will be explained in more detail along the rescue chain and its process steps.

6.10.1.1 Securing the Accident Scene and Personal Safety

After an accident, occupants and witnesses to the incident often experience shock or fall into excitement and panic. There is uncertainty about what to do next. It is precisely at this point that targeted action saves the lives of the injured and helps to prevent further damage.

After a road accident, the first step is to secure the accident scene. Initial steps include stopping the vehicle and turning on the *hazard warning lights*. For personal protection during further securing of the accident scene, *safety vests* should be worn. Occupants of crashed vehicles should definitely wait outside the vehicle in a safe place for the arrival of rescue forces. Flowing traffic, especially on expressways, often leads to subsequent accidents involving disoriented and shocked individuals. In a next step, the flowing traffic is *signaled with a warning triangle, hazard warning light* and also by waving to slow down. Arriving *first responders* must also turn on the hazard warning lights to secure their own vehicle and illuminate the accident scene in the dark.

The flowing traffic forms a *rescue lane* to make room for rescue forces. Obstructions to rescue measures or inappropriate behavior by onlookers at the accident scene are sanctioned.

The measures essentially also apply to the securing of a *breakdown vehicle* or in an emergency without a collision.

6.10.1.2 Emergency Call

The reporting of an accident to a rescue control center (also *emergency call center)* is the initial information for dispatching rescue forces for the care of injured people and further securing the accident site. High-quality information about the emergency or accident situation is a prerequisite for the targeted alerting of rescue services.

Control centers receive emergency calls via various communication channels. They receive both calls and automatically transmitted reports from involved vehicles or even from the smartphones of the occupants. Variants of emergency call systems, such as reports of an accident event to a local responsible control center, are presented below.

Callers can reach local emergency centers via a single emergency number. The *European emergency number* **112** *(Euro emergency call)* provides a free connection via landline or mobile phone to the control center. The caller's phone number and location data are automatically transmitted. Using the "**5 W**" (**1:** Where is the accident? **2:** What happened? **3:** How many injured? **4:** Who is calling? and **5:** Wait for questions!) the control center tries to gather information to initiate the necessary rescue measures and provides first aid advice until the arrival of the rescue forces.

Hearing and speech impaired individuals or visitors with limited language skills can reach rescue services via a barrier-free emergency call app (e.g., Germany: *nora emergency call app*) without having to speak. Until now, relay services were primarily required for emergency calls from hearing impaired people, which enable the emergency call using sign language.

Vehicles can also report a traffic accident to the Emergency call center using automatic emergency call systems. In Europe, the installation of *eCall systems* has been mandatory since 2018 (Chap. 5 and 7). The emergency call is automatically initiated, for example, after airbags are triggered. Afterwards, a voice connection can be established between the emergency call center and the vehicle. Other road users can also manually start an emergency call via the emergency call button in their vehicle and are connected to the vehicle manufacturer's call center (concierge service). It then establishes a connection to the emergency call center. In emergencies other than a traffic accident, an emergency call can also be manually activated via the emergency call button in the vehicle.

Nowadays, smartphones or smartwatches also have functions for accident detection and automatic emergency calls. The sensing of various traffic accident scenarios (frontal, side, rear impact and rollover) is carried out both via the acceleration sensors and through the analysis of characteristic acoustic signals [19]. If an emergency pass with medical advice is stored, this data is shared with the rescue control center and can be transmitted to the rescue forces.

The rescue control center determines the *indication for use* based on the information received and decides on the dispatch of the available and necessary rescue services from its point of view, including ambulances, emergency doctors, fire brigade with rescue vehicle and rescue device or also fire brigade, police or breakdown service.

Furthermore, if necessary, information and warnings of the remaining traffic can be recorded and distributed via the traffic warning service and the national registration offices.

6.10.1.3 First Aid

Everyone is obliged to provide assistance if they would not put themselves in immediate danger by doing so. A failure to provide *first aid* in traffic accidents is sanctioned.

However, rescuing vehicle occupants from the vehicle can be difficult for first responders if doors cannot be opened or the occupants are trapped. Immediate life-saving measures must then initially be provided from the outside until the vehicle can be opened. Here, rescue and recovery equipment must first be deployed by the fire brigades.

6.10.1.4 Securing and Recovery by the Fire Brigades

If the fire brigade has been alerted to rescue trapped accident victims, they will reach the accident site with a fire engine and additional equipment for technical assistance or also with further *rescue* and *salvage devices*.

The tactical approach is determined based on the assessment of the danger situation. The *immediate rescue* without further delay, for example, occurs in case of fire hazard or risk of the vehicle falling. Otherwise, the *patient-oriented rescue* is carried out with the aim of avoiding further consequential damage.

If the vehicle needs to be cut open to rescue the occupants, the fire brigade can quickly determine a suitable approach for the operation based on the vehicle sketch of the correct vehicle type with vehicle-specific information such as the installation locations of airbags, seat belt tensioners, airbag control unit, batteries, fuel tank or pressurized gas container. These sensitive systems can pose a danger to both the occupants and the rescue forces during rescue measures. For rescue forces operating at the accident site, this information is published by the vehicle manufacturers in the form of *rescue data sheets* (also: *rescue card*, rescue guide). Rescue data sheets are standardized [55] and can already be accessed online for several thousand vehicle types. They can also be displayed with specific software for rescue forces [23] (Fig. 6.143).

Rescue forces can determine the specific vehicle type by querying the license plate at the *National Vehicle Register* (ZFZR) of the Federal Motor Transport Authority (KBA). Since the rescue data sheets do not list vehicle-specific modifications (e.g., gas system), these vehicle-specific data can only be determined by querying the license plate. The connection to the *eCall* system can be used to further accelerate the query process.

Rescue card QR codes, which are affixed to the vehicle, also serve to provide rapid information, as it can be difficult to determine the exact type of a heavily damaged vehicle after an accident.

While in the past, information about the pyrotechnic occupant protection systems was the main motivation for the development of rescue data sheets, it is now also information about the type of drive of the vehicle as well as fuels, batteries, gas storage or gas lines.

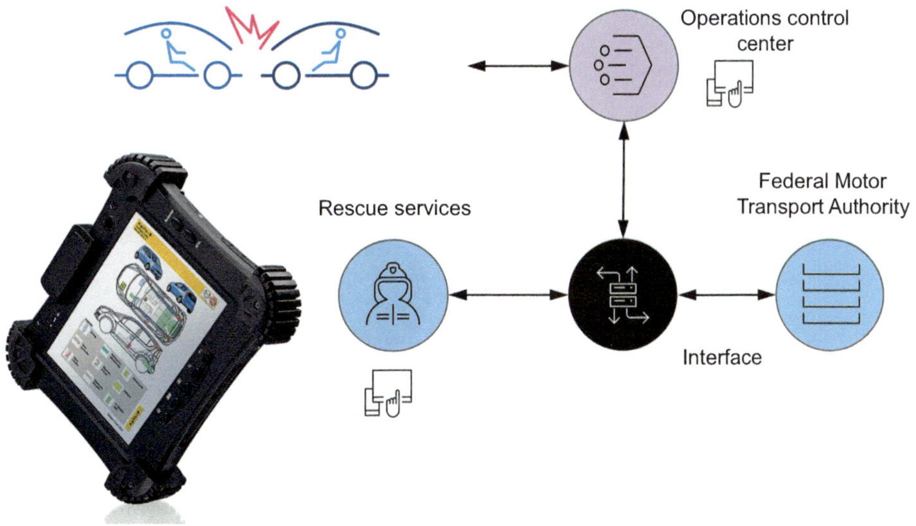

Fig. 6.143 Online rescue data sheet system for fire brigades. (*Source*: DAT)

6.10.1.5 Emergency Care by Rescue Services

The injuries of traffic accident victims range from minor lesions to life-threatening injuries to multiple parts of the body (polytrauma). The rescue team determines the procedures for preclinical initial care and adjusts them depending on the situation on site and further course.

For emergency patients, the principle of "*treat first what kills first*", applies to ensure survival in the next few minutes or hours by securing vital functions. For the structured examination and care of emergency patients, rescue workers therefore follow guidelines (*rescue schemes*) such as the *ABCDE-schema* [123] or cABCDE-schema, which sets priorities for life-threatening conditions: **c:** Critical bleeding (life-threatening external bleeding), **A:** Airway, **B:** Breathing (respiratory rate), **C:** Circulation, **D:** Disability (neurological deficits), **E:** Exposure (exploration, includes body temperature, fractures, environment).

Nowadays, comprehensive communication of rescue workers from the accident site with the control center and the clinic can coordinate further care for the injured. Such a system is referred to as an emergency assistance system and has been implemented in Germany, Austria, and Switzerland as an *emergency, information and documentation assistant* (NIDA) in many regions.

Ambulances and emergency doctor vehicles are equipped with mobile devices such as tablets for online connection via mobile data connections (Fig. 6.144), through which they are connected to their clinic and the rescue control center and can handle voice, video, and data communication throughout the entire period of the operation. Further

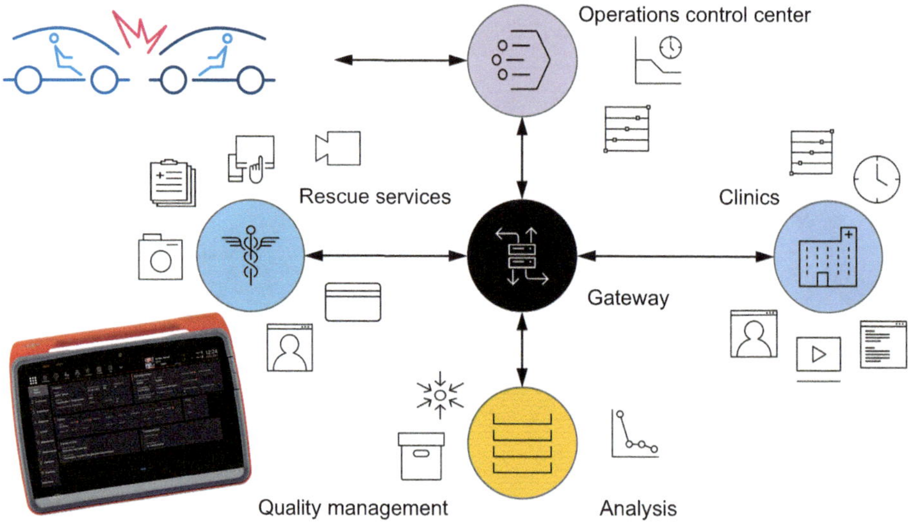

Fig. 6.144 Information chain after the accident event for the deployment of the rescue service. (*Source*: ZTM, medDV)

information for preparing the operation can be transmitted on the way to the accident site—for example, when the control center has learned new details.

For the further course, it must be decided which clinic is suitable for the care of the injured and currently has the necessary capacities. This is particularly true for severely injured people who need special care. Here, the emergency assistance system can support and accelerate the coordination process. Once the emergency care is completed and the patient is stabilized, transport to the designated clinic begins.

The data from the emergency assistance system support the handover to the emergency room in the clinic. The emergency doctor's initial diagnosis, patient data (history, electrocardiogram, vital parameters) as well as images or videos can be sent to the emergency room from the accident site. In the clinic, the accident situation, the accident kinematics (Chap. 3) of the occupants as well as probable stresses and injuries of the injured (e.g., *acceleration-deceleration trauma)* can be assessed much better. Preparations can already be made in the emergency room before the ambulance arrives.

After the emergency operation, the systematic documentation of preclinical measures is carried out according to standardized and legally secure procedures. For this purpose, for example, the *emergency operation protocol* [26] and the *minimal emergency data set* (MIND) [27] of the German Interdisciplinary Association for Intensive and Emergency Medicine (DIVI) are used [85] Fig. 6.145). This documentation is also an element of quality management in rescue services and thus also contributes to the identification of further potential for improvement.

Rescue chain and integral safety in the post-crash phase

A selection of further information offers and topic pages on the technologies of integral safety in the post-crash phase and the rescue chain for traffic accident victims is listed in the following list:

- ADAC—www.adac.de—Rescue sheets
- Deutsche Automobil Treuhand (DAT)—www.datgroup.com—Rescue card database
- German Interdisciplinary Association for Intensive Care and Emergency Medicine (DIVI)—www.divi.de—DIVI-emergency operation protocol, minimum emergency data set MIND4.0
- Euro NCAP—www.euroncap.com—Rescue and extrication
- International Association of Fire and Rescue Services—www.ctif.org
- ISO—www.iso.org—ISO 17840—Road vehicles—Information for first and second responders
- Center for Telemedicine (ZTM)—www.ztm.de—NIDA

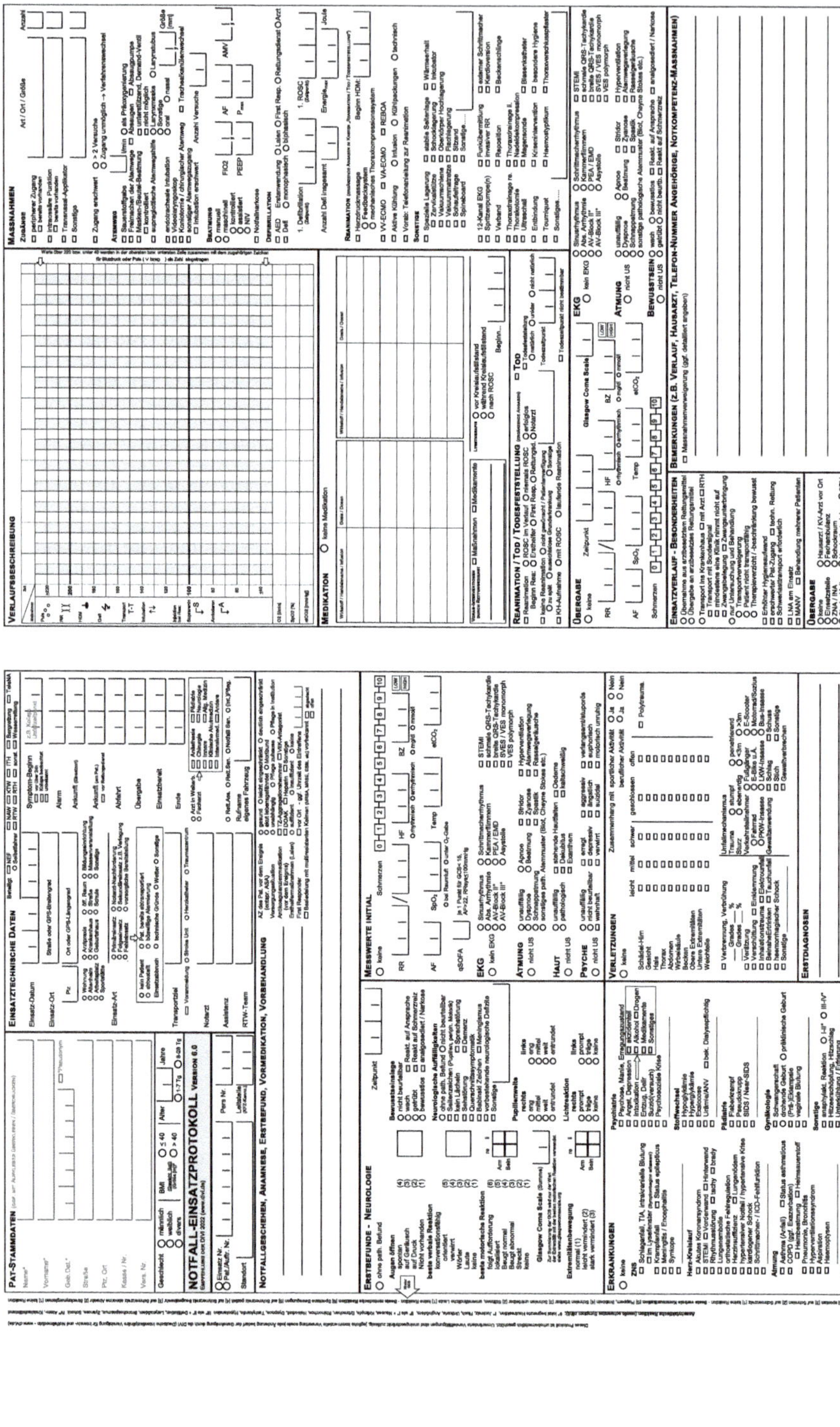

Fig. 6.145 Standardized recording form DIVI-emergency operation protocol [26]. (*Source:* ZTM)

6.10.2 Restoration of Traffic Safety

In addition to the emergency care of injured persons by the rescue services, further tasks have to be fulfilled at the accident site in order to quickly restore traffic safety at the accident site and release the road for flowing traffic.

This includes the recovery of vehicles, towing or loading and transporting damaged and no longer roadworthy vehicles, road cleaning, securing and blocking measures, hazardous substance removal, road repair. These activities are carried out by breakdown services, fire brigades, the road maintenance department, municipal services or the technical relief agency.

After the accident vehicles have been moved to a storage area, they are secured against possible subsequent damage, e.g. by an undetected short circuit or a smoldering fire. For this purpose, the power supply of the low-voltage batteries is interrupted. It may also be necessary to drain any operating fluids remaining in the vehicle. Special measures and regulations apply to the parking of vehicles with high-voltage batteries and alternative drives.

6.10.3 Fire Protection

Vehicle fires as a sole accident event or as a result of a collision pose a great danger to occupants, other persons in the traffic area and rescue forces. A vehicle fire can also severely damage infrastructure such as roads, bridges, tunnels or parking garages due to heat development, residues of extinguishing agents, water ingress or smoke gases. Therefore, fire protection in motor vehicles includes a bundle of measures, which will be discussed in detail at this point.

A common cause of vehicle fires is leaking fuel, which ignites on hot components of the engine and drivetrain. Vehicle fires as a cause or as a result of a traffic accident occur much less frequently.

6.10.3.1 Fire Protection through Portable Fire Extinguishers

Incipient fires can be effectively combated with portable *fire extinguishers* carried in the vehicle. For example, a cable fire can be extinguished at an early stage, preventing the vehicle from burning out. Car fire extinguishers are only mandatory in a few countries so far.

With *ABC powder*, fires of the three *fire classes*—fires of solid materials (A), liquid fires (B), and gas fires (C)—can be extinguished. Since ABC extinguishers are also durable and cost-effective, this variant is the most widespread universal extinguishing agent. However, the elaborate removal of the powder after the fire is a major disadvantage and can cause high consequential damage.

Alternative portable fire extinguishers are foam extinguishers, which are suitable for fire classes A and B. The extinguishing agent is a water-foam mixture.

6.10.3.2 Vehicle-Based Fire Detection and Extinguishing Systems

As an additional facility, *automatic fire detection-* and *extinguishing systems* are primarily installed in motorsport or are available as aftermarket products.

Especially in commercial vehicles such as buses, special vehicles or work machines, fire protection is continuously evolving. However, standard installation is still rather rare here, as extinguishing systems in particular require additional maintenance during the life cycle. After activation by the fire alarm system, occupants and the environment are warned by visual and acoustic signals and an emergency call can be sent to a control center. The fire extinguishing system releases the extinguishing agent.

The basic structure of a fire alarm system for combustion vehicles includes heat sensors installed in the engine compartment, which activate the fire extinguishing system from a certain trigger temperature, or detection or activation cables laid in the engine compartment, which trigger when they come into contact with an open flame. The extinguishing agent then reaches the engine compartment and the fuel tank via a pipe system and distributor nozzles from the extinguishing agent container. Depending on the operating conditions, ABC powder, foam extinguishing agents or carbon dioxide (CO_2) are used as extinguishing agents in a vehicle-installed fire extinguishing system. Another variant is an aerosol extinguishing system. By decomposing potassium bicarbonate into potassium carbonate, water vapor, and carbon dioxide, the fire is deprived of thermal energy through an endothermic reaction. In addition, the resulting potassium compounds release radicals that interrupt the combustion process by neutralizing flame-propagating free radicals. This enhances the extinguishing effect. Aerosol extinguishing systems are offered for various vehicle applications for fire protection in engine compartments, for the protection of work machines, but also for fire protection in installed high-voltage battery systems.

6.10.3.3 Fire Protection Measures for Vehicles with Internal Combustion Engines

In vehicles with petrol or diesel drive, the fuel tank and the fuel system are sensitive components. Hot engine or engine-mounted parts, the exhaust system or hot components of the drive train also pose fire risks.

The *fuel tank* is placed in a location in the vehicle where there is the least risk of deformation, damage, or leakage causing a fire in the event of an accident. It is located, for example, under the rear seat bench in front of or above the rear axle.

6.10.3.4 Fire Protection in Vehicles with High-Voltage Batteries

Vehicles with high-voltage battery and power supply systems like plug-in hybrid vehicles (PHEV) and battery electric vehicles (BEV) as well as vehicles with a hydrogen-powered fuel cell (FCEV) require adjusted measures for fire protection..

Vehicles with high-voltage systems have several safety systems that reduce the risk of a fire after a collision due to a short circuit or the risk of electric shock for both occupants and first responders and rescue workers.

If the airbag control unit detects an accident, it triggers restraint systems such as airbags and seat belt pretensioners. In addition, the power supply of the high-voltage system is now also permanently and directly interrupted at the high-voltage battery and also at the charging system. Here, *pyrotechnic disconnectors* (also: *pyro-fuse*) are used as actuators. With the help of a pyrotechnic propellant, a bolt or wedge permanently (non-reversible) disconnects the flow of electric current. The power shutdown is achieved within a few milliseconds. Thus, the high-voltage network in the vehicle is already separated from the battery before damage and a short circuit as a possible cause of fire can occur due to the collision. In other emergency situations, the pyrotechnic disconnector can also be activated by the *battery management system* (BMS).

6.10.3.5 Fire Protection in Vehicles with Hydrogen Drive

Vehicles with hydrogen drive will gain a larger market share in the future—especially in the commercial vehicle sector. Practical experience will lead to further developments of technical solutions and the development of fire-fighting strategies for the various hydrogen applications; this process is already underway. The following specific challenges of hydrogen technology must be taken into account:

- Hydrogen molecules are the smallest molecules and have a low viscosity. Since hydrogen is odorless and colorless, leaks are difficult for humans to detect. Pressurized gas containers, lines and connections are therefore designed to be leak-proof.
- The flame of burning hydrogen emits little infrared or visible light, so it is hardly perceived as heat by humans. On the other hand, the flame emits UV radiation detectable by UV sensors or UV-IR hybrid detectors.
- Escaping hydrogen can form a gas-air mixture that burns explosively.
- The small size of hydrogen molecules can cause hydrogen embrittlement, reducing ductility and potentially leading to cracking. This must also be taken into account when designing the supply chain and pressure gas containers.

Standard operating procedures have been designed for fire departments, which are currently being discussed and further developed.

References

1. AS Autoglas: Airbag und Windschutzscheibe. https://www.as-autoglas.de/wissenswertes/airbag-und-windschutzscheibe/ (2023)
2. Adomeit, H.-D.: Neue Bewertungsgrößen für die Frontalaufprallsicherheit des gurtgesicherten Insassen. Dissertation. https://www.tib.eu/de/suchen/id/TIBKAT%3A1009385046 (1980)
3. Appel, H., Kramer, F.: Biomechanik und Kraftfahrzeugsicherheit. Lecture manuscript, Institut für Fahrzeugtechnik, Technische Universität Berlin (1987)

4. Autoliv AB: Cyclist Head Injuries: A Persistent Challenge. https://www.autoliv.com/sites/autoliv/files/pr/202207074695-1.pdf?ts=1657519220 (2022, September)
5. Autoliv AB: Airbag technology can make the difference – A motorcycle helmet with an integrated airbag. https://www.autoliv.com/sites/default/files/2022-11/A5%20Airoh%20Flyer.pdf (2022, November)
6. Autoliv AB: Autoliv performs first crash test of an e-scooter airbag. https://www.autoliv.com/press/autoliv-performs-first-crash-test-e-scooter-airbag-1769461 (2020, Januar 31)
7. Autoliv AB: Autoliv and the Piaggio Group join forces to develop a scooter and motorcycle airbag. https://www.autoliv.com/press/autoliv-and-piaggio-group-join-forces-develop-scooter-and-motorcycle-airbag-1961244 (2021, November 4)
8. Ford Motor Company: Automotive Industry's Only Five-Row Side-Curtain Airbag Debuts In All-New Ford Transit: News. https://media.lincoln.com/content/fordmedia/fna/us/en/news/2014/06/25/five-row-side-curtain-airbag-debuts-in-all-new-transit.html (2014, Juli 25)
9. Bass, C. R., Crandall, Jeff. R., Pilkey, Walter D.: Out-of-Position Occupant Testing (OOPS3 Series) (S. 118). University of Virginia, Automobile Safety Laboratory. https://www.nhtsa.gov/sites/nhtsa.gov/files/oopreport_0.pdf (1998)
10. Bieler, D., Lechleuthner, A.: Entwicklung des Notarztwesens in der Präklinik: Zum 25. Geburtstag der Notfall + Rettungsmedizin. Notfall + Rettungsmedizin **25**(3), 156–158. https://doi.org/10.1007/s10049-022-01001-3 (2022)
11. Bohlin, N.I.: Safety Belt (United States Patent Office Patent Nr. US-3043625-A). (1962). https://patents.google.com/patent/US3043625A/en
12. Bonin, S.J., DeMarco, A.L., Siegmund, G.P.: The effect of MIPS, headform condition, and impact orientation on headform kinematics across a range of impact speeds during oblique bicycle helmet impacts. Ann. Biomed. Eng. **50**(7), 860–870 (2022). https://doi.org/10.1007/s10439-022-02961-w
13. Bubb, H., Bengler, K., Grünen, R.E., Vollrath, M.: Automobilergonomie. Springer Fachmedien, Wiesbaden (2015). https://doi.org/10.1007/978-3-8348-2297-0
14. Bundesanstalt für Straßenwesen (BASt): Optimierung von Kinderschutzsystemen im Pkw. Fahrzeugtechnik, Fachthemen. https://bast.opus.hbz-nrw.de/frontdoor/deliver/index/docId/241/file/F67.pdf (2023)
15. Bundesministerium für Gesundheit: Leistungsfälle bei Rettungsfahrten und Krankentransporten der Versicherten der gesetzlichen Krankenversicherung (Gesundheitsberichterstattung des Bundes). Bundesministerium für Gesundheit (BMG). https://www.gbe-bund.de (2021)
16. Burczyk, C., Merz, U.: Beltbag – Erhöhung der Sicherheit und des Komforts im Fond. ATZ – Automobiltechnische Zeitschrift **115**(10), 750–757 (2013).https://doi.org/10.1007/s35148-013-0266-1
17. Burdi, A.R., Huelke, D.F., Snyder, R.G., Lowrey, G.H.: Infants and children in the adult world of automobile safety design: Pediatric and anatomical considerations for design of child restraints. J. Biomech. **2**(3), 267–280 (1969). https://doi.org/10.1016/0021-9290(69)90083-9
18. Bustos, A., Schultz, D.: Beifahrerairbag im Dach. ATZ – Automobiltechnische Zeitschrift, 114(4), 334–339. https://doi.org/10.1007/s35148-012-0336-9 (2012)
19. Butler, R. D.: Mobile emergency attack and failsafe detection (Patent and Trademark Office (USPTO) Patent Nr. US-20140066000-A1). https://image-ppubs.uspto.gov/dirsearch-public/print/downloadPdf/20140066000 (2015)
20. Class, U., Seyffert, M., Seewald, A.: Active seat belt buckle for comfort and safety. ATZ Worldwide **115**(6), 24–28 (2013). https://doi.org/10.1007/s38311-013-0070-2
21. Class, U.: Die Verbindung von passiver und aktiver Sicherheit mit dem Gurtaufroller ACR-2. ATZ – Automobiltechnische Zeitschrift **107**(10), 888–892 (2005). https://doi.org/10.1007/BF03221743

22. Costas Muñoz, D., Majzel, M.: Improving collapsibility robustness of an EPS-CD by means of simulation and six sigma. Adv. Model. Simu. Eng. Sci. **7**(1), 7 (2020). https://doi.org/10.1186/s40323-020-0144-9
23. DAT: SILVERDAT 3 PRO. Deutsche Automobil Treuhand GmbH (DAT). https://www.dat.de (2022)
24. DIN:DIN EN 1078:2023–02, Helme für Radfahrer; Deutsche und Englische Fassung prEN_1078:2023. Beuth Verlag GmbH. https://doi.org/10.31030/3409307 (2023)
25. DIN: DIN EN 1080:2023–02, Stoßschutzhelme für Kleinkinder; Deutsche und Englische Fassung prEN_1080:2023. Beuth Verlag GmbH. https://doi.org/10.31030/3408891 (2023)
26. DIVI: DIVI-Notfalleinsatzprotokoll 6.1. Deutsche Interdisziplinärer Vereinigung für Intensiv- und Notfallmedizin (DIVI). https://www.divi.de/component/edocman/service/mind-notfalleinsatzprotokoll (2022)
27. DIVI: Minimaler Notfalldatensatz MIND 4.0 (S. 8). Deutsche Interdisziplinärer Vereinigung für Intensiv- und Notfallmedizin (DIVI). https://www.divi.de/component/edocman/service/mind-notfalleinsatzprotokoll (2021)
28. Destatis: 23621: Personal im Gesundheitswesen, Gesundheitspersonalrechnung 2000–2021 (EVAS-Nr. 23621; Gesundheitspersonalrechnung, GPR). Statistisches Bundesamt, Destatis. https://www-genesis.destatis.de/genesis/online?sequenz=statistikTabellen&selectionname=23621 (2023)
29. Deutsches Patent und Markenamt (DPMA): Dreipunktgurt. https://www.dpma.de/dpma/veroeffentlichungen/meilensteine/erfindungenmitgeschichten/dreipunktegurt/index.html (2022, November 3)
30. EU-MOVE: EU Road Safety – Results conference 2022. https://road-safety.transport.ec.europa.eu/eu-road-safety-results-conference-2022-2022-12-08_en (2022, August 12)
31. Eckstein, L: Strukturentwurf von Kraftfahrzeugen: Bauweisen und Package; Fertigungs- und Fügeverfahren; Leichtbau und Betriebsfestigkeit; Passive Sicherheit; Fußgängerschutz; Vorlesungsumdruck Strukturentwurf (4. Aufl.). fka, Forschungsgesellschaft Kraftfahrwesen (2014)
32. Euro NCAP: Euro NCAP – Test Protocol – VRU (Test Protocol Version 9.0.2; S. 56). Euro NCAP. https://www.euroncap.com/en/for-engineers/protocols/vulnerable-road-user-vru-protection/ (2023)
33. Euro NCAP: Euro NCAP – Assessment Protocol – VRU (Assessment Protocol Version 11.2.2; S. 30). Euro NCAP. https://www.euroncap.com/en/for-engineers/protocols/vulnerable-road-user-vru-protection/ (2023)
34. Euro NCAP: Compatibility Assessment (TB 027; 1.0). https://cdn.euroncap.com/media/41786/tb-027-compatiblity-assessment-v10.201811161304153907.pdf (2018)
35. Euro NCAP: Child Occupant Protection (COP) (v8.0) [Assessment Protocol]. Euro NCAP. https://www.euroncap.com/en/for-engineers/protocols/child-occupant-protection/ (2021)
36. European Commission (EC): Horizon 2020 – Future Occupant Safety for Crashes in Cars (OSCCAR). https://cordis.europa.eu/project/id/768947 (2018, Juni 1)
37. European Commission: Weißbuch – Die europäische Verkehrspolitik bis 2010: Weichenstellungen für die Zukunft (S. 152). Europäische Kommission (EU). https://www.europarl.europa.eu/meetdocs/committees/rett/20020219/com(2001)370de.pdf (2001)
38. Evers, C.: Gurte, Kindersitze, Helme und Schutzkleidung – 2021. https://www.bast.de/DE/Publikationen/DaFa/2022-2021/2022-02.html (2022, Februar)
39. Fang, X.: Anforderungen an die Karosserie durch die passive Sicherheit. In X. Fang, Karosserieentwicklung und -Leichtbau, S. 233–318. Springer, Berlin, Heidelberg (2023). https://doi.org/10.1007/978-3-662-67118-4_6

40. Fédération Internationale de l'Automobile (FIA).: Safety Harness (FIA Standard Nr. 8854/98). https://www.fia.com/fia-standard-885498-safety-harnesses-updated-28092012 (2012)
41. Gasch, B.: Struktur des Rettungswesens in Deutschland. In F. Lasogga & B. Gasch (Hrsg.), Notfallpsychologie (S. 397–402). Springer Berlin Heidelberg. https://doi.org/10.1007/978-3-540-71626-6_24 (2008)
42. Gehre, C., Kramer, S., Schindler, V.: Seitenairbag und Kinderrückhaltesysteme. Carl Ed. Schünemann KG. https://www.nw-verlag.de/seitenairbag-und-kinderrueckhaltesysteme.html (2004)
43. Girstmair, J., Machens, K.-U., Witfeld, E., Ober, F.: Noise of Seatbelt Retractors A Real Challenge. ATZ Worldwide **119**(7–8), 42–45 (2017). https://doi.org/10.1007/s38311-017-0061-9
44. Gross, D., Hauger, W., Schröder, J., Wall, W.A.: Technische Mechanik 3: Kinetik. Springer, Berlin Heidelberg. (2015). https://doi.org/10.1007/978-3-642-53954-1
45. Gwehenberger, J., Bende, J., Bogenberger, T., Langieder, K:. ECBOS Work-Package 1, Task 1.3 – Database Integration [Studie]. https://www.dsd.at/ (2004)
46. HDLI: Convertibles versus coupes (Vol. 37, No. 4; HLDI Bulletin, S. 13). Highway Loss Data Institute. https://www.iihs.org/media/5c8ec650-fecd-479b-8191-788fcb30642b/qIk1Wg/HLDI%20Research/Bulletins/hldi_bulletin_37-04.pdf (2020)
47. Halldin, P., Gilchrist, A., Mills, N.J.: A new oblique impact test for motorcycle helmets. Int. J. Crashworthiness **6**(1), 53–64 (2001). https://doi.org/10.1533/cras.2001.0162
48. Halldin, P., Gilchrist, A., Mills, N. J.: A new oblique impact test for motorcycle helmets. IRCOBI Conference Proceedings 2001, 3. http://www.ircobi.org/wordpress/downloads/irc0111/2001/Session6/6.4.pdf (2001)
49. Hetrick, J. W.: Safety cushion assembly for automotive vehicles (United States Patent and Trademark Office (USPTO) Patent Nr. US-2649311-A). https://image-ppubs.uspto.gov/dir-search-public/print/downloadPdf/2649311 (1953)
50. Hornquist, N., Aderum, T.: APG-1, Autoliv's New Generation of „green" and clean Airbag Inflators. 11th International Symposium and Exhibition on Sophisticated Car Occupant Safety Systems. Airbag 2012, Karlsruhe. https://www.tib.eu/de/suchen/id/TIBKAT%3A734786883 (2012, März 12)
51. Hu, J., Reed, M. P., Rupp, J. D., Fischer, K., Lange, P., Adler, A.: Optimizing seat belt and airbag designs for rear seat occupant protection in frontal crashes. 2017–22–0004. (2017). https://doi.org/10.4271/2017-22-0004
52. Hu, J., Rupp, J., Reed, M. P., Fischer, K., Lange, P., Adler, A.: Rear Seat Restraint Optimization Considering The Needs From a Diverse Population (DOT HS 812 248; S. 76). National Highway Traffic Safety Administration (NHTSA). https://www.nhtsa.gov/sites/nhtsa.gov/files/812248-rearseatscarab_report.pdf (2016)
53. ISO.: DIN EN ISO 14451–5:2013–08, Pyrotechnische Gegenstände für Fahrzeuge – Teil 5: Anforderungen und Kategorisierung von Airbaggasgeneratoren (DIN EN ISO 14451–5:2013–08). Beuth Verlag GmbH. (2013). https://doi.org/10.31030/1935086
54. ISO.: Road vehicles – Anchorages in vehicles and attachments to anchorages for child restraint systems – Part 1: Seat bight anchorages and attachments – Amendment 1: CRF reduced height specification (ISO 13216–1:1999/Amd 1:2006). ISO. https://www.iso.org/standard/36592.html (2006)
55. ISO 17840–1: Road vehicles – Information for first and second responders (ISO 17840–1: 2022; S. 15). International Organization for Standardization (ISO). https://www.iso.org/standard/78461.html (2022)
56. Imam, S. A.: Evaluation Of far-side occupant safety based On numerical modeling [Wayne State University]. https://digitalcommons.wayne.edu/oa_dissertations/2416 (2020)

57. Janssen, E. G., Huijskens, C. G., Beusenberg, M. C.: Reduction in seat belt effectiveness due to misuse. International IRCOBI Conference on the Biomechanics of Impacts. Verona (Italy) (1992)
58. Knecht, T.: Der Sicherheitsgurt – wer hat ihn erfunden? Tom's Rückspiegel. https://saabblog.net/2020/12/10/der-sicherheitsgurt-wer-hat-ihn-erfunden/ (2020, Dezember 10)
59. Kompetenzzentrum – Das virtuelle Fahrzeug Forschungsgesellschaft mbH: Future Occupant Safety for Crashes in Cars (OSCCAR). https://www.osccarproject.eu/ (o. J.)
60. Kontrollierter Sturz mindert Verletzungen. Gute Fahrt, 10 (1986)
61. Kramer, F., Franz, U., Lorenz, B., Remfrey, J., Schöneburg, R.: Integrale Sicherheit von Kraftfahrzeugen: Biomechanik – Simulation – Sicherheit im Entwicklungsprozess. Springer Fachmedien, Wiesbaden (2013). https://doi.org/10.1007/978-3-8348-2608-4
62. Kramer, F., Appel, H., Adomeit, H.-. D.: Der Einfluß von Strukturparametern auf die Belastung und Kinematik von PKW-Insassen. Internationaler Kongreß der VDI-Gesellschaft Fahrzeugtechnik „Berechnung im Automobilbau" (1990)
63. Kramer, F., Kramer, M., Lüder, M., Herpich, T., Class, U.: Die Bedeutung von OoP-Situationen im Straßenverkehr hinsichtlich künftiger Auslegung von Insassenschutz-Systemen. VDI-Tagung „Innovativer Kfz-Insassen- und Partnerschutz" (2003)
64. Kramer, F., Scholpp, G., Otte, D.: Das Verletzungsrisiko in unterschiedlich schweren Fahrzeugen und daraus abgeleitete Erfordernisse für verbesserte Rückhaltesysteme. VDI/BMW-Gemeinschaftstagung „Sicherheit im Straßenverkehr". München (1993)
65. Kurt, M., Laksari, K., Kuo, C., Grant, G.A., Camarillo, D.B.: Modeling and Optimization of Airbag Helmets for Preventing Head Injuries in Bicycling. Ann. Biomed. Eng. **45**(4), 1148–1160 (2017). https://doi.org/10.1007/s10439-016-1732-1
66. König, P. & Hochschule Trier: Adaptives Fahrzeug-Rückhaltesystem. https://www.hochschule-trier.de/hauptcampus/forschung/forschung-am-hauptcampus/projekte-entdecken/adaptives-fahrzeug-rueckhaltesystem (2023)
67. Kübler, L., Gargallo, S., Elsäßer, K.: Bewertungskriterien zur Auslegung von Insassenschutzsystemen. ATZ – Automobiltechnische Zeitschrift **111**(6), 426–433. (2009). https://doi.org/10.1007/BF03222079
68. Kühn, M., Fröming, R., Schindler, V.: Fußgängerschutz: Unfallgeschehen, Fahrzeuggestaltung, Testverfahren. Springer Berlin Heidelberg. https://doi.org/10.1007/978-3-540-34303-5 (2007)
69. Laakmann, F.: Integrated Safety – Enabling Technologies for Future Vehicles. Airbag 2018, 14th International Symposium and Accompanying Exhibition on Sophisticated Car Occupant Safety Systems. Airbag 2018, Pfinztal. https://publica.fraunhofer.de/handle/publica/404813 (2018, November 27)
70. Laakmann, F: Integrated Safety. From ADAS to Automated Driving. ntegrated Safety, Detroit (2018, Oktober 9)
71. Laakmann, F., Seyffert, M., Herpich, T., Saupp, L., Ladwig, S., Kugelmeier, M., Vollrath, M.: Benefits of Tactile Warning and Alerting of the Driver through an Active Seat Belt System. 27th International Technical Conference on the Enhanced Safety of Vehicles (ESV), 24. https://www-esv.nhtsa.dot.gov/Proceedings/27/27ESV-000216.pdf (2023)
72. Langwieder, K: Heckunterfahrschutz bei Nutzfahrzeugen. Automobiltechnische Zeitschrift (ATZ, Nr. 103) (2001)
73. Linderer, W.: Einrichtung zum Schutze von in Fahrzeugen befindlichen Personen gegen Verletzungen bei Zusammenstößen (Deutsches Patent und Markenamt (DPMA) Patent Nr. 869312). https://www.dpma.de/docs/postergalerieneu/patentschriften/028airbag-walterlinderer.pdf (1953)

74. Lippert, H., Herbold, D., Lippert-Burmester, W:. Anatomie: Text und Atlas (10., erweiterte Auflage). Urban & Fischer Verl/Elsevier. (2017) https://shop.elsevier.de/anatomie-9783437261831.html
75. Lohs, T.: Qualitätsbericht Rettungsdienst Baden-Württemberg (Berichtsjahr 2021; S. 145). Stelle zur trägerübergreifenden Qualitätssicherung im Rettungsdienst Baden-Württemberg (SQR-BW). https://www.sqrbw.de/fileadmin/SQRBW/Downloads/Qualitaetsberichte/SQRBW_Qualitaetsbericht_2021_web.pdf (2022)
76. Lutz, H., Machens, K.-U., Hoffmann, S.-Y., Kübler, L.: SPR4 – Gurtstraffer der Nächsten Generation. ATZ – Automobiltechnische Zeitschrift **112**(12), 886–891. https://doi.org/10.1007/BF03222217 (2010)
77. MAN Truck & Bus SE: Integralbauweise Buskarosserie. Glossar. https://www.mantruckandbus.com/de/unternehmen/glossar/integralbauweise-buskarosserie.html (2023)
78. Machens, K.-U., Kuebler, L.: Dynamic Testing with Pre-Crash Activation to Design Adaptive Safety Systems. 27th International Technical Conference on the Enhanced Safety of Vehicles (ESV), 24. https://www-esv.nhtsa.dot.gov/Proceedings/27/27ESV-000067.pdf (2023)
79. Malczyk, A., Koch, M.: Injuries to heavy truck occupants – Severity, patterns and causation. IRCOBI Conference 2019, 14. http://www.ircobi.org/wordpress/downloads/irc19/pdf-files/83.pdf (2019)
80. Malczyk, A., Koch, M.: Verletzungsgeschehen von Insassen schwerer LKW (Nr. 102; Unfallforschung kompakt). Gesamtverband der Deutschen Versicherungswirtschaft e. V. (GDV). https://www.udv.de/resource/blob/79256/656f0f885e4e2c46ecafb3aa3e347851/102-verletzung-insassen-lkw-data.pdf (2020)
81. Mercedes-Benz: Das Experimental-Sicherheits-Fahrzeug ESF 2009 – Belt Bag: Clevere Kreuzung aus Gurt und Airbag. Technologie & Innovationen. http://www.daimler.com (2012, Oktober 21)
82. Mercedes-Benz Group AG: 30 Jahre Beifahrer-Airbag, 20 Jahre Windowbag. Innovationen, Meilensteine. https://www.mercedes-benz.com/de/innovation/meilensteine/airbag-jubilaeum/ (2018)
83. Mercedes-Benz Group: Experimental-Sicherheits-Fahrzeug (ESF) 2019 – Neue Sicherheitsideen für eine neue Mobilität. Technologie. https://group.mercedes-benz.com/innovation/esf-2019.html (2023)
84. Mertz, H. J.: Restraint Performance of the 1973–76 GM Air Cushion Restraint System. 880400. (1988) https://doi.org/10.4271/880400
85. Messelken, M., Schlechtriemen, T., Arntz, H.-R., Bohn, A., Bradschetl, G., Brammen, D., Braun, J., Gries, A., Helm, M., Kill, C., Mochmann, C., Paffrath, T.: Der Minimale Notfalldatensatz MIND3. Der Notarzt **27**(05), 197–202 (2011). https://doi.org/10.1055/s-0031-1276903
86. NHTSA: Takata recall spotlight. National Highway Traffic Safety Administration (NHTSA). https://www.nhtsa.gov/equipment/takata-recall-spotlight (2023)
87. National Highway Traffic Safety Administration, NHTSA.: Seat Belts. Risky Driving. https://www.nhtsa.gov/risky-driving/seat-belts (2023)
88. National Highway Traffic Safety Administration, NHTSA: Seat Belt Use in 2022 – Overall Results (DOT HS 813 407; S. 4). National Highway Traffic Safety Administration, NHTSA. https://crashstats.nhtsa.dot.gov/Api/Public/ViewPublication/813407 (2023)
89. National Inventors Hall of Fame (NIHF): Nils I. Bohlin – Safety Belt. Inducties. https://www.invent.org/inductees/nils-i-bohlin (2023)
90. Niederer, P. F.: Der Stoßgürtel – Ein Sicherheitskonzept für Leichtfahrzeuge. Leichtmobil-Symposium. Wildhaus (CH) (1992)

91. Niederer, P., Walz, F., Zollinger, U.: Adverse effects of seat belts and causes of belt failures in severe car accidents in Switzerland during 1976. 21st Stapp Car Crash Conference, New Orleans, LA. https://doi.org/10.4271/770916 (1977, Februar 1)
92. Nissan Motor: President (1990: JHG50). Nissan Heritage Collection. https://www.nissan-global.com/EN/HERITAGE/president.html (2023)
93. Pohle, M., Maier, R.: Identifikation von unfallauffälligen Stellen motorisierter Zweiradfahrer innerhalb geschlossener Ortschaften. Fachverlag NW in der Carl Schünemann Verlag GmbH. https://bast.opus.hbz-nrw.de/files/1658/V269_Anhang.pdf (2016)
94. Research Council for Automobile Repairs (RCAR). A Procedure for Evaluating Motor Vehicle Head Restraints (RCAR, S. 14) [Protocol]. Insurance Institute for Highway Safety (IIHS). https://www.iihs.org/media/c9d1f323-11bd-464d-be63-88cae2423de3/5of9kQ/Ratings/Protocols/archive/rcar-iihs_evaluation_procedure_v3_0308.pdf (2008)
95. Richter, B., Appel, H., Hoefs, R., Langwieder, K.: Entwicklung von PKW im Hinblick auf einen volkswirtschaftlich optimalen Insassenschutz. Abschlußbericht des Forschungsprojekts TV 8036 des Bundesministers für Forschung und Technologie (1984)
96. Robert Bosch GmbH, Simon, B.: Geschichte der Airbag-Steuergeräte. History Blog. https://www.bosch.com/de/stories/geschichte-der-airbag-steuergeraete/ (2023)
97. Rokosch, U: Airbag und Gurtstraffer (2., überarb. Aufl). Vogel-Buchverl, Würzburg (2011)
98. Röper, J.W.A.: Kosten der hubschraubergestützten Notfallversorgung: Innovationsbasierte Szenarioanalyse und Empfehlungen zur Gestaltung von Luftrettungssystemen. Springer Fachmedien, Wiesbaden (2022). https://doi.org/10.1007/978-3-658-38301-5
99. SAE USCAR: USCAR24–2 – Inflator Technical Requirements and Validation (Nr. USCAR24–2). SAE International. https://doi.org/10.4271/USCAR24-2 (2013)
100. SFI Foundation: Advanced Motorsport Restraint Assemblies (SFI SPECIFICATION 16.6). https://www.sfifoundation.com/wp-content/pdfs/specs/Spec_16.6_042018.pdf (2018)
101. Scania CV AB: Side curtain airbags keep drivers safe. Newsroom. https://www.scania.com/group/en/home/newsroom/news/2019/side-curtain-airbags-keep-drivers-safe.html (2019, Februar 27)
102. Schaffner, T., Girstmair, J., Neumann, J., Witfeld, E., Melzer, W.: Prediction of Eigenfrequencies and Eigenmodes of Seatbelt Retractors in the Vehicle Environment, Supporting an Acoustically Optimal Retractor Integration by CAE. 2018–01–1543. https://doi.org/10.4271/2018-01-1543 (2018)
103. Schlott, S.: Airbag: Die zündende Idee beim Insassenschutz. Verl. Moderne Industrie. https://www.tib.eu/de/suchen/id/TIBKAT:09005511X (1996)
104. Schmiedel, R., Behrendt, H: Leistungen des Rettungsdienstes. 2016/17. Fachverlag NW in Carl Ed. Schünemann KG. https://bast.opus.hbz-nrw.de/files/2319/M290.pdf (2016)
105. Schöneburg, R., Hart, M., Feese, J., Mücke, S., Richert, J.: ESF 2019 – Experimental Safety Vehicle Meets Automated Driving Mode. 26th International Technical Conference on the Enhanced Safety of Vehicles (ESV), 8. https://www-esv.nhtsa.dot.gov/Proceedings/26/26ESV-000042.pdf (2019)
106. Seeck, A: Einfluss der Bodenfreiheit der EEVC-Barriere im Vergleich zum Fahrzeug/Fahrzeug-Seitenaufprall. Bundesanstalt für Straßenwesen (BASt). https://www.bast.de (1994)
107. Springer Fachmedien Wiesbaden: Inflatable seatbelts: New dimension to safety. Auto Tech. Rev. **1**(8), 64–64 (2012). https://doi.org/10.1365/s40112-012-0113-x
108. StVZO: § 21 – Betriebserlaubnis für Einzelfahrzeuge, StVZO § 21, Bundesrepublik Deutschland, StVZO. https://www.gesetze-im-internet.de/stvzo_2012/__21.html

109. StVZO: § 21a – Anerkennung von Genehmigungen und Prüfzeichen auf Grund internationaler Vereinbarungen und von Rechtsakten der Europäischen Gemeinschaften, StVZO § 21a, Bundesrepublik Deutschland, StVZO. https://www.gesetze-im-internet.de/stvzo_2012/__21a.html
110. StVZO: § 35a – Sitze, Sicherheitsgurte, Rückhaltesysteme, Rückhalteeinrichtungen für Kinder, Rollstuhlnutzer und Rollstühle, StVZO § 35a, Bundesrepublik Deutschland, StVZO. https://www.gesetze-im-internet.de/stvzo_2012/__35a.html
111. StVZO: § 35f – Notausstiege in Kraftomnibussen, StVZO § 35f, Bundesrepublik Deutschland, StVZO: https://www.gesetze-im-internet.de/stvzo_2012/__21.htmlhttps://www.gesetze-im-internet.de/stvzo_2012/__35f.html
112. Standardization Administration of the People's Republic of China (SAC): Compulsory Product Certification Implementation Rules – Car Seat Belts (GB 14166–2013 (CNCA-C11–04)). https://openstd.samr.gov.cn/bzgk/gb/index (2014)
113. Standardization Administration of the People's Republic of China (SAC): The protection of the occupants in the event of an off-set frontal collision for passenger car (GB/T 20913–2007). https://openstd.samr.gov.cn/bzgk/gb/index (2007)
114. Standardization Administration of the People's Republic of China (SAC): The protection of the occupants in the event of a lateral collision (GB 20071–2006). https://openstd.samr.gov.cn/bzgk/gb/index (2007)
115. Statista: Durchschnittliches Leergewicht neu zugelassener Personenkraftwagen in Deutschland von 2012 bis 2022 (Verkehr & Logistik, Fahrzeuge & Straßenverkehr). Statista. https://de.statista.com/statistik/daten/studie/12944/umfrage/entwicklung-des-leergewichts-von-neuwagen/ (2023)
116. Statistisches Bundesamt, Destatis: 46241. Statistik der Straßenverkehrsunfälle (EVAS-Nr. 46241; Statistik der Straßenverkehrsunfälle). Statistisches Bundesamt, Destatis. https://www-genesis.destatis.de/genesis//online?operation=table&code=46421-0052 (2022)
117. Statistisches Bundesamt, Destatis: 46421–0052. Luftfahrtbewegungen, Flugart 2019–2021 (EVAS-Nr. 46421–0052; Verkehrsleistungsstatistik im Luftverkehr). Statistisches Bundesamt, Destatis. https://www-genesis.destatis.de/genesis//online?operation=table&code=46421-0052 (2022)
118. Stigson, H., Rizzi, M., Ydenius, A., Engström, E., Kullgren, A:. Consumer testing of bicycle helmets. 2017 IRCOBI Conference, 9. http://www.ircobi.org/wordpress/downloads/irc17/pdf-files/30.pdf (2017)
119. Subbaiah, V. M:. Investigation of Takata Inflator Ruptures (1600478.000–0471; S. 34). Exponent, Inc. https://www.nhtsa.gov/document/exponent-research-summary (2016)
120. Teske, L., Goßmann, H., Timm, H., Plath, A., Pecho, W., Derks, M., Veith, S., Reinstettel, M., Rauber, C., Thomer, K. W. Karosserie. In: S. Pischinger & U. Seiffert (Hrsg.), Vieweg Handbuch Kraftfahrzeugtechnik, S. 169–260. Springer Fachmedien, Wiesbaden (2021) https://doi.org/10.1007/978-3-658-25557-2_4
121. The Automotive Hall of Fame (AHF): Nils Bohlin. https://www.automotivehalloffame.org/honoree/nils-bohlin/ (2023)
122. The Automotive Hall of Fame (AHF): Allen K. Breed. https://www.automotivehalloffame.org/honoree/allen-k-breed/ (2023)
123. Thim, T., Krarup, G., Rohde, L.: Initial assessment and treatment with the Airway, Breathing, Circulation, Disability, Exposure (ABCDE) approach. Int. J. Gen. Med. **117** (2012) https://doi.org/10.2147/IJGM.S28478
124. UN Regulation No. 29: – Cabs of commercial vehicles, Nr. UN ECE R29, United Nations, Rev. 2. https://unece.org/transport/vehicle-regulations-wp29/standards/addenda-1958-agreement-regulations-21-40 (2012)

125. UN Regulation No. 66: – Strength of superstructure (buses), Nr. UN ECE R66, United Nations, Rev. 1, Corr. 3. https://unece.org/transport/vehicle-regulations-wp29/standards/addenda-1958-agreement-regulations-61-80 (2007)
126. UN Regulation No. 80: – Strength of seats and their anchorages (buses), Nr. UN ECE R80, United Nations, Rev. 1. https://unece.org/transport/vehicle-regulations-wp29/standards/addenda-1958-agreement-regulations-61-80 (2012)
127. UN Regulation No. 93: – Front underrun protective devices (FUPDs), Nr. UN ECE R93, United Nations, Rev. 1. https://unece.org/transport/vehicle-regulations-wp29/standards/addenda-1958-agreement-regulations-81-100 (1994)
128. UN Regulation No. 58: – Rear underrun protective devices (RUPDs), Nr. UN ECE R58, United Nations, Rev. 3. https://unece.org/transport/vehicle-regulations-wp29/standards/addenda-1958-agreement-regulations-41-60 (2017)
129. UN Regulation No. 73: – Lateral protection devices (LPD), Nr. UN ECE R73, United Nations, Rev. 1. https://unece.org/transport/vehicle-regulations-wp29/standards/addenda-1958-agreement-regulations-61-80 (2011)
130. UN Regulation No. 14 – Safety-belt anchorages, Nr. UN ECE R14, United Nations, Rev. 6. https://unece.org/un-regulations-addenda-1958-agreement (2020)
131. UN Regulation No. 16: – Safety-belts, Nr. UN ECE R16, United Nations, Rev. 9. https://unece.org/un-regulations-addenda-1958-agreement (2018)
132. UN Regulation No. 95: – Lateral collision protection, Pub. L. No. UN ECE R-94, Rev. 2. https://unece.org/un-regulations-addenda-1958-agreement (2014)
133. UN Regulation No. 137: – Frontal impact with focus on restraint systems, Nr. UN ECE R137, United Nations, Rev. 1. https://unece.org/un-regulations-addenda-1958-agreement (2017)
134. UN Regulation No. 94: – Frontal collision protection, Pub. L. No. UN ECE R-94, Rev. 3. https://unece.org/un-regulations-addenda-1958-agreement (2017)
135. UN Regulation No. 137: – Frontal impact with focus on restraint systems, Pub. L. No. UN ECE R-137, Rev. 1. https://unece.org/un-regulations-addenda-1958-agreement (2017)
136. UN Regulation No. 135: – Pole Side Impact (PSI), Pub. L. No. UN ECE R-135, 43, Rev. 1. https://unece.org/un-regulations-addenda-1958-agreement (2016)
137. UN Regulation No. 44: – Child Restraint Systems (CRS), Nr. UN ECE R44, United Nations, Rev. 3. https://unece.org/un-regulations-addenda-1958-agreement (2014)
138. UN Regulation No. 129: – Enhanced Child Restraint Systems (ECRS), Nr. UN ECE R129, United Nations, Rev. 3. https://unece.org/un-regulations-addenda-1958-agreement (2013)
139. UN ECE: Global Technical Regulation (GTR) No. 6 – Safety glazing. UN ECE. https://unece.org/fileadmin/DAM/trans/main/wp29/wp29wgs/wp29gen/wp29registry/ECE-TRANS-180a6a1app1e.pdf (2011)
140. UN Regulation No. 43: – Safety glazing, Nr. UN ECE R43, United Nations, Rev. 4. https://unece.org/un-regulations-addenda-1958-agreement (1995)
141. United States: Federal Motor Vehicle Safety Standard (FMVSS) 208 – Occupant crash protection. https://www.ecfr.gov/current/title-49/subtitle-B/chapter-V/part-571/subpart-B/section-571.208 (2004)
142. United States: Federal Motor Vehicle Safety Standard (FMVSS) 201 – Occupant protection in interior impact. https://www.ecfr.gov/current/title-49/subtitle-B/chapter-V/part-571/subpart-B/section-571.201 (2004)
143. United States:. Federal Motor Vehicle Safety Standard (FMVSS) 214 – Side impact protection. https://www.ecfr.gov/current/title-49/subtitle-B/chapter-V/part-571/subpart-B/section-571.214 (o. J.)
144. United States: Federal Motor Vehicle Safety Standard (FMVSS) 226 – Ejection Mitigation. https://www.ecfr.gov/current/title-49/subtitle-B/chapter-V/part-571/subpart-B/section-571.226 (o. J.)

145. Volvo AB (Volvo Trucks): Volvo trucks in 2000s. Safety. https://www.volvotrucks.com/en-en/about-us/who-we-are/our-heritage/2000s.html (2023)
146. Volvo AB (Volvo Trucks): Toward zero – Safety. Safety. https://www.volvotrucks.us/our-difference/safety/ (2023)
147. Volvo Cars: Volvo Cars airbag celebrates 20 years. Mai 24, 2007. (2007, Mai 24)
148. Wikipedia: Fahrradhelme – Fahrradhelmpflicht. https://en.wikipedia.org/wiki/Bicycle_helmet (2023, Mai 10)
149. World Health Organization (WHO):. Global plan – Decade of action for road safety 2021–2030 (S. 36). World Health Organization (WHO). https://cdn.who.int/media/docs/default-source/documents/health-topics/road-traffic-injuries/global-plan-for-road-safety.pdf (2021)
150. World Health Organization: Global status report on road safety 2018. World Health Organization (WHO). https://apps.who.int/iris/handle/10665/276462 (2018)
151. Zander, O., Lorenz, B., Seeck, A., Langner, T: Sicherheitsbewertung von airbagbasierten Kopfschutzsystemen für Fahrradfahrer (S. 27). Bundesanstalt für Straßenwesen (BASt). https://bast.opus.hbz-nrw.de/files/2175/Sicherheitsbewertung_von_airbagbasierten_Kopfschutzsystemen.pdf (2016)
152. Ziegenfuß, T.: Notfallmedizin. Springer, Berlin Heidelberg. https://doi.org/10.1007/978-3-662-52775-7 (2017)
153. Zodiac Automotive: Inflatable Tubular Structure (ITS). Products. http://www.zodiacautomotive.com/products/side-impact-rollover-protection/inflatable-tubular-structure-its/ (2023)

Sensor Technology for Accident Detection

7

Stephan Zecha, Johannes Clemm, Andreas Forster and Marc Menzel

Since the 1980s, passive safety has been standard equipment in most automobiles; in addition to the safety passenger compartment, this also includes seat belt and airbag systems. These occupant protection systems are now unthinkable without the necessary electronics for sensing and diagnosis as well as for timely control. The decision whether the restraint systems such as the pretensioner and load limiter of the seat belt system as well as driver -, passenger -, head - and thorax airbags should be activated in the event of an accident depends on the signals detected by the sensors used. Both in the airbag control units and in the supporting front and side sensors, micromechanical acceleration sensors are usually used today. However, pressure sensors installed in the vehicle doors are also used in the event of side collisions.

The development of modern occupant protection systems follows the generally observed trend towards increased use of individually controlled safety devices, such as multi-stage airbag modules or variable vent openings in the airbag. The first systems, for example, were equipped with single-stage gas generators, and in further development, two-stage gas generators, so-called dual stage inflators, were used. The latest

S. Zecha (✉)
Continental Safety Engineering International GmbH, Alzenau, Germany
e-mail: stephan.zecha@continental-corporation.com

J. Clemm
Alzenau, Germany

A. Forster
Alzenau, Germany

M. Menzel
Alzenau, Germany

R. Schöneburg (ed.), *Integral Vehicle Safety*,
https://doi.org/10.1007/978-3-658-48100-1_7

development trend is to use multi-stage gas generators or to develop variable generators, where the generator performance can be continuously adjusted within certain limits. For the effective activation of these new systems, there will be more sensors and more sensor networks in motor vehicles in the future compared to today's systems. When selecting suitable sensors, it is crucial in which phase of the accident the sensing should take place: acceleration and pressure sensing is exclusively limited to sensing within the crash phase. However, if sensing is to take place in the pre-crash phase, appropriate predictive environmental sensing must be used.

Today's occupant protection systems are undergoing a massive change: The development of safety systems is focusing on an improved restraint function, with the protection of vulnerable road users increasingly coming to the fore. This requirement necessitates a new generation of sensors for occupant detection and classification, in the pre-crash phase and for environmental detection (Fig. 7.1). There is an increasing flow of information between the pre-crash and in-crash phases in order to meet the increasing need for safety.

7.1 Development of Sensor Technology

The introduction of various types of airbag systems necessitates further demands for different types of crash sensors. For the first airbag systems, which consisted only of driver and front passenger airbags, central crash sensing was sufficient. The electronic modules are equipped with a single accelerometer and in cases where an offset crash is to be detected, with two sensors. These sensors are oriented at a 90° angle to each other and often mounted in a ±45° orientation relative to the vehicle's longitudinal axis. This arrangement is used for most airbag systems.

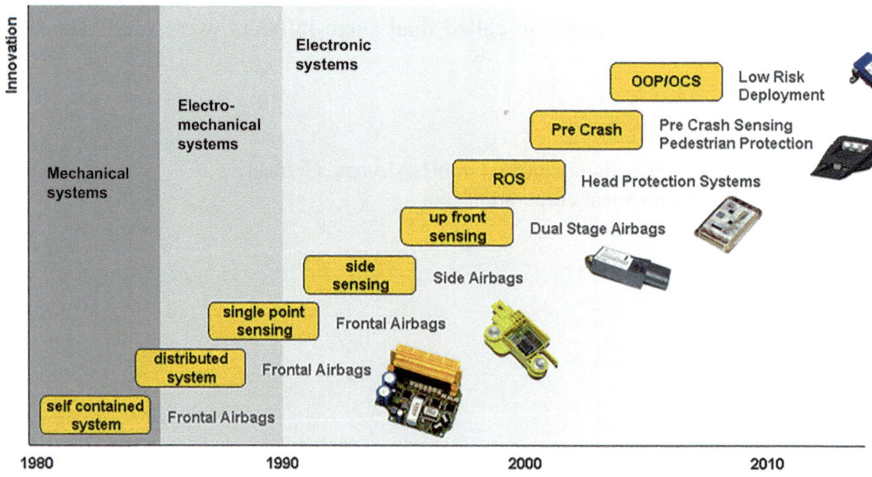

Fig. 7.1 Development of occupant protection systems and suitable sensors (from [7])

7.1.1 Mechanical Sensors

Only for the sake of completeness, it should be mentioned here that the first airbag systems used mechanical sensors; they are based on the impact igniter principle. In these solutions, the gas generators are equipped with an ignition cap that is activated by an impact igniter: When a sufficiently large acceleration overcomes a defined mass, it strikes the ignition cap, which in turn ignites the propellant of the gas generator. These systems can only be tuned to the accident situation to a very limited extent. Another disadvantage of mechanical sensors is that the triggering cannot be synchronized. In extreme cases, it can happen that in a frontal impact in the lower speed range only one of the airbags is triggered. For these reasons, such systems were only used to a very limited extent and were replaced by systems with electromechanical sensors.

7.1.2 Electromechanical Sensors

The introduction of electrical igniters was accompanied by the use of electromechanical crash sensors. Two fundamentally different types of electromechanical sensors are distinguished: In the first airbag systems, they are directly integrated into the ignition circuit and directly involved in the triggering of the airbags; they are referred to as crash sensors. With the advent of micromechanical acceleration sensors and the increasing use of electronics, the electromechanical sensors in the front area of the vehicle were eliminated and used within the airbag electronics as so-called safing or trigger sensors.

The most well-known **crash sensors** are ball and rolling mass sensors, which are mounted on the wheel arches in the front area of the vehicle and are therefore referred to as "front crash sensors". Another sensor of the same design is installed in the control unit (SDM: sensing and diagnostic module) for safety and plausibility. These sensors are based on the spring-mass principle: upon impact, the sensors close the electrical contacts of a circuit and activate the airbags. A graded evaluation of the crash signal is not possible with this type of sensor.

In the case of the **ball sensor** (ball-in-tube sensor) from BREED (today: KEY SAFETY SYSTEMS), a steel ball is housed in a precisely manufactured cylinder (Fig. 7.2). The steel ball represents the sensor mass; it is held in its resting position by a permanent magnet. In the event of a defined deceleration sufficient to overcome the holding force of the magnet, the ball leaves its resting position and closes the electrical contacts. It is noteworthy that the ball moves within the exact guide against an air cushion, thus exhibiting a low-pass characteristic. High-frequency accelerations dampen the movement of the ball in this way, and the sensor is largely insensitive to vibrations in the vehicle. When the circuit is closed, the airbags are ignited and the occupant protection systems are triggered. For better conductivity and to avoid corrosion, the steel ball and contacts are gold-plated. The sensor housing consists of a hermetically welded steel

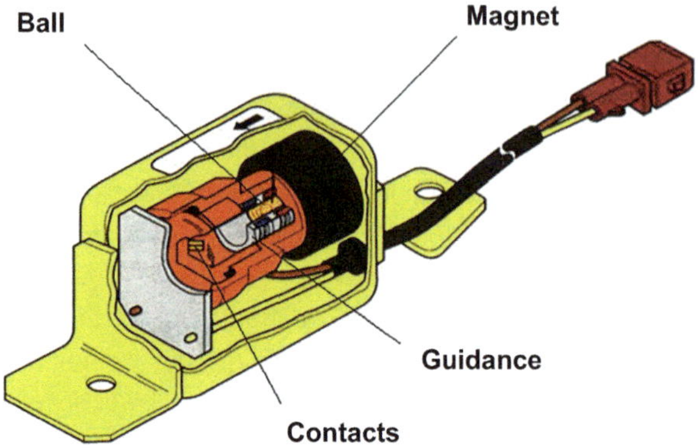

Fig. 7.2 Ball sensor (ball-in-tube sensor), from [18]

housing and is installed in the direction of travel, indicated by an arrow on the top of the housing.

In the case of the **rolling mass sensor** (Rolamite from ZF), a mass designed as a roller is held and guided in its initial position by a band spring. In the event of a defined deceleration, the roller moves over a curved running surface and in doing so runs over a contact spring, which closes the circuit to activate the igniter (Fig. 7.3). The sensor can

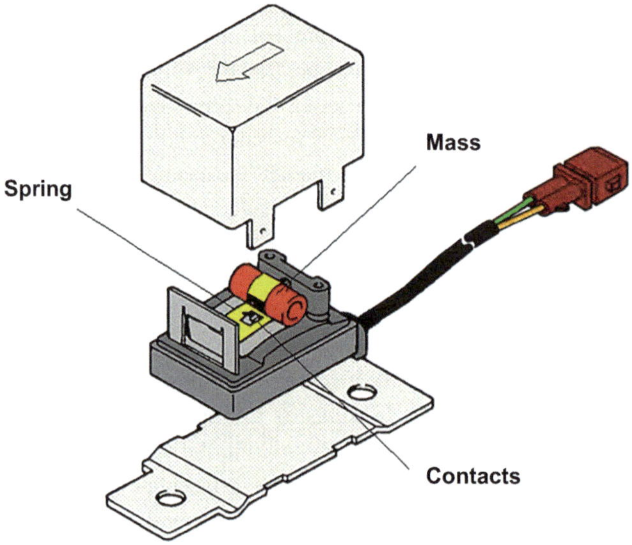

Fig. 7.3 Rolling mass sensor (Rolamite sensor), from [26]

be set to a specific threshold or response value via an adjustment screw. This sensor also realizes a low-pass characteristic through the band spring stretched over the curved surface: high-frequency accelerations, which for example originate from vibrations, are thus kept away from the roller and its movement option is restricted. As with the ball sensor, the housing of the rolling mass sensor is hermetically sealed, its installation direction is indicated by a marking arrow on the steel housing.

Since the use of micromechanical accelerometers, the requirement for a second, independent confirmation of the deceleration signal for triggering various occupant protection systems has been formulated. The sensors used for this are referred to as safing sensors, they are also based on the spring-mass principle. An early design of these safing sensors used in airbag electronics consists of a glass body filled with mercury, hence **mercury safing sensor**. The inclined glass body is mounted on the circuit board in the direction of travel using a holder, at the upper end of which contacts are attached (Fig. 7.4). If a sufficiently high deceleration acts on the sensor, the mercury moves within the inclined glass body due to its inertia and closes the contacts at the upper end. Due to the undefined switching behavior during a collision, but also for environmental protection reasons, this type of safing sensor was only used for a short time.

In **rollover sensors**, such as those installed in convertibles, several spirit levels filled with liquid (alcohol) are used as an alternative. Here, the position of the air bubble in the spirit level is monitored with a light barrier. If a rollover occurs, the position of the air bubble changes and the rollover protection devices are activated. Such sensors are still used today in some rollover systems for convertibles. In Fig. 7.5, the two liquid spirit levels can be seen, the air bubble is at the top end, between the light barrier. To the right of it is a G-sensor, which additionally closes an electrical contact in the event of a rollover. The electronics are mounted vertically in the vehicle, so the G-sensor acts in the z-axis, i.e., in the vertical vehicle direction. The sensor signals are read by

Fig. 7.4 Design of a mercury safing sensor

Fig. 7.5 Rollover sensor with liquid spirit level

a microcontroller and evaluated with an algorithm tailored to the vehicle. The six large capacitors serve as an energy reserve for controlling the electromagnetic unlocking system for the roll bars. The **G-sensor** in its simplest form is a mass that is movably mounted in a housing. At the top end of the housing are contacts that are closed by the mass as soon as the sensor is turned over (Fig. 7.6); due to gravity, the mass is pressed against the contacts.

The **safing sensor with reed contact** consists of a guide sleeve on which a round magnet and a spiral spring made of stainless steel are attached; magnetic particles are mixed into the round magnet, which is made of plastic by injection molding. A reed contact is integrated as a switch in the guide sleeve, its contacts are hermetically sealed in a glass body filled with protective gas (Fig. 7.7). An externally effective magnetic field closes the contact, allowing a current to flow. When a defined deceleration occurs on the safing sensor, the mass of the ring magnet is moved against the spring and the circuit for activating the occupant protection systems is closed. The required static deceleration for switching the safing sensor is approximately 2 to 3 g.

In order to achieve a sufficiently long closing time of the contacts, the special design shown in Fig. 7.8 (left illustration) is used: The magnet is guided in a helical groove in the guide sleeve and the mass of the magnet is converted into a rotational movement above the switching point of the reed contact offset in the event of a deceleration in the longitudinal direction. This achieves the longer closing time of the safing sensor.

Safing sensors are used in both standing and lying configurations. In the lying configuration, connection flags of different lengths are used. The long connection flags offer the advantage of a greater distance to the circuit board, so that additional components can be placed under the sensor; they thus allow a higher packing density. With safing sensors, the otherwise common integrated resistor is omitted. The connections of the contacts are duplicated for diagnostic reasons, so that a test current can monitor the

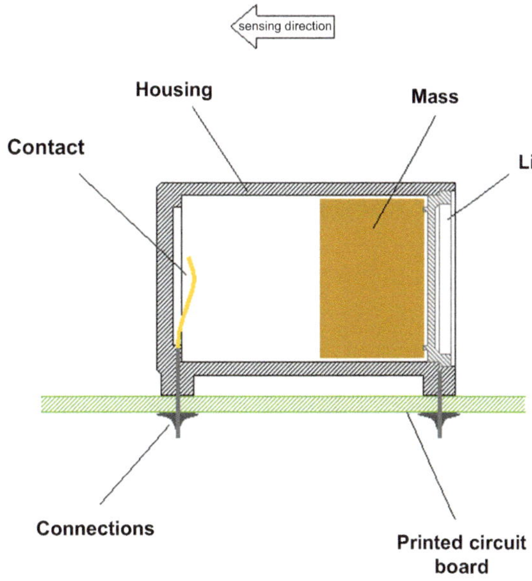

Fig. 7.6 Design of a G-sensor (from [7])

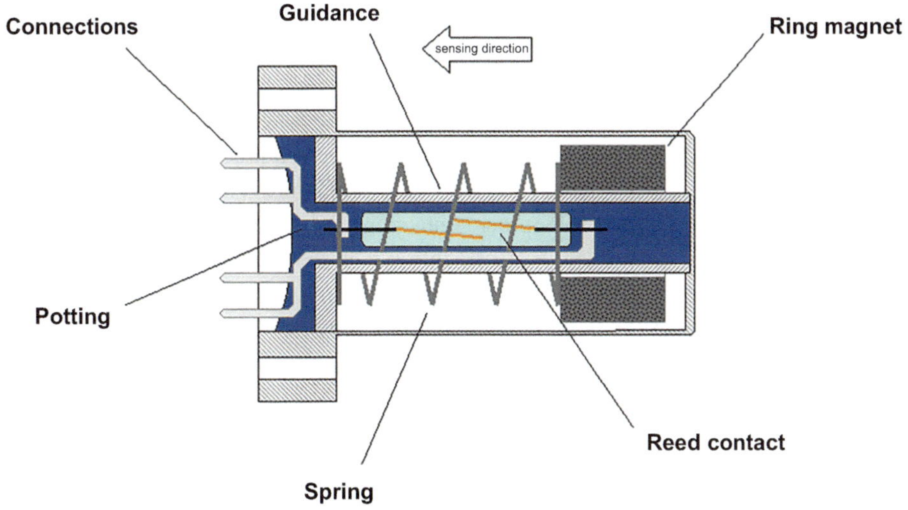

Fig. 7.7 Design of a safing sensor with reed contact

proper contacting of the sensors on the circuit board. The housing of the safing sensor is made of plastic, the guide sleeve is pressed in with the integrated reed contact, the stainless steel spring and the ring magnet and is potted with a PU material. A direction of travel arrow on the top of the housing clearly indicates the installation direction of the sensor.

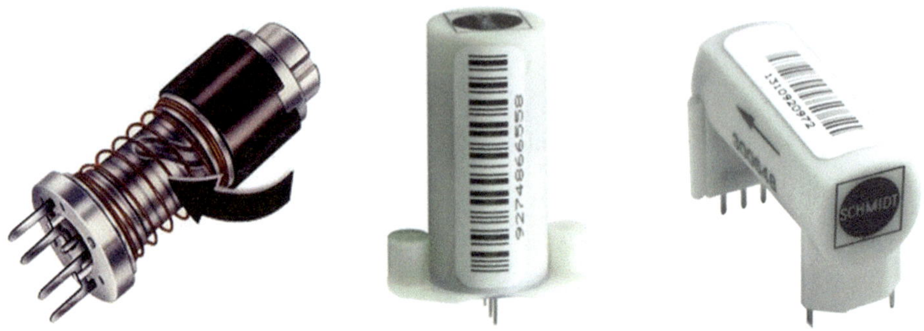

Fig. 7.8 Different designs of safing sensors with reed contact

Due to the complex occupant protection systems, the goal is to achieve the longest possible closing time, which cannot be sufficiently realized with the directly switching safing sensors. This led to the development of the **trigger sensors**, with the help of which the signal for triggering the restraint systems is artificially extended by an electronic circuit. This circuit is usually a configurable and programmable component, a so-called monoflop, with which the switching duration can be adjusted according to the various requirements.

The spring/mass principle is also used in trigger sensors: The mass is designed as a steel ball that presses against a spiral spring made of stainless steel. Behind the spiral spring is a pair of contacts that is closed by the spring when there is sufficient acceleration, thus providing the trigger signal (Fig. 7.9). The required static deceleration for switching the trigger sensor is about 1.5 to 2.0 g. For diagnostic reasons, the connections of the contacts are duplicated. The proper contacting of the trigger sensor can be checked with a test current. The complete housing, into which the contacts are injected and the spring and ball are inserted, is made of plastic and is closed with a lid. It is potted with a PU material for safety. A direction of travel arrow on the top of the housing indicates the installation direction of the trigger sensor.

7.1.3 Electronic Sensors

The next development step in airbag control units was seen in the introduction of the so-called single-point sensing systems (see Fig. 7.1). This new generation was only made possible by the provision of electronic acceleration sensors. The first variants of these electronic sensors were piezoelectric sensors, which were replaced by so-called "bulk-micro-machined" and "surface-micro-machined" crash sensors (Fig. 7.10). This achieved improved detection of the crash signal and enabled targeted analysis of the deceleration.

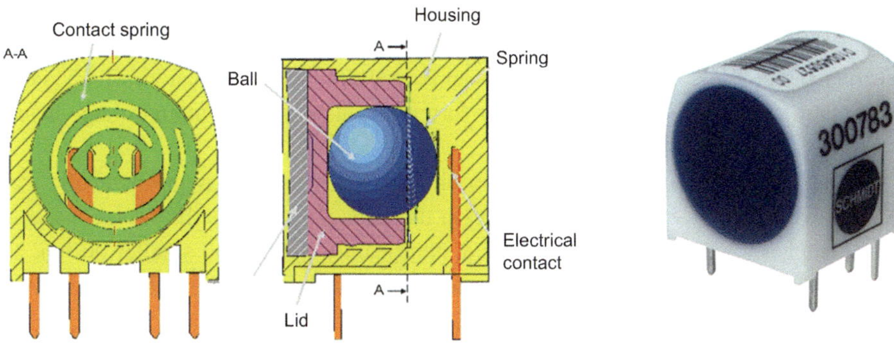

Fig. 7.9 Design and realization of a trigger sensor

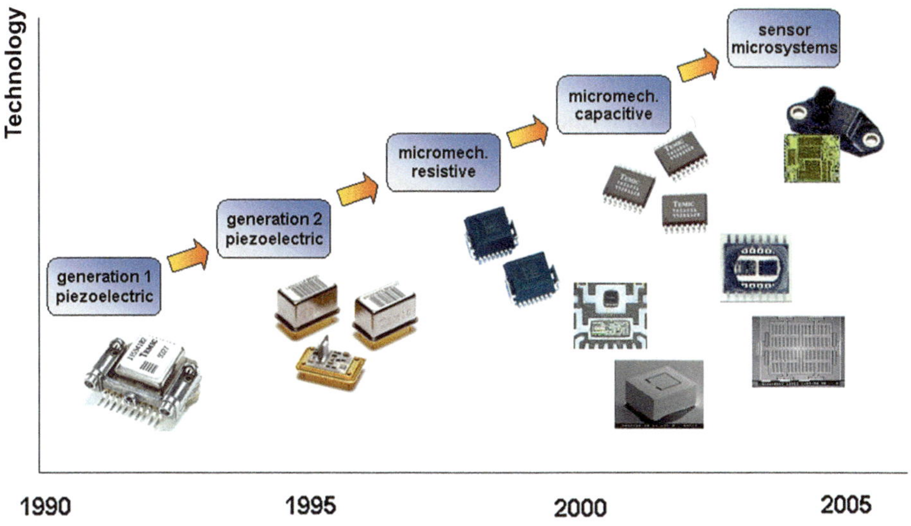

Fig. 7.10 Development stages of electronic sensors (from [25])

Electronic sensors have a measuring range of approximately 30 to 50 g. An analog or a digital output signal is available for evaluating the airbag electronics. In analog systems, the output voltage is about 40 mV/g and thus corresponds to a measuring range of ±50 g at an operating voltage of $U_b = 5$ V. To ensure that both deceleration and acceleration can be measured, the zero point of the output voltage is usually set at $0.5 \cdot U_b$. In digital systems, the typical data width is 8 bits, thus resulting in a resolution of the measuring range from −128 to +127 values.

In **piezoelectric sensors**, the piezo effect is used: A piezo-ceramic element is mounted on a carrier and provides an electrical voltage. If a deceleration acts on the piezo element, it is deformed and there is an increase in the piezo tension, which is available as an analog value at the output for further evaluation. The output voltage behaves proportionally to the deceleration.

In the first generation of piezoelectric sensors, the housing was mechanically stable and screwed for better pulse transmission in the airbag electronics. In the second generation, emphasis was placed on realizing a significantly lighter and smaller design. This made it possible to solder the sensors onto the circuit board while still ensuring pulse transmission.

In Fig. 7.11, the left illustration shows a piezoelectric sensor in which the piezo element is set up vertically in the holder in order to be able to realize sensing in the horizontal direction. The right illustration shows a sensor with a lying element; the sensing direction is perpendicular to the horizontal plane. To provide a hermetically sealed housing, the Al_2O_2 ceramic is soldered onto the housing base and connected to the connections in the form of glass/metal feedthroughs. The housings of the sensors are made of nickel-plated steel, with the lid welded to the housing base.

To manufacture **micromechanical sensors**, economical and rational manufacturing processes are used, which are offered by the modern semiconductor industry; this is the so-called MEMS technology (MEMS: micro-electrical mechanical systems). The sensor element and the corresponding evaluation are integrated in an electronic component. In contrast, the first micromechanical sensors consisted of the sensor element and an externally constructed signal processing. An example of this was the SA20 sensor from SensoNor, which was used in large quantities in the airbag electronics.

A distinction is made between single-chip and dual-chip design in micromechanical sensors: The single-chip design is usually used in so-called surface-micro-machined sensors, while the dual-chip design is used in bulk-micro-machined sensors. In Fig. 7.12, a

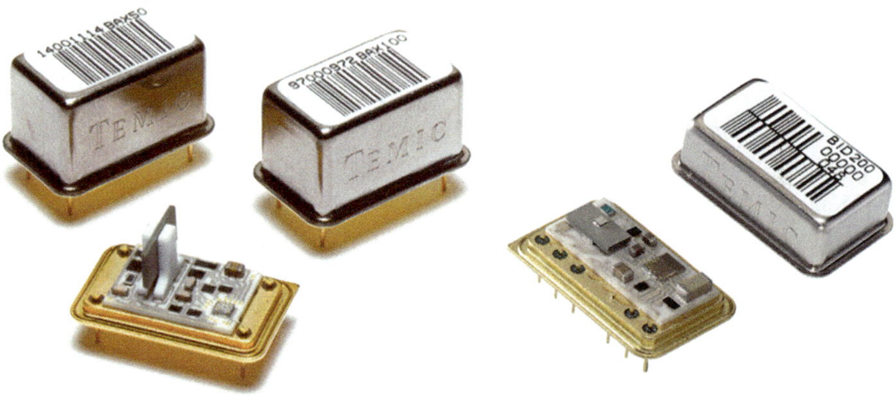

Fig. 7.11 Different designs of piezoelectric sensors (from [7])

Fig. 7.12 Micromechanical dual chip sensor with sensor element and evaluation circuit (from [7])

dual-chip sensor is shown: The two chips form the sensor element and the required evaluation circuit, usually based on ASIC (ASIC: application specific integrated circuit). The signals from the sensor element are amplified and filtered by the evaluation circuit and are finally available at the output of the airbag electronics for further processing.

In the upper part of the left illustration in Fig. 7.12, the sensor element is shown, below that the evaluation circuit (ASIC) for signal evaluation. Both chips are mounted on a common frame and electrically connected to each other via wires. The complete sensor is encapsulated in plastic in a special transfer molding process, with the connections preformed for a so-called "chip-on-board assembly".

Figure 7.13 also shows a dual chip sensor, which is based on the principle of capacitance. The assembly of the two chips and their electrical connection are comparable to the sensor shown in Fig. 7.12. On the right side of the image, the *g*-cells of the sensor based on the capacitive measurement method can be seen, magnified by SEM (scanning electronic microscope). The meandering structure serves for the flexible suspension of the seismic mass. Accordingly, the sensing direction of this sensor is perpendicular to the horizontal plane.

In the production of **micromechanical sensor elements**, two basic methods are distinguished: the bulk-micro-machining and the surface-micro-machining process. In the bulk-micro-machining process, the seismic mass is etched out as a three-dimensional structure from a relatively thick, up to 500 µm thick silicon material. To protect it against environmental influences and other impurities, the mass is enclosed between two thinner, approximately 200 to 300 µm thick silicon layers. In the sensor elements, the measurement is carried out either by a resistance, a capacitance, or a frequency change, which is caused by an acceleration or a deceleration.

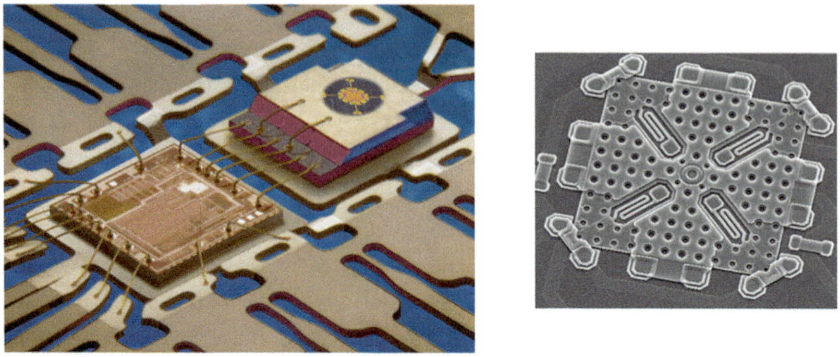

Fig. 7.13 Micromechanical capacitive dual chip sensor (from [7])

Figure 7.14 shows a sensor element produced in the bulk-micro-machining process; the three-dimensional structure can be seen, but the silicon layer intended as the upper cover has not yet been applied. To measure the acceleration, a mass element is moved in such a sensor element, resulting in a resistance change, as the resistances are located in the connecting elements between the seismic mass and the frame. They are connected as a Wheatstone bridge (Fig. 7.15), with the mass connections being led out separately from the sensor element for diagnostic reasons. For diagnosis, an additional resistance is used, which simultaneously covers the entire structure. If a break occurs in the structure, the resistance becomes high-impedance. If the sensor is based on the capacitive principle, the seismic mass is designed as a capacitor plate. The upper and lower cover of the sensor is insulated from the carrier by a glass layer, which serves as another capacitor plate. This makes it possible to detect the deceleration through a capacitive differential measurement system. In these sensors, an additional electrode is attached, which allows the sensor element to be deflected and detected electrostatically by applying a tension.

Fig. 7.14 Bulk-micro-machined sensor element (from [12])

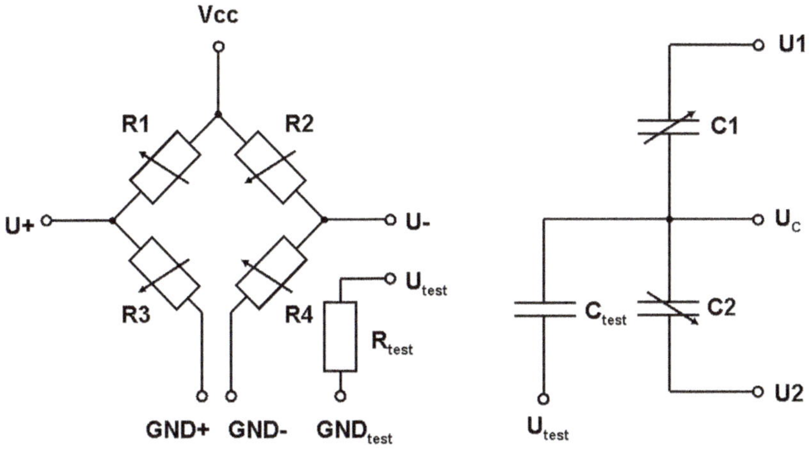

Fig. 7.15 Equivalent circuit diagram for resistance- and capacitance-based sensors

In a sensor manufactured using the surface-micro-machining process, significantly less silicon is required than in the bulk-micro-machining process; a significant cost advantage is the result. These sensors are typically based on the capacitive measurement principle. The seismic mass of the sensor, shown in Fig. 7.16, is designed as a comb structure and is flexibly mounted on the carrier. The carrier itself is also designed as a comb structure, but is rigidly mounted and engages with the comb structures of the seismic mass. The interlocking teeth of the two combs act as capacitor plates, between which the capacitance is detected according to the differential measurement principle. A significant advantage of the surface-micro-machining process is that the complete signal processing can be accommodated on the chip. This allows for the production of extremely cost-effective sensors, so-called single-chip sensors.

The use of **structure-borne sound sensors** is based on the following considerations: The term "structure-borne sound" is used for vibrations in solid bodies, they are in the range of human hearing (20 to 20,000 Hz). The acceleration sensors used in today's airbag systems are only suitable for detecting the bending wave that occurs during a collision due to their frequency response (see Fig. 7.17). This wave propagates at an average speed of 300 m/s. However, to enable faster detection of the deceleration signal, sensors are required that detect occurring density waves, so-called longitudinal waves. This wave form occurs in infinitely extended media, meaning that the extent of the medium is large in all directions compared to the length of the longitudinal wave. The average propagation speed of the longitudinal waves is 5,000 m/s, allowing for a significantly shorter sensing time. This principle is used in structure-borne sound sensors, which are designed for use in airbag electronics. They are capable of capturing a signal up to a frequency of 20 kHz at the location. However, it should be taken into account that the construction of

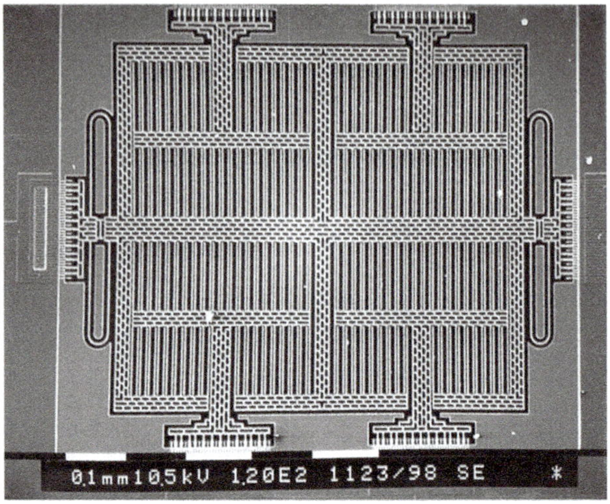

Fig. 7.16 Surface-micro-machined sensor element (from [7])

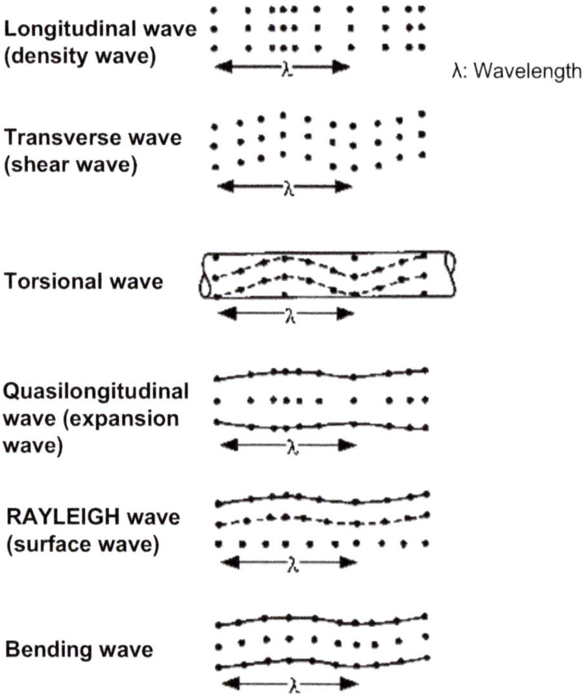

Fig. 7.17 Waveforms of structure-borne sound

Fig. 7.18 Roll rate sensor

the vehicle has a significant influence on the signal transmission. The characteristic of pulse transmission depends on the materials used, the number and type of connections.

By using a structure-borne sound sensor, it is possible to dispense with the use of additional external sensors for timely plausibility checking without restricting the system requirements. In the near future, the use of a structure-borne sound sensor in external side and up-front sensors appears feasible.

In order to detect the rollover of automobiles (e.g., convertibles, off-road vehicles, and luxury sedans) at an early stage and to activate existing safety measures such as head protection systems and seat belt pretensioners in a timely manner, **rollover or roll rate sensors** (Fig. 7.18) are used. These sensors also use micromechanical sensor systems. They are mostly based on the Coriolis effect: The micromechanical structure, e.g., a tuning fork or a rotationally oscillating micromechanical structure, is set into vibration for this purpose. If the structure is now subjected to an external rotation, the Coriolis force becomes effective and a corresponding counterforce is established, which can be measured and represents a measure of the roll rate. This roll rate corresponds to a rotational speed and is given in degrees per second; such sensors typically have a measuring range of 200 to 300 °/s. Rollover sensors are usually integrated into control units, but they can also be used as separate modules in which the rollover sensor technology is integrated, providing a signal for further processing.

Finally, to conclude this section, a summary of the sensors used in vehicle safety today will be provided. **MEMS sensor** (MEMS: micro-electrical, mechanical system), however, without claiming completeness, is shown. It is listed in Table 7.1.

Table 7.1 Overview of MEMS sensors

Sensor	Manufacturer	Type	Axis	Measurement range
Acceleration	ADI	ADXL78	x	± 35 to ± 70 g
		ADXL278	x/y	± 35 to ± 70 g
	Bosch	SMB050/052	x	± 35 to ± 50 g
		SMB060/062	x/y	± 35 to ± 50 g
		SMB065/067	x/-x	± 35 to ± 50 g
		SMB180	x	± 50 to ± 100 g
		SMB190	x	± 200 g
		SMB25x	x	± 35 to ± 200 g
		SMB26x	x/y	± 35 to ± 200 g
	Freescale	Hera	x/y	± 20 to ± 100 g
Low g-acceleration	ADI	ADXL103	x	± 1.7 g
		ADXL203	x/y	± 1.7 g
	Bosch	SMB220	x	± 4.7 g
	Kionix	KXM60–1120	x	± 1 to ± 6 g
		KXM60–1130	x/y	± 1 to ± 6 g
	Memsic	MXR7150V	x/y	± 2 g
	Freescale	Colossus	z	± 1.5 to ± 10 g
	ST-M	LIS2L02AL	x/y	± 2 g
	VTI	SCA610	x	± 0.5 to ± 1.7 g
		SCA1000	x/y	± 1.7 g
		SCA1020	z/y	± 1.7 g
Angular rate sensors	ADI	ADXRS150	x	± 150°/s
		ADXRS300	x	± 300°/s
	BAE [1]	CSR05		± 50 to ± 200°/s
	BEI [2]	MicoGyro		± 75 to ± 300°/s
	Bosch	SMG040	x	± 250°/s
		SMG060	x	± 240°/s
		SMG061	x	± 240°/s
		SMG070	x	± 187°/s
	Freescale	Newton	x	± 300°/s
		Galilée	z	± 300°/s
	Honeywell	*		
	Kionix	KGF01–300	x	± 300°/s
	Melexis	*		
	muRata	GYROSTAR	x	± 300°/s

(continued)

Table 7.1 (continued)

Sensor	Manufacturer	Type	Axis	Measurement range
	Panasonic	EWTS62	x	± 300°/s
	SensoNor	SAR10	x	± 250°/s
Multiple Sensors[3]	ADI	*		
Pressure Sensors	Bosch	SMD085	-	60 to 115 kPa
	Infineon	KP106	-	60 to 130 kPa

Notes:
[1] Silicon sensing system
[2] Systron Donner Automotive Division
[3] Combined sensor e.g.: angular rate and low g-sensor
* Custom development, no data available

7.2 Sensor Applications

The sensors currently in use are to be differentiated below into the sensor technology used in frontal, side and rollover accidents, with a distinction being made between acceleration, angular velocity, pressure and shock wave sensors. In addition, sensors for the detection of occupants and external road users as well as traffic situations are explained. Finally, the assistance systems that are already used in many cars today are presented.

7.2.1 Up-Front Sensors

The up-front sensors (Fig. 7.19) are mounted in the front area of the automobile; their competence and responsibility is to detect a collision at the earliest possible stage. The aim is to achieve a safety-oriented gradation of various crash scenarios, but especially to detect the severity of an accident. The detected signal is digitally transmitted to the airbag electronics and evaluated via a suitable crash algorithm. This enables a distinction to be made between fire/no-fire accidents and, depending on the accident severity, to activate so-called adaptive occupant protection systems in a timely and graduated manner.

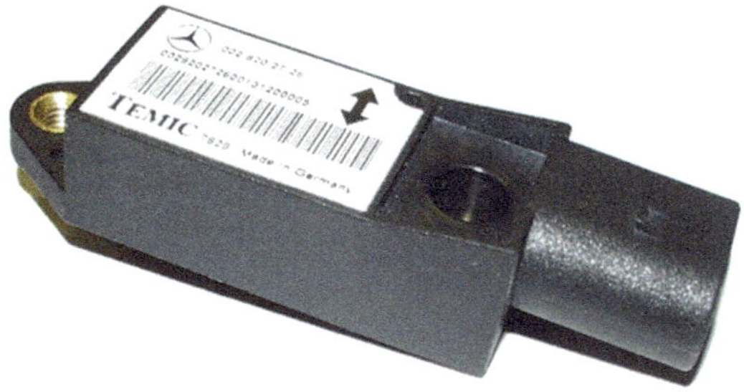

Fig. 7.19 Up-front sensor (from [7])

Fig. 7.20 Side sensor with integrated signal preprocessing (from [24])

Due to their exposed installation in the front area of the vehicle, up-front sensors must be protected against splash water and have waterproof connectors. They are also designed for a higher temperature range, up to 105 °C.

7.2.2 Side Sensors

With the introduction of side airbags, additional sensors were required to detect a lateral collision in time. However, due to the signal transit times in the vehicle body structure, timely detection of an impending collision was not possible with conventional sensors. The solution was provided by external sensors, known as side sensors.

The first side sensors consisted of an **accelerometer** and a microcontroller, which performed signal preprocessing using an algorithm (Fig. 7.20). The signal required for airbag deployment was transmitted as a PWM signal (PWM: pulse width modulation) in three stages to the airbag electronics: Here, the pulse/pause ratio corresponded to the severity of the accident or impact. Two additional stages were used once as a "life" signal and another as an error signal, so that a total of five different stages could be realized for signal transmission.

The currently used side sensors are equipped with a digital communication unit, which transmits the crash signal directly to the airbag control unit. The signal is then evaluated using a crash algorithm according to predefined criteria, and the restraint systems are activated.

The side sensor shown in Fig. 7.21 consists of the micromechanical sensor element and the integrated, application-specific ASIC unit (ASIC: application specific integrated circuit). The acceleration signal is made digitally available to the airbag electronics.

Fig. 7.21 Side sensor with digital data transmission (from [7])

A different principle for detecting side collisions is applied with **pressure-sensors**: These sensors are installed in the vehicle door and measure the pressure increase in a side impact using a pressure transducer. They are also equipped with a digital communication unit, which transmits the current pressure profile to the airbag electronics. The signal is processed and evaluated for the activation of the occupant protection systems using a crash algorithm. The pressure sensor shown in Fig. 7.22 has the micromechanical sensor element with the necessary inlet opening on the right. The ASIC circuit is used for signal processing and transmission of the pressure signal to the airbag control unit.

7.2.3 Systems for Occupant Detection

In the field of occupant detection, there has been significant development since around 2020. This is due to two market movements:

Fig. 7.22 Pressure sensor with digital communication unit (from [7])

- Safety ratings, e.g. by Euro NCAP with:
  - child presence detection
  - driver drowsiness warning
- Increasing automation of driving functions and the necessary assessment of the driver's ability to take over.

Systems for so-called "driver monitoring" are often covered by camera systems. The use case "child presence detection" is intended to prevent children from being left sleeping in a vehicle and suffering health damage due to uncontrolled heating of the vehicle interior by sunlight.

To understand the possibilities for harnessing these market trends, a brief consideration of the status quo of legislation is initially useful.

In order to keep the injury risk of occupants as low as possible in the event of an airbag deployment, US safety legislation requires a low risk airbag deployment and/or an airbag suppression in OoP (out-of-position) situations or in the presence of occupants (see Fig. 7.23). It is important to consider the different design of airbag systems in Europe and the USA, as in the USA, unlike in Europe, the airbag systems must also be designed for the unbelted occupant. The consequence of this is that airbags must be filled with a larger gas quantity within the same inflation time. The airbags designed for the North American market are therefore more aggressive than those used in Europe; this could lead to OoP situations resulting in severe, sometimes even fatal injuries to the occupants. Since 2015, after a 10-year ramp-up phase, it can be assumed that 100 % equipment quota according to FMVSS 208 is in place, i.e. all vehicles sold in the USA contain a mechanism for low risk airbag deployment and are tested according to the following scheme:

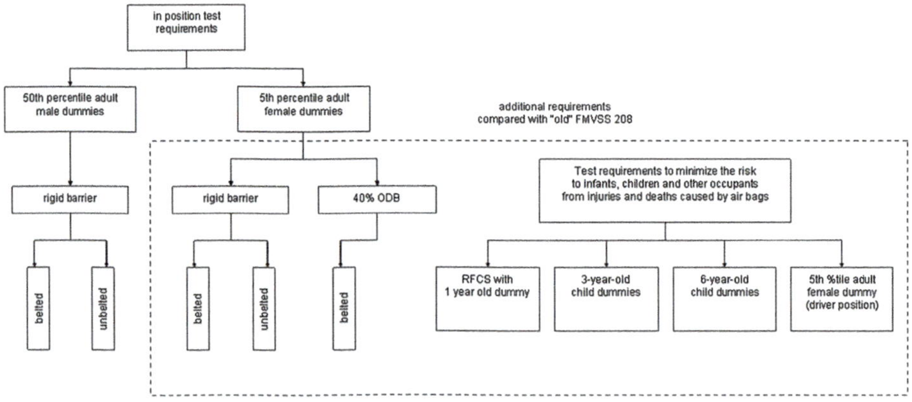

Fig. 7.23 Test requirements according to FMVSS 208 (from [7])

	Driver side	Passenger side		
Dummy:	Hybrid III 5% female	1 year old child	3 year old child	6 year old child
Configuration:	1. Chin on module 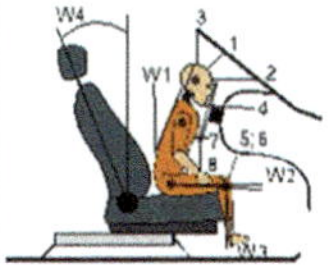 2. Chin at rim 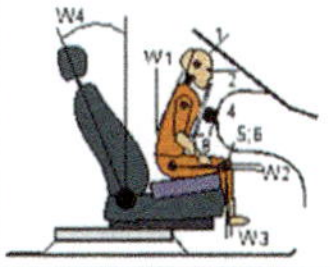	Dummy in 23 different child safety seats 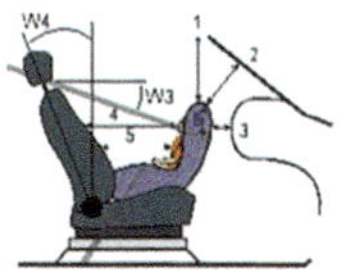	1. Chest at IP 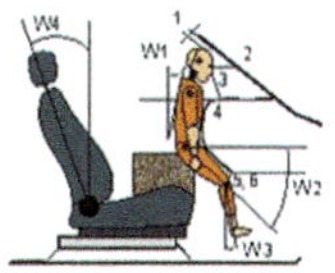 2. Head at IP 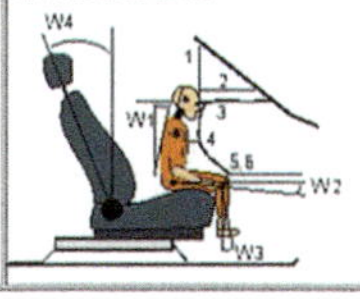	1. Chest at IP 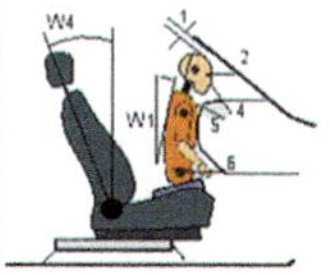 2. Head at IP 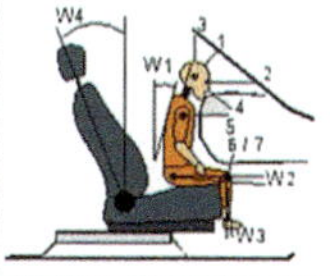

Fig. 7.24 Test requirements according to FMVSS 208 (low risk deployment), from [7]

The low risk tests during an airbag deployment include tests where the passenger side is occupied by a 12-month-old child dummy in child seats (various seating positions), with a 3-year-old child dummy (with and without booster seat) and with a 6-year-old child dummy. For tests to demonstrate the reduced risk of the driver's airbag, the driver's side must be occupied by a HIII-5F (5% female dummy); the tests are carried out in various test configurations (Fig. 7.24).

To demonstrate a low risk airbag deployment and/or to justify airbag deactivation in out-of-position (OoP) situations or in the presence of occupants, the airbag deployment must be adapted to the occupant or based on seat occupancy. For this purpose, in today's systems, the seat occupancy is determined by measuring the weight on the seat or the seat imprint of the child seat device and a corresponding classification is made.

In addition to the automatic deactivation of the passenger airbag when **using child seats**, some vehicle manufacturers in Europe also install a manually operated switch, an **airbag-off switch**, in the vehicle. With this switch, which is operated with the ignition key, the front passenger airbag can be activated and deactivated by the user. A suitable display in the instrument panel indicates the respective system state. The activation and deactivation of the system depend on the responsibility and attention of the driver in this manual solution.

As an alternative to the airbag-off switches, the use of **child seat transponders** for detection of child seats was discussed. In this system, an antenna is integrated into the surface of the vehicle seat; communication is via a high-frequency field with the transponder located in the child seat. This can determine the presence, but also the orientation of the child seat, so that it can be detected whether the child seat is installed forward or backward on the vehicle seat. The disadvantage of this system is that this measure only works with child seats that have a specially attached transponder. These systems have not established themselves in the market and as of today play no role in the design.

Table 7.2 Overview of typical weight classes

Class	Meaning	Mass class
0	Empty or child seat	0 to ≤ 15 kg
1	Occupied with child seat or with child up to 6 years	> 15 to ≤ 30 kg
2	Occupied with 5th percentile woman or larger	> 30 kg

The seat-bound occupant detection systems used in today's vehicles are used to detect the presence and, if positive, the weight of the passenger. Depending on the measured mass, a classification is made to distinguish between children and adult occupants. For this purpose, the classification summarised in Table 7.2 is usually used.

However, the mass classes shown in Table 7.2 only provide reference values; the actual values used depend on the requirements of the vehicle manufacturer and the occupant detection system used. It should also be noted that the American legislation and Euro NCAP occasionally use new child seat products for the tests. For example, a list of 7 seats, which are "significantly heavier than average", is included in an update planned for 2025 in the USA. These come very close to the threshold with an "empty weight" of 14.7 kg (e.g. a pram-child seat combination with foldable wheels) and can also approach close to 30 kg when occupied. Two fundamentally different principles are applied in the classification: the seat surface-integrated and the seat structure-integrated detection system.

The **seat surface-integrated detection system** involves seat mats that are located between the seat cushion and the cover. The most common systems are the seat mats from IEE and Aptiv. The IEE seat mats have a certain number of pressure-dependent resistors integrated into a film (Fig. 7.25). The shape of the seat mat and the number of measuring points are adapted to the respective seat surface. If the number of measuring points is large enough, the profile of the occupant on the seat can be determined with sufficient accuracy. The Aptiv seat mat, referred to as a bladder, uses a silicone-filled cushion that is connected to a pressure sensor via a hose. The load exerted by the occupant on the seat results in a defined pressure increase in the cushion, which is measured with the pressure sensor. A major challenge of the systems integrated in the seat surface is the high sensitivity to mechanical and thermal stresses in the seat surface, which occur during long-term use of a vehicle. When using seat surface-integrated systems, additional electronic equipment is required to evaluate the actual detection signal and transmit the result to the airbag control unit directly or via the vehicle CAN connection (CAN: controller area network).

An alternative to this can be seen in the **seat-structure-integrated detection systems**. Weight sensors are used to measure the occupant's weight acting on the seat structure (Fig. 7.26). The weight measurement is taken at the seat attachment points. Various

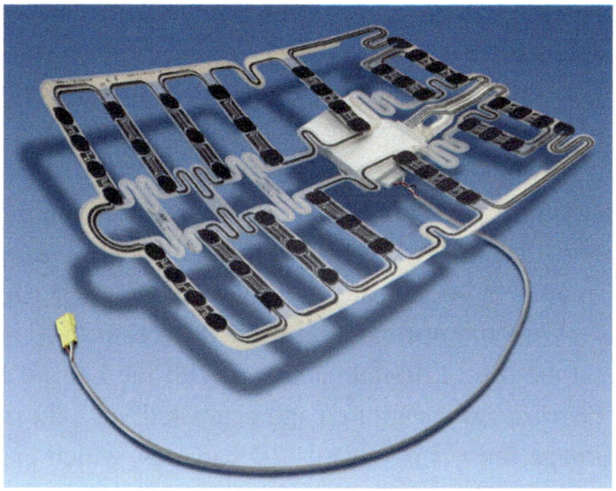

Fig. 7.25 Seat mat as seat surface-integrated detection system (from [7])

technologies are used for the weight sensors: capacitive measurement methods, the moving coil principle, Hall sensors or strain gauges are used. The weight sensors are either individually or connected to the airbag electronics via a data bus system. However, this technology has taken a back seat to the bladder mat in recent years.

The weight-based detection systems, whether integrated in the seat surface or in the seat structure, have a common, serious disadvantage: A child seat fastened to the seat with great force to avoid belt slack is interpreted as high weight when the belt retractor is simultaneously locked. To compensate for this misinterpretation and thus to be able to exclude a confusion between child seat and an adult occupant, additional seat belt force sensors can be used to ensure a clear detection.

The market trends outlined at the beginning, towards more **occupant detection systems** due to other use cases, lead in the area of passive safety firstly to manufacturers striving to depict the legal case through the new systems. In other words: to gradually replace the mentioned weight-based systems while maintaining the safety requirements with other systems. And secondly, the additional information provided by novel occupant detection systems can be used for a more finely tuned restraint effect.

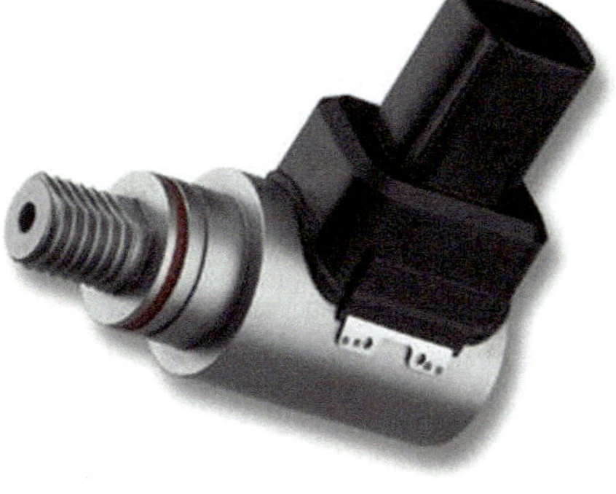

Fig. 7.26 Weight sensor as a seat-structure-integrated detection system (from [21])

The general objective of future occupant detection systems is to detect the presence of occupants and to classify them in the positive case in order to realize an occupant protection system adapted to the respective occupant. Both the physical characteristics of the occupants such as weight, size and seating position, as well as the current accident situation in terms of accident severity, direction of impact and affected vehicle body structure are taken into account. In addition, a further focus of recent sensor development is to detect a risky forward displacement of the occupants, i.e. a possible OoP situation, in order to specifically control or even prevent the airbag deployment.

With the **optical occupant detection systems**, the current development is focusing on the use of digital cameras. With the help of one or more such cameras, the scene to be assessed is captured and evaluated via a downstream image processing. The trend for image sensors is towards 2D cameras, which enable a 3D capture of the vehicle interior using modern evaluation algorithms and deep learning.

Alternatively, three-dimensional cameras (3D cameras) can be used. These determine the depth information either from two individual images or from the time-of-flight of the light for each pixel (pixel-oriented). Other 3D camera systems use modulated light and determine the phase shift between the emitted and the reflected light. This phase shift provides an exact runtime determination and thus the distance of the collision object for each pixel.

In Table 7.3, currently used sensors and further developments of seat occupancy systems are presented in a summary without claiming to be complete. This is intended to provide an overview of the state of the art of occupant detection systems.

7.2.4 Pre-Crash Sensing

The increasing requirements for modern occupant protection systems require a new generation of sensors. Traditional protection systems are only activated when the collision has already occurred. In contrast, future safety measures should be designed to be activated before the impact to avoid decelerations or to make safety devices, so-called reversible systems, prophylactically effective. This requires sensors that can detect and assess an impending or even inevitable collision in advance, i.e., before the actual impact, the beginning of the in-crash phase. These are pre-crash sensors, which are used to activate necessary protection systems early. These sensors must be able to detect collision opponents in the immediate vicinity of the vehicle in terms of both time and space.

The first generation of such sensors work in the close range as optical pre-crash sensors. They enable the distance, velocity, and direction of a collision opponent in the direction of the collision object to be recognized. This allows an impending collision to be detected early on, what kind of accident is to be expected, and which stages of the occupant protection systems need to be activated. The information provided by the pre-crash sensor is used to control occupant protection devices such as reversible seat belt pretensioners and adaptive load limiters.

Table 7.3 Overview of today's common occupant detection systems [20]

Type	Technical principle	Installation	Possible manufacturers
Occupant classification and passenger occupancy detection	Pressure distribution with liquid bladder	Passenger seat surface	Aptiv
Occupant classification and passenger occupancy detection	Capacitive measurement	Passenger seat surface	IEE
Occupant classification and passenger occupancy detection	Weight measurement	Passenger seat structure	Aisin
Occupant classification and occupancy detection	Camera-based detection	Interior	Camera Aptiv
Occupancy detection	Pressure distribution with film/contact switch	Seat surfaces	IEE, Scherdel, Bauerhin

Pre-crash sensors are preferably mounted in the upper area of the windshield near the rearview mirror. This allows for seamless coverage of the monitoring area in front of the vehicle. The principle of such sensors is based on a runtime measurement. Pulsed infrared laser light with a wavelength of 905 nm is emitted and reflected by objects within the sensor's range. The decoder calculates the runtime between the emitted and reflected laser pulse. The sensor has a multiple beam system, so that the environment is scanned at a defined angle range in front of the vehicle (Fig. 7.27). With the help of the multiple beam system, the direction of the approaching object is determined and the speed is calculated from repeated distance measurements. These sensors can also be equipped with secondary functions such as rain and brightness sensors.

With the information provided by a pre-crash sensor, it is not only possible to activate reversible seat belt systems early, for example, but they also allow the development of additional actuators to enable extended seating options in the vehicle. The future pre-crash sensor also provides an approach speed of the collision opponent that is independent of the direction of the collision object, thus allowing a much more precise pre-determination of the airbag deployment required under given circumstances than is possible with conventional systems.

Another possible application for pre-crash sensors is seen in the activation of pedestrian protection systems, which have been demanded and introduced in Europe since mid-2005. With such a sensor, the motion sequence of a collision opponent can be analyzed before an impending impact (Fig. 7.28) and allows a distinction between "hard" and "soft" objects after evaluating the sensor information. If a "soft" object, which characterizes the pedestrian, is recognized and further selection criteria for parameters such

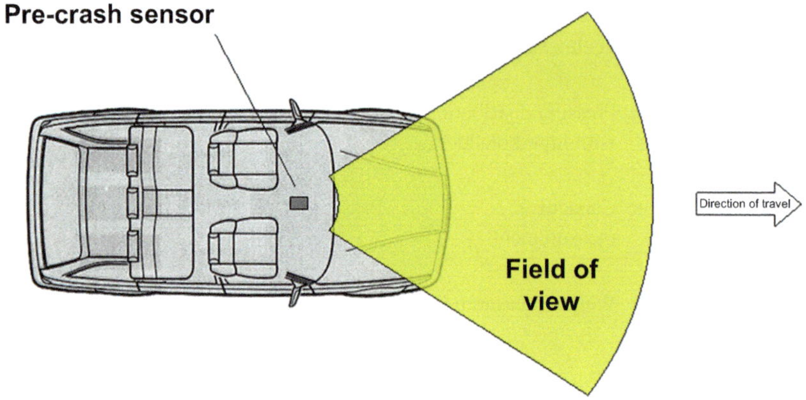

Fig. 7.27 Pre-crash sensor and the frontal sensing area (from [21])

as object geometry and approach speed are met, the corresponding pedestrian protection systems can be specifically activated via the algorithm integrated in the control unit.

7.2.5 Environment Sensors for Integral Vehicle Safety–RADAR

For integral and active vehicle safety, environment sensors are used that also support a multitude of driver assistance functions. A particularly suited sensor for vehicle safety is the RADAR sensor due to its robust, reliable measuring properties.

Radar is a measurement method for the contactless detection and localization of objects in the vicinity of the radar system. The radar principle is based on the emission of electromagnetic waves in the high and ultra-high frequency range, which are reflected by the objects and received again by the radar antenna. The position of the detected object is determined by runtime and direction measurement. In addition, the relative speed of the measured object can be determined via the doppler effect. The following measurement data are provided by an automotive radar to the vehicle:

Fig. 7.28 Pre-crash sensor for pedestrian protection (from [7])

- Object detection and classification
- Radial distance to the object
- Azimuth (elevation) angle to the object
- Doppler relative speed
- Radar backscatter cross-section

The radar principle and the underlying radar basic equation, which describes the basic characteristics of wave propagation, is shown in Fig. 7.29 and formula 7.1 [27].

$$P_e = \left(\frac{P_a G}{4\pi d^2}\right) \cdot \left(\frac{\sigma}{4\pi d^2}\right) \cdot \left(\frac{G l^2}{4\pi}\right) \ldots \sim \frac{1}{d^4} \tag{7.1}$$

Measurement methods for determining distance and relative speed to the target object:

To determine the distance to the target object, it is necessary to imprint a specific marker on the electromagnetic wave. A continuous wave would not contain any information about the transmission and reception time and thus would not allow for runtime measurement.

This marking is done by modulating the wave. Three parameters of the wave can be modulated:

- Amplitude A(t): e.g. for pulse modulation (Fig. 7.30)
- Frequency f(t): e.g. linear frequency modulation
- Phase phi(t): e.g. binary phase modulation

After demodulation, the runtime and thus the distance to the object is determined.

In addition to a robust measurement of the distance to objects, radar also offers the possibility of measuring the relative speed to the target objects. This is done by using the Doppler effect. The Doppler effect arises from the relative movement of the measured object and the associated compression (when approaching) or expansion (when moving away) of the electromagnetic waves. Physically, the relative movement causes a phase shift of the signal. This phase shift and thus the relative speed can be determined by mixing the reflected signal with the transmitted signal, as shown in Fig. 7.31:

The Doppler frequency is calculated as follows, Eq. 7.2:

$$f_{doppler} = \frac{\frac{\Delta\varphi}{2\pi}}{\Delta t} = 2 \cdot v_{rad,rel} \cdot \frac{f_{rf}}{c} \tag{7.2}$$

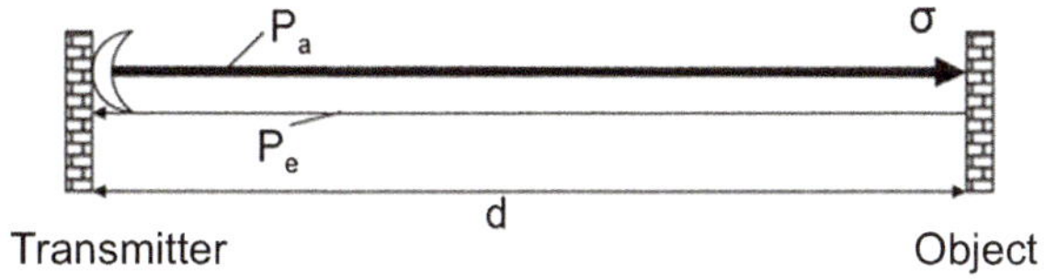

Fig. 7.29 Radar basic principle [27]

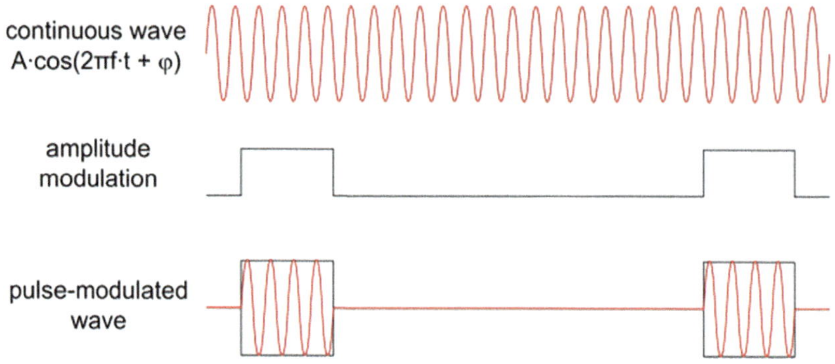

Fig. 7.30 Runtime measurement with pulse-Doppler method [9]

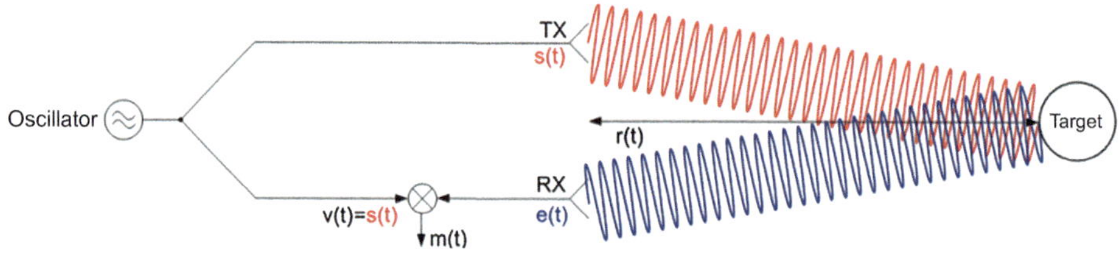

Fig. 7.31 Determination of the Doppler frequency by mixing the input and output signals [9]

The Doppler formula shows that the Doppler frequency is directly proportional to the transmission frequency. This means that with an increasing transmission frequency, a significantly better Doppler signal is available, and thus a significantly better resolution of the relative speed.

The trend towards high-frequency radar systems in the range of 77–79 GHz has thus improved the speed estimation of target objects.

Various methods have been used in radar development to determine the angular offset of an object. These methods, as they were also chronologically introduced into automotive radars, are:

- Mechanical scanning of the radar beams (Fig. 7.32)
- Switching between "single antennas" that radiate in different directions (Fig. 7.33)
- Electronic scanning by electronically "tracking" the radar beams
- Digital beamforming (Fig. 7.34).

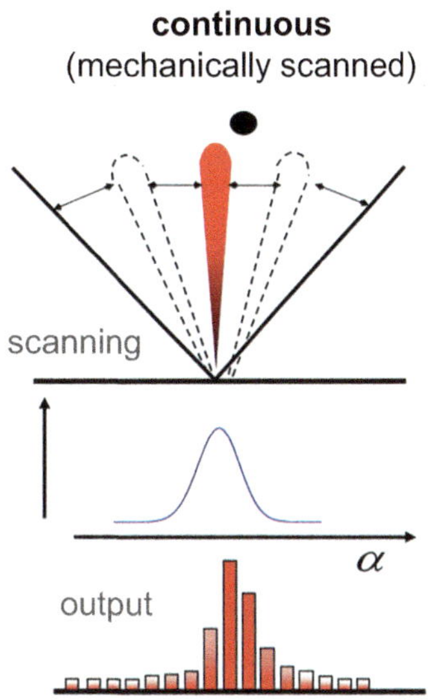

Fig. 7.32 Mechanical scanning

In digital beamforming, a multi-patch antenna is used, in which several individual antenna patches are placed equidistantly next to each other (see Fig. 7.34).

The measurement concept is based on the fact that the electromagnetic wave reflected by the object arrives at the adjacent antenna elements with a small time offset in the case that the object has a lateral angular offset. This time offset means that the wave arrives at the parallel antenna elements with a phase offset. This phase offset allows the angle to the target object to be calculated using formula 7.3:

$$\alpha_{AZ} = \arcsin\left(\frac{b}{d}\right) \tag{7.3}$$

Where:

$$b = (\varphi_{RX1} - \varphi_{RX2})\frac{\lambda}{2\pi} \tag{7.4}$$

The phase difference is determined by summing the phase-shifted signals and subsequent processing steps.

The typical measurement and detection range of a modern front radar is shown in the following Fig. 7.35.

Depending on the available radar cross-section, passenger cars can typically be detected and "tracked" up to a distance of 250 m. Pedestrians, with a significantly smaller radar cross-section, can usually only be detected up to a distance of 100 m.

The detection range of a radar shown in Fig. 7.35 of the 4th generation has the peculiarity that it has a significantly larger azimuthal detection range up to 60 m in the near range. This allows critical cross-traffic situations to be well detected.

Fig. 7.33 Switching between antennas

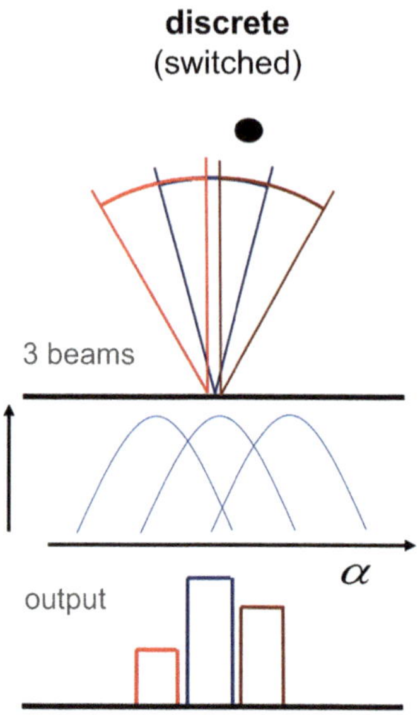

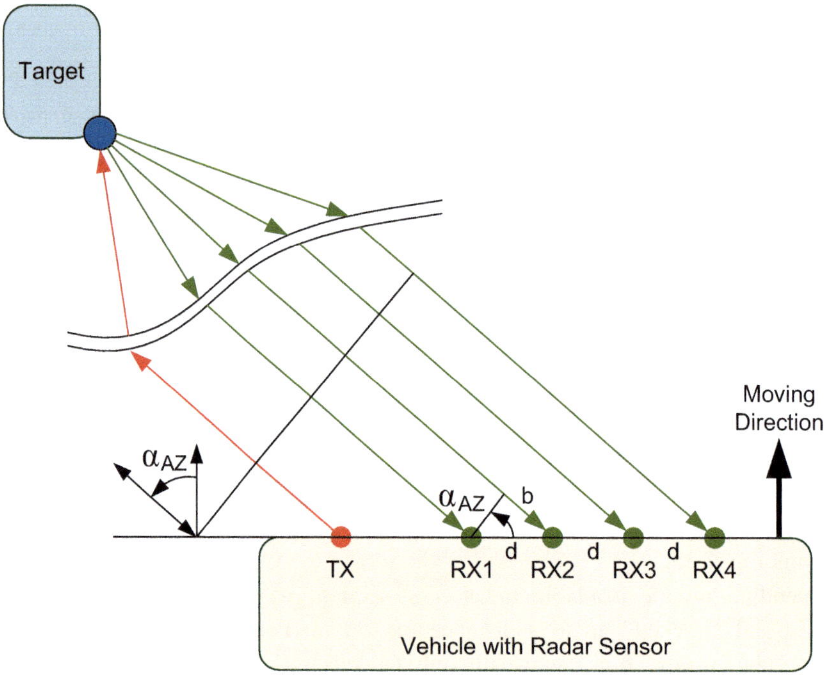

Fig. 7.34 Digital beamforming concept [9]

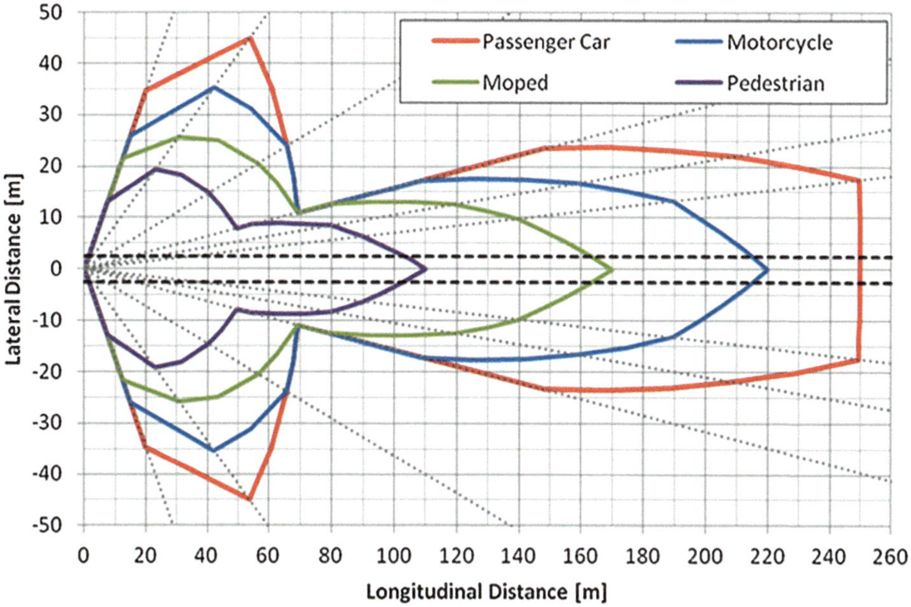

Fig. 7.35 Detection range of a modern front radar [8]

The performance spectrum of such a radar is shown in Table 7.4. In addition to the radar's detection range, the speed range and the resolution capability in radial and azimuthal direction are of essential importance.

7.2.6 Assistance Systems

The biggest risk factor when driving is still the human being: In Germany, according to the accident figures of the Federal Statistical Office [22] in 2011, about 87 % of all recorded accidents with personal injury were due to the driver's misconduct. In the INVENT report [16], the cause of accident in about 85 % of all accidents with personal injury is attributed to the driver. In contrast, it is stated in [5] that only 36 to 50 % of all accidents can be explained by human failure. Regardless of the actual proportion of accident causes, driver assistance systems can provide a remedy, supporting the driver in demanding, dangerous situations, but also offering scope for relief in monotonous driving with decreasing attention.

Almost every driver knows the helpless feeling when a car skids straight ahead with locked wheels during heavy braking and is no longer steerable. Effective here is an interval braking, which however only few drivers can master in an emergency situation. In such situations, the **anti-lock braking system** (ABS) automatically intervenes and prevents the wheels from locking when braking: The vehicle remains steerable and stable

Table 7.4 Performance data of a 5th generation radar sensor [27]

		ARS540	
Distance	Range	0.2 - 300 m	
	Accuracy	0.1 - 0.3 m	*dependent on v_ego (**)*
	Resolution	0.4 m	*dependent on v_ego (**)*
Speed	Range (unambiguous)	-400 to +200 km/h	*negative value: oncoming car*
	Accuracy	±0.1 km/h	
	Resolution	0.35 km/h	
Azimuth	Range (for max. misalignment)	±60°	
	Accuracy	±0.1°... ±0.5°	*±0.1° within ±15°, ±0.2° @50°, ±0.5° @60° linear interpolation in between*
	3dB beamwidth	1.2°...1.68°	*1.2°@0°...±15°, 1.68°@±45°*
Elevation	Range (for max. misalignment)	±4°... ±20°	
	Accuracy	±0.1°	
	3dB beamwidth	2.3°	
	Antenna channels	12 x 16	*TX x RX*
	Output Power (EIRP, average)	<= 37 dBm	*(@ -40°C ambient temp.); reduction for countries with lower limits available (see HCC section)*
	Auto alignment	±4° ±6°	for azimuth for elevation
	Cycle Time	50 ms	

(**) thresholds at 115/110 kph

even during strong braking. This allows the driver to avoid an obstacle without having to release the brake again.

This is made possible by wheel speed sensors attached to the wheels, which send signals to a control unit. The control unit regulates the brake pressure and can prevent individual wheels from locking. Automatic release and application of the brake alternate, ensuring that the vehicle remains steerable despite the driver's full braking.

The **drive slip control** (ASR or DSC, DTC, TCS or TRACS) or traction control is based on the ABS sensors, evaluates the speed of the individual wheels and thus detects spinning of individual or several wheels when starting or accelerating. The function of drive slip control is comparatively simple: If the system detects excessive slip during acceleration, it reduces the engine power. To avoid stalling the engine in tricky situations, the control allows a minimum slip (up to 20 %). Nowadays, the functions of the drive slip control are mostly integrated into the driving dynamics control (ESC). However, for commercial vehicles or sports cars, manufacturers can still be found who offer a pure ASR function and deactivate ESC functions.

In addition to reducing engine power, some vehicle manufacturers also work with brake interventions. This has the advantage that, for example, if a wheel is spinning, it is slowed down and more torque is supplied to the wheel with grip via the differential, so that the vehicle can be accelerated. Depending on the design of the system, this is also referred to as an electronic brake differential, because ASR with brake function can partially take over the function of a differential lock. ASR systems that also take into account the steering angle are already working close to the limit of ESC.

In reversal of the ASR, some manufacturers also offer an engine drag torque control (MSR). If the driver suddenly takes off the gas or engages too quickly, the drive wheels can block and the vehicle can become unstable due to the engine's drag torque on slippery surfaces. MSR prevents this tendency to block by regulating the gas only as far back or, in the case of too fast engagement, as high as necessary to prevent the effect from occurring.

In the case of the **brake assist** (BAS), there are two variants: the brake assist that derives an emergency braking situation solely from the driver's reaction and does not incorporate environmental information, and the brake assist that additionally incorporates environmental information (usually the distance to the obstacle) into the function.

First introduced by Mercedes-Benz in the S-Class in 1996, it senses the actuation speed of the brake pedal (via corresponding pedal travel sensors). The brake assist (BAS or iBrake, active city stop, predictive safety system or PRE-SAFE® brake) ensures maximum braking pressure regardless of the braking force applied by the driver. Numerous studies show that most drivers do brake in an emergency situation, but do not press the pedal far enough to build up maximum pressure and thus the highest possible deceleration, or release it too early. The brake assist recognizes from the characteristic pedal movements—time from moving the foot from the gas to the brake pedal, time when releasing the gas pedal and speed at which the brake pedal is actuated—that an emergency situation exists and builds up braking pressure independently of the driver.

The brake assist achieves its effectiveness through a brake booster with a partially evacuated chamber, which is subjected to atmospheric air pressure during emergency braking. This increases the brake boost if necessary. An increase in brake pressure is also achieved in some systems via a hydraulic unit. The brake assist is triggered via an electronic control. To prevent unwanted wheel lock, the brake assist is coupled with ABS and/or ESC. Unlike the emergency brake assist, however, it only operates after the brake pedal has been actuated by the driver and cannot autonomously trigger deceleration.

The **emergency brake assist** (EBA) goes far beyond the functionality of a brake assist: Such an EBA system allows a brake intervention completely autonomously, without any driver reaction having occurred.

The brake intervention occurs autonomously when the vehicle is heading towards an obstacle and there is an immediate risk of collision, but the driver has so far shown no

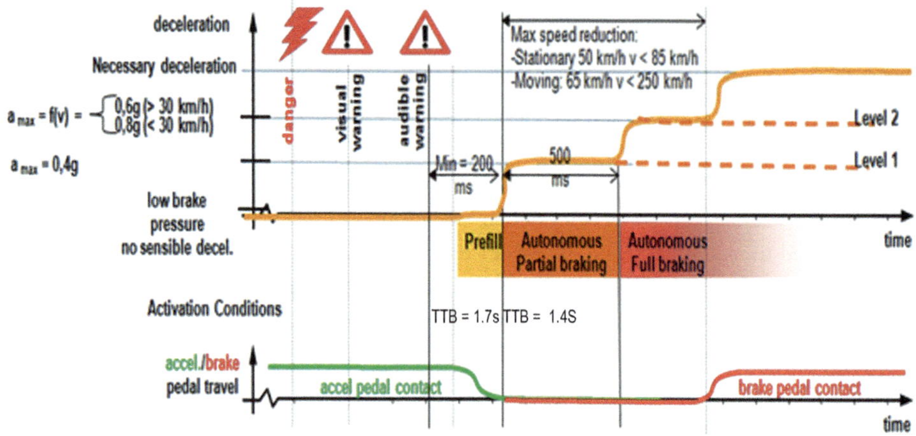

Fig. 7.36 Triggering sequence of an autonomous emergency braking system [10]

reaction to intervene in the braking process. The obstacle and the impending risk of collision are detected by a highly reliable sensor. In the computing unit of this sensor or in a directly connected safety control unit, the position of the detected object is compared with the current driving path. If a critical collision expectation value is exceeded, braking is triggered autonomously. A deceleration request is forwarded to the brake control unit. Depending on the current driving speed, decelerations of 6 m/s^2 to 8 m/s^2 are requested. A cascaded triggering sequence of an autonomous emergency brake is shown in Fig. 7.36.

The sensor predestined for this function is radar. The measuring principle of radar allows a very precise determination of the distance and relative speed to the potential collision object. Based on these highly reliable measurement data and the existing driving dynamics, the risk of collision can be reliably estimated. The functional principle of front radars used for this purpose is described in Sect. 7.2.5.

Through the continuous further development of radars, the performance of obstacle detection has been gradually improved. Accordingly, a distinction is made between 4 EBA generations (Table 7.5):

The requirements for emergency braking systems are contained in the standards ISO 15623, ISO 22839 and also the NHTSA regulation on “collision imminent braking”. Since a mistakenly triggered emergency braking can cause serious collateral damage, a significant focus in the design of these EBA systems is on the functional safety design of the system in accordance with ISO standard 26262. An EBA sensor is designed according to the potential hazard to an ASIL-B safety level.

The **driving dynamics control:** (ESC: electronic stability control, also known as ESP®: electronic stability program, as well as DSC, VSC, PSM, among others) is a

Table 7.5 Further development of the autonomous emergency braking system

EBA 1st generation Emergency braking on a **moving** (slower) object Triggering **up to 30 km/h**	
EBA 2nd generation emergency braking on a **stationary** object	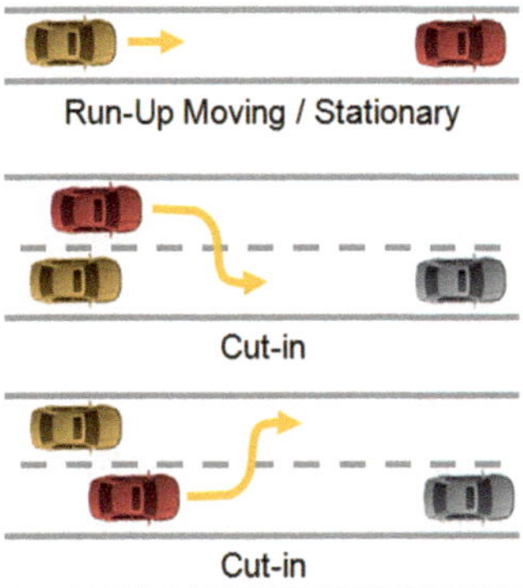
EBA 3rd gen: Emergency braking on pedestrians and activation of the autonomous brake **up to 60 km/h**	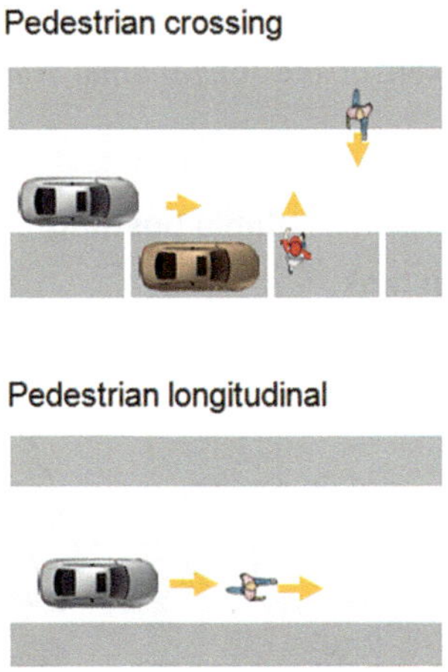
EBA 4th gen: Emergency braking for all typical partners in road traffic (vehicles, pedestrians, cyclists) Triggering of the autonomous brake **up to 100 km/h**	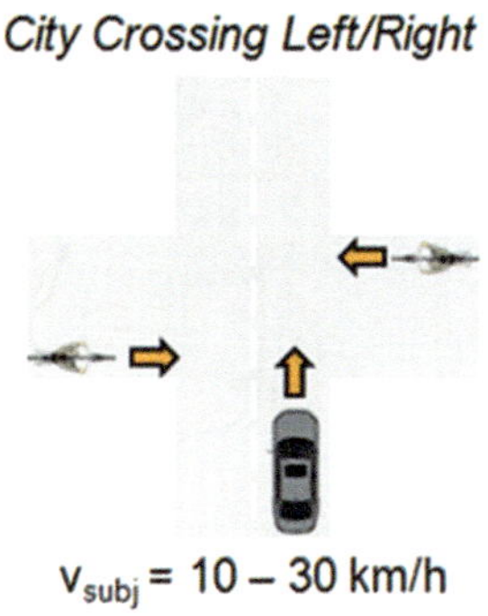

combination of the anti-lock braking system ABS, the drive slip control ASR, the electronic brake force distribution EBV, and the yaw moment control GMR (Fig. 7.37). By targeted braking interventions in individual wheels and in the engine control (in a disassembled system also in the steering system), a skidding of the vehicle can be prevented within physical limits. The electronic driving stability control ESC combines the functions of the wheel slip controls (ABS, EBV, ASR) with the yaw moment control (GMR).

Using a physical model, ESC calculates the vehicle behavior intended by the driver from the wheel speeds, the steering wheel angle, and the master cylinder pressure possibly controlled by the driver. ESC records the actual driving behavior using the yaw rate and the lateral acceleration. Especially with very fast steering movements, a vehicle can no longer convert the steering wheel lock into the expected change of direction. This leads to either understeering or oversteering, in extreme cases up to "skidding" (Fig. 7.38). The yaw moment control detects the deviation between real and targeted driving behavior and intervenes in a supportive and stabilizing manner.

Understeering is primarily corrected by the yaw moment control by braking the inner rear wheel of the curve, oversteering by braking the outer front wheel of the curve. This selective braking builds up unilaterally acting longitudinal forces and thus the desired yaw moment. A supportive effect is created by the targeted reduction of lateral guiding forces as a result of the longitudinal forces built up through braking moments. If

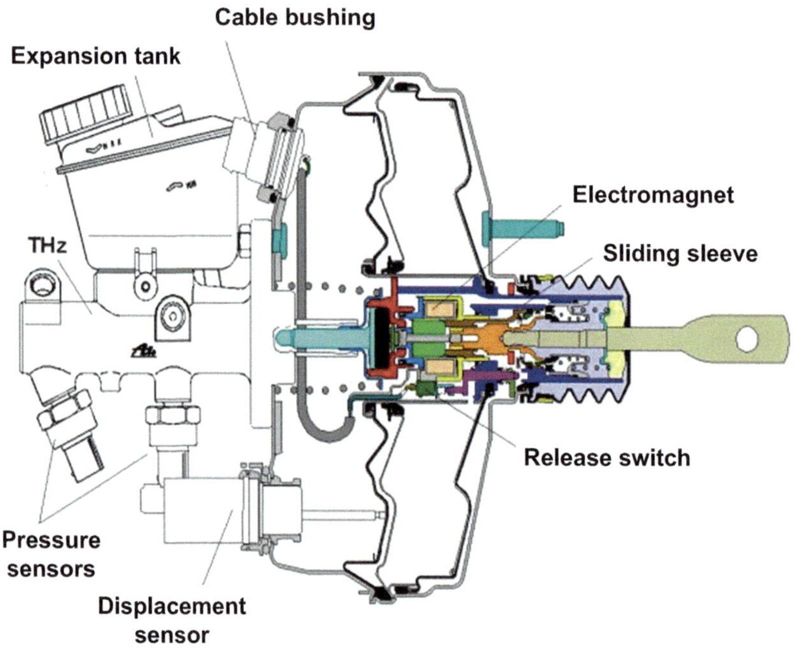

Fig. 7.37 Brake assist

necessary, ESC reduces excessive drive torque by intervening in engine management. The extended ABS/ASR hydraulics with the integrated electronic controller is the core of the ESC. This hydraulics allows the selective build-up of brake pressure on each wheel independently of the operation of the brake pedal. The ability of the ESC hydraulics to build up pressure in the wheel brake cylinders independently of pedal operation is used for another additional function, the so-called "hydraulic brake assist" (HBA); the HBA uses the existing ESC sensor technology.

Pressure sensor signals serve the brake controller to detect an extremely fast operation of the brake pedal in emergency situations. If a parameterizable critical pressure gradient is exceeded, the ASR separation valves are closed, the electric suction valves are opened, and the pump is activated. This increases the pressure introduced via the pedal to blocking pressure level. In this pressure control, the pressure in the wheel brake cylinders follows that of the tandem master cylinder, which allows modulation of the wheel brake pressures within the HBA mode. If a minimum pressure is fallen below (release of the brake pedal), the HBA function switches off again. The hydraulic brake assist is an example of systems in which the hydraulic control unit (HCU: hydraulic control unit) supports the function of the vacuum brake booster.

The **emergency evasion function** ESA (emergency steer assist) assists the driver in performing a high-dynamic evasive maneuver in front of an obstacle. The ESA function detects the critical collision situation with predictive sensors and determines optimal evasion trajectories around the potential collision object. As soon as the driver initiates an

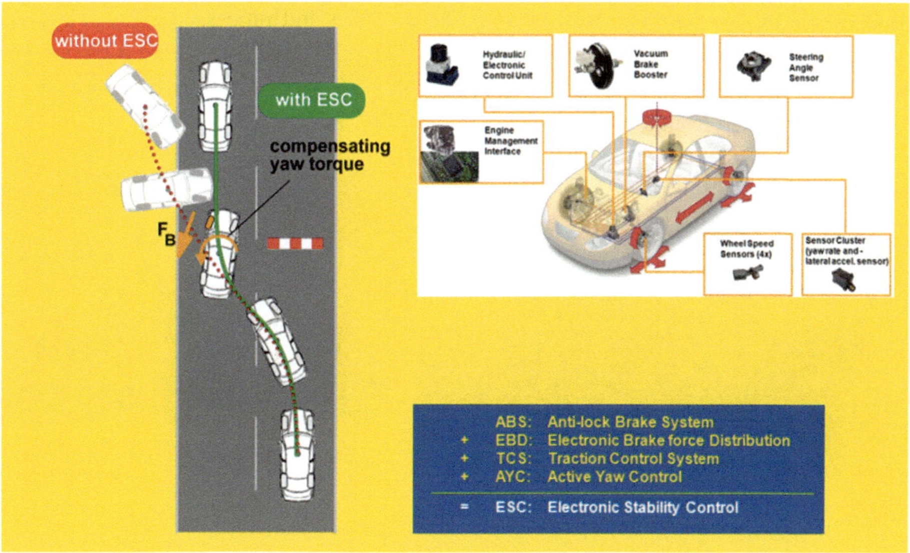

Fig. 7.38 Components and function of the electronic stability control (ESC)

evasive maneuver, the ESA function switches on and supports the evasion steering with an active steering intervention to drive the optimal evasion trajectory. This allows the vehicle to avoid the pedestrian or the vehicle in its direction of travel without collision, as shown in Fig. 7.39. If necessary, an emergency braking can also be superimposed on the emergency evasion function. The autonomous steering intervention can be overridden by the driver at any time.

An evasion maneuver that takes place in the own lane, as shown in the right part of Fig. 7.39, allows a fully autonomous evasion process of the vehicle. If it can be ensured via appropriate sensor technology that there is no obstacle in the potential evasion area, the critical object can be bypassed by an autonomous vehicle intervention without the driver's intervention.

The **adaptive cruise control** ACC is an extension of the known cruise control; its area of application is primarily the highway. A distance sensor in the front of the vehicle monitors the environment in front of the vehicle (Fig. 7.40). Due to the speed profile to be processed, distances up to 200 m and distance changes up to 200 km/h have to be measured, a requirement that requires the use of radar technology. If the vehicle threatens to collide with a vehicle ahead at the set speed, the adaptive cruise control reduces the gas by intervening in the engine control and, if necessary, brakes independently. It determines the distance to the vehicle in front and keeps it constant. The length of the distance can be set by the driver in most systems in five steps. If he makes no choice, the system regulates the distance depending on the speed based on the legally prescribed distance.

The first generation ACC systems were based on a still simple algorithm, with the help of which only moving objects could be detected and processed. The object relevant for the distance control (called target) was determined, usually the next object in the lane. Newer ACC generations regulate up to standstill and, at the driver's request, also after the vehicle in front starts moving again.

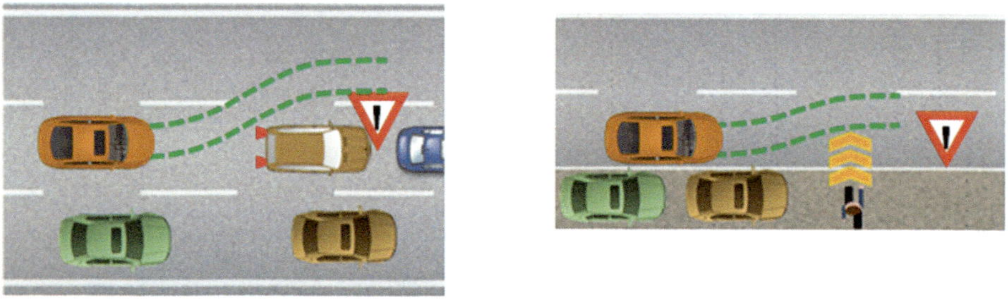

Fig. 7.39 ESA-1: Functional principle of the emergency evasion function [10]

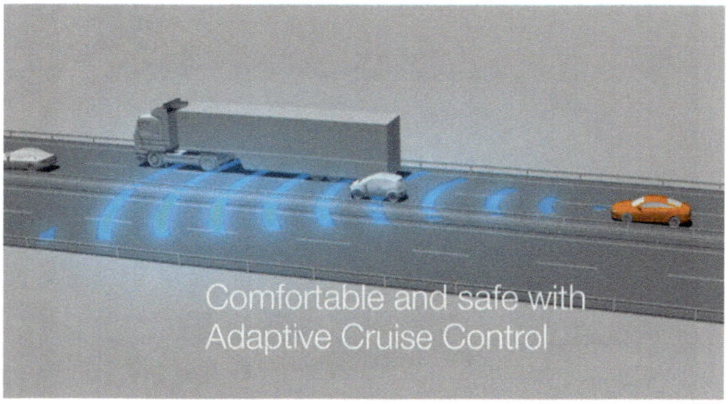

Fig. 7.40 Adaptive cruise control

7.2.7 Telematics Functions for Accident Prevention

All currently in vehicles used environment sensors, such as radar, camera or LIDAR, are sensors that require a direct line of sight between the sensor and the detected object. With the information thus obtained about the vehicle environment, a large number of accident scenarios can be detected and accidents can be avoided—but not all. In [11] it was shown that typical ADAS sensors can theoretically avoid about 55% of all scenarios, but the precise evaluation of the GIDAS data from 2015-2020 shows that in over a third of all accidents the lack of visibility of the sensor to the accident opponent is a problem because it is detected too late. In city traffic with 50 km/h, a threshold of two seconds is critical, because for physical reasons, even with optimal road and weather conditions, a crash with an emergency brake is no longer avoidable from 1.6 s warning time. If you include the driver's reaction time, the critical limit even shifts to three seconds.

The solution to this problem can be vehicle communication. Figuratively speaking, after the ADAS sensors, which represent "eyes", the vehicle is now given "mouth" and "ears" to communicate with other vehicles and infrastructure facilities. This vehicle communication is usually referred to as the abbreviation "V2X" (vehicle-to-X), where the X is supposed to indicate that many types of communication partners are possible, such as vehicles, traffic lights, networks, etc. (Fig. 7.41).

The field test sim^{TD} [4] showed that V2X can not only theoretically but also actually significantly reduce accident numbers. Here, the so-called intersection collision warning achieved an efficiency of >50 %, and in relation to accidents with severe personal injuries (MAIS > 2), even an efficiency of >65 %. By including intelligent infrastructure, e.g. at intersections, which also detects road users without V2X equipment through sensor systems and sends appropriate messages on their behalf, the effectiveness of V2X systems in the vehicle can be significantly increased. Especially the effectiveness in interaction with the equipment of the infrastructure also dispels concerns that V2X can only take effect when a large part of the vehicle fleet is equipped. It can even be shown that a significant reduction in fatal accidents can be expected after just a few years if a general equipment of vehicles and infrastructure units is decided [13].

Fig. 7.41 V2X concept overview

7.2.7.1 Safety Functions

Two examples will show how traffic safety can be increased with V2X.

The first example is the **electronic emergency brake light** (EEBL), which is an extension of the classic emergency brake light that signals to the following traffic via the brake lights that the vehicle is not just performing a normal braking operation, but is braking sharply and thus there is a risk of collision. The extension of the EEBL function becomes clear in the picture below (Fig. 7.42). It shows a typical convoy drive, where the leading vehicle has to brake sharply. The first vehicle can usually brake without problems, but due to the delay of the driver's reaction, the reaction time for those driving further back in the convoy becomes shorter and shorter, so that a collision becomes more and more likely. This delay cascade is broken by V2X, as the vehicle driving at the back can warn its driver at the same time as the brake light warns the directly following driver. To actually get more reaction time for the driver, the latency time for the entire function chain, from the detection of sharp braking in the sending vehicle to the display of the warning in the receiving vehicle, must be significantly less than one second. Since detection and warning are already optimized for other vehicle functions, the latency of the communication technology is of central importance to avoid accidents. Usually, 100-200 ms are assumed as the maximum time for communication. This time includes all processes described further below. The pure data transmission on the access layer (see below) must take place significantly faster than 100 ms.

These latency time requirements are mainly met by communication technology with direct communication between the final communication partners (ITS-G5, LTE-V2X see below). Other communication systems like 4G or 5G, which use base stations and a network, are useful for functions with longer warning time, as they also offer the necessary

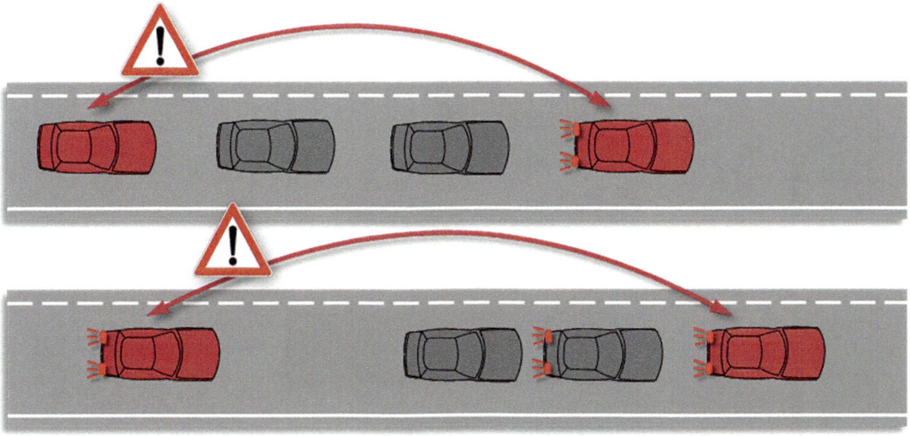

Fig. 7.42 Electronic emergency brake light (EEBL)

range. Since the accident avoidance potential of the directly communicating systems is highest, these are described in detail below.

The effectiveness of EEBL was investigated in the sim^{TD} project. It was found that the "time-to-collision", i.e., the vehicle distance, actually increases during braking in columns, thus reducing the risk of collision (see Fig. 7.43). For a vehicle that directly follows the braking vehicle, however, the TTC decreased, as the warning on the HMI apparently was more of a disturbance than a support in the critical emergency braking situation. A fact that illustrates the crucial importance of the type of communication integration into the integral safety system.

Technically, the EEBL function is implemented via an event message (DENM—see Sect. 7.2.7.3). The DENM ensures maximum communication range and, above all, prioritized channel access, so that latency can be minimized. This message is sent out as soon as the "triggering conditions" from [6] are met. These standardized triggering conditions are important so that the receiver knows exactly how to interpret the received data and the results are recognized and processed equally by all manufacturers. The reaction on the receiver side is not standardized in order to optimally integrate the function into the operating and warning logic of the respective vehicle.

A second important V2X function is the "**intersection collision warning**"[1] (ICW). It serves to warn drivers of a potential collision with another vehicle. Both vehicles regularly send out status messages (CAM, see below) several times per second, which contain their position, direction of travel, and speed. With this information, the receiving

[1] In some documents, ICW is also referred to as "intersection movement assist" (IMA), in sim^{TD} this function was referred to as "cross traffic assistant".

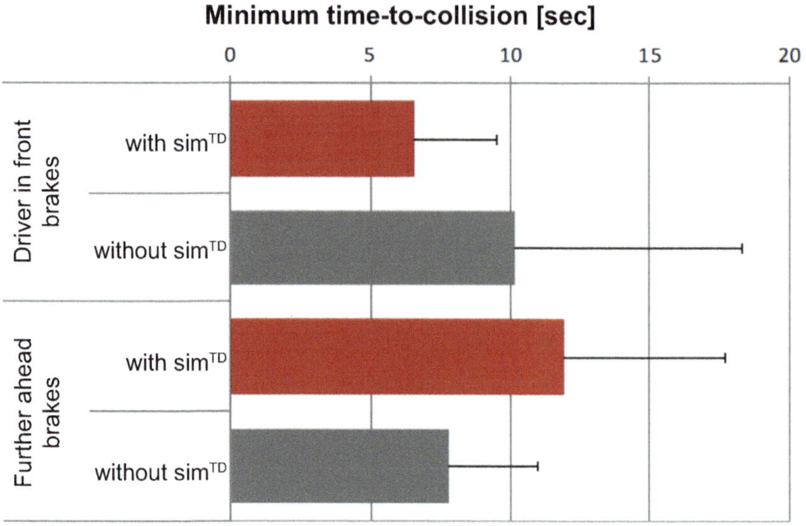

Fig. 7.43 TTC for EEBL. *Source* sim^{TD} final presentation

vehicle extrapolates the trajectory of the sending vehicle and checks whether it has a collision with its own trajectory. In the event of a collision risk, a warning is displayed to the driver. This is particularly helpful for intersections where one or both vehicles come from a cover and cannot be detected by either the drivers or ADAS sensors (see Fig. 7.44). The ICW is very effective, as studies in sim^{TD} have shown. 450 cross-traffic collisions from the GIDAS accident database were re-simulated, once with and once without V2X warning for the drivers. The driver's reaction to V2X warnings was simulated according to a driver model that was obtained from sim^{TD} experiments. The result of these simulations was that about half of the accidents were completely avoided and in many other situations the accident severity could be significantly reduced, so that a reduction of the MAIS > 2 injured by about 2/3 was achieved. Since cross-traffic accidents, due to the small "crumple zone", pose a particularly high injury risk, the ICW functions are particularly relevant to reduce the number of people injured in road traffic.

Both EEBL and ICW only take effect when many vehicles are equipped with a compatible V2X technology, so functions are needed for a successful market launch that also work with few compatible vehicles. Ideal are functions that use infrastructure information. This consideration led to the rollout of the "road works warning" (RWW) function in Germany and Austria in lane blocking trailers of the highway maintenance departments parallel to the introduction of the ITS-G5 V2X technology in Europe. This function sends the position and geometry of the construction site to the vehicles, which then display corresponding information in the HMI. This function uses DENM (see Sect. 7.2.7.3) to achieve maximum range.

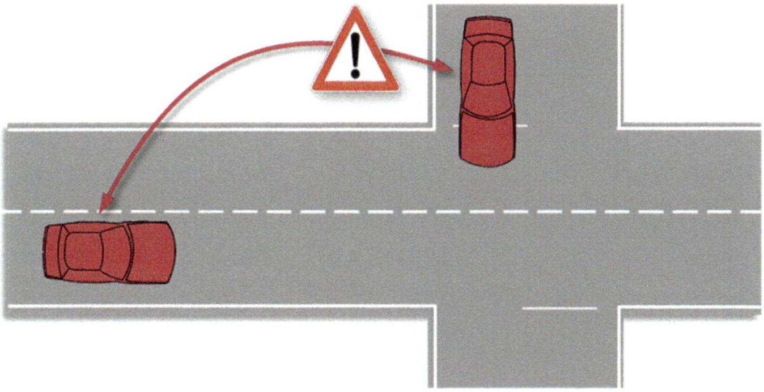

Fig. 7.44 Intersection collision warning (ICW)

In addition to the functions explained, there are many more V2X functions (Table 7.6).

It is expected that the upcoming Euro NCAP specifications will also include V2X functions, so that the accident prevention potential of this technology will also become visible to consumers in the form of additional evaluation points.

7.2.7.2 V2X Architecture

The technical basis for the described functions consists of standardized data exchange between vehicles. There are specific standards for this in the various world regions. In Europe, it is the family of ETSI ITS-G5 standards [1], in the USA it is SAE "WAVE" and in China it is the family of CSAE C-V2X standards.

Europe

- ETSI EN 302 637-2 Intelligent transport systems (ITS)—Vehicular communications; Basic set of applications; Part 2: Specification of cooperative awareness basic service
- ETSI EN 302 637-3 Intelligent transport systems (ITS)—Vehicular communications; Basic set of applications; Part 3: Specifications of decentralised environmental notification basic service
- ETSI TS 103 301 Intelligent transport systems (ITS)—Vehicular communications; Basic set of applications; Facilities layer protocols and communication requirements for infrastructure services
- ETSI EN 302 665 Intelligent transport systems (ITS); Communications architecture
- ETSI TS 102 894-2, Intelligent Transport Systems (ITS); Users and applications requirements; Part 2: Applications and facilities layer common data dictionary
- ETSI EN 302 636-4-1, Intelligent Transport Systems (ITS); Vehicular Communication; Geonetworking; Part 4: Geographical addressing and forwarding

Table 7.6 V2X functions

Name	Abbreviation	Explanation
Emergency Vehicle Warning	EVW	Warns of approaching emergency vehicles
Wrong Way Driving Warning	WWDW	Warns of wrong-way drivers
Stationary Vehicle Warning	SVW	Warns of broken down vehicles
Blind Spot Warning	BSW	Warns of overtaking vehicles
Do Not Pass Warning	DNPW	Warns of overtaking vehicles
Control Loss Warning	CLW	Warns when ABS or ESC have been activated (black ice/oil slick etc.)
Hazardous Location Notification	HLN	Warns of dangerous road sections (e.g. obstacles, potholes, tight curves, …)
Traffic Jam Warning	TJW	Warns of traffic jams
Redlight Violation Warning	RVW	Warns of a vehicle that ignores a red traffic signal
Lane Change Assistance		Subfunction of maneuver coordination
Pre-crash Sensing Warning		Exchange of crash-relevant data to prepare passive safety systems for a crash
Co-operative Forward Collision Warning	FCW	Warns of rear-end collisions
Co-operative Glare Reduction		Automatic dimming of high beams
Left Turn Alert/Right Turn Alert	LTA	Warns of collisions when turning left/right
Vulnerable Road User Warning		Warning of pedestrians
Motorcycle approaching indication	MAI	Make motorcycles more visible
Decentralized Floating Car Data	FCD	Obtain traffic flow data from V2X messages
Greenlight Optimal Speed Advisory	GLOSA	Optimize speed to achieve a "green wave"
In-Vehicle Signage	IVS	Display traffic signs in the vehicle
Co-operative Adaptive Cruise Control	CACC	Optimize ACC with V2X by including more vehicles in the calculation
Co-operative Highway Platooning		Link vehicle platoons via V2X so that only the first vehicle is controlled by a driver
Adverse Weather Warning	AWW	Warning of weather hazards

for point-to-point and point-to-multipoint communications; Sub-part 1: Media-Independent Functionality
- ISO/TS 19091, Intelligent transport systems—Cooperative ITS—Using V2I and I2V communications for applications related to signalized intersections
- ETSI EN 302 663, Intelligent Transport Systems (ITS); Access layer specification for Intelligent Transport Systems operating in the 5 GHz frequency band
- ETSI TS 102 687, Intelligent Transport Systems (ITS); Decentralized Congestion Control Mechanisms for Intelligent Transport Systems operating in the 5 GHz range; Access layer part
- ETSI TS 102 792, Intelligent Transport Systems (ITS); Mitigation techniques to avoid interference between European CEN Dedicated Short Range Communication (CEN DSRC) equipment and Intelligent Transport Systems (ITS) operating in the 5 GHz frequency range, V1.2.1 (2015-06)
- ETSI TS 102 724, Intelligent Transport Systems (ITS); Harmonized Channel Specifications for Intelligent Transport Systems operating in the 5 GHz frequency band
- ETSI EN 302 571 V2.1.1, Intelligent Transport Systems (ITS); Radiocommunications equipment operating in the 5 855 MHz to 5 925 MHz frequency band; Harmonised Standard covering the essential requirements of article 3.2 of Directive 2014/53/E
- ETSI EN 302 636-5-1, Intelligent Transport Systems (ITS); Vehicular Communications; GeoNetworking; Part 5: Transport Protocols; Sub-part 1: Basic Transport Protocol
- ETSI TS 103 248, Intelligent Transport Systems (ITS); GeoNetworking; Port Numbers for the Basic Transport Protocol (BTP)
- ETSI EN 302 931, Vehicular Communications; Geographical Area Definition
- ETSI TS 102 636-4-2, Intelligent Transport Systems (ITS); Vehicular Communications; GeoNetworking; Part 4: Geographical addressing and forwarding for point-to-point and point-to-multipoint communications; Sub-part 2: Media-dependent functionalities for ITS-G5
- ETSI TS 103 097, Intelligent Transport Systems (ITS); Security; Security Header and Certificate Formats
- ISO 8855, Road vehicles—Vehicle dynamics and road-holding ability—Vocabulary
- ETSI TS 103 175, Intelligent Transport Systems (ITS); Cross Layer DCC Management Entity for operation in the ITS G5A and ITS G5B medium
- ISO/TS 19321, Intelligent transport systems—Cooperative ITS—Dictionary of in-vehicle information (IVI) data structures
- ISO 3166-1, Codes for the representation of names of countries and their subdivisions—Part 1: Country codes
- ISO 14816, Road transport and traffic telematics; Automatic vehicle and equipment identification; Numbering and data structure
- ISO/TS 14823, Intelligent transport systems—Graphic data dictionary

- IEEE 80211, IEEE Standard for Information technology—Telecommunications and information exchange between systems, local and metropolitan area networks—Specific requirements, Part 11: Wireless LAN Medium Access Control (MAC) and Physical Layer (PHY) Specifications
- Car2Car Consortium—Basic Standards Profile (V1.6.3 or later)

USA

- SAE J3161, LTE Vehicle-to-Everything (LTE-V2X) Deployment Profiles and Radio Parameters for Single Radio Channel Multi-Service Coexistence
- IEEE 1609.2, IEEE Standard for Wireless Access in Vehicular Environments—Security Services for Applications and Management Messages
- IEEE 1609.3, IEEE Standard for Wireless Access in Vehicular Environments (WAVE)—Networking Services
- IEEE 1609.12 IEEE Standard for Wireless Access in Vehicular Environments (WAVE)—Identifier Allocations
- SAE J2735, Dedicated Short Range Communications (DSRC) Message Set Dictionary

China

- YD/T 3400 Overall technical requirement for LTE-based vehicular communication
- YD/T 3707 Technical requirements of network layer of LTE-based vehicular communication
- TCSAE 53_CH "Cooperative intelligent transportation system; vehicular communication; application layer specification and data exchange standard"
- YD/T 3709 Technical requirement of message layer for LTE-based vehicular communication
- YD/T 3594 General technical requirements of Security for Vehicular Communication based on LTE
- GB/T 37376 Transportation- Digital certificate format
- YD/T3340 Technical requirement of air interface for LTE-based vehicular communication
- YD/T 2018-0176 Technical requirement of sidelink-enabled on-board unit for LTE-based vehicular communication

The architectures of these standard families can be traced back to a common basis, which is a modification of the usual ISO-OSI communication architecture (open systems interconnection model) (Fig. 7.45): The lowest layer represents wireless communication and includes the physical layer and the definition of media access (e.g., CSMA/CA for IEEE802.11 based systems or TDMA/FDMA for LTE-V2X). Furthermore, this layer defines measures for coexistence with other radio technologies, such as CEN DSRC, and the handling of channel overload situations (decentralized congestion control).

This basic information exchange layer is then used by the network layer, which ensures that information is sent to defined recipients or areas or as a broadcast. A special capability of the ETSI ITS-G5 network layer is that it offers so-called "multihop" communication, so that messages are carried over several stations to increase the range of communication. In addition, the ETSI ITS-G5 offers various addressing types, in which geographical areas and not as usual stations are addressed. This geoaddressing is optimized for V2X functions, as V2X warnings should apply to defined areas that may be further away than the average radio range. This network layer is then used to send or receive the actual V2X messages.

In addition to the actual message generation, essential additional functions, such as time synchronization or station positioning, are summarized and specified as a separate layer of the architecture—the facility layer.

At the top of the V2X stack are the actual warning functions, which on the sender side detect events and initiate the sending of the appropriate messages. On the receiver side, the functions analyze the received messages to see whether the reported events or the situation pose a relevant danger to the received vehicle. The receiving functions may also initiate the warning of the driver.

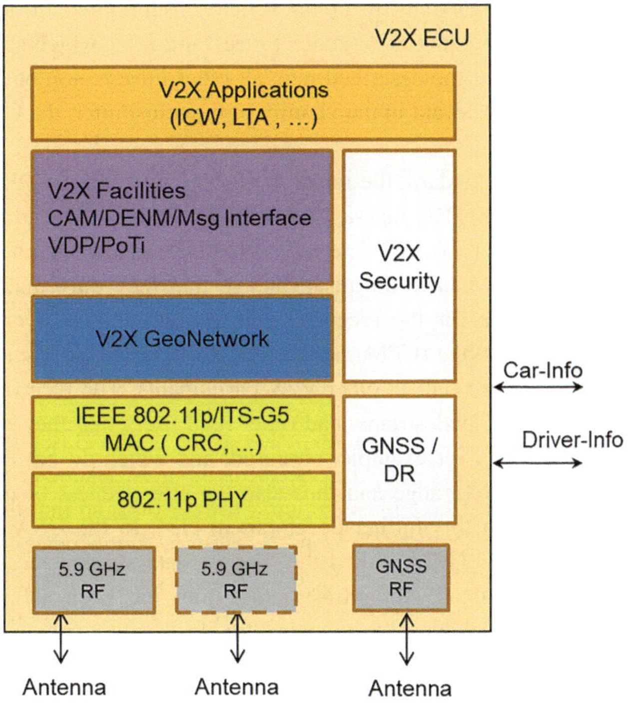

Fig. 7.45 Overview: V2X stack architecture

Other important architectural elements are security functions, which are necessary to provide the necessary trust in the transmitted data. Details on this are described in Sect. 7.2.7.5.

7.2.7.3 V2X Messages

The most important messages in the ETSI ITS-G5 system are the "cooperative awareness message" (CAM) and the "decentralized environmental notification message" (DENM). The DENM (Table 7.7) is sent out when events in road traffic are to be reported. Examples of such events are sharp braking, weather events, a broken down vehicle or an active emergency vehicle. Since the DENM-triggering events are relevant for a larger radius around the sender or also for further defined areas, DENM are sent as "geonetwork multihop message". The exact conditions when a DENM with which data content is to be sent are defined in the so-called "triggering conditions" of the Car-to-Car Communication Consortium [6].

The CAM is always sent when the vehicle has moved by 4 m, has changed its direction of travel by 5°, or has experienced a change in driving speed of more than 0.5 m/s, but at least once per second, and at maximum 10 times per second. The purpose of this message is to inform the surroundings about the presence, status, position, and direction of movement of a vehicle, so that other vehicles can check whether the transmitting vehicle can become dangerous to the receiving vehicle. To minimize the bandwidth used, only the data of the "HighFrequency" container (see Table 7.8), which contains the driving dynamics data, is sent at the described rate, all other information only once per second. Since a CAM is only relevant in the vicinity of the transmitter, the CAM is sent as a "single hop" broadcast.

In the US and Chinese standard, the function of the CAM and the DENM is summarized in the so-called "basic safety message" (BSM). In these V2X standards, the driving dynamics data, the vehicle status, and detected event flags are always sent at 10 Hz.

CAM, DENM, and BSM have the disadvantage that they only make vehicles that send themselves detectable for the receiver. This shortcoming is overcome with the cooperative perception message (CPM), as object data detected by a sender with radar, camera, or Lidar sensors are sent to other V2X participants. The receivers thus receive information about vehicles, pedestrians, and other road users that they themselves cannot detect because they are, for example, obscured, and which do not themselves send V2X messages. The effective range and thus also the effectiveness of the V2X system increases significantly with this further development [13]. In the USA, a similar V2X message is defined in SAE J3224, called the "sensor data sharing message" (SDSM). In China, a message specifically for infrastructure units has been defined, which also reports detected objects to vehicles, the so-called "roadside safety message" (RSM).

The data from the CPM can be used similarly to CAM data and improve functions such as "intersection collision warning" or "left turn alert".

Specifically for the coordination of vehicles, e.g., in lane change situations, the "maneuver coordination message" (MCM) is currently being defined. It is particularly

Table 7.7 Data of the CAM

Container	Data
StationType	StationType
ReferencePosition	Latitude
	Longitude
	PosConfidenceEllipse
	Altitude
BasicVehicleContainerHighFrequency	Heading
	Speed
	DriveDirection
	VehicleLength
	VehicleWidth
	LongitudinalAcceleration
	Curvature
	CurvatureCalculationMode
	YawRate
	AccelerationControl (Opt.)
	LanePosition (Opt.)
	SteeringWheelAngle (Opt.)
	LateralAcceleration (Opt.)
	VerticalAcceleration (Opt.)
	PerformanceClass (Opt.)
	CenDsrcTollingZone (Opt.)
RSUContainerHighFrequency	ProtectedCommunicationZonesRSU (Opt.)
BasicVehicleContainerLowFrequency	VehicleRole
	ExteriorLights
	PathHistory
PublicTransportContainer	EmbarkationStatus
	ptActivation (Opt.)
SpecialTransportContainer	SpecialTransportType
	LightBarSirenInUse
DangerousGoodsContainer	DangerousGoodsBasic
RoadWorksContainerBasic	RoadworksSubCauseCode (Opt.)
	LightBarSirenInUse
	ClosedLanes
RescueContainer	LightBarSirenInUse
EmergencyContainer	LightBarSirenInUse
	CauseCode (Opt.)
	EmergencyPriority (Opt.)

(continued)

Table 7.7 (continued)

Container	Data
SafetyCarContainer	LightBarSirenInUse
	CauseCode (Opt.)
	TrafficRule
	SpeedLimit

Table 7.8 DENM Data

Container	Data
ActionID	StationID
	SequenceNumber
ReferenceTime	TimestampIts
DetectionTime	TimestampIts
Termination (Opt.)	Termination
ReferencePosition	Latitude
	Longitude
	PosConfidenceEllipse
	Altitude
RelevanceDistance (Opt.)	RelevanceDistance (Opt.)
RelevanceTrafficDirection (Opt.)	RelevanceTrafficDirection (Opt.)
ValidityDuration	ValidityDuration
TransmissionInterval (Opt.)	TransmissionInterval (Opt.)
StationType	StationType
InformationQuality	InformationQuality
CauseCode(eventType)	CauseCodeType
	SubCauseCodeType
CauseCode(linkedCause) (Opt.)	CauseCodeType
	SubCauseCodeType
EventHistory	EventPoint (1 … 23)
Speed (Opt.)	SpeedValue
	SpeedConfidence
Heading (Opt.)	HeadingValue
	HeadingConfidence
Traces	PathHistory (0 … 7)
Roadtype (Opt.)	Roadtype
LanePosition (Opt.)	LanePosition

(continued)

Table 7.8 (continued)

Container	Data
ImpactReductionContainer (Opt.)	HeightLonCarrLeft
	HeightLonCarrRight
	PosLonCarrLeft
	PosLonCarrRight
	PositionOfPillars
	PosCentMass
	WheelBaseVehicle
	TurningRadius
	PosFrontAx
	PositionOfOccupants
	VehicleMass
	RequestResponseIndication
Temperature (Opt.)	Temperature
RoadWorksContainerExtended (Opt.)	LightBarSirenInUse (Opt.)
	ClosedLanes (Opt.)
	RestrictedTypes (Opt.)
	SpeedLimit (Opt.)
	CauseCode (Opt.)
	ItineraryPath (Opt.)
	DeltaReferencePosition (Opt.)
	TrafficRule (Opt.)
	ReferenceDenms (Opt.)
PositioningSolutionType (Opt.)	PositioningSolutionType
StationaryVehicleContainer (Opt.)	StationarySince (Opt.)
	Causecode (Opt.)
	DangerousGoodsExtended (Opt.)
	NumberOfOccupants (Opt.)
	VehicleIdentification (Opt.)
	EnergyStorageType (Opt.)

interesting for automated vehicles, as they lack the ability to coordinate with other drivers via non-verbal communication. An extensive elaboration and demonstration of MCM technology was carried out in the IMAGinE research project [17].

In addition to these messages, which are mainly sent from vehicles, there are a number of messages that are specifically sent by infrastructure facilities such as traffic signal systems. These are "signal phase and timing" (SPAT) together with the "intersection MAP message" (MAP). The MAP provides information about the topology of an intersection and which traffic light is assigned to which lane. The SPAT provides the status of the traffic signals and change times. With both messages together, a receiving vehicle can calculate which signal is relevant, what status it has, and can adjust its driving strategy with the change time information.

The "in vehicle information message" (IVIM) provides information about traffic signs to vehicles, which is particularly interesting for variable traffic signs. Details can be found in [6].

7.2.7.4 Facility Layer Functions

The most important information that a V2X vehicle emits is the position of the vehicle and, since the vehicle is moving, the time stamp of this position. In order for the receiver of the V2X message to correctly interpret the position, it must know what time the sender used. Therefore, it has been determined that all ETSI ITS-G5 stations must be synchronized to TAI (Temps Atomique International, international atomic time) with an accuracy of ± 20 ms. Stations that are not synchronized are not allowed to transmit. Since normal, quartz-based clocks only have an accuracy of a few ppm, they quickly fall out of the required accuracy band, especially when the ambient temperature rises or falls sharply. Accordingly, V2X systems are always synchronized using GNSS (global navigation satellite system—GPS/Galileo/etc.).

The GNSS are mainly used for positioning the vehicles in the V2X system. GNSS-based positions are always subject to considerable noise, which is why requirements for position accuracy are given as a radius in which 95 % of the position indication lies around the respective true value, this radius is referred to as confidence interval. For moving vehicles, the position accuracy is often different along and across the direction of travel, which turns the confidence circle into an ellipse. Since the currently estimated confidence interval is part of the V2X messages (PosConfidenceEllipse), the respective receiver can decide whether the position indication is accurate enough to use the message in its function. In [6] maximum confidence intervals are required for different driving situations. The radii range from 5 m for unrestricted GNSS reception to 15 m for positions in an "urban canyon". Other standards only require compliance with a confidence interval for "open sky". All standards have in common that their requirements aim to enable the positioning of a vehicle in a lane.

Since GNSS positions are either not available at all (e.g., in a tunnel) or fluctuate greatly in situations where reception is impaired (shadowing from buildings or mountains), GNSS positions are improved with vehicle sensor signals. This inertial navigation uses, for example, vehicle speed, accelerations, and rotation rates to determine vehicle movement and extrapolate the position. This extrapolated position is then used to smooth the GNSS position. As a by-product of these algorithms, the position confidence interval is also estimated.

The collection of vehicle data for the messages is also considered a function of the "facility layer", but is naturally vehicle-specific and therefore not further standardized.

The same applies to the interface to the warning functions, for which there is a standard, but this is of subordinate importance, as the functions themselves are not standardized and the interface must be adapted to the functions implemented.

7.2.7.5 V2X Security

For V2X functions to be useful to a driver, he must be able to rely on the transmitted data. To this end, the messages must be signed before sending so that the recipient can verify that they are unchanged and, above all, trustworthy. The digital signature of V2X systems is based on the ECDSA algorithm (elliptic curve digital signature algorithm) [2] and uses cryptographic algorithms based on elliptic curves over finite fields. For V2X, 256-bit long keys are used to provide a level of security that meets current and medium-term requirements. ECDSA is an asymmetric method that uses public keys for verification and private keys for signing. Since in situations with dense traffic and many V2X stations several hundred messages are received and each message has to be verified, a high computing power must be provided for the ECDSA calculation, usually via special hardware accelerators.

The public keys are attached as part of a certificate to the V2X message, while the private keys are stored in specially protected storage areas of the sending system and do not leave these areas. The certificates in turn are signed by a "public key infrastructure" (PKI), which thus confirms the trustworthiness of the sender. Before a vehicle can have certificates signed, it must identify itself to the PKI and only known and registered vehicles receive signed certificates. To limit the damage that can be done with potentially stolen certificates, these are time-limited (~1 week) and can be restricted for certain functions. The vehicles use a so-called "enrollment certificate" (EC), which they receive during the production of the vehicle or at the first commissioning, as identification. Additional security measures are necessary to secure the enrollment and prevent unauthorized persons from receiving certificates.

Vehicles always use several certificates at the same time and change these and all identifiers used in the communication protocols to obscure the path of a vehicle. A typical certificate change time is about 10 minutes, as this achieves a high probability that different certificates will be used at the beginning and end of a journey, so a local observer cannot easily draw conclusions about the identity of a vehicle. On the other hand, the 10-min constant identification of the vehicle ensures that safety functions can track a vehicle through longer-lasting danger situations. Since the certificates are supposed to obscure the identity of the vehicles, they are sometimes also referred to as "pseudonym certificates". In the ITS-G5 standards, they are referred to as "authorization tickets" (AT). To separate enrollment and the identity of a vehicle from the issuance of certificates, different "security authorities" are set up, each of which in turn ensures their authenticity via a common "root certification authority" (RCA). The public key of the RCA is securely stored in the vehicles so that the vehicles can also check that received messages and the certificates contained therein come from a trustworthy authority, i.e., were accepted by the same RCA. An overview of the relationships in the V2X PKI is given in Fig. 7.46.

7.2.7.6 V2X Communication Technology

V2X systems, which are based on "standard mobile communication networks", use all available mobile communication technologies that are supported by the respective vehicle and the corresponding telecommunications contract. The functions in the vehicle use IP communication protocols (TCP/UDP/http/MQTT/TLS/…), to connect the vehicle functions with the respective service server. More detailed information on these technologies can be found in [14].

Direct vehicle-to-vehicle communication uses the 5.9 GHz band and either IEEE 802.11 (WLAN) based radio technology or LTE based PC5 sidelink communication. Currently, IEEE 802.11 based ITS-G5 is used in Europe, and LTE-V2X in China and USA. Detailed information about LTE and WLAN technologies can be found in [23].

Typical V2X implementations with ITS-G5 or LTE-V2X operate with a transmission power of approx. 23 dBm, which achieves a communication range of approx. 800 m in free-field conditions and approx. 150–200 m in typical city conditions. Due to the comparatively high frequency, the positioning of the antennas on a vehicle is important so that the vehicle itself does not interfere with the wave propagation, e.g. through a curved roof that disturbs the emission in the direction of travel. Many vehicles therefore use two antennas, e.g. a roof antenna and an antenna in the area of the front camera. To ensure interference-free transmission operation with two antennas, the transmission module must support "cyclic delay diversity" (CDD).

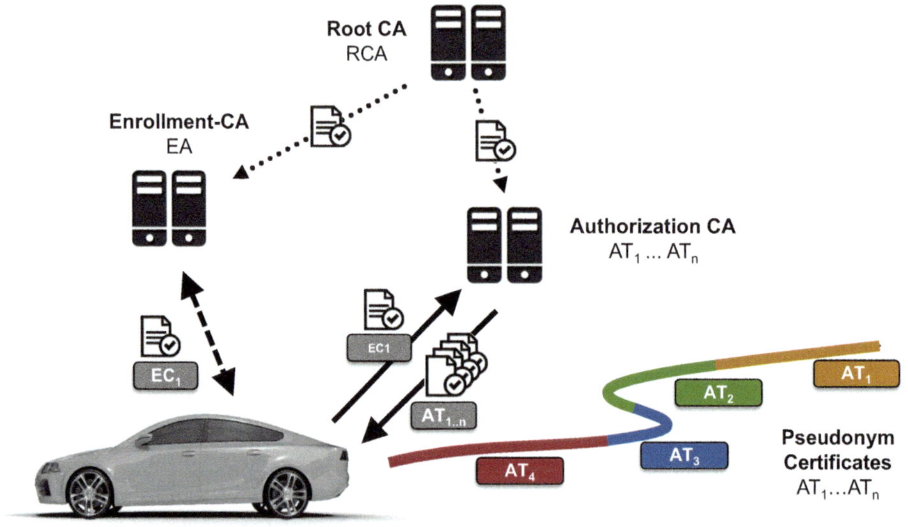

Fig. 7.46 V2X public key infrastructur (PKI)

Both ITS-G5 and LTE-V2X use "orthogonal frequency division multiplexing" (OFDM) as a modulation method to optimally utilize the available bandwidth. Due to the difficult communication conditions in road traffic (large Doppler shift due to high self-speed of the vehicles, changing multi-path propagation of the signal due to many reflections on house walls etc.), the modulation is usually limited to quadrature phase-shift keying (QPSK) or 16QAM, otherwise packets with a few hundred bytes of payload would too often not be able to be decoded and thus the effective communication range would be too short.

ITS-G5 systems use one or more 10 MHz wide channels and offer a very short maximum latency of 4 ms. ITS-G5 uses the IEEE 802.11 CSMA/CA method and sends its message as soon as the radio channel is recognized as free.

LTE-V2X uses a 20 MHz wide channel and a maximum latency of typically 100 ms, as LTE-V2X operates with "semi-persistent scheduling". In this case, each station selects a free time and frequency slot and uses it for a few seconds. The messages are sent every 100 ms in the selected frequency/time resource block. For longer messages, several blocks must be reserved. Since the individual stations must very precisely maintain the time slots in order not to disturb the messages of the neighboring stations, the stations must synchronize the internal clocks to ±391 ns. A GNSS is used as the synchronization source.

For both radio technologies, there is an upper limit for the number of messages that can be sent per second on the radio channel. Depending on how the worst-case scenario is designed (number of vehicles with high/low speed/ number of roadside units/ number of events and message lengths), this number is different. Generally, about 1,000 messages per second are considered a realistic upper limit, with ITS-G5 reaching about 50 % channel load. To grant all vehicles within radio range access to the channel despite this upper limit, there is the so-called "decentralized congestion control", which defines rules from which channel load the transmission frequency of the messages is reduced. Particularly urgent messages (e.g. DENM for EEBL) are exempt from this rate limit.

7.2.7.7 Implementation Examples

Telematics functions are implemented in almost all modern vehicles, but rarely as a true safety function in the sense of the preceding sections. The following examples are not an exhaustive market overview, but just a few prominent examples that are currently (March 2023) available. There are many more implementations by almost all vehicle manufacturers, who continuously bring new V2X systems to the market.

Since the end of 2019, a European OEM has been delivering V2X based on ITS-G5 as a standard component in various vehicle models. According to European registration figures, at the end of 2022, there were approximately 800,000 vehicles equipped with this technology active in road traffic. More vehicle models are to be equipped with V2X technology.

The following V2X functions are available in these vehicles:

- Stationary vehicle warning
- Traffic jam warning
- Road works warning
- Emergency vehicle warning
- Control loss warning
- Electronic emergency brake light (only sending)

These systems can receive the DENM of the lane blocking trailers of the German Autobahn GmbH, which are equipped with ITS-G5 transmitters. In addition, various telematics services are available via mobile radio, but these do not have a safety aspect.

Increasingly, warning information is also sent to the driver via mobile phone-based telematics systems. For example, in the USA, warnings of wrong-way drivers are spread via the "HaaS Alert" network and used by various OEMs to sensitize drivers.

Vehicles of another European OEM have been using the mobile phone systems since 2013 not only to receive warnings, but also to send data about traffic hazards to a server. This server aggregates and validates the data and, if necessary, sends warnings to other vehicles of the OEM, which are in the danger area or are approaching it.

Depending on the equipment of the vehicle, the following functions are offered, with the designations adapted to the terms used so far:

- Stationary vehicle warning
- Road works warning
- Control loss warning
- Adverse weather warning
- General warning (hazard warning lights active)

V2X technology is in a rapid development phase. For example, new radio technologies have been standardized, such as IEEE 802.11bd or 3GPP 5G NR V2X, but also new architectures, such as "mobile edge computing", where the V2X servers are no longer hosted by the OEMs, but by the telecommunications providers. In addition, work is being done to extend the radio technology with object location functions, so that the communication system becomes an environment sensor. Examples can be found in [19].

V2X-based functions are also becoming increasingly diverse and will require an extension of the protocols to data elements in the future that enable a functionally safe implementation of the functions. Corresponding standardizations were started in 2023, so that the second generation of V2X systems can take over significantly more safety functions.

7.3 Boundary Conditions

The boundary conditions necessary for a functioning and effective sensor system include the electronics, represented here for the airbag deployment, the power supply with energy reserve, the safety requirements as well as the data transmission. These essential conditions are presented and explained in detail below.

7.3.1 Airbag Control Unit

The airbag control unit include the power supply, sensing, signal evaluation and information storage, as well as the activation of the safety measures required for the protection of occupants and pedestrians (Fig. 7.47).

The tasks of the airbag control unit can be summarized as follows:

- Continuous monitoring of the entire airbag system
- Error display via warning lamp,
- Full functionality during diagnosis,
- Autonomous power supply for a defined period,
- Data exchange with occupant detection systems,
- Activation or deactivation of passenger protection systems depending on the status of the occupant detection system,
- Recognition and classification of the present input information,
- Impact and rollover detection,
- Activation of the protective devices,
- Non-volatile storage of operational, status, error and collision information,
- Notification of an impact or rollover event and the status of other control units via bus or discrete connections and finally
- Disposal ignition for end-of-life vehicle recycling.

In sensing, a distinction is made between internal and external sensors. The evaluation of the sensor signals is carried out using an algorithm tailored to the respective vehicle. If several external sensors are used, they are connected and connected by additional circuits. This creates the possibility of using additional acceleration or pressure sensors. A corresponding number of ignition circuits is required for the ignition of the safety systems; these are usually two, four or even six separate ignition circuits that can be connected together. The airbag control unit shown in Fig. 7.48 is a variant with a two-channel accelerometer, as well as a trigger sensor. The three recognizable capacitors serve the energy reserve of the airbag system.

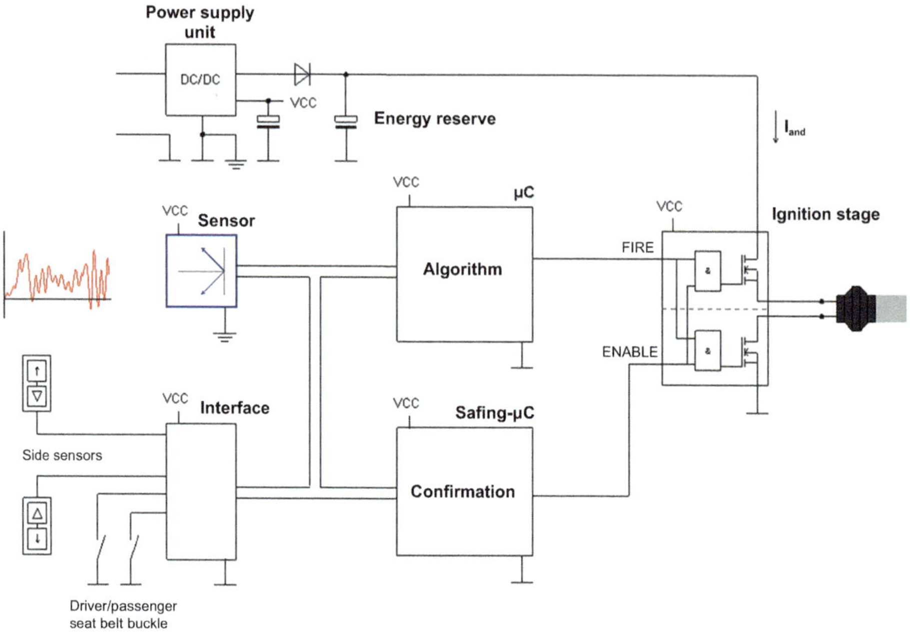

Fig. 7.47 Schematic of airbag control units (from [7])

7.3.2 Power Supply and Energy Reserve

The power supply (see Fig. 7.47) serves to provide the required voltage, it is fed from the vehicle's electrical system. It is designed as a switching or linear regulator and protects the airbag electronics from reverse polarity, overvoltage, electrostatic discharge (ESD) and electromagnetic interference (EMC: electromagnetic compatibility). A voltage of 3.3 to 5.0 V is provided for the supply of the electronics, the sensors, the semiconductors and other components; the charging voltage for the energy reserve is usually 20 to 40 V. This energy reserve is used to activate the restraint systems and is usually designed so that the airbag system is still supplied with the required voltage for about 150 ms after a voltage drop, which can be caused by a short circuit or by the loss of the vehicle battery during the accident.

7.3.3 Sensors, Control and Monitoring Units

One or more sensors, depending on the type of protection systems, are integrated into the airbag control unit. The **external sensors** are connected to the airbag control unit via

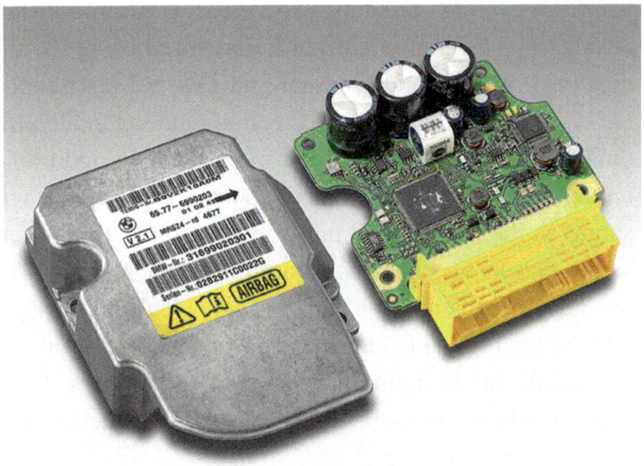

Fig. 7.48 Airbag control unit (from [7])

interfaces in order to be able to import the signals. In addition, the **internal sensors** are responsible for monitoring the status of seat belt buckles as well as seat occupancy and seat position detection. They also serve to gather and provide information from occupant detection systems and effective belt forces.

Another key component is the **microcontroller** (μC), also referred to as the master-μC. It controls the entire process within the electronics and checks software-implemented functions such as operating system, communication, diagnostics, data acquisition, crash algorithm, plausibility check, and finally the control of the ignition stages. The inclusion of additional requirements for occupant detection and classification or for pedestrian protection constantly increases the need for memory (RAM: random access memory; ROM: read only memory) and higher clock frequency. In addition, future tasks need to be tackled, such as the implementation of the event data-storage (EDR: event data recorder) or the disposal ignition as a recycling requirement for the disposal of pyrotechnic restraint systems (e.g., airbags, seat belt pretensioner) when decommissioning the vehicle.

The hardware-based, independently executed monitoring circuit is realized using a special small microcontroller, the so-called **safing-μC**. With it, the correct system clock of the software running sequentially in the microcontroller is monitored, the sensor signals are checked, and the start or enable signal for the ignition circuits is electronically generated.

7.3.4 Ignition Device

The **ignition stages** are usually power switches contained in integrated circuits (ASIC: application specific integrated circuit) and also include diagnostic and communication circuits. The power switches are used to control the igniters, which are located in the generators of the protection systems. The ignition stages are monitored for possible errors and the ignition current is measured using the diagnostic circuit. The tests of the diagnosis include both internal and external test scopes. Internally, the power switches are checked for their correct function, and externally, the correct resistance values of the igniters including the lines are monitored. In addition to possible open lines and short circuits, the checks also detect short circuits of the ignition lines to vehicle ground or supply voltage. Electrical faults in the igniters or gas generators can also be detected. For safety reasons, a physical separation of the high-side and low-side switches is made in the design of the ignition stages. This is realized by separate integrated circuits, so-called ASICs, for high-side and low-side or by using a third switch, which externally enables the supply of the ignition voltage of the circuits. Increasingly, a single-chip design is used, in which the high-side and low-side switches are separated by a sufficiently wide silicon trench on the chip.

To ensure safe disposal of pyrotechnic restraint systems (e.g., airbags, seat belt pretensioners) when decommissioning a vehicle, ISO-26021 (road vehicles—end-of-life activation of in-vehicle pyrotechnic devices) was published in 2008. It represents a manufacturer-independent technical specification for a **disposal ignition**. The specification has been available in version 2 since 2022 and defines the interfaces that airbag control units should offer for disposal ignition, for example the CAN Identifier 0 × 7F1 and 0 × 7F9 for request and response of the UDS communication protocol. A suitable device is connected to the airbag electronics via a vehicle-side installed diagnostic unit (OBD: onboard diagnostic connector) and activates the disposal mode via a communication interface. The proper ignition of the airbags, the seat belt pretensioners, and the other pyrotechnically activated components of restraint systems is automatically logged and documented at the end of the disposal.

7.3.5 Storage of Event Data

In 2012, the storage of event data (EDR: event data recorder) was legally anchored in the USA. In the following years, legislation on EDR was passed in numerous countries worldwide, and the EU Parliament also decided to introduce an event data recorder as part of the "General Safety Regulations" GSR 2 in 2019 (Regulation 2019/2144).

The technical design of this event data recorder was carried out within an international working group (IWG) under the umbrella of the UN-ECE WP.29 and can be found in UN Regulation R 160. This regulation standardized the existing requirements from the USA and China and took into account the increasing automation of driving functions through driver assistance systems. UN-R 160 became mandatory in the EU for new type approvals on July 6, 2022 and will be replaced in July 2024 by the first revision of UN-R 160 ("Series 01") with additional event data on assistance functions.

7.3.6 Safety Requirements for Airbag Control Units

Today, safety requirements must be imposed on modern airbag control units, which must be taken into account during development. Specifically, these are requirements regarding

- system reliability,
- system availability, even in the event of a power failure,
- a reliable ignition decision based on two independent signal sources,
- 360° sensing and plausibility checking, and finally,
- individual errors must not lead to unintentional triggering, but also must not prevent intentional triggering.

Over the past few years, airbag control units have been significantly developed and improved, as the mentioned safety requirements could not be realized or only insufficiently realized with the first airbag systems. However, with the use of modern electronics, it has been possible to increasingly meet the mentioned requirements.

With the first, purely **mechanical systems**, there was naturally neither independent sensing nor a check of plausibility. An exact evaluation of the accident signal as well as an accurate calculation of the trigger time were not possible. A weak point of the mechanical systems, an example of which is shown in Fig. 7.49, is that the system only has a single triggering path; moreover, the system state cannot be checked and displayed.

With the use of **electromechanical systems**, the safety against false triggers has been significantly increased, and the system state could be monitored. Should an intended ignition occur, in addition to the safing sensor installed in the control unit, at least one of the crash sensors installed in the front area must be closed (Fig. 7.50). This safety sensor also prevents unintentional triggering in the event of line faults. In the event of a loss of supply voltage during an accident, the system is also kept ready to trigger by its own energy reserve. To reduce sensitivity to electromagnetic radiation, twisted lines are used in the connections. The system can be specifically monitored through the resistors installed in the sensors; even the igniters conduct a weak diagnostic current for checking the ignition circuits. If an error still occurs, a warning lamp is activated, and if this warning lamp fails, a buzzer integrated in the SD module (SDM: sensing and diagnostic module) of the control unit sounds. The system also has a fuse that is deactivated in the event of a serious system error. This allows the system to be permanently switched off to safely prevent unintentional triggering. In the ignition circuit of the SD module, a low-resistance resistor is also integrated, which continuously monitors the voltage drop of the microcontroller. If an ignition current flows, the voltage drop increases, and the system state is stored in a permanent memory for possible review.

In the first generation of **airbag control units**, a **safing sensor** is integrated in series into the ignition circuit (Fig. 7.51). The response threshold of the safing sensor is relatively low and fulfills two basic requirements in newer systems: On the one hand, it is

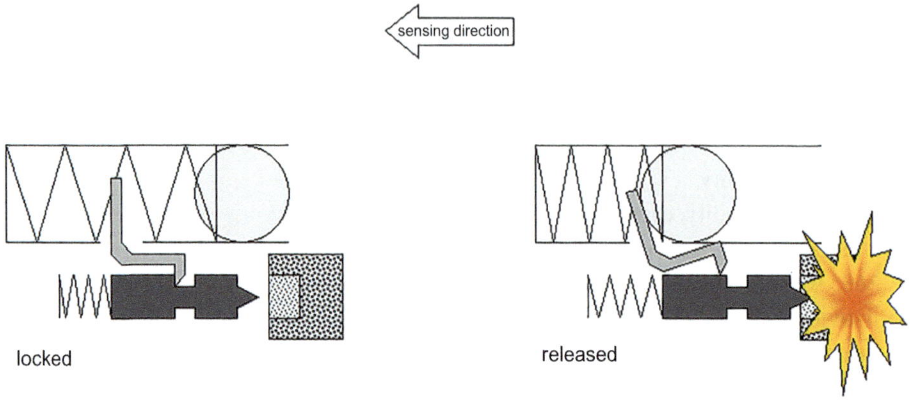

Fig. 7.49 Mechanical system: locked on the left and triggered on the right

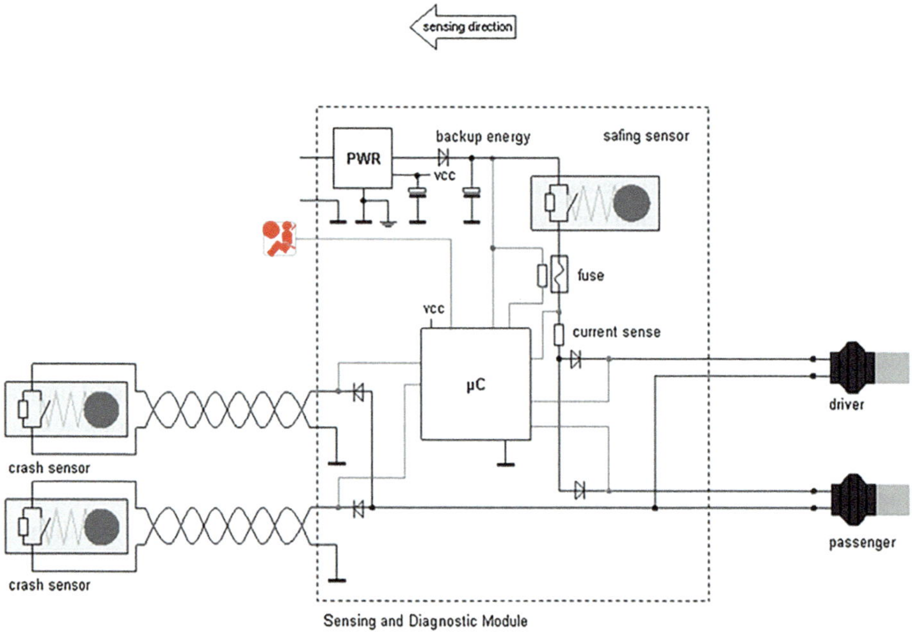

Fig. 7.50 Circuit diagram of an electromechanical system with safing sensor

used as an independent input for confirming the triggering decision, and on the other hand, it prevents unintentional triggering in the event of an error. With such electronic systems, it was possible for the first time to achieve a comprehensive diagnosis of the system state. All lines as well as accelerometers and igniters are checked for their defined resistance values. Furthermore, the energy reserve is monitored in terms of capacity and charge. The entire diagnosis is carried out by the diagnostic cycle of the airbag control unit and is run through in a millisecond cycle. The only component that cannot be checked in the diagnostic cycle is the safing sensor. This disadvantage has been tried to be compensated by deflecting the magnet using an electromagnetic field and thus closing a reed contact. However, due to the high power consumption, the required installation space, and the weight, this arrangement has not been established in series. Here too, a warning lamp is activated by the airbag electronics in the event of a system error.

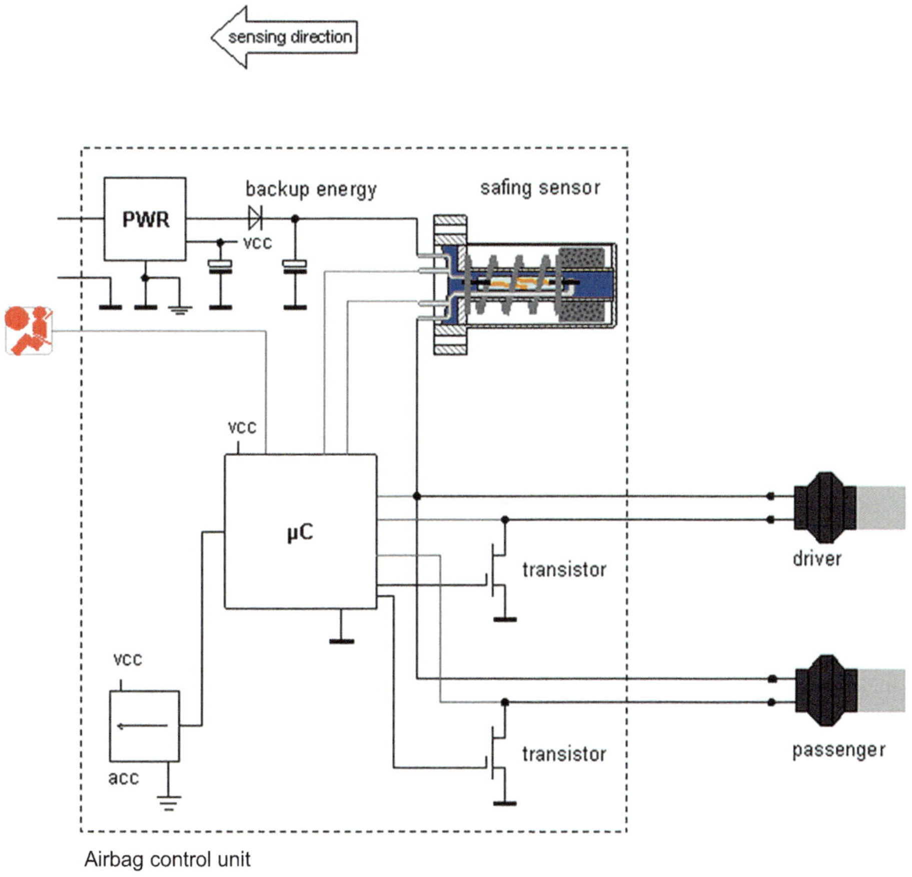

Fig. 7.51 Circuit diagram of an airbag control unit with safing sensor

When integrating the safing sensor in the ignition circuit, on the one hand, the number of restraint systems to be controlled is limited by the current that can flow over the reed contact. On the other hand, the closing time is limited, which can have a negative effect when controlling multi-stage airbag systems. For this reason, the safing sensor was replaced by a **trigger sensor** (Fig. 7.52). Unlike the safing sensor, it is not integrated into the ignition circuit. It is also significantly smaller and provides a switching signal that can be electronically extended. This means that the signal for plausibility checking is available over the entire accident time span.

Airbag control units, in which roll rate sensors are used, require additional sensors for plausibility checks that record relatively low accelerations in the *z* direction; these are so-called low *g z*-sensors. This sensor technology allows a rollover to be reliably distinguished from extreme driving maneuvers. The use of such sensors allows for early, safe, and reliable activation of the necessary restraint systems.

With the constant improvement of passive safety in motor vehicles, the number of different occupant protection systems is also increasing. The continuous development towards comprehensive protection of vehicle occupants requires a 360° confirmation of the crash signal. Since the safing and trigger sensors used so far only have one direction of action, they are unsuitable for this. The only solution is the use of a direction-independent safing microcontroller, the so-called **Safing-μC** (Fig. 7.53). It is also in the

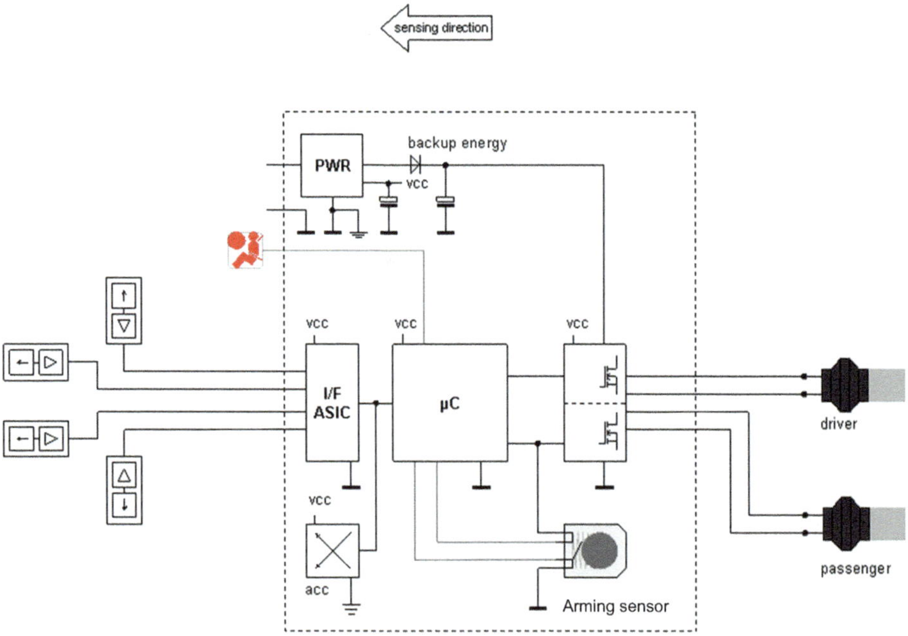

Fig. 7.52 Circuit diagram of an airbag control unit with trigger sensor

location to process the signals of the acceleration sensors and to monitor the system clock and the sequential program run of the master controller. Any irregularities that may occur can thus be reliably detected. Since there are no longer any electromechanical components in the airbag electronics, it is possible to diagnose the system completely and comprehensively electronically.

To meet the increasing requirements for reliability, modularity, space requirements, and costs in modern airbag control units, new partitions or submodules of the components used are introduced. In the following airbag control unit, the power supply, the safing function, the watchdog unit, the charge pump for the energy reserve, and a number of interfaces for external sensors are integrated in a common circuit, the **safing-IC** (Fig. 7.54). If additional external sensors need to be connected, the system can be expanded with additional interface circuits (ASICs). The ignition circuits are now activated by the safing-IC and the algorithm in the master controller. If more are needed, the system can be expanded almost indefinitely in this way.

Vehicle manufacturers today require a so-called SIL classification (SIL: safety integrity level) for airbag electronics according to [15]. This **integrity level for reliability** indicates in various stages the value ranges of the failure probabilities of a safety function; SIL 1 represents the lowest level and SIL 4 the highest level. In [15], two operating modes for safety functions are described: one with a low demand rate (low demand

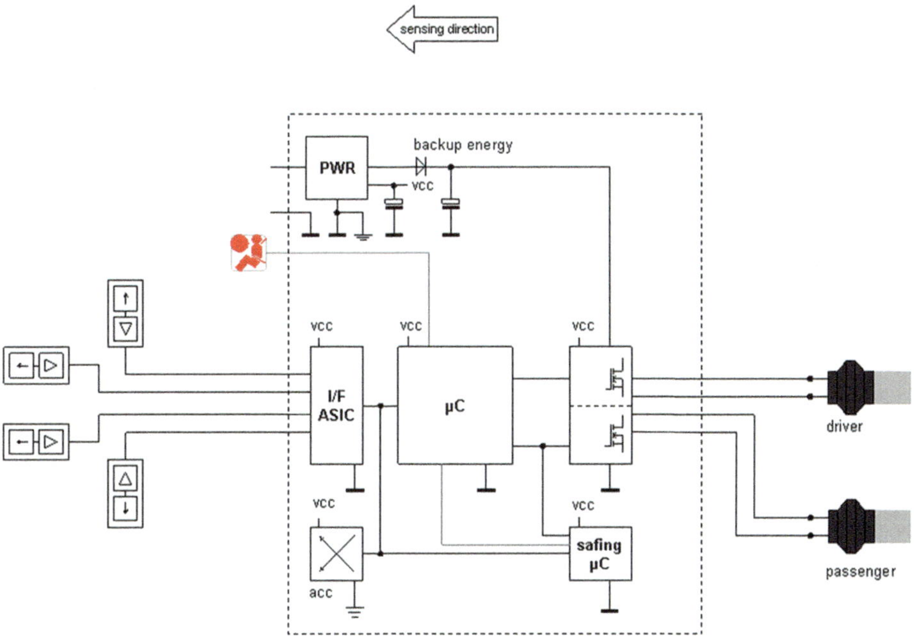

Fig. 7.53 Schematic of an airbag control unit with safing-μC

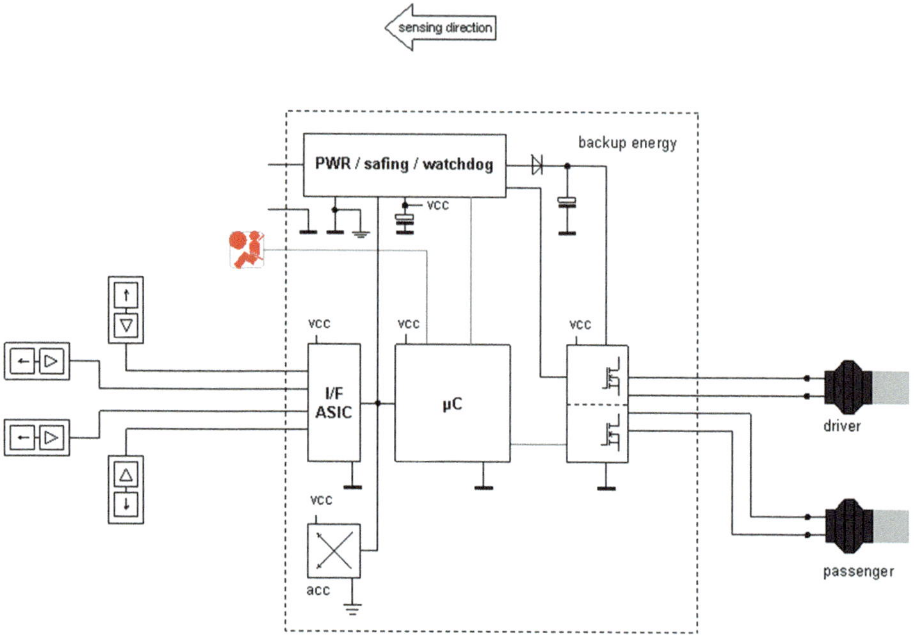

Fig. 7.54 Schematic of an control unit with safing-IC

mode) and one with a high or continuous demand rate (high demand or continuous mode). The formal definitions and terminology are given in [15].

A safety function operating in demand mode is only executed upon request and brings the system to be monitored (EUC: equipment under control) into a defined, secured state. The safety-related system that executes this exact safety function has no influence on the monitoring system as long as no demand is made on the safety function. This applies, for example, to airbag systems. A low demand rate is present when the demand on the safety-related system is made no more than once a year and the frequency is less than twice the frequency of the repeat test. In contrast, a high or continuous demand rate is present when the demand on the safety-related system occurs more than once a year or is greater than twice the frequency of the repeat test. A safety function operating in continuous mode always keeps the system to be monitored in a normal, secured state. The safety-related system thus continuously monitors the EUC monitoring system. A failure of this system therefore leads directly to a hazard, unless other safety-related systems become effective or external measures to avert danger are taken. In Table 7.9, the safety integrity levels for modes of operation with low and high or continuous demand rates are summarized.

The conditions mentioned are generally applied to the entire safety-related system according to the standard [15]. For airbag control units, a safety integrity level SIL 3 is usually aimed for in the mode of operation with a low demand rate; this corresponds to an average failure probability in the range of 10^{-4} to $10-^{3}$.

Finally, for the sake of completeness, the standard according to IEC/EN 61508 [15] should be discussed. It consists of a total of seven parts, with only the first three parts containing normative requirements:

Part 1: General requirements,
Part 2: Requirements for safety-related electrical/electronic/programmable electronic systems,
Part 3: Software requirements,
Part 4: Definitions and abbreviations,
Part 5: Examples for determining the level of safety integrity levels (SIL),
Part 6: Guidelines for the application of Part 2 and 3, and
Part 7: Overview of techniques and measures.

7.3.7 Data Transmission

The data transmission from external sensors, for example the side sensors, was carried out in earlier systems with a PWM signal (PWM: pulse width modulation). In this signal, five different T_i/T_p ratios were chosen for data transmission; however, a six-bit code was also used—depending on the manufacturer's requirement. In both application cases, the acceleration information in the external side sensors was evaluated with an algorithm and thus preprocessed, so that only the ignition decision was transmitted to the airbag control unit. In the current data transmission from the external sensors, e.g. the side or up-front sensors, to the airbag control unit, only a digital raw data transmission is used today. This is predominantly the PSI-5 protocol (PSI: peripheral sensor interface) developed jointly by Autoliv, Bosch and Continental. It was published in 2008 in version 1.3 based on the previously developed PAS-3 and PAS-4 protocols (PAS: peripheral acceleration sensor) by Bosch and the "Pegasus" protocols used by Siemens VDO (later Continental). By 2012, version 2 of the protocol with its own sub-standard for airbags with transmission rates of 125 kbps and 189 kbps was mature and has since been the

Table 7.9 Safety integrity level SIL for different modes of operation

Safety integrity level (**SIL**)	Mode of operation with low demand rate (average for failure probability on demand)	Mode of operation with high or continuous demand rate (failures per hour)
1	10^{-2} to 10^{-1}	10^{-6} to 10^{-5}
2	10^{-3} to 10^{-2}	10^{-7} to 10^{-6}
3	10^{-4} to 10^{-3}	10^{-8} to 10^{-7}
4	10^{-5} to 10^{-4}	10^{-9} to 10^{-8}

standard for sensor communication for airbag control units. PSI-5 is the result of consortium work and can be used license-free by all consortium members. These various protocols use the Manchester-coded data transmission method (see Fig. 7.55).

In the PAS-3 protocol, the data width is 8 bits and in the PAS-4 protocol it is 10 bits. In contrast, the PSI-5 protocol has a variable data frame from 8 to 24 bits. This allows additional information, such as manufacturer code, sensor type and characteristics, to be transmitted. The transmission of this additional information to the airbag control unit takes place during the deployment phase of the airbag system. The specification for the PSI-5 protocol v1.3 29.07.2008 (see Fig. 7.56) will be presented below:

- Two-wire current interface
- Manchester-coded data transmission
- 8 to 24 bit data word length
- 0 to 19 mA current consumption during data pause
- Transmission of manufacturer code, sensor type and characteristics
- Synchronous and asynchronous operating mode, adjustable
- Sensor cluster and semi-data bus communication capable

A consistent continuation of the digital data transmission is the development of the **bus systems** for the application of occupant protection systems. While there have been corresponding proposals and individual developments from various manufacturers in the past, which could not prevail due to the diversity in the market, the interest of vehicle manufacturers in uniform, standardized solutions was met, and there was a worldwide merger of manufacturers active in the field of passive safety. The efforts for standardization led to a specification with the title "Safe-by-Wire Plus—Automotive Restraint System Bus

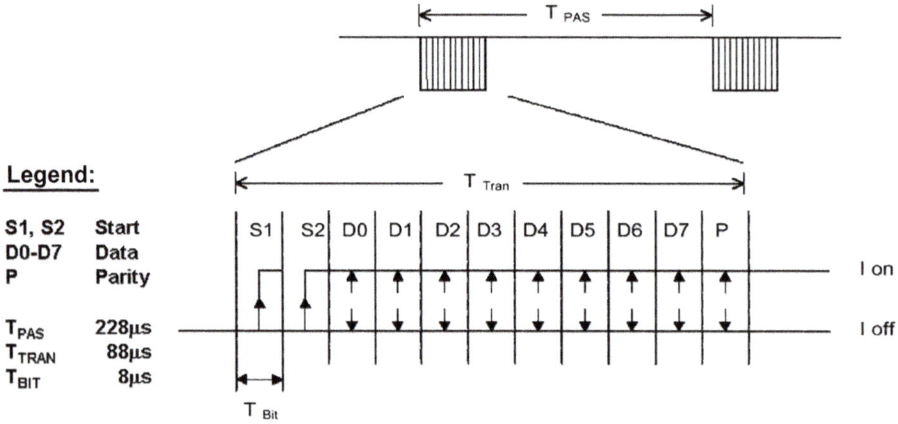

Fig. 7.55 Data transmission according to the PAS-3 protocol

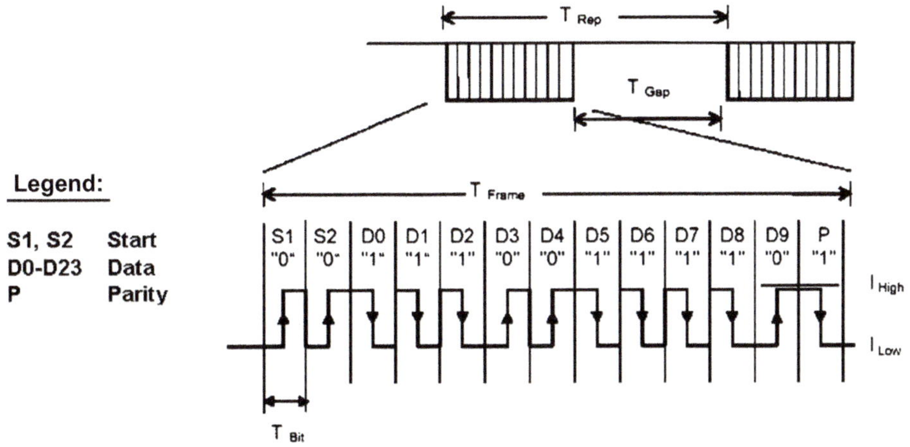

Fig. 7.56 Data transmission according to the PSI-5 protocol (from [21])

Specification 2.0" (see Fig. 7.57), which was submitted for standardization, the ISO standard (ISO 22896:2006).

The following properties are specified for the standard "Safe-by-Wire Plus":

- Bus for ignition and sensor applications:
  - Bus for igniters and sensors in occupant protection systems
  - Sensor bus for dynamic and static sensors
  - Combined sensor and ignition bus
- Bi-directional two-wire bus with integrated power supply:
  - Master/Slave operation
  - Priority control for ignition commands from the master
  - Interrupt capability for "smart" sensors
  - Optional multi-master capability
- Variable bus speed through dynamic slaves:
  - 20, 40, 80 or 160 kbps ± 13%
- Flexible bus topology: star, ring or mixed structure
- Enabling of daisy-chain, parallel or mixed systems:
  - 20, 40, 80 or 160 kbps ± 13%
  - Bus length for parallel system up to 40 m
  - Bus length for daisy-chain system up to 25 m
- Recovery from bus errors within the specified limits in real time
- Communication error detection by sender and by receiver:
  - CRC check
- Multiple safeguards against unintentional triggering:

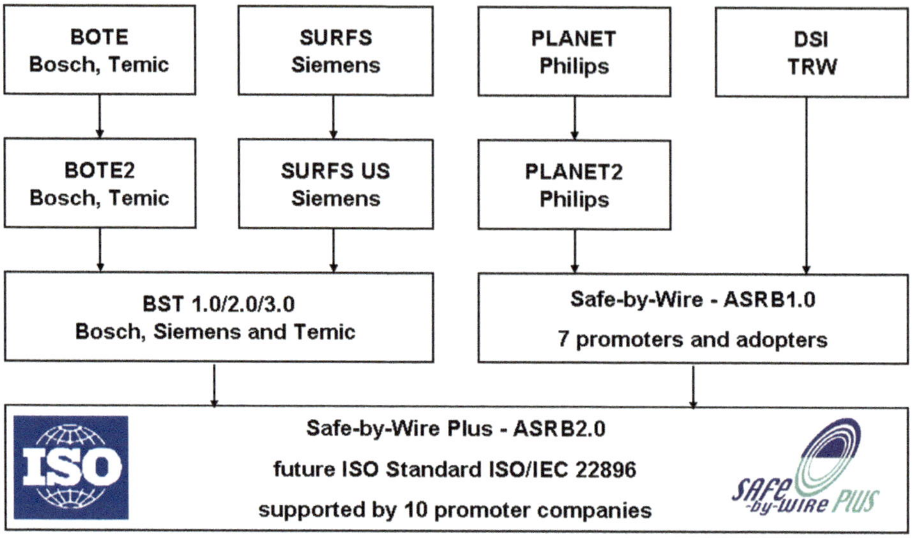

Fig. 7.57 Standardization initiative for the data bus system "Safe-by-Wire Plus"

  - Bit error detection by master and slave
  - Reserved bus voltage levels for ignition command (analog safing)
  - Ignition only when charged energy reserve in the igniter
  - Ignition only after previous ignition release command
  - Support of the high-side and the low-side switches in the igniter, which can be controlled separately over the bus
- Use of support capacitors in the slaves possible

Taking these properties into account allows for a flexible system design: In addition to parallel systems, symmetric and asymmetric daisy-chain systems are also possible (Fig. 7.58). Furthermore, there is the possibility to combine the different systems. By using daisy-chain slaves at appropriate nodes, branches that are designed as a parallel system can be decoupled from the rest of the system in case of a failure.

For the integration of the **bus-capable igniter**, special designs are required that, in addition to the normal pyrotechnics, must have a suitable electronic circuit in order to enable their operation in a data bus system at all. This circuit initially includes a bus interface as well as the addressing and decoding logic, the diagnostic device, a power supply with energy reserve and the circuit breaker for controlling the ignition bridge and is conveniently designed as an ASIC circuit (Fig. 7.59). Since the igniters only have a small energy reserve, so-called low-energy ignition bridges are used. The energy required for ignition (with the parameters bridge resistance 1.1 to 10 Ω, ignition capacitor 0.3 to 1 μF and ignition voltage 22 V) is about 70 to 240 μJ depending on the design. In comparison, a conventional bridge wire igniter has an energy requirement of about 1 to 3 mJ. Due to the pyrotechnics contained in the igniter, the requirements regarding electrostatic charging (ESD: electrostatic discharge) are very demanding; they must be able to withstand an ESD-voltage of 30 kV (150 Ω; 150 pF) without damage.

The circuit of the bus-capable igniter is applied on an Al_2O_3 substrate with a size of 6.0 x 8.0 mm. For contact with a connector, the circuit has connector pins with standard contacts of 1 mm diameter, which are accommodated by a glass/metal base. On the top side, shown in Fig. 7.59 top right, the capacitors C1 and CZ as well as the resistor R1 are applied, with the resistor R1 being printed directly onto the substrate and located under the capacitor CZ for space reasons. The capacitors are standard components that are contacted with high-temperature solder. In addition, the two contact surfaces serve for contacting the ignition bridge, a so-called HfHx element. On the underside of the substrate, shown in Fig. 7.59 bottom right, the ESD protection circuit with the suppressor diodes D1 and D2, as well as the evaluation circuit is applied. The contacting of the individual components is carried out using a special adhesive press connection process, the so-called die-bond process.

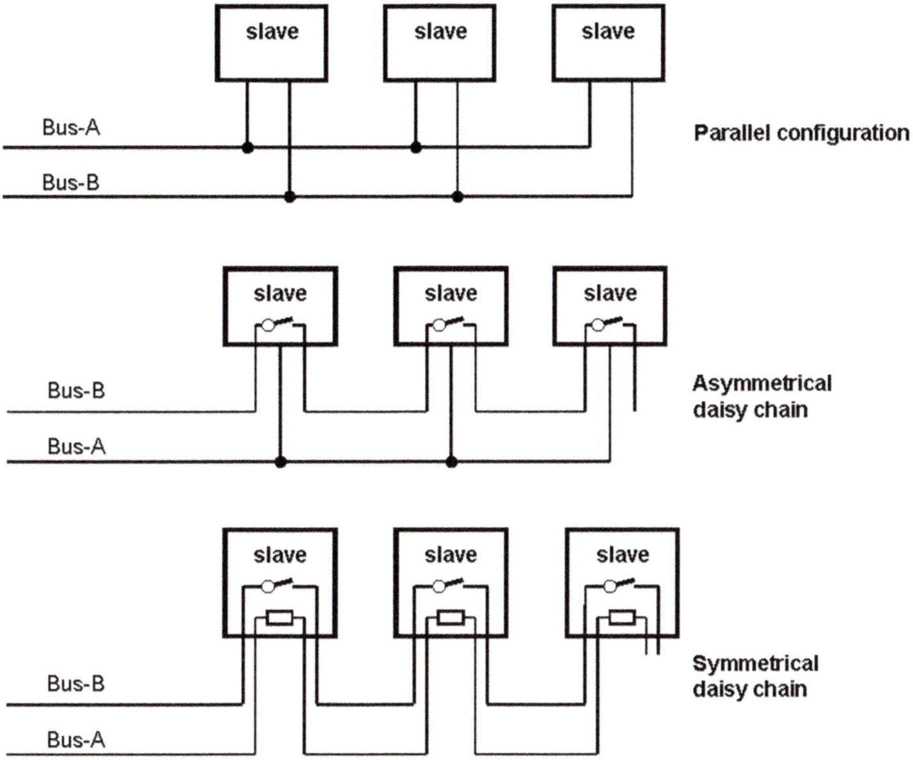

Fig. 7.58 Flexible system design through different bus structures (from [7])

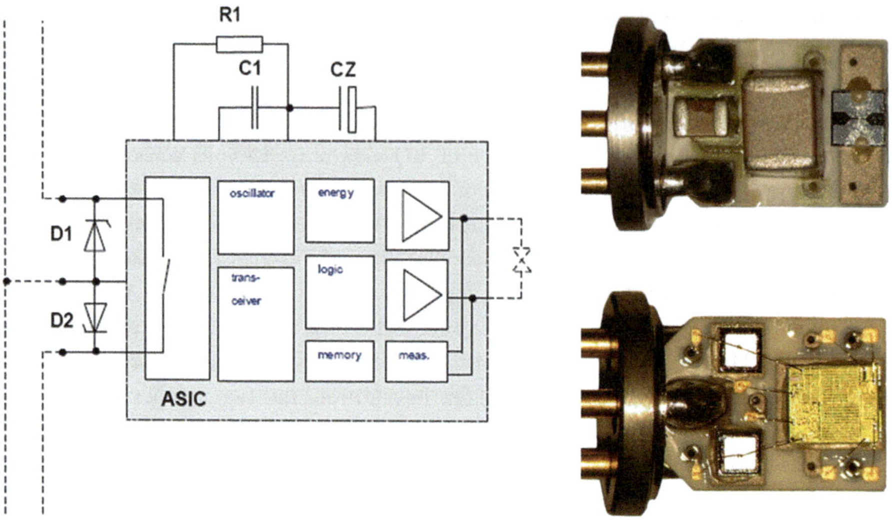

Fig. 7.59 Schematic and realization of an integrated bus-capable igniter (from [7])

7.4 System Integration with Regard to Active and Passive Safety

The enormous potential that the use of electronics has unlocked in functions, previously performed mainly mechanically, as well as the development of completely new functions made possible by it, have led to a triumph of electronics throughout the vehicle. The average vehicle has a value share of electronics of over 30% and on average over 30 control units in the areas of powertrain, interior, and chassis. The next big step involves networking these individual electronic systems into extended functional blocks. The integral safety, i.e., the combination of active and passive safety functions (see Fig. 7.60) and the expansion of assistance systems, helps the driver to better master tricky traffic situations than the classic approach of separate active and passive safety.

The occupant protection systems used so far in automobiles have been undergoing a significant change since around 2015 due to information generated in the temporal area of active safety: They were only activated when the accident had actually occurred. The current state of the art is the improvement of the functionality and effectiveness of the protection systems for vehicle occupants, and increasingly also for the protection of external road users, whose safety needs are increasingly coming to the fore. These requirements open up the possibility of developing a new generation of sensors, such as occupant detection systems, pre-crash and environment sensors, to decide before the collision occurs which safety systems can offer an optimum level of protection in the event of an impact. Passive safety benefits, so to speak, from the information that can be provided and used from active safety. In this sense, there is a link, albeit a one-sided one, between active and passive safety in order to meet increasing safety requirements.

Active safety is expected to handle critical situations in the form of a failed driving task in order to avoid an impending accident. Depending on the spatial and temporal range of the sensors used, the information generated with them is used to warn and support the driver. However, this information can also be used to activate occupant protection systems, which can be controlled earlier and more precisely than the systems available today. This can achieve a significant improvement in safety, Fig. 7.60. The necessary sensors are being consistently further developed for use in the area of accident prevention, i.e., active safety. Market-introduced examples of the advanced driver assistance systems (ADAS) include the adaptive cruise control ACC, the lane departure warning device LDW, and the lane keeping support system LKS. These devices enable the driver to be warned of an unintentional lane change, or to keep the vehicle automatically in the lane in conjunction with an electrically assisted steering system. Due to legal requirements and consumer rating programs (e.g., Euro NCAP), these assistance systems are increasingly supported by emergency brake assistants. The collision probability calculated for braking has a direct influence on passive safety.

From road traffic accident statistics, it is known that restraint systems are an effective measure to minimize or even avoid injury consequences in accidents. From the same

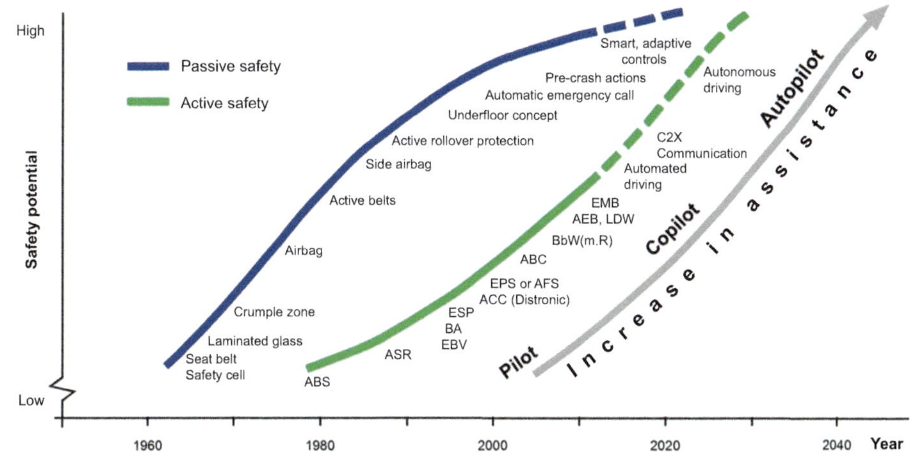

ABS	anti-lock braking system	BA	brake assist system	ABC	active body control
ASR	traction control	ACC	adaptive cruise control	EMB	electromech. brake
EBV	electronic brake force distribution	BbW	brake-by-wire	DbW	drive-by-wire
ESP	electronic stability program	(m.R)	(with mechanical fallback level)	C2X	car-to-X communication
		CA	collision avoidance		

Fig. 7.60 Development of driver assistance systems using the example of active and passive safety

accident data material, however, it can also be derived that situations occur in which an improvement in the effectiveness of restraint systems would be necessary and appears possible. Likewise, the accident data show that there is a need for protection also for vulnerable road users, e.g., for pedestrians, motorcyclists, or cyclists, who have not been sufficiently considered so far. Thus, the European Union demands considerable efforts aimed at protecting pedestrians and initially incorporated into the voluntary commitment of European car manufacturers, grouped together in the ACEA (European Automobile Manufacturer Association). The agreement reached there represented a major step towards passive, but also towards active safety. In the meantime, pedestrian protection has been legally regulated in the EU, see Chap. 4.

Future challenges for the restraint systems in modern vehicles will concern the architecture, the multifunctional application, and the increasing number of sensors. In addition, intelligent networking and the reduction of costs, weight, and assembly effort are inevitable. Sophisticated, powerful bus systems will be used, which will handle the data exchange between the individual sensor and control modules. Finally, systems will be used that are capable of completely replacing mechanical and hydraulic components, so-called X-by-wire systems, which are unthinkable without powerful and reliable bus systems (Fig. 7.61).

The integral safety concept—Just as the chassis subsystems are growing together in a global control network, the integration of active and passive safety systems is also in full progress, see Sect. 1.3. In a first stage, Mercedes introduced a system of integrated safety under the name PRE-SAFE® in 2002. Newer systems, such as the subsequent generations of PRE-SAFE® or the system called "Pre-Crash Safety Monitor" by Continental, additionally incorporate environmental information into a graded accident avoidance and injury protection strategy [3].

These approaches assume different driving phases in connection with accidents and work on reducing the consequences of accidents for each phase with appropriate priority (Fig. 7.62):

Normal driving (Information and relief)—During normal driving and without an acute danger, the driver is relieved of routine actions by assistance functions (e.g., adaptive cruise control, ACC); the so-called lane keeping assist (LKA) assists in lane keeping. The driver is supported in maintaining a safe distance to the vehicle ahead, the lane is better maintained and haptically reported back using camera information and steering intervention, the driving light is regulated according to weather conditions and time of day, speed limits are displayed, traffic information arrives via radio or navigation.

Dangerous situation (Warning and support in accident prevention)—The electronic brake system and the electric steering system (EPS) are key technologies for all intervening safety functions. Anticipatory brake assist functions, ranging from preconditioning the brake to automatic emergency braking, are activated in critical rear-end situations. The driver is also supported in performing an evasive maneuver to keep the vehicle safely on the road. If the accident is still inevitable, the following three driving phases play an important role in reducing the consequences of the accident.

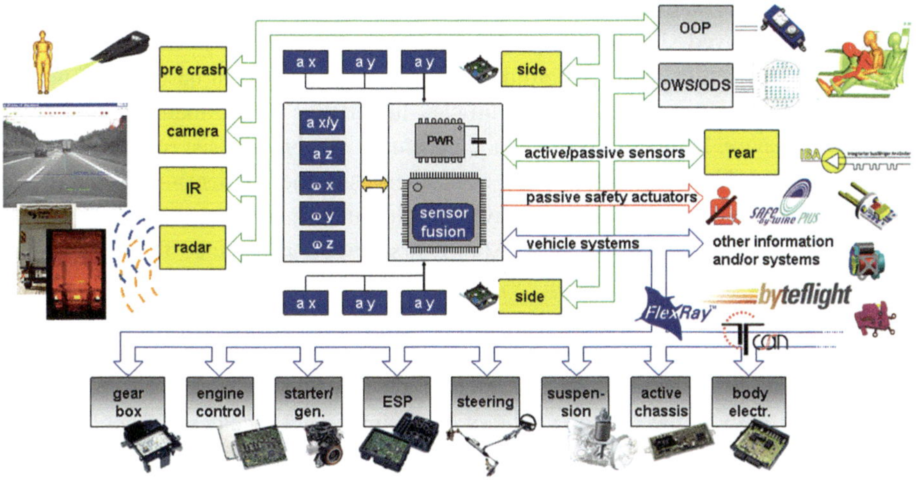

Fig. 7.61 System integration active/passive safety systems (from [7])

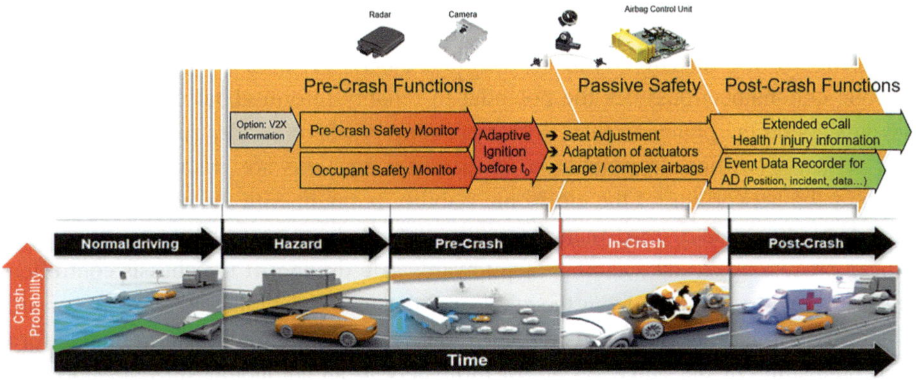

Fig. 7.62 Networking of active and passive safety functions for integral safety. *Source* Continental

Before the accident (Preparation)—The reduction of the impact speed by an emergency brake assistant can significantly reduce the injury risk. Before an impact, additional protective measures (e.g., activating seat belt pretensioners, positioning the occupants, pre-conditioning the other restraint systems, position of seat and headrests) that prepare the vehicle and its occupants for an accident are initiated. Depending on the evaluated information about the accident severity and the accident constellation, a decision to ignite the non-reversible restraint systems is made that is adapted to the situation, so that the vehicle occupants experience the greatest possible and reliable protection.

During the accident (Impact protection)—To ensure maximum occupant protection, airbags are deployed situationally based on data from pre-crash sensors. Technologies such as rollover sensors and the optimization of airbag deployment based on body sound (crash impact sound sensor) are of significant importance due to their fast and precise response behavior. Even during an impact, the emergency brake assist remains active.

After the accident (Minimizing the consequences of the accident)—Shortly after the first impact, measures are initiated to avoid further impact or try to reduce the accident severity and also alert rescue forces. After a first impact is detected by the ignition of the airbags, the electronic braking system automatically brakes the vehicle to a standstill. The life-saving "eCall" is an automatically generated emergency call. The location of the vehicle is given and the so-called "minimum set of data" (MSD) is transmitted. This information is important for the rescue forces so that the vehicle can be correctly located, especially in the case of a solitary (nighttime) single accident without other parties involved.

7.5 Future Developments and Expectation Horizon

The long-term goal is seen in **accident prevention**. The already widespread dynamic navigation with connection to the standardized mobile network (GSM: global system for mobile communications) has the potential to reliably inform about current traffic jams or dangers on the planned route. With increasing precision of electronic road maps and the vehicle's position determination, further new functions are gradually implemented, such as the networking of the electronic stability program with the adaptive cruise control ACC. It evaluates road markings using extended sensor technology (camera with image processing) to take another step towards safe driving. If the chassis controller of such a vehicle recognizes from the additional navigation data that the car is too fast for the upcoming hairpin curve, this can lead to driver warning, reduction of engine torque and preloading of the brake system or ultimately to an autonomously initiated braking process.

With increasing reliability of traffic information, which is fed into the chassis computer via the car radio (traffic message channel) or digital communication, the vision of active accident prevention through networked electronic components can be further elaborated: If there is a risk of black ice on the section of the route to be passed soon or if the road is blocked due to an accident or a construction site, the electronics can warn the driver or even slow down the vehicle by active interventions in the steering and braking systems. Drive-by-wire becomes a technological prerequisite (enabling technology) that teaches the car to "see" and "hear", cf. Sect. 7.2.7. Bearing in mind the demands that such a complex networking of chassis systems with each other and with the outside world places, it is a long way to go before accidents can be completely avoided by cars. The achievement of "Vision Zero" (Sect. 1.4) will take decades. But with chassis control devices (global chassis control), environment-information supported integral safety technology, and extensive vehicle/vehicle or vehicle/environment communication V2X, a partial solution is conceivable and manageable. In addition to powerful sensors, particularly powerful actuators are required. The advanced braking system, which covers all future requirements, is of utmost importance (Fig. 7.62).

Outlook: Autonomous driving

Automated driving is one of the mega trends in vehicle development. The possibility of relieving the driver temporarily or completely from the driving task not only gives the driver a substantial recovery phase, but also allows him to pursue other activities during this time. Thus, the vehicle becomes an "IoT device" (internet of things) for a pleasant pastime.

In the 2010s, the step from assisted driving to fully automated driving appeared as a development step that could be implemented in the near future. Accordingly, a 5-stage model for the transition from driver-led vehicle to fully autonomous driving was defined (Fig. 7.63).

Automation is implemented at various stages with increasing degrees of automation:

- Level L 1: Assisted driving in one dimension
  The driver is supported by individual driver assistance systems. Assistance systems such as an adaptive cruise control or a lane keeping system are used here. They intervene in the driving process independently of each other. The driver retains responsibility for the driving behavior at all times.
- Level L 2: Coordinated assisted driving
  The driver assistance systems known from level 1 are used here in a combined and thus extended form. Here, several driver assistance functions are executed in a coordinated manner. An example of this is the traffic jam assistant, which combines the longitudinal guidance of the vehicle with the lateral guidance. This means that an adaptive cruise control is operated in combination with a lane keeping system. The control task is significantly more complex and the support of the driver is significantly more extensive. Even at level 2, the driver retains responsibility for the vehicle at all times.
- Level L 2+: Accompanied driving
  Here, the driver receives complete support for all driving tasks necessary for long-distance driving. Under certain conditions, the driver can completely take his hands off the steering wheel during the drive. The vehicle drives and steers completely independently. However, the driver must still keep an eye on the driving situation. Even at level L 2+, the responsibility for the driving behavior remains with the driver.
- Level L 3: Partially automated driving
  At this level, the driver is relieved of the driving task for the first time for a certain period of time. The responsibility for the driving behavior is transferred to the vehicle manufacturer. This expansion stage not only requires the use of highly reliable driver assistance systems for temporary fully automatic driving behavior, but also the application of driver monitoring systems. These "monitoring" systems are needed to

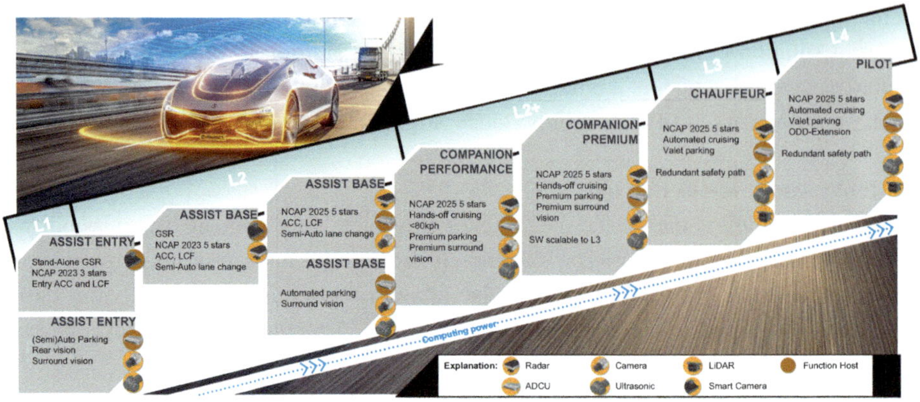

Fig. 7.63 Classification of the automation levels of a vehicle according to SAE

capture the current seating and attention status of the driver. Based on this information, the driver must be specifically informed that he must take over the driving task again. This handover must take place within a few seconds so that the journey does not have to be interrupted by a "safe stop" because the competence and responsibility of the L3 system cannot yet be managed and the driver has not yet taken over the leadership task. This targeted handover of the driving order to the driver represents a substantial challenge for the development of a serial-ready L3 system. With this expansion stage, the primacy of the driver assistance systems to avoid vehicle reactions to "false positive" objects is transferred to the primacy of avoiding any "false negative" situations. This requires a substantial increase in the reliability of object detection by the environment sensors.

- Level L 4: Automated driving (with the possibility of driver intervention)
  At this automation level, the driver is relieved of the driving task for the entire journey. The vehicle systems completely take over the driving task from the handover point to the destination or until an interruption caused by the driver. At this automation level, the driver still has the option to intervene in the driving process and drive the vehicle independently.
- Level L 5: Autonomous driving
  This level corresponds to the degree of automation of level L4. The difference to L4 is that manual driving is no longer possible in these vehicles. There are also no control components provided in these vehicles that would allow manual driving. Typical examples of L5 vehicles are the so-called robo-taxis.

The central challenge of automated driving is to ensure a highly reliable automated driving function. Vehicle automation must enable at least as safe driving as is currently guaranteed by an average driver. This challenge is directly related to providing scientifically recognized evidence of available reliability. Known methods for empirically proving sufficient reliability from the world of driver assistance systems lead to an enormous number of test kilometers, which cannot be implemented on an economically viable scale. A broad introduction of automated vehicles will only be possible when a feasible proof method has been found.

References

1. ETSI Standards: https://www.etsi.org/deliver/etsi_en/—https://www.etsi.org/deliver/etsi_ts/—https://www.etsi.org/deliver/etsi_tr/
2. ANSI X9.62. Public Key Cryptography for the Financial Services Industry: The Elliptic Curve Digital Signature Algorithm (ECDSA). (2005)
3. Bader, D., Forster, A., Heiner, J, Peters, B.: Umfassende Schutzfunktion für Mensch und Batterie. ATZ 05/2023 (2023)
4. BMBF/BMWi Forschungsprojekt simTD. Deliverable D5.5 – Teil B -1 A: https://www.eict.de/fileadmin/redakteure/Projekte/simTD/Deliverables/simTD-TP5-Abschlussbericht_Teil_B1-A_GIDAS-Wirkgradanalyse_V10.pdf. Sindelfingen (2013)
5. Bundesvereinigung der Straßenbau- und Verkehrsingenieure BSVI: Daten zum Straßen- und Verkehrswesen. Publikation, Hannover (1988)
6. Car2Car Communication Consortium: Basic System Profile inkl. Triggering Conditions. https://www.car-2-car.org/documents/basic-system-profile. (2023)
7. Continental Automotive Technologies GmbH: https://www.continental-automotive.com/de.html. Regensburg (2023)
8. Continental Engineering Services: Produktbeschreibung Langreichweitenradare. https://conti-engineering.com/areas-of-expertise/components/radar-sensor (2023)
9. Waldschmidt, C., Hasch, J., Menzel, W.: Automotive Radar—From First Efforts to Future Systems. IEEE J. Microw. (2021)
10. Continental Automotive Technologies GmbH: Beschreibung Fahrerassistenzsysteme. https://www.continental-automotive.com/de/solutions/sicherheitstechnologien/fahrerassistenzsysteme (2023)
11. European Horizon 2020 research project PROSPECT: Deliverable D2.3. https://ec.europa.eu/inea/en/horizon-2020/projects/H2020-Transport/Safety/PROSPECT. Zugegriffen: 28. Nov. 2023 (2019)
12. Freescale Semiconductor Inc.: https://services.austintexas.gov/edims/document.cfm?id=243075. Austin, TS (USA) (2005)
13. Feifel, H. et al.: REDUCING FATALITIES IN ROAD CRASHES IN JAPAN, GERMANY, AND USA WITH V2X-ENHANCED-ADAS. Paper 23–0082, 27. ESV-Conference, Japan (2023)
14. Hromkovic, J.: Lehrbuch Informatik. ISBN 978-3-8348-0620-8. Vieweg+Teubner, Wiesbaden (2008)
15. DIN EN 61508-1 VDE 0803-1:2011-02 Funktionale Sicherheit sicherheitsbezogener elektrischer/elektronischer/programmierbarer elektronischer Systeme. VDE-Verlag (2011)

16. INVENT-Projekt: Vorausschauende Aktive Sicherheit (VAS) – Ergebnisbericht des Projekts Intelligenter Verkehr und nutzergerechte Technik 1998–2003. Kirchheim unter Teck (2003)
17. IMAGine Projekt: https://www.imagine-online.de/en/findings-publications (2023)
18. Key Safety Systems Inc.: Sterling Heights, MI (USA) (2005)
19. Forschungsinitiative Ko-FASt: www.ko-fas.de. Großwallstadt (2014)
20. Mercedes-Benz: https://www.mercedes-benz.de/. Sindelfingen (2023)
21. Robert Bosch GmbH: Kraftfahrtechnisches Taschenbuch, 27. Auflage. Vieweg-Teubner, Plochingen (2011)
22. Statistisches Bundesamt Wiesbaden (Hrsg.) Verkehrsunfälle 2021. Fachserie 8, Reihe 7. https://www.destatis.de/DE/Publikationen/Thematisch/TransportVerkehr/Verkehrsunfaelle/VerkehrsunfaelleJ.html (2021). Accessed: 1 Nov 2023
23. Sauter. M.: Grundkurs Mobile Kommunikationssysteme – 5G New Radio und Kernnetz, LTE-Advanced Pro, GSM, Wireless LAN und Bluetooth. ISBN 978-3-658-36962-0. Springer Nature, Wiesbaden (2022)
24. Infineon: Pressure Sensors for Side Crash Detection (SAB). https://www.infineon.com/cms/en/product/sensor/pressure-sensors/pressure-sensor-for-side-crash-detection-sab/#!support. Accessed: 27 Nov 2023
25. Schmidt Technology: St. Georgen (2005)
26. ZF Automotive Germany GmbH: https://www.zf.com/mobile/de/technologies/integrated_safety/integrated_safety.html.
27. Knoll, P.: Prädiktive Fahrassistenzsysteme. Script KIT_ITE (2017)

Integral Safety in the Vehicle Development Process

8

Rodolfo Schöneburg, Ulrich Franz, Andre Haufe, Florian Kramer and Adrian Zlocki

The vehicle development process has changed significantly over the past decades due to external influences, competitive pressure, and the need for timely and customer-oriented product placement. In the past, vehicles were developed over a period of seven to nine years, but today the development time has been reduced to less than three to four years. This marks a trend that will continue to evolve in order to better meet market and customer-specific requirements (Fig. 8.1).

In the past, the approach of engineering departments was characterized by sequential processing, as well as a manageable product range and vehicle complexity. Today, vehicles and the integrated systems are more complex and interconnected, the number of variants is higher, the available series and special equipment are more numerous, and the available development time is shorter. Such environment required a parallel working method with the help of computer-aided development tools and efficient processes.

Fast and high-quality product development is therefore not only dependent on the competence of the involved engineers and suppliers, but is primarily ensured by efficient and controlled processes. All involved development areas must be closely aligned in the product development process and engineers need clear interim goals for targeted development work. The following will illustrate a product development process as

R. Schöneburg (✉)
RSC Safety Engineering, Hechingen, Germany
e-mail: rsc@rschoeneburg.de

U. Franz · A. Haufe
Stuttgart, Germany

A. Zlocki
Aachen, Germany

F. Kramer

R. Schöneburg (ed.), *Integral Vehicle Safety*,
https://doi.org/10.1007/978-3-658-48100-1_8

yesterday	Development factors	today
Little competition in the premium segment	**Market development**	Strong competition in the premium segment
Few model variants and optional extras	**Complexity**	Many model and country variants, niches and optional extras
Proven technology	**Technology**	Short innovation cycles
Seven years and more	**Development time**	Four years and less
Low networking, manageable	**Mode of operation**	High level of networking, many stakeholders
Low degree of tension	**Costs**	High degree of tension

Fig. 8.1 Vehicle development yesterday and today using the example of the premium segment [95]

an example, with a particular focus on the scope of passive safety. The focus is on the "development tools" experimental and computational simulation, but also innovation management and the integration of accident research into the development process using the example of a vehicle manufacturer will be explained.

8.1 Process Targets and Development Organization

Business process design and management in vehicle development has the following objectives:

- Description and optimization of the product development process,
- Synchronization of workflows through defined interfaces,
- Transparency of parallel processes to recognize the effects of delays,
- Clear representation of interim goals to check the achievement of objectives and quality, and finally
- Definition of tasks, competencies and responsibilities.

At most vehicle manufacturers, the development organization is set up in a matrix organization for this purpose (Fig. 8.2). The project organization defines, organizes, controls,

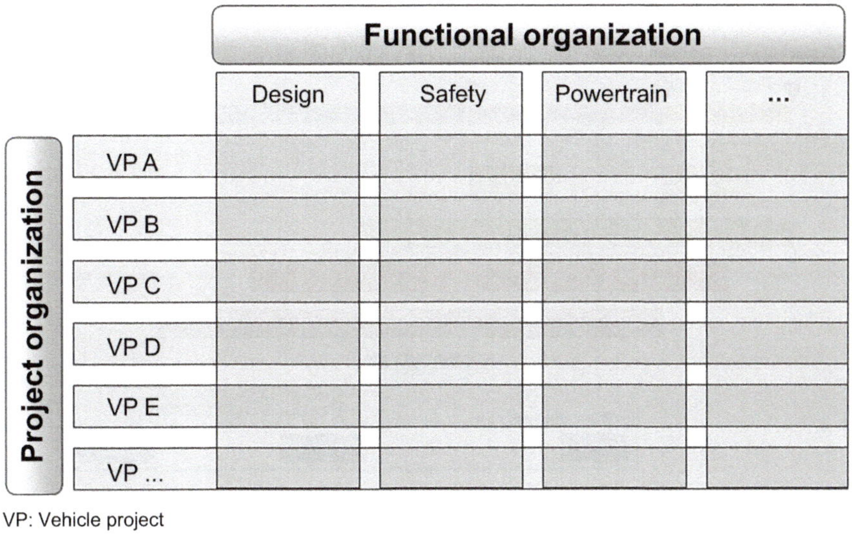

Fig. 8.2 Organizational structure of a vehicle manufacturer (matrix organization)

and represents the respective vehicle projects. It also coordinates work packages between purchasing, production, and sales. The functional organization, on the other hand, develops, realizes, and qualifies the vehicle contents and researches and integrates new vehicle characteristics, here for example the product characteristics for achieving integral vehicle safety.

8.2 Development Process and Development Phases

The typical development process of a vehicle is divided into different phases by almost all manufacturers. Figure 8.3 shows a highly schematic, generic representation of a vehicle development process, often also referred to as product development process PEP, with a focus on the contents relevant to passive safety [95].

The process can be divided into a definition or strategy phase, a validation phase, a realization phase, and a series phase. The various development phases are divided by milestones with the aim of synchronizing parallel work processes, such as the design process, construction, digital development, hardware setup, and testing.

The transition from one phase to another is characterized by the review of the development progress using quality criteria in the context of so-called "quality gates". The application of this quality principle, also in internal customer/supplier relationships, serves as the foundation for ensuring product quality. Figure 8.4 shows the general contents of each development phases. The individual work processes are again described within these phases to ensure an integrated way of working.

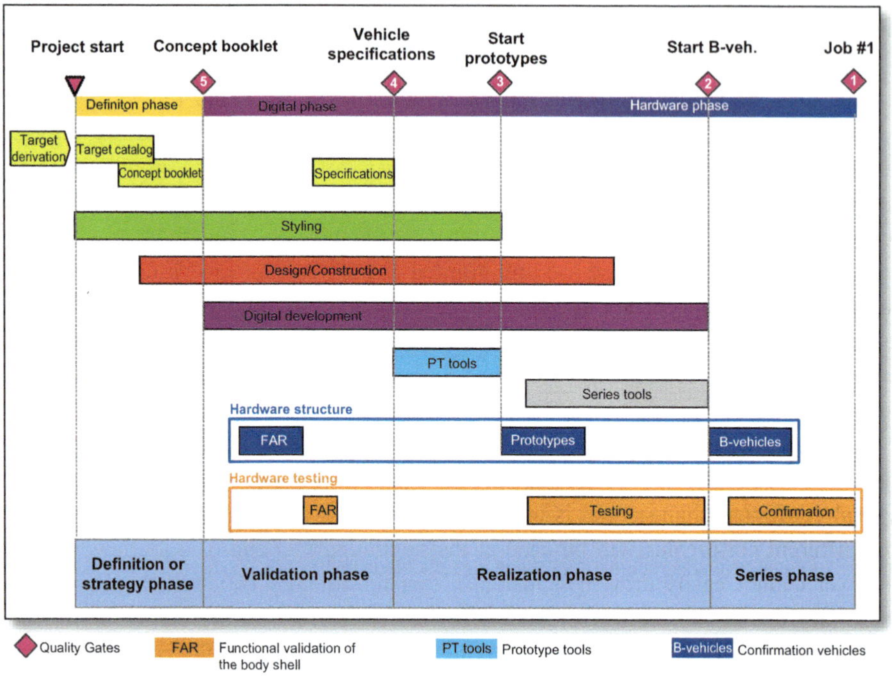

Fig. 8.3 Schematic vehicle development process for vehicle safety according to [95]

Definition or Strategy phase	Validation phase	Realization phase	Series phase
Definition of vehicle characteristics and positioning Input variables: • Market research • Corporate strategy • Customer feedback • Predecessor model ⇒ **Concept booklet**	Decision on vehicle variants and components Proof of feasibility ⇒ **Vehicle specifications**	**Series development** of components and variants Ensuring function and durability	**Vehicle and system release** Preparation for production Production of series vehicles **Production start-up**

Fig. 8.4 Content and sequence of the development phases

In each development phase, passive safety has various interactions with regard to the components and sub assemblies but also their properties in the vehicle. Knowledge of these interactions is of utmost importance, so that conflicts of objectives and influencing parameters can be recognized in time and included in the vehicle design. Passive safety must be considered in the early development phase in the brand strategy, in the positioning of the vehicle with regard to competition, and in the sales requirements (markets, service, customer requirements). Already at this point, both the external legal and consumer protection requirements as well as the internal safety requirements for the vehicle to be developed, are defined in the form of test limit values.

8.3 Requirements for Passive Vehicle Safety

In an early vehicle development phase, it is first necessary to compile all objectives that can be derived from the experiences of the predecessor model from the real accident occurrence and all requirements of the legislators and consumer protection organizations (Fig. 8.5).

This compilation serves the first basic preconditions of vehicle body structure and vehicle equipment. It should be noted that these requirements must be met or maintained at the beginning of market introduction, but also throughout the entire life cycle of a model series. Generally, a period of about 15 years is assumed for this, which is significantly longer than the current average age of cars in Germany, namely 9.8 years in 2021 [51]. However, this long time frame for consideration carries substantial risks, particularly due to the steadily growing safety requirements in global markets.

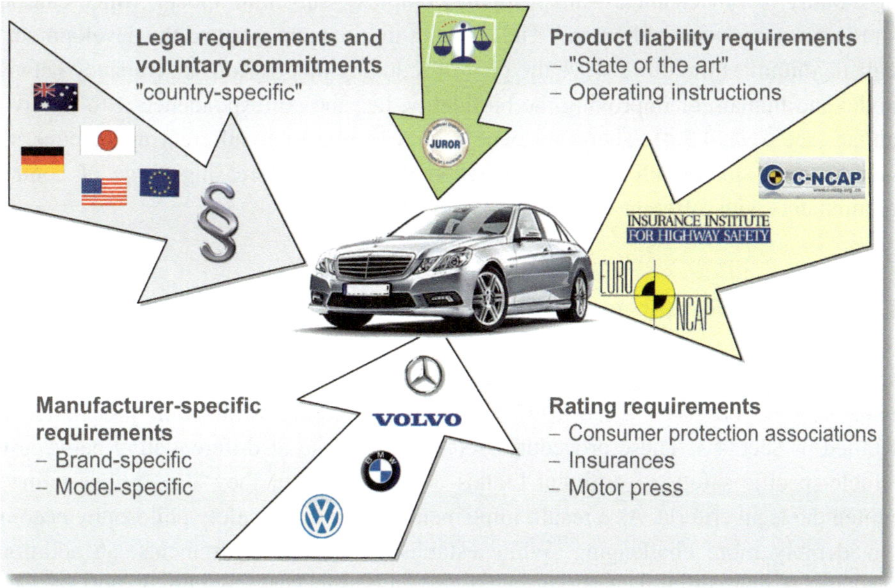

Fig. 8.5 Consideration of all product requirements [95]

8.3.1 Legal Requirements

Based on past experience, it must be considered that national safety standards are constantly evolving and that new regulations may be introduced or existing ones amended on short notice. These innovations must then—sometimes with considerable effort—be retroactively incorporated into an already started vehicle development program or integrated into an existing vehicle concept. The increase in these regulations further exacerbates the already existing conflict of goals between the different regulations of individual countries.

The efforts of a car manufacturer to introduce a vehicle that meets the requirements of all markets, i.e., to build a "world vehicle", are thus increasingly restricted. He is therefore left with the decision to either not serve certain markets or to produce expensive country variants. There are efforts to harmonize safety requirements, but they are progressing very slowly (see Sect. 4.2).

An example of the divergence of requirements are the test configurations for lateral collision [62]: Both EU and US law undoubtedly pursue the objective of improving occupant protection in side impacts, but the methodology and criteria are very different (Fig. 4.6). However, the goal of the vehicle manufacturer is to protect the "human being", who is subject to the same biomechanical requirements everywhere in the world, albeit with some variation, and not the respective dummy type. Another disadvantage, which however affects all such test procedures, is seen in the model building. Because the dummy is a mechanical—albeit highly complex—substitute model, which can only approximate an "average human". The research in biomechanics and the development of realistic dummies therefore serve the goal of reducing this obvious discrepancy between model and human, or improving the biofidelity, i.e., the correspondence with the living human (see Sect. 4.2.4). There are even efforts to represent different age groups. Car manufacturers are therefore forced to reconcile an ever-increasing range of complex requirements with different objectives for a vehicle.

8.3.2 Rating Requirements

Analogous to the legal regulations, a clear increase in different consumer rating requirements can be observed since 1997 [7, 107], the current evaluation procedures are detailed in Sect. 4.3. These procedures serve the purpose of differentiating between the vehicle-specific safety of different brands and models; but they also further refine or tighten the legal criteria. As a result, implementing a uniform safety philosophy becomes considerably more challenging. While establishing these requirements, an additional challenge is that additions and modifications can be introduced within a short period of time — for example, in as little as twelve months. In contrast, legal regulations are typically introduced with a lead time of three to four years, or at least accompanied by transition periods. Consequently, the vehicle-specific development process lacks sufficient

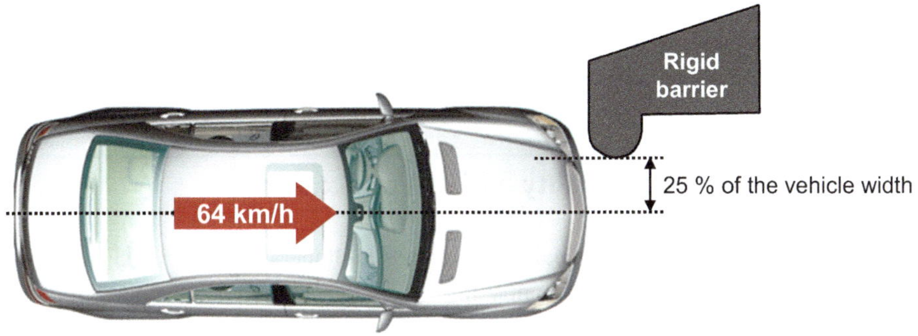

Fig. 8.6 Configuration of the frontal impact according to IIHS SOB (small overlap barrier)

lead time for proper implementation. Figure 8.6 shows as an example the configuration of the 'small overlap' frontal impact, which was first introduced by the American Insurance Institute for Highway Safety (IIHS) in 2012.

8.3.3 Product Liability Requirements

The safety features of a passenger car have high relevance with respect to product liability. If customers are injured in an accident in a vehicle, the manufacturer must be able to prove at all times that the product complies with the latest state of science and technology. In addition, he must be able to prove that it can withstand the expected loads and meets the customer expectations possibly raised by the manufacturer. When selecting the test spectrum and the test criteria, it is necessary to take into account the specific user behavior, i.e., the type of use of the vehicle and the respective environmental conditions. And finally, during the development process, the testing, computational optimization, and later compliance with quality standards must be documented in detail. Product liability plays a very important role for these reasons throughout the entire development process, from the definition phase to the series phase.

8.3.4 Manufacturer-Specific Requirements

In addition to the multifaceted requirements that cannot be influenced by the manufacturer, companies often have their own requirements for the safety features of a vehicle. The motivation for this can vary greatly: Often, own requirements serve to underline and expand a specific brand profile, which is expressed, for example, in an image evaluation "builds safe vehicles" [11]. Manufacturer-specific requirements can be reflected, for example, in the focus on innovation. But they can also show that a vehicle manufacturer offers safety features as standard and not as optional equipment. In addition, it is possible to document the focus with particularly high requirements in the fulfillment of rating tests. Depending on the brand profile, however, a manufacturer may also aim to focus only on mandatory requirements.

8.4 Development Quality and its Verification

The demand for high product quality, which is simultaneously associated with low warranty and goodwill costs, significantly influences the methods and criteria for securing and confirming quality in vehicle development. For the area of passive safety, in which the safety components are subject to documentation, high demands are placed on complete and consistent documentation of the design and test results due to increasingly strict policies in the area of product liability worldwide.

The starting point for quality assurance of any vehicle development are the concept booklet and the vehicle specifications, in which the immediate requirements and expectations for a planned vehicle are formulated, Fig. 8.7. These are, for example, legal and manufacturer-specific requirements for the safety of the entire vehicle in various accident scenarios as well as for the behavior of vehicle components during an accident, i.e., the accident situations are defined as "load cases". The contents of the vehicle specifications must be described in a conflict-free, complete, and unambiguous manner. Since the overall vehicle can be composed of different components, individual component requirements are derived from the vehicle specifications.

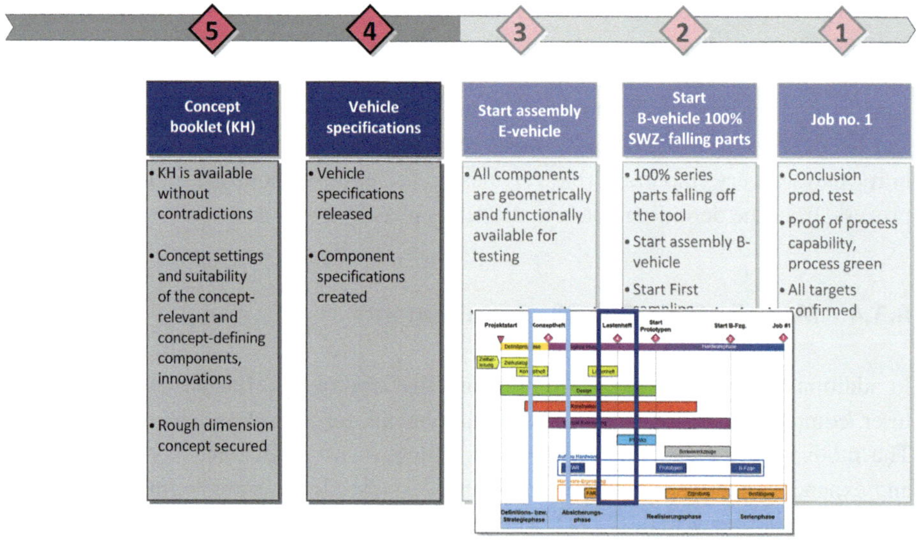

Fig. 8.7 Development milestones concept booklet and vehicle specifications

The success of a new product is crucially dependent on the consideration of experiences from the development of predecessor models, both positive and negative. One way to systematically capture these experiences is referred to as "lessons learned". This is a method in which the experiences of all project participants are solicited and reflected upon after the completion of a vehicle project. The surveys to collect the experiences usually take place in workshops using questionnaires or in oral interviews. To sustainably improve the quality of the follow-up project, it is also important to address these experiences correctly so that they can be fully taken into account at the project start in the strategy phase. The decision about a product or even just a concept can consequently only be made when the proof of feasibility has been carried out, taking into account the systematically prepared experiences. The decision whether a concept is suitable for the planned application and whether it can be developed and produced in sufficient quantities, at acceptable costs and acceptable quality, is prepared by an interdisciplinary working group staffed with people from the areas of design, testing, production, purchasing, sales, and quality. Essential for a viable decision-making process is a risk analysis to be carried out on the mentioned criteria. Once the decision for a product has been made, the phase of development (design) and functional validation (computational simulation and testing) begins.

A central and recognized quality management tool in the automotive industry is the failure mode and effects analysis **(FMEA),** Fig. 8.8. With the help of this analysis technique, errors that could occur should be detected at an early stage and avoided through appropriate measures. The increase in product maturity and quality achievable with FMEA reduces the risk of product errors in a later phase of development, which would then require expensive and time-critical changes or, in the worst case, would occur at the customer's site. In addition, the application of FMEA offers further advantages, which

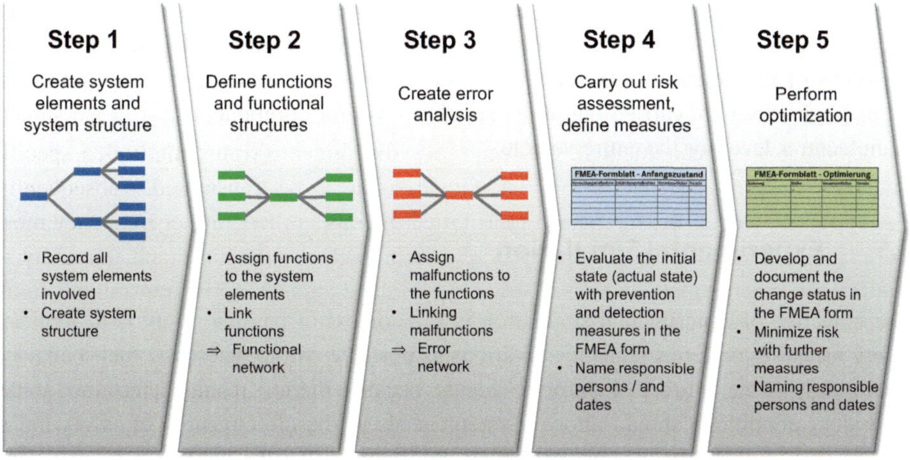

Fig. 8.8 Basic process of quality assurance using FMEA

can be seen in the bundling of cross-disciplinary competence, in the structured, complete documentation of possible errors and the measures to avoid them, as well as in the increase in the quality awareness of the employees.

Usually, a distinction is made between three types, the system, design, and process FMEA, which build on each other but are completely different in terms of their scope of consideration. The respective processing is basically carried out in interdisciplinary working groups: The system FMEA in passive safety considers the interaction of the components that can influence the protection potential in different accident scenarios. The goal of the **system FMEA** is therefore to ensure the optimal design of the overall system based on the vehicle specifications by avoiding errors and checking safety and functionality. With the help of the **design FMEA**, possible errors on individual components of the system, e.g., on the airbag or the seat belt, are to be recognized and avoided by constructive or manufacturing measures. This allows the effectiveness of all constructive measures to be confirmed in prototype testing even before the series tools are commissioned. The engineer responsible for the component, referred to as the "driver", is responsible for the design FMEA. The **process FMEA** is based on the results of the design FMEA and uncovers potential errors in the individual process and production steps within manufacturing and assembly. The process planning is evaluated before ordering manufacturing facilities. The management of the process FMEA is taken over by an employee from production planning. After the suitability for series production has been proven at the component level based on extensive tests, the system and vehicle approval can be started at the full vehicle level.

The **release** serves to prove the functional and legal provisions defined in the specifications. It is carried out according to the country-specific requirements in the form of a self-certification, in the presence of a technical service approved for vehicle certification, or under the supervision of state representatives of the respective country. Only after successful certification, the so-called type approval, the delivery of the vehicles can start. The manufacturer is responsible for ensuring that suitable precautions are taken in production and test procedures are defined to ensure that all produced vehicles comply with the status of this release. In the area of passive safety, conformity tests (CoP: conformity of production) are carried out at the supplier level for components and at the vehicle manufacturer level for the entire vehicle.

8.5 Experimental Simulation

The objective of experimental simulation in the context of passive safety is to replicate safety measures under as realistic conditions as possible and to determine their behavior. Reality means the entirety of traffic accidents, because the test results determined under laboratory conditions should allow a statement about the effectiveness of safety measures in an accident. To this end, occurring deformations on the vehicle body and damage to the occupant protection systems are examined; the forces, accelerations, and other loads acting on the test dummies used serve as a criterion for the severity of the expected injuries. However, since not every single accident can be replicated due to the multitude of possible accident constellations, accident situations are usually replicated according to internationally agreed protocolls that represent a certain accident group and behave similarly in terms of accident and injury mechanics, such as in frontal, side, and rear collisions. Depending on the question and requirement, however, subsystems can also be derived, the examination of which is sufficient to investigate and prove the functionality and effectiveness of safety components; availability and costs, especially in the development phase of motor vehicles, also play a decisive role.

When selecting and conducting tests to check and evaluate passive safety, it is important to consider the necessary boundary conditions, which can be expressed with the help of the "three big Rs":

- Relevance with regard to the accident event,
- Representativeness with regard to the accident and injury mechanics, and
- Reproducibility with regard to the measurement results.

For passive safety, experimental simulation means using experimental crash mechanics simulations to observe and analyze structural deformation behavior and occupant loading during a collision. This is done using the crash test facility and the associated peripheral facilities, with the measurements of physical quantities and the filming of the movement behavior serving the analysis and finally the documentation of the test results (Fig. 8.9).

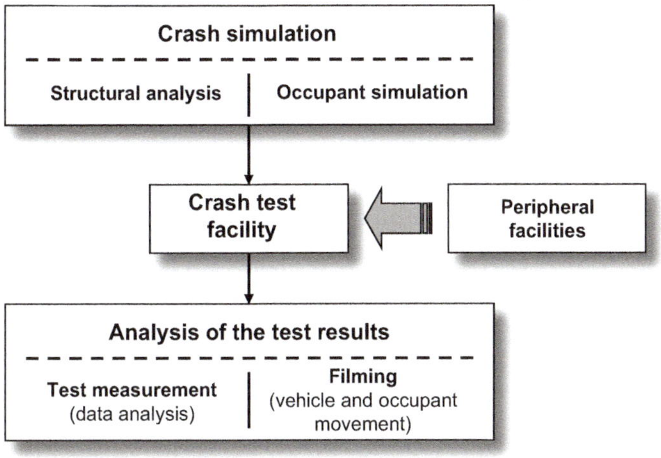

Fig. 8.9 Crash test facility and the pursued objective

8.5.1 Types of Tests

The accident occurrence is too diverse to be simulated in its entirety. Therefore, it is differentiated into representative types of tests, which will be presented in the following. They represent a certain section of the accident world, but differ in simulation quality, i.e., they allow a statement about the extent to which the respective accident is replicated. In addition, the objective for the experimental simulation is of decisive importance, which can be seen, on the one hand, in the functionality of protection system components under accident conditions to be checked; on the other hand, the research objective is to demonstrate the effectiveness of protective measures with regard to a specific accident constellation.

8.5.1.1 Vehicle Tests

The highest simulation quality is achieved in tests with complete vehicles, especially with regard to their motion- and deformation behavior. The required use of anthropomorphic test devices, so-called dummies, as occupants and pedestrians serves on the one hand the motion and on the other hand the load simulation, i.e., both the impact and contact points are determined as well as the "injury" probability is replicated. Sensors are installed inside the dummies for this purpose, so loads in form of accelerations, contact and deformation forces, torques and deformations or displacements can be measured (see Fig. 8.10).

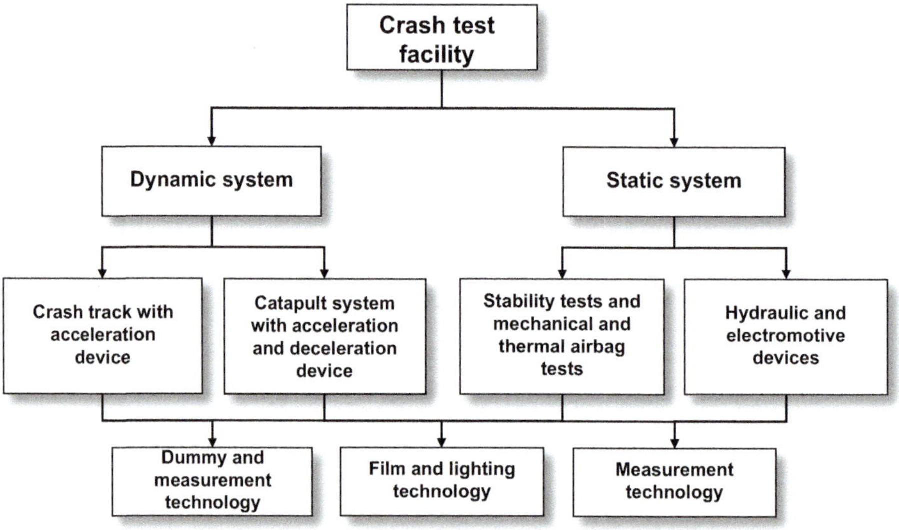

Fig. 8.10 Crash test facility and associated facilities

Table 8.1 Selection of test methods for frontal collisions

Scope of application	Barrier	Overlap	Velocity [km/h]	Objective
Europe	deformable	40 % on the driver's side	56	Structure, ISS, dummy load
Euro NCAP	deformable	40 % on the driver's side	64	Structure, ISS, dummy load
USA, Canada	0°, rigid or deformable	100 or 40 %	48... 53	Structure, ISS, dummy load
USA	30°, rigid	100 %	48... 53	Structure, ISS, dummy load
NCAP (USA)	0°, rigid	100 %	56	Rating test
IIHS (USA)	0°, rigid	25% on both sides	64	Rating test
Europe	10°, rigid	40 % on the driver's side	15	Rep. friendliness, sales insurance classification
USA	rigid pendulum	lateral, center	(2.5 mph) 4.0	No damage permitted on the bumper
Canada	rigid pendulum	lateral, center	(4.8 mph) 8.0	No functional impairment permitted

In **frontal impact vehicle tests**, the partially roadworthy vehicles are accelerated on the crash track using an acceleration device (Fig. 8.10) from a standstill to the intended speed and crash against a defined obstacle, which is rigid or deformable and has full or only partial overlap. The following summary (Table 8.1) presents, without claiming to be exhaustive, the multitude of test constellations for frontal collisions that are common today.

The tests essentially aim to verify the vehicle's deformation behavior and compartment structure as well as the restraint systems; the dummy load values serve as the evaluation criterion. In this context, the multitude of additional, manufacturer-specific test configurations, e.g., the frontal crash test of two cars proposed in [114] to verify structural compatibility, must be disregarded. Similarly, the frequently conducted tests for external safety, such as the simulation of pedestrian accidents or accidents involving motorized two-wheelers and bicycles, are not represented. However, the tests for ease of repair in crash and parking accidents as well as for the initial classification of comprehensive (VK-) insurance are included in Table 8.1 [110]; a change introduced in 2003 should be noted here [83]: The impact area was tilted by 10° (Fig. 8.11). In the past, the European rating system Euro NCAP has been controversial not only in terms of its evaluation methodology but also its impact speed: a test speed of $v = 64$ km/h was introduced, which, compared to the velocity of the ECE crash test with $v = 56$ km/h, corresponds to an approximately 30 % higher impact energy. In this context, it is often argued that a higher test speed would correspond to a correspondingly higher safety. However, accident analytical facts, such as the speed distribution in frontal collisions, the Δv disadvantage of smaller vehicles in vehicle/vehicle collisions as well as the probability of side collisions and the higher risk to laterally impacted occupants associated with a stiffer front structure in lateral accidents, are all too often overlooked. Due to the complexity of the accident event, it is therefore to be feared that an increase in test speeds for frontal

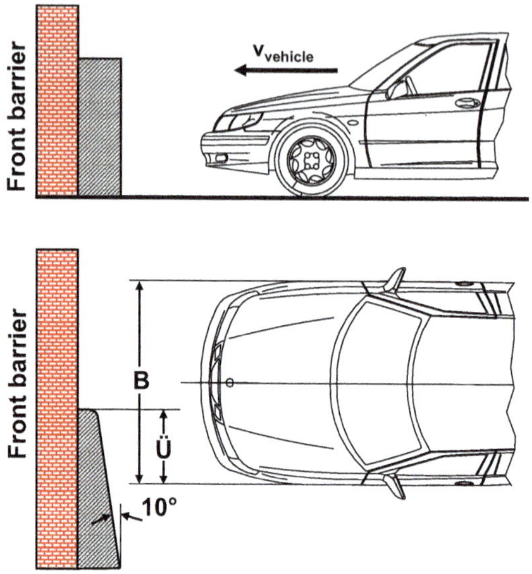

Fig. 8.11 Test setup for frontal impact to verify ease of repair; $v = 15^{+1}$ km/h, 40 % overlap, 10° tilt [52]

impact could reduce the injury risk of occupants in frontal accidents, but at the expense of occupant load in side collisions.

In the execution of **side impact tests**, the vehicle under investigation usually stands still, while a moving deformable barrier (MDB) is brought to the test speed from a standstill using an acceleration device and collides with the stationary vehicle. To take into account the speed component of the vehicle moving longitudinally in the accident, the stationary vehicle, rotated by an angle of 27°, is impacted by a diagonally moving barrier according to the US regulation FMVSS 214. Accordingly, the initial impact impulse does not act exactly in the lateral direction, i.e., perpendicular to the direction of travel, but at an angle of 27° to the rear. The test configurations for side collisions according to European and US safety standards are compared in Fig. 4.6. In the pole impact test, both movement options can be used: once the laterally moving vehicle against the pole fixed to the crash barrier and on the other hand a pole mounted on the moving barrier against the stationary vehicle. In most cases, however, the vehicle is moved against the rigid pole today, which has a diameter of 254 mm (10 in.). To take into account the stiffness of the front deformation structure resulting from the rigid wall impact, a lateral collision with two identical cars is carried out in the compatibility test; first proposed in [3]. This is intended to prevent the front structure from being designed too stiffly to protect the occupants in frontal collisions, but otherwise acting more aggressively in a lateral collision and leading to higher dummy load values. In Table 8.2, common test methods for side collision tests, the essential test conditions, and the objectives are summarized.

The primary goal of the **rear impact tests** is to check the integrity of the fuel supply system according to UN-R 34 or FMVSS 301. The mobile impact barriers used for this purpose are shown in Fig. 8.12. According to US regulation, an impact barrier with a

Table 8.2 Selection of test methods for side collisions, ISS: injury severity score

Scope of application	Barrier	Direction	Velocity [km/h]	Objective
Europe	deformable	90°	50	Side structure, ISS, dummy load
Euro NCAP	deformable	90°	60	Side structure, ISS, dummy load
Euro NCAP	pole, ∅ 254	90°	32	Side structure, ISS, dummy load
USA	deformable	27°	(33.5 mph) 54	Side structure, ISS, dummy load
USA	rigid	90°	(20.0 mph) 32	Integrity of the fuel system
SI-NCAP (USA)	deformable	27°	(38.0 mph) 61	Rating test
IIHS (USA)	deformable	90	60	Rating test
Manufacturer-spec.	pole, ∅ ≥250	90°	≤35	Side structure, ISS, dummy load
Manufacturer-spec.	car front	90°	50	Compatibility for side collisions

Fig. 8.12 Mobile barrier for rear impact tests, according to UN-R 34 (left) or FMVSS 208/301 (right); according to [65]

deformable element, a 70 % overlap, and a speed of 80 km/h hits the stationary vehicle. Immediately after the test, the vehicle is placed on a rollover rotary platform and rotated around its longitudinal axis to specific positions (a total of 360 degrees); any fuel that may leak out must not exceed a defined leak quantity. In addition to the leak test, the rear impact test allows for the examination of the deformation behavior of the rear structure, ensuring that doors can be opened, the seat attachment and backrest strength, as well as occupant protection according to FMVSS 208.

As with frontal collisions, low-speed tests in rear collisions serve to assess the ease of repair after collisions or parking accidents as well as the classification for the fully comprehensive insurance [110]; here too, the impact area has been rotated by 10° since 2003 [83] (Fig. 8.13). A summary of the test methods for rear collisions is shown in Table 8.3.

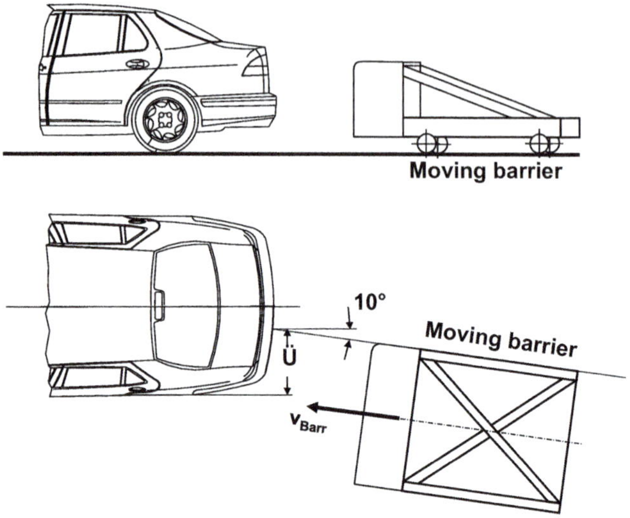

Fig. 8.13 Test setup for rear impact to check the ease of repair; $v = 15^{+1}$ km/h, 40 % overlap, 10° inclination [52]

Table 8.3 Test methods for rear collisions

Scope of application	Barrier	Overlap	Velocity [km/h]	Objective
Europe	180°, deformable	100 %	50	Fuel system integrity, structural behavior
USA	180°, deformable	70 %	80	Fuel system integrity, structural behavior, seats
Europe	170°, rigid	40% on the driver's side	15	Ease of repair, insurance classification
USA	rigid pendulum	lateral, center	(2.5 mph) 4.0	No damage permitted on the bumper
Canada	rigid pendulum	lateral, center	(2.5 mph) 4.0	No functional impairment permitted

Ensuring the structural integrity of the **survival space of the passenger compartment** can be checked in various ways, namely statically or dynamically. In the static method, the roof frame is laterally loaded with a certain force. For dynamic testing, the vehicle can be made to roll over by driving up a one-sided ramp and suddenly turning the steering wheel. However, it is common to move the vehicle sideways on a 23° inclined sled (rollover sled); it is dropped by initiating a strong sled deceleration using deformation elements, causing the vehicle to roll over several times laterally (rollover or rollover test). The test setup is shown in Fig. 8.14.

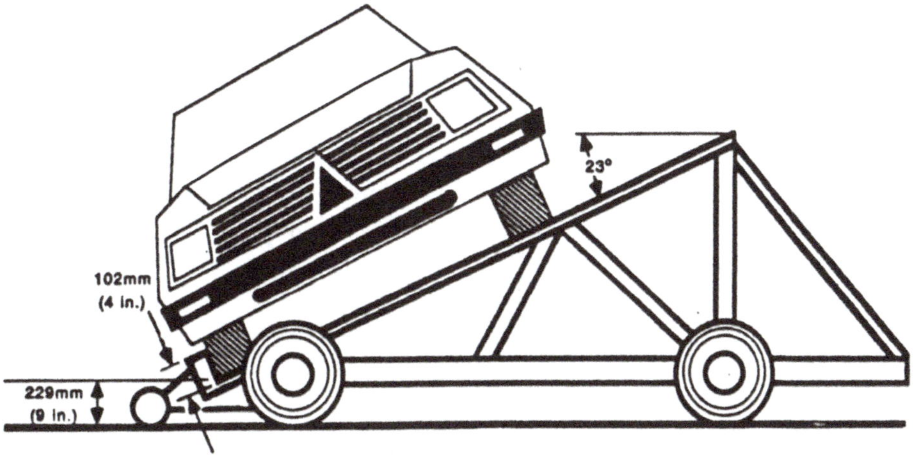

Fig. 8.14 Setup for the rollover test using a rollover sled [25]

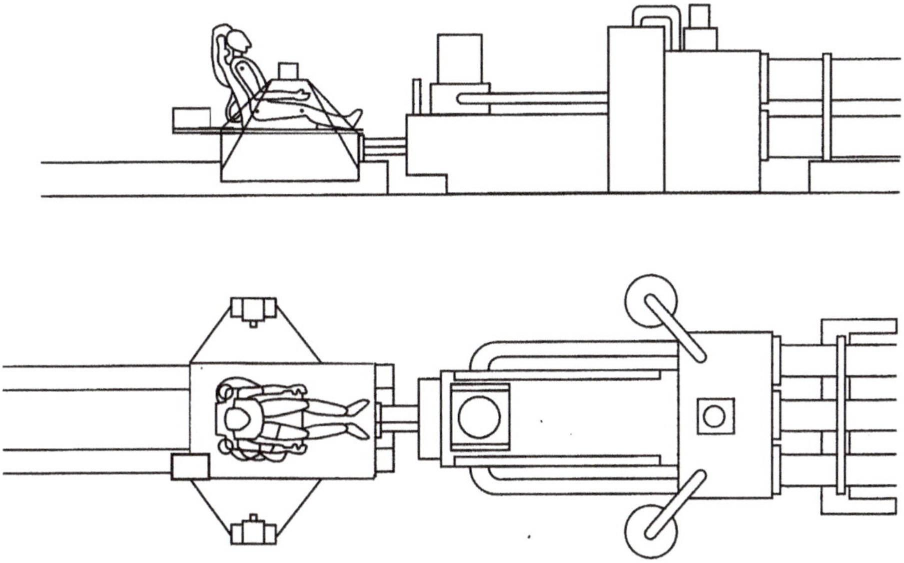

Fig. 8.15 Acceleration sled with hydraulic drive [57]

8.5.1.2 Sled Tests

With the help of full vehicle tests, passive safety is validated taking into account the interaction between the deformation structure and the restraint system, while sled tests primarily serve the design of restraint system components. The main advantage of sled tests is that the effect of occupant protection systems can be examined without destroying a complete vehicle. To accommodate the occupant protection system, the sled is often equipped with a partial body (sled test fixture), but this is so stiffened that it survives several tests (up to 50 trials) without being deformed. Therefore, injury-relevant intrusions, such as the penetration of the steering wheel, the dashboard, and the footwell, can only be replicated with great effort and therefore only in the rarest cases.

Two different principles are used for the **simulation of frontal collisions**: the acceleration and the deceleration sled (see Fig. 8.10). The **acceleration sled** (catapult sled) is accelerated from rest against the direction of travel, where the variable temporal progression of the acceleration corresponds to the deceleration of the vehicle during rigid wall impact (Fig. 8.15). The sled is accelerated backwards, during which the dummy load is measured and is brought to a standstill again by means of a brake. During the acceleration phase, the sled is driven by a hydraulic piston with a force of up to 1,600 kN. A correspondingly pre-stressed gas storage, which is separated from the hydraulic oil by a membrane, releases the necessary energy. Several differently sized, multi-stage servo valves are used to control the oil flow ($\leq$ 90,000 l/min) and thus the acceleration signal ($\leq$ 55 g) [57]. Another possibility for oil flow supply is the shaping of the control piston, which opens different cross-sections in certain positions when opening the hydraulic oil valve. However, with this control principle, the control piston must be replaced to realize a certain acceleration profile [4]. This Bendix principle, named after the manufacturing company, has largely been replaced by the fast and precise servo valve control.

The **deceleration sled** is brought from rest to a defined test speed in the direction of travel using a tow rope (propulsive phase); the rope drive is powered by energy storage in the form of a falling weight, a hydraulic, electric or combustion engine, or pre-stressed rubber ropes (bungee cords). When the test speed is reached, the rope drive is disconnected from the sled, so that the sled is in a non-accelerated state (free-running phase). The deceleration of the sled (deceleration phase) takes place due to the lack of vehicle deformation structure by means of a stationary deceleration device (see Fig. 8.10), in which the kinetic energy is converted into deformation work or flow energy. Plastic tubes, which are expanded by a mandrel attached to the sled during penetration, serve as deformation elements [27]. The possibilities for variation to reproduce a given deceleration profile are extremely limited, but they are sufficient to maintain the deceleration/time corridor defined, for example, in [27].

Another principle of deceleration is realized by the parallel arrangement of several, possibly stepped, deformation tubes. With this principle, the increase can be reproduced, but the decrease of a given deceleration/time profile can only be reproduced to a limited extent.

Today, band bending brake devices are usually used because of the possibilities for variation, in which a "nose" attached to the sled deforms commercially available flat steel ligaments by bending, Fig. 8.16. These are fixed by rollers and built up in several stages arranged one behind the other.

Another deceleration device is based on the principle of displacement of fluids, in which a nose mounted on the sled and designed as an inclined plane displaces a hydraulic piston perpendicular to the sled movement, which in turn displaces hydraulic oil

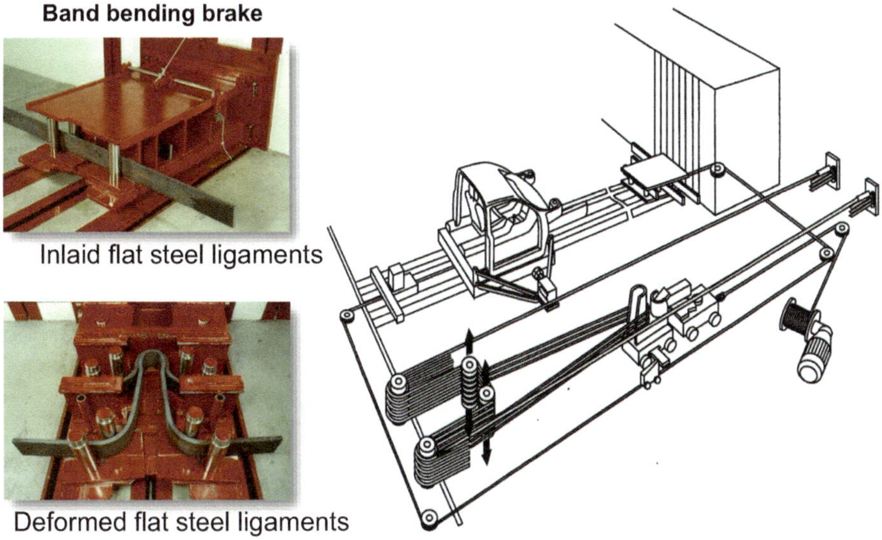

Fig. 8.16 Deceleration sled with deceleration device (taken from [16])

through defined valve cross-sections, thus realizing the required deceleration curve [72]. With these modern hydraulic brake sled, deceleration force curves of up to 3,500 kN can be applied.

HyperG sled systems, in which a cylinder rod is preloaded via a large air piston with pressure chamber, are another variant. The deceleration of the sled takes place after the acceleration phase through a hydraulic, servo valve-controlled brake.

Finally, electromagnetic sled systems are increasingly being used, in which large magnetic fields in the sled guide move the structure highly dynamically. Here, both acceleration and deceleration profiles can be represented highly dynamically and very precisely.

The measurement of the load on the dummy starts at the beginning of deceleration. Figure 8.16 shows a sled with a stiffened car body and the deceleration device designed as a band brake.

Due to the more complex injury mechanics, and depending on the expectations regarding the simulation quality, a **sled facility for side collisions** is much more complicated, as the injury-inducing, penetrating side door must be brought from a standstill to the defined penetration speed in order to realistically reproduce the motion quantities prevailing in side collisions, especially the contact speed between door and occupants. In the sled facility shown in Fig. 8.17, the passenger compartment, including the seated dummy and the side door attached to the door frame with deformation elements, is at rest. The impact sled, brought to collision speed by a cable drive, is equipped with a locking device that locks the door with the sled at the time of contact. In the further course, the side door penetrates into the passenger compartment at the speed imposed

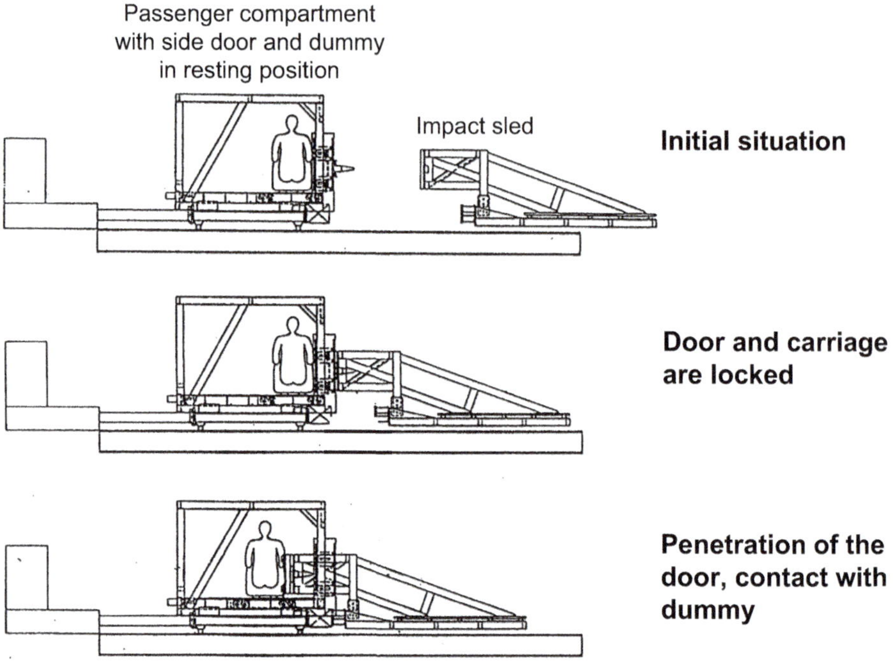

Fig. 8.17 Sled facility for the simulation of side collisions according to [53]

on it by the sled and contacts the dummy still at rest. The loads determined in this way largely correspond to the values measured in vehicle tests. For less elaborately designed sled facilities for side collision tests, more or less large compromises have to be made with regard to load and the temporal course of the motion quantities in terms of simulation quality.

8.5.1.3 Component Tests

With the help of component tests, investigations on the functionality of subsystems of passive safety are carried out. These can refer to the deformation and intrusion behavior of external or internal structures, but they can also aim at the function and mode of operation of components of the restraint system. For this purpose, body shells, assemblies such as bumpers with longitudinal beams, steering device with front end and subsystems such as seat, instrument panel, seat belt system and airbag are prepared and provided for the respective investigations. A large part of these component tests is defined in terms of type and criteria in legal regulations, standards or manufacturer-specific requirements.

To investigate the stiffness and deformation conditions on **body shells and assemblies**, for example, the following tests are conducted:

- Static roof crush tests (Fig. 8.18) according to FMVSS 216 to ensure survival space in the event of rollover accidents,

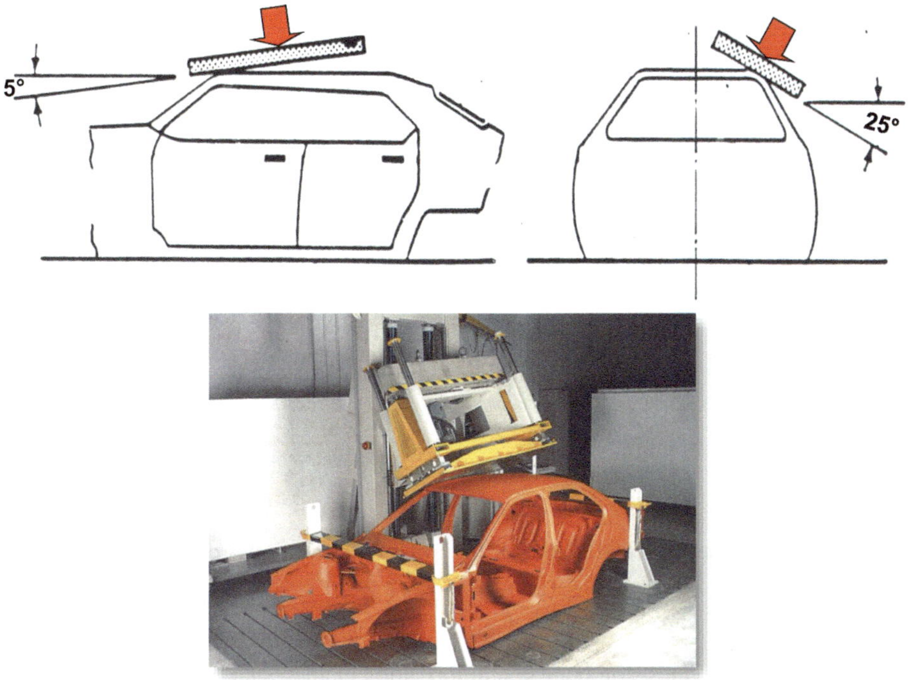

Fig. 8.18 Test setup for the static roof crush test to check the roof stiffness (taken from [113])

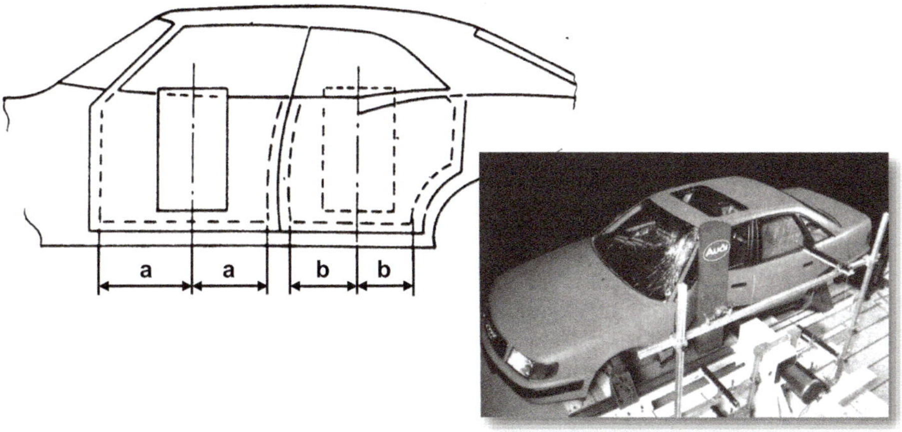

Fig. 8.19 Setup for the static door crush test to check the door stiffness (taken from [77])

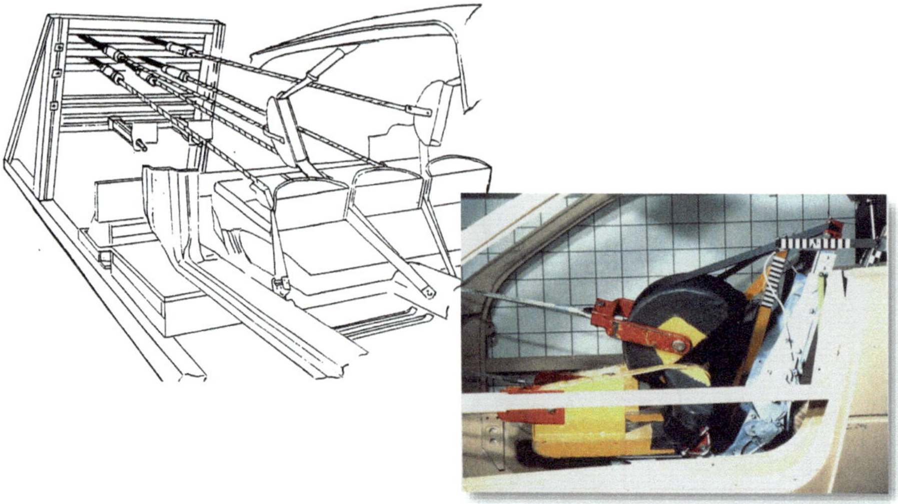

Fig. 8.20 Setup for the tensile test of the seat belt anchorage points to check the body strength (taken from [67])

- Static and dynamic tests on bumpers, deformation elements and longitudinal beams in accordance with FMVSS 215 to check the compliance and energy absorption capacity in parking and minor accidents,
- Static door crush tests (Fig. 8.19) according to FMVSS 214a to ensure door stiffness in side collisions—replaced by the dynamic side impact test according to FMVSS 214,
- Static tensile tests at the seat belt anchorage points (Fig. 8.20) according to FMVSS 210 to ensure body strength at the corresponding attachment points on the B- and C-pillars,
- Dynamic impact tests according to FMVSS 204 using a torso block to check the compliance and energy absorption capacity during steering wheel impact, and
- Dynamic impact tests using a free-flying head form (free motion head form) according to FMVSS 201 to check the compliance and energy absorption capacity of the contact points (lining on the roof frame as well as on A-, B-, C- and possibly D-pillars) in the interior.

These are carried out by applying a defined quasi-static or dynamic load. The individual test setups and execution regulations are described and illustrated in detail in the annually published, revised and expanded SAE manual (e.g., [68, 25, 113, 77, 67]).

Subsystems are used to investigate the functionality of occupant protection systems under accident conditions, which must be ensured over their operating time (initially 12, then 15 years, now indefinitely). This involves the use of artificial aging effects through time-lapse with regard to various environmental influences such as vibrations, temperature, sunlight, humidity, and other long-term effects. The following US tests can be named as examples of functional examinations on subsystems without claiming to be exhaustive:

- Static and dynamic tests on seats, seat backs and headrests according to FMVSS 201 and 202 to check the strength and energy absorption capacity in frontal and rear collisions,
- Dynamic impact tests on instrument panels using a free-flying head form according to FMVSS 201 to check the compliance and energy absorption capacity of the contact points as well as checking sharp edges through minimum radii,
- Static and dynamic testing on child restraint systems according to FMVSS 210 to check the strength, function and operability,
- Static and endurance tests on seat belts, fastening fittings and retractor devices as well as environmental simulation according to FMVSS 209 and 210 to check the strength, abrasion resistance, function and durability due to environmental influences,
- Dynamic tests on retractor- or lock-side mounted seat belt pretensioners to check the strength and function, and
- Deployment and static tests with driver, passenger and side airbags to check the shape, function and effectiveness on adult and child dummies in OoP (out-of-position) situations (according to FMVSS 208) including the covers and trim parts; these OoP tests are sometimes mistakenly referred to as "static" tests, which however ignores the fact that these are highly dynamic processes.

The list of test types mentioned here is essentially based on legally prescribed or standardized test arrangements and procedures. However, these regulations are usually only minimum requirements for effectiveness and functionality of protective measures. It is therefore obvious that vehicle manufacturers and suppliers of restraint systems, especially in the development stage, must carry out a much more extensive range of tests and trials. Test program details are confidential information of each vehicle manufacturer. In addition, functional tests are subject to constant change, which is oriented towards the development of new protective measures before the tests can be standardized.

8.5.2 Test Facilities and Equipment

The execution of the types of tests mentioned in the preceding subchapter takes place in test laboratories and at crash test facilities, which depending on the task and requirement, take place in test halls and—albeit to a lesser extent today—outdoors. The **outdoor facilities** are used for simulating vehicle collision accidents (commercial vehicles, buses and passenger cars) in oncoming traffic and at certain angles, and are characterized by the space required for the approach path and the necessary run-out. The main disadvantage lies in the dependence on the weather (temperature, wind, rain or snow and light conditions) and is therefore limited to the bare minimum of cases that cannot be carried out in the test hall due to space constraints.

The vast majority of vehicle and sled tests take place at **crash** test facilities in test halls. These include the actual crash track (up to 160 m long), at the end of which is the crash block made of concrete. On the front side of the crash block, a mounting plate is embedded, to which a wooden plate for frontal impact tests or various extensions (force measuring wall, angled or offset barriers with and without deformation elements or with pole extensions) can be attached. The crash track can be divided into the acceleration, calming, and deceleration or crash sections. The length of the deceleration section is composed of the deformation and run-out path and the structural length of the extensions; it is about 3 m. The length of the acceleration section depends on the collision speed to be achieved (up to 90 km/h) and the maximum permissible vehicle or sled acceleration (≤ 0.3 g) and thus requires approximately 160 m. Since the vehicle or sled should hit the obstacle without acceleration, there is a calming section of about 2 m length between the acceleration and deceleration sections. The total length of the crash track is therefore approximately 170 m. If higher velocities are to be achieved, either the total length must be designed accordingly larger or the maximum permissible vehicle acceleration must be increased.

To accelerate the vehicle or test sled, a centrally guided endless steel cable is used in a shaft below the crash track, which is driven by a pulley powered by a hydraulic, electric or combustion engine. The drive is regulated by a target/actual comparison of the cable speed in such a way that the impact speed envisaged at the end of the acceleration phase can be realized reliably enough, with a deviation of $\pm$ 0.1 km/h. Today, other types of acceleration are increasingly rare in the use of a falling weight or a rubber cord drive. The intended speed results from a certain extension length of the falling weight or the rubber cords. The vehicle or sled is brought into a certain position against the falling weight or the rubber cord tension applied via the pull cord by means of a winch and locked on the crash track. The winch's cable is then released from the vehicle or sled, which are then in a pre-tensioned state in the rest position. The run-up is initiated by the release, and the vehicle or sled moves towards the crash block with increasing speed.

Installation rooms for the catapult system and for vehicle or sled preparation as well as laboratories for component tests, for measurement technology and for dummy calibration are usually located on the side of the crash track.

For obvious reasons, almost all automobile manufacturers and suppliers of occupant protection systems such as seat belt and airbag systems, seats, child restraint systems, etc., operate crash test facilities of the types described here; but also university institutes and monitoring institutions (DEKRA, TÜV in Germany etc.) and increasingly also innovative system development companies and engineering service providers are equipped with such facilities.

8.5.3 Anthropomorphic Test Devices (Dummies)

The essential criteria for the effectiveness of safety measures are the expected probability of injury and the injury severity of road users in accidents. For obvious reasons, the use of humans in experimental simulation is prohibited, as the risk of injury cannot be ruled out. Therefore, in the context of biomechanics research, the best possible correlation between typical accident injuries and measurable loads has been and continues to be established with the help of volunteers, human corpses, animals, mechanical and mathematical models. This approach allows for the experimental determination of load values in order to make statements about a possible injury risk. Mechanical human models are used in the experimental simulation, which largely correspond to the human body in terms of size and mass, mass distribution, and movement behavior. They are equipped with sensors for measuring mechanical quantities, i.e., load quantities, from which load values can be determined. These anthropomorphic test manikins or dummies usually consist of a skeleton made of metal or plastic, whose body parts are joined with each other. Muscles and soft tissues are replicated by elastic plastic materials, and the body surface consists of a plastic skin. The main objectives of the design, equipment, and composition are maximum reproducibility and traceability of the measurement results under the same test conditions.

8.5.3.1 Requirements

Most dummies are used in frontal, lateral, or rear impact vehicle crash tests or sled tests. The crucial requirement for the design of a dummy type is therefore initially the **load direction**. Projects aiming to develop a universally applicable dummy for all use cases have so far failed [48]. However, the use of complex dummies is not necessary in all cases; rather, for some component tests, simplified body part replacement models are sufficient, such as the torso block for the body block test to check the compliance of the steering system, the femur test body for the bumper test, or the free-flying skull cap to check the front hood and interior. The reason for such **simplifications** lies in the possible ambiguity of the test results, the higher probability of errors, and the expensive and complex calibration of a complex dummy with its extensive instrumentation.

In safety tests, the 50th percentile dummy is most commonly used, i.e., 50% of the male population is smaller than the 50th percentile value, which represents the average value of the frequency distribution of an anthropometric series examination [86]. To investigate extreme situations regarding the **anthropometry** (size and mass) of the occupants, two other dummy sizes are used: a 5th percentile female dummy and a 95th percentile male dummy for small or large occupants respectively. The 5th percentile dummy represents size and weight of the data of the population where 5% of all women are smaller, and the 95th percentile dummy represents the anthropometric data of the population where 95% of all men are smaller. For these two dummies, the joint distances, the body part masses and their centers of gravity, as well as the compliances of the body parts under load, were derived from the 50th percentile dummy using a scaling approach from similarity mechanics [64].

The most challenging task in dummy development is certainly the replication of the mechanical characteristics of human limbs and body parts; this attribute is referred to as **simulation quality** or **biofidelity**. Thus, the dummy used in tests must exhibit the same movement behavior and the same stiffness and damping properties as the human occupant during a collision in an accident. This means that the dummy in the test must deform the contact structure in the same way as it is damaged by the human in the accident. If this is not the case, the dummy will either not hit the contact structure or hit it at a wrong angle or cause a different form of deformation; its flexibility and compliance would be inaccurately replicated. The dummy is exposed to a situation in the test where the human body would be injured or limbs could break. In contrast, the dummy or its limbs should not break apart, but the impact must be detectable in such a way that the degree of the corresponding overload, which leads to fractures in human body parts, is measurable and can thus be evaluated.

The initially demanded **reproducibility** in the use of dummies means that the same dummy leads to the same measurement results in repeatedly conducted, identical impact tests. On the other hand, it must be ensured that the same measurement results are obtained when using different dummies in otherwise identical tests. The crucial prerequisite for this is the calibration of dummies, which is to be carried out on the basis of measurements under defined test conditions and visual inspections. Since calibration is time-consuming and costly, the dummy must withstand a larger number of tests, e.g., 15 tests without a seat belt or airbag system, without showing any damage. In addition, the defined calibration limit curves, which guarantee reliable function, must be adhered to. However, a dummy is usually subjected to calibration after about three to ten tests, even if no externally visible damage occurs to the dummy, to ensure the reproducibility of the dummy measurement values.

8.5.3.2 Instrumentation

Injury criteria for the various body regions of the living human are transferred into protection criteria on the dummy, which are defined in legal regulations and safety standards. Both criteria are expressed in physical quantities that can be measured on the dummy. The tolerance threshold corresponds to the test limit values that are to be aimed for in legally prescribed tests in order not to exceed the protection criteria. The sensors installed in the dummy enable the measurement of these physical quantities.

Today, the common method for assessing injuries of the **head** involves measuring the resultant of the triaxially measured translational accelerations in the center of gravity of the head and calculating the head injury criterion (HIC; Eq. 3.4) from it. The measuring range of these piezoresistive sensors ranges from 200 to 2,000 g. Since the biomechanical basis for the HIC value results from direct head impact, the HIC calculation is limited to an interval of typically 15 or 36 ms. In experiments, peak values of up to 200 g are sometimes measured in the dummy head. For the calculation of the GAMBIT value (cf.

Eq. 3.20), in addition to the translational acceleration components, the orthogonal components of the rotational accelerations are required. Their measurement is carried out by deriving the signals from triaxial angular velocity sensors with a measuring range of up to 100 krad/s^2.

At the **neck and lumbar spine**, bending moments, bending angles, compressions, and shear forces are measured as protection criteria. The measurement of forces and moments using piezoresistive sensors is easily possible. They are either designed uniaxially or have multiple components.

With direct load of the **thorax**, many injuries occur due to extensive chest deformations. On the other hand, injuries result from the inertia-related reactions of internal organs, leading to overstretching and tearing. Such injuries occur without major rib deformations. Therefore, in modern dummies, such as HIII, THOR, ES-2, or WorldSID, both accelerations and indentations are measured in the thorax. To determine the thoracic trauma index (TTI; Eq. 3.15), maximum values of accelerations at the chest spine (T12) and at the lateral rib are measured in the case of lateral load on the US-SID-Dummy, formerly prescribed in the US standard FMVSS 214. The combination of deformation speed and relative thorax indentation leads to the calculation of the viscous tolerance criterion (VC; Eq. 3.12), where the velocity is calculated from the measured indentation. The indentations are recorded using rotary potentiometers (Fig. 8.21) or capacitive or

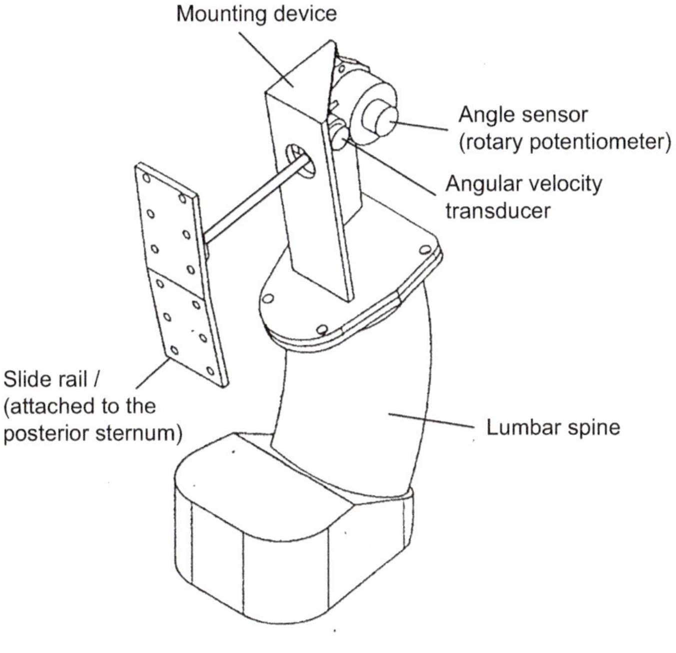

Fig. 8.21 Arrangement of the rotary potentiometer for the recording of chest indentation [112]

inductive displacement sensors, and the accelerations are recorded with piezoresistive acceleration sensors.

In frontal collision tests, the penetration of the pelvic belt leads to injuries in the **abdominal area**. As a submarining criterion, therefore, pelvic angles are integrated from rotational speeds or strain gauge force sensors are used on both sides of the pelvic crest for measurement. In lateral load, indentations and forces are measured.

Frontal loads on the **pelvis** are initiated by knee and thigh forces and occasionally lead to fractures of the joint socket, while lateral loads can lead to fractures of the bony pelvis. It therefore seems sufficient to measure three-axial accelerations in the dummy pelvis, for which, however, no test limit values are prescribed in frontal collision tests. Only in FMVSS 214 is a peak value of up to 130 g permissible for lateral loads. In the European regulation UN-R 95, on the other hand, a force measurement in the pubic symphysis with a limit value of up to 10 kN is permitted.

The injury mechanism of the **lower extremities** include fractures of the bone structure, ruptures and dislocations of the joints, and damage to the ligament structure. For the knee—femur—pelvis area, the femur longitudinal force, measured with force transducers, is established as a protection criterion. For the lower leg, the tibia index has primarily established itself as a protection criterion, more on this in Chap. 3. In addition, longitudinal forces and bending moments on the lower legs as well as shear forces in the knee joint are measured.

8.5.3.3 Dummies used and their Application

It has already been explained that the use of living humans in the context of experimental simulation can only be limited to exceptional cases—volunteer trials in the significantly subcritical load range—due to the expected load. Therefore, dummies were initially developed for frontal collisions (SAM, VIP, HYBRID I, II and III, THOR) since the late 1940s and for side collisions (US-SID, Bio-SID, EuroSID or ES-2 and WorldSID) since the late 1970s. To investigate the increasingly important cervical spine distortions, the BioRID (biofidelic rear impact dummy) was developed at Chalmers University of Technology in Gothenburg (S) since the late 1990s, distributed since 2002 and used in tests. It is particularly characterized by a realistic reproduction of the spine [50]. A rough overview of the timeline of **dummy developments** is shown in Table 8.4.

When determining the hip point to determine the seat reference point (SRP) that is crucial for the development of restraint systems, a metallic, anatomical articulated doll, **H-point manikin** or H-point machine (HPM), is used, which helps simulating the seating forces on the seat. Once the seat reference point is fixed, it must be ensured that a dummy takes the corresponding position; i.e., the H-point machine is then used to adjust the seating position and backrest inclination. In this greatly simplified dummy (Fig. 8.22), torso, buttock, femur, and lower leg weights are placed in a frame consisting of articulated seat and back shape and suspended thighs, so that the loads of the seat cushion and backrest can be adjusted.

Table 8.4 History of dummy developments for vehicle safety technical application (selection, according to [50, 18, 9, 23, 45])

Designation	Developer	Start o development	Explanations
ALDERSON	Alderson Res.Lab. & Sierra Engineering	1946	Mechanical replicas of human limbs
SAM	Sierra Engineering	1949	First dummy developed for the US Air Force
GARD DUMMY	Grumman-Alderson	1952	Research dummy
FBP dummy	Alderson Res.Lab.	1954	Further development of the GARD DUMMY for use in the US Air Force and automotive engineering
ASP dummy	Alderson Res.Lab.	1956	Dummy for APOLLO space project
VIP	Alderson Res.Lab.	1962-1971	Dummies in seated position for vehicle technology (50th -, 5th -, 95th percentile) and child dummies (3 and 6 years)
	Holliman Air Force Base	1963	Comparative study between SIERRA SAM and ARL VIP dummy and volunteers in the test; result: VIP dummy more realistic than SIERRA SAM
VIP 50 A and B	Alderson Res.Lab.	1964-69	Improvements; basis: VIP 50th percentile dummy; Further development of dummies for testing ejection seats for the US Navy and the US Air Force
ACTD	ARL & SIERRA	1950-70	Development of the crash test dummy for vehicle safety (ACTD: Automotive Crash Test Dummy); 50th and 95th percentile dummy (male) and 5th percentile dummy (female) based on GARD DUMMY, VIP and SAM
HYBRID I	General Motors (GM)	1971	Standardization of the ARL and SIERRA 50th percentile dummy (male) to one dummy; head of SIERRA
HYBRID II	General Motors (GM)	1972	Improvement of the HYBRID I with regard to head, neck, shoulder, vertebral column and knee. Detailed documentation
Part 572	NHTSA	1973	Introduction of the HYBRID II dummy in Part 572 of the US Code of Federal Regulations for use in FMVSS 208
ATD 502	General Motors (GM)	1973	Anthropomorphic test device 50th percentile 2nd generation: Improvement of the HYBRID II with regard to head, neck, joints, ribs; new design of spine and knees; improvement of posture to meet NHTSA requirements
Pinocchio	TNO (Netherlands)	1973	Prototype of a toddler dummy
CAMI dummy	Civil Aeronautical Medical Institute of the US Federal Aviation Administration	1973	Prototype of an 18-month child dummy for the evaluatio of child seats
HYBRID III	GM	1974	Further development of the ATD 502: New design of the neck and thorax; more transducers. Can be used optionally for HYBRID II as standard in FMVSS 208
P-Series	TNO (Netherlands)	1979-1994	TNO together with child seat experts from the UN Working Group: Child dummies (newborns to 10 years old child) for later use in the UN-R44

Table 8.4 (continued)

US-SID	NHTSA & University of Michigan Transportation Research Institute (UMTRI)	1979	Side Impact Dummy (SID) based on the HYBRID II; new thorax design for use in side collisions. Later standard forFMVSS 214
EuroSID	EEVC-WG9 Þ TRL, INRETS, APR, BASt & TNO	1983	New development of a dummy (50th percentile (male)) for Side collisions; later use in tests according to UN-R 95
HYBRID III	Humanetics & SAE	1987	Derivation of the 95th percentile dummy (male) and the 5th percentile dummy (female) based on HYBRID III 50th percentile dummy
BioSID	GM, FTSS & SAE	1988	Development of a more realistic dummy for side collisions (BioSID: Biofidelic Side Impact Dummy) based on the HYBRID III
CRABI 6 months and HYBRID III 3 years old	First Technologies Safety Systems (FTSS)	1992	Child dummies for US legislation, later supplemented by other age groups
SID-IIs	Humanetics	1994	Developed to meet the requirements of improved side impact protection - especially side airbags. The dummy is based on the Hybrid III 5th percentile (female) ATD, but also largely corresponds to the anthropometry of a small teenager.
THOR	NHTSA & GESAC, later Humanetics	1994	Start of the development of the "Test Device for Human Occupant Restraint", a new generation of dummies for frontal collisions, which should take into account previous findings in dummy development
Q-Series	Various EU projects	1996-2012	Development of the Q-series child dummies (6 weeks to 10 years) in various EU projects (CREST, CHILD, EPOCh) as the successor to the P-series, additional measurement options
BioRID	Chalmers University (Sweden)	1997	Biofidelic Rear Impact Dummy (50th percentile (male)) specifically for low speed rear-end collisions to assess the risk of (cervical) spinal injuries
World SID	International Standardization Organization (ISO)	1997	Project to develop a worldwide harmonized side impact dummy as a joint project of the ISO dummy experts
EvaRID	Chalmers University (Sweden), Humanetics	2010	Dummy for the 50th percentile (female) specifically for Rear-end collisions, based on the BioRID
THOR-AV	Humanetics	2018	Modified THOR dummies for 5th percentile (female) and 50th percentile (male) for alternative seating positions in future autonomous vehicles, developed as part of the AVOS consortium (Autonomous Vehicles Occupant Safety)

The seating procedure is specified in UN-R 14; the insertion is difficult because the penetration depth depends on the penetration speed due to the visco-elastic behavior of the seat cushion.

In the early 1960s, the simplified **TNO-10-Dummy** was developed by the TNO Road Vehicle Research Institute for the verification of seat belt system design (UN-R 16) and introduced in 1970. It consists of a head, neck, torso, thighs, but only one lower leg (Fig. 8.23). The joints between the individual body parts can be adjusted. Its size is 174 cm and its weight can vary between 70 and 82 kg. No acceleration or force measurements can be made on the dummy itself. Rather, in the sled test prescribed by UN regulation 16, where the seat belt must not tear and no components shall break, only the limit of the dummy forward displacement is defined.

With the introduction of UN Regulation 44 on child restraint systems in motor vehicles in 1980, TNO presented a parallel developed **child dummy series** that represent nine months ($P\frac{3}{4}$), three (P3), six (P6) and ten years (P10) old children. In 1984, a dummy representing a newborn child was added. To ensure realistic movement behavior,

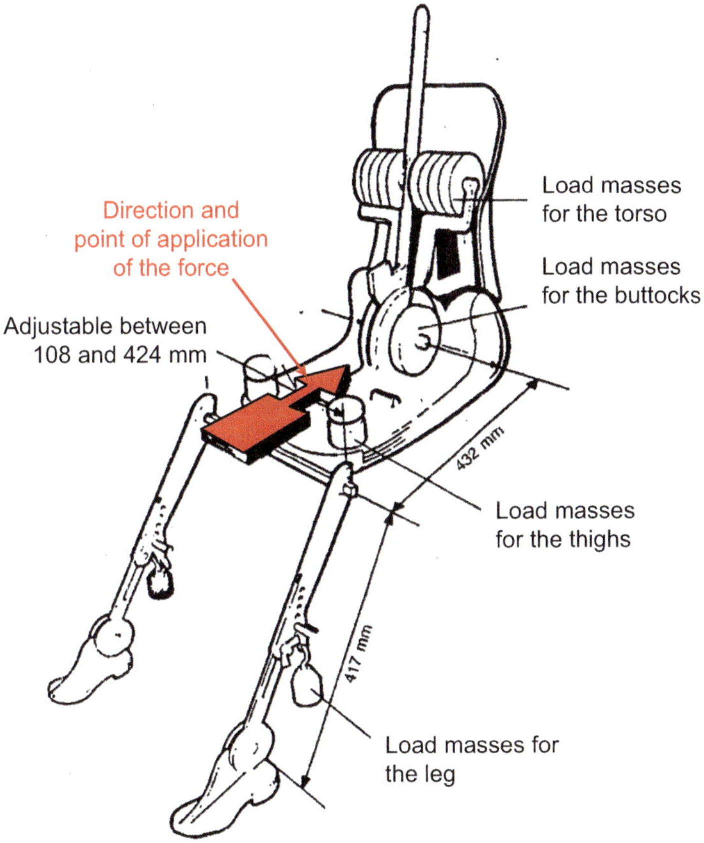

Fig. 8.22 H-point machine (H-point manikin) for determining the seat reference point (from UN-R 14)

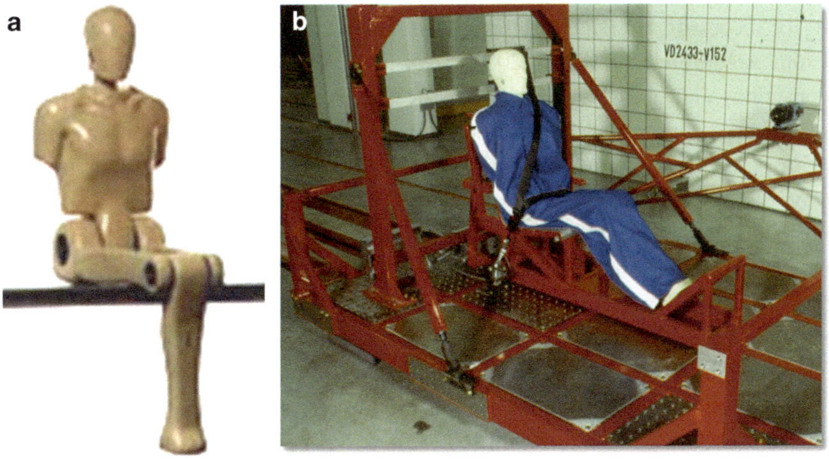

Fig. 8.23 Structure of the TNO-10-Dummy (**a**) and used in a sled test (**b**)

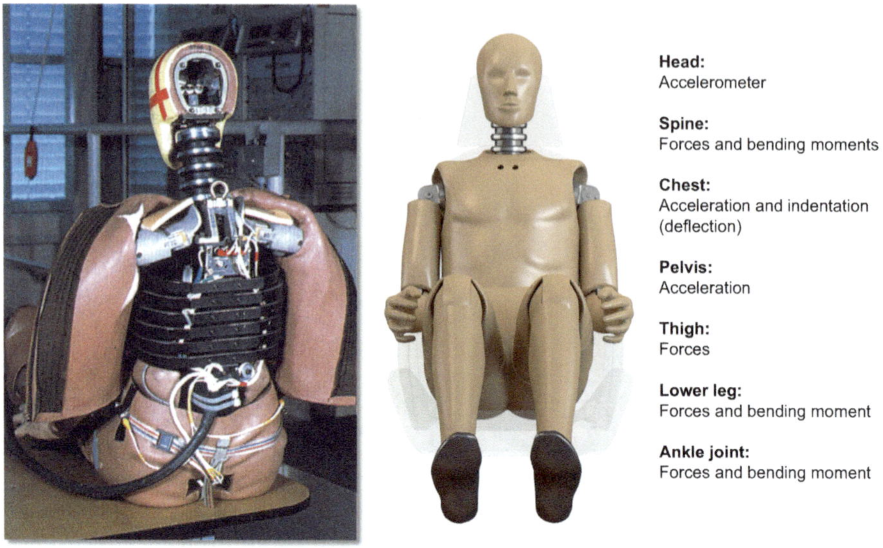

Fig. 8.24 Dummy HYBRID III: Structure and measurement points (according to [23, 45])

shoulder and hip joints are designed as ball joints, while elbows and knees are designed as hinge joints. The head can be rotated around its vertical axis and at the level of the two upper cervical vertebrae (atlas and axis) around its transverse axis. The abdominal area contains a belly insert made of soft foam. Accelerations can be measured in the head and chest with three-axial sensors, and sensors are also used to measure neck forces and moments. From 2013, the outdated dummies of the P-family (P $\frac{3}{4}$, P1 $\frac{1}{2}$, , P6 etc.) were replaced by child dummies of a new generation, the so-called Q-Series.

For the experimental verification of complete motor vehicles, highly developed dummies are required. The **HYBRID II** originally developed by General Motors and especially its successor **HYBRID III** are used worldwide as standard dummies in frontal collisions according to FMVSS 208 or UN-R 94 and UN-R 137, in consumer protection programs and in all research activities in this area (Table 8.4 and Fig. 8.24). They are also used as pedestrian and motorcyclist dummies. The 5th percentile (female) and the 95th percentile dummy (male) were derived from the HYBRID-III-50th percentile dummy.

Up to 60 measurement channels can be used with the HYBRID-III dummy. Since a few years, the dummies of the HYBRID-III generation have been replaced by the new dummies of the **THOR** series, where THOR stands for "**T**est device for **H**uman **O**ccupant **R**estraint", Fig. 8.25. A significant feature of the new dummy generation is the significantly improved sensor equipment. Up to about 200 measurement channels allow even more precise statements about the injury risk of vehicle occupants in traffic accidents. The 50th percentile dummy of THOR has been used in the European new car

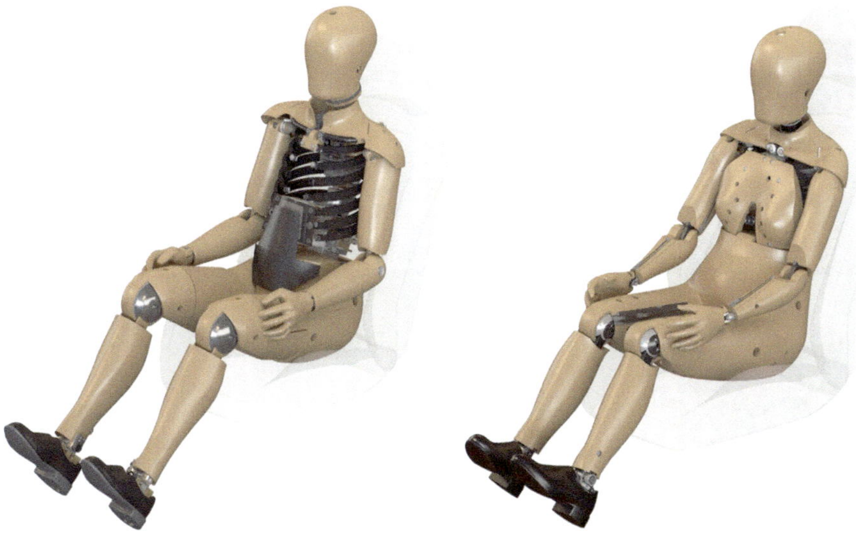

Fig. 8.25 Dummies of the THOR series: 50th (left) and 5th percentile (right) [45]

assessment program Euro NCAP and other consumer protection programs since 2020 and is expected to replace the corresponding HYBRID III in the next years. The 5th percentile dummy (female) of this dummy generation, the so-called THOR-5F, is the first dummy based exclusively on anthropometric data from women. The market launch is currently being prepared.

In the review of **side collisions**, there are two different test procedures: the impact of a movable barrier, which represents another vehicle, on the test vehicle (FMVSS 214 or UN-R 95) and the impact of the test vehicle on a pole (FMVSS 214 or UN-R 135 and GTR 14). For this, there are specific dummies: the US-SID 50th percentile, the so-called ES-2 (the second generation of the EuroSID, Fig. 8.26), the SID-IIs and the WorldSID.

The **US-SID**, a Hybrid-III dummy 50th percentile (male) from the USA modified for side impact, is now hardly used anymore. More modern is the **ES-2**, a 50th percentile dummy (male), Fig. 8.26, which was originally developed in Europe, but is also prescribed for tests according to FMVSS 214 in a slightly modified variant and for many rating tests. The **SID-IIs** is a small dummy that represents a female occupant of the 5th percentile. The **WorldSID** represents the current generation of new, worldwide harmonized dummies. The WorldSID-50M, a 50th percentile dummy (male) is already used for legally prescribed tests according to GTR 14/UN-R 135 and in safety rating programs, Fig. 8.27, the female 5th percentile variant WorldSID-5F is in the late phase of development. All dummies for side collisions have in common the arrangement of the sensors so that the load of the human body can be optimally evaluated in a side impact. Up to about 100 measurement channels are available for this with the most modern dummies.

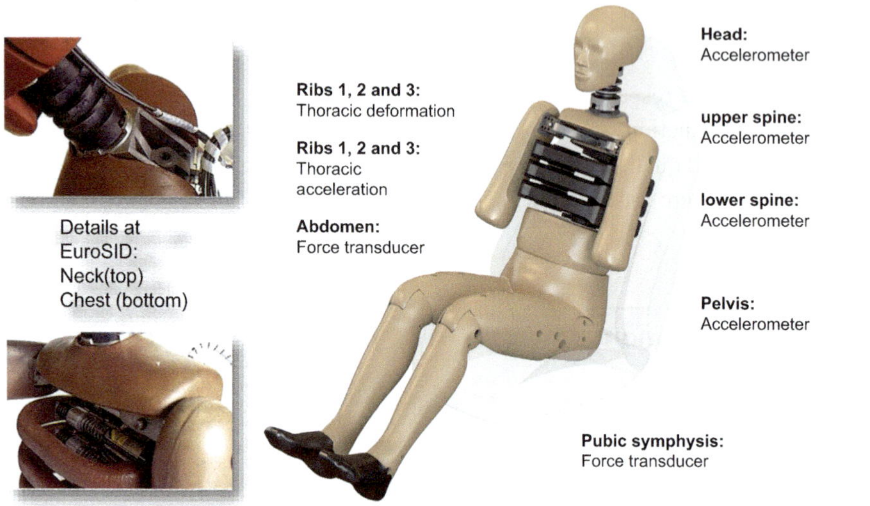

Fig. 8.26 The European dummy for side collisions (EuroSID) with details of the neck and chest (according to [23, 45])

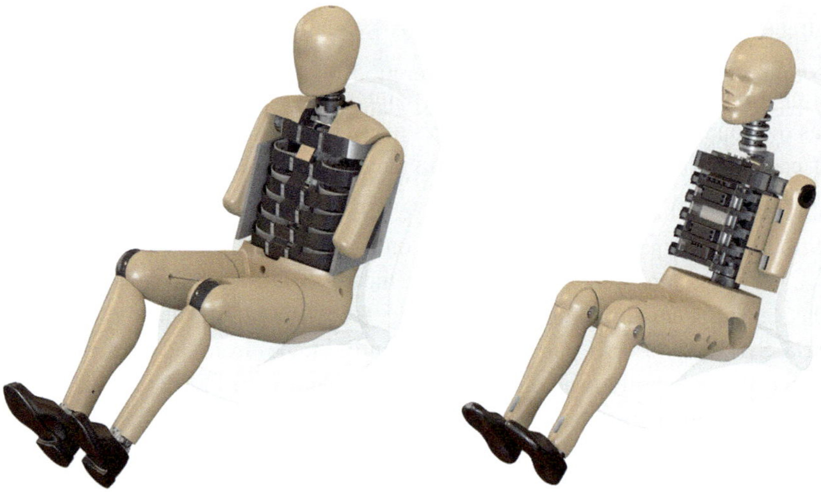

Fig. 8.27 Dummies for side impact: WorldSID-50M (left) and SID-IIs (right) [45]

Special dummies have also been developed in the past for the experimental simulation of injury risks in rear-end collisions. The so-called **BioRID-II,** the biofidelic rear impact dummy of the 2nd generation, Fig. 8.28, has become established. This dummy was specifically developed to assess the risk of cervical spine whiplash injuries in low-speed rear-end collisions and is used in the GTR 7 as well as the UN-R 17 and in safety rating tests. A female variant of this male 50th percentile dummy, the so-called EvaRID,

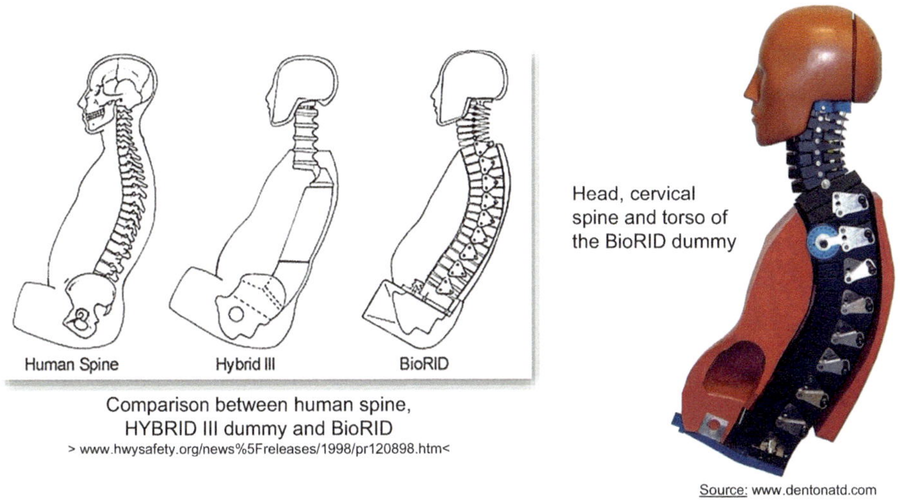

Fig. 8.28 BioRID (biofidelic rear impact dummy) in comparison and in reality [23]

was developed in Sweden around 2010 and represents a 50th percentile woman. A decision on future use has not yet been made.

In addition to the dummies that represent adult vehicle occupants, there are also special child dummies. These have been developed in the USA and Europe since the early 1970s. The most common dummies today are the child dummies of the HYBRID-III series, the CRABI dummies, the so-called P-dummies and the Q-dummies. The **HYBRID III** 3 years old, the H III 6 years old and the H III 10 years old are dummies based on the HYBRID-III series and represent a 3-year-old, a 6-year-old and a 10-year-old child. In addition to the HIII dummies, there are **CRABI** dummies (child restraint and airbag interaction) for the 6-month-old, the 12-month-old and the 18-month-old infant. The HYBRID-III and the CRABI dummies are used in US regulations (FMVSS 208, FMVSS 213) to test the effectiveness of vehicle restraint systems for children and to evaluate child seats. The **P-Series** consists of dummies in the age groups newborn, 9-month-old and 18-month-old infant as well as 3-year-old, 6-year-old and 10-year-old child. Initially, these dummies were developed by TNO in the Netherlands and prescribed in UN regulation 44. They are used both for the evaluation of child seats and for the evaluation of restraint systems in vehicles. The most modern series of child dummies is the **Q-Series**, which has been developed in various European research projects since the early 2000s. The Q-dummies represent the same age classes as the P-Series, but are characterized by a more realistic anatomy and mass distribution. In addition, plastics are predominantly used for the dummy design. Figure 8.29 shows the equipment of the Q-dummies with measuring channels. The Q-dummies are used in UN-R 129 as well as in various consumer protection programs. The Q-dummy series is complemented by the

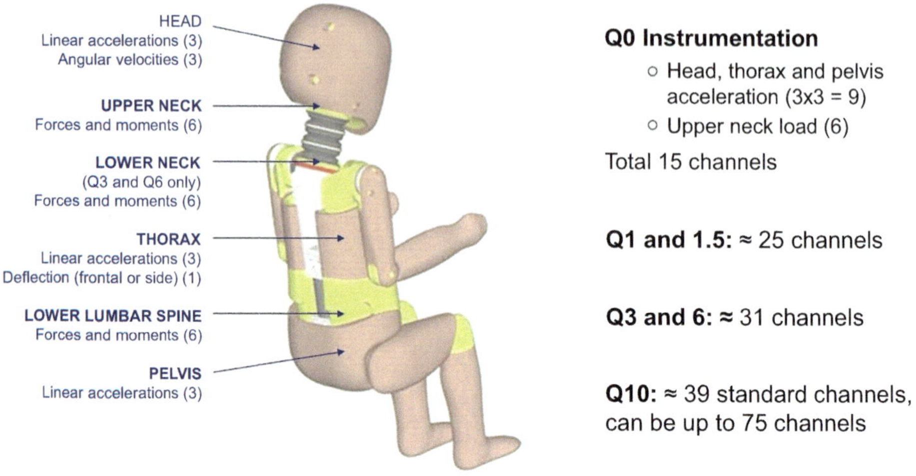

Fig. 8.29 Instrumentation of the Q-dummies with measuring channels [45]

Q3s, a Q3 dummy that has been further optimized for use in side collisions. This was prescribed by the NHTSA in 2020 for use in FMVSS 214.

8.5.4 Measurement Technology

The load quantities measured on the dummy serve as a criterion for the severity of injuries to persons expected in the event of an accident, while the quantities measured on the vehicle technically represent a measure of the accident severity. These mechanical quantities are recorded using sensors and converted into electrical signals, which can be converted, analyzed, and stored or registered in analog or digital form. The following will first discuss the general structure of measurement chains. Subsequently, principles of transducers or sensors used in safety tests will be explained and the recording and processing of measurement data will be presented.

8.5.4.1 Measuring Chain

The term measuring chain refers to the entirety of all facilities used to determine the measurement value of a measured variable [80]. The structure of such a measuring chain is essentially characterized by three subsystems: the transducer, the transmission part, and the storage or registration. The measurement task determines the transmitter properties, the type of display or registration devices, and the transmission behavior of the entire measuring chain. The following criteria are important when choosing the measuring chain:

- Measurement range, sensitivity, and error limits,
- Lower and upper limit frequency,

- Space requirements, i.e., distance between the measurement object (transmitter) and measurement periphery (storage and registration device), and
- Feedback on the measurement object due to mechanical, climatic, chemical, and electromagnetic influencing factors during measurement operation.

8.5.4.2 Transducer

Transducers convert a physical, usually non-electrical quantity (acceleration, displacement, force, etc.) into an electrical quantity; this often happens via a non-electrical intermediate stage. In the true sense, the transducer consists of a measurement variable sensor (e.g., acceleration on the measurement object and on the transducer housing—seismic mass—beam bending) and a converter that converts a mechanical quantity into an electrical measurement signal (detection of the deflection of the bending beam using strain gauges or piezoresistive elements). The relationship between the electrical output signal and the mechanical input or measurement variable is characterized by the calibration curve, which can be linear, non-linear, and/or discontinuous. In the case of a (largely) linear relationship, the calibration curve simplifies to a calibration value, which is referred to as the transducer constant.

For the **acceleration measurement**, sensors are used that have a vibration-capable suspended mass (so-called seismic mass). The mass is connected to the housing via a spring. The relative movement of the mass against the housing resulting from the acceleration or inertial force is used for measurement by attaching piezoresistive elements (formerly: strain gauges), which change their electrical resistance according to the mechanical load, to the cantilevered bending beam and electrically connecting them as a *Wheatstone* bridge. The deflection of the bending spring that occurs when acceleration is effective causes a voltage change in the bridge circuit, which is proportional to the introduced acceleration. Compared to the initially common strain gauges, piezoresistive measuring elements have only a small, negligible mechanical hysteresis and are orders of magnitude more sensitive. Quasi-static measurements ($f = 0$ Hz) can be carried out with piezo accelerometers. Figure 8.30 shows a piezoresistive measuring element (left illustration) and three orthogonally arranged accelerometers (right illustration) for measuring the three fixed acceleration components in x-, y- and z-direction, which are used in the

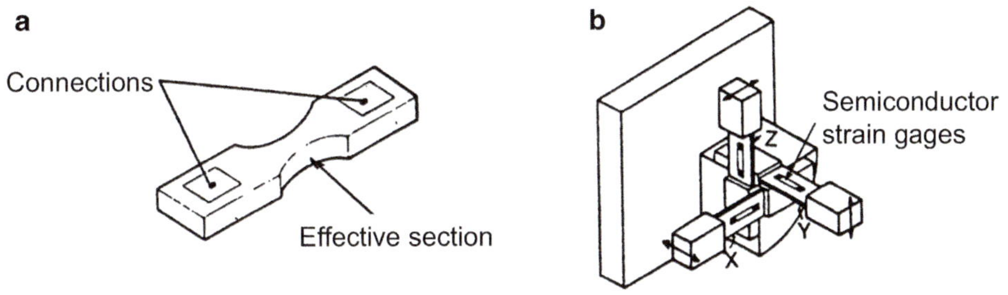

Fig. 8.30 **a** Piezoresistive measuring element made of silicon and **b** three-axial accelerometer.

dummy. Another possibility for acceleration measurement is to arrange the mass separately from the housing by piezo elements or carbon sheets, which are electrical resistances. The inertial force effective during acceleration results in a pressure change, which causes a voltage or resistance change (bridge circuit) and can be tapped off [79]. Silicone oil filled in the housing serves to dampen natural vibrations.

The **force measurement** is carried out via a displacement or strain measurement on an elastically deformable body. The strain gauges arranged in a bridge circuit, made of wire or piezoresistive measuring elements, result in resistance changes, which in turn result in a change in the bridge voltage. With piezoelectric force transducers, voltage changes occur that are proportional to the force being measured. However, force measurements are also possible using the moving coil system. The force to be measured is opposed by an electrically generated reaction force (compensation measurement). By readjusting, a force equilibrium is established. The current required for this is proportional to the reaction force and can be tapped off for measurement.

In the case of the **pressure measurement**, the principle of force measurement is applied, as it is carried out by measuring the force that occurs through the deformation of a membrane on which strain gauges are attached. Newer techniques use the application of thick-film resistors.

In the case of the **angle measurement**, slide potentiometers are usually used, with the help of which the analogy between the length of a wire or layer resistance and its resistance value is used for measurement. The tapped wire length results constructively from the relative angle to be measured.

Indentations and deformations are determined using the **distance measurement**. The change in distance is converted into electrical analog or digital signals in such a way that proportional changes in the capacity or inductance of a capacitor or a magnetic field result from the changes in distance to the object being measured. However, indentations can also be determined from the measurement of velocity or acceleration, followed by one or more integrations. In the case of chest deformation on the dummy, a rotary potentiometer is used, with the aid of a compensation circuit to take into account the angle function (see Fig. 8.21).

Integrated sensors include transducers and signal electronics directly at the measuring point to relieve the control unit. The data transmission can be done with digitized signals, making it significantly more interference-proof. If necessary, the signal electronics can consist of a signal processing unit, a digital/analog converter and/or a microcomputer.

8.5.4.3 Measurement Data Acquisition and Processing

As the description of the measuring chain focused on three subsystems (transducer, transmission part and storage or registration), this section will discuss the acquisition of measurement data and its processing at the measurement object, which ultimately leads to the analysis of the measurement results. Measurement data acquisition, as shown in Fig. 8.31, includes the generation of measurement data using the transducer, the

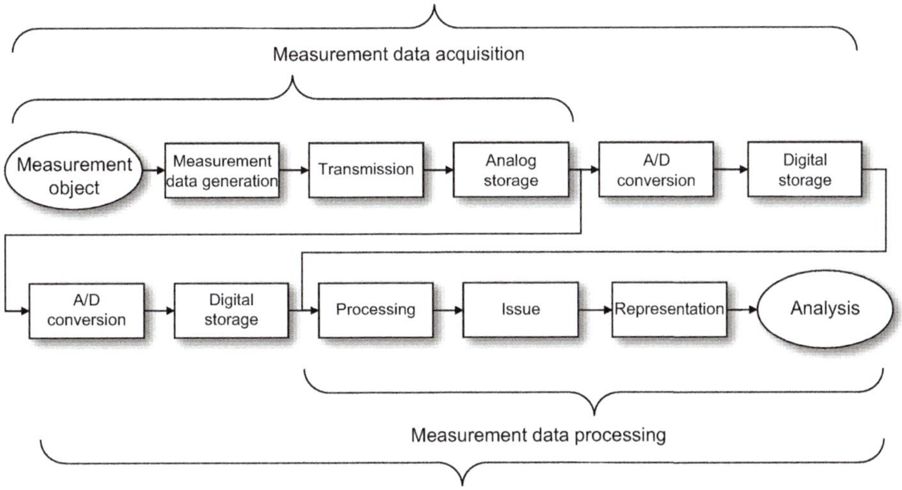

Fig. 8.31 Tasks and terms of measurement data acquisition and processing

transmission, the analog or digital storage, and the necessary conversion before digital storage. Measurement data processing generally includes the conversion of measurement data for digital storage and the storage itself, numerical processing, and the output of data in memory and/or the display of measurement data and results. The conversion and digital storage of measurement data can thus take place both in the context of acquisition and in the processing of measurement data.

In the case of the **generation or production** of measurement data, the physical quantity to be measured is converted into a technically detectable quantity, e.g., acceleration is converted into a beam deflection, and into an electrical measurement signal (detuning of the bridge voltage). The electrical measurement signal must be adapted to the expected physical quantity at the measurement object in terms of measurement range and sensitivity and, if necessary, amplified.

The **transmission** of measurement data includes commutation, modulation, analog/digital (A/D) conversion, and multiplexing of the measurement signal. Here, the **modulation**, a temporary transformation of the measurement signal, primarily serves transmission security. The reversal is referred to as demodulation. Occasionally, transformations are also referred to as modulation, in which the number of transmitted signals is increased without increasing transmission security, while dispensing with the continuity of the signals (e.g., in the case of PAM). The most important types of modulation are according to [87]:

- AM: Amplitude modulation (time and amplitude continuous),
- FM: Frequency modulation (time and amplitude continuous),
- PM: Phase modulation (time and amplitude continuous),

- PLM: Pulse-length modulation (time continuous, amplitude discrete),
- PPM: Pulse-phase modulation (time continuous, amplitude discrete),
- PFM: Pulse-frequency modulation (time continuous, amplitude discrete),
- PAM: Pulse-amplitude modulation (time discrete, amplitude continuous) and
- PCM: Pulse-code modulation (time and amplitude discrete).

Often there is a need to transmit multiple measurement signals over only one channel. The greater the distance between the measurement object and the storage system, the more relevant this becomes. It does not matter whether the transmission is via a cable or by telemetry, i.e., wirelessly. This is the multiplex transmission, for which two different methods exist:

The essential feature in the **time division multiplex method** is seen in the commutation and the decommutation. Here, all signals present at the commutator are cyclically tapped, and a transmission signal is created, consisting of a time-sequenced chain of segments from the different signals. These are later decomposed by the synchronously running decommutator, resulting in discrete representations of the originally continuous signals to be transmitted. In the **frequency division multiplex method,** a specific frequency range is assigned to each individual signal to be transmitted. Subsequently, the modulated signals are mixed and transmitted directly or after further frequency modulation. The advantages of frequency modulation, such as continuity and immunity to interference, are retained, and yet several measurement signals can be transmitted simultaneously over one channel.

Analog measurement data can be stored using a magnetic tape recording device, but this is no longer state of the art today and is therefore limited to special cases. However, analog signals can be excellently recorded with an x/y writer, a direct writer, a light beam oscillograph, or an oscilloscope. Analog data is usually converted for **storage** using the A/D converter and fed to a central control unit, while digital measurement data is transmitted directly to the control unit for storage. From the central unit, the measurement data is transmitted to a digital storage. Digital magnetic tapes, magnetic disks, magnetic disk stacks, and compact discs (CDs) are used as media. A frequently encountered digital storage is the transient recorder for high speed events, which can record the digitized dummy and vehicle measurement data directly on the vehicle or on the sled. It is shockproof and can therefore be exposed to the accelerations occurring during the crash test without damage.

Measurement values are often overlaid by unwanted electrical signals. To separate the measurement values from these interference signals, **filters** are used, which "cut off" the signal in the amplitude spectrum towards higher (low-pass filter) and occasionally towards lower frequencies (high-pass filter). For measurement signals in impact tests, a channel frequency class (CFC: channel frequency class) for the various applications is defined by specifying the filter frequency class (e.g., CFC 1000) according to [47]. It indicates a permissible range in which the amplitude spectrum of the measurement transducer must lie (Fig. 8.32).

In Table 8.5, typical applications are listed and, in addition to the frequency class (F_H), the corner and upper cut-off frequency (F_N or $2 \cdot F_H$) are indicated. The drop of

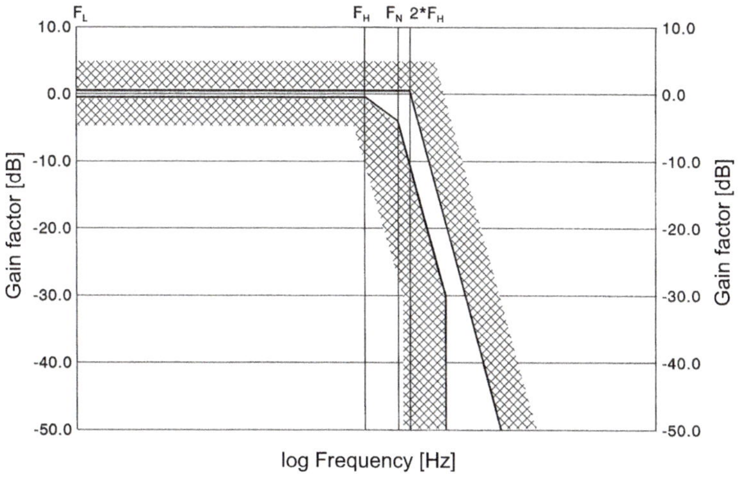

Fig. 8.32 Permissible range of the amplitude spectrum for measurement signals in safety tests (from channel frequency class in [47])

Table 8.5 Frequency classes in safety tests (according to [47])

Application	Frequency class FH [Hz]	Corner Frequency FN [Hz]	Cut-off Frequency 2 · FH [Hz]
Acceleration on vehicle structures and test facilities:			
Vehicle cell or carriage deceleration (experimental and computational)	60	100	120
Integration to determine the speed and distance	180	300	360
Structural components	600	1,000	1,200
Force measurement at the barrier	60	100	120
Forces on the seat belt system	60	100	120
Loads on the steering system	600	1,000	1,200
Test dummy:			
Head accelerations (translational and rotational) and neck forces	1,000	1,650	2,000
Accelerations of the head dome	1,000	1,650	2,000
Neck moments, chest indentation, thigh forces and moments	600	1,000	1,200
Spinal accelerations	180	300	360
Rib and sternum accelerations	1,000	1,650	2,000
Lumbar spine forces and moments	1,000	1,650	2,000
Pelvic accelerations, forces and moments	1,000	1,650	2,000

the boundary lines is −24 dB. The most commonly used filters are the *Butterworth*, the *Chebyshev*, and the *Bessel* filter. While the former is usually used for continuous signals, the *Chebyshev* filter has two charakteristics that make it unsuitable for some applications: the frequency response is not flat and the phase response starts at a very low frequency. The advantage is undoubtedly the high filter slope. *Bessel* filters are characterized by the linear phase shift depending on the frequency and the good damping properties. In digital filtering, the measurement signal is digitized and processed in such a way that the returned analog signal is attenuated at certain frequencies. As with any digitization process, the so-called alias effect, i.e., the error that occurs during digital sampling due to non-observance of the sampling theorem, must be avoided. This is achieved by preceding analog filtering (anti-aliasing) or by a sampling rate that is at least three times the value of the highest signal frequency [79].

The **numerical processing** of the measurement data, for example for the calculation of the resultant from the individual acceleration components or the HIC value, as well as the overall evaluation of the test results, such as in the form of the safety index described in Chap. 4 or a star ranking in NCAP tests [37], is usually carried out using commercially available software packages, occasionally supplemented and extended by user programs, on electronic computers. Measurement signal curves and numerical results that can be displayed on the screen, can be printed or stored in document management systems for **documentation purposes**. Initial results from crash tests are available a few minutes after the test has been carried out; however, a complete evaluation takes several hours.

In newer crash facilities, a multitude of channels, up to 600 today, are recorded and stored simultaneously. Therefore, for efficient execution of experiments, a system for **centralized timing and synchronization** is required for sensor calibration, for the warning and safety devices (locking to block crash track access, horn and warning light), for switching on the lighting device, for starting the sled or vehicle run-up, for starting up the high-speed cameras, and for data recording, storage, and processing of measurement data.

8.5.5 High-Speed Camera and Lighting Technology

For the analysis of deformation behavior of vehicle structures as well as for motion and contact point analysis in the interaction between dummy body parts and interior parts, the **high-speed film technology** is used in experimental simulation. Here, a distinction is made between stationary and sled or vehicle-mounted, traveling film cameras. For many years, primarily **high-speed video cameras** with 1,000 fps (frames per second), for special applications up to 36,000 fps, have been used because they are available immediately after the experiment and allow a simpler and faster photogrammetric evaluation. Traveling video cameras with a frame rate of usually 1,000 fps are shock-resistant, i.e., acceleration-resistant, up to accelerations of 60 g. They are mounted on vibration-stable booms on the sled or on the vehicle. The control unit and the power supply are usually also mounted on the sled or the vehicle. They are started by a central control, e.g., via trailing cables. The necessary synchronization between measurement signals and film recordings is achieved by simultaneously triggering the measurement and high-speed camera equipment. With the help of the trigger signal, flash lamps are activated and the crash-resistant timers are started. They are positioned so that they are visible in the film section.

The required **lighting system** for high-speed recording technology usually consists of two movable lighting units, each equipped with a series of metal halide lamps (HMI), mounted on both sides above the object under observation. The illuminance is up to 200,000 lx (lx) and produces a dazzlingly bright light, flicker-free up to a frame rate of 36,000 fps [16]. Such lighting systems have a lifespan of up to 500 h or 1,600 on/off cycles. An innovative lighting technology that has increasingly established itself can be seen in LED headlights, which were developed by Messring [65] under the name M = LIGHT LED, among others, and offer significant advantages in terms of lifespan, energy consumption, and light output (Fig. 8.33). So-called on-board lighting units are used to illuminate the passenger compartment, which, like the cameras, must be designed to be crash-resistant.

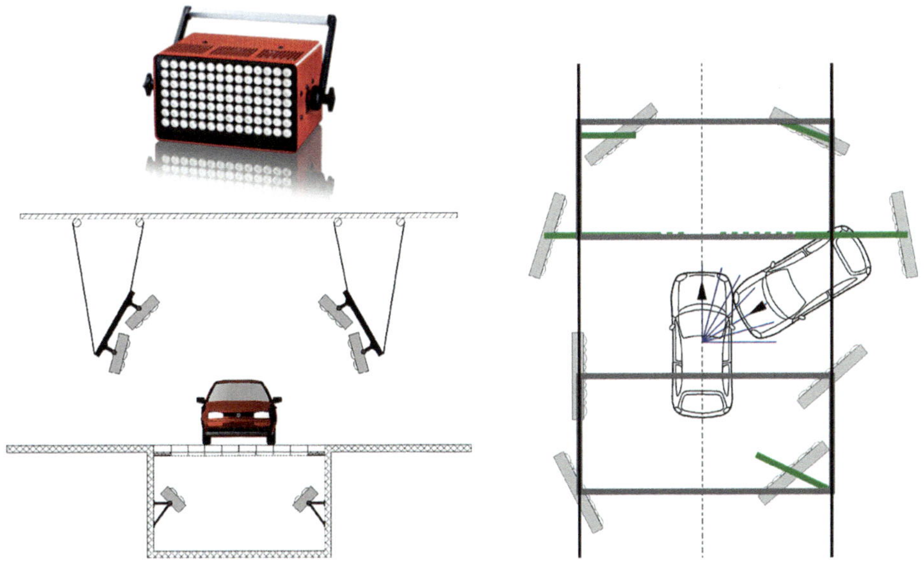

Fig. 8.33 Concept for lighting devices for frontal collisions (left) and for oblique angle collisions (right) as well as individual headlights with LED technology [65]

8.5.6 Scenario-Based Testing of Active Safety and Automated Driving

For the evaluation of **active safety**, scenario-based testing has been established. The test scenarios are derived from the analysis of accident and traffic events and thus represent statistically representative scenarios. In addition to regulatory requirements for scenarios (e.g., for automated lane keeping systems (ALKS) in UN-Regulation 157 [111]), scenario-based testing is particularly carried out by consumer protection programs (see, for example, Euro NCAP [89]).

During the development and market launch of a new motor vehicle, testing is initially carried out in computational simulation. Common simulation tools include vehicle models, environment models and driver models. These can be parameterized according to specific scenarios. Active safety functions, such as automated emergency braking, can also be well represented in the calculation. This allows a large number of scenarios with many parameter variations to be simulated.

When real vehicles are available, scenario-based testing is carried out on test tracks. For this purpose, a scenario is prepared similar to a script, in which the starting conditions are defined and adhered to. The sequence then depends on the performance of the safety system. Different elements are available for designing the scenario. These include, for example, dynamic and static objects, such as balloon cars, pedestrian dummies, traffic lights or traffic signs. To evaluate the performance, a reference system, the so-called ground truth, must record the sequence so that function-independent measurement data are available. Many scenarios are driven with steering and braking robots to ensure reproducibility and comparability.

Against the background of automated driving, the safety of highly complex sensors and algorithms can no longer be demonstrated by simple test cases and statistically representative practical tests. The approval of automated driving functions can only be justified if they promise at least a reduction in damage in the sense of a positive risk balance compared to human driving performance [24]. The scenario-based evaluation is carried out by dividing entire journeys into individual scenarios, which represent more or less extensive related traffic sequences. The scenarios are stored in scenario databases and are available similar to accident data depending on the specification and application domain (ODD, operational design domain). The advantage of scenario-based evaluation compared to field testing is that the system can be specifically evaluated for relevant traffic situations. For example, to evaluate the safety of a system, scenarios can be specifically considered that contain situations requiring collision avoidance. This is intended to structure the test effort for the safety assessment of automated driving functions and make it more efficient.

The term "scenario" is used in many ways and is interpreted differently. Scenarios must be defined as events that are, for example, maneuver-based and combined through causal chains. In the context of validating active safety functions and automated driving, different levels of abstraction have been established:

1. Functional scenarios: Functional scenarios represent an abstract level of description that is described in natural language at a conceptual level. They represent intuitively readable expert knowledge and generally do not specify specific physical values.
2. Abstract scenarios: Abstract scenarios are formalized, declarative descriptions, derived from functional scenarios.
3. Logical scenarios: Logical scenarios are described with the inclusion of parameters, with the values of some of the parameters defined as ranges.
4. Concrete scenarios: Concrete scenarios can be derived from logical scenarios and represent the most diverse form of scenarios. These are represented with explicit parameter values that describe physical attributes. This form of scenarios can be used for testing in simulation or on the test site.

The structured description and storage of scenarios requires a complete concept in which all parameters can be mapped. As part of the German PEGASUS project, the 6-layer model was introduced to structure the elements of a scenario in order to establish a universal description of relevant elements [10].

The 6-layer model (Fig. 8.34) is capable of categorizing all aspects of a scenario comprehensively, so that all relevant parameters can be incorporated into a corresponding simulation environment. Layer 1 describes the road network including its geometries and topologies. This layer includes, among other things, the description of curve radii, cross and longitudinal inclinations, as well as other factors. Furthermore, permanent objects

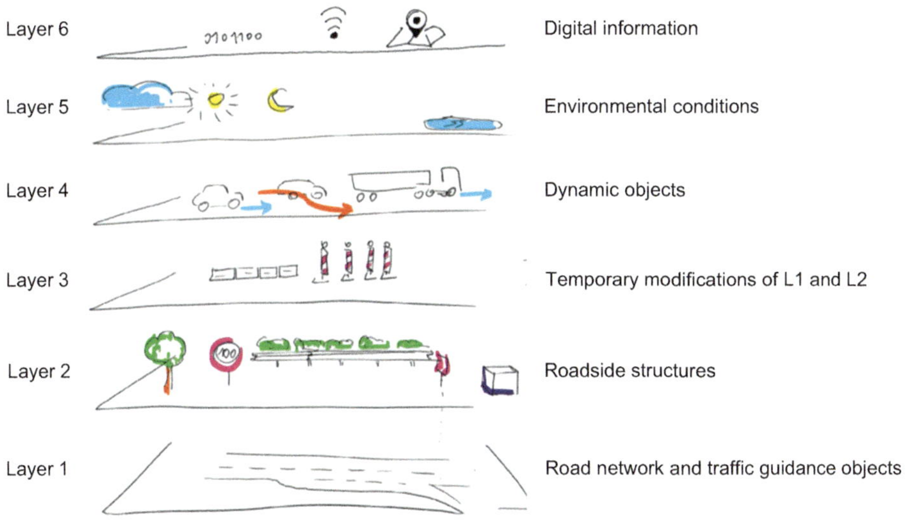

Fig. 8.34 6-layer model according to PEGASUS [88]

for traffic guidance are defined. These include markings and traffic signs according to the traffic sign catalog. The roadside environment is described on layer 2. This includes, for example, the description of safety structures (e.g., guardrails) and other static environmental objects such as trees. Layer 3 includes temporary modifications of layers 1 and 2. If necessary, bottlenecks and road expansions as well as construction sites, which are marked, for example, by cones or yellow markings on the road, are added. Layer 4 introduces the description of road users and other dynamic objects including their interactions. This layer can be filled using a basic scenario concept, as was developed, for example, in [117] for the federal highway driving domain. While the first three layers define static objects, the fourth layer includes all potentially movable objects. Layer 5 defines the environmental conditions, such as the weather and the time of day. In addition, influences of these on layers 1 to 4 are described, as for example snow can change the road condition in terms of the friction coefficient. Layer 6 includes digital communication. This includes both digital map data and corresponding localization, as well as vehicle-to-everything (V2X) communication and the status of traffic signal systems and variable traffic signs.

To test active safety systems and automated driving of different automation levels, various tools must be combined. In particular for computational simulation, there are already a large number of tool chains for individual components (sensors, algorithms, etc.), which ideally can be linked together.

The central elements are the relevant scenarios, which were previously covered by high mileage for each implementation of a new system. The storage of the relevant scenarios in a so-called scenario database makes these available in various phases of the validation process, see Fig. 8.35.

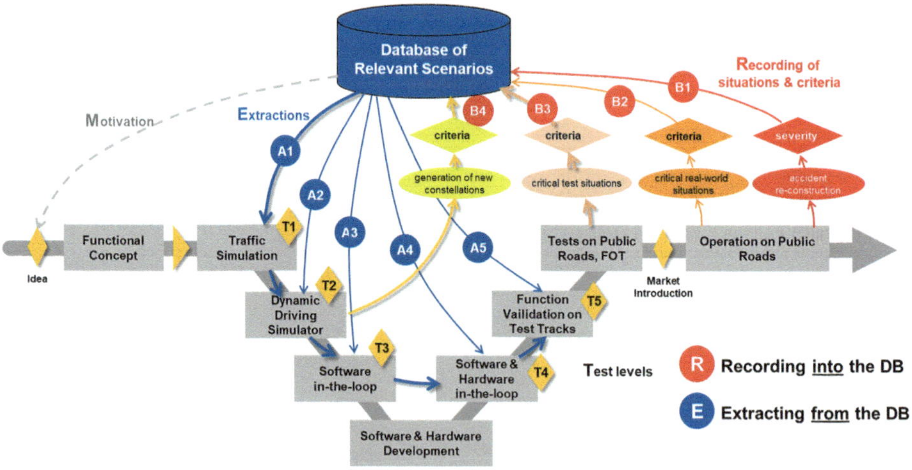

Fig. 8.35 Concept for the database of relevant scenarios according to [28]

A major advantage of such a database is the selection of scenarios with relevance for a specific function. Using the function description and the application domain (ODD, operational design domain), it is possible to play out exactly those scenarios that cover the corresponding application area. Changes in the function scope can therefore be matched with new scenarios with which the automated driving functions are confronted.

Another advantage is the possibility to consider the 'state of the art' of relevant scenarios in the validation process. While in field tests there is a certain statistical probability of encountering or missing certain relevant scenarios, the database allows the function to be tested against targeted or all scenarios that have been imported into the database so far. Of course, it is necessary to continuously fill the database with relevant scenarios to take into account changes in the traffic system that arise, for example, from the introduction of new technologies. Therefore, the proposed approach (see Fig. 8.35) allows both the extraction of scenarios from the database and the inclusion of scenarios in the database, creating the so-called "circle of relevant scenarios".

The original set of scenarios in the database is captured from real traffic through the analysis of accident data, e.g., from accident databases, and traffic scenarios that were recorded in public traffic. In addition, the results of field tests and new relevant driving scenarios from driving simulator studies are fed into the database.

Extractions or assignments (A1 to A5 in Fig. 8.35) of scenarios from the database are fed into the various development and validation testing tools during the process. This allows for a re-evaluation of the impacts of an automated driving system in a previously recorded scenario that is classified as relevant for the specific system. The effects of the automated driving function in these scenarios can be analyzed in the most suitable testing environment. Traffic simulations in various standardized software tools are capable of analyzing a large number of scenarios, supplemented by virtual variations of the traffic scenarios. With the help of a dynamic driving simulator, it is also possible to estimate whether the introduction of innovative functions results in new relevant driving scenarios that can be included in the database. Traffic scenarios are also assigned to software- or hardware-in-the-loop tests (A3 and A4) of components.

The introduction of scenario databases is taking place among vehicle manufacturers, but is also being discussed at national and international levels.

8.6 Computational Simulation

For vehicle type approval, full vehicle tests are internationally agreed and legally prescribed. Experimental simulation often reveals unexpected weaknesses in design or function. A disadvantage, however, is the fact that the test objects to be examined must first be available as prototypes or samples in order to be experimentally tested. This not only means high production costs, but also a high expenditure of time, which contradicts the ever-shortening development cycles of new vehicle types and platforms. For this reason, computational simulation has developed into a decisive and recognized development tool over the past decades. The application ranges from the concept phase to series development and is characterized by a high degree of reliability, accuracy and forecasting capability. This applies both to the static and dynamic calculation of vehicle behavior and components of the occupant protection system as well as to the simulation of the movement and load behavior of occupants.

Within crash simulation, computational methods are used in which the components to be examined are assumed to be deformable individual bodies or coupled rigid bodies. Accordingly, programs from the field of the finite element method (FEM) or programs for multi-body systems (MKS) are used. The following is a brief outline of the individual mathematical methods after a summary of the historical development of the computational method. This is followed by a description of the types of models used in the different computational disciplines. Finally, the necessary criteria for evaluating a computational simulation are discussed and the possibilities for optimizing systems using simulation are explained.

8.6.1 The History of Computational Simulation

If one uses MKS modeling, this leads to systems of ordinary differential equations. A group of algorithms for solving initial value problems of ordinary differential equations that is still relevant today was published at the beginning of the 20th century, i.e., before the development and spread of computers. The methods are named after the developers, the German mathematician and physicist Carl David Tolmé *Runge* (1856–1927), who taught as a professor in Göttingen, and the German mathematician Martin Wilhelm *Kutta* (1867–1944), who taught in Stuttgart. However, the *Runge–Kutta* methods are not yet capable of effectively solving the equations that occur in occupant safety without additional modifications. It was not until the mid-20th century that Charles *Curtiss* and Joseph *Hirschfelder* published studies on so-called stiff differential equation systems. The methods based on this for solving a stiff differential equation system, such as *Runge–Kutta–Rosenbrock* or *Runge–Kutta–Nyström,* now allow the tasks to be solved numerically efficiently.

ZF LIFETEC
PIONEERING AUTOMOTIVE SAFETY

One of the first two-dimensional occupant simulation models was established in 1963 in the USA by *McHenry* at the Cornell Aeronautical Laboratory (C.A.L., later renamed Calspan Corp.) under the name CAL-2D and continuously developed further. This led to the creation of programs such as ROS (revised occupant simulation) by *Segal* in 1971, MODROS (modified revised occupant simulation) by *Danforth* and *Randall* in 1972, and PSOS (program for simulation and optimization of seat belts) by *Niederer* in 1977. Parallel to the *Calspan*- model, *Robbins* developed the model known as MVMA-2D, which was published by him and other co-authors in 1970. A passenger crash mechanics computational model ICMF [54] developed by *Kramer* at the Technical University of Berlin in the early 1980s was intensively used in research projects, but was never commercially distributed; its origins lie in a program developed by *Anselm* in 1975. TNO in the Netherlands developed the MADYMO-2D program, the description of which was first published by *Bacchetti* and *Maltha* in 1978.

The development of three-dimensional occupant simulation models began as early as 1970 with the HSRI (Highway Safety Research Institute, University of Michigan, MI, USA) model by *Robbins*, which was modified by *King* in 1972, and has rapidly progressed in the following years. At the same time (published in 1970), a similar model was also created at the Texas Transportation Institute (TTI, USA) by *Young*. Also in 1970, *Furusho* and *Yokoya* expanded their flat, two-dimensional computational model, whose formulation was very similar to the HSRI model, in the third dimension. *Bartz* also developed a spatial model at *Calspan* in 1972. A modification of the same, in which the occupant can be described by a tree structure of up to 20 mass elements, was made by *Fleck* in 1975. Finally, in 1974, *Huston* et al. expanded their UCIN-CRASH (University of Cincinnati, OH, USA) model into a twelve-body part occupant simulation model for frontal collisions.

In Europe, essentially two spatial occupant models were developed, the generally formulated, three-dimensional multi-mass system by *Wittenburg* in 1977, which is no longer used today, and the MADYMO-3D program, which was first published by *Maltha*, *Bacchetti* and *Heijer* in 1977.

The above summary of the development history of occupant simulation programs based on the MKS approach is abbreviated and does not claim to be complete, the interested reader is referred to the detailed documentation presented in [54].

In contrast to the multi body system approach (MKS) based on rigid body interaction, the finite element method (FEM) is used in crash simulation to determine deformations up to the failure of components. The first investigations into the finite element method surprisingly took place at a time when the necessary computer architecture—and thus the computing power—was not yet available. The *Ritz* method was investigated by the Swiss mathematician and physicist Walter *Ritz* (1878–1909), who worked in Zurich and Göttingen, and the mathematician Richard *Courant* (1888–1972), who emigrated from Göttingen to New York in 1943. However, the work of the two mathematicians had no practical significance due to the lack of available computing power. The first relevant application of the FE method did not occur until 1956 at the aircraft manufacturer *Boeing* in Seattle for the calculation of swept aircraft wings. With the advent of increasingly powerful computers, the FE method underwent a rapid development in structural mechanics in the following years, which was essentially driven by applications in engineering sciences. The work of the University of California in Berkeley, shaped by R. W. *Clough,* Edward *Wilson* and R. L. *Taylor,* the University of Stuttgart with J. *Argyris* and the University of Wales in Swansea with O. C. *Zienkiewicz* had a great influence. The first publications on nonlinear finite element methods were made in 1965 and 1967 by *Argyris, Marcal* and *King*. One of the first commercial programs in the nonlinear FEM area was *Marc,* named after the founder's name *Marcal,* from 1968. From 1972, the program *Wrecker* was available, with which for the first time investigations into nonlinear, 3-dimensional as well as transient (time-dependent) problems of structural mechanics could be carried out. An important step in the development of FEM programs based on explicit time integration methods was the first version of the DYNA-2D program of the Lawrence Livermore National Laboratories (LLNL) in California in 1976. Since 1980, three commercial programs have been derived from the DYNA program, which today dominate the crash simulation market.

8.6.2 Computational Methods

In the following, a few theoretical aspects of the individual methods will be highlighted. This is done with the aim of explaining certain terms that appear in the application of a corresponding program system. For detailed descriptions of the methods, the reader is referred to the sources mentioned below.

8.6.2.1 Multi-Body Systems with a Focus on "Occupant Safety"

In multi-body systems, a system is considered that consists of rigid bodies connected by joints; it is therefore often referred to as a rigid body system. For each of these bodies, a translational and rotational movement is permitted, provided that the connections to the adjacent bodies allow this. Occupant models consist of individual body elements that can be combined into a chain of links: They thus represent a kinetic system of rigid bodies. The principles of analytical mechanics required to set up the equations of motion essentially contain energy statements, e.g., based on virtual displacements along an imagined (fictitious) movement.

With the *Lagrange*'s equations of motion, which are derived in detail in [54], a set of exactly f equations for the f independent movements of the occupant model is obtained; f denotes the number of degrees of freedom. The reaction forces that do not contribute to the movement, which, for example, have to be set up in elementary calculation methods and then eliminated again in the equations of motion, do not appear in the calculation from the outset. This is the main advantage of formulating the equations of motion according to *Lagrange*. In the mathematical sense, the f equations represent a nonlinear, inhomogeneous, and ordinary differential equation system of the second order, which resists a closed analytical treatment. However, it can be solved using a numerical approximation method, with solutions of almost any desired accuracy being achievable: Using, for example, the *Runge-Kutta-Nyström* method, the displacement and velocity coordinates are calculated approximately. For this purpose, averages are determined from a total of four preliminary function values (one at the initial time value t_i, two at half the step size $t_{i+1/2}$ and finally one at the end of the step t_{i+1}), from which the final approximation values at the end of the step, i.e., at the point ($t + \Delta t$), are calculated. This numerical integration method works without iteration and start-up calculation, but determines the approximate solution by extrapolation. In terms of its accuracy, it is a fourth-order method, i.e., the errors enter with the fifth order. This means that the error decreases very rapidly when the step size is reduced. Since the method does not indicate an exceeding of the permissible step size by itself through large differences in results (stability) and noticeably slow or no convergence, correct and automatic step sizing is of crucial importance [54].

In Fig. 8.36, the procedure for solving the differential equation system is shown. Starting from the provision of the equation system—using the contact model to determine the forces and moments—the equation in matrix form is solved for the acceleration vector. The numerical integration method provides the displacements and velocities for the next time step, from which the occupant displacement and the contact forces are determined. This results again in a differential equation system that is solved and integrated again. This loop is run through until the specified time limit (for example, t_{max} = 200 ms) is exceeded [55].

8.6.2.2 Multi-Body Systems with a Focus on "Accident Reconstruction"

In accident reconstruction, a variety of programs and methods are used that are based on the laws of classical mechanics, such as collision theory with different supplementary hypotheses, the center of mass and angular momentum theorem, the impulse and angular impulse theorem, and the law of conservation of energy, with elasto-plastic material laws also being used depending on the field of application. These software tools are crucial for forensic accident analysis to determine collision speeds [52, 14]. The reconstruction

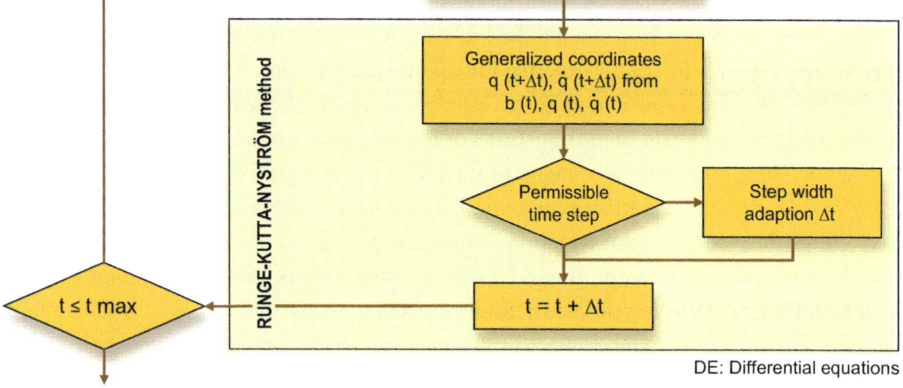

Fig. 8.36 Setup and solution of the differential equation system (from [55])

methods most commonly used until the mid-1990s are summarized under the term "backward calculation" and were applied in a graphical-manual manner. Later, they were implemented and used in calculation programs (detailed explanations in [70, 71]). In this type of reconstruction, the collision and ultimately the driving speed are deduced from the final positions of the involved vehicles under as clearly defined assumptions as possible.

In contrast, with calculation programs for "forward calculation", the sequence of motion of a vehicle/vehicle or a vehicle/obstacle collision is considered from any point in time before the start of the collision until the vehicle or vehicles come to a standstill. For this purpose, a theoretical vehicle model is used as a basis, for which the equations of motion are set up to describe the sequence of motion during the in-crash and post-crash phase.

8.6.2.3 Finite Element Method

In Sect. 8.6.2.1, systems of ordinary differential equations were introduced, which result from the motion of rigid bodies. The analysis of the motion of deformable bodies leads to second-order partial differential equations. These partial differential equations are classified into three groups: elliptic, parabolic, and hyperbolic differential equations, with the latter two describing transient problems. In crash analysis, equation solving is typically done using software applications based on the finite element method. In the literature, different equivalent approaches to the finite element method (FEM) can be found. In mathematics, the FEM is often derived from the weak formulation of the differential equation and a suitable choice of trial functions [12, 49]. In mechanics, on the other hand, the FEM is derived from the principle of virtual displacements or from the principle of minimal potential energy. Numerous textbooks on the diverse applications of the finite element method with an engineering background are available, with only the following being listed as examples for the field of structural mechanics [6, 19, 8, 44, 93].

All formulations have in common that the solution of the problem is approximated by a suitable summation of functions, each of which is non-zero only on a small area; these functions are called trial or shape functions. For the finite element method, the geometry, i.e., the body or component to be described, is represented by small (finite) contiguous elements (beam, shell, or volume elements); this is also referred to as spatial discretization or meshing of the geometry. The points defining the geometry of these elements, such as corner points for linear elements, are called nodes. The shape functions are automatically determined by the nodes, i.e., for linear elements, functions are obtained that have the value one at a node and linearly decrease to zero at the neighboring nodes of the surrounding elements. With the FEM, taking into account the material behavior and possibly further boundary conditions (imposed boundary conditions, loads, contact boundary conditions), an approximate solution is found that results from the suitable weighted summation of the shape functions defined by the nodes.

A task that leads to an elliptical differential equation, for example, is the calculation of the stress distribution in a component under static load. In dynamic, transient problems, such as crash simulation, differential equations of hyperbolic character are solved. For transient problems, the finite element method is used to describe the spatial problem, resulting in an equation that only contains time derivatives (semi-discretization). These can be solved numerically by treating the remaining time variables with a suitable time integration method. In practice, different time integration methods are used depending on the application, but all are based on the finite difference method. After time discretization, the problem is prepared to the extent that the solution can be determined with the help of a computer. A schematic representation of the procedure is given in Fig. 8.37.

An important distinguishing feature of different time discretization methods is seen in the necessity of whether large, (possibly non-) linear systems of equations need to be solved within the algorithm or not. If such a large system of equations, which originates from the required equilibrium at the end of the considered time step, needs to be solved, this is simplistically referred to as **implicit methods**, if this is not necessary, i.e. equilibrium conditions are not per se fulfilled, this is referred to as **explicit methods**. This difference is of great practical importance, as solving a large system of equations imposes specific requirements on computer architectures. If a computer system is to be used for explicit time integration methods, the most important requirement for the computer hardware is the processor speed. The requirements for the size of the working memory and the I/O (input/output) performance are significantly lower than those for computer systems optimized for implicit methods. An advantage of explicit methods is also that many non-linear phenomena, such as contact conditions between two components, can be represented algorithmically very simply using penalty methods. The numerical advantage over implicit methods, which consists in not having to solve a system of equations, is offset by the disadvantage that the time step size depends on the spatial and physical discretization, i.e. on the so-called characteristic element edge length and the material properties to be defined. The stiffer the material and the smaller the element edge length, the smaller the

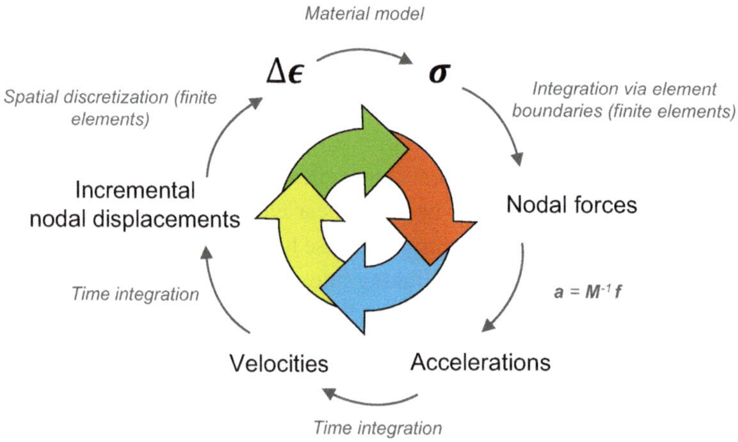

Fig. 8.37 The process of a time step of a transient, nonlinear finite element calculation. In implicit time integration, an iteration loop is provided to ensure equilibrium. In explicit time integration, the time steps are much smaller; only the energy balance is checked

time step must be chosen. Thus, only processes with a relatively short duration are usually considered for a simulation. This restriction does not apply to implicit methods. Here, the individual time step, especially for linear or weakly non-linear tasks that are expected to yield a smooth solution, can be very large. In particular, questions with complex contact situations in very short time ranges, as typically occur in impact, crash- and occupant simulation, are therefore a domain of explicit time integration methods.

The different requirements and possibilities that arise from time discretization are the reason why commercial providers of FEM software originally focused either on the implicit or the explicit time integration method. Only in the recent past has the other method been additionally implemented in the commercial program systems. This means that the advantages of both methods can now be used in one software environment. A combination of both methods for solving technical problems is discussed in Sect. 8.6.2.7.

8.6.2.4 Brief Introduction to Explicit Time Integration

As occupant simulation usually involves highly nonlinear, dynamic processes of short duration, the finite element method with explicit time discretization, also known as explicit finite element method, is commonly used. Commercial implementations are available to the user, which do not necessarily require detailed knowledge of the theory, but a basic understanding is still advisable. For further reading on crash calculation and occupant simulation, the following standard reference by Belytschko et al. [8] is particularly recommended.

Firstly, it should be noted that the basic spatial discretization, as already mentioned, is carried out with finite elements. This method is now a common and indispensable development tool in the engineering field and is used millions of times a day across all application disciplines for linear problems in structural mechanics. The extension of the equation set for transient, nonlinear problems, in discretized form for the time t is:

$$\boldsymbol{M}\,\ddot{\boldsymbol{u}} + \boldsymbol{C}\,\dot{\boldsymbol{u}} + \boldsymbol{K}\boldsymbol{u} = \boldsymbol{P} \tag{8.1}$$

Herein, $\boldsymbol{M}$ is the mass matrix, $\boldsymbol{C}$ is the damping matrix and $\boldsymbol{K}$ is the stiffness matrix already known from the static application of the FE method. $\boldsymbol{u}$ denotes the vector of node displacements, $\dot{\boldsymbol{u}}$ the vector of node velocities and $\ddot{\boldsymbol{u}}$ the vector of node accelerations in the available spatial coordinates of the corresponding finite element; $\boldsymbol{P}$ represents the vector of possible external forces.

It should also be noted that this equation must be solved for a defined time step when using either an implicit or explicit time integration method. For explicit time integration, which is advantageous in the present case of occupant or crash calculation, as high degrees of nonlinearity result from material properties due to large deformations and especially from various contact situations, the equation can be rewritten considering the approaches of a central difference method with the time step size Δt as follows.

For the velocities at time step n, the following applies:

$$\dot{\boldsymbol{u}}_n^{forw} = \dot{\boldsymbol{u}}_{n+1/2} = \frac{\boldsymbol{u}_{n+1} - \boldsymbol{u}_n}{\Delta t} \quad \text{und} \quad \dot{\boldsymbol{u}}_n^{backw} = \dot{\boldsymbol{u}}_{n-1/2} = \frac{\boldsymbol{u}_n - \boldsymbol{u}_{n-1}}{\Delta t} \tag{8.2}$$

Here, the forward difference quotient and the backward difference quotient are each denoted by $\dot{\boldsymbol{u}}_n^{forw}$ and $\dot{\boldsymbol{u}}_n^{backw}$ respectively. The central difference quotient thus results in

$$\dot{\boldsymbol{u}}_n = \frac{1}{2}\left(\dot{\boldsymbol{u}}_n^{forw} + \dot{\boldsymbol{u}}_n^{backw}\right) = \frac{\boldsymbol{u}_{n+1} - \boldsymbol{u}_{n-1}}{2\Delta t}. \tag{8.3}$$

The accelerations correspondingly apply

$$\ddot{\boldsymbol{u}}_n = \frac{1}{\Delta t}\left(\dot{\boldsymbol{u}}_n^{forw} - \dot{\boldsymbol{u}}_n^{backw}\right) = \frac{\dot{\boldsymbol{u}}_{n+1/2} - \dot{\boldsymbol{u}}_{n-1/2}}{\Delta t} = \frac{\boldsymbol{u}_{n-1} - 2\boldsymbol{u}_n + \boldsymbol{u}_{n+1}}{\Delta t^2}. \tag{8.4}$$

Thus, Eq. 8.1 can be written for the time step n

$$\boldsymbol{M}\ddot{\boldsymbol{u}}_n + \boldsymbol{C}\dot{\boldsymbol{u}}_n + \boldsymbol{K}\boldsymbol{u}_n = \boldsymbol{P}_n \tag{8.5}$$

to be

$$\boldsymbol{M}\frac{\boldsymbol{u}_{n-1} - 2\boldsymbol{u}_n + \boldsymbol{u}_{n+1}}{\Delta t^2} + \boldsymbol{C}\frac{\boldsymbol{u}_{n+1} - \boldsymbol{u}_{n-1}}{2\Delta t} + \boldsymbol{K}\boldsymbol{u}_n = \boldsymbol{P}_n \tag{8.6}$$

or, after further rearranging the terms, to

$$\left(\boldsymbol{M} + \frac{\Delta t}{2}\boldsymbol{C}\right)\boldsymbol{u}_{n+1} = (\boldsymbol{P}_n - \boldsymbol{K}\boldsymbol{u}_n)\Delta t^2 + \boldsymbol{C}\frac{\Delta t\, \boldsymbol{u}_{n-1}}{2} - \boldsymbol{M}(\boldsymbol{u}_{n-1} - 2\boldsymbol{u}_n). \tag{8.7}$$

Thus, on the left side are the unknown displacement quantities; on the right, all quantities are known. By inverting the term $\boldsymbol{M} + \frac{\Delta t}{2}\boldsymbol{C}$, the sought displacements $\boldsymbol{u}_{n+1}$ and the corresponding accelerations can therefore be determined by further substitution. It is important to note that for reasons of execution speed, the mass matrix $\boldsymbol{M}$ is typically used as a so-called concentrated mass matrix. This is only populated on the diagonal with the node masses, so inversion is trivial. Furthermore, damping may be completely omitted or assumed to be mass-proportional. Both together allow a runtime-optimized calculation of the new state variables at the time t_{n+1}.

The central finite difference procedure is second order accurate, but only conditionally stable and requires a start procedure at the beginning of the calculation, as obviously the displacements u_{n-1} for $n = 0$ need to be determined. The stability condition is known under the term Courant-Friedrichs-Levy and describes the critical time step size over the maximum eigenfrequency of the discretized system: $\Delta t_n \leq \Delta t_{crit} = 2/\omega_{max}$. The latter is defined over the smallest characteristic element edge length l_{char} (when discretizing with finite elements) in the system, as well as the material properties stiffness and density. For a discretization with i solid elements, a maximum time step for the discretized system is therefore exemplarily given by:

$$\Delta t_n \leq \Delta t_{crit} = \min_i\left(l_{char}\sqrt{\frac{\rho(1+\nu)(1-2\nu)}{E(1-\nu)}}\right) \tag{8.8}$$

Solving the further technical task now requires the description of the geometry (spatial domain over finite elements), the material properties, the boundary and initial conditions, and the external loads.

8.6.2.5 Application Example

The following is intended to illustrate, using a simple example, the steps necessary to prepare a problem for an FEM calculation. The impact of a dummy head on a steel profile using the finite element method is to be processed. The objective of this task is to calculate the resulting head acceleration or deceleration. To do this, the CAD geometry of the body or component is imported to a pre-processor and meshed with finite elements. This step divides the geometry into segments and simple volumes, the so-called elements, in which the trial functions for displacements and stresses are stored. This subdivision of the geometry is commonly referred to as meshing. In addition to volume elements, element types with structural properties are available: e.g., shell, plate, beam, and rod elements. Typically, modern finite element software offers different element formulations for each element type, with differences arising from different underlying assumptions, as discussed in classical shell theory, or from different properties of the trial functions. Those elements that are connected by common nodes are considered as a continuum. Since the solution results from the linear combination of the shape functions, which are defined over the division of the geometry, the geometry should be finely divided, i.e., finely meshed, at those locations where larger, locally occurring deformations due to loads, geometry, or material are expected. For example, in the present case, it makes sense to mesh the scalp more finely in the area of the impact point.

In addition to the geometric properties, the material properties must also be taken into account, which significantly determine the component behavior. These properties are described by material laws, i.e., the relationship between strain and the associated stress state. For the steel profile, a very simple relationship, the so-called *Hooke's Law* is to be used as an example. This is a linear-elastic material law, for whose description only the modulus of elasticity and the *poission ratio* (or lateral strain ratio) are needed. The skin of the dummy head is made of vinyl, which should be modeled with a visco-elastic material law. The user must decide which of the many material laws offered in the program he wants to use. Depending on this choice, he can use material parameters from the literature or he must carry out material tests to identify the respective material parameters based on these. The vinyl skin of the head is stretched over a hollow aluminum body, which represents the skull. The aluminum body is not supposed to deform during the planned load and can therefore be assumed to be rigid in the calculation. However, a local separation of the scalp from the aluminum skull should be possible. The assignment between the material properties and the geometry is done via the discretization of the individual elements.

To further describe the problem, the boundary conditions must be defined. In the present example, this is the fixed mounting of the steel profile at the edge of the surrounding body structure. The impact speed of the head is given as an initial condition. Finally, the contact definition with the physical contact parameters (e.g., the friction coefficient) for the contact between the scalp and the aluminum body as well as the steel profile must be made. Before the calculation can start, it must be specified which real process period should be considered for the impact problem. With this, all necessary inputs have been made, and the simulation run can be initiated. After the calculation, the post-processing takes place, in which the results are visualized. Figure 8.38 shows the model of the head and Fig. 8.39 a sequence of the head impact.

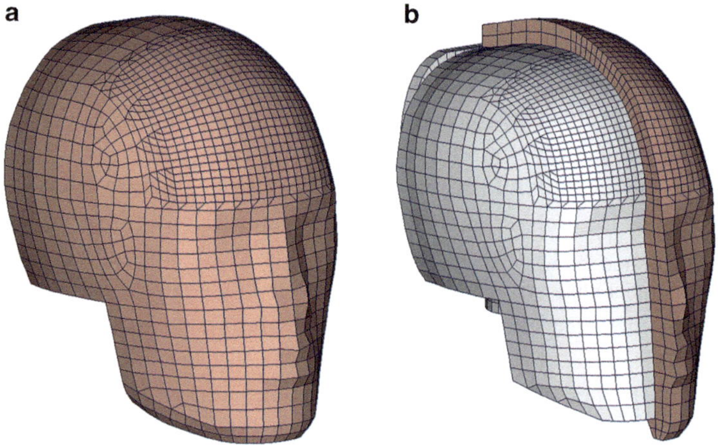

Fig. 8.38 Simple finite element model for the head impact according to FMVSS 201, **a** Scalp made of vinyl, **b** Aluminum core (model from [46])

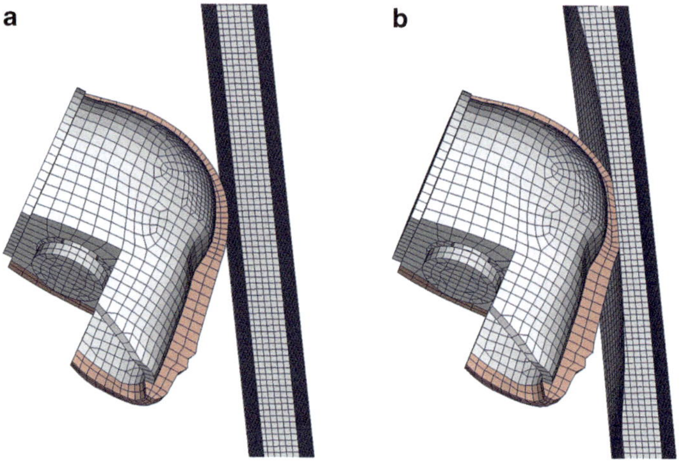

Fig. 8.39 Sequence of images during the head impact on the deformable steel profile, with the right half of the vinyl skin of the head being hidden in each case

The commercial FEM programs used in technical design for passive safety have many specific, sometimes simplified models in addition to the purely classical FEM functions. These allow, for example, complex processes in specific components to be depicted in a physically reduced form. For example, the mechanical effect of a seat belt retractor can be described in a substitute model with curve definitions as well as ignition and deceleration times. A detailed modeling of the mechanical structure of the retractor is therefore not necessary. Other such substitute models exist, for example, for spot welds (or other joining techniques in general), joints, joint stops, spring-dampers, seat belts, seat belt retractors, seat belt load limiters, seat belt pillar loops, and acceleration sensors.

8.6.2.6 Control Volume and Flow Simulation

A crucial component of a vehicle's restraint system is the airbag. The unfolding process of the airbag, which is triggered by the ignition of the gas generator, is a highly complex physical process. The pressure conditions in the airbag are strongly dependent on the temporal course of the unfolding and the resulting gas volume and vice versa. They are also influenced by the inflowing gas mass and its temperature, as well as by the gas losses at the permeable airbag fabric or at the airbag vents. Furthermore, the gas flow, which can reach local supersonic speed, is deflected and obstructed by deflector plates on the generator and airbag tethers in the airbag cushion. And finally, the exact physical description of the orthotropic, nonlinear airbag fabric is extremely difficult. Therefore, in the past, many simulations were satisfied with a greatly simplified modeling of the spatial characteristics of the airbag gas volume. This is based on a scalar balance equation that balances inflowing and outflowing masses. The method is based on a publication by *Wang* and *Nefske* [116] that appeared at the end of the 1980s. Over the years, the method has been expanded several times to model effects that occur during complex unfolding through additional parameters. Thus, for example, *jetting effects,* temperature influences or complex fabric permeabilities can now be depicted. However, all these extensions have in common that the pressure in the airbag is not resolved locally; therefore, they are also referred to as so-called control volume- or uniform pressure airbags. The latter is an indication that the pressure for airbag deployment in the numerical model is identical at all inner surfaces of the airbag at all times. Another common simplification concerns the modeling of fabric properties: During folding, but also during the unfolding process itself, many very small folds occur, for the modeling of which an extremely fine finite element mesh of membrane elements would be required. This, in turn, would lead to models with a very large number of degrees of freedom. In order to enable modeling with a manageable number of elements, the membrane elements can be simplified by being compressed without tension in the membrane plane. This allows the geometric property of materials (or membranes in general) to buckle and form folds under in-plane compressive stresses, i.e., in the membrane plane, to be simplified by modifying the material model properties.

These and other simplifications require one or even several attempts to adjust the model when creating a model of an airbag system. This inclusion of experimental results is referred to as validation of the model. For this purpose, a similar or even identical, folded airbag is ignited and the impulse upon impact on a pendulum or a ball is measured, and the unfolding kinetics are recorded with digital high-speed cameras. The model is then adjusted by sensibly adjusting various parameters in the airbag definition and the structural properties of the material model so that it delivers accelerations of the impactor comparable to the experiment and/or unfolding kinetics as in the high-speed video. Once the simulation model has been adjusted in this way, it is used for calculation in more complex environments for the design of the restraint system. Figure 8.40 shows a typical experimental setup for airbag validation. In load cases where the interaction of the occupant with the airbag takes place at a time when the airbag is already deployed, very good results can be achieved with the method described above and a relatively low computational effort.

Recent regulations, such as the safety standard FMVSS 208, and complex airbag constructions increasingly require exact simulation even in the early deployment phase. Here, the simplified models are often no longer sufficient, as they do not allow for the resolution of local flow conditions in the airbag. One difficulty arises from the very small geometric details of the generator housing, possible deflection plates for gas flow, and possibly existing supply lines. In addition, the airbag in the folded state has many small "chambers" between different layers and folds as well as extensive flat layers through which the gas must "find" its way during deployment. This requires a very precise spatial resolution of the flow phenomena. To complicate matters further, the location of the airbag folds changes spatially extremely quickly, thus presenting a very challenging technical question on the fluid-structure interaction.

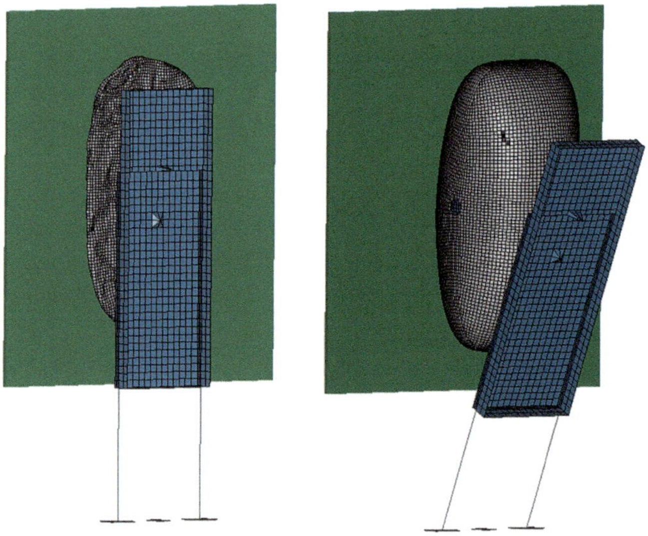

Fig. 8.40 Finite element model of a simple experiment with a stationary pendulum for the validation of an airbag model (with kind permission of Volkswagen AG [115])

To simulate the gas flowing in at supersonic speed, different approaches exist in the available commercial programs: Work is done besides the finite point-set method (FPM) with a special type of ALE coupling (ALE: arbitrary Lagrangian–Eulerian) [73] or even the coupling with existing programs for flow simulation (computational fluid dynamics, CFD). Another very common method is the corpuscular method, in which a larger number of gas molecules are combined and their interaction in the flow process is examined [74]. All these methods are capable of resolving the state of the gas locally, but still work with simplifications of gas modeling and geometric details, as an exact treatment of the complicated physical processes is not yet possible with a reasonable computational effort. Despite the many simplifications presented, it is now possible to realistically depict the deployment process of an airbag in all deployment phases and thus significantly reduce the development times for restraint systems with numerical simulation.

8.6.2.7 Coupled Solvers

If a single software program does not have all the necessary properties to solve a physical problem, the programs can possibly be coupled together. A coupling can take place once, by incorporating the result of one program into the other. However, a coupling can also occur simultaneously, i.e., at the same time, when two solvers are working on a problem. In this case, a regular exchange of information (partial results) is carried out between the various applied software programs during a calculation run. An example can be the coupling of an MKS program, which calculates the global dummy movements and delivers these via a file interface or API to an explicit FEM solver, which then calculates the deformation of the contact environment and thus the vehicle geometry. Figure 8.41 shows an MKS dummy model in an FEM-discretized vehicle environment. Here, two software programs were coupled for the calculation. A calculation program coupling can combine the advantages of several programs, but requires the respective licensing and sufficient knowledge of each program. If the programs are to be used on distributed computer systems (clusters), there are often strong restrictions for the efficient use of the hardware due to a possibly insufficiently coordinated parallelization of all involved software programs.

The simultaneous coupling of different numerical methods in a program system, on the other hand, has significantly fewer disadvantages. Figures 8.42 and 8.43 show applications of simultaneous couplings of different methods, such as *computational fluid dynamic* (CFD) or *arbitrary Lagrangean Eulerian* (ALE), each discretized with explicit finite elements in a toolchain and also widely used on distributed computing systems.

In a sequential coupling, results from one method are used as input data for another computational method. For example, implicit and explicit time integration methods can be combined for calculation tasks. An implicit FEM calculation is ideal for determining the positioning and initial position of a dummy model, as the implicit method can solve the slow adjustment process, which also exhibits only minor nonlinearities, much more efficiently than an explicit algorithm. Figure 8.44 shows a dummy model positioned by implicit time integration. Following this pre-calculation, the highly non-linear problem

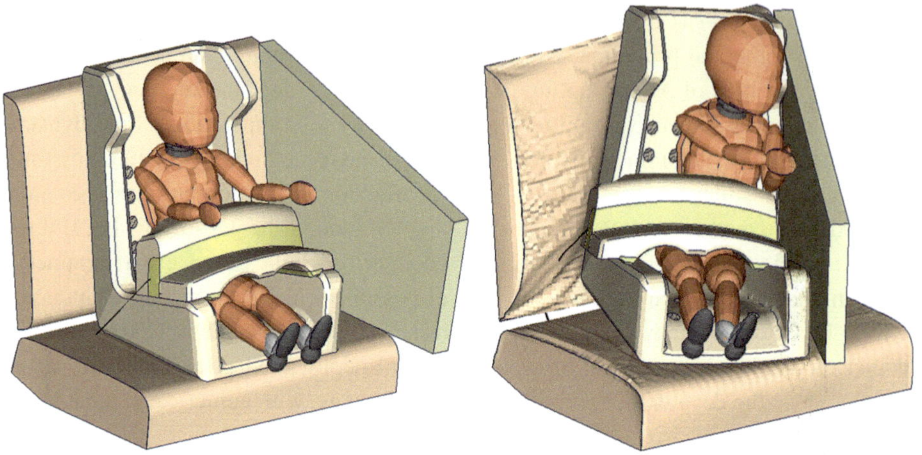

Fig. 8.41 Coupling of MADYMO and LS-DYNA for the investigation of test scenarios of a child restraint system in side impact (from [91])

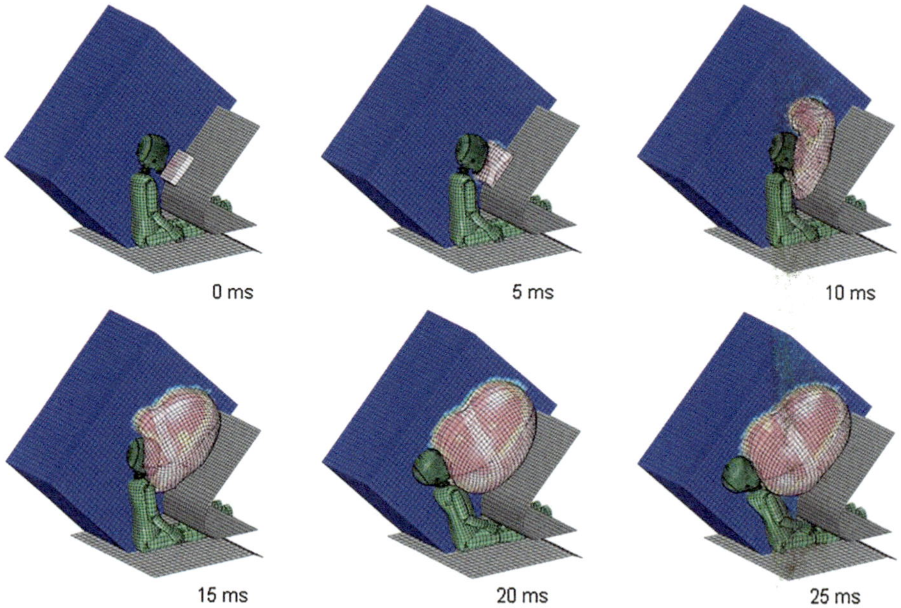

Fig. 8.42 Coupling of the finite element method with the ALE method in a program system (from [41])

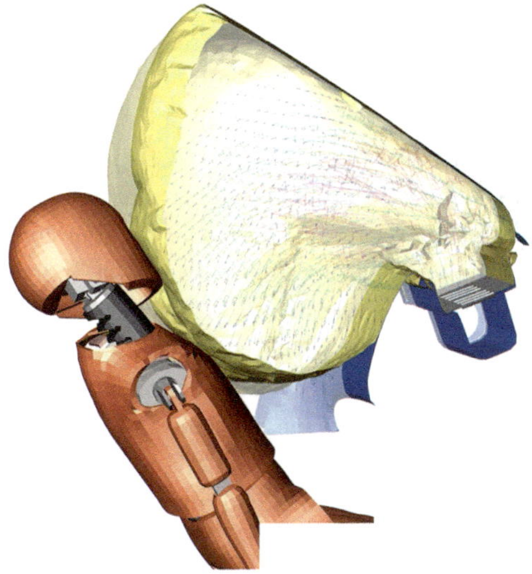

Fig. 8.43 Out-of-position-simulation with coupled flow and FEM simulation in a program system [109]

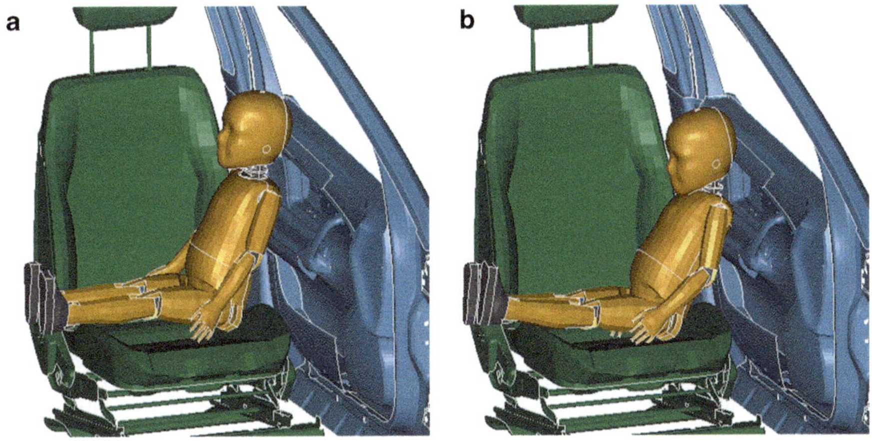

Fig. 8.44 Sequential coupling of various computational methods, **a** Dummy after positioning by a pre-processor, **b** Dummy in static rest position after implicit calculation (from [32]). Subsequently, the simulation for the coupled fluid-structure airbag simulation, similar to Fig. 8.42, can be started

of airbag deployment and interaction with the dummy can be calculated using explicit time integration. Further examples of sequentially switched calculations are used when, for example, the influences of the manufacturing process on individual components in a crash simulation are taken into account or the elastic springback of the components after a crash is to be determined.

8.6.2.8 Hardware Architectures

While calculations were carried out on large vector computers in the 1980s and 1990s, cluster systems have generally been used since the 2000s. These clusters use components such as CPUs with a multitude of computing cores, memory or graphics chips, which are also used in commercial PCs (personal computers). The FEM programs break down the simulation problem into several sub-problems for calculation, which are then solved separately on the individual computing cores in the cluster. During the calculation, the exchange of information between the individual processors takes place via the network and/or the motherboard. With cluster technology, it is possible to achieve very high computing power at an affordable price. A further significant performance boost could also be achieved for specific applications through the use of GPUs (graphics processing units). As an operating system, on cluster systems Linux is still mainly used, while the number of Windows installations on workstations has been increasing in recent years. There is also a clear trend towards central simulation services in dedicated data centers (*cloud computing*, both *in-house* administered and via *simulation-as-a-service* on third-party hardware) with smaller and mobile workstations (very powerful laptop computers) for the CAE development work of engineers.

8.6.2.9 Commercial Program Systems

The methods described above are implemented in some program systems: The most widely used representative of the MKS method for occupant protection is MADYMO [60]. As an MKS program for accident reconstruction, in addition to PC-CRASH [76], the program CARAT (computer aided reconstruction of accidents in traffic) [30] is most commonly used in Europe. For design with explicit FEM programs, ANSYS LS-DYNA [58] and ESI's Virtual Performance Solution (ESI-VPS, formerly PAM-CRASH) [75] are the most widely used. Other competitors are RADIOSS [82], as well as the explicit program part of ABAQUS [1], which has also allowed complex explicit calculations for occupant simulation for several years. All the programs mentioned above have been greatly expanded in recent years to also cover the adjacent calculation methods, especially coupled multi-physics issues in the field of short-term dynamics. For example, MADYMO was supplemented with an FEM program part or LS-DYNA was expanded with an integrated implicit, thermal and electromagnetic program part.

8.6.3 Calculation Models

The goals of computational simulation in vehicle development in the field of passive safety lie in achieving the highest possible predictive accuracy. This includes

- the modelling of real vehicle/vehicle and vehicle/obstacle collisions (contact points and motion parameters such as accelerations, deformations and deformation speeds),

- the design of deformation and contact structures as well as occupant protection systems on or in the vehicle with a focus on stiffness and energy absorption capacity,
- the prediction of loads on various body regions of occupants and vulnerable road users and their evaluation with regard to the body part-specific protection criteria, and
- the analysis of injury mechanisms on simulation models of human body parts (head, cervical spine, chest, pelvis andabdominal area as well as extremities).

The toolchains of software applications used in this process employ models with varying levels of detail. These are shown as an overview in Fig. 8.45 and will be described in the following.

8.6.3.1 Simulation Models for Accident Reconstruction

In the most commonly used calculation programs for *forward computation* for accident reconstruction, there are two possibilities for determining the collision (in-crash phase): Either the impact equations are applied in an alternative form of backward calculation, or a continuous calculation is made over the approximate impact duration (force calculation). In the force calculation, stiffness-based models in the form of contact ellipsoids [22] or dynamic meshes [39] are used to calculate the contact forces. For the analysis of the run-out movements (post-crash phase), mathematical models for the vehicles are used in the current programs. The input data set for the in-crash and post-crash phases consists of the vehicle geometry (length, width, center of gravity position, wheelbase

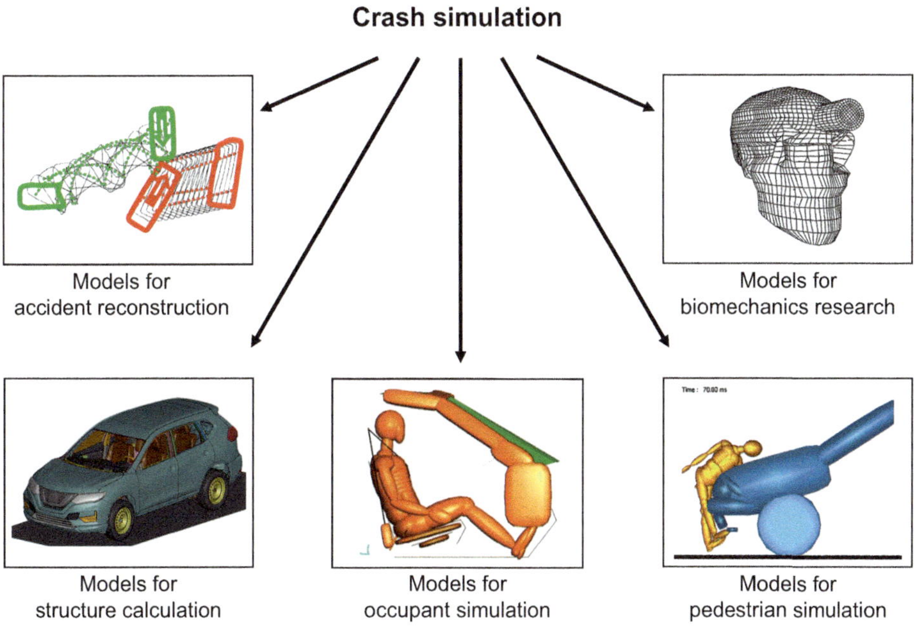

Fig. 8.45 Types of models and areas of application in crash simulation

and track width), the kinetic parameters (mass and moment of inertia, tire circumference forces, friction and rolling resistance coefficients, translational and rotational speeds) and the structural rigidities. The program output consists of the geometric motion sequence of the involved vehicles (Fig. 8.46) without or with calculated damages at the contact points, the movement of the wheel contact points and the center of gravity as locus curves, as well as diagrams with the motion quantities depending on time. The determination of the penetration depths is either done in the impact models by evaluating photos of the damaged vehicles or in the force calculation using the respective contact model in more or less accurate approximation. Based on the size of the penetration and the given rigidities, the deformation forces are determined. As an example, the illustration in Fig. 8.46 shows the collision position at first contact and the run-out movements along the secured tracks in a car/car collision with a velocity of 70 km/h for the car moving from left to right (Skoda) and with 50 km/h for the car approaching from the right (VW Golf). The example was calculated with the program CARAT [30, 40].

8.6.3.2 Computational Structural Mechanics

In computational structural mechanics, the finite element method is primarily used, for example, to investigate the static properties of car bodies or to analyze vibrations and noises. From the mid-1980s, the calculation was extended to the behavior of body structures in accident situations, with the first studies focusing on the calculation of networked assemblies such as longitudinal beams of different cross-sections and materials.

Fig. 8.47 shows the structure of a VW Polo (1986) consisting of approximately 6,000 shell and beam elements with a total of 5,100 nodes [42]. Despite the (from today's perspective) very simple model due to computational capacity reasons, the deformations showed decent agreement with the experimentally determined results (Fig. 8.48). In addition, for the first time, a computationally created balance of deformation energy in the front vehicle area could be presented, which at that time would have been possible

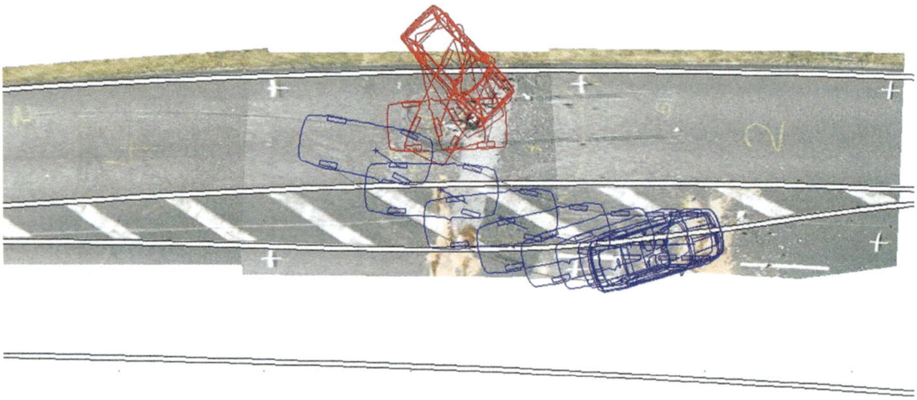

Fig. 8.46 Result of the reconstruction program CARAT including the locas curves for the wheel contact points in a car/car head-on collision with 70 and 50 km/h

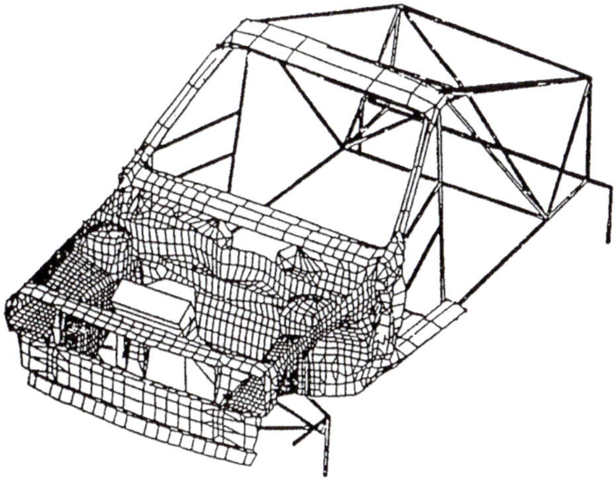

Fig. 8.47 FEM model from 1986 for calculating the front structure of a VW Polo consisting of 5,555 shell, 106 beam elements and 5,100 nodes (from [42])

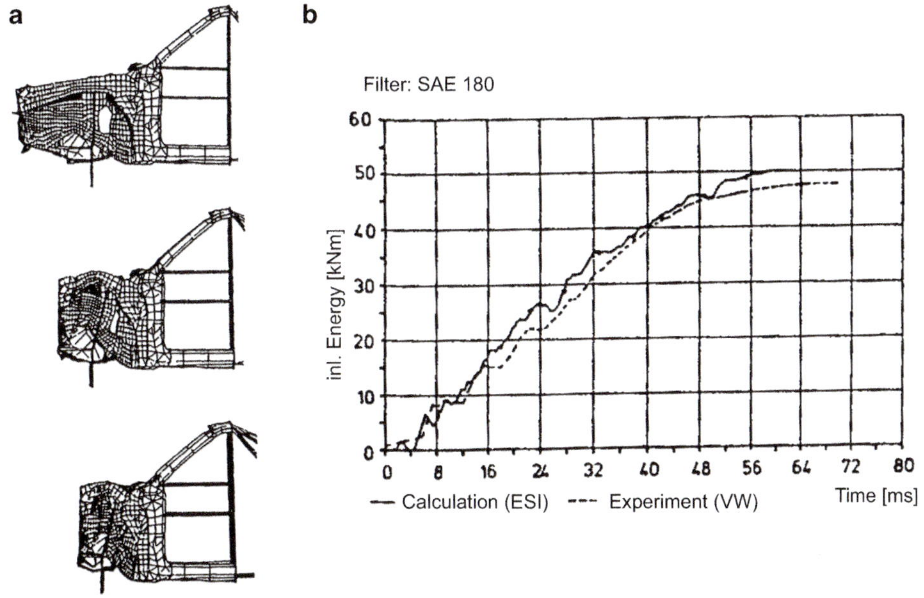

Fig. 8.48 Computationally simulated deformation behavior of the front structure at 0, 30 and 40 ms **(a)** and comparison of deformation energy between calculation and experiment **(b)** (from [4, 42])

experimentally only with a disproportionately high effort. Accordingly, the deformation energy is distributed to approximately 70 % on the longitudinal beam and floor group, and the remaining share of 30 % on the wheel arches, the front cross beam (frontend), the bulkhead, and the compartment with the rear structure; doors and flaps as well as the cladding of the wheel arches (fenders) were not modeled.

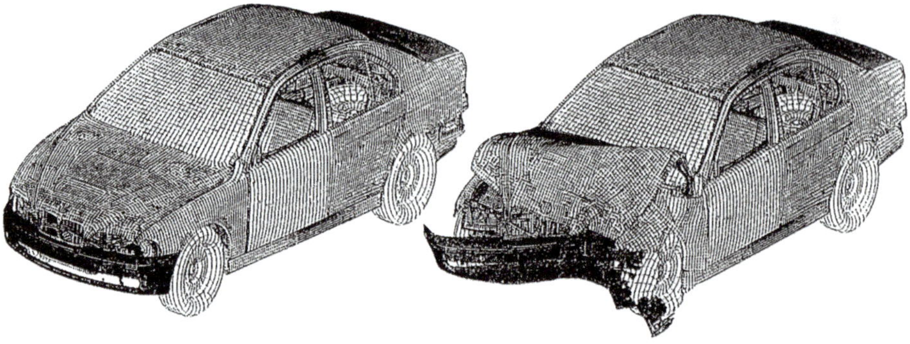

Fig. 8.49 FEM model from 1995 for calculating the structural deformation of a BMW (5 series), consisting of approximately 250,000 elements [29]

With the increase in computational capacity and CPU performance, the level of detail of the FEM models in structural mechanics computation of the entire vehicle also grew.

The model used for the VW Golf in 1987 consisted of approximately 20,000 elements, the model of the VW Passat (1990) only three years later already consisted of 30,000 and the benchmark model from BMW (1995) for calculating a 40% offset crash even consisted of approximately 61,000 elements [85, 59]. In the year 2000, models often still had a level of detail of under 500,000 elements. The representations in Fig. 8.49 show a FEM model of an older BMW vehicle, which comprises about 250,000 elements for structural mechanics computation in a 40% offset crash at 64 km/h. Figure 8.50 from the year 2005 shows a model of the Opel Astra, which has about 2.5 million elements. In 2012, vehicle models had a size of about 3–8 million elements, ten years later, in 2022, the model size is regularly at 15 to 20 million elements, with some vehicle manufacturers also building and calculating much finer and thus larger models for detailed investigations. The development towards ever finer models is inevitable due to the high demands on forecast quality and reliability of the calculations. Compared to earlier years, today's simulations contain much more detailed models for describing material behavior and material failure or for connection technology, such as for spot welds, rivet and screw connections or for adhesive connections. These highly detailed material models typically require a sufficiently fine mesh. In addition, for many load cases today, occupant simulation and structural simulation are carried out simultaneously with one model. This means that in addition to the model of the vehicle and the barrier, a detailed model of the vehicle interior, the seats, the restraint systems and possibly several dummy models are included in a calculation.

Depending on the component and its function in the vehicle, the requirements for describing the material behavior are very high, so that a realistic description of the material behavior using complex material models requires a large number of parameters that must be calibrated. A rough distinction is made between elastic-plastic (steel and

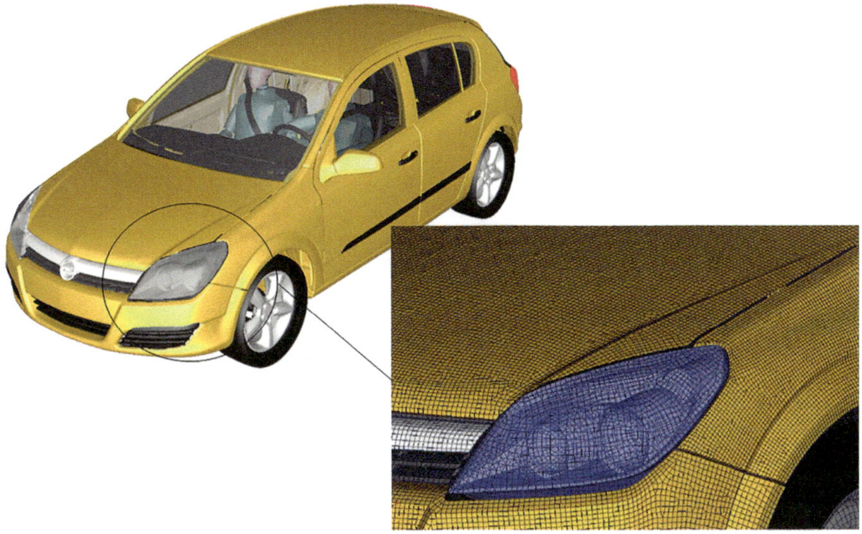

Fig. 8.50 FEM model for calculating the crash safety of an Opel Astra, consisting of approximately 2.5 million elements from the year 2005 [34]

lightweight alloys), visco-elastic or visco-plastic (plastics, reversible foams) and hyper-elastic (rubber) behavior. For parameter identification of these models, not only static but also dynamic characteristics are needed. For all crash-relevant or injury-relevant components, it is now indispensable to correctly represent the load path in the model, which necessitates the use of progressive damage and failure models. The requirements for the predictive quality of these material models are also very high with different discretizations (i.e., element edge length and element type) and require very extensive material and component tests. Today, only a few freely accessible or commercially available material databases exist for use in crash calculation. The knowledge of how to model the phenomena occurring with the material models available in the crash programs used has been developed in individual companies over many years and is rarely openly shared with third parties. However, in recent years, the sharing of so-called calibrated material cards with development partners in encrypted form has become established. For structures that are repeatedly used in crash tests and prescribed in hardware by homologation authorities, such as barriers, there are validated models that can be accessed if needed. Here too, only a few models are freely available. The currently most extensive library is stored at George Mason University (VA, USA) and can be downloaded from the internet [17]. Figure 8.51 shows a freely available vehicle model. The deformation of a commercial barrier model for a validation load case is shown in Fig. 8.52 [5].

Fig. 8.51 Freely available FEM model of a vehicle in isometric and cut view. The model consists of more than 3.2 million deformable elements and 1,280 components. It thus ranks at the lower end of the currently commercially used vehicle models for crash calculation [17]

8.6.3.3 Occupant Simulation with Dummy Models

The most frequently studied load cases are simulations of the occupant in frontal and side collisions. The scenarios examined mainly relate to load cases that are required by law for approval (e.g., according to UN-R and FMVSS), those developed for vehicle comparison by consumer rating associations (e.g., Euro NCAP), and those used by insurance companies for insurance classification (e.g., IIHS: Insurance Institute for Highway Safety). In addition, manufacturer-internal load cases are often examined. Accordingly, the objectives and thus the number of different calculations are diverse.

Various models of dummies and barriers are available on the market, developed by software manufacturers or dummy manufacturers, and can be used with the vehicle-specific models of occupant- or pedestrian protection. For the validation of the dummy models, the manufacturers make a great effort. This involves static and dynamic tests to model the material properties as well as the behavior of dummy components and the complete dummy model in a simplified vehicle environment. This ensures that the dummy model used delivers results that match the measurements obtained from the real test dummy. Since the development of the models is very complex, in Germany, the Research Association for Automotive Technology e. V. (FAT) and the Partnership for Dummy Technology and Biomechanics (PDB) have been promoting a manufacturer-independent development of the models through appropriate industry working groups for many years. In these working groups, representatives of the car manufacturers and many suppliers work together to ensure a model development according to their ideas; details on the development are summarized in [33]. In addition, the development of models is being advanced in European or internationally funded projects; a selection of dummy models is shown in Figs. 8.53, 8.54, 8.55, 8.56.

Dummy models are used either in whole vehicle models or in submodels; in the latter, the model boundaries are guided by corresponding displacement boundary conditions. Since in frontal collisions the movement of the occupant has a negligible influence on the vehicle deformation, this submodel technique can be applied very well: The results

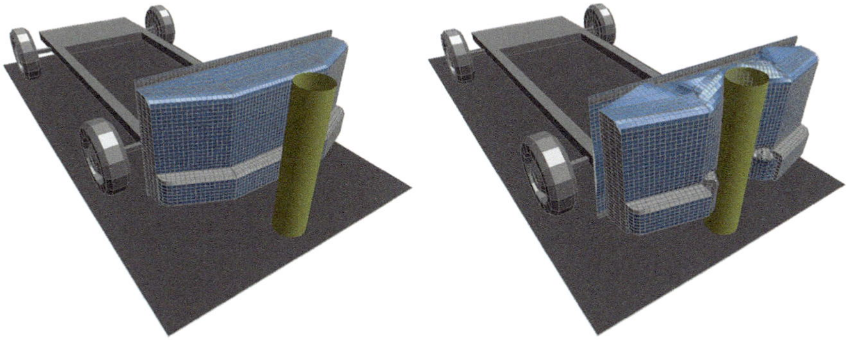

Fig. 8.52 Finite element model of the barrier of the Insurance Institutes for Highway Safety (IIHS) for a validation load case [5]

Fig. 8.53 Faceted MKS and FEM model of the HYBRID III 50% dummy [109]

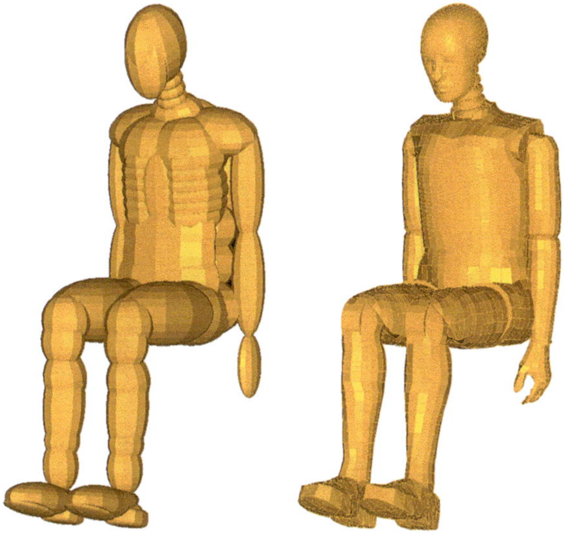

of a whole vehicle calculation are used to impose them as boundary conditions on a more finely discretized submodel. This can consist, for example, of the dummy, the seat, the steering wheel, the steering column, the instrument panel (I-panel), the restraint system, and all other parts with which the dummy can come into contact. The information about the movement of the seat rail, the I-panel, and the seat belt attachment points is taken from the previous calculation of the whole vehicle. In the submodel, for example, the influence of airbag variants or the different ignition times of the airbag, the seat belt pretensioner, or the retractor can be examined with greatly reduced computational effort. The level of detail of such submodels is adapted to the respective requirements and can therefore vary considerably. In investigations with submodels, MKS models and MKS

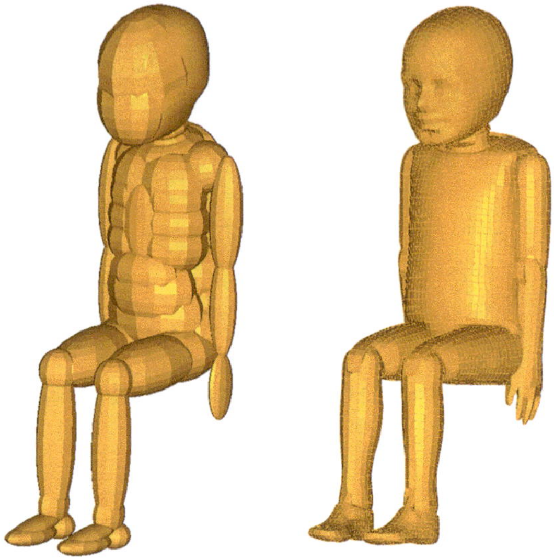

Fig. 8.54 MKS and FEM model of the 3-year-old HYBRID III dummy [109]

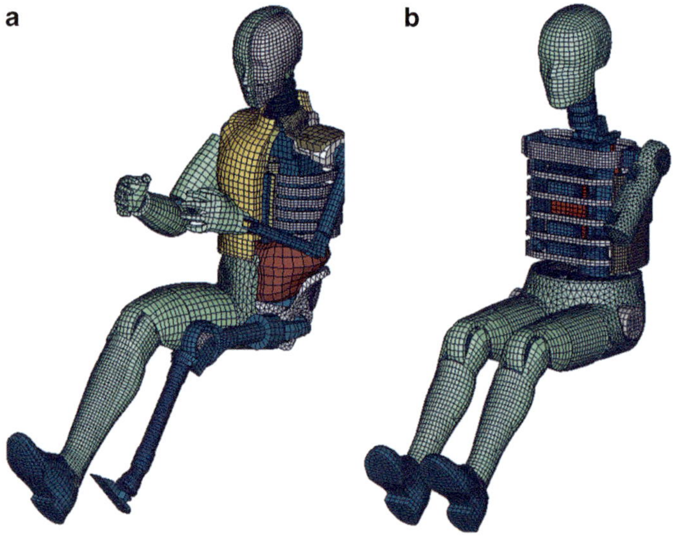

Fig. 8.55 (**a**) FEM models of the HYBRID III 50th dummy and (**b**) of the side impact dummy SID IIs (from Humanetics Inc. [46])

solvers are also used, as this allows a very high number of variants to be examined with even less computational effort.

For many years, an important research topic was, for example, the distortion of the cervical spine (HWS) in rear-end collisions. Here, submodels were also used: The dummy loads are created by interaction between the dummy and the seating system, at

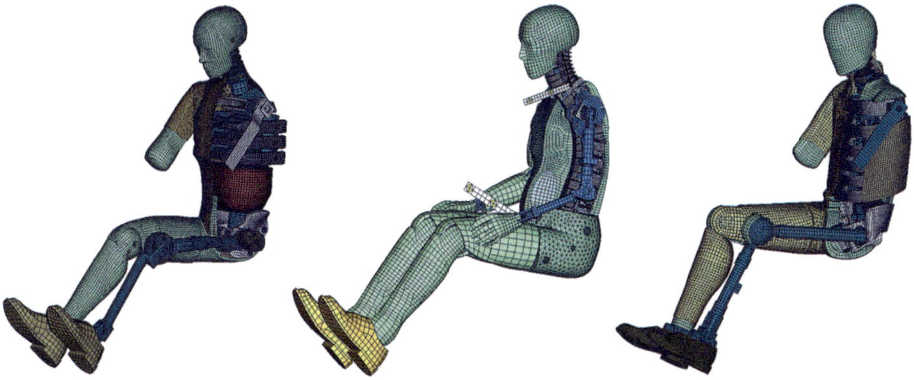

Fig. 8.56 FEM models of the dummy ES-2, BioRID-2 and the WorldSID 50% with partially hidden components (from DYNAmore GmbH [20])

whose attachment points the vehicle acceleration, i.e., the crash pulse, acts. The influence of the dummy's own movement on the crash pulse at the seat base is of only minor importance and can therefore be neglected. Thus, the crash pulse, which is essentially shaped by the deformation behavior of the rear structure, can be determined once with the structural mechanics computation on the vehicle during the rear impact and then finds its way as simulation input into the submodel, with which the effects of changes to the seat construction on the dummy load can now be examined in isolation. This approach is also suitable, for example, for investigations of out-of-position (OoP) situations, in which the effect of the deploying airbag on the dummy is analyzed. In Figs. 8.42, 8.43 and 8.44, child dummies are shown in such typical OoP seating positions.

Not all aspects of occupant protection can be investigated using the submodel method. For example, deformations on the door during side impact are also influenced by the occupant, and it is extremely difficult to determine a suitable model for this. Figure 8.57 shows in section the deformations of the dummy model of the US-SID compared to the model of the ES-2re, a further development of the European EuroSID dummy, at certain times during the simulation of a lateral collision.

The main disadvantage of submodels is the potentially considerable effort required to update them in the event of design changes, which must be incorporated into both the main model and the submodels. If the vehicle with occupant protection system and dummy is considered in a complete model, this is referred to as "integrated simulation"; an example of this is shown in Fig. 8.58.

In structural mechanics computation, questions about modeling the correct initial state, such as pre-damage from the manufacturing of components, are frequently raised. These discussions also naturally occur when using dummies, which can, for example, exhibit pre-stresses [32]. The causes of pre-stresses are diverse and arise from manufacturing and joining processes and from the positioning of different components relative to

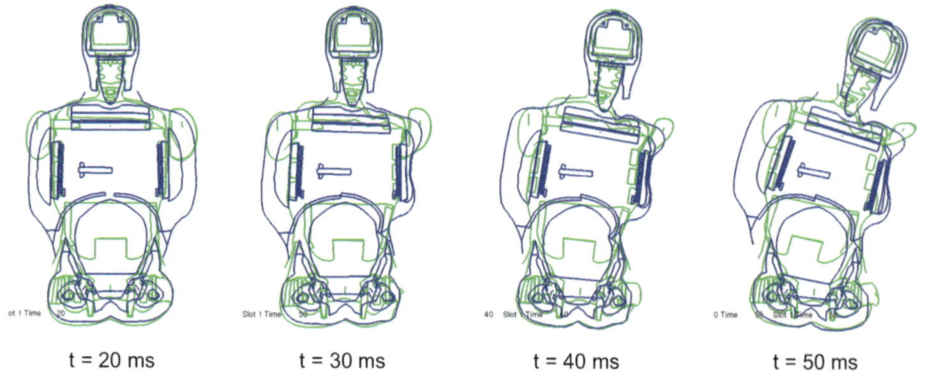

Fig. 8.57 Analysis of dummy models in lateral collision: cross-section through US-SID *(blue)* and ES-2re (green) [92]

Fig. 8.58 Integrated simulation at Mercedes-Benz AG [21]

each other. While these stresses adjust themselves in the experiment, the calculation usually starts with a stress-free state of all components. If necessary, however, the pre-stress can be calculated by a pre-simulation, and the results are transferred to the model at the start of the actual crash calculation. For example, in a rear-end crash, the pre-stress in the seat and in the dummy spine has a significant influence on the dummy load values of the BioRID-II dummy and cannot be neglected.

8.6.3.4 Simulation of Pedestrian and Two-Wheeler Collisions

With the help of pedestrian and two-wheeler models, collisions between cars and pedestrians or two-wheelers (with so-called riders) are described. The objective of such simulations is to determine mechanical load values of the modeled vulnerable road user and comparing them with the protection criteria. This can provide indications for improvements in passive safety through kinematically more favorable movement sequences and/or with the help of suitable measures, such as the constructive increase in the energy absorption capacity of contact structures and protection systems. By varying geometric conditions and kinematic input data, the influence of certain external accident parameters, the vehicle contour, and the structural properties on the load of the human during the accident can be determined. Furthermore, the simulation can be used to design and optimize a sensor system in the vehicle, for example for activating an transient pedestrian protection system, like an active hood or a windshield airbag.

Models for simulating pedestrian accidents are multi-body systems with up to 15 body elements and 32 degrees of freedom, which in earlier years were two-dimensional, but are now typically designed in three dimensions. Figure 8.59 shows a two-dimensional simulation model from the early days of pedestrian simulation to determine the influence of shape-forming parameters of the vehicle front [36]. The validation of the pedestrian models is based on tests with both dummies and cadavers. For example, the motion sequences in Fig. 8.60 show a good correlation between a two-dimensional MKS pedestrian model and a male cadaver of the same body size and mass [35]. For some time now, FEM models have also been available for pedestrian simulation. Figure 8.61 shows the impact of an MKS pedestrian model with an FEM submodel of the vehicle front end.

The main objective is to analyze the kinematics of pedestrians depending on external accident parameters (collision speed, contour of the car's front area) and to determine the contact points or regions as well as the impact speed of various body parts (lower leg,

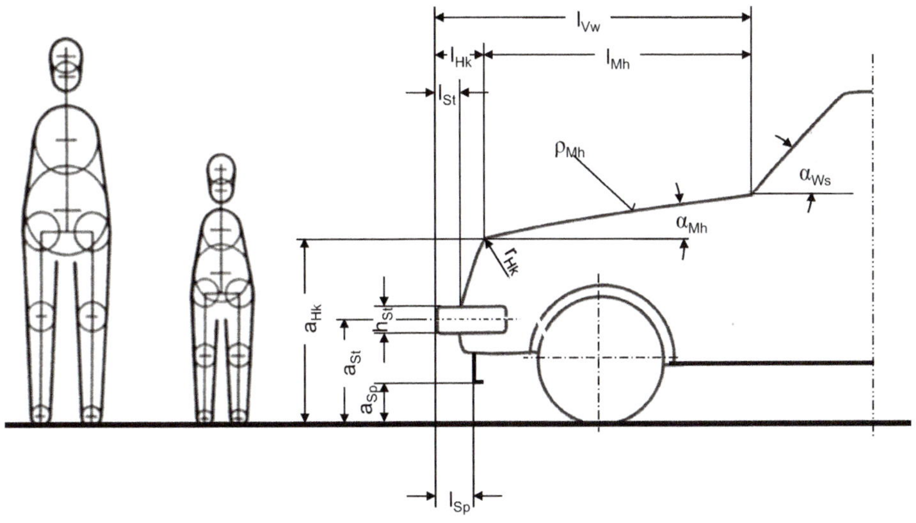

Fig. 8.59 Two-dimensional pedestrian models and variation parameters for describing the car exterior shape (acc. to [36])

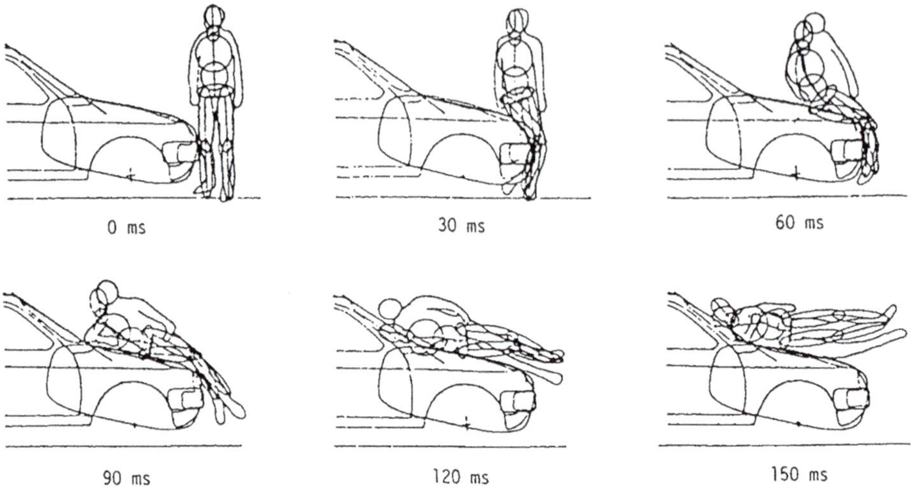

Fig. 8.60 Comparison of the motion sequence of a pedestrian simulation with that of a cadaver test (from [35])

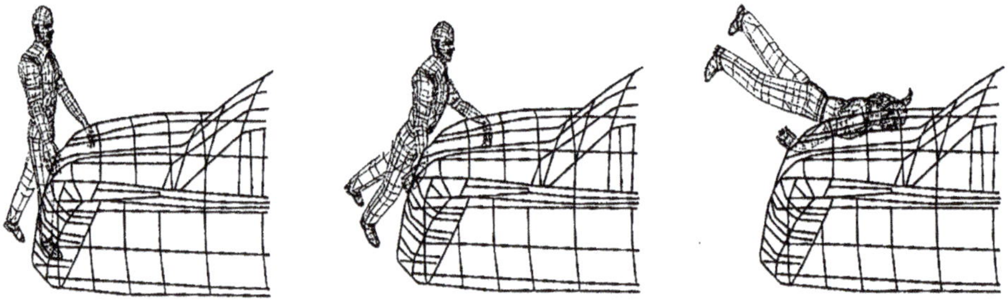

Fig. 8.61 Motion sequence of a three-dimensional MKS pedestrian model when impacting the FEM structure of a car front end (from [78])

pelvis, shoulder region, head) when in contact with the corresponding vehicle structures. For such parameter studies, pedestrian models are used for both adults and children (Fig. 8.59). Subsequently, critical contact points on the vehicle can be specifically defused by shaping, stiffness adjustment, and improvement of energy absorption. In Fig. 8.61, three phases of a car/pedestrian collision with a three-dimensional MKS model for early conceptualization of a pedestrian-friendly vehicle front-end, which is designed as an FEM model, are shown. Figure 8.62 shows the finite element pedestrian model *THUMS* (*total human model for safety,* [84]), as well as the recently newly developed human body model (HBM) *HANS*.

For the computational simulation of pedestrian protection tests using head, hip and leg components, according to the relevant regulations (UN-R 127 and Euro NCAP),

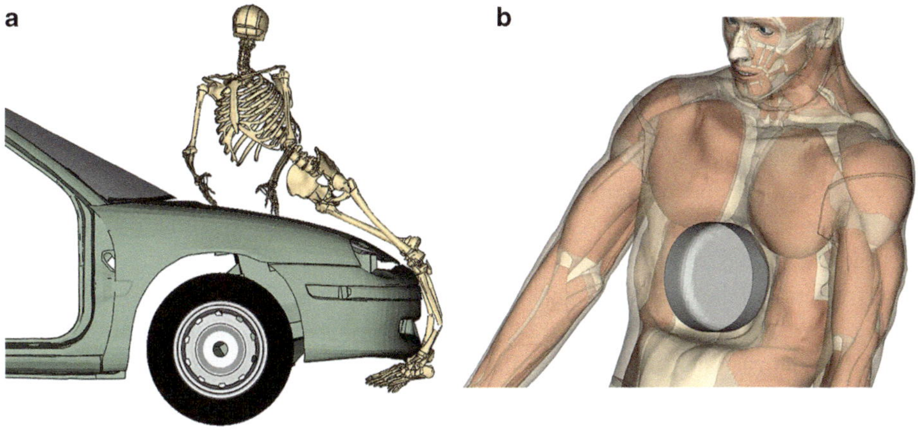

Fig. 8.62 Motion sequence of the FEM pedestrian model THUMS [84] upon impact on a car front-end **(b)**; as well as validation calculation of the newly developed human model *HANS* with circular impactor **(a)** (see [121])

validated models for the used impactors naturally also exist. In Fig. 8.63, the models from different software manufacturers are shown.

The simulation of motorcycle collisions against cars is much more complex than that of pedestrians, as on the one hand the motion sequence is more complicated and on the other hand the interaction of the motorcyclist not only takes place with the car structure, but also contact with the motorcycle itself occurs. Models of motorcyclists are usually multi-body systems, which consist of 15 and more body elements. The motorcycle is simplified as a rigid disc in two-dimensional models and usually described as a chain of

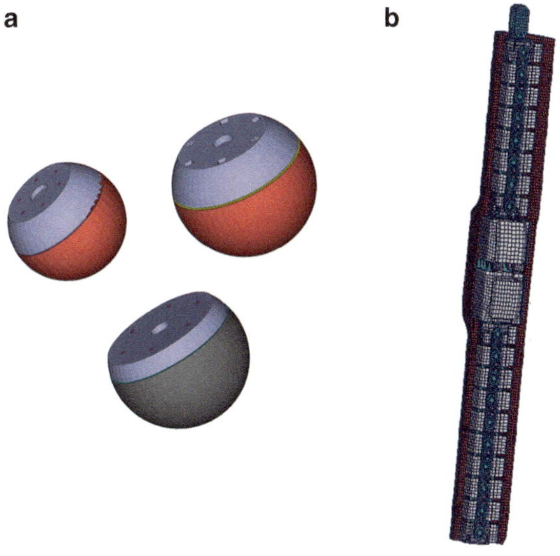

Fig. 8.63 FEM models for pedestrian protection; **(a)** three head models according to ECE [31], **(b)** FlexPLI-GTR leg model [26]

links, consisting of a total of six elements (main frame, fork, front wheel and suspension, rear wheel and suspension), in three-dimensional models. Initial studies with simulation models aimed to compare the kinematics of the modeled driver with actual accidents and with replicated tests [108]. Further analyses were carried out, for example, with the objective of improving the motion sequence when separating between driver and motorcycle through support measures on the motorcycle. In addition, for reasons of energy management, the deformation behavior of the front motorcycle structure, the motorcycle airbag [118, 56] as well as the energy absorption capacity especially of the lateral contact structures on the car (A-, B- and C-pillars as well as the roof frame) were examined.

Figure 8.64 shows the sequence of motion at certain points in time during a simulated, two-dimensional motorcycle collision against the side of a stationary car, as well as the trajectory of the pelvic center of gravity in comparison between experiment and calculation. In the case of an oblique collision, however, a two-dimensional model is not sufficient for realistic simulation. The complexity of the motion sequence of both the

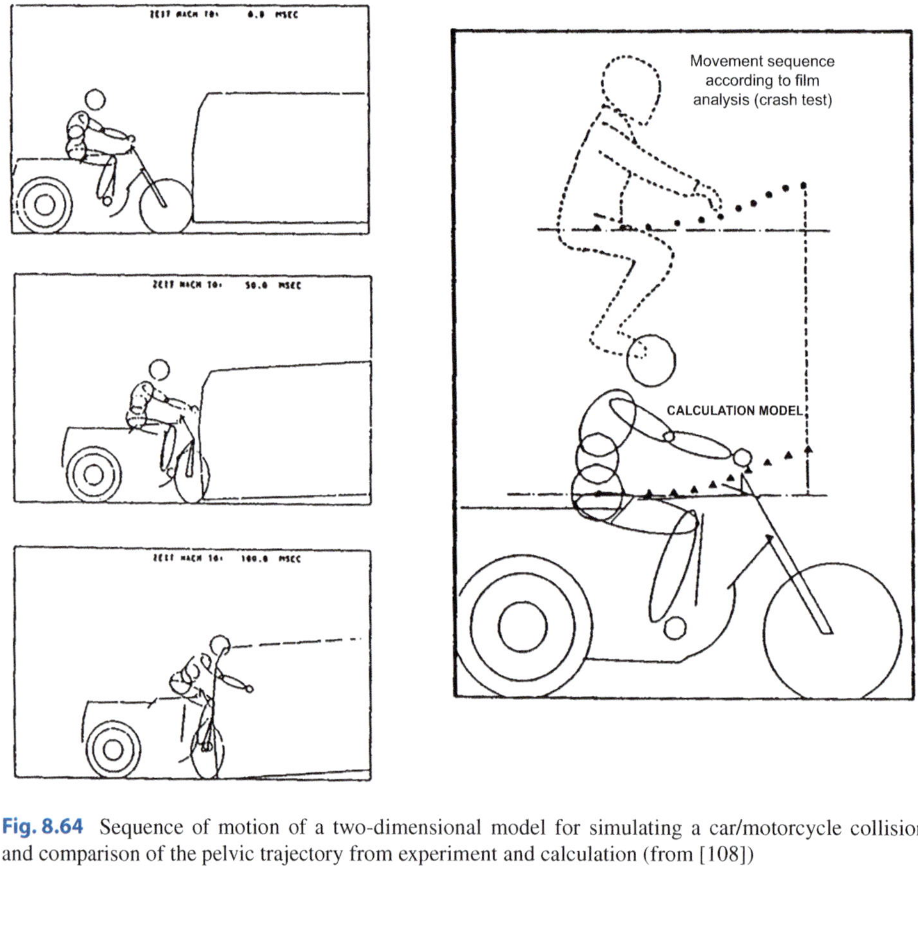

Fig. 8.64 Sequence of motion of a two-dimensional model for simulating a car/motorcycle collision and comparison of the pelvic trajectory from experiment and calculation (from [108])

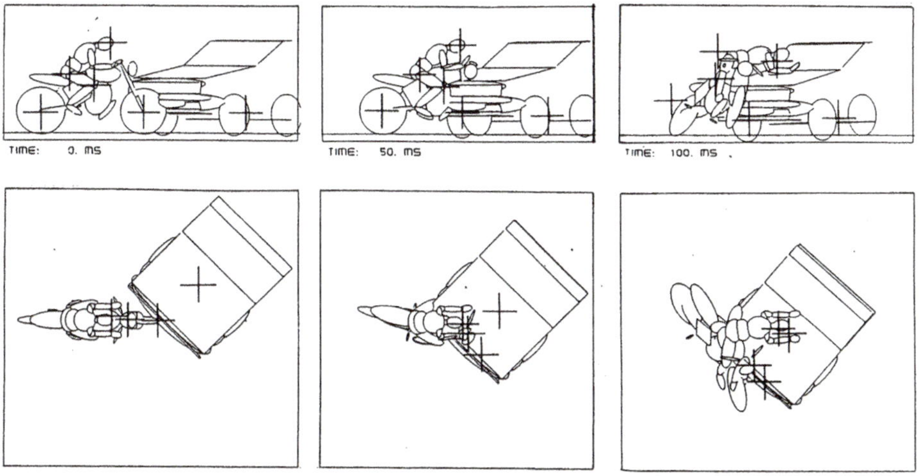

Fig. 8.65 Sequence of motion using a three-dimensional model to simulate a 45° car/motorcycle collision (shown from the front and from above); in each case $v = 32$ km/h (from [118])

motorcycle and the driver can be understood from the computational simulation of a 45° collision against a car front shown in Fig. 8.65. Recent research takes into account possible impact scenarios for two-wheelers in complex transient explicit finite element simulations, which also include the deployment of possible airbags of the two-wheeler (see Fig. 8.66, [61]).

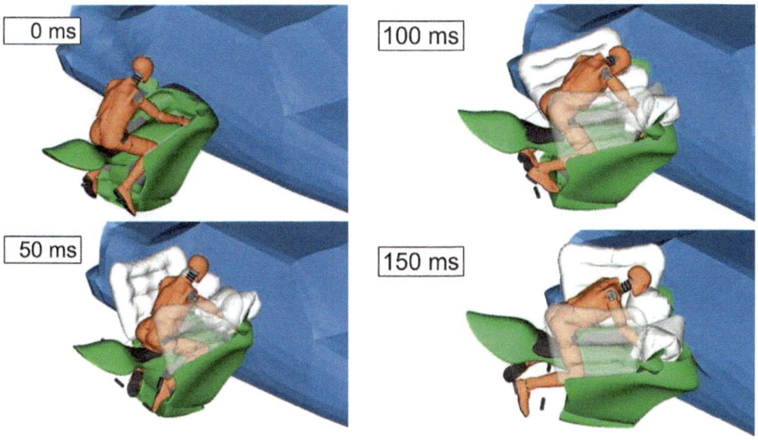

Fig. 8.66 Accident scenario according to ISO-13232-3 ($v_{Motorcycle} = 48$ km/h, $v_{auto} = 24$ km/h, impact angle = 90°)

8.6.3.5 Simulations with Human Body Models

In Fig. 8.62, the simulation of a human in a collision with a vehicle using an FEM model was shown. For such numerical simulations, the data to describe the human body are much more complicated and also much more difficult to obtain than shown in the examples above.

The endeavor to develop and apply FEM models for human body parts, limbs and ultimately for the entire body, taking into account size, mass distribution, gender, and constitution including age, may seem utopian, but many steps have already been taken in this direction, and large investments are being made to achieve this goal. Various objectives are pursued: to identify injury mechanisms and protection criteria, to support the development of dummies, and to provide a human-like FEM model for occupants and vulnerable road users (pedestrians, cyclists, and motorcyclists), with the help of which safety measures can be computationally developed, tested, and approved for traffic.

The necessary techniques for mathematical modeling and numerical treatment, the necessary information about injury mechanical relationships from biomechanics experiments and accident reports, as well as the provision of "material" data are, at least in part, available. However, further efforts are required for comprehensive realization and professional application; according to Haug [43], they refer to the development and provision of:

- Geometric data for biomechanical human models (so-called bio-models) for the skeleton, the ligaments, the muscles, the tendons and all organs considered relevant to the load path (see the illustrations in Fig. 8.67),

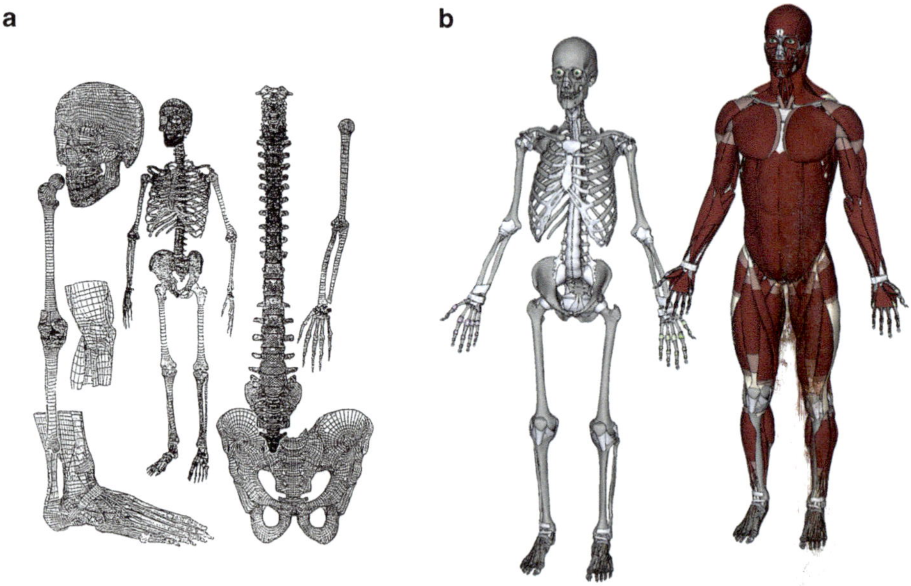

Fig. 8.67 (**a**) Discretized surfaces of human body parts for the application of biomechanical FEM models (from [43]); (**b**) as well as recent developments in the field of human body modelling [121]

- FEM models, in particular the discretization of surfaces and the generation of shell or volume elements, and
- Mechanical properties of human tissues and structures, such as stiffnesses at different load directions, velocities and other influencing factors up to crack formations and ruptures.

The available models originate from a Brite/Euram project, called HUMOS (human model for safety), or are developments in collaboration with the Toyota Central Rentity-value Labs Inc. (Fig. 8.68). In the recent past, there has been an increase in development efforts in creating more finely discretized, robust, efficient, and predictive human models. As an example, *HANS* [121] is mentioned.

Given the diverse, yet not fully understood mechanical behavior of vastly different tissues and structures of the human body, the use of reliable and predictively applicable human models on a broad basis seems to lie in the future. However, simulation results are already available today, compared with results from cadaver tests under accident conditions, from which further material properties can be experimentally determined and used as a dataset in the FEM model. There is a wide field for parameter studies and sensitivity analyses to process usable information for different individuals and accident-relevant impact conditions. An example of an application is presented in [63]. With the use of human models, simulation can contribute to vehicle development, which cannot be contributed experimentally. Because a human model would enable a much more accurate analysis of injuries and their origin. Figure 8.69 shows the pelvis and the lower spine of an ES-2 dummy and a human model. The time required to enable reliable injury

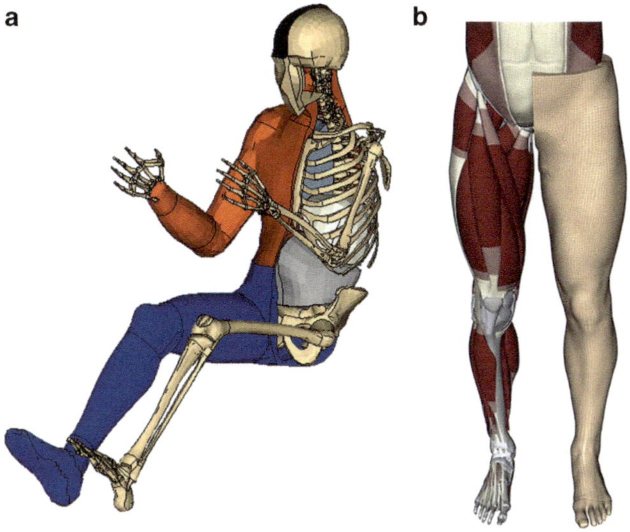

Fig. 8.68 Models **(a)** *THUMS* and **(b)** *HANS* for the computational simulation of the human body; tissues and other body parts partially hidden ([84, 121])

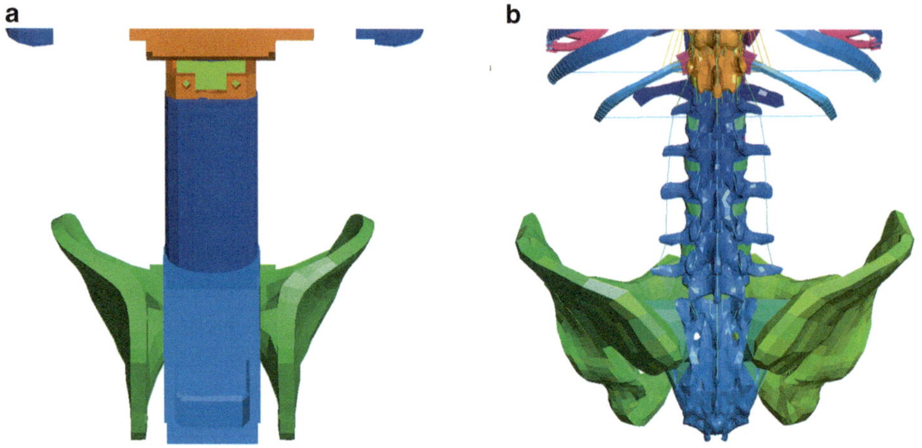

Fig. 8.69 (**a**) Pelvis and lower spine in the FEM model of the ES-2 dummy and (**b**) in the FEM human model (from [81])

prediction with the help of human occupant models will depend on the further goal-oriented cooperation of international public and private institutions and their financial resources.

8.6.4 Aspects of Crash Simulation

The application of the presented methods and models in the vehicle development process allows for quick solutions with significantly reduced effort for physical tests. This results in a significant time and cost advantage. The following describes questions and solutions that arise when using FEM in an industrial environment.

8.6.4.1 Computational Evaluation

The physical processes during an accident are very complex and diverse. Even though commercial simulation programs suggest through their realistic visualization that an accident can be exactly replicated, it should always be remembered that simplifications had to be made at many points in order to carry out the calculation within an acceptable timeframe. Furthermore, many detailed problems are still the subject of research and the modeling compared to many other technical applications is very coarse due to the coupling to the time step in explicit time integration. It is also advisable to check the result for plausibility after each calculation. An important criterion is the development of the **energy balance** during the deformation process. The total energy of the calculation model should not increase and the conversion of kinetic energy into deformation and internal energy should be plausible according to the load case. Special attention should be paid to the occurring contact energies and the stabilization energy (e.g., hourglass components), which should be small in relation to the total energies. If the energy balance is satisfactory, further simple checks for consistency of the result are recommended, such as the resulting airbag pressure or the seat belt restraint force progression depending on time. For the interpretation of the results, it should be considered how accurately and comprehensively the input data (i.e., geometry data, material data, etc.) were determined and taken into account. The computational accuracy depends on the exact description of the initial state and the material behavior. If the geometry is only roughly or inadequately meshed or if material models are used that only simplify the material behavior, this will of course only provide conditionally reliable results.

8.6.4.2 Computational Optimization in the Field of Passive Safety

If suitable computational models are available for simulation, variants can be examined with very little effort. A goal-oriented parameter variation allows the engineer to analyze the structure precisely and improve it relatively quickly. This could mean, for example, that the component under investigation becomes lighter, that it absorbs a higher deformation energy in the event of a crash, or that it can be manufactured more cost-effectively. In addition to improving the solution through selected simulations, there is also the possibility of automating the optimization process. To do this, the variables to be varied, the design variables, and the quantification of the goal using a target function and possible constraints (restrictions) must be defined. For example, in the head impact example discussed in Sect. 8.6.2.5, design variables could be the sheet thickness and the bending radius. The goal of the calculation could be to develop a component that is as light as possible, for which constraints of a maximum head acceleration and a minimum bending and torsional stiffness of the steel profile are defined. Using an optimization program, the design variables can now be determined in such a way that the target function—in this case the component mass—is minimized and the mentioned restrictions regarding head acceleration and profile stiffness are met. The determination is made through an iterative process in which the optimization program changes these geometry- and material-related, so-called design variables in a suitable manner and transfers them to the computational program. The results are then evaluated with the intention of readjusting the design variables to the target setting.

A difficulty for many tasks in the context of crash calculation arises from the geometrically and physically nonlinear behavior of vehicle structures. This means that the dependence of the behavior on changed parameters (typically the geometry and the material) can only be estimated with difficulty. For this reason, methods that require significantly more computational runs are used for optimization in the field of crash safety than methods for linear problem situations require. Since a single crash calculation on a modern computing cluster typically takes several hours and several hundred variants may need to be calculated for an optimization, optimization based on FEM solutions requires extensive computing resources. For basic optimizations with many parameters, MKS solutions are therefore also used. Several optimization examples with MKS and FEM models are described in [90]. Figure 8.70 shows a pilot study by Audi AG on optimizing the restraint system for several load cases in frontal collisions.

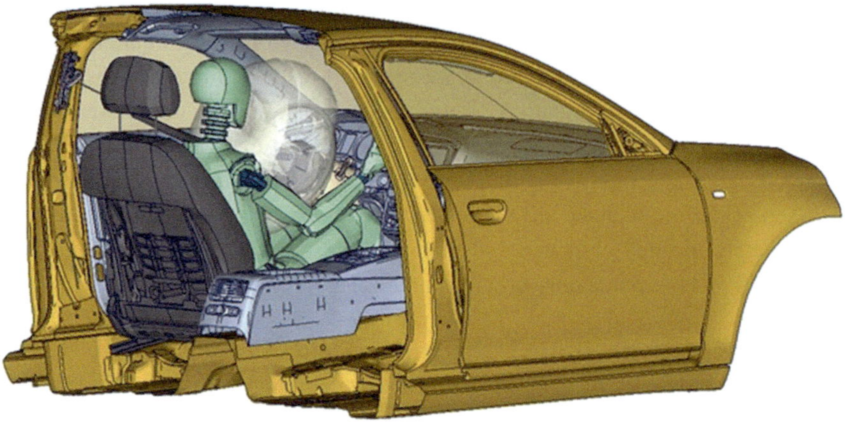

Fig. 8.70 Submodel for investigating the restraint system in a head-on collision (from [120])

In addition to optimizing the model, many commercial programs also offer the possibility to evaluate the robustness of a structure variant. In a **robustness analysis**, the system response is examined depending on small variations of defined input parameters. Figure 8.71 shows a robustness investigation on a vehicle longitudinal beam. All structural parameters—such as component thickness, geometry, and material properties—can vary due to manufacturing tolerances and also due to aging. In addition, crash simulations involve highly nonlinear problems. Therefore, robustness analyses are a crucial part of any reliable numerical simulation.

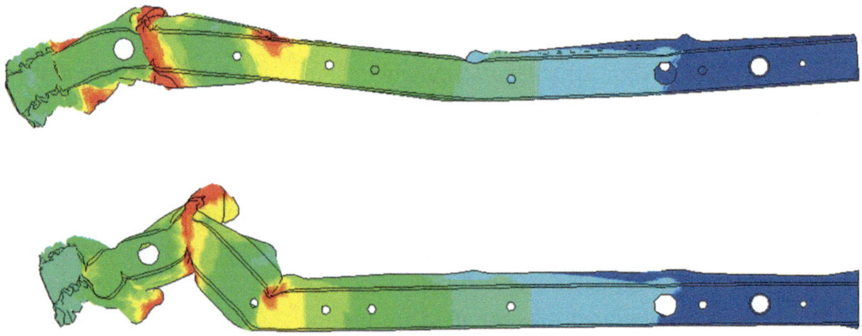

Fig. 8.71 Stochastic investigation on a longitudinal beam; the variance of the displacements is color-coded [69]

8.6.4.3 Simulation Data Management

For the development of vehicles with the aim of a high level of protection through passive safety, as explained, a large number of load cases need to be examined. Each of these load cases requires detailed, validated, and predictive models. Some of the required data were presented in the previous chapters. Another complexity arises from the fact that different people work on the individual load cases or the development of the input data. Therefore, it is extremely difficult to incorporate changes at a certain point in time into all models of a vehicle.

The following example should explain the complexity: It is assumed that the material of a sheet metal part should be changed so that it can be manufactured more easily or cost-effectively. For all materials used in the calculation, so-called material cards are necessary, i.e., validated parameters for the material model associated with the material, which are available in the respective input format of the software, describing the elastic and plastic (irreversible) behavior and the damage and failure development during deformation. An introduction to the complex matter of failure models for crash calculation can be found in [2]. If the modified component is joined, further material cards are necessary, which describe the behavior of the spot weld and possibly the heat-affected zone depending on the joining partners. If the considered sheet metal part is used in several vehicles or vehicle variants, an update must be made for all calculation models and for all vehicle variants. The safeguarding of such a process is very complex. To complicate matters, changes are typically made at many points of a calculation model by different people at the same time. For this reason, there is increasing use of simulation data management systems (SDM), which allow simultaneous work on these models. Such systems are able to automate the structure of the calculation models and initiate simulations for different load cases. The evaluation and visualization of the calculation is also automated. Figure 8.72 shows the SDM system Loco2 used at Audi. Another innovative step is seen in the automated linking of the calculation results with the test data for the respective, defined "milestones" of the development process.

There is currently no comprehensive solution that covers all development areas, but there are already solutions for many sub-areas that are increasingly being linked. With an SDM system, it is then possible to utilize computing clusters with over 15,000 computing cores. For the safety-related design of the structure of the Audi A3 (8 V), for example, 10,000 crash calculations were carried out [66].

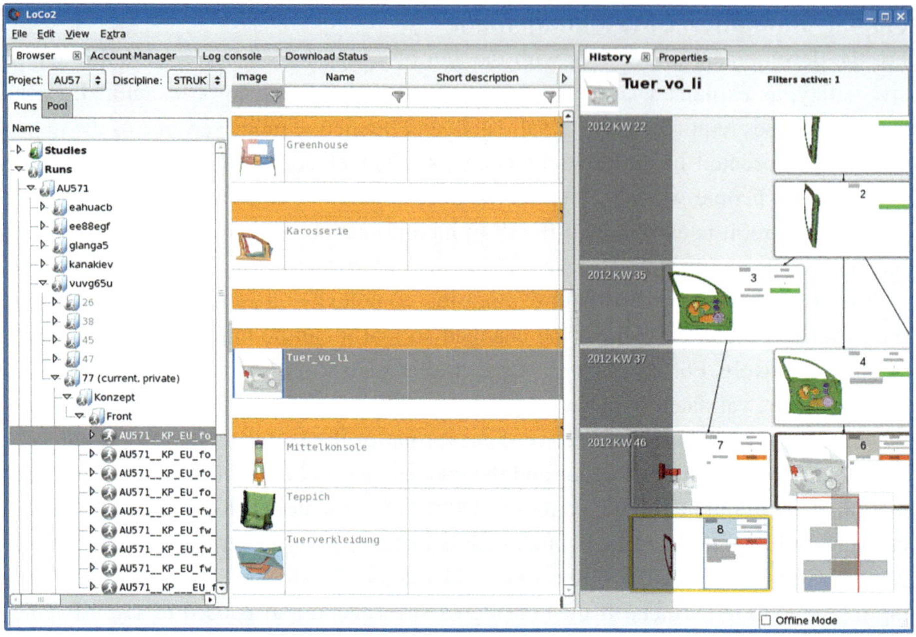

Fig. 8.72 SDM system Loco2 at Audi AG

8.7 Networked Vehicle Development

For efficient development of measures in the field of passive safety, an optimal networking and integration of design, computer simulation, prototype construction, and testing is ensured by a well-structured process model, see Fig. 8.73 [95].

The first process step in the development of a vehicle begins with CAD design (CAD: computer aided design) and the creation of a virtual or digital prototype (VP or DPT), which is secured by calculation, as detailed in Sect. 8.6. Subsequently, a second virtual prototype can be constructed and calculated with optimization measures from the first loop. If the maturity level is good enough, the construction of a so-called mule or vehicle for functional validation of the body shell (FAR) is carried out in parallel. However, this step is only necessary today for completely new concepts or materials due to the high forecasting capability of digital development tools.

The early digital development steps and resulting measures ensure that necessary long-runner tools for body-in-white and interior parts can be commissioned in a timely manner. In the same way, if necessary with corresponding change loops, the prototypes (PT) are developed and whole vehicle tests (Sect. 8.5) are carried out.

The process model is structured in such a way that simultaneous development (SE: simultaneous engineering) takes place, i.e., parallel work, so that feedback from computer simulation and testing can be incorporated into the design for the next development phase in a timely manner. This process must be networked with a multitude of

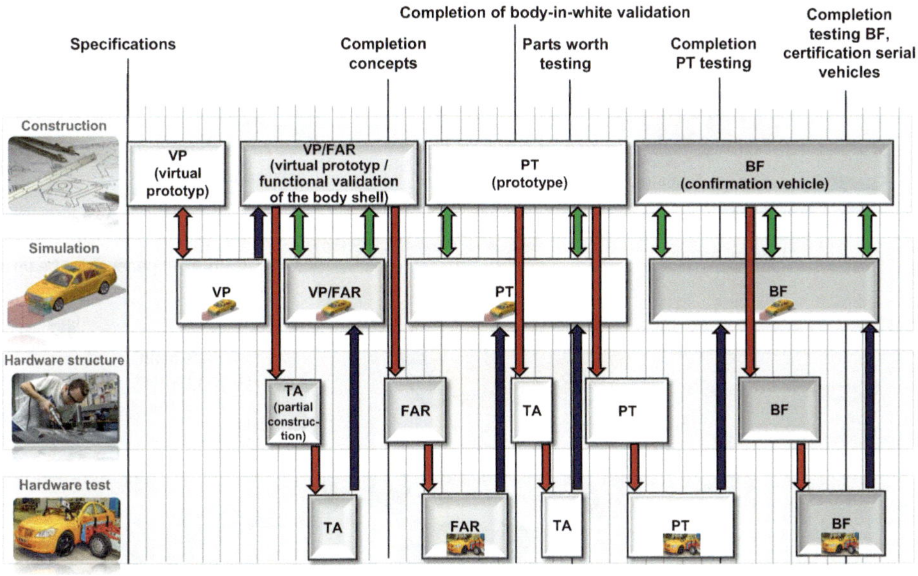

Fig. 8.73 Collaboration model between design, computational simulation, and testing [95]

development steps for all subtasks of passive safety, such as structural design, occupant protection, pedestrian protection or vehicle compatibility, as well as with other vehicle functions, for example for the realization of durability or vehicle acoustics. Their influences, changes and optimizations can indeed lead to contradictory measures that need to be coordinated.

At the end of the development process, confirmation vehicles (BF) are built with parts and components from series tools and final tests are carried out. The sum of the results and their positive evaluation ultimately lead to the development approval and certification of the vehicle.

8.7.1 Use of Development Tools

As part of the development process, the individual vehicle statuses are functionally verified. This is done either purely through tests (Sect. 8.5) or with the help of computational simulation (Sect. 8.6) and the subsequent experimental validation. In the case of CAE simulation, commercial programs that are based on the method of finite elements and an explicit time integration method are used (see Sect. 8.6.2.3).

The body-in-white is modeled with a very high level of detail in order to capture as many effects and interactions as possible. Similarly, the various aggregate parts as well as components and their space requirements, the so-called package, are represented in the **CAE models** (Fig. 8.74). For the different materials, modern material descriptions are used that take into account the strength, energy absorption capacity, elasticity, and

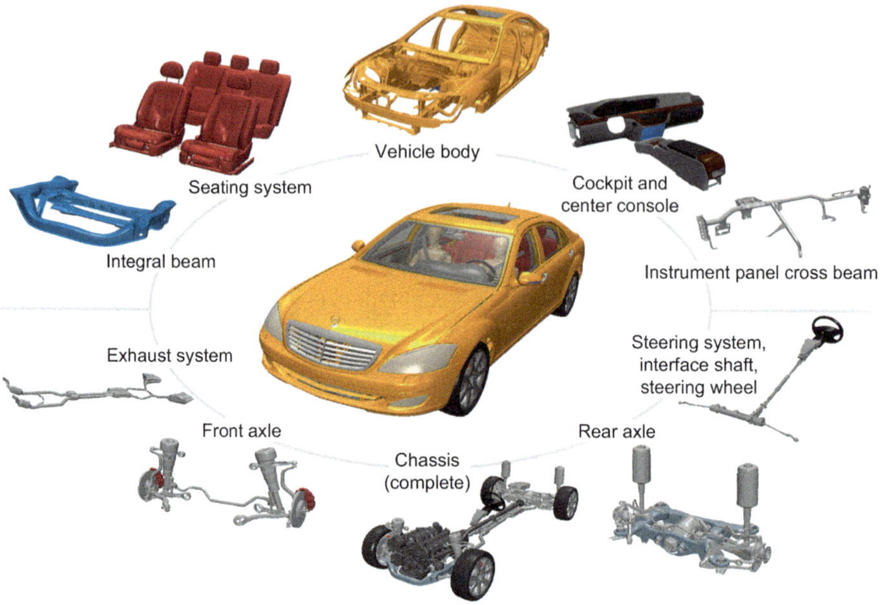

Fig. 8.74 Example of the level of detail of a CAE simulation model [95]

viscosity of the materials, i.e., the speed-dependent deformation behavior, to a sufficient extent. Additional challenges are modeling of connections, joining techniques, and plastic materials [38, 96]. In addition to the structural mechanics computation, the occupant protection system is designed at a very early stage in the development process using computational simulation [97–99]; some examples are shown in Fig. 8.75. Only the coordinated, optimal interaction of structure and occupant protection system components, such as seat belt system and airbag, meets the demand for a high level of passive vehicle safety.

The **experimental simulation** and the certification process (Sect. 8.5) make use of a whole range of different testing facilities: Testing begins initially with component testing for structurally relevant components and then only moves on to system testing facilities, e.g., the sled facility for designing occupant protection systems. If these results are promising, numerous configurations for the overall test of the vehicle are experimentally tested on the crash facility (Fig. 8.76), in order to provide the metrological proof for the requirements formulated at the beginning of the development.

In all the tests to be carried out, a very complex **measurement technology** is used, starting with anthropomorphic test devices, through a multitude of standard to special measuring points for acceleration, force and distance measurements. The measurement data is of very high importance, because the load sizes determined from it can be of decisive importance for decisions on changes during the development process and for vehicle

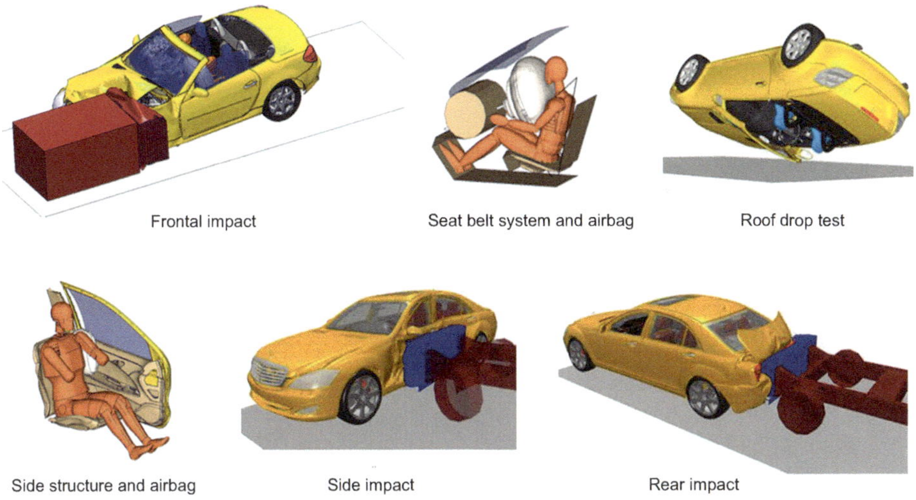

Fig. 8.75 Examples from structure and occupant crash simulation [95]

Fig. 8.76 Selection of different configurations for crash tests with the complete vehicle [95]

optimization. The individual testing facilities and the necessary equipment are described in detail in Sect. 8.5 and are shown in an overview in Fig. 8.77.

The procedure for the functional release, according to the defined requirements, is usually carried out through development loops with the **engineering departments**. The basic procedure can be seen from the flow diagram, shown in Fig. 8.78. The development process starts with an initial design status within the process. This status is led to the functional release of the corresponding milestone in collaboration between calculation and testing. The first phase, the virtual prototype, is hardware-free except for parallel component tests, i.e., the development is digital. Consequently, the functional

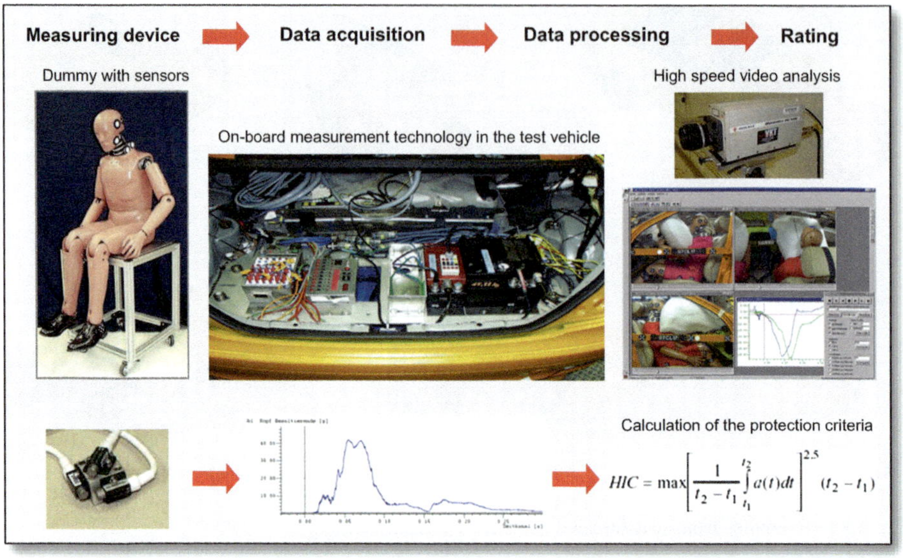

Fig. 8.77 Crash measurement technology and evaluation of the measurement data [95]

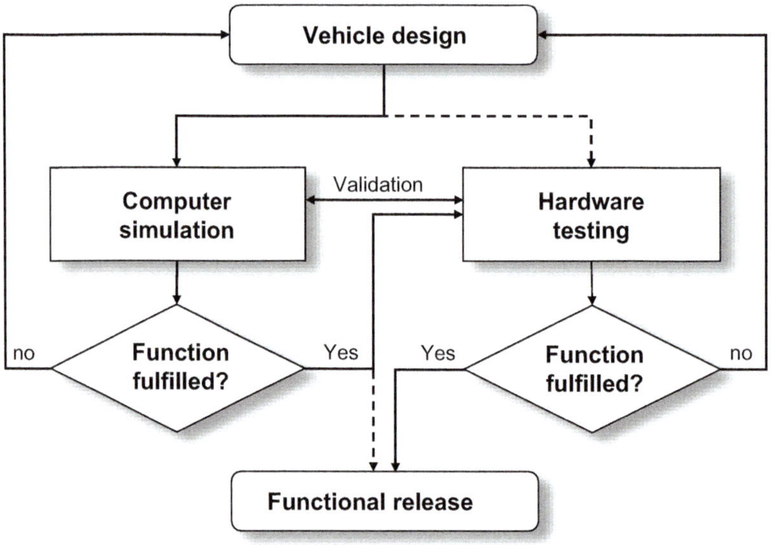

Fig. 8.78 Development loops for achieving the functional release

release is also digital. In this phase, after the design data has been received in the numerical simulation, the computer model is built. Based on the determined calculation results regarding the functions crash safety, stiffness, durability, vibrations, acoustics, etc., the functional release takes place when the function criteria are met, and the achieved design status is transferred unchanged into the next development phase. In case of non-fulfillment, a modified status is worked out together with testing and design departments. These necessary modifications can range from simple material and sheet thickness variations, to reinforcements, or even completely different design concepts. The design status thus achieved is then recalculated and evaluated together with the test engineers. This loop repeats until the defined criteria for all requirements are met.

During the run-through of the development loop, recurring conflicts of objectives between passive safety and other functions are often unavoidable. This will be explained using the example of an integral carrier, which must be designed as stiff as possible for reasons of stiffness, durability, and vibration. On the other hand, according to the requirement for crash safety, it serves as a deformation element and therefore must have a targeted deformation behavior in the vehicle's longitudinal direction. These conflicting objectives require detailed optimization taking into account all functions. The deformation of a computationally optimally designed integral carrier is shown in Fig. 8.79. A frontal impact according to Euro NCAP was considered, in which a vehicle collides with a deformable barrier at a velocity of 64 km/h and an overlap of 40 % based on the vehicle width.

In some cases, additional **component tests** must be carried out for the computational forecasting capability. This is particularly necessary when new materials or joining concepts are to be introduced and secured. These component tests are then computationally represented and the calculation results are compared with those determined experimentally. This process is referred to as validation. Further calculations are then only carried out with the models validated in this way.

While in the early, purely digital development phase the computational release takes place and the necessary test samples (hardware parts) are limited to structural components, the scope of development of the test increases significantly from the middle phase in the development process. Here, vehicles for verifying the body and aggregates are no longer built exclusively virtually, but also in reality in the form of individual parts. Depending on the load case, a computational or hardware-oriented safeguarding can be economically sensible. As a rule, however, a combination of computational and experimental verification is considered advantageous.

As development progresses, vehicle components are increasingly being manufactured as **prototypes** and made available for testing; hence, this phase of the process is called the "prototype phase". Up to certification and series release, the computational simulation in the development loop steadily decrease, while the testing activities increase (Fig. 8.80).

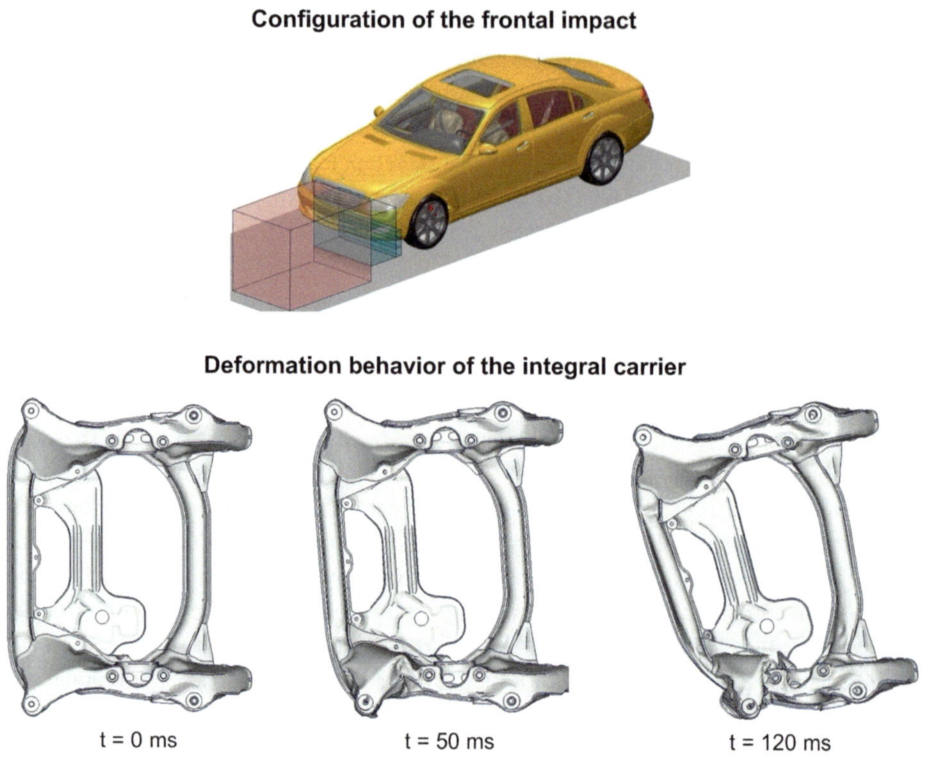

Fig. 8.79 Deformation behavior of an integral carrier during a computationally simulated frontal impact [95]

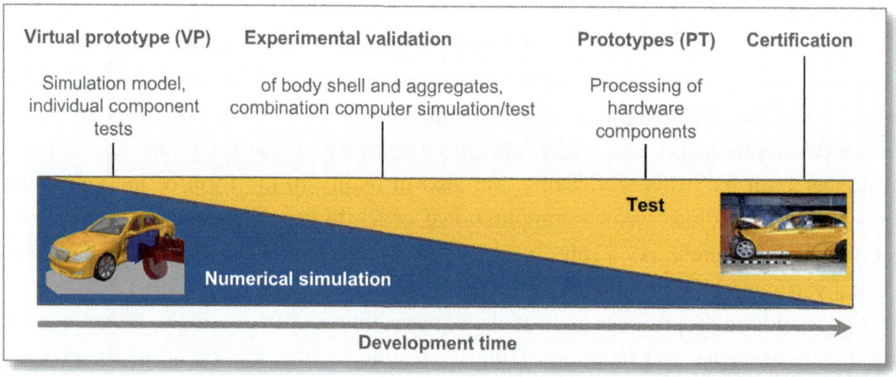

Fig. 8.80 Use of development tools computational and experimental simulation in the course of the development process

The **certification** is currently based solely on experimental verifications. However, certification based on computational simulation results is the subject of current discussions. There are initial examples of digital certification in connection with bumper load cases, but also in insurance classifications.

To increase the forecast accuracy, the computational simulation results are always compared with the test results. This continuous validation with results from whole vehicle tests provides valuable insights for subsequent development projects.

8.7.2 Qualification Process of New Technologies and Materials

If new technologies or materials are used within the development process, early experimental validation is necessary for the validation of the computational simulation (Fig. 8.81). Tensile tests are usually the first step in evaluating new materials, and depending on the field of application, further load directions such as pressure, shear or bending are considered. Such tests are carried out at various loading speeds until material failure, as almost all materials exhibit a certain viscosity and a specific fracture behavior dependent on the strain rate. After the material behavior has been experimentally determined, the tests are modeled via computational simulation to validate the material laws under model conditions, see Sect. 8.6.3. These generated material descriptions are used, if necessary, for component tests for validation. The same procedure is followed when new technologies are intended to be used. Only when the validation

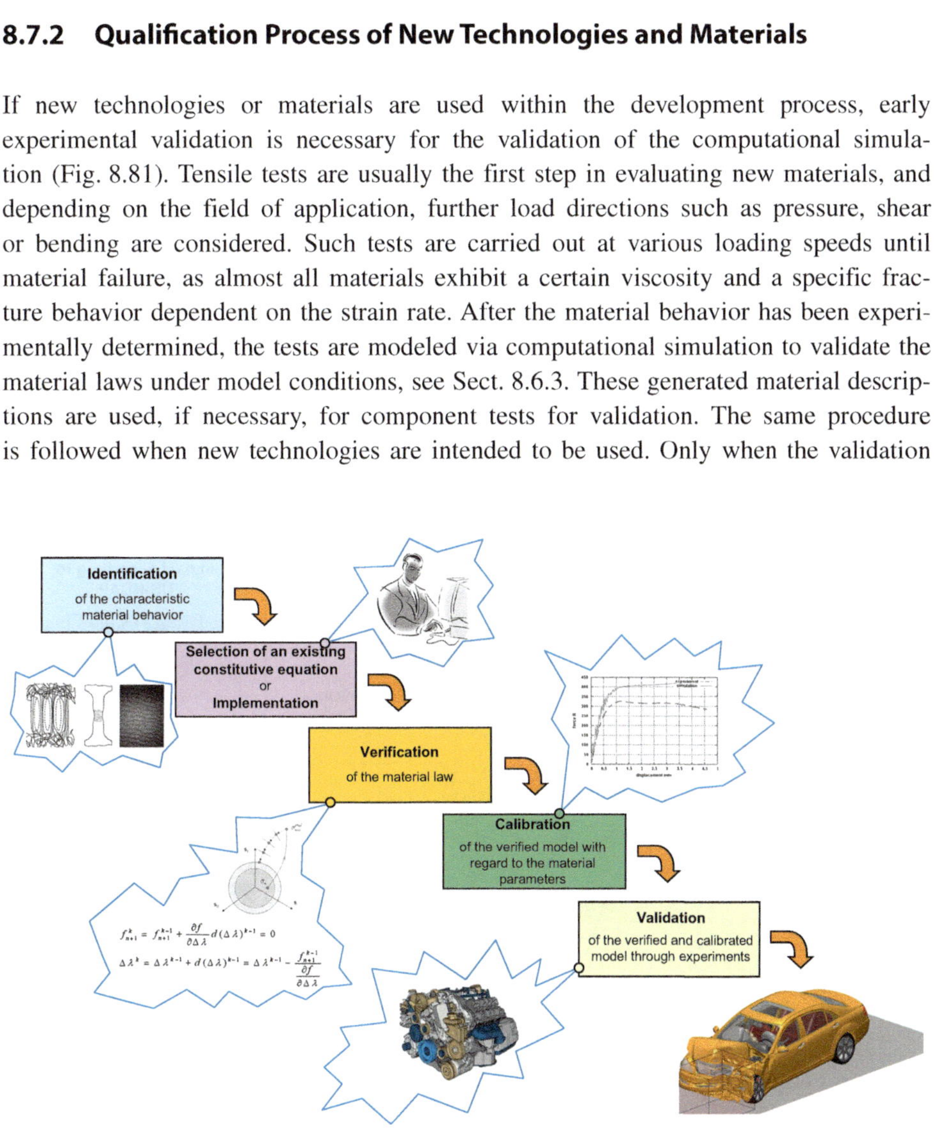

Fig. 8.81 Example of a validation process

phase has been successfully completed and a sufficiently high forecasting capability can be considered secured, the use in the whole vehicle model is carried out and thus integrated into the development.

8.7.3 Opportunities and Limitations of Simulations

The high level of detail of today's simulation models allows very accurate forecasts regarding the structural behavior of body-in-white and assembly parts. Based on the a/v/s diagram shown in Fig. 8.82, named after the time-dependent motion variables acceleration, velocity and displacement, the **forecasting capability** of the computational simulation compared to test results is to be clarified. The depicted variables result from a frontal impact test according to US-NCAP specifications, in which a vehicle collides with a rigid barrier at a speed of 56 km/h and an overlap of 100 % (full overlap).

A complete match between test and computational simulation cannot be achieved due to the method of finite elements, which is by definition an approximation method, due to the complex deformation behavior, the scattering of material characteristics and manufacturing-related tolerances, as was explained in detail in Sect. 8.6. However, such deviations also occur in measurement results of identical tests.

A significant advantage of numerical simulation is that it provides detailed insights into the design during the simulated load process. However, a model only exactly represents what was considered during model building. Therefore, it has limitations.

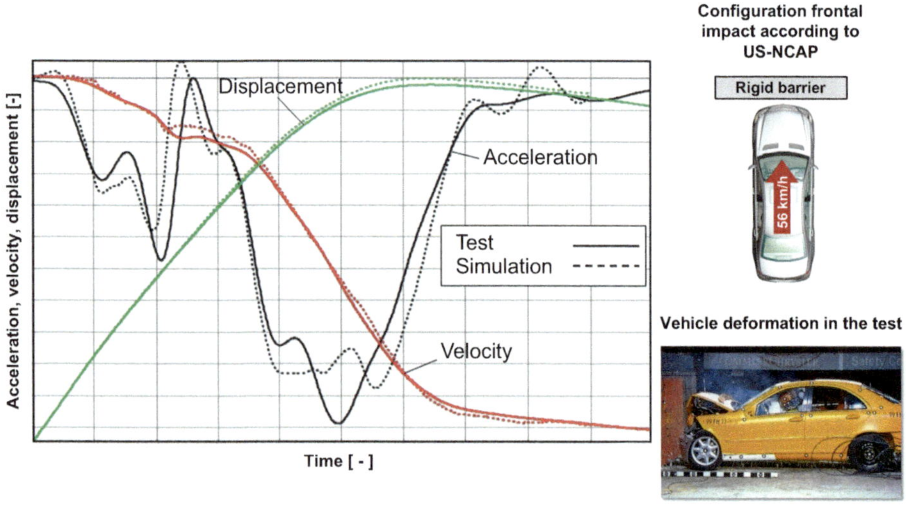

Fig. 8.82 Comparison between test and numerical simulation based on the vehicle motion variables [95]

Idealizations and discretizations limit the informative value. This is particularly evident in specific applications of crash simulation, for example in

- material models for composite materials,
- consideration of manufacturing influences,
- models for reliable prediction of component failure (e.g., in the case of crack formation),
- models for describing crack progression, and
- modeling of joining techniques (adhesives, spot welds, etc.) including failure prediction.

However, suitable approaches have been developed for the mentioned problem areas, which allow an approximate prediction and can be used for initial design. Moreover, not only are there currently unavoidable limits in computational simulation, but experimental simulation also comes with certain uncertainties, such as

- idealizations in early vehicle development phases,
- prototype-specific components and joining techniques,
- status of control units and software,
- variations in prototypes, in the experimental setup, and in the experimental procedure, and finally
- reproducible positioning of dummies.

Today, a goal-oriented development of passive safety can only be optimally and efficiently operated when the opportunities and limits of all available tools of computational and experimental simulation are known and optimally coordinated.

In the future, the use of numerical simulation will significantly expand to reduce the cost-intensive production of test samples and their testing. However, the remaining confirmation tests will become more complex, as comprehensive statements about the complete vehicle are expected from the results of the tests with a reduced number of prototypes and confirmation vehicles. In addition, future testing will increasingly target accident occurrence to generate insights from the interaction between active and passive safety or compatibility phenomena [15, 100].

The development of the passive safety of an entirely new vehicle, including certification without real prototypes and confirmation vehicles, is not yet conceivable today, as the overall functionality, including crash sensors, airbag flap tearing behavior, control units, door opening forces, fuel system integrity, CAN bus, etc., would have to be functionally represented in the simulation. From today's perspective, however, the creation of a holistic virtual model of a vehicle with all its functionalities is too complex and thus too costly and time-consuming.

The task of the development engineer will continue to lie in critically examining the computational simulation and testing results and constantly checking their informative value, so that the product reaches a high maturity level as early as possible and a completely safe vehicle reaches market maturity.

8.7.4 Challenges within Project Work

The networking between passive safety and other disciplines in the vehicle development process becomes clear using the example of body development: The body, as a significant component of passive safety, is an integral part of the vehicle. It interfaces with many components of the vehicle (e.g., restraint systems, doors, interior trim, bumpers, cockpit, seats, axles, electrical systems, etc.) and it must fulfill a multitude of other customer-relevant properties (e.g., corrosion protection, durability, stiffness, vibration comfort). These requirements can sometimes conflict with each other and demand optimal solutions in terms of cost, weight, quality, and time.

Another example of the interaction of disciplines is the increasing networking between system components of active and passive safety. Current attempts to develop and evaluate crash safety are standardized and conducted in the laboratory. For reasons of reproducibility of test results and to evaluate development progress, the same test conditions, such as velocities, vehicle overlaps, seating positions, occupant weights and sizes, are always specified and adhered to. Based on the results, vehicles are evaluated and developed with a focus on their protection capabilities during accidents. The prescribed limit or target values set the direction of development, with the degree to which the measured values are achieved being determined in regular standardized tests and serving as a benchmark for development progress.

But how does this relate to vehicle safety in actual accident occurrence? As is well known, vehicle speed and impact configuration (type of impact, angle, overlap, etc.) are essential parameters for accident severity. As part of active safety, systems for accident severity reduction or accident prevention are developed, such as anti-lock braking systems (ABS), electronic stability programs (ESP), brake assistants or environment recognizing systems, which have a significant influence and great potential to increase the overall safety of a vehicle. Mercedes-Benz redefined integral safety with the introduction of PRE-SAFE® [101, 102] in 2002. By targeted use of sensors and active safety systems, such as assistance systems, the impact of real accidents on the persons involved is reduced. However, this interaction of systems in terms of requirements and functionality demands completely new, networked work structures and flexible processes in development. Previously unconnected development departments must now reorganize their content and development programs and redesign the resulting sub-processes. Increasingly, electrical comfort systems, such as seat adjustments, sunroofs, window lifters but also rear seat systems and reversible seat belt systems, are integrated into preventive occupant protection. This in turn places higher demands on the failure and signal security of the bus systems and at the same time increases the requirements for the durability and availability of the actuators. For the so-called electrical/electronic topology, this means a further increase in system networking to take into account the actuator control (see Sect. 7.3). As a result, new requirements, such as minimum speeds in signal transmission, must be ensured and documented. Further challenges arise from questions about standardization of components as well as communication protocols and signal processing, as systems from different suppliers are integrated into the vehicle and must work reliably together there. The task of vehicle manufacturers now consist in creating overarching opportunities for simplification and standardization.

The number of people and departments involved in the development of such complex systems has skyrocketed in recent years. The processes required to manage the extensive system task must therefore be clearly described and mastered by the participants, but at the same time offer enough flexibility and openness for the introduction of innovations to keep pace with both customer wishes and technological changes.

8.8 Safety Innovations in the Development Process

Fundamentally, new solutions and concepts can significantly influence the development process. Depending on the degree of novelty and scope of a solution or concept, innovations in the development process represent a risk factor. During the course of development, there is a lack of experience, and there can always be unknown, unexpected problems. Therefore, the development process must be flexible enough to lead to a mature product despite difficulties and obstacles. Of course, it is particularly important to ensure careful preliminary development for innovative solutions and concepts. In case a car manufacturer aims for first-to-market innovations, the available development time is limited. If a manufacturer takes too much time in development, they run the risk of being second to market.

The goal is therefore to carefully weigh the development risks and introduce an innovation at the right time into a suitable, flexible overall development process. The flexibility of the process may mean that optimization loops need to be planned and more resources need to be used, or in the worst case, so-called fallback solutions need to be available. This makes it clear that depending on the type and complexity of innovations, the development processes must be designed differently. Accordingly, shorter processes are possible for more evolutionary standard products than for complex, revolutionary, innovative products.

The extent to which innovations influence the process and at what stage of overall development they need to be introduced depends on the concept relevance of a solution. The following categories are distinguished (Fig. 8.83):

- **Vehicle concept-defining safety innovations**: Innovations that significantly shape the design, the overall vehicle package, the body structure, the interior design, or the electrical/electronic architecture. These must be considered at the start of the project in the definition or strategy phase.

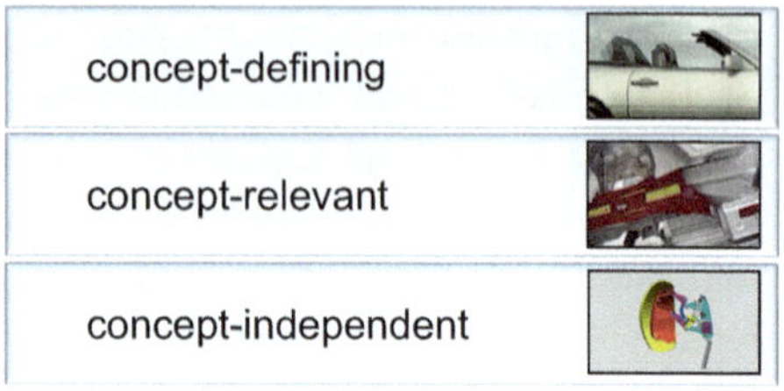

Fig. 8.83 Relevance of innovations for the vehicle concept [95]

- **Vehicle concept-relevant safety innovations**: Innovations that place significant requirements on the design, the overall vehicle package, the body structure, the interior design, or the electrical/electronic architecture. These must be incorporated into the concept booklet.
- **Vehicle concept-independent safety innovations**: Innovations that either do not influence or only slightly influence the design, overall vehicle package, body structure, interior design, or the electrical/electronic architecture, and are therefore easily integrated into existing concepts. These can possibly still be considered after the "vehicle specifications" milestone in the series development or realization phase of the vehicle.

Vehicle concept-defining safety innovations could, for example, be measures for pedestrian protection that have a significant influence on the design and silhouette of a vehicle. Concept-relevant innovations are, for example, structural measures that significantly influence axle geometry and package. Also conceivable are safety systems that place very specific requirements on an electrical/electronic architecture. Vehicle concept-independent innovations influence the development process only to a minor extent, as they have a clearly defined and limited scope. Such innovations can be integrated relatively late into a process and are also suitable as facelift measures, as they can be implemented into an existing series vehicle.

Like all innovations, safety innovations must also be assessed regarding their quality in respect to the status of the vehicle development process, i.e., they must pass so-called quality gates. To be considered in a concept booklet for a new vehicle, it must be demonstrated on substructures or test carriers that the principle of this innovation works and produces the desired effects in terms of expected customer benefit. This proof is called **maturity of design principle**.

For the concept booklet, proof of the **maturity of design concept** is required. This means that the function is ensured in the specific vehicle environment and, in addition, initial risk analyses are available. Furthermore, at this point, evidence for manufacturing and ensuring future quality goals must be present.

For the milestone "vehicle specifications", the decisive hurdle is the comprehensive proof of the **maturity of series production process**. With the maturity of series production process, the function in connection with further innovations of the target vehicle must be proven, i.e., the achievement of target weight and target costs is ensured,

3D-CAD data are updated, and all target conflicts are resolved. To reduce risks in the development process, vehicle concept-defining innovations should already have proven the maturity of series production process at the "concept booklet" milestone.

At the end of the validation phase, i.e., at the quality gate "vehicle specifications" (see Fig. 8.3), the development responsibility usually shifts from research/pre-development to series development. This handover is traditionally difficult and a stumbling block for many innovations, as there is often a different understanding of the maturity level for a series production process in pre-development and in series development. Practice shows that it seems sensible to realize a gradual transition of responsibilities by having responsible project managers switch to series development with the innovation or by deploying a series developer in the project ahead of the handover.

8.9 Integration of Accident Research into the Development Process Using the Example of Mercedes-Benz

Many vehicle safety innovations are generated from the observation of real accident events. Since 1969, engineers from Mercedes-Benz have been specifically investigating road traffic accidents. The goal pursued is to learn from accidents to improve vehicle safety. The work of the Mercedes-Benz accident researchers was only made possible by a decree of the Baden-Württemberg Ministry of the Interior in Germany on April 29, 1969. To date, the researchers have investigated about 5,000 accidents with injured occupants. A working group consisting of several engineers primarily conducts these investigations in the greater Stuttgart area. For interesting cases, the survey area is also expanded. Cooperation is mainly with the police, towing companies, and car dealerships. Whether an investigation is carried out depends on the following criteria:

- Injuries to the occupants,
- Deformations of the vehicle, and
- Activation of the restraint systems

The tasks and procedures of the accident researchers are described in detail in Chap. 2. Various information goes into the evaluation of an accident, which relates to the investigation of the accident vehicles, the measurement of the accident site, and the knowledge about the occupant injuries. From this information, data are derived that subsequently enable a detailed reconstruction of the accident. The reconstruction is done computationally using computer programs that allow an animation of the accident sequence. An example is shown in Fig. 8.84.

The so-called collision analysis, i.e., the calculation of forces and moments during the impact, is crucial for the achieved result, see Sect. 8.6.3. All boundary conditions such as hooking or sliding on the collision opponent are taken into account [13]. The EES value (EES: energy equivalent speed—see Sect. 2.2.2) plays an important role in this

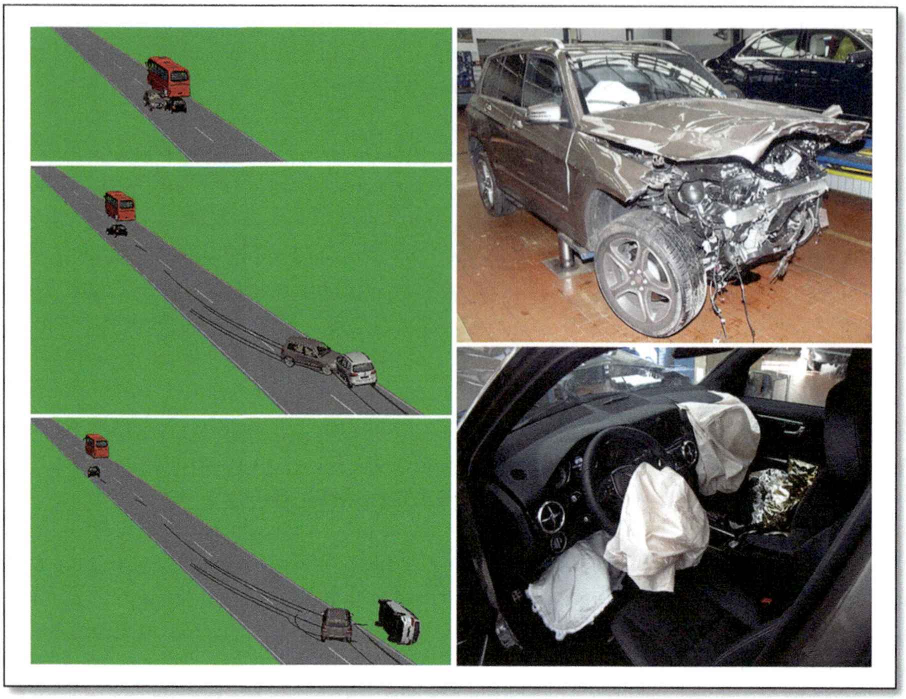

Fig. 8.84 Example of an accident reconstruction [95]

calculation as a control variable. It is a value that represents the energy input that is converted into deformation, expressed as velocity in km/h. It is determined from crash tests and represents a measure of the accident severity [119].

The occupant injuries are recorded in detail and classified and coded using the AIS scale (abbreviated injury scale—see Sect. 3.2.2). Each individual injury to the body is evaluated. Finally, the MAIS value is determined, which then summarises the maximum occurring body part-specific single injury severity of the occupant.

Figure 8.85 shows the essential relationships in the accident and the execution of an evaluation. With the help of the data and information, an overall picture of accident severity, cause of injury, and injury severity can be formulated. From these findings, ideas and concrete concepts for future safety measures then emerge.

With the help of the findings from the accident analysis, numerous improvements to the vehicle and new test procedures, e.g., the "offset crash" characterised lateral offset frontal impact, derived in the 1970s. The PRE-SAFE® system, which is active before the accident, also results from the analysis of road traffic accidents [11, 102]. A new

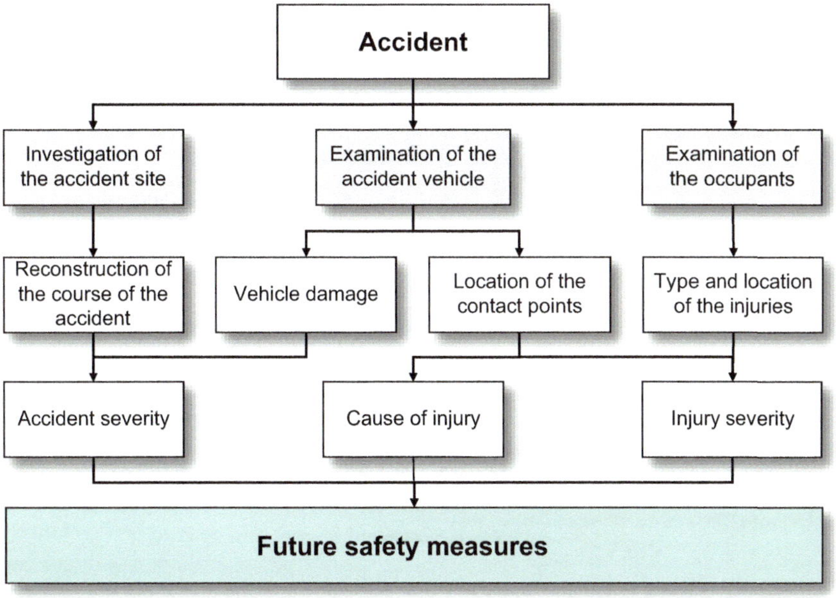

Fig. 8.85 Causalities in the accident and derivation of safety measures

direction that can also be derived from accident research is seen in the individualization of occupant protection systems. Here, depending on occupant parameters such as body size and weight, gender, etc., more precise adjustments of the restraint systems will be made in the future using sensor devices in the vehicle. Initial developments show significant potential for possible adjustments to the restraint systems [103].

After the series introduction of new technologies, these can be evaluated again in accident occurrence. This "accident research control loop", which is fed by the accident data as an input variable, is shown in Fig. 8.86. Meaningful test configurations can be derived and in the end, a safety prediction for the respective vehicle types or an evaluation of systems already on the market is available.

However, accident research is no longer just focused on protection in collisions themselves, but vehicle safety and therefore accident research must be viewed holistically. Every accident begins with a critical situation and ends with the rescue, Fig. 1.7. In particular, accident root cause analysis is becoming increasingly important. Mercedes-Benz accident research is indispensable from today's development process and is a pioneer of new, evolving safety requirements and many innovations [94, 104–106].

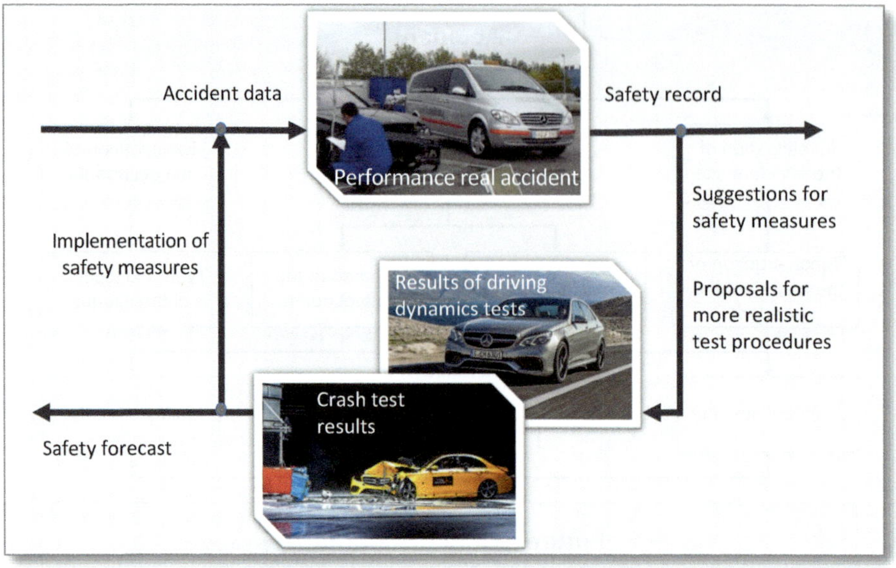

Fig. 8.86 Accident research control loop [95]

References

1. ABAQS: Dassault Systems, Suresnes (F). https://www.3ds.com
2. Andrade, F. X. C., Feucht, M., Haufe, A., Neukamm, F., An incremental stress state dependent damage model for ductile failure prediction. Int. J. Fract. https://doi.org/10.1007/s10704-016-0081-2
3. Appel, H., Kramer, F., Glatz, W., Lutter, G. et al.: Quantifizierung der passiven Sicherheit für PKW-Insassen. Technische Universität Berlin. Bericht zum Forschungsprojekt 8517/2 der Bundesanstalt für Straßenwesen (Eds.). Bergisch Gladbach (1991)
4. Appel, H., Kramer, F.: Biomechanik und Kraftfahrzeugsicherheit. Umdruck zur gleichnamigen Vorlesung am Institut für Fahrzeugtechnik an der Technischen Universität Berlin, Berlin (1987)
5. Arup Ltd., Solihull (UK): https://www.arup.com/dyna
6. Bathe, K.J.: Finite Element Procedures. Prentice Hall, Englewood Cliffs (1996)
7. Baumann, K.-H., Schöneburg, R.: Manufacturer's Demand on Crash Test Programs. TÜV Symposium Worldwide Harmonization of Crash Test Programs. Köln (1999)
8. Belytschko, T., Liu, W. K., Moran, B., Elkhodary, K.: Nonlinear Finite Elements for Continua and Structures, 2nd. edn, ISBN: 978-1-118-63270-3 (2013)
9. Beusenberg, M.C., Janssen, E.G., Schreuder, J.J.H.: Five Years Experience of Using EUROSID-1 in Sled and Car Tests. Report Paper. Bd. 94 S6 O 09. TNO Crash-Safety Research Centre, Delft (1994)
10. Bock, J., Krajewski, R., Eckstein, L., Klimke, J., Sauerbier, J., Zlocki, A.: „Data basis for scenario-based validation of HAD on highways", 27. Aachener Kolloquium (2018)
11. Breitling, T., Mellinghoff, U., Metzler, H.-J., Schöneburg, R.: Das Experimental-Sicherheits-Fahrzeug ESF 2009. ATZ Automobiltechnische Zeitschrift **111**(07–08), 532–541 (2009)

12. Brenner, S.C., Scott, L.R.: The mathematical theory of finite element methods. Springer, New York (2002)
13. Burg, H., Moser, A. (Hrsg.): Handbuch Verkehrsunfallrekonstruktion. Friedr. Vieweg & Sohn Verlag/GWV Fachverlage GmbH, Wiesbaden (2007)
14. Burg, H., Moser, A.: Handbuch Verkehrsunfallrekonstruktion – Unfallaufnahme, Fahrdynamik, Simulation, 2nd. edn. Vieweg + Teubner GWV Fachverlage GmbH, Wiesbaden (2009)
15. Busch, D., Cakmak, M., Schöneburg, R., Zobel, R.: Evaluation of Crash Compatibility of Vehicles with the Aid of Finite Element Analysis. 15th ESV-Conference, Paper Number 96-S4-W-25. Melbourne (AUS) (1996)
16. Böhmler, K.: Beschreibung der TRW Repa-Crashanlage, Alfdorf (1990)
17. Center for Collision Safety and Analysis, George Mason University (USA). https://www.ccsa.gmu.edu/models/2020-nissan-rogue. Accessed June 19, 2023
18. Crash Test Dummy Development History: Newsline of the First Technology Safety Systems, Inc. Plymouth (MI), USA (1993)
19. Crisfield, M.A.: Non-Linear Finite Element Analysis, Vol. 2: Advanced Topics. Wiley, New York (1997)
20. DYNAmore GmbH, Stuttgart, (D): https://www.dynamore.de
21. Daimler AG, Stuttgart, (D): https://www.daimler.com
22. Day, T.D., York, A.R. II: Validation of DyMesh for vehicle vs Barrier Collisions. SAE Technical Paper Series 2000-01-0844. International Congress and Exposition. Detroit (MI), USA (2000)
23. Denton Products & Service: Presentation Robert A. Denton Inc, Rochester Hills MI (USA) (2004)
24. Di Fabio, Udo et. Al., „Bericht der Ethik-Kommission Automatisiertes und Vernetztes Fahren" Bundesministerium für Verkehr und digitale Infrastruktur, available online at https://bmdv.bund.de/SharedDocs/DE/Publikationen/DG/bericht-der-ethik-kommission.pdf?blob=-publicationFile. Accessed May 3, 2023
25. Dolly Rollover Recommended Test Procedure. SAE J2114 APR93, SAE Recommended Practice. 2003 SAE Handbook, Vol. 3 On-Highway Vehicles and Off-Highway Machinery. Society of Automotive Engineers Inc. Warrendale (PA), USA (2003)
26. Dutton, T.: Finite Element Models for European Testing: Side Impact Barrier to WG13 and Pedestrian Impactors WG17, 4th European LS-DYNA Conference, Ulm, Germany, 2003. https://www.dynalook.com/conferences/european-conf-2003/finite-element-models-for-european-testing.pdf/@@download/file/LS-DYNA_ULM_C-II-29.pdf
27. ECE-R 16: Einheitliche Bedingungen für die Genehmigung der Sicherheitsgurte und Rückhaltesysteme für erwachsene Personen in Kraftfahrzeugen. Bremseinrichtung, Anhang 6
28. Eckstein, L. "Safety, Efficiency & Driving Experience – from Vision to Reality", 16. VDA Technical Congress 20.-21.3.2014, Hannover, Germany (2014)
29. Finsterhölzl, H.: Strukturberechnung am 5er BMW (E 39) bei einem 40 %-igen Off-set-Crash mit 64 km/h unter Verwendung einer deformierbaren Barriere (IIHS-Test). Verwendung der Bilder und Erläuterungen mit freundlicher Genehmigung der Fa. BMW, Abteilung Karosserieberechnung. (1998)
30. Fittanto, D.A., Ruhl, R.A., Southcombe, E.J., Burg, H., Burg, J.: Overview of CARAT-4, a Multi-body Simulation an Collision Modelling Program. SAE Technical Paper Series 2002-01.1566. International Congress and Exposition. Detroit (MI), USA (2002)
31. Frank, T., Kurz, A., Pitzer, M., Söllner, M.: Development and Validation of Numerical Pedestrian Impactor Models, 4th European LS-DYNA Conference, Ulm, Germany, 2003. https://www.dynalook.com/conferences/european-conf-2003/development-and-validation-of-numerical-pedestrian.pdf/@@download/file/LS-DYNA_ULM_C-II-01.pdf

32. Franz, U., Schuster, P., Stahlschmidt, S.: Influence of Pre-stressed Parts in Dummy Modeling – Simple Considerations, 9. International LS-DYNA Konferenz. Detroit (MI), USA (2004). https://www.dynalook.com/conferences/international-conf-2004/05-4.pdf
33. Franz, U., Stahlschmidt, S., Schelkle, E., Frank, T.: „15 Years of Finite Element Dummy Model Development Within the German Association for Research on Automobile Technology (FAT)", JRI Japanese LS-DYNA Conference, Nagoya, (JP) (2008)
34. Frik, S., Gosolits, B., Böttcher, C.-S.: 20 Years of Crash Simulation at Opel – Experiences for Future Challenge, 4th LS-DYNA-Forum. Bamberg, Germany (2005)
35. Gibson, T.J., Hinrichs, R.W., McLean, A.J.: Pedestrian Head Impacts: Development and Validation of a Mathematical Model. International IRCOBI Conference on the Biomechanics of Impacts. Zürich, (CH) (1986)
36. Glöckner, H.: Fußgängerschutz am PKW – Ergebnisse mathematischer Simulation. Dissertation an der Technischen Universität Berlin, Berlin (1982)
37. Hackney, J.R., Kahane, C.J.: The New Car Assessment Program: Five Star Rating System and Vehicle Safety Performance Characteristics. SAE Technical Paper Series 950888. International Congress and Exposition. Detroit (MI), USA (1995)
38. Haldenwanger, H.-G., Liman, U., Reim, H., Schöneburg, R.: Die Notwendigkeit von Computersimulationen bei der Entwicklung von Kunststoffbauteilen. VDI-K, Kunststoffe im Automobilbau. VDI-Verlag, Düsseldorf, S. 87–116 (1995)
39. Handbuch PC-Crash 10.0: A Simulation Program for Vehicle Accidents. DSD, Dr. Steffan Datentechnik, Linz, Austria (2013)
40. Handbücher Carat-3 und Carat-4 - A Multi-body Simulation and Collision Modeling Program. IbB-Informatik GmbH, Mülheim/Mosel, Germany (2002)
41. Haufe, A., Franz, U.: On the Simulation of Out-of Position Load Cases with the ALE-Method. Airbag 2004, 7th International Symposium and Exhibition on Sophisticated Safety Systems. Karlsruhe (2004)
42. Haug, E., Scharnhorst, T., Dubois, P.: FEM-Crash-Berechnung eines Fahrzeugfrontalaufpralls. VDI-Tagung „Berechnung im Automobilbau". VDI-Berichte. Vol. 613. VDI, Würzburg (1986)
43. Haug, E.: Biomechanical models in vehicle accident simulation. PAM'95 Fifth European Workshop on Advanced Finite Element Simulation Techniques. Bad Soden (1995)
44. Hughes, T.J.R.: The finite element method: Linear static and dynamic finite element analysis. Dover Publications, Mineola, NY (2000)
45. Humanetics ATD Crash Test Dummies: https://www.humaneticsgroup.com
46. Humanetics Innovative Solutions Inc., Huron, (OH) (USA). https://www.humaneticsatd.com
47. Instrumentation for Impact Test – Part 1 – Electronic Instrumentation. SAE J211/1 MAR95, SAE Recommended Practice. 2003 SAE Handbook, Bd. 3 On-Highway vehicles and off-highway machinery. Society of Automotive Engineers Inc. Warrendale (PA), USA (2003)
48. Janssen, E.G.: Mechanical human body simulators. TNO Industrial Report. TNO Road-Vehicles Research Institute, Delft, NL (1993)
49. Johnson, C.: Numerical solution of partial differential equations by the finite element method. Cambridge University Press, Cmabridge (1995)
50. Kelly, J.R.: BioRID-Iic Rear Impact Crash Test Dummy. Robert A. Denton Inc, Rochester Hills (MI, USA) (2003)
51. Kraftfahrt-Bundesamt – Fahrzeugalter – Durchschnittsalter der Personenkraftwagen wächst, https://www.kba.de. Accessed June 19, 2023
52. Kramer, F.: Unfallrekonstruktion. Vorlesungsskript zur gleichnamigen Lehrveranstaltung an der Hochschule für Technik und Wirtschaft (HTW) Dresden (2010/11)
53. Kramer, F.: Significance of Restraint Systems for the Secondary Safety of Motor Vehicles. 6th KIA International Academic Seminar. Seoul, Korea (1993)

54. Kramer, F.: Insassen-Crashmechanik-Rechenmodell für Frontalkollisionen – Programmbeschreibung, Verifikation und Validierung. Forschungsbericht Nr. 324/89. Institut für Fahrzeugtechnik, Technische Universität Berlin, Berlin (1988)
55. Kramer, F.: Schutzkriterien für den Fahrzeug-Insassen im Falle sagittaler Belastung. Dissertation an der Technischen Universität Berlin. Fortschritt-Berichte, VDI-Reihe 12. Vol 137. (1989)
56. Kramer, F., Hönig, M., Leithold, L.: Airbag für Motorräder. Berichte und Informationen der Hochschule für Technik und Wirtschaft (HTW) Dresden **15**(2) (2007)
57. Krämer, J.: Crash Simulationsanlagen – Technische Dokumentation. Vertriebsinformation Produktbereich P der Fa. SCHENCK AG. (1994)
58. LS-DYNA: LSTC Inc., Livermore (CA), (USA), https://www.lstc.com , https://www.dynamore.de . Jüngst übernommen durch Ansys Inc., https://www.ansys.com
59. Lonsdale, G. et al.: Experience with Industrial Crashworthiness Simulation Using the Portable, Message-passing PAM-CRASH Code. PAM'95 Fifth European Workshop on Advanced Finite Element Simulation Techniques. Bad Soden (1995)
60. MADYMO: TNO, Eindhoven (NL), https://plm.sw.siemens.com/en-US/simcenter/mechanical-simulation/madymo
61. Maier, S., Helbig, M., Hertneck, H. & Fehr J.: Characterisation of an energy absorbing foam for motorcycle rider protection in LS-DYNA, 13th European LS-DYNA Conference 2021, Ulm, Germany
62. McNeill, A., Haberl, J., Holzner, M., Trautenhahn, U., Schöneburg, R., Strutz, T.: Current Worldwide Side Impact Activities – Divergence versus Harmonisation and the Possible Effect on Future Car Design. 19th ESV-Conference, Paper Number 05-0077-O. Washington D.C. (VA/USA) (2005)
63. Meister, M.: Finite Element Simulation of Human and Dummy Kinematics in Side Crash Scenarios. Europam Conference, Mainz (2003)
64. Mertz, H.J., Irwin, A.L., Melvin, J.W., Stalnaker, R.L., Beebe, M.S.: Size, Weight and Biomechanical Impact Response Requirements for Adult Size Small Female and Large Male Dummies. SAE Technical Paper Series 890756. International Congress and Exposition. Detroit (MI), USA (1989)
65. Messring. Prüfeinrichtungen – Daten-Akquisition – Beleuchtung. https://www.messring.de. Accessed Juni 13, 2023
66. Mlekusch, B., Danzel, M.: Erst Rechnen, dann Crashen. Audi Magazin (01/2012)
67. Motor Vehicle Seat Belt Anchorages – Test Procedure. SAE J384 JUN94, SAE Recommended Practice. 2003 SAE Handbook, Vol. 3 On-Highway Vehicles and Off-Highway Machinery. Society of Automotive Engineers Inc. Warrendale (PA), USA (2003)
68. Moving Rigid Barrier Collision Tests. SAE J972 DEC88, SAE Recommended Practice. 2003 SAE Handbook, Bd. 3 On-Highway Vehicles and Off-Highway Machinery. Society of Automotive Engineers Inc. Warrendale (PA), USA (2003)
69. Müllerschön, H., Günther, F.C., Roux, W.: Robustness Study of a Front Impact Crash Model Regarding Uncertainties in Material Properties and Sheet Thicknesses. Crashworthiness of Light-Weight Automotive Structures, NTNU, Trondheim, (NOR) (2004)
70. Nagel, U.: Entwicklung eines Programms zur Rekonstruktion von Fahrzeugunfällen. Technische Universität Berlin, Studienarbeit (1989)
71. Nagel, U.: Rechnergestützte Unfallrekonstruktion – Programmsystem auf der Basis der Stoß- und der Kraftrechnung. Diplomarbeit, Technische Universität Berlin, Berlin (1991)
72. News 2005, Crash Test – Facility Design – Crash Components – Data Analysis: Präsentation der Firma Messring Systembau MSG GmbH, München (2005)

73. Olovsson, L.: On the Arbitrary Lagrangian-Eulerian Finite Element Method. Division of Solid Mechanics, Department of Mechanical Engineering, Linköpings universitet. Linköping (Schweden) (2000)
74. Olovsson, L.: Corpuscular Method for Airbag Deployment Simulations in LS-DYNA, Report R32S-1 IMPETUSafea AB, ISBN 978-82-997587-0-3 (2007)
75. PAM-CRASH: ESI-Group, Paris (F). https://www.esi-group.com
76. PC-CRASH: DSD GmbH, Linz (A). https://www.dsd.at
77. Passenger Car Door System Crush Test Procedure. SAE J367 JUN80, SAE Recommended Practice. 2003 SAE Handbook, Vol. 3 On-Highway Vehicles and Off-Highway Machinery. Society of Automotive Engineers Inc. Warrendale (PA), USA (2003)
78. Peter, W.: Anteil der Berechnung am Entwicklungsprozess eines Automobils. VDI-Tagung „Berechnung im Automobilbau“. VDI-Berichte. Bd 613. VDI, Würzburg (1986)
79. Petzsche, T.: Handbuch der Schock- und Vibrationsmesstechnik – Ein Leitfaden für die praktische Anwendung. Endevco, Heidelberg (1992)
80. Profos, P., Pfeifer, T.: Grundlagen der Meßtechnik. R. Oldenbourg Verlag, München, Wien (1992)
81. Pyttel, T., Floss, A., Thibaud, C.: Realitätsnahe Simulationsmodelle für Airbag und Mensch – neue Möglichkeiten und Grenzen der FE Simulation. VDI Fahrzeug und Verkehrstechnik, Internationaler Kongress. Berlin (2005)
82. RADIOSS: Altair Engineering Inc., Michigan, Troy, USA. https://www.altair.com
83. RCAR Research Committee for Automobile Repairs: News Release: November 2003 – Low Speed Crash Testing. https://www.rcar.org. Accessed June 19, 2023
84. Recent advances in THUMS: Development of Individual Internal Organs, Brain, Small Female and Pedestrian Model, 4th European LS-DYNA Conference, Ulm, Germany, 2003. https://www.dynalook.com/documents/4th_European_ls-dyna/LSDYNA_ULM_C-I-01.pdf
85. Schettler-Köhler, R.: Numerische Simulation von Crashvorgängen in der Fahrzeugentwicklung in Rechenmethoden in der Fahrzeugentwicklung. In: Dirschmid, W. (eds.): Fortschitte in der Fahrzeugtechnik. vol 12. Vieweg & Sohn Verlagsgesellschaft, Braunschweig, Wiesbaden (1992)
86. Schneider, L.W., et al.: Development of Anthropometrically Based Design Specification for an advanced adult anthropomorphic dummy family. The University of Michigan Transportation Research Institute, Ann Arbor (MI), USA (1983)
87. Schneider, H.-J.: Lexikon der Informatik und Datenverarbeitung. R. Oldenbourg Verlag GmbH, München (1983)
88. Scholtes, M., et al.: 6-Layer Model for a Structured Description and Categorization of Urban Traffic and Environment. IEEE Access **9**, 59131–59147 (2021). https://doi.org/10.1109/ACCESS.(2021)
89. Schram, R.: “Euro NCAP’s first step to assess automated driving systems” Paper Number 19-0292, ESV Conference (2019)
90. Schumacher, A.: Optimierung mechanischer Strukturen – Grundlagen und industrielle Anwendungen. Springer-Verlag, Berlin (2004)
91. Schuster, P., Franz, U., König, C.: Simulation von Kinderrückhaltesystemen im Seitencrash. Update-Tag Dummy-Modelle, Stuttgart (2004)
92. Schuster, P., Franz, U., Stahlschmidt, S., Pleschberger, M., Eichberger, A.: Comparison of ES-2re with ES-2 and US-SID Dummy – Considerations for ES-2re model in FMVSS Tests. 3. LS-DYNA Forum. Bamberg (2004)
93. Schwarz, H.R.: Methode der Finiten Elemente. Teubner Verlag, Wiesbaden (1991)
94. Schöneburg, R.: Auf dem Weg zur elektronischen Knautschzone. ATZ Automobiltechnische Zeitschrift **110**, 953 (2008)

95. Schöneburg, R.: Integrale Fahrzeugsicherheit. Vorlesungsmanuskript Technische Universität Dresden und Hochschule für Technik und Wirtschaft Dresden, Dresden (2019)
96. Schöneburg, R., Baumann, K.-H., Timmel, M., Bürkle, H.: Integrale Sicherheit bei Mercedes-Benz: Kunststoffe – Erfahrungen und zukünftige Anforderungen. VDI-K Internationaler Kongress, Kunststoffe im Automobilbau, Düsseldorf (07. April 2011)
97. Schöneburg, R.: Einsatz numerischer Simulationsverfahren im Fahrzeugsicherheitsversuch. Bag & Belt, 2nd International Akzo Symposium on Occupant Restraint Systems, pp. 21–31. Köln (1992)
98. Schöneburg, R.: Numerische Simulation in der Airbagentwicklung – Stand der Technik, Entwicklungen und Tendenzen. Airbag 2000, Int. Symposium on Sophisticated Car Occupant Safety Systems, pp. 8–1–14. Karlsruhe (1992)
99. Schöneburg, R.: CAE in the New Development Process MDS Excellence. 9. EDM CAE Forum, Stuttgart (12./13. Juli 2011)
100. Schöneburg, R.: Kompatibilität von Personenwagen, wesentliches Kriterium für sichere Fahrzeuge. VDA Technischer Kongress. Rüsselsheim (2004)
101. Schöneburg, R, Breitling, T.: Enhancement of Active & Passive Safety by Future PRE-SAFE®Systems. 19th ESV-Conference, Paper Number 05-0080-O. Washington D.C. (VA/USA) (2005)
102. Schöneburg, R., Baumann, K.-H., Fehring, M.: The Efficiency of PRE-SAFE Systems in Prebraked Frontal Collision Situations. 22th ESV-Conference, Paper Number 11-0207-O, Washington D.C. (USA) (2011)
103. Schöneburg, R.: Individual Safety – Potentiale für die weitere Erhöhung der Insassensicherheit im PKW. 15. automobil-forum. Stuttgart (2004)
104. Schöneburg, R., Baumann, K.-H.: Auf dem Weg zur virtuellen Knautschzone – Möglichkeiten des präventiven Energieabbaus und Auswirkungen auf Fahrzeug und Insassen in der Vorunfallphase. 11. Symposium AAET (Automatisierungs-, Assistenzsysteme und eingebettete Systeme für Transportmittel), Braunschweig (2010)
105. Schöneburg, R., Hart, M. , Feese, J. , Muecke, S. , Richert, J.: ESF 2019 – Experimental Safety Vehicle meets Automated Driving. 26th ESV-Conference, Paper Number 19-0042-O, Eindhoven (Netherlands) (2019)
106. Schöneburg, R.: Elektromobilität bei Mercedes-Benz – Auswirkungen auf die Fahrzeugsicherheit und die Entwicklungsabläufe beim Crashversuch. 37. Tagung Werkstoffprüfung, Deutsche Gesellschaft für Materialkunde e.V. (DGM), Neu-Ulm (2019)
107. Schöneburg, R., Baumann, K.-H.: Passive Safety – the Influence of new Legislation and Rating Systems on Vehicle Development. 1st Sicherheit im Automobil. München (2001)
108. Sporner, A.: Experimentelle und mathematische Simulation von Motorradkollisionen im Vergleich zum realen Unfallgeschehen. Dissertation an der Technischen Universität München, München (1982)
109. TNO Automotive Germany GmbH, Stuttgart, Deutschland, https://www.tno-automotive.de . Jüngst umformiert zu TASS, https://tass.plm.automation.siemens.com
110. The Insurance Crash Test Parameters, Appendix 1 in Vehicle Design Features for Optimum Low Speed Impact Performance. RCAR Research Committee for Automobile Repairs. Conference The RCAR Meeting 1995. Rotherwick, UK (1995)
111. UN Regulation No. 157 "Uniform provisions concerning the approval of vehicles with regard to Automated Lane Keeping SystemsEuro NCAP", ECE/TRANS/WP.29/2020/81 (2021)
112. Uriot, J., Page, M., Tarriere, C., Bendjellal, F. et al.: Measurement of Submarining on HYBRID III 50° & 5° Percentile Dummies. 14th ESV Conference. München (1994)
113. Vehicle Roof Stength Test Procedure SAE J374 MAY91, SAE Recommended Practice. 2003 SAE Handbook, Vol. 3 On-Highway Vehicles and Off-Highway Machinery. Society of Automotive Engineers Inc. Warrendale (PA), USA (2003)

114. Vieweg, C.: Alles über die Mercedes-Benz A-Klasse. Daimler-Benz Service, Stuttgart (1997)
115. Volkswagen AG, Wolfsburg, https://www.volkswagen.de
116. Wang, J.T., Nefske, D.J.: A New CAL§D Airbag Inflation Model. SAE Technical Paper Series 880654. International Congress and Exposition. Detroit (MI), USA (1988)
117. Weber, H. et al., "Grundszenarien für die Fahrdomäne Bundesautobahn", BASt Bericht Heft F 149 (2022)
118. Wismans, J., Goudswaard, A.P., Nieboer, J.J.: Motorcycle Airbag Systems. International AKZO Symposium „Bag & Belt" on Occupant Restraint Systems. Köln (1992)
119. Zeidler, F.: Die Bedeutung der Formänderungsenergie für die Unfallforschung und das EES-Unfallrekonstruktionsverfahren. Verkehrsunfall und Fahrzeugtechnik. Verlag Information Ambs GmbH, Kippenheim (1984)
120. van den Hove, M., Mlekusch, B., Müllerschön, H., FE-Simulation Based Optimization of an Adaptive Restraint System Considering Multiple Front-Crash Load Cases using LS-OPT. 5th LS-DYNA Forum, Bamberg, 2005. https://www.dynamore.de/download/af05/papers/A-I-79.pdf
121. https://www.dynamore.de/en/products/models/dynamore-human-body-model. Accessed June 19, 2023

GPSR Compliance

The European Union's (EU) General Product Safety Regulation (GPSR) is a set of rules that requires consumer products to be safe and our obligations to ensure this.

If you have any concerns about our products, you can contact us on ProductSafety@springernature.com

In case Publisher is established outside the EU, the EU authorized representative is:

Springer Nature Customer Service Center GmbH
Europaplatz 3
69115 Heidelberg, Germany

Batch number: 10739813

Printed by Printforce, the Netherlands